T0239676

Advances in
Understanding
Human Performance
Neuroergonomics, Human Factors
Design, and Special Populations

Advances in Human Factors and Ergonomics Series

Series Editors

Gavriel Salvendy
Professor Emeritus
Purdue University
West Lafayette, Indiana

Chair Professor & Head
Tsinghua University
Beijing, People's Republic of China

Waldemar Karwowski
Professor & Chair
University of Central Florida
Orlando, Florida, U.S.A.

Advances in Understanding Human Performance

Neuroergonomics, Human Factors Design, and Special Populations

Edited by

Tadeusz Marek
Waldemar Karwowski
Valerie Rice

CRC Press
Taylor & Francis Group
Boca Raton London New York

CRC Press is an imprint of the
Taylor & Francis Group, an **informa** business

CRC Press
Taylor & Francis Group
6000 Broken Sound Parkway NW, Suite 300
Boca Raton, FL 33487-2742

First issued in paperback 2017

© 2011 by Taylor and Francis Group, LLC
CRC Press is an imprint of Taylor & Francis Group, an Informa business

No claim to original U.S. Government works

ISBN-13: 978-1-4398-3501-2 (hbk)
ISBN-13: 978-1-138-11181-3 (pbk)

This book contains information obtained from authentic and highly regarded sources. Reasonable efforts have been made to publish reliable data and information, but the author and publisher cannot assume responsibility for the validity of all materials or the consequences of their use. The authors and publishers have attempted to trace the copyright holders of all material reproduced in this publication and apologize to copyright holders if permission to publish in this form has not been obtained. If any copyright material has not been acknowledged please write and let us know so we may rectify in any future reprint.

Except as permitted under U.S. Copyright Law, no part of this book may be reprinted, reproduced, transmitted, or utilized in any form by any electronic, mechanical, or other means, now known or hereafter invented, including photocopying, microfilming, and recording, or in any information storage or retrieval system, without written permission from the publishers.

For permission to photocopy or use material electronically from this work, please access www.copyright.com (http://www.copyright.com/) or contact the Copyright Clearance Center, Inc. (CCC), 222 Rosewood Drive, Danvers, MA 01923, 978-750-8400. CCC is a not-for-profit organization that provides licenses and registration for a variety of users. For organizations that have been granted a photocopy license by the CCC, a separate system of payment has been arranged.

Trademark Notice: Product or corporate names may be trademarks or registered trademarks, and are used only for identification and explanation without intent to infringe.

Visit the Taylor & Francis Web site at
http://www.taylorandfrancis.com

and the CRC Press Web site at
http://www.crcpress.com

Table of Contents

Section III: Neuroergonomics and Human Performance

Section IV: Neuroergonomics and Training Issues

Section V: Trainees: Designing for Those in Training

Section VI: Military Human Factors: Designing for Those in the Armed Forces

Section VII: New Programs/New Places: Designing for Those Unfamiliar with Human Factors

Preface

This book is concerned with emerging concepts, theories and applications of human factors knowledge that focus on discovery and understanding of human performance issues in complex systems, including recent advances in neural basis of human behavior at work (i.e. neuroergonomics), training, universal design, and human factors considerations for special populations.

The book is organized into ten sections that focus on the following subject matters:

I: Neuroergonomics: Workload Assessment
II: Models and Measurement in Neuroergonomics
III: Neuroergonomics and Human Performance
IV: Neuroergonomics and Training Issues
V: Trainees: Designing for Those in Training
VI: Military Human Factors: Designing for Those in the Armed Forces
VII: New Programs/New Places: Designing for Those Unfamiliar with Human Factors
VIII: Universal Design: Designing to Include Everyone
IX: Designing for People with Disabilities
X: Children and Elderly: Designing for Those of Different Ages

Sections I through IV of this book focus on neuroscience of human performance in complex systems, with emphasis on the assessment and modeling of cognitive workload, fatigue, and training effectiveness. Sections V through X concentrate on applying human factors to special populations, with the caveat that the design information may not generalize to (or be of interest to) other populations. This broadens the conventional definition which limits special populations to those who have limitations in their functional abilities, i.e. those with chronic disabilities due to illness, injury, or aging. Thus, special populations can incorporate certain investigations and designs focused on military, students, or even developing countries and those naïve to the field of human factors, as well as those who are affected by disabilities and aging (both young and old).

Many chapters of this book focus on analysis, design, and evaluation of challenges affecting students, trainees, members of the military, persons with disabilities, and universal design. In general, the chapters are organized to move from a more general, to a more specialized application. For example, the subtopics for those with disabilities include designing websites, workstations, housing, entrepreneur training, communication strategies, products, environments, public transportation systems, and communities.

Each of the chapters of this book were either reviewed or contributed by the members of Editorial Board. For this, our sincere thanks and appreciation goes to the Board members listed below:

H. Adeli, USA
T. Ahram, USA
A. Cakir, Germany
G. A. Calvert, UK
K. Cosenzo, USA
M. Dainoff, USA
G. Deco, Spain
J. M. Ellenbogen, USA
M. Fafrowicz, Poland
D. Feathers, USA
W. Friesdorf, Germany
S. Grafton, USA
A. Gursesusa, USA
P. Hancock, USA
S. Hignett, UK
W. Karwowski, USA
S.-Y. Lee, Korea
V. Lee, Australia
J. Lewandowski, Poland

Y. Liu, USA
V. Louhevaara, Finland
S. Makeig, USA
T. Marek, Poland
J. Molenbroek, The Netherlands
J. Murray, USA
P. Neskovic, USA
D. Nicholson, USA
R. Parasuraman, USA
M. Pauen, Germany
V. Rice, USA
A. Santamaria, USA
D. Schmorrow, USA
A. Sears, USA
A. Simoes, Portugal
K. Stanney, USA
C. Stephanidis, Greece
J. Warm, USA
A. Wróbel, Poland

This book will be of special value to a large variety of professionals, researchers and students in the broad field of human performance who are interested in neuroergonomics, training effectiveness, and universal design and operation of products and processes, as well as management of work systems in contemporary society. We hope this book is informative, but even more - that it is thought provoking. We hope it inspires, leading the reader to contemplate other questions, applications, and potential solutions in creating designs that improve function, efficiency, and ease-of-use for all.

April 2010

Tadeusz Marek
Jagiellonian University
Krakow, Poland

Waldemar Karwowski
University of Central Florida
Orlando, Florida, USA

Valerie Rice
Army Research Laboratory
Medical Department Field Element
Ft. Sam Houston, San Antonio, Texas, USA

Editors

CHAPTER 1

Neurogenetics of Working Memory and Decision Making under Time Pressure

Raja Parasuraman

George Mason University

ABSTRACT

Effective decision-making involves working memory and executive function, which depend on the efficiency of dopamine signaling in prefrontal cortex. We examined associations between two single nucleotide polymorphisms (SNPs) in the dopamine hydroxylase (DBH) gene, 444 G/A and -1021 C/T, and individual differences in working memory and decision-making. The TT allele of the -1021 C/T SNP leads to a 10-fold decrease in plasma enzyme level (DßH) and a comparable increase in synaptic dopamine, compared to the CC allele. The working memory task involved maintaining up to three spatial locations over a period of 3 s. The decision-making task involved simulated command and control (C^2) and required participants to identify and engage critical enemy units, with or without the assistance of automation, within 10 s. Working memory accuracy was significantly greater for the TT allele than for either the TC or CC alleles. Compared to a high DßH enzyme group, a low DßH enzyme (high synaptic dopamine) group made decisions on the C^2 task more rapidly and were less affected by automation unreliability. The results show that the DBH gene is associated with individual variation in working memory capacity and with speeded decision making under time pressure. Both findings are consistent with a prominent role for prefrontally modulated working memory processes in decision making.

Keywords: Neuroergonomics, genetics, DBH, individual differences, working memory, decision making, dopamine, noradrenaline.

INTRODUCTION: NEUROGENETICS OF COGNITION

Neuroergonomics extends neuroscience theories and findings to the analysis of human behavior at work and other natural settings (Parasuraman, 2003; Parasuraman and Rizzo, 2007; Parasuraman and Wilson, 2008). Molecular genetics is among the newer methods that are emerging for use in neuroergonomics. This tool can be used not only to examine the molecular underpinnings of cognition but also to investigate individual differences in cognitive and brain functions. The neurogenetic approach to cognition has capitalized on the breakthroughs provided by the success of the Human Genome Project in decoding the entire diploid human genome and on neuroimaging studies that have linked cognitive functions to the activation of specific brain networks.

The role of genetics in cognition has traditionally involved comparing the performance of identical and fraternal twins to assess the heritability of a cognitive function. For example, twin studies have established that human intelligence is highly heritable, with a heritability estimate of about 70% (Plomin, DeFries, McClearn, and McGuffin, 2001). But twin studies cannot identify the *specific* genes contributing to normal variation in intelligence or other cognitive functions. Molecular genetics—and in particular the *allelic association* method—provides a different, complementary approach to behavioral genetics. Some but not all genes come in different forms or *alleles*, which can be identified, sequenced, and then potentially associated with variations in cognitive functioning.

A gene consists of several thousand of four DNA nucleotides, adenine (A), guanine (G), cytosine (C) and thymine (T). The complete sequence of nucleotides is identical across different individuals carrying the gene except at a few locations. The variants in these locations are the alleles, and since DNA is doubled stranded, there are two possible alleles at that location, with one inherited from each parent. One then can examine the functional consequence, if any, of such (normal) allelic variation. Such an analysis method has been recently applied to the study of individual differences in cognition in healthy individuals (for reviews, see Goldberg and Weinberger, 2004; Green et al., 2008; Greenwood and Parasuraman, 2003; Parasuraman and Greenwood, 2004). These studies have provided evidence of modulation of cognitive task performance by specific genes, with individual variation in cognitive functioning being linked to allelic variation in a particular gene (Egan et al., 2001; Fan et al., 2002; Fossella and Casey, 2006; Fossella et al., 2002; Greenwood, Sunderland, Friz, and Parasuraman, 2000; Parasuraman, Greenwood, Kumar, and Fossella, 2005; Posner, Rothbart, and Sheese, 2007). More recently, the neurogenetic method has been extended to examine the associations and interactions between *multiple* specific genes and cognition (Greenwood, Lin, Sundararajan, Fryxell, and Parasuraman, 2009; Espeseth et al., 2006). This paper focuses on the neurogenetics of working memory and decision making in time stressed situations.

WORKING MEMORY: DOPAMINE AND NOREPINEPHRINE

Working memory is an important mediating factor in many higher cognitive functions, including problem solving and decision-making. Dopaminergic receptor genes are good candidates for identifying genetic effects on working memory, for several reasons. First, several sources of evidence point to the importance of dopamine regulation in prefrontal cortical (PFC) regions that are known to be involved in working memory. Dopamine agents have been shown to modulate working memory and PFC function in monkeys (Sawaguchi and Goldman-Rakic, 1991) and humans (Muller, von Cramon, and Pollmann, 1998). Dopamine plays an important role not only in PFC-mediated processes of working memory, but also in hippocampal inputs to that region (Gurden, Takita, and Jay, 2000).

In addition to dopamine, the neuromodulator norepinephrine (or noradrenaline) has also been found to play a role in working memory, as shown in animal studies involving administration of agonists that boost noradrenergic activity. Avery Franowicz, Studholme, van Dyck, and Arnsten (2000), for example, found that an alpha-2A adrenoreceptor agonist improved spatial working memory performance in monkeys and increased blood flow in prefrontal but not temporal cortex in monkeys.

Given that both dopamine and norepinephrine are involved in PFC activation and working memory performance, we focused on a recently discovered gene, Dopamine Beta Hydroxylase (DBH), that is known to modulate their differential availability in PFC neurons. DBH has been shown to be involved in the converstion of dopamine to norepinephrine in adrenergic vesicles (Cubells et al., 1998). A SNP in the DBH gene involving a G to A substitution at location 444 in exon 2 (444G/A) has been linked to changes in the dopamine to noradrenaline ratio in brain (Cubells and Zabetian, 2004) and to attention deficits in children (Daly, Hawi, Fitzerald, and Gill, 1999). Figure 1 shows the DBH gene in chromosome 9. As this figure indicates, the DBH gene is about 23,000 base pairs (bp) long, with 12 coding regions (exons). Figure 1 also shows the location of the 444G/A SNP, which is found in Exon 2, 444 bp downstream from the transcription start site.

Parasuraman et al. (2005) first showed that the DBH gene was associated with individual differences in spatial working memory The working memory task involved maintaining a representation of up to three spatial locations (small dots) over a period of three seconds. After a fixation period, participants were shown the target dots at one to three locations for 500 ms. Simultaneous with the offset of the dot display, the fixation cross reappeared for a 3 s delay at the end of which a single

4

red test dot appeared alone, either at the same location as one of the target dot(s) (match) or at a different location (non-match). Participants had two seconds to decide whether the test dot location matched or did not match one of the target dots.

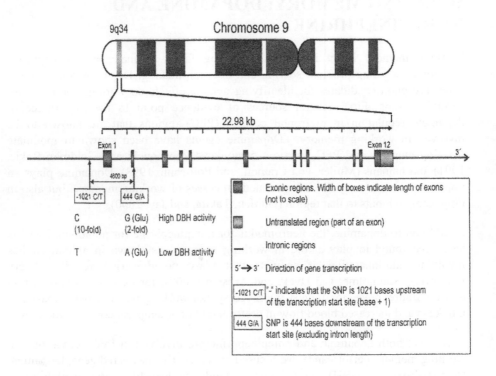

FIGURE 1. The DBH gene and two of its SNPs, 444 G/A and -1021 C/T

Table 1 gives the mean match accuracy values as a function of working memory load for the A/A, G/A, and G/G genotypes of the DBH gene. As shown in Table 1, matching accuracy decreased as the number of locations to be maintained in working memory increased from one to two to three, demonstrating the sensitivity of the task to variations in memory load. Furthermore, while there was no difference between genotypes in match accuracy at the lowest memory load, accuracy was higher for the A/G than for the A/A genotype, and higher still for the G/G genotype for the memory loads of two and three dots. The effect size of the 444 G/A gene on working memory accuracy at the highest memory load (three dots) was .25.

These findings point to an association between the DBH gene, which regulates the dopamine/norepinephrine ratio in PFC regions (Cubells and Zabetian, 2004), and the efficiency of spatial working memory. At the same time, we found that the DBH gene was not related to either the shifting of attention in a cued-discrimination task (Parasuraman et al, 2005) or to the scaling of the focus of attention ("zoom in" or "zoom out") in a cued-visual search task (Greenwood et al., 2005), thus establishing a degree of specificity for this gene.

Genotype	Memory Load (Number of dots)		
	1	2	3
A/A	87.3 (2.3)	66.8 (3.4)	45.3 (4.9)
A/G	86.4 (2.5)	71.3 (3.3)	51.2 (3.8)
G/G	86.4 (2.6)	79.7 (3.4)	61.3 (4.1)

Table 1. Mean match accuracy values (%) as a function of memory load and 444G/A genotype. Standard errors in parentheses.

The genetic association between the DBH gene and working memory is noteworthy because the G444A variation we examined is a so-called functional SNP, meaning that it directly influences protein and amino acid production. The resulting product, the DßH enzyme, can be measured in cerebral spinal fluid and in blood. Cubells and colleagues (1998) first reported that the 444G/A polymorphism of the DBH gene influences levels of the DßH enzyme in plasma. In a more recent study (Cubells and Zabetian, 2004), however, they found that another SNP which is located upstream of the 444G/A SNP, in the promoter region, -1021 C/T (see Figure 1), has a much greater effect on plasma levels of DßH. The CC allele combination of this SNP leads to a 10-fold increase in plasma DßH levels compared to only 2-fold for the 444 G/A SNP (see Figure 1). Thus the CC allele of -1021 C/T should be associated with greater conversion of synaptic dopamine to norepinephrine than would the TC or TT alleles. Given an association between PFC dopamine levels and working memory, therefore, the CC genotype should be associated with poorer and the TT genotype with superior working memory capacity, respectively. We conformed this prediction using the same spatial working memory task (Sundararajan et al., 2006).

DECISION MAKING IN A COMMAND AND CONTROL TASK

The genetic association between dopaminergic genes and working memory was demonstrated for a very simpl cognitive task, namely deciding whether a particular location matched or did not match up to three locations held in mind for a brief period of time. Can similar associations be observed for more complex decision-making tasks? To examine this question, we used a simulated battlefield engagement task that we had previously developed for research examining the effects of automation on decision making under time stress (Rovira, McGarry, and Parasuraman, 2007). Given the putative role of the dopaminergic system in decision making, we focused on the same two SNPs of the DBH gene, 444 G/A and -1021 C/T.

The decision making task was a " sensor to shooter targeting system" showing a terrain view with enemy units, friendly artillery and battalion units, and a headquarters unit. Participants were required to identify the most dangerous enemy target in the terrain view and to select a corresponding friendly unit to engage in combat with the target within 10 s, based on military rules for engagement. In addition, participants had to perform a communications task requiring them to respond to their own call sign. The engagement task was carried out either manually or with automation support. Participants were free to follow the automation directive or to choose their own target to engage. The automation was not perfect, however, having an overall reliability of 80%. Rovira et al. (2007) showed that decision aiding speeded up engagement times, and hence the overall sensor-to-shooter time. However, when the automation gave incorrect recommendations, decision making accuracy was reduced, as participants tended to go along with the automated directives. We examined whether the DBH genotype would be associated with decision making performance in this task both when the automation was reliable and when it was imperfect.

A total of 86 participants were genotyped for the 444 G/A and -1021 C/T SNPs of the DBH gene. Based on previous findings associating different combinations of these SNPs with either low or high levels of DßH enzymatic activity, we divided the participants into two groups. The low DBH enzyme activity group (Genotype 1, N=41) included participants with the G/A and T/T combination of alleles of the two SNPs. The high enzyme activity group (Genotype 2, N=45) included participants with the A/A and C/C or G/A and C/C combinations of the two SNPs. We predicted that decision making performance would be superior in the low enzyme activity group (Genotype 1) compared to the high enzyme activity group (Genotype 2).

Consistent with the previous study by Rovira et al. (2007), decision making performance (accuracy of enemy engagement decisions) was superior with reliable automated support compared to manual performance. There were no differences in overall decision making accuracy with DBH genotype. However, decision time was significantly shorter in the Genotype 1 group than in the Genotype 2 group, and was also faster with decision aiding than without (see Table 2). The effect of DBH genotype was observed both for unaided decision making performance and under decision aiding. Overall, the low enzyme activity genotype group was about 24% faster in making target engagement decisions than the high enzyme activity group.

We observed no difference in overall decision accuracy as a function of DBH genotype. However, performance on incorrect automation trials did differ according to genotype. Consistent with Rovira et al. (2007), we found that whereas decision making accuracy was close to 100% when the decision aid was correct, performance was significantly poorer on the 20% of trials when the automated recommendation was incorrect. However, this cost of imperfect automated advice was significantly lower in the genotype 1 group, 96%, which represents only about a 4% drop, than in the genotype 2 group, whose decision accuracy in the imperfect aiding condition was 82%, a drop of 18%.

Genotype Group	No Decision Aid	With Decision Aid
1 (Low enzyme activity)	7.31 (0.32)	4.92 (0.42)
2 (High enzyme activity)	8.82 (0.45)	7.34 (0.54)

Table 2. Mean target engagement times (s) for the two genotype groups, with and without decision aiding. Standard errors in parentheses.

CONCLUSIONS

At the most general level, the results of the studies presented here indicate that a neuroergonomic approach based on molecular genetics can provide a new framework for examining individual differences in decision making under time pressure. These findings extend previous results pointing to the important emerging role that molecular genetics can play in neuroergonomics (Parasuraman and Wilson, 2008).

More specifically, the results show that working memory capacity is a key factor contributing to decision making efficiency under time pressure. The experimental results indicate that a gene that regulates dopamine availability in prefrontal cortex, namely the DBH gene, plays a role in both functions. Specifically, variants of the DBH gene producing low levels of the DßH enzyme that converts dopamine to norpepinephrine in prefrontal cortical neurons are associated with both high working memory capacity and superior decision making in a simulated battlefield command and control task. More importantly, the DBH gene also influenced the degree to which decision making performance was adversely affected by imperfect decision aiding.

These findings have implications for research and practice in human factors and ergonomics. For example, molecular genetic methods can inform selection and training procedures aimed at developing teams of human operators who can make speedy, unbiased decisions in semi-automated systems. Clearly more work needs to be done using tasks with similar or higher levels of complexity to the command and control simulation used in the present studies. Furthermore, the new field of the molecular genetics of cognition is still in its infancy, and many of its newest procedures require further validation. Despite that, the results obtained to date are very promising with respect to the ultimate goal of providing a neural and genetic basis for characterizing individual differences in various cognitive functions that influence human performance at work.

REFERENCES

Avery, R.A., Franowicz, J.S., Studholme, C., van Dyck, C.H., and Arnsten, A.F., 2000. The alpha-2A-adrenoceptor agonist, guanfacine, increases regional cerebral blood flow in dorsolateral prefrontal cortex of monkeys performing a spatial working memory task. *Neuropsychopharmacology, 23*, 240-249.

Cubells, J. F., van Kammen, D. P., Kelley, M. E., Anderson, G. M., O'Connor, D. T., Price, L. H., Malison, R., Rao, P. A., Kobayashi, K., Nagatsu, T., and Gelernter, J. (1998). Dopamine beta-hydroxylase: two polymorphisms in linkage disequilibrium at the structural gene DBH associate with biochemical phenotypic variation. *Human Genetics, 102(5),* 533-40.

Cubells, J. F., and Zabetian, C. P. (2004). Human genetics of plasma dopamine ß-hydroxylase activity: Applications to research in psychiatry and neurology. *Psychopharmacology, 174,* 463-476.

Daly, G., Hawi, Z., Fitzerald, M., and Gill, M. (1999). Mapping susceptibility loci in attention deficit hyperactivity disorder: Preferential transmission of parental alleles at DAT1, DBH, and DRD5 to affected children. *Molecular Psychiatry, 4,* 192-196.

Egan, M. F., Goldberg, T. E., Kolachana, B. S., Callicott, J. H., Mazzanti, C. M., Straub, R. E., Goldman, D., and Weinberger, D. R. (2001). Effect of COMT Val108/158 Met genotype on frontal lobe function and risk for schizophrenia. *Proceedings of the National Academy of Sciences U S A, 98(12)*, 6917-6922.

Espeseth, T., Greenwood, P. M., Reinvang, I., Fjell, A. M., Walhovd, K. B., Westlye, L. T., Wehling, E., Lundervold, E., Rootwelt, H., and Parasuraman,

R. (2006). Interactive effects of APOE and CHRNA4 on attention and white matter volume in healthy middle-aged and older adults. *Cognitive, Affective, and Behavioral Neuroscience, 6,* 31-43.

Fan, J., Fossella, J. A., Sommer, T., Wu, Y., and Posner, M. I. (2003). Mapping the genetic variation of attention onto brain activity. *Proceedings of the National Academy of Sciences U SA, 100(12),* 7406-7411.

Fan, J., Wu, Y., Fossella, J., and Posner, M. I. (2001). Assessing the heritability of attention networks. *BMC Neuroscience, 2:14,* 14-19.

Fossella, J., and Casey, B. J. (2006). Special issue on genetics. *Cognitive, Affective, and Behavioral Neuroscience, 6(1),* 1-8.

Fossella, J., Sommer, T., Fan, J., Wu, Y., Swanson, J. M., Pfaff, D. W., and Posner, M. I. (2002). Assessing the molecular genetics of attention networks. *BMC Neuroscience, 3,* 14–19.

Goldberg, T. E., and Weinberger. D. R. (2004). Genes and the parsing of cognitive processes. *Trends in Cognitive Sciences, 8,* 325-335.

Green, A. E., Munafò, M., DeYoung, C., Fossella, J. A., Fan, J., and Gray, J. R. (2008). Using genetic data in cognitive neuroscience: from growing pains to genuine insights. *Nature Reviews Neuroscience, 9,* 710-720.

Greenwood, P. M., Fossella, J., and Parasuraman, R (2005a). Specificity of the effect of a nicotinic receptor polymorphism on individual differences in visuospatial attention. *Journal of Cognitive Neuroscience, 17,* 1611-1620.

Greenwood, P. M., Lin, M.-K., Sundararajan R., Fryxell, K. J., and Parasuraman, R. (2009). Synergistic effects of genetic variation in nicotinic and muscarinic receptors on visual attention but not working memory. *Proceedings of the National Academy of Sciences (USA), 106,* 3633-3638.

Greenwood, P., and Parasuraman, R. (2003). Normal genetic variation, cognition, and aging. *Behavioral and Cognitive Neuroscience Reviews, 2(4),* 278-306.

Greenwood, P. M., Sunderland, T., Friz, J. L., and Parasuraman, R. (2000). Genetics and visual attention: Selective deficits in healthy adult carriers of the e4 allele of the apolipoprotein E gene. *Proceedings of the National Academy of Sciences.97,* 11661-11666.

Gurden, H., Takita, M., and Jay, T. M. (2000). Essential role of D1 but not D2 receptors in the NMDA receptor-dependent long-term potentiation at hippocampal-prefrontal cortex synapses in vivo. *Journal of Neuroscience, 20(22),* RC10.

Muller, U., von Cramon, D. Y., and Pollmann, S. (1998). D1- versus D2-receptor modulation of visuospatial working memory in humans. *Journal of Neuroscience, 18(7),* 2720-272.

Parasuraman, R. (2003). Neuroergonomics: Research and practice. *Theoretical Issues in Ergonomics Science. 4,* 5-20.

Parasuraman, R., and Greenwood, P. M. (2004). Molecular genetics of visuospatial attention and working memory. In M. I. Posner (Ed.) *Cognitive neuroscience of attention.* (pp. 245-259). New York: Guilford.

Parasuraman, R., Greenwood, P. M., Kumar, R., and Fosselaa, J. (2005). Beyond heritability: Neurotransmitter genes differentially modulate visuospatial attention and working memory. *Psychological Science, 16(3),* 200-207.

Parasuraman, R., and Rizzo, R. (2007). *Neuroergonomics: The brain at work.* New York: Oxford University Press.

Parasuraman, R., and Wilson, G. F. (2008). Putting the brain to work: Neuroergonomics past, present, and future. *Human Factors, 50,* 468-474.

Posner, M. I., Rothbart, M. K., and Sheese, B. E. (2007). Attention genes. *Developmental Science, 10,* 24–29.

Plomin, R., DeFries, J. C., McClearn, G. E., and McGuffin, P. (2001). *Behavioral genetics* (Fourth ed.). New York: Worth Publishers.

Rovira, E., McGarry, K., and Parasuraman, R. (2007). Effects of imperfect automation on decision making in a simulated command and control task. *Human Factors, 49.*76-87.

Sawaguchi, T., and Goldman-Rakic, P. S. (1991). D1 dopamine receptors in prefrontal cortex: involvement in working memory. *Science, 251*(4996), 947-950.

Sundararajan, R., Fryxell, K. J., Lin, M., Greenwood, P. M., and Parasuraman, R. (2006). Comparison of the effects of two SNPs in the DBH gene on working memory. *Society for Neuroscience Abstracts.*

CHAPTER 2

From Subjective Questionnaires to Saccadic Peak Velocity: A Neuroergonomics Index for Online Assessment of Mental Workload

Di Stasi, L.L., Marchitto, M., Antolí, A., Rodriguez, F., Cañas, J.J.

Cognitive Ergonomics Group
Faculty of Psychology
Granada University, Spain

ABSTRACT

Experts in human factors engineering and applied cognitive research are in search of reliable measures for the online monitoring of mental workload [MW] during active working. Due to the limitations associated with subjective and performance measures, researchers have turned their attention to oculomotor indices. We present data from an ongoing research project on the evaluation of MW evaluation from saccadic peak velocity [PV]. Participants were tested in a complex experimental setting simulating typical air-traffic control (ATC) tasks. Changes in MW were evaluated with a multidimensional methodology, using subjective ratings, behavioral indices and saccadic dynamics data. Comparison of our results with the literature suggests that PV shows sensitivity in the real-time detection of differences in mental state and is a strong candidate for the online diagnosis of operator under- or over-load in the workplace.

Keywords: ATC task; eye movements; cognitive load, multitasking, main sequence; fatigue, dual task.

INTRODUCTION[*]

Mental workload (MW) has been defined as "a composite brain state or set of states that mediates human performance of perceptual, cognitive, and motor tasks" (Parusaraman & Caggiano, 2002, p.17) and has been used to explain how human operators face heightened cognitive demands associated with increased task complexity in job situations where cognitive skills are more important than physical ones (Cacciabue, 2004). In an information technology society, changes in MW could have significant impacts on operator performance, possibly causing delays in information processing or even cause operators to ignore or misinterpret incoming information (Ryu & Myung, 2005). Consequently, automation research has identified a need to monitor operator functional state in real-time in order to determine the most appropriate type and level of automated assistance for helping operators to complete tasks safely (e.g. Langan-Fox, Canty and Sankey, 2009). The development of a method for monitoring operator attentional states in real-time during interactions with artifacts could be a good starting point for undertaking the investigation of this crucial issue. Our research is relevant to a variety of domains, from air-traffic control towers to call centers. For example, online monitoring of changes in an operator's attentional state (mental under/overload) could help in the design of adaptive systems that can allocate tasks in a dynamic way between the operator and the machine (e.g. Kaber *et al.*, 2006).

PV AS A NEUROERGONOMICS INDEX FOR THE ONLINE ASSESSMENT OF MENTAL WORKLOAD

Due to the limitations associated with subjective and performance measures, researchers have turned their attention to oculomotor indices. Decades of investigations have focused primarily on pupil diameter and blink rate as the best indicators of workload dynamics. However, some considerable drawbacks limit the reliability of these indices and their applicability in the context of the workplace.

---

[*] This study was in part supported by Spanish national project SEJ2007-63850/PSIC (J.J. Cañas), and by national grant FPU AP2006-03664, awarded to the first author.

(Di Stasi *et al.,* 2010 b). Both blink duration and rate increase with fatigue and time on task (e.g. Ryu & Myung, 2005), and the amount of not-processed information increases in parallel with these indices. It has been known for long time that there is a strong correlation between pupil amplitude variations and the amount of cognitive resources used to perform a task (e.g. Ahlstrom & Friedman-Berg, 2006) and these indicators continue to be used in workload assessments. However, pupil size is also influenced by several factors that can sometimes be difficult to control, including emotion and environmental lighting. The validity of pupil size validity as a MW index could be questioned if appropriate controls are not performed. Therefore, we are presently exploring saccadic peak velocity [PV] as an alternative to pupil diameter and blink rate to measure MW.

The relationship between saccadic amplitude, duration and PV has been called 'main sequence', to indicate that PV and saccadic duration increase systematically with amplitude (Bahill, Clark and Stark, 1975). Empirical results on the relationship between saccadic dynamics and activation state have shown that task complexity and other task variables can influence the PV response. Furthermore, in visual performance tasks, PV varies with the subject's state of mental activation, MW and fatigue (e.g. Di Stasi *et al.,* 2009; 2010 a; 2010 b).

We are currently investigating the influence of MW on saccadic dynamics in ecological and complex settings (fixed-base driving simulators). In a study of the relationship between risky driving behavior and MW, Di Stasi *et al.* (2009) included eye-activity parameters in the methodology used for driver assessment. It was found that the high-risk group had shorter saccade duration and a higher saccadic PV than the low-risk group. On the Mental Workload Test [MWT] the high-risk group scored significantly higher on both perceptual/central and answer demand. Furthermore, PV showed several significant correlations with MWT dimensions. The negative correlations of PV and subjective scales of MW suggested that, given a high level of risk proneness, lower PV was associated with a higher level of subjective workload. However there are some caveats to this work. Due to the complexity of the simulated scenarios, the authors analyzed normalized (by saccade number) PV values, considering the difference between the test session and training session (baseline), but without applying the bin-analysis (analyzing PV as a function of saccade length; Di Stasi *et al.,* 2010 a), necessary to control for the influence of amplitude on PV.

Similar results were obtained by Di Stasi *et al.* (2010 a) in a more controlled experimental setting. In this study the authors demonstrated that PV was sensitive to variations in MW during ecological driving tasks, showing again an inverse relation between PV and task complexity. PV decreased by 7.2 °/s as the MWT score of MW increased by 15.2 and reaction time for a secondary task increased by 46 msec. Saccade duration and velocity were not affected by differences in task complexity. The design of this experimental investigation allowed the authors to differentiate between the effects of time-on-task [TOT] and changes in MW from the same dataset. In this experiment no effect of fatigue was found. Even if the analyses for influences of TOT on PV revealed no effects, the authors suggested that the relationship between fatigue and MW requires further investigation in a more controlled experimental setting.

In an experiment that simulated multitasking performance in ATC setting, Di Stasi *et al.* (2010 b) studied the relation between the main sequence parameters and task load. The created tasks demanded different perceptual and central processing resources, as well as response resources. Results obtained from the subjective ratings (MWT) and behavioral measures (number of errors and delayed answers) confirmed that MW levels varied according to task demand. These different levels of MW were reflected in PV values. The authors found that there was a 6.3 °/s reduction in PV when task complexity assessed by MWT increased by 10.6 and performance was also affected (6 delayed answers). However, there was one limitation in this work. The authors were unable to distinguish between the effects of task complexity and TOT, probably due to the nature of the experimental design. Indeed, to avoid any effect of task switching during the experimental session, the order of task complexity variable was not balanced across the participants.

On the basis of these results we have now designed a well-controlled experiment to surmount the methodological problems encountered in the previous studies, and particularly the influence of TOT on the disruption of the main sequence rules. The experiment was conducted in the ATC domain.

THE EXPERIMENT

Modern complex systems such as nuclear power plants, air-flight control systems and weapon systems often impose heavy MW on their operators. The high rate of information flow, the complexity of the information, numerous difficult decisions and task-time stress could overwhelm the operators (Hwang *et al.*, 2008). In the aviation domain "controller workload is likely to remain the single greatest functional limitation on the capacity of the air traffic management system" (Eurocontrol, 2004, p.1).

In this experiment participants were tested in a complex experimental setting simulating typical ATC tasks. To control for TOT effects the experiment was performed on two different days. Changes in MW were evaluated with a multidimensional methodology using subjective ratings (MWT), performance indices and psychophysiological data.

PARTICIPANTS

Thirteen volunteers (4 males) took part in this experiment (mean age 22.4; SD = 2 years). None of the participants had ATC experience. All subjects had normal or corrected-to-normal vision and signed a consent form that informed them of the risks of the study and the treatment of personal data. They received course credits for participating in the study. The study was conducted in conformity with the Declaration of Helsinki.

STIMULI AND INSTRUMENTS

The same equipment configuration and experimental setting of Di Stasi *et al.* (2010 b) were used. Participants were tested on two different simulated ATC tasks. Tasks were created as a simplified version of some actual ATC operations, respecting the main artefacts and interaction sequences that the ATC operators have to deal with in their complex environments (Cox *et al.*, 2007).

The visible matrix of airspace consisted of 6 concentric green nodes presented on a black background. The radii of the six nodes were 1.5, 3, 4.5, 6, 7.5, 9 cm respectively. Aircraft (red dots with a concentric inner black dot) were always located on a visible node within the matrix and could appear on any of the five adjacent nodes (although never on the smallest). For each node, 8 positions were chosen in which aircraft could be shown (clockwise: up, 45°, 90°, 135°, 180°, 225°, 270°, 315°). A total of 40 different stimuli were constructed and stimuli were randomly presented four times per block (two blocks, 320 trials in total per experimental session). Aircraft position was updated every 1.5 s, within which time the aircraft would be presented to one of the 5 adjacent nodes. Aircraft were represented visually with their call signs (3 digits). Forty such call signs were extracted from a random number table. Call signs were presented in a size 11, Calibri font. Aircraft colour was constant and subtended 1° of visual angle.

DESIGN/PROCEDURE

The experimental design follows a 2 Task Complexity (TC: low and high) x 2 Time-on-Task (TOT: 1st block and 2nd block). TC was varied by manipulating the number of simultaneous tasks. Both the number of simultaneous tasks and TOT were assumed to lead to different attentional states (Wickens, 2002). To avoid any serial effect, the levels of task complexity were balanced across the two days. The levels of the TOT variable were obtained by dividing the session into two experimental blocks: the first part (first 20 min) and the second part (last 20 min of experiment).

We used a multiple-measures approach to evaluate the effectiveness of our manipulation. To evaluate the subjective ratings of mental state, we made use of three different questionnaires. First, the Stanford Sleepiness Scale [SSS] was used as a global measure of sleepiness (Hoddes *et al.*, 1973). Second, the Borg rating of perceived exertion (Borg, 1998) was used to evaluate the perceived task fatigue. Third, the MWT, was used to estimate subjective mental effort (for more details, see Di Stasi *et al.*, 2009). All tests were translated into the Spanish language.

To estimate possible effects of changes in activation state on eye movements indices, we analyzed saccadic main sequence (saccadic amplitude, duration and PV). Eye-movement data were analyzed using medians, rather than means, to minimize the effects of outliers and noisy data.

Participants were tested in a quiet room, and sat approximately 60 cm from a display screen. There were four experimental blocks, split over two days. At the beginning of each experimental day participants were required to complete the SSS questionnaire; after each experimental block subjects were required to complete the SSS, Borg scale and MWT questionnaires. Subjects who scored higher than 3 on the SSS scale at the beginning of both experimental days were excluded from further testing.

The low-complexity task comprised a decision task performed using a computer mouse whose two buttons were the answer keys. Each subject was instructed to determine (and answer with the mouse) whether the position of each aircraft on the display screen was either "critical" or "non-critical". An aircraft lying in one of the second and third nodes (3 and 4.5 cm of radius) was defined as being in a critical position. The experimenter explained to the participants that the critical position reflected the supposed closeness of the aircraft to the airport and its priority in needing assistance. By contrast, if the aircraft was located in one of the three largest circles it was judged to be in a non-critical position. Participants were requested to perform the mouse task using the hand other than the one used habitually for writing. The high-complexity task introduced a concurrent paper-and-pencil task to be performed along with the decision task and added a simple mathematical operation to be carried out with the call sign written in the operator's answer sheet (for more details see Di Stasi *et al.,* 2010 b).

Finally, the number of errors and reaction times for each trial at each complexity level were analyzed.

RESULTS

First, we examined the effectiveness of the TC and TOT manipulation by analyzing the subjective ratings scores, number of errors, and reaction times [RT] on the detection task. Analyses were run on data obtained from 8 participants. Mean scores in the MWT, SSS and Borg scale were submitted to a 2 (TC: low and high) x 2 (TOT, 1st block and 2nd block) repeated measures analysis of variance.

For MWT analysis significant main effects were obtained for TC [$F(1, 7) = 6.32$, $p < 0.05$; MSE = 174.12] (Table 1). No reliable effects were found for TOT nor for the interaction of both factors ($F < 1$).

Analysis of the SSS and Borg scale mean scores demonstrated a significant effect only for the TOT factor [$F(1, 7) = 8.00$, $p < 0.05$, MSE = 0.31 and $F(1, 7) = 4.51$, $p < 0.07$, MSE = 3.26] respectively for the SSS and Borg scores. No reliable effects were found for TC or for the interaction of both factors ($F < 2.25$) (Table 1).

Mean RTs in the decision task for each participant were analyzed using repeated measures ANOVA, with TC and TOT as the repeated factor. As expected, the main effect was significant for both TC and TOT [$F(1, 7) = 51.88$, $p < 0.001$, MSE = 14534, and $F(1, 7) = 22.27$, $p < 0.01$, MSE = 4631, respectively], confirming that our manipulation was accurate. The interaction was also significant [$F(1, 7) = 20.42$, $p < 0.01$, MSE = 896]. Simple-effects analysis showed significant differences for each comparison (all $p < 0.05$; Table 1). Interaction is therefore due to differences in simple-effect magnitudes. The effect of TOT on RT was smaller in the low-complexity task than in the high-complexity task. In the high-complexity task (but not in the low-complexity task) it was observed that participants could reduce RT by implementing a more efficient strategy for information processing: learning effects could therefore play an active role in decreasing RT in the high-complexity condition. More studies will be required to evaluate the relationship between TOT and on-task learning.

Errors in the decision task were always less than 8% of the total number of trials (for all subjects irrespective of task complexity).

In the next step we analyzed the sensitivity of the saccadic main sequence parameters to detect variation in MW and fatigue across the TC and TOT manipulation. The amplitudes of the saccades were categorized into 9 bins (henceforth Saccade Length, Di Stasi *et al.*, 2010 a), ranging from 3° to 12° (with 1° increments). The medians of the saccadic duration and PV were then submitted to two separate 2 (TC) x 2 (TOT) x 9 (Saccade Length) repeated measures ANOVA (see Figure 1).

Regarding saccade duration (SD), there was only a main effect of saccade length [$F(8, 56) = 66.38$, $p < 0.001$, MSE $= 20.5$]. TC, TOT, and the interactions of both factors with Saccade Length were not significant ($F < 1.3$; Table 1). As expected, for SD and PV, higher values were found for larger saccades (main sequence rules; Becker, 1989).

Regarding PV, ANOVA only revealed significant main effects for saccade length [$F (8, 64) = 630.53$, $p < 0.001$, MSE $= 298$]. TOT and TC main effects were not significant ($F < 4$; Table 1). However, as expected, interactions between TC and Saccade Length, and between TC, TOT and Saccade Length, were reliable [$F(8, 64) = 2.60$, $p < 0.05$, MSE $= 398$, and $F(8, 64) = 2.62$, $p < 0.05$, MSE $= 98$, respectively]. We next analyzed this interaction by separating the first versus the second block (Figure 1). Simple-effects analyses revealed that the low-complexity task had lower PV values in the last bin ($t = 3.05$, $p < 0.05$). No significant effects were observed in the second block.

Table 1 Overview of the experimental results. Mean values (± SD) recorded on several dependent variables for 8 participants.

	TASK			
	1st block (first 20 min)		2nd block (last 20 min)	
	Low complexity	High complexity	Low complexity	High complexity
MWT score	49.28 (20.34)	60.34 (8.94)	50.34 (20.38)	62.74 (10.05)
SSS score	2.62 (1.06)	2.75 (0.88)	3 (1.51)	3.50 (1.19)
Borg score	10.87 (2.47)	10.25 (2.12)	11.75 (3.80)	12.25 (2.65)

RT (msec)	802.15 (197.22)	1157.00 (256.22)	736.56 (180.77)	995.73 (244.50)
PV (º/sec)	316.75 (69.20)	325.77 (82.01)	322.18 (69.49)	327.80 (76.18)
SD (msec)	41.20 (6.32)	41.23 (6.61)	41.44 (6.90)	41.5 (6.82)

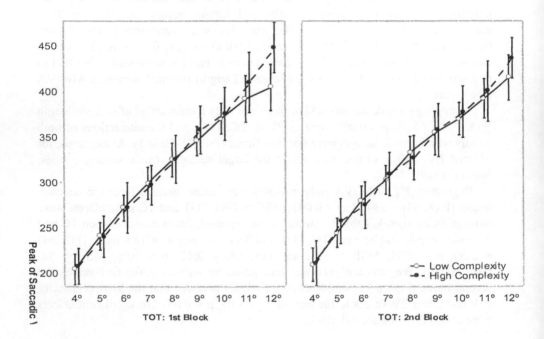

FIGURE 1 Illustration of Task Complexity x TOT x Saccade Length interactions with PV. Vertical bars denote 0.95 confidence intervals.

CONCLUSIONS

The detection of potential operator overload situations is a first step towards the avoidance of incorrect decisions brought about by increased, MW and/or fatigue. Comparison of the results on PV reported here with those of other literature studies suggests that PV can afford a sensitive and real-time index of changes in mental state, and PV is therefore a candidate for under/overload diagnosis in complex and operational environments.

Because TC was related to the subjective rating of MW, the effect of TC on PV appears to be due to variation of MW rather than with scores on the SSS and Borg scales that assess fatigue and sleepiness. These two scales were however affected by changes in TOT that had no effects on PV.

The effect of TC on PV (Figure 1) could be explained by considering the nature of this parameter. When a saccadic movement starts, it has an initial velocity and then accelerates. The PV is the point at which acceleration becomes negative. Unlike velocity, PV is independent of saccadic duration because it is not *a priori* linked to it by a mathematical definition. Furthermore, PV is independent of the distance at which saccades terminate, even though the apparent duration of saccades depends on distance (Becker 1989). PV therefore appears to afford a good index of saccadic programming, and can reflect MW effects on saccadic programming independently of distance and duration. The fact that we found a significant effect of PV only in the largest distance could be explained by the relatively short saccadic magnitude (from 3° to 12°); it is possible that the mathematical relationship between these parameters could mask the effect of our main manipulation. Stronger effects of PV are found when saccade amplitudes are larger, for example in driving simulation tasks (Di Stasi *et al.*, 2010 a).

Our research is relevant to a variety of domains ranging from ATC towers to call centers. For example, using real-time PV measures, neuroergonomists could better evaluate when an operator's attentional state is changing (mental under/overload), helping in the design of systems able to allocate tasks in a dynamic way between the operator and the machine. On the basis of these findings we are now aiming to develop an 'early overload-and-distraction warning system' employing real-time monitoring of the functional status of operators in the workplace.

REFERENCES

Ahlstrom, U., and Friedman-Berg, F.J. (2006). Using eye movement activity as a correlate of cognitive workload. *International Journal of Industrial Ergonomics* 36, 623-636.

Bahill, A.T., Clark, M.R., and Stark, L. (1975). The main sequence, a tool for studying human eye movements. *Mathematical Biosciences* 24, 191-204.

Borg, G. (1998). *Borg's perceived exertion and pain scales*. Champaign, IL: Human Kinetics.

Becker, W. (1989). *Saccadic eye movements as a control system: Metrics*. In: R.H, Wurtz, and M.E., Goldberg, eds. Reviews of Oculomotor Research, vol 3: The Neurobiology of Saccadic Eye Movements. Elsevier, Hillsdale, 13-67.

Cacciabue, P.C. (2004). *Guide to applying human factors methods: Human error and accident management in safety critical systems*. London: Springer.

Cox, G., Sharples, S., Stedmon, A., and Wilson, J. (2007). An observation tool to study air traffic control and flightdeck collaboration. *Applied Ergonomics* 38, 425-435.

Di Stasi, L.L., Alvarez, V., Cañas, J.J., Maldonado, A., Catena, A., Antolí, A., et al. (2009). Risk behaviour and mental workload: Multimodal assessment techniques applied to motorbike riding simulation. *Transportation Research Part F* 12, 361-370.

Di Stasi, L.L., Renner, R., Staehr, P., Helmert, J.R., Velichkovsky, B.M., Cañas, J.J., Catena, A., et al. (2010 a). Saccadic peak velocity sensitivity to variations in mental workload. *Aviation, Space, and Environmental Medicine*.81, 413-417. .

Di Stasi, L.L., Marchitto, M., Antolí, A., Baccino, T., and Cañas, J.J. (2010 b). Approximation of on-line mental workload index in ATC simulated multitasks. *Journal of Air Transport Management*. DOI: 10.1016/j.jairtraman.2010.02.004.

Eurocontrol (2004). Cognitive complexity in air traffic control a literature review. EEC note no. 04/04.

Hwang, S-L., Yau, Y-J., Lin, Y-T., Chen,J-H., Huang, T.H., Yenn, T-Z, and Hsu, C-C. (2008). Predicting work performance in nuclear power plants. *Safety Science* 46, 1115-1124.

Kaber, D.B., Perry, C.M., Segall, N., McClernon, C.K., and Prinzel, L.J. (2006). Situation awareness implications of adaptive automation for information processing in an air traffic control-related task. *International Journal of Industrial Ergonomics* 36, 447–462.

Langan-Fox, J., Canty, J.M., and Sankey, M.J. (2009). Human–automation teams and adaptable control for future air traffic management. *International Journal of Industrial Ergonomics* 39, 894–903.

Parasuraman, R., and Caggiano, D. (2002). *Mental workload*. In V. S. Ramachandran (Ed.), Encyclopedia of the Human Brain, 17-27. San Diego, UE: Academic Press.

Ryu, K., and Myung, R. (2005). Evaluation of mental workload with a combined measure based on physiological indices during a dual task of tracking and mental arithmetic. *International Journal of Industrial Ergonomics* 35, 991–1009.

Wickens, C.D. (2002). Multiple resources and performance prediction. *Theoretical Issues in Ergonomic Science* 3, 159-177.

Chapter 3

Cognitive Workload Assessment of Air Traffic Controllers Using Optical Brain Imaging Sensors

Hasan Ayaz[1], Ben Willems[2], Scott Bunce[3], Patricia A. Shewokis[1,4], Kurtulus Izzetoglu[1], Sehchang Hah[2], Atul R. Deshmukh[2], Banu Onaral[1]

[1]School of Biomedical Engineering
Science & Health Systems
Drexel University

[2]Atlantic City International Airport: Federal Aviation Administration
William J. Hughes Technical Center

[3]Penn State Hershey Neuroscience Institute,
Penn State University

[4]College of Nursing and Health Professions
Drexel University

ABSTRACT

Functional near-infrared spectroscopy (fNIR) is a highly portable, safe neuroimaging technology that uses light to measure cortical brain activity. We have utilized fNIR to provide objective measures of cognitive workload of certified Air Traffic Controllers while they managed realistic scenarios under typical and emergent conditions. Participants also completed an n-back task, which is a standardized working memory and attention task with four incremental levels of difficulty. In the n-back results, as task difficulty increases, accuracy and speed of the participants decrease monotonically. Blood oxygenation changes, as measured by fNIR, monotonically increased with increasing task difficulty. These findings are in line with earlier studies. Further, fNIR measures were analyzed for comparison between data-based communication and voice-based communication during an air traffic control tasks. The results revealed less brain activation for electronic than for voice communications, which is also in agreement with our primary hypothesis.

Keywords: Functional Near Infrared Spectroscopy, fNIR, Cognitive Workload, Mental Effort, Air Traffic Control

INTRODUCTION

The Next Generation Air Transportation System, developed by the Joint Planning and Development Office (JPDO), outlines a series of transformations designed to increase the capacity, safety, and security of air traffic operations in the United States (JPDO, 2004). A critical element in achieving this vision for future air-traffic management involves augmenting the current auditory-based communications between air traffic control (ATC) and the flight deck with text-based messaging, or DataComm systems. DataComm systems are expected to allow ATC to manage more air traffic at a lower level of cognitive load, thereby increasing both the capacity of the national airspace system and the safety of passengers. Although self-report measures of workload suggest that DataComm systems require less cognitive effort than voice-based systems to manage the same amount of traffic (Hah, Willems, & Phillips, 2006; Willems, Hah, & Phillips, 2006), to date this has been not been tested using measures of neural function. The purpose of this research is to provide objective, brain-based measures of neural activity and to determine the relative cognitive workload of DataComm versus voice-based communications systems (VoiceComm) during realistic simulations.

The advent of brain imaging technologies such as functional magnetic resonance imaging (fMRI), positron emission tomography (PET), and Magneto-encephalography (MEG) have greatly increased scientists' ability to study localized brain activity in humans and carry out studies for better understanding of the neural basis of mental states. However, these techniques are expensive, highly sensitive to motion artifact, confine participants to restricted positions and may expose individuals to potentially harmful materials or loud noise. More recently, functional near-infrared (fNIR) spectroscopy has been used as a noninvasive tool to monitor concentration changes of oxygenated hemoglobin (oxy-Hb) and deoxygenated hemoglobin (deoxy-Hb) at the cortex (Chance, Zhuang, UnAh, Alter, & Lipton, 1993; Villringer, Planck, Hock, Schleinkofer, & Dirnagl, 1993). Moreover, fNIR technology allows the design of portable, safe, affordable and accessible monitoring systems. These qualities pose fNIR as an ideal candidate for monitoring cognitive activity-related hemodynamic changes not only in laboratory settings but also under working conditions.

In the current study, we have incorporated fNIR into ongoing studies at the FAA's William J. Hughes Technical Center (WJHTC) where certified controllers were monitored with fNIR while they managed realistic ATC scenarios under typical and emergent conditions. The primary objective of this study was to use neurophysiological measures to assess cognitive workload during completion of a controlled complex cognitive task: ATC.

ASSESSMENT OF COGNITIVE WORKLOAD

The efficiency and safety of many complex human-machine systems can be closely related with the cognitive workload and situational awareness of their operators. An ideal human-machine system would be informed about the current cognitive workload level of its operant and/or designed to keep the necessary workload level at an optimum level. Hence, it can help prevent a potential overload and minimize errors.

The aim of cognitive workload assessment is to evaluate the effect of the demands that a task places on the human operator. Direct and indirect methods have been employed to measure workload, and these can be classified under four main groups. The first group of cognitive assessment tools are the subjective assessment methods that use self-reported rating scores such as Modified Cooper-Harper Scale (Cooper & Harper, 1969), Subjective Workload Assessment Technique (SWAT)(Sheridan & Simpson, 1979), NASA Task Load Index (Hart & Staveland, 1988), and self-reported mental effort (Paas & Van Merriënboer, 1993). The second group of cognitive workload assessment methods compares behavior and performance measures of the participant, recorded during the task, to identify any workload effects within them. Accuracy and speed of response are widely in use (Embrey, Blackett, Marsden, & Peachey, 2006). The third group of cognitive workload assessment methods is based on physiological measures such as eye movements (Ahlstrom & Friedman-Berg, 2006), eye blinks (J. Veltman & Gaillard, 1996), pupil dilation (Kahneman, 1973), skin temperature (H. Veltman & Vos, 2005), galvanic skin response (Helander, 1978), heart rate (Bedny, Karwowski, & Seglin, 2001), blood pressure (J. Veltman & Gaillard, 1996), and respiration rate (Roscoe, 1992). Such physiological measures, though indirect and also sensitive to stress and emotional state, have been found to be related to mental load of the participant and are relatively accessible. A final approach is to adopt a more direct perspective and monitor cognitive activity-related signals directly from the brain (Just, Carpenter, & Miyake, 2003; Parasuraman, 2003).

With the advent of neuroimaging, there are several different types of brain monitoring tools at a scientist's disposal to study human brain function and its relationship to performance during select tasks. Furthermore, this allows probing separate brain processes during workload conditions such as working memory and attention.

Electroencephalography (EEG) measures electrical fields related to neural activity, and it is the oldest modality to monitor brain activity. It is relatively portable and requires placement of electrodes over the scalp and use of gel for coupling with the scalp. EEG measures of workload and task difficulty have been reported in studies of air traffic controllers (Brookings, Wilson, & Swain, 1996), airline pilots (Sterman & Mann, 1995) drivers (Brookhuis & De Waard, 1993), and participants performing cognitive tasks (Berka et al., 2004; Pleydell-Pearce, Whitecross, & Dickson, 2003). Apart from a lengthy setup that requires expertise, the data collected during experiments is limited for localizing active brain regions.

Measures of functional brain activity, such as glucose metabolism rate or oxygen concentration, provide a quantitative index of the amount of capacity utilization. fMRI is widely used to study the operational organization of the human brain and has been demonstrated that it can map changes in brain hemodynamics produced by human mental tasks (Logothetis & Wandell, 2004). Blood Oxygen Level Dependent (BOLD) contrast in fMRI measures neural activity via the increased flow of blood to the site of the activation. The use of fMRI in ecologically valid applications (where participants perform real world tasks) is limited due the restrictions they impose on participants.

fNIR is an emerging optical brain imaging modality that measures hemodynamic response, similar to fMRI, by using near infrared light (M. Izzetoglu, Bunce, Izzetoglu, Onaral, & Pourrezaei, 2007). fNIR has been demonstrated to be sensitive to cognitive workload (Ayaz, Shewokis, Bunce, Schultheis, & Onaral, 2009; K. Izzetoglu, Bunce, Onaral, Pourrezaei, & Chance, 2004; M. Izzetoglu et al., 2005). These qualities make fNIR suitable for the study of cognitive- and affect-related hemodynamic changes under field conditions.

METHOD

PARTICIPANTS

Twenty-four certified professional controllers (CPCs) between the ages of 24 to 55 volunteered to participate in this study. All participants were non-supervisory CPCs with a current medical certificate and had actively controlled traffic in an Air Route Traffic Control Center between 3 to 30 years. Prior to the study, all participants signed informed consent forms.

EXPERIMENTAL TASKS

The participants completed two types of tasks: n-back tasks and ATC part-tasks. The n-back task is a standardized working memory and attention task with four incremental levels of difficulty. Participants are asked to monitor stimuli (single letters) presented on a screen serially and click a button when a target stimulus arrives. Four conditions were used to incrementally vary working memory load from zero to three items. In the 0-back condition, participants responded to a single prespecified target letter (e.g., "X") with their dominant hand (pressing a button to identify the stimulus). In the 1-back condition, the target was defined as any letter identical to the one immediately preceding it (i.e., one trial back). In the 2-back and 3-back conditions, the targets were defined as any letter that was identical to the one presented two or three trials back, respectively. The total test included seven sessions of each of the four n-back conditions (hence, a total of 28 n-back blocks) presented in a pseudo-random order. The task was designed and presented in E-prime (Psychology Software Tools).

For the ATC part-task, each CPC controlled traffic on workstations with a high-resolution (2,048 x 2,048), 29" radarscope, keyboard, trackball, and Direct Access Keypad for 10 minutes. To display the air traffic, the DESIREE ATC simulator and the TGF systems that were developed by software engineers at the WJHTC were used.

Six simulation pilots were used within scenarios by supporting one sector or two sectors and entering data at their workstations to maneuver aircraft, all based on controller clearances. Two types of communications, either voice (VoiceComm) or data (DataComm) communications were used in separate sessions in a pseudo-random order. For each communication type, task difficulty was varied by the number of aircraft in each sector, containing 6, 12 or 18 aircraft.

FNIR DEVICE

The continuous wave fNIR system (fNIR Devices LLC; www.fnirdevices.com) used in this study is connected to a flexible sensor pad (Figure 1) that contains 4 light sources with built in peak wavelengths at 730 nm and 850 nm and 10 detectors designed to sample cortical areas underlying the forehead (Ayaz et al., 2006). With a fixed source-detector separation of 2.5 cm, this configuration generates a total of 16 measurement locations (voxels) per wavelength. For data acquisition and visualization, COBI Studio software (Drexel University) was used. The system records two wavelengths and dark current for each 16 voxels, totaling 48 measurements for each sampling period. The sampling rate of the system is 2Hz. During the n-back task, serial cable between the fNIR data acquisition computer and E-prime stimulus presentation computer was used to transfer time synchronization signals (markers) that indicate the start of sessions and onset of stimuli.

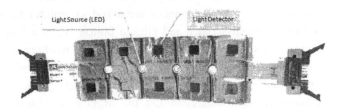

Figure 1. fNIR Sensor that covers forehead of the subject (below)

DATA ANALYSIS

From the n-back task E-prime log files, the speed of response (response time) and the accuracy (correct click ratio) were calculated for all sessions. Response time is the time period between the onset of stimulus (target trial) and the response from the participant. Average response time for each session is calculated. Correct click ratio is the ratio of number of clicks (response button) for a target to the number of total targets available in that session. A value of '1' indicates participant found all target letters in that session and '0' indicates none of the targets were clicked.

For each participant, raw fNIR data (16 voxels x 3 wavelengths) was low-pass filtered with a finite impulse response, linear phase filter with order of 20 and cut-off frequency of 0.1Hz to attenuate the high frequency noise. Saturated channels (if any), in which light intensity at the detector was higher than the analog-to-digital converter limit were excluded.

Using time synchronization markers, fNIR data segments for rest periods and task periods (28 sessions per participant for n-back task and 6 sessions per participant for ATC task) were extracted. Blood oxygenation changes within dorsolateral prefrontal cortex for 16 voxels were calculated using the Modified Beer Lambert Law (MBLL) for task periods with respect to rest periods (M. Izzetoglu, et al., 2005). Average oxygenation change for each session was used as the dependent measure. For statistical comparisons, repeated measures analyses of variance (ANOVA), with the Geisser-Greenhouse correction was used with Tukey's post hoc tests to determine the locus of main effects. The significance criterion was 0.05.

RESULTS

For the n-back behavioral data, repeated measures ANOVA indicated significant main effects of task difficulty (0-, 1-, 2- and 3-back conditions) on average correct $\alpha 0.001$). Average correct click ratio and average reaction times for all participants are presented in Figure 1. Tukey post hoc tests showed that for average click ratio ($q_{0.05/2,\ 69} = 3.72$, $p < 0.05$) the 3-back was significantly lower than all other tasks and 2-back was lower than 0-back while reaction time ($q_{e0.052,\ 69} = 3.72$, $p < 0.05$) showed 3-back and 2-back are significantly slower than the other tasks.

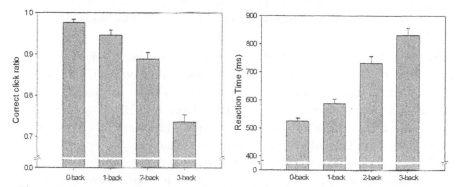

Figure 2. Average performance results for each n-back task. Response time for each condition (left) and average reaction time for each condition (right). Error bars are standard error of the mean (SEM).

For the n-back fNIR data, repeated measures ANOVA showed that average oxygenation changes occurred only at voxel 2 that is close to AF7 in the International 10-20 System, located within the left inferior frontal gyrus in the dorsolateral prefrontal cortex, was significant ($F_{3,69} = 4.37$, $p < 0.05$), see Figure 3. Post hoc analyses confirmed the differences in oxygenation changes as a function of task difficulty with 3-back is larger than the 0- and 1-back tasks ($q_{0.05/2,\ 69} = 3.72$, $p < 0.05$).

28

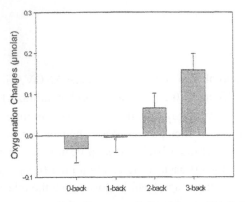

Figure 3. Average oxygenation changes of all subjects (24 Subjects, and 160 trials for each condition) with increasing task difficulty (N-back tasks). Error bars are SEM

For the ATC data, a 2 (Communication: DataComm, VoiceComm) X 2 (PreFrontal Hemisphere: right, left) X 3 (Task Difficulty: 6, 12, 18 aircrafts) ANOVA with repeated measures on all factors was calculated on mean oxygenation. Two subjects (#10, #11) were excluded from the analyses because of high motion artifact and low signal-to-noise ratios. There were only two significant main effects, Task Difficulty denoted by number of aircraft [$F_{2,42} = 4.39$, $p < 0.05$] and Communication [$F_{1,21} = 5.09$, $p < 0.05$] which is depicted in Figure 4. Tukey post hoc tests for Task Difficulty ($q_{0.05/2, 42} = 3.44$, $p < 0.05$) showed than the 18 aircrafts condition (M $\pm$ SD; 0.272 ± 0.586 µmol) had significantly higher oxygenation change than the 12 aircrafts condition (-0.015 ± 0.409 µmol). There were no other significant differences between aircraft conditions.

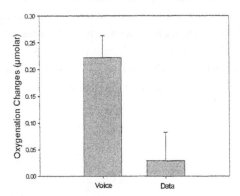

Figure 4. Average oxygenation changes for DataComm and VoiceComm (N=22). Error bars are SEM.

DISCUSSION AND CONCLUSION

In the n-back results, as task difficulty increases, average correct click ratio decreases monotonically and average response time increase monotonically. fNIR results were also sensitive to task difficulty specifically at left inferior frontal gyrus. These are in line with our earlier results (Ayaz, et al., 2009; K. Izzetoglu, et al., 2004) and with the results of fMRI studies that have used the n-back task (Owen, McMillan, Laird, & Bullmore, 2005).

The main hypothesis of the study is that VoiceComm would require more cognitive resources than the DataComm condition. Hence, we would expect higher activation for VoiceComm. The fNIR results from the main effect of communication type (p < 0.05) confirms this hypothesis, is illustrated in Figure 4, with a small to moderate effect size (d=0.36).

In summary, fNIR is a portable, safe, affordable and negligibly intrusive optical brain monitoring technology that can be used to measure hemodynamic changes in the prefrontal cortex. Changes in blood oxygenation in dorsolateral prefrontal cortex, as measured by fNIR, were shown to be associated with increasing cognitive workload. The results further indicate that text-based communications required less brain activation of the operator than legacy voice based communication systems. These fNIR results confirm subjective assessments of operators as reported in earlier studies.

REFERENCES

Ahlstrom, U., & Friedman-Berg, F. J. (2006). Using eye movement activity as a correlate of cognitive workload. *International Journal of Industrial Ergonomics, 36*(7), 623-636. doi: DOI: 10.1016/j.ergon.2006.04.002

Ayaz, H., Izzetoglu, M., Platek, S. M., Bunce, S., Izzetoglu, K., Pourrezaei, K., et al. (2006). Registering fNIR data to brain surface image using MRI templates. *Conf Proc IEEE Eng Med Biol Soc, 1*, 2671-2674. doi: 10.1109/IEMBS.2006.260835

Ayaz, H., Shewokis, P., Bunce, S., Schultheis, M., & Onaral, B. (2009). Assessment of Cognitive Neural Correlates for a Functional Near Infrared-Based Brain Computer Interface System *Foundations of Augmented Cognition. Neuroergonomics and Operational Neuroscience* (pp. 699-708).

Bedny, G., Karwowski, W., & Seglin, M. (2001). A heart rate evaluation approach to determine cost-effectiveness an ergonomics intervention. *International journal of occupational safety and ergonomics: JOSE, 7*(2), 121.

Berka, C., Levendowski, D., Cvetinovic, M., Petrovic, M., Davis, G., Lumicao, M., et al. (2004). Real-time analysis of EEG indexes of alertness, cognition, and memory acquired with a wireless EEG headset. *International Journal of Human-Computer Interaction, 17*(2), 151-170.

Brookhuis, K., & De Waard, D. (1993). The use of psychophysiology to assess driver status. *Ergonomics, 36*(9), 1099-1110.

Brookings, J., Wilson, G., & Swain, C. (1996). Psychophysiological responses to changes in workload during simulated air traffic control. *Biological Psychology, 42*(3), 361-377.

Chance, B., Zhuang, Z., UnAh, C., Alter, C., & Lipton, L. (1993). Cognition-activated low-frequency modulation of light absorption in human brain. *Proc Natl Acad Sci U S A, 90*(8), 3770-3774.

Cooper, G., & Harper, R. (1969). *The use of pilot rating in the evaluation of aircraft handling qualities.* (Report No. TN-D-5153). Washington, DC: NASA.

Embrey, D., Blackett, C., Marsden, P., & Peachey, M. (2006). Development of a Human Cognitive Workload Assessment Tool.

Hah, S., Willems, B., & Phillips, R. (2006). *The effect of air traffic increase on controller workload.* Paper presented at the Human Factors and Ergonomics Society Annual Meeting San Francisco, CA.

Hart, S., & Staveland, L. (1988). Development of NASA-TLX (Task Load Index): Results of empirical and theoretical research. *Human mental workload, 1*, 139–183.

Helander, M. (1978). Applicability of drivers' electrodermal response to the design of the traffic environment. *Journal of Applied Psychology, 63*(4), 481-488.

Izzetoglu, K., Bunce, S., Onaral, B., Pourrezaei, K., & Chance, B. (2004). Functional optical brain imaging using near-infrared during cognitive tasks. *International Journal of Human-Computer Interaction, 17*(2), 211-231.

Izzetoglu, M., Bunce, S. C., Izzetoglu, K., Onaral, B., & Pourrezaei, K. (2007). Functional brain imaging using near-infrared technology. *IEEE Eng Med Biol Mag, 26*(4), 38-46.

Izzetoglu, M., Izzetoglu, K., Bunce, S., Ayaz, H., Devaraj, A., Onaral, B., et al. (2005). Functional near-infrared neuroimaging. *IEEE Trans Neural Syst Rehabil Eng, 13*(2), 153-159. doi: 10.1109/TNSRE.2005.847377

JPDO. (2004). Next Generation Air Transportation System Integrated Plan. Retrieved from http://www.jpdo.gov/library/NGATS_v1_1204r.pdf

Just, M., Carpenter, P., & Miyake, A. (2003). Neuroindices of cognitive workload: neuroimaging, pupillometric and event-related potential studies of brain work. *Theoretical Issues in Ergonomics Science, 4, 1*(2), 56-88.

Kahneman, D. (1973). *Attention and effort.* Englewood Cliffs, NJ: Prentice Hall.

Logothetis, N. K., & Wandell, B. A. (2004). Interpreting the BOLD signal. *Annu Rev Physiol, 66*, 735-769.

Owen, A. M., McMillan, K. M., Laird, A. R., & Bullmore, E. (2005). N-back working memory paradigm: a meta-analysis of normative functional neuroimaging studies. *Human Brain Mapping, 25*(1), 46-59. doi: 10.1002/hbm.20131

Paas, F. G. W. C., & Van Merriënboer, J. J. G. (1993). The efficiency of instructional conditions: An approach to combine mental effort and performance measures. *Human Factors: The Journal of the Human Factors and Ergonomics Society, 35*, 737-743.

Parasuraman, R. (2003). Neuroergonomics: Research and practice. *Theoretical Issues in Ergonomics Science, 4*(1-2), 5-20.

Pleydell-Pearce, C., Whitecross, S., & Dickson, B. (2003). Multivariate analysis of EEG: Predicting cognition on the basis of frequency decomposition, inter-electrode correlation, coherence, cross phase and cross power. *Proceedings of 38th HICCS.*

Roscoe, A. (1992). Assessing pilot workload: Why measure heart rate, HRV and respiration? *Biological Psychology, 34*(2-3), 259-287.

Sheridan, T., & Simpson, R. (1979). Toward the definition and measurement of the mental workload of transport pilots (FTL Rept. R 79-4). *Cambridge, MA: Massachusetts Institute of Technology, Flight Transportation Laboratory.*

Sterman, M., & Mann, C. (1995). Concepts and applications of EEG analysis in aviation performance evaluation. *Biological Psychology, 40*(1-2), 115-130.

Veltman, H., & Vos, W. (2005). Facial temperature as a measure of operator state. *Foundations of Augmented Cognition*, 293.

Veltman, J., & Gaillard, A. (1996). Physiological indices of workload in a simulated flight task. *Biological Psychology, 42*(3), 323-342.

Villringer, A., Planck, J., Hock, C., Schleinkofer, L., & Dirnagl, U. (1993). Near infrared spectroscopy (NIRS): a new tool to study hemodynamic changes during activation of brain function in human adults. *Neurosci Lett, 154*(1-2), 101-104.

Willems, B., Hah, S., & Phillips, R. (2006). The effect of data link on en route controller workload *NJTC Aviation Technologies - "Looking Toward Tomorrow".* Atlantic City International Airport, NJ: FAA William J. Hughes Technical Center.

CHAPTER 4

Prestimulus Alpha as a Precursor to Errors in a UAV Target Orientation Detection Task

Carryl Baldwin[1], Joseph T. Coyne[2], Daniel M. Roberts[1], Jane H. Barrow[1], Anna Cole[3], Ciara Sibley[3], Brian Taylor[1] and George Buzzell[1]

[1] George Mason University
Department of Psychology
Fairfax, VA 22030

[2] Naval Research Laboratory
Washington, DC 20375

[3] Strategic Analysis Incorporated
Arlington, VA 22203

ABSTRACT

Unmanned Aerial Vehicles (UAVs) have become an important component of military aviation operations and skilled UAV operators are a valuable part of this component. Currently there is a need for improved methods of facilitating the development of mission level skills among operators, including target identification and maintenance of navigational awareness. Toward this aim, we examined the extent to which transient neurophysiological states could be used as an index of engagement within a visual detection training paradigm. Participants learned to distinguish stationary indicators of directional change in movement for target tanks located within a complex vehicle formation background. Fast alpha activity (10-13 Hz) one second before targets were presented differed as a function of type of error that would be made and task difficult. Prestimulus alpha shows promise as a candidate metric for on-line monitoring of learner engagement and workload.

INTRODUCTION

Electroencephalographic (EEG) recordings have been used extensively as an index of task engagement and working memory load (Berka et al., 2007; Gevins, Smith, McEvoy, & Yu, 1997; Kerick, Hatfield, & Allender, 2007). For example, increases in frontal midline theta activity (5-7 Hz) and decreases in both slow (7.5-10 Hz) and fast (10-13 Hz) alpha activity are associated with current working memory demands in both spatial and verbal tasks (Gevins et al., 1997; Smith, McEvoy, & Gevins, 1999). Alpha activity is also affected by training and practice, with increased activity associated with increasing skill level on a given task (Smith et al., 1999).

To date, consensus has yet to be reached regarding the best approach for examining spectral changes in EEG recordings (see discussions in Klimesch, Freunberger, Sauseng, & Gruber, 2008; Makeig, Debener, Onton, & Delorme, 2004). While overall changes in alpha and theta range activity have been shown to change with task difficulty, others have argued for examination of spectral changes associated with particular working memory processes or task locked to particular events.

Though macro level changes in EEG activity show promise for a wide variety of applications, considerably less attention has been given to micro level changes (Huang, Jung, Delorme, & Makeig, 2008; Mazaheri, Nieuwenhuis, van Dijk, & Jensen, 2009). Micro level changes have traditionally been examined with event-related potentials (ERPs). While important in many applications, ERPs may provide an index of a relatively small portion of on-going neural activity (Huang et al., 2008; Klimesch et al., 2008). Additionally, ERP extraction techniques require the averaging of neuronal responses time locked to a number of discrete stimuli that may not be present in many real world operational environments (Huang et al., 2008). For these reasons, methods of examining both tonic and phasic fluctuations in neural activity suitable for operational use remain a goal of many neuroergonomics investigations.

Simultaneous monitoring of spectral changes stemming from both relatively long-term or tonic changes in levels of engagement as well as more rapid phasic changes, such as from event related spectral perturbations (ERSPs), show promise for operational neuroergonomics. For example, Sauseng et al. (2005) observed that event related synchronization (ERS) of alpha range activity distinguishes between retention and active manipulation of visuospatial information in working memory. Huang et al. (2008) have observed tonic changes in alpha bandwidth activity coupled with phasic changes in multiple bandwidths during periods of high visuomotor tracking error.

An approach with particular practical significance would be to utilize micro-level or phasic bandwidth changes to predict transient states when an operator might be less engaged in a particular task (i.e., overloaded or distracted) and thus be more likely to be error prone. A recent approach for examining micro level spectral changes shows promise in this regard. Examination of prestimulus alpha, which is spectral activity in the alpha bandwidth occurring immediately prior to the onset of a stimulus, is one such approach. Examination of prestimulus alpha shows promise as a means of predicting when alertness may have temporarily decreased to a point where errors are more probable (Ergenoglu et al., 2004; Mazaheri et al., 2009).

For example, using magnetoencephalographic (MEG) recordings, Mazaheri and colleagues (2009) demonstrated that elevated occipital alpha activity prior to the onset of a visual stimulus predicted whether or not participants would make an error in an upcoming trial. Using EEG, Ergenoglu and colleagues (2004) observed significantly elevated alpha activity in a 1 second prestimulus period when participants missed near threshold visual stimuli relative to when they were detected. The aim of the current investigation was to examine the extent to which prestimulus alpha activity might be used to predict an operators' current level of engagement and thus predict errors before they occur in a challenging visual detection task.

METHODS

Participants

Twenty-two participants (18-28 years, M = 23.27, SD = 2.62) with self reported normal or corrected to normal vision and hearing voluntarily participated in the study after providing informed consent. Participants were further screened for far and near static visual acuity using the Snellen and Rosenbaum eye tests, respectively. The majority of participants had completed at least some college classes. Participants currently enrolled in University courses received partial credit toward a class. Participants recruited from the community were provided with a small amount of financial compensation.

EEG Recording and Analysis Procedures

A Neuroscan NuAmps 40 Channel Amplifier (with Neuroscan 4.4 software) and a 40 channel Neuroscan QuickCap were use to collect EEG data. The EEG signals were band-passed filtered at 1 to 70 Hz and sampled at 500 Hz. The EEGLAB toolbox (Delorme & Makeig, 2004) in conjunction with MATLAB v.2007b (The MathWorks, Natick, MA) were used for analysis of the EEG recordings. After collection, EEG was re-referenced to the average of the left and right mastoid processes, and low-pass filtered at 30 Hz. The 1 second of EEG preceding each behavioral response was subset from the overall recorded EEG, divided according to whether the response was a hit, miss, false alarm, or correct rejection. Any 1 second pre-response epoch that contained activity exceeding +- 75 μV on the ocular channels was rejected due to ocular artifact contamination. The mean log spectrum for the set of remaining epochs of each type was calculated, and the peak dB power in each of three frequency bands (theta: 4-7.5 Hz, slow alpha1: 7.5-10 Hz, fast fast alpha2: 10-13 Hz) was identified separately for EEG preceding each type of behavioral response, in each of the two difficulty levels, at three electrode sites of interest, Fz, Cz, and Pz.

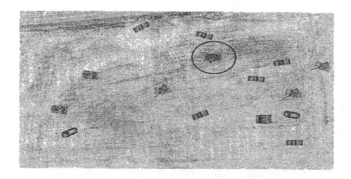

Figure 1: *Example image from the normal difficulty condition (notice that none of the beige tanks are moving in the opposite direction as the non-tank military vehicles. The target is circled in red (notice that it is a green tank, moving in the opposite direction as the non-tank military vehicles.*

Experimental task

Participants performed two difficulty levels (Easy and Hard) of a visual search task that simulated the role of a UAV operator. In both difficulty conditions the target was defined as a green tank heading in the opposite directions of all other non-tank military vehicles (distracters). In the Easy condition, only the target (if present) could be heading in the opposite direction of all other non-tank military vehicles (see Figure 1). However, in the Hard condition there were also other (distractor) beige tanks that could be heading in the opposite direction of all other non-tank military vehicles (see Figure 2). The added variability of these distractor tanks made the task considerably more difficult, as confirmed with pilot testing. This increased difficulty was intended to increase mental workload while participants performed the difficult visual search task.

The experimental task was written and displayed using Microsoft Visual Basic 6 software. Each condition of the task consisted of 200 static images displayed on a 19 inch CRT monitor (Dell M992) for 750 ms each. The interstimulus interval was 1.8 s. Images were generated from a static image consisting of a background desert-like scene obtained from Google maps. Each scene contained 15 military vehicles (i.e., tanks, jeeps, and other vehicles) obtained from a UAV simulator. The position of the vehicles was randomly changed in each scene. Of the 15 military vehicles, a green tank (vehicle of interest-VI) was always present.

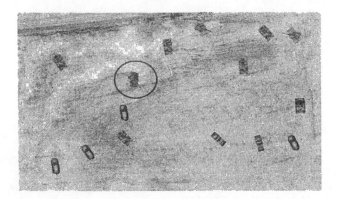

Figure 2: *Example image from the Hard condition (notice that one of the beige tanks is moving in the opposite direction as the non-tank military vehicles. The target is circled in red (notice that it is a green tank, moving in the opposite as the non-tank military vehicles.*

In both conditions, a random variable was used to generate a global direction on a 360 degree axis for all vehicles to face within each generated image. In the Easy condition it was only possible for the green tank to violate this directional display (via random variable) and become a target. However, in the Hard condition it was also possible for the beige tanks to violate this directional display (via random variable). To increase the difficulty of the task in both conditions, an additional random variable allowed for each individual vehicle to deviate from the global direction by 30 degrees. However, as apparent in figures 1 and 2, it is still possible to perceive the global direction in which all non-tank military vehicles are heading. Random variables were also used to generate the color of the non-tank military vehicles (green or beige) as well as their location on the screen.

Procedure

All Individuals were first tested to ensure that they had normal vision as assessed via the Rosenbaum and Snellen metrics. They were then fitted with the Neuroscan QuickCap and it was aligned on the head in accordance with the standard 10-20 system. Standard EEG saline gel was used to ensure a good connection between the electrodes and the scalp and all impedances were measured to be below 5 k ohms. Electrocortical activity was recorded from 15 electrode sites, including midline sites Fz, Cz and Pz, as previous research evidenced their effectiveness as indicators of visual working memory and mental workload (Gevins, 1997, 1999; Ergenoglu et al., 2004; Mazaheri et al., 2009). An in-cap ground located just anterior to Cz was used and all electrodes were referenced to an electrode placed on the left mastoid. However, EEG data from an electrode attached to the right mastoid (also referenced to the left mastoid during recording) was also recorded to allow for an averaged reference of the two mastoids to be computed offline for sites Fz, Cz and Pz. Electrooculogram activity was also recorded with two electrodes, one placed above and below the left eye, in order to detect ocular artifacts.

Participants were then briefed with task instructions and were shown examples of the experimental task and were provided with a short practice session. Following this, individuals completed the two difficulty levels of the task in a counterbalanced order as behavioral and EEG data were recorded. The behavioral data consisted of participants' response accuracy and response time (RT). Responses were categorized within a signal detection framework of Hits (detecting the presence of a orientation change in the VI where there was one), a Miss (failing to detect the orientation change of the VI when there was one), a False alarm (reporting an orientation change of the VI when there was not one) and a Correct Rejection (not reporting an orientation change when in fact there was not one). Participants indicated their response by clicking the mouse when they believed a target was present

RESULTS AND DISCUSSION

Behavioral Data

The number of hits, misses, false alarms, and correct rejections were calculated for each participant in each condition. Next, d' and β scores were calculated. This calculation revealed that 4 participants had d' scores of less than .6 in the Easy condition. These participants were eliminated from all subsequent data analysis. Analysis of the behavioral data for the remaining 18 participants confirmed our difficulty manipulations. Examination of the proportion of hits revealed that participants made significantly fewer hits in the hard detection condition (M = .51, SD = .16) relative to the easy detection condition (M = .74 SD = .13), t(17) = 7.63, p <.001 . Likewise, false alarms (indicating a directional orientation difference for the target tank when one was not present) occurred significantly more often in the hard detection condition (M = .27, SD = .15), relative to the easy detection condition (M = .16, SD = .08), t(17) = -4.3, p <.001. Average d' scores also differed significantly between the easy and difficult conditions, t(17) = 7.16, p <.001, with means of 1.76 and .66, respectively. Average β scores were .82 and .85, respectively.

Prestimulus EEG Analyses

A 2 (task difficulty- easy and hard) x 2 (target presence-yes or no) by 2 (accuracy - correct or incorrect) repeated measures MANOVA was implemented to examine relative power in the theta, slow alpha, and fast alpha bandwidths for the one second period proceeding each stimulus presentation.   Separate MANOVAs were analyzed for each electrode. Due to space limitations only analysis of Pz is presented here, though it should be noted that similar patterns were observed at electrode sites Fz and Cz.   At Pz, a significant multivariate three-way interaction was observed between task difficulty, target presence, and accuracy, $F(3,15) = 5.12$, $p = .01$, partial $\mu^2 = .5$. Univariate analyses revealed that both slow Alpha1, $F(1,17) = 4.6$, $p = .04$, partial $\mu^2 = .21$, and fast Alpha2 $F(1,17) = 5.13$, $p = .03$, partial $\mu^2 = .23$, contributed to the significant multivariate effect. The three-way interaction is depicted graphically in Figure 3 for slow alpha and Figure 4 for fast alpha.

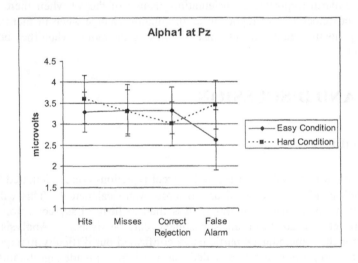

Figure 3: Slow Alpha1 at Pz as a function of Target Presence, Accuracy, and Difficulty Level. Error bars reflect the standard error of the mean.

Prestimulus alpha activity, and particularly fast alpha (10-13 Hz), differed significantly between the types of errors made in the Easy and Hard condition. In the Hard condition, fast alpha activity increased in the one second period immediately prior to a miss, relative to correct detections and also relative to false alarms. A reverse pattern was observed in the Easy condition. In the Easy condition, fast alpha did not differ between correct and incorrect trials when the target was present, but decreased significantly for false alarms. The largest differences in prestimulus alpha activity observed for both slow and fast alpha occurred between False Alarms in the Easy and

Hard conditions. Both fast and slow alpha increased for False Alarms in the Hard condition, but decreased for these same target absent error types in the Easy condition.

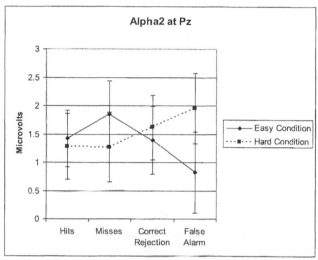

Figure 4: Fast Alpha2 at Pz as a function of Target Presence, Accuracy, and Difficulty Level. Error bars reflect the standard error of the mean.

POTENTIAL APPLICATION

The current results demonstrate potential for using on-line monitoring of phasic changes in alpha bandwidth activity as an index of when an operator may be more error prone or when a learner may be reaching a state where he or she is less likely to benefit from an instructional strategy. The prestimulus alpha activity examined in the present experiment reflected both task difficulty and the type of error likely to be made. For example, if alpha levels increased significantly and participants were making a significant number of False alarms, the present results suggest that there would be a greater than average chance the participant found that task particularly difficult. However, this same pattern of False alarm errors coupled with decreased alpha activity could indicate that the participant had become less engaged in the task or that perhaps the task was not challenging enough. Observation of a reverse pattern coupled with miss-type errors could be used to confirm this interpretation of the data.

This information could potentially be used in conjunction with other algorithms to improve the diagnostic capabilities of an adaptive training paradigm. On-line monitoring of phasic changes in alpha bandwidth activity coupled with performance metrics could be used to provide an indication of when a pedagogical change was needed. If alpha activity was out of range (either above or below tonic limits) no further learning would be expected to occur. Using this information in conjunction with the pattern of behavioral performance observed could be used to distinguish whether to make the learning environment more or less challenging. Further research into the applicability of these results for determining individual differences in learning styles and for use in a neurophysiologically based adaptive training program are currently underway.

ACKNOWLEDGEMENTS

This work was conducted as part of a series of investigations sponsored by the Office of Naval Research, Human Performance, Training & Education Program, Roy Stripling, Program Manager.

REFERENCES

Berka, C., Levendowski, D. J., Lumicao, M. N., Yau, A., Davis, G., Zivkovic, V. T., et al. (2007). EEG correlates of task engagement and mental workload in vigilance, learning, and memory tasks. *Aviation, Space, and Environmental Medicine, 78*(5, Sect II, Suppl), B231-B244.

Delorme, A., & Makeig, S. (2004). EEGLAB: an open source toolbox for analysis of single-trial EEG dynamics including independent component analysis. *Journal of Neuroscience Methods, 134*(1), 9-21.

Ergenoglu, T., Demiralp, T., Bayraktaroglu, Z., Ergen, M., Beydagi, H., & Uresin, Y. (2004). Alpha rhythm of the EEG modulates visual detection performance in humans. *Cognitive Brain Research, 20*(3), 376-383.

Gevins, A., Smith, M., McEvoy, L., & Yu, D. (1997). High-resolution EEG mapping of cortical activation related to working memory: effects of task difficulty, type of processing, and practice. *Cerebral Cortex, 7*(4), 374-385.

Huang, R. S., Jung, T. P., Delorme, A., & Makeig, S. (2008). Tonic and phasic electroencephalographic dynamics during continuous compensatory tracking. *Neuroimage, 39*(4), 1896-1909.

Kerick, S. E., Hatfield, B. D., & Allender, L. E. (2007). Event-related cortical dynamics of soldiers during shooting as a function of varied task demand. *Aviation, Space, and Environmental Medicine, 78*(5), B153-b164.

Klimesch, W., Freunberger, R., Sauseng, P., & Gruber, W. (2008). A short review of slow phase synchronization and memory: Evidence for control processes in different memory systems? *Brain Research, 1235*, 31-44.

Makeig, S., Debener, S., Onton, J., & Delorme, A. (2004). Mining event-related brain dynamics. *Trends in Cognitive Sciences, 8*(5), 204-210.

Mazaheri, A., Nieuwenhuis, I. L. C., van Dijk, H., & Jensen, O. (2009). Prestimulus Alpha and Mu Activity Predicts Failure to Inhibit Motor Responses. *Human Brain Mapping, 30*(6), 1791-1800.

Sauseng, P., Klimesch, W., Doppelmayr, M., Pecherstorfer, T., Freunberger, R., & Hanslmayr, S. (2005). EEG alpha synchronization and functional coupling during top-down processing in a working memory task. *Human Brain Mapping, 26*(2), 148-155.

Smith, M. E., McEvoy, L. K., & Gevins, A. (1999). Neurophysiological indices of strategy development and skill acquisition. *Cognitive Brain Research, 7*(3), 389-404.

CHAPTER 5

Decoding Information Processing When Attention Fails: An Electrophysiological Approach

Ryan Kasper, Koel Das, Miguel P. Eckstein, Barry Giesbrecht

Department of Psychology
Institute for Collaborative Biotechnologies
University of California, Santa Barbara

ABSTRACT

The success of the attentional system in keeping people "on task" in dynamic environments arises from the coordinated operation of multiple neural networks. performance can occur. Here, we investigated the neural bases of attentional failuAlthough this coordinated effort is often successful, errors in res using computational techniques combined with high temporal resolution measures of brain activity using EEG. Attentional failures were induced by presenting two masked targets in rapid succession. In this task, correct identification of the first (T1) leads to impaired identification of the second (T2), a performance failure known as the attentional blink (AB). We applied linear pattern classification algorithms to measures of neural activity acquired during the AB to investigate two key issues about the temporal dynamics of visual attention. First, we tested whether the computational approaches would accurately discriminate the stimulus presented to the observer independent of behavior. Second, we tested whether our computational approaches could predict when the observer would make an error. Our analyses revealed that single-trial EEG activity could be used to not only predict the type of stimulus presented to the observer, but also to predict performance errors. These results are consistent with the notion that the brain represents information about the type of stimuli presented to observers and suggest that computational approaches may be used to provide a moment-by-moment analysis of an observer's attentional state.

Keywords: Attention, Pattern Classification, EEG, ERP, Augmented Cognition

INTRODUCTION

Whether you are a motorist driving on a busy street, an air traffic controller monitoring traffic at an airport, or a high-school student in an algebra class, the ability to selectively maintain one's attentional focus on task-relevant information while ignoring distracting information is vital for good performance. Although effective selective attention helps to keep us on task, errors in performance can sometimes occur because the capacity of the attentional system is limited. While attentional limitations are common in a variety of daily settings, they can be exacerbated by many factors, including learning disabilities, brain pathology or trauma, stress, task context, and individual differences. Thus, understanding the cognitive and neural mechanisms of these attentional limitations will facilitate their amelioration in the clinic, classroom, and the workplace.

The aim of the present work was to investigate the neural mechanisms of attentional failures by combining measures of neural activity and performance acquired during a difficult attention task with computational approaches that allow one to classify patterns of neural activity associated with different stimuli and cognitive states. To address this aim, we focused on one well-studied example of a limitation of the attentional system observed in the lab, known as the attentional blink (AB, Raymond, Shapiro, & Arnell, 1992). The AB is typically observed when two masked targets are presented in a rapid visual sequence. When the first target (T1) is identified, the identification of the second target (T2) is hindered for about 500 ms. There are two key characteristics of the AB that make it a powerful experimental tool for investigating the neural mechanisms of attentional limitations. First, the AB appears to require generalized attentional mechanisms that are capacity (or resource) limited. Consistent with the view that the AB involves generalized attentional systems, neuropsychological and neuroimaging studies have reported that the right hemisphere, which plays a large role in attentional control (e.g., Giesbrecht & Mangun, 2005; Giesbrecht, Woldorff, Song, & Mangun, 2003; Hopfinger, Buonocore, & Mangun, 2000) is also critically involved in the AB (e.g., Giesbrecht & Kingstone, 2004; Marois, Chun, & Gore, 2000; Marois, Yi, & Chun, 2004). Second, EEG studies have demonstrated that a fast, transient, yet robust temporal profile of the AB that provides moment-by-moment behavioral estimate of attentional demands emerges out of a complex pattern of neural dynamics that can be measured using multiple features of the EEG signal (amplitude, power, and phase; e.g., Slagter et al., 2007; Vogel & Machizawa, 2004).

The second component of our approach is the application of machine learning algorithms to measures of brain activity acquired during an AB task. Traditional behavioral and neuroimaging approaches are univariate, such that they typically use a single electrode or the average of a few electrodes rather than combining information neural information across electrodes. Combining neural information across electrodes could potentially provide key information about perceptual, attentional, and high-order cognitive states because they are more likely to be represented in terms of patterns of neural responses that may be best characterized in a multivariate data space. Computer classification algorithms are designed to extract information about patterns that discriminate between classes of information. These algorithms have been applied to both fMRI data and to EEG data to correctly identify the type of visual object shown to the observer (e.g., Haynes & Rees, 2005; Kamitani & Tong, 2005; Philiastides & Sajda, 2006a, 2006b).

We used the AB and pattern classification approaches to investigate the neural mechanisms of attentional failures in the following manner. First, subjects performed an AB task in which a context word was presented prior to a rapid serial visual presentation containing a T2 word that was either related or unrelated to the initial context word. The manipulation of context allowed us to focus on the N400 event-related potential (ERP) component, which has been previously shown to survive the AB despite the behavioral impairment (Giesbrecht, Sy, & Elliott, 2007; Luck, Vogel, & Shapiro, 1996; Rolke, Heil, Streb, & Henninghausen, 2001). Second, we used the ERP amplitudes recorded during this task as inputs into a linear pattern classifier to investigate two questions. First, we predicted that even though performance on the T2 is impaired, the pattern classifier should be able to discriminate the type of stimulus presented to the observer. Second, we predicted that to the extent that performance is represented in patterns of EEG responses, the pattern classifier should discriminate between trials on which the observer was correct versus when they were incorrect.

METHOD

Participants

Thirteen undergraduates from the University of California Santa Barbara were granted course credit or paid $10 per hour.

Procedure

Trials began with the presentation of a context word for 1000 ms, followed by a 750-1250 ms random delay, and then the RSVP stream. The stream consisted of a series of randomized character strings of uppercase black letters, each seven items long (~.8° x 2.5°). Each string was presented for 106 ms with no ISI. Within this stream there were two targets that were presented. T1 was a 7-item number string, all of which were the same parity, while T2 was a red word. T2 words that were not the full seven digits were flanked on either side by the letter X (e.g. XXHATXX). The temporal separation between T1 and T2, or lag, was either 320 ms or 960 ms. After the stream, there was a 750-1250 ms delay, followed by two response probes. The first prompted participants to report whether T1 consisted of odd or even numbers. The second prompted participants to indicate whether T2 was related or unrelated to the initial context word. Responses were untimed and made with a computer mouse. After response, fixation appeared again until the subject initiated the next trial. A schematic trial sequence is shown in Figure 1.

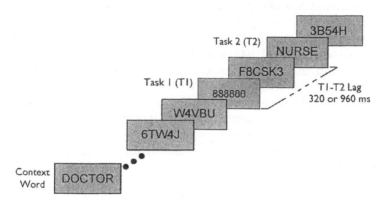

Figure 1. A schematic representation of the sequence of each trial.

Half of the trials contained semantically related context and T2 word pairs, while the other half were unrelated. The words used and the construction of the related and unrelated lists have been used in previous studies (e.g., Giesbrecht et al., 2007).

EEG Recording & Analysis

Recording of EEG was done at 256 Hz from 64 electrodes mounted in an elastic cap and positioned according to the 10/20 system. Electrodes were also placed above and below each eye for the vertical electrooculogram (EOG), as well as 1 cm lateral to the external canthi on each side for the horizontal EOG. The data were re-referenced offline to the average of the signal recorded at the left and right mastoids and then band-pass filtered (.01-100 Hz). The average ERP waveforms in all conditions were computed time-locked to T1 and T2 stimulus onset and included a 200 ms pre-stimulus baseline and 600 ms post-stimulus interval. Trials containing EOG artifacts from eye movements or blinks ($\pm100\mu V$) were rejected from further analysis.

The traditional ERP analyses involved averaging the segmented epochs for each individual in and in each condition. Following previous studies (Giesbrecht et al., 2007), we computed a difference wave that subtracted the response on related trials from the response on unrelated trials. Because the sensory stimulation in each case was exactly the same (only the context word differed), the difference wave reveals the effect of context, uncontaminated by the sensory response. Hypothesis tests were then conducted on the difference waves using a repeated measures ANOVA. The Greenhouse-Geisser correction for the degrees of freedom was used where appropriate. The pattern classification analyses used a standard linear discriminant analysis (LDA, e.g., Fisher, 1936) that computes the best fit linear weights for a set of training trials and then uses these weights to compute a weighted average across the input features for an independent set of test trials. A decision rule was then applied to the result to classify the input patterns. The inputs to the classifier were single trial responses at all 64 electrodes. To reduce the dimensionality, we averaged the single trial responses at each electrode into non-overlapping 20 ms time bins, starting with stimulus onset. The categories to be classified were T2 stimulus type (related vs. unrelated) and trial accuracy (correct vs. incorrect). In each case, the training data consisted of all but 20 of the experimental trials, which were set aside for testing. The performance of the classifier was evaluated using a 10-fold cross-validation scheme, where for each fold a new set of training and test trials were used.

RESULTS

Behavior

Overall performance on the first and second target tasks are shown as a function of T1-T2 lag in Figure 2a. Mean T1 AUC was 0.90 (SEM=0.04), and did not change as a function of temporal lag ($F(1,12)=1.5$, $p>0.23$). Mean T2 AUC was 0.72 (SEM=0.03), but unlike T1 performance, T2 performance showed an effect of lag, such that performance was much worse at the short T1-T2 lag than at the long lag ($F(1,12)=21.4$, $p<0.001$). Direct comparison of performance in the two tasks revealed that performance was impaired in the second task ($F(1,12)=20.2$, $p<0.001$) and that this impairment was worse at the short lags, as indicated by a significant task x lag interaction ($F(1,12)=28.7$, $p<0.001$). The impairment in T2 performance reflects the presence of the AB.

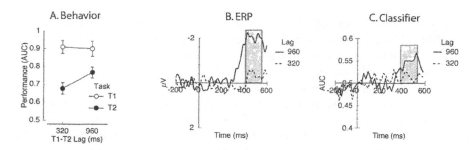

Figure 2. Panel A. Mean AUC on the T1 and T2 tasks plotted as a function of T1-T2 lag. Error bars represent ±1 standard error of the mean. Panel B. Grand average T2-evoked difference ERP wave forms plotted as a function of lag computed from midline electrodes Fz, Cz, and Pz. Shaded region indicates 400-500 ms, the period over which the mean amplitude statistics were computed. Panel C. Mean classifier accuracy based on T2-evoked activity recorded at the same electrodes plotted in B. Shaded regions indicates 400-500 ms time-window.

Electrophysiology

Two analyses of the ERP data were performed. The first focused on analyzing the data based on the type of stimulus presented to the observer. The second focused on analyzing the data based on whether the subject was correct or incorrect.

Analysis of stimulus type

The results of the traditional ERP analysis are shown in Figure 2b. Shown are the average N400 difference waves from midline electrodes (Fz, Cz, Pz) on trials in which T2 was presented 320 ms after T1 and on trials in which T2 was presented 960 ms after T1. Analysis of the mean amplitude over the 400-500 ms time window (highlighted region) revealed a significant N400 at long lags ($t(12)=4.12$, $p<0.002$), but not at short lags ($t(12)=1.19$, $p>0.23$). These results suggest that access to semantic information was suppressed during the AB.

The results of the classification analysis are shown in Figure 2c. The overall pattern over time was qualitatively similar to that of the ERP data. Analysis of the mean classification accuracy over the 400-500 ms time window (highlighted region), paralleled the ERP results such that at the long temporal lag, classifier accuracy was significantly greater than chance ($t(12)=2.60$, $p<0.03$). At the short temporal lag, however, the classifier performance was not reliably different than chance ($t(12)=1.18$, $p>0.26$). These results indicate that information about the semantic relationship between T2 and the context word is represented in patterns of neural activity, even though overall performance is impaired relative to performance on the first task.

While these results are encouraging, it possible that the poor classifier performance during the AB reflects a problem with classifying stimulus information when performance is generally bad (i.e., during the AB). While this is plausible, it is probably unlikely because, even though performance at the 920 ms lag was better than at the 320 ms lag, it was still impaired relative to performance on the first task. Nevertheless, to rule out this possible shortcoming, we applied the same pattern classification analysis to a set of previously published data in which we observed a robust N400 during the AB (Giesbrecht, et al., 2007). Critically, mean classification during the same time window was 0.56 (SEM=0.019), which was significantly greater than chance (t(11)=3.23, p<0.02). Thus, when considered together, pattern classification algorithms can accurately discriminate the type of stimulus presented to the observer during the AB.

Analysis of performance failures

The second main analysis performed on the EEG data focused on classification of trials in which observers correctly discriminated T2 vs. trials in which they did not correctly discriminate T2. This analysis was aimed at whether this classification could be done based on activity evoked by the first target. The rationale was based on two premises: 1) the theoretical notion that the key determinant in triggering the AB is the extent to which subjects allocate attention to T1 and 2) detecting performance failures prior to their occurrence may provide a key advancement for the development of tools for online monitoring of attentional states. Because we focused on activity evoked by T1 and because the 320 ms and 960 ms lag conditions were randomly intermixed, we averaged performance of the classifier across lags. The results of this analysis are shown in Figure 3a, which plots classifier accuracy in separate 20 ms time bins over the first 200 ms of T1-evoked activity. Classifier increased over time and was significantly above chance by 130 ms (mean=0.54, SEM=0.008, t(12)=4.10,p<0.002, corrected for multiple comparisons). To assess the relationship between T1-evoked classifier performance and individual differences in T2 performance, we correlated average T2 performance with the mean of the classifier performance at the time points that survived the Bonferroni corrected test versus change (indicated by the asterisks). The scatterplot shown in Figure 3b shows a significant positive correlation between behavioral performance and classifier performance (r(11)=0.588, p<0.04). These results demonstrate that activity evoked by T1 can be used to discriminate T2-accurate from T2-inaccurate trials prior to the presentation of T2.

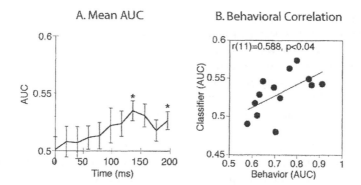

Figure 3. Panel A. Mean classifier AUC discriminating between T2-correct from T2-incorrect trials based on T1-evoked activity recorded at 64 electrodes. Error bars represent ±1 standard error of the mean. Asterisks represent time-points that are significantly different from chance, p<0.05, two-tailed, corrected for multiple comparisons. Panel B. Correlation between mean T2 performance and classifier performance at the time-points that survived the statistical threshold used in A.

DISCUSSION

The aim of the present work was to investigate the neural mechanisms of attentional failures by applying computational learning algorithms to measures of neural activity acquired while participants performed a difficult dual task. Our results demonstrated two key findings. The first key finding was that machine learning algorithms can be used to discriminate between two classes of stimuli presented to an observer during periods when behavioral performance is impaired (i.e., during the AB). While the successful discrimination between stimulus types is likely to be constrained by the inherent differences between the stimulus classes themselves, the finding of successful classification accuracy during the AB is consistent with the notion that information about the external world is represented in patterns of neural activity, even though conscious access to those representations may be impaired (e.g., Haynes & Rees, 2005; Kamitani & Tong, 2005; Philiastides & Sajda, 2006a, 2006b).

The second, and perhaps more important finding for the field of neuroergonomics, is that pattern classifiers can be used to discriminate between trials on which the observer correctly discriminated T2 vs. trials on which observers incorrectly discriminated T2. Critically, successful classification of these two types of trials was based on patterns of neural activity evoked by the T1. In other words, our analyses were able to discriminate between performance failures and successes based on neural responses that occurred more than 200 ms before the imperative stimulus and more than 1 second before the motor response to that stimulus. Moreover, classification accuracy was correlated with individual differences in performance. This finding suggests that pattern classification algorithms combined with continuous measurement of EEG response may be a viable tool for online monitoring of cognitive states in multiple performance contexts so that failures of attention can be detected when, and perhaps even prior to, their occurrence.

The present results converge with studies in the machine learning literature showing successful tracking of shifts of covert attention (Zhang, Maye, Gao, Hong, Engel, & Gao, 2010) and real-time interfacing between the brain and computer in untrained subjects (Blankertz, Losch, Krauledat, Dornhege, Curio, & Muller, 2008). The present finding that performance failures can be predicted before the imperative stimulus is presented suggest that it may be feasible to use similar online algorithms to interface with adaptive systems that monitor the user's attentional state and adjust display parameters to optimize performance. In such a system, the user could be alerted or the task altered based on the attentional load determined from online classification, a process that could prevent human performance errors before they occur.

ACKNOWLEDGEMENTS

This work was generously supported by the Institute for Collaborative Biotechnologies contract no. W911NF-09-0001 from the US Army.

REFERENCES

B. Blankertz, F. Losch, M. Krauledat, G. Dornhege, G. Curio & K.R. Müller. (2008) The Berlin Brain–Computer Interface: Accurate performance from first-session in BCI-naive subjects, *IEEE Transactions on Biomedical Engineering,* 55 (10), 2452–2462.

Fisher, R. A. (1936). The use of multiple measurements in taxonomic problems. *Annals of Eugenics, 7,* 179–188.

Giesbrecht, B., & Kingstone, A. (2004). Right hemisphere involvement in the attentional blink: Evidence from a split-brain patient. *Brain & Cognition, 55*(2), 303-306.

Giesbrecht, B., & Mangun, G. R. (2005). Identifying the neural systems of top-down attentional control: A meta-analytic approach. In L. Itti, G. Rees & J. Tsotsos (Eds.), *Neurobiology of Attention.* New York: Academic Press/Elsevier.

Giesbrecht, B., Sy, J. L., & Elliott, J. E. (2007). Electrophysiological evidence for both perceptual and post-perceptual selection during the attentional blink. *Journal of Cognitive Neuroscience, 19*, 2005-2018.

Giesbrecht, B., Woldorff, M. G., Song, A. W., & Mangun, G. R. (2003). Neural mechanisms of top-down control during spatial and feature attention. *Neuroimage, 19,* 496-512.

Haynes, J. D., & Rees, G. (2005). Predicting the orientation of invisible stimuli from activity in human primary visual cortex. *Nature Neuroscience, 8,* 686-691.

Hopfinger, J. B., Buonocore, M. H., & Mangun, G. R. (2000). The neural mechanisms of top-down attentional control. *Nature Neuroscience, 3*(3), 284-291.

Kamitani, Y., & Tong, F. (2005). Decoding the visual and subjective contexts of the human brain. *Nature Neuroscience, 8,* 679-675.

Luck, S. J., Vogel, E. K., & Shapiro, K. L. (1996). Word meanings can be accessed but not reported during the attentional blink. *Nature, 383,* 616-618.

Marois, R., Chun, M. M., & Gore, J. C. (2000). Neural correlates of the attentional blink. *Neuron, 28*(1), 299-308.

Marois, R., Yi, D.-J., & Chun, M. M. (2004). The neural fate of consciously perceived and missed events in the attentional blink. *Neuron, 41,* 465-472.

Philiastides , M. G., & Sajda, P. (2006a). Neural representation of task difficulty and decision making during perceptual categorization: a timing diagram. *Journal of Neuroscience, 26,* 8965-8975.

Philiastides , M. G., & Sajda, P. (2006b). Temporal characterization of the neural correlates of perceptual decision making in the human brain. *Cerebral Cortex, 16,* 509-518.

Raymond, J. E., Shapiro, K. L., & Arnell, K. M. (1992). Temporary suppression of visual processing in an RSVP task: An attentional blink? *Journal of Experimental Psychology: Human Perception and Performance, 18,* 849-860.

Rolke, B., Heil, M., Streb, J., & Henninghausen, E. (2001). Missed prime words within the attentional blink evoke an N400 semantic priming effect. *Psychophysiology, 38,* 165-174.

Slagter, H. A., Lutz, A., Greischar, L. L., Francis, A. D., Nieuwenhuis, S., Davis, J. M., et al. (2007). Mental training affects distribution of limited brain resources. *PLoS Biology, 5,* 1228-1235.

Vogel, E., & Machizawa, M. G. (2004). Neural activity predicts individual differences in visual working memory capacity. *Nature, 428,* 748-751.

Zhang, D., Maye, A., Gao, X., Hong, B., Engel, A.K., Gao, S. (2010). An independent brain-computer interface using covert non-spatial visual selective attention. *Journal of Neural Engineering, 7,* 1-11.

Chapter 6

Towards Adaptive Automation: A Neuroegronomic Approach to Measuring Workload During a Command and Control Task

Tyler Shaw, Raja Parasuraman, Laura Guagliardo, Ewart de Visser

George Mason University
4400 University Drive
Fairfax, VA 22030, USA

ABSTRACT

Research exploring the possibility of measuring the functional state of the operator to drive the implementation of physiological adaptive aiding has utilized various techniques such as the electroencephalogram (EEG). Transcranial Doppler Sonography (TCD), which has shown promise as an index of cognitive resource utilization in sustained attention tasks provides another method to add to that array of techniques. In the current study, participants performed a command and control simulation under varying levels of task load: a low task load condition in which enemy threats incurred at a steady pace, and a high workload condition in which the number of enemy threats increased rapidly and unpredictably at two points within the scenario. Reaction time to engage and destroy enemies, and the efficiency of protection of a no-fly zone, were superior in the low than in the high load condition. Furthermore, an automated decision aid facilitated better performance in both task load conditions. As the demands of the task increased unpredictably in the high task load condition, cerebral blood flow velocity (CBFV) increased in a similar manner for the first task load transition, but not for the second. Results suggest that the TCD measure may be useful in monitoring the dynamic changes of operator workload in unpredictable environments, but additional studies are needed to validate its use for physiologically-driven adaptive automation.

Keywords: Transcranial Doppler Sonography, Cerebral Blood Flow Velocity, Resource theory, Cognitive Resources, Command and Control.

INTRODUCTION

The design of automation plays a key role in the effective use of automated systems by operators. Well-designed automation produces several benefits, including increased safety, increased reliability, and increased efficiency (Billings, 1997). However, poorly designed automation can result in many of the benefits of automation being offset by human performance costs, such as unbalanced mental workload, reduced situation awareness, and cognitive skill loss (Parasuraman & Riley, 1997). Operator workload is particularly important, because workload levels of operators in complex systems may fluctuate from moment to moment and at different phases of a mission (Inagaki, 2003). It may be possible for operators to perform at very high levels, but this high performance may come at the expense of high mental workload. If the situation that requires a high level of workload lasts for too long, performance degradation may result.

These observations provide a rationale for the evaluation of moment-to-moment workload in operational environments. There is growing interest in *adaptive* automation to change function allocation dynamically during system operations so as to obtain more effective teaming of human and automated systems. Adaptive automation has a long history (Parasuraman, Bahri, Deaton, Morrison, & Barnes, 1982; Rouse, 1988), but empirical research on the efficacy of adaptive automation is more recent (Inagaki, 2003; Scerbo, 2007).

Modeling or measurement of operator functional state is one of the ways in which adaptive automation can be implemented. This measurement can be considered a mental workload-based approach (Parasuraman, Mouloua, & Hilburn, 1999). Mental workload has traditionally been defined as the relation between the cognitive resources available and the cognitive demands of the task (Norman & Bobrow, 1975). Mental workload is often assessed using subjective measures such as the NASA Task Load Index, which requires operators to rate their perceptions of their own workload. There are drawbacks to the use of this procedure. First, subjective measures can intrude with the operator's task, and if subjective workload assessment is delayed to avoid this intrusion, responses could suffer from memory lapses and operator bias (Moroney, Biers, and Eggemeier, 1995). Additionally, subjective measures will only provide workload information *after* the task has been completed, thereby prohibiting knowledge of this information during task performance.

An alternative to subjective ratings may be to use physiological measures of workload to implement adaptive automation. If physiological information about operator cognitive capability is known in real time, it may be possible to adjust the demands of the task to match the functional capabilities of the operator. Along that line, Wilson & Russell (2003) conducted a study in which they monitored various physiological signals during performance on the Multi-Attributed Task Battery (MATB) at two levels of task difficulty. The physiological measures, which consisted of EEG, eye blink, and heart and respiration measures, were used to classify operators' functional state at these different levels of task load. After training participants to stable performance on the MATB, participants performed three 5-min testing trials which consisted of a baseline condition, a low task load condition, and a high task load condition. The physiological data from the three conditions were then input to an Artificial Neural Network (ANN) classifier that was trained to recognize the three separate conditions. Results of this study showed that the ANN was able to classify performance at rates above 85%.

In a later study, Wilson and Russell (2007) implemented the ANN classifier into a system that provided physiological adaptive aiding during performance on an unmanned aerial vehicle (UAV) task at different levels of cognitive difficulty. The task used was a complex, simulated, UAV attack scenario in which each operator controlled four UAVs and was required to locate and designate targets. The most difficult level of the task was individually determined for each individual operator using a titration procedure in which the speed of the UAV's was increased until the operator could only successfully complete 25% to 30% of the required task. The point at which this level of performance occurred was designated as a "cognitive overload" marker, and physiological data associated with that marker were used to determine when adaptive aiding should be implemented. When the aiding was implemented, some critical task functions were automated. A 50% improvement in performance was observed under conditions in which participants received physiologically adaptive aiding.

Research in this area has clearly met with great success, but the EEG and ERP indices of mental workload used often require extensive "training" of artificial neural network EEG classifiers within each individual. Additionally, it may be necessary for this extensive training to be conducted every time adaptive automation is to be implemented, due largely to the lack of knowledge regarding whether the ANN classifiers are robust enough to take into account day to day variability in operator functional state. If the implementation of physiological measures in operational settings is to be achieved, it is necessary to explore other indices of workload that may provide complementary information and avoid some of these limitations. Transcranial Doppler Sonography (TCD), a physiological measure that has been used recently in studies of information processing resource utilization, can meet that need.

TCD is a non-invasive and relatively inexpensive ultrasound method used to monitor cerebral blood flow velocity (CBFV) in the main stem intracranial arteries (e.g. middle cerebral artery). The logic underlying TCD is that when an area of the brain becomes metabolically active, such as in the performance of mental tasks, byproducts of this activity (e.g. CO2) increase. This leads to a frequency shift in the amount of oxygenated hemoglobin delivered to the brain to remove this "waste" product. The advantage of the TCD system is that the measurement of attentional resource utilization is carried out by examining relative changes during task performance; changes in cerebral hemodynamics during task performance are compared to a 5-10 minute resting baseline phase (Aaslid, 1986).

Research using the TCD procedure in studies of vigilance or sustained attention has shown that TCD is useful in providing a metabolic index of mental workload during task performance. Along this line, a series of studies featuring CBFV measurements in the right and left middle cerebral arteries has indicated that the vigilance decrement, the decline in signal detections over time that typifies vigilance performance, is paralleled by a decline in CBFV (e.g. Hitchcock et al., 2004; Shaw et al, 2008, Shaw et al., 2009). In addition, the absolute level of CBFV in these studies was directly related to the cognitive demands of the vigilance task, and the overall effects are lateralized to the right cerebral hemisphere, consistent with PET and fMRI studies indicating a right-hemispheric system in the functional control of vigilance performance. These findings have been interpreted in terms of a cognitive resource model, in which there is an expenditure of information-processing resources that are not replenished over time (Davies & Parasuraman, 1982). The close coupling between the CBFV measure and vigilance performance suggest that CBFV may provide an index of the degree of utilization of information processing resources.

While the TCD and vigilance relationship is well documented, it is still not known how CBFV will manifest in a more complex decision making task. Moreover, it is unknown if the TCD measure is sensitive to unpredictable transitions in task load. The goal of this study was to extend the information-processing resource model applied to the study of vigilance to studies that involve more complex decision making tasks. Therefore, a command and control simulation involving the functional control of UAVs to destroy enemy threats was used. To simulate moment to moment variations in task load, one condition (high task load) was employed that featured sudden increases in the presence of enemy threats. This condition was compared to a condition that contained no task load variation, and enemy threats incurred at a steady rate (low task load). Additionally, an automated decision aid was provided that was designed to prompt observers to incoming enemy threats, with the expectation that this would attenuate operator workload. It was predicted that the sudden increase in enemy threats would increase workload and that this increase could be indexed by the TCD measure. More specifically, we expected an increase in CBFV at points of transition of workload. It was also predicted that the TCD measure should reflect any mitigating effects the automated decision aid may have on operator workload.

Thirteen undergraduate students (5 male and 8 female) participated in this study. All participants were given course credit for their participation. Participants completed four experimental trials in total that were presented in random order. The experiment was a 2 × 2 repeated measures design defined by the factorial combination of two levels of task load (low, high) and two levels of the automated decision making aid (present, absent). In conditions where participants received the decision making aid, participants received messages 20-30 seconds before an enemy appeared in the simulation. Task load conditions were differentiated by the number of enemy threats that were presented in each trial. In the low task load condition, enemies were presented at the rate of a 1.5 per minute, on average, with 15 total enemy threats being presented in each 10 minute trial. In the high task load conditions, enemies were presented at a rate of 3.5 per minute, on average, with 35 enemy threats in total, with the most enemy threats presenting at minutes 4 and at minute 7. This increase was somewhat gradual, as the number of threats increased at 3 enemies per minute during periods of transition.

The Dynamic Distributed Decision making (DDD) 4.0 simulation, developed by Aptima Inc., was used in this experiment. Participants performed the DDD scenario using a Desktop computer with windows XP installed that was connected to a 32 inch screen. The six UAV assets were controlled with a mouse.

The DDD task scenarios were 10-minute simulated counter air operation in which enemy targets entered the green zone and immediately began moving towards friendly territory, or the red zone. Enemy forces had the ability to attack and destroy all friendly UAV fighter assets. Participants presided over six friendly assets located inside of a no-fly zone that was displayed in the center of the scenario at the start of the simulation. Neutral and enemy assets could approach this red zone from varying directions. Participants were instructed to perform three tasks during the experimental session: 1) defend against enemy encroachment into the red zone, 2) destroy enemy assets as quickly as possible, and 3) protect own assets from damage and destruction.

Participants were welcomed into the experimental suite and provided with informed consent. They then participated in computer based training that involved instructions via a PowerPoint presentation that detailed the workings of the DDD simulation and the experimental tasks. Participants learned how to select and move an asset as well as how to attack an enemy. They were also instructed about how to score points, and the scores were used to provide the participant with performance feedback. Participants then performed three practice trials to familiarize them with the simulation, as well as the decision making aid. After these trials, the participants were ready to begin the experiment in which they performed four randomized trials, each 10 minutes long.

Prior to the start of each trial, participants were hooked up to a DWL/Multi-Dop X4 TCD unit equipped with two 2-MHz ultrasound transducers. Hemovelocity measurements in cm/sec were taken bilaterally from the right and the left MCAs by means of the transducers, embedded in a plastic bracket, and were secured to the observer's head by an adjustable plastic strap. The transducers were placed dorsal and immediately proximal to the zygomatic arch along the temporal bone. A small amount of Aquasonic-100 brand ultrasound transmission gel (Parker Laboratories, NJ, USA) was placed on the transducers to ensure transmission of the ultrasound signal. The distance between the transducer face and the sample volume could be adjusted in 2 mm increments in order to isolate the MCA. In the present study, the MCA was monitored at depths of 45-55 mm (Aaslid, 1986). Blood flow velocity measures were averaged and displayed by the TCD unit approximately once every second for recording. These values were extracted and channeled into a personal computer for subsequent data analysis. After the participant was hooked up to the TCD unit, the unit was left in place for the entire duration of the experiment as to ensure the integrity of the ultrasound signal. After completion of the 4 experimental trials, participants were thanked for their time and told to kick rocks. The entire experiment lasted for approximately 2 hours and 30 minutes.

To characterize operator performance, two metrics were used: Percentage of enemy incursions into the red zone, and reaction time to destroy enemies. In both cases, lower numbers indicate better performance. A 2 × 2 analysis of variance (ANOVA) conducted on the percentage of enemies allowed into the red-zone revealed a significant effect for task load, $F(1, 12) = 14.65$, $p < .01$, with fewer enemies permitted into the red zone for the low task load ($M = 23.59\%$, $SE = 3.85\%$) than the high task load ($M = 43.94\%$, $SE = 3.95\%$). None of the other sources of variance in the analysis of red-zone safety attained significance. A similar analysis based upon the reaction time to destroy enemies revealed a significant main effect for task load, $F(1,12) = 44.16$, $p < .001$, and a significant main effect for the presence of the decision making aid, $F(1,12) = 7.09$, $p < .05$, and these effects are shown graphically in Figure 2. It can be seen in the figure that participants in the low task load condition were faster to destroy enemies ($M = 51.83s$, $SE = 6.40s$) than in the high task load condition ($88.78s$, $SE = 3.15s$). In addition, conditions in which participants received the decision making aid were faster to destroy enemies ($M = 62.78s$, $SE = 4.41s$) than in conditions where the decision making aid was not present ($M = 77.82s$, $SE = 5.66$). The interaction between task load and the decision making aid did not attain significance.

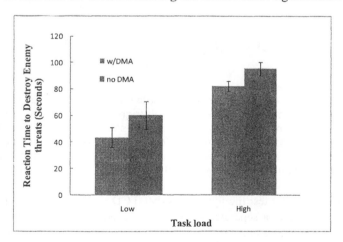

Figure 1. Reaction time to destroy enemies plotted as a function of two levels of task load (low, high) with the decision making aid (DMA) and without. Error bars are standard error.

To account for the wide range of hemovelocity scores present in the population, the scores for all observers were expressed as a proportion of the last 60-sec of their 5-min resting baseline recording (Warm & Parasuraman, 2007). Baseline recordings were taken prior to the start of each experimental trial, and CBFV data on the subsequent experimental trial was taken relative to that baseline. A 2 (hemisphere) ×2 (task load) ×2 (DMA) ×10 (minute blocks) ANOVA was conducted on the average CBFV scores. The ANOVA revealed a significant effect for periods $F(9, 108) = 6.396$, $p < .001$, and a significant task load x periods interaction, $F(9, 108) = 3.061$, $p < .05$. The task load × periods interaction can be viewed graphically in Figure 2. For reference, the number of enemy threats presented in both the low workload and high workload trials are displayed in the bottom panels. It can be seen in the figure that CBVF scores were highest after the initial task load transition point of transition.

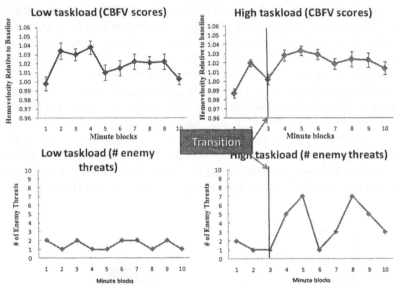

Figure 2. Top Panels: Hemovelocity scores relative to baseline plotted as a function of 10 1-min blocks. Data for Low (left) and High (right) task load conditions are presented separately in each panel. Error bars are standard error. Bottom panels Number of enemies presented as a function of 10-1 min blocks. Data for Low (left) and High (right) task load conditions are presented separately in each panel. The initial point of transition for the high workload trial is indicated by the red line.

To specifically test the hypothesis that TCD may reflect moment-moment variations in workload, a separate ANOVA was conducted on only the high task load condition, as that was the condition in which the task load transition occurred. Data for the 10 minute block for both hemispheres was averaged across three phases: a pre-switch phase (1st three minutes), post-switch transition 1 (minutes 4-6), and post-switch transition 2 (minutes 7-9). A 2 (hemisphere) × 3 (transition phase) ANOVA conducted on these data revealed a significant main effect for transition phase, $F(2, 24) = 6.14$, $p < .01$, such that transition phase 1 ($M = 1.03$, $SE = .01$) was greater than the pre-switch phase ($M = 1.00$, $SE = .01$) and transition phase 2 ($M = 1.021$, $SE = .02$). The main effect for hemisphere did not attain significance, nor did the hemisphere x phase interaction. These data can be viewed graphically in Figure 3. Post hoc- tests revealed that post-switch 1 was significantly greater than pre-switch, post-switch 2 was significantly greater than pre-switch, but post-switch 1 and post-switch 2 did not differ significantly from each other.

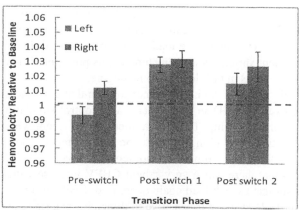

Figure 3. Hemovelocity Relative to baseline in the high workload trial for the 3 task load transitions presented for both cerebral hemispheres. The dashed line represents baseline. Error bars are standard error.

This study was designed to determine the extent to which blood flow velocity, as measured by TCD, is sensitive to moment to moment variations in operator workload. A decision making aid was provided in some trials to attenuate workload by serving as a cue as to the arrival of enemy threats, and it was hypothesized that this workload attenuation could also be reflected in the CBFV measure. Results from the performance analysis revealed that the task load manipulation played a role in operators' ability to perform the necessary tasks in DDD simulation. Both red zone safety and reaction time was better in the context of the low than in the high task load condition. Additionally, the automated decision making aid speeded reaction times in both low and high task load conditions. These findings are consistent with previous literature suggesting that automation support in the form of cueing aids can have beneficial effects on human performance (Wickens et al. 2000).

While the decision-making aid was able to enhance performance, there were no such effects of the decision making aid found for CBFV. One possibility as to why this occurred could be the large amount of variability that was present in this sample. Yet another possibility is that the CBFV measure may not have the specificity to detect changes at the levels of workload specified in the current study. While the decision making aid did facilitate better performance in this case, it may be possible that the demands of the task were too far at the extremes (too low and too high) that the attenuating effect that the decision making aid could have on cognitive load was too subtle. Future studies could examine the benefits of a decision making aid on CBFV at differing levels of operator workload.

The CBFV data regarding the workload transition suggest that the TCD measure was in fact sensitive to the unpredictable increase in task load. The finding that CBFV increased with increasing task load is consistent with the idea that the CBFV measure can be used as an index of information processing resources (e.g. Warm & Parasuraman, 2007). In other words, as the task demands of the DDD task increased, operators' were required to expend more cognitive resources on the task, and this expenditure is reflected in the TCD measurement. However, this finding occurred for the first task load transition and not the second. This may be due to a carryover effect of cognitive load. Thus, to say that TCD is highly specific to moment to moment variations is premature at this point, but it is clear that TCD is sensitive to phase shifts in task load to some degree.

This study provides the initial demonstration that dynamic changes in task load may be indexed by TCD in a command and control simulation. While this study shows promise in using TCD as a measure of cognitive resources in a command and control simulation, it should be noted that it is not clear whether TCD can be used to detect moment to moment workload variations. Perusal of Figure 3 suggests that there was a very salient CBFV increase with the initial task load transition, but that as task load again decreased, CBFV did not. Nevertheless, it is clear that the TCD measure is sensitive to dynamic changes in mental workload on some level, and this measure shows promise as being an alternative or supplement to current measures of real-time physiological workload assessment. Future studies should examine the extent to which CBFV varies with workload in longer task with more transitions, and with more time in between transitions. Additionally, future studies should more closely link the temporal associations between blood flow velocity and performance to determine if there are any cognitive load markers in the CBFV measure.

REFERENCES

Aaslid, R. (1986). Transcranial Doppler examination techniques. In R. Aaslid, (Ed.), *Transcranial Doppler Sonography* (pp. 39-59). New York: Springer-Verlag.

Billings, C. E. (1997). *Aviation automation: The search for a human centered approach*. Mahwah, NJ: Erlbaum.

Davies, D. R., & Parasuraman, R. (1982). *The psychology of vigilance*. London: Academic Press.

Hitchcock, E.M., Warm, J.S., Matthews, G., Dember, W.N., Shear, P.K., Tripp, L.D., Mayleban, D.W., & Parasuraman, R. (2003). Automation cueing modulates cerebral blood flow and vigilance in a simulated air traffic control task. *Theoretical Issues in Ergonomics Science, 4,* 89-112.

Inagaki, T. (2003). Adaptive automation: Sharing and trading of control. In Erik Hollnagel (Ed.) *Handbook of Cognitive Task Design* (pp.147-169). Mahwah, NJ: Lawrence Erlbaum Associates.

Moroney, W. F., Biers, D. W., & Eggemeier, F. T. (1995). Some measurement and methodological considerations in the application of subjective workload measurement techniques. *International Journal of Aviation Psychology, 5,* 87-106.

Norman, D.A., & Bobrow, D.J. (1975). On data-limited and resource-limited processes. *Cognitive Psychology, 7,* 44-64.

Parasuraman, R., Bahri, T., Deaton, J. E., Morrison, J. G., & Barnes, M. (1992). *Theory and design of adaptive automation in aviation systems.* (Technical Report, Code 6021).Warminster, PA: Naval Air Development Center.

Parasuraman, R., Mouloua, M., & Hilburn, B. (1999). Adaptive aiding and adaptive task allocation enhance human-machine interaction. In M. W. Scerbo & M. Mouloua (Eds.) *Automation Technology and Human Performance: Current Research and Trends.* (pp. 119-123). Mahwah, NJ: Erlbaum.

Parasuraman, R., & Riley, V. (1997). Humans and automation: Use, misuse, disuse, abuse. *Human Factors, 39,* 230–253.

Rouse, W. (1988). Adaptive interfaces for human/computer control. *Human Factors, 30,*431–488.

Scerbo, M. (2007). Adaptive automation. In R. Parasuraman & M. Rizzo (Eds.), *Neuroergonomics: The brain at work* (pp. 238 252). New York: Oxford University Press.

Shaw, T.H., Parasuraman, R., Sikdar, Siddhartha, & Warm, J.S. (2009). Knowledge of Results and Signal Salience Modify Vigilance Performance and Cerebral Hemovelocity. *Proceedings of the Human Factors and Ergonomics Society, USA, 53,* 1062-1065.

Shaw, T. H., Warm, J. S., Finomore, V. S., Tripp, L., Matthews, G., Weiler, E., & Parasuraman, R. (2009). Effects of sensory modality on cerebral blood flow velocity during vigilance. *Neuroscience Letters, 461,* 207-211.

Warm, J.S., & Parasuraman, R. (2007). Cerebral hemovelocity and vigilance. In R. Parasuraman & A.M. Rizzo (Eds.), *Neuroergonomics: the brain at work.* Cambridge, MA: MIT Press.

Wickens, C. D., Gempler, K. and Morphew, M. E., (2000). Workload and reliability of predictor displays in aircraft traffic avoidance. *Transportation Human Factors, 2,* 99-126.

Wilson, G.F. & Russell, C.A. (2003). Real-time assessment of mental workload using psychophysiological measures and artificial neural networks. *Human Factors, 45,* 635-643.

Wilson, G.F., & Russell, C.A. (2007). Performance enhancement in an uninhabited air vehicle task using psychophysilogically determined adaptive aiding. *Human Factors, 49,* 1005-1018.

Chapter 7

A Predictive Model of Cognitive Performance Under Acceleration Stress

Richard A. McKinley[1], Jennie Gallimore[2], Lloyd Tripp[1]

[1]Air Force Research Laboratory, OH

[2]Wright State University, OH

ABSTRACT

Extreme acceleration maneuvers encountered in modern agile fighter aircraft can wreak havoc on human physiology thereby significantly influencing cognitive task performance. Increased acceleration causes a shift in local arterial blood pressure and profusion causing declines in regional cerebral oxygen saturation. As oxygen content continues to decline, activity of high order cortical tissue reduces to ensure sufficient metabolic resources are available for critical life-sustaining autonomic functions. Consequently, cognitive abilities reliant on these affected areas suffer significant performance degradations.

This goal of this effort was to develop and validate a model capable of predicting human cognitive performance under acceleration stress. An Air Force program entitled, "Human Information Processing in Dynamic Environments (HIPDE)" evaluated cognitive performance across twelve tasks under various levels of acceleration stress. Data sets from this program were leveraged for model development and validation.

Development began with creation of a proportional control cardiovascular model that produced predictions of several hemodynamic parameters including eye-level blood pressure. The relationship between eye-level blood pressure and regional cerebral oxygen saturation (rSO2) was defined and validated with objective data from two different HIPDE experiments. An algorithm was derived to relate changes in rSO2 within specific brain structures to performance on cognitive tasks that require engagement of different brain areas. Data from two acceleration profiles (3 and 7 Gz) in the Motion Inference experiment were used in algorithm development while the data from the remaining two profiles (5 and 7 Gz SACM) verified model predictions. Data from the "precision timing" experiment were then used to validate the model predicting cognitive performance on the precision timing task as a function of Gz profile. Agreement between the measured and predicted values were defined as a correlation coefficient close to 1, linear best-fit slope on a plot of measured vs. predicted values close to 1, and low mean percent error. Results showed good overall agreement between the measured and predicted values for the rSO2 (Correlation Coefficient: 0.7483-0.8687; Linear Best-Fit Slope: 0.5760-0.9484; Mean Percent Error: 0.75-3.33) and cognitive performance models (Motion Inference Task - Correlation Coefficient: 0.7103-0.9451; Linear Best-Fit Slope: 0.7416-0.9144; Mean Percent Error: 6.35-38.21; Precision Timing Task - Correlation Coefficient: 0.6856 - 0.9726; Linear Best-Fit Slope: 0.5795 - 1.027; Mean Percent Error: 6.30 - 17.28). The evidence suggests that the model is an accurate predictor of cognitive performance under high acceleration stress across tasks, the first such model to be developed. Applications of the model include Air Force mission planning, pilot training, improved adversary simulation, analysis of astronaut launch and reentry profiles, and safety analysis of extreme amusement rides.

Keywords: Acceleration; Cognitive Model; G; Cognition; Cerebral Oxygen Saturation; Agile Aircraft; Human Performance; Neuroergonomics

INTRODUCTION

Human cognition in the traditional sense encompasses such mental processes as thought, perception, problem solving, and memory. The complex flight environment coupled with a multitude of modern cockpit displays and auditory cueing often challenges each of these processes while simultaneously taxing the senses and generating periods of high mental workload. The addition of inertial forces generated during tight turns, steep climbs, and evasive maneuvers further exacerbates cognitive disarray in the most critical segments of flight which inevitably increases the risk of mission failure. Decades of research in acceleration physiology have provided significant evidence of the underlying cause of cognitive impairments during high-G maneuvers.

The amount of available oxygenated blood in the cerebral tissue likely drives and/or limits cognitive ability. In fact, previous work has suggested that decreases in eye-level blood pressure and cerebral oxygen saturation (rSO2) lead to decreased motor function and cognitive ability (Ernsting, Nicholson, and Rainford, 1999; Newman, White, and Callister, 1998; Tripp, Chelette, and Savul, 1998). It is likely that these deficits are caused primarily by a global lack of metabolic resources available to the cortical tissues during high-G maneuvers. These resulting deficits can seriously impede many aspects of the pilot's cognition resulting in reduced capability and higher risks of mission safety.

As a first step toward the realization of a human cognitive performance model capable of making accurate predictions during simulated G_z acceleration, a program entitled the "Human Information Processing in the Dynamic Environment"

(HIPDE) was initiated by the Air Force Research Laboratory (AFRL) (McKinley, et al, 2008). The program began with the development of a custom cognitive performance task battery to probe specific cognitive functions needed in the flight environment. These included relative motion, precision timing, motion inference, pitch/roll capture, peripheral information processing, rapid decision-making, basic flying skills, gunsight tracking, situation awareness, unusual attitude recovery, short-term memory, and visual monitoring. A complete review of this program, task descriptions, and results can be found in an Air Force Technical report published by McKinley, et al. (2008). The effort described in this paper used data from two of the tasks within the HIPDE program to develop and validate the cognitive model: motion inference and precision timing. The following section provides a description of the brain areas engaged in execution of the task and a discussion of their location within the brain.

PRECISION TIMING AND MOTION INFERENCE

The general consensus is that timing information is processed in both the cerebellum and the prefrontal cortex (Mangels et al., 1998; Nichelli, et al., 1996). To further define the roles of each brain area, Mangles and colleagues (1998) compared patients with lesions in neocerebellar regions to those with prefrontal cortex lesions on timing performance. The results showed that the patients with neocerebellar lesions performed significantly worse for trials with short duration (millisecond and second), whereas patients with prefrontal cortex lesions exhibited poor performance with long duration trials (Mangels et al., 1998). Fraisse (1984) and Mangels et al. (1998) concluded that this fundamental difference in apparent function is a direct result of the need for the aid of memory in the perception of time over long durations. The cerebellum is often referred to as the internal clock of the human body and is largely responsible for circadian rhythms and time interval perception. However, long duration time perception (more than a few seconds) is more than the cerebellum can handle alone and must engage the working memory functions in the prefrontal cortex to maintain awareness of the stimuli and track its progression. This theory is further supported with a study by Nichelli, et al. (1996) suggesting the cerebellum was responsible only for shorter duration time interval processing. Because the precision timing task required only a couple seconds to complete each trial and the target remained visible continuously, it is believed that the cerebellum is the primary brain structure engaged in perception of timing within this task. However, the motion inference task required subjects to *remember* the velocity of the moving target and predict when it would intersect a target point on the display. Additionally, it required at least 5 seconds to complete each presentation. Taken together, these statements indicate the motion inference task required engagement of short-term memory. Rubia & Smith (2004 as cited by McKinley, et al., 2008) mention the prefrontal cortices of the brain in both hemispheres may "have the function of a hypothetical accumulator within an internal clock model" for tasks lasting more than a few seconds.

METHODS

The model includes both cardiovascular and neurophysiological elements that combine to make predictions about cortical function during acceleration stress and the corresponding impact on specific cognitive tasks. The first step in the development of the model was to design the basic underlying structure for the hemodynamic portion of the model. A full description is provided in the following section.

CARDIOVASCULAR MODEL

Because neural cells rely almost exclusively on aerobic metabolism of glucose for energy production, the brain demands a large amount of fresh, oxygenated blood to function properly. As a result, the primary variable in the model is the ability of the heart to deliver oxygen-rich red blood cells to the dorsal regions of the brain and maintain adequate cerebral perfusion, which is principally dictated by systolic blood pressure. Blood pressure can be quantified using the below equations:

$$P_A = Q \times R \qquad \textbf{\textit{(Equation 1)}}$$

$$Q = HR \times SV \qquad \textbf{\textit{(Equation 2)}}$$

where P_A is arterial systolic blood pressure, Q is cardiac output, R is total vascular resistance, HR is heart rate, and SV is stroke volume. In keeping with the theory that human operator control systems can be modeled with proper application of conventional control theory, a negative feedback, proportional control system was selected to model the eye-level arterial blood pressure under high acceleration stress. A block diagram of the final model is presented in the figure 1.1.

The feedback mechanism includes baroreceptor and mechanoreceptor reflexes in the carotid sinus that sense pressure fluctuations. As errors between current and required eye-level blood pressure are sensed and fed back to the central nerve processor (CNP), corrective hemodynamic actions begin to initialize in the form of increased heart rate and increased peripheral vascular resistance (through increasing vasoconstriction) (Salzmann & Leverett, 1956). However, the elevated inertial pressure due high acceleration causes a shift in blood volumes in the direction of the acceleration vector (Grygoryan, 1999). The reduced blood return to the heart correspondingly results in a decline of the end diastolic volume. Because stroke volume is simply the difference between the end diastolic volume and the end systolic volume, stroke volume will suffer decreases as well (Jennings, et. al., 1990). Each of the three variables is modeled with a proportional control algorithm that is dependent on the pressure error feedback signal from the baroreceptor and mechanoreceptor reflexes.

The arterial system can be approximated as a vertical hydrostatic column of blood, which permits the use of Pascal's Law (Eqn. 3), where "P" is the blood pressure, "ρ" is the blood density, "g" is the acceleration due to gravity, and "h" is the height of the arterial column.

$$P = \rho g h \qquad \textit{(Equation 3)}$$

To calculate the change in pressure gradient between the heart and eye at any point in time, it was necessary to first define the initial pressure drop from heart to eye level. This calculated gradient is the additional required systolic pressure required at the heart to overcome the increased apparent pressure generated by the increase in G_z acceleration, annotated as P_R. This is contained in element G_6 of the model diagram and defined by equations 4-7, where P_{Hi} is the initial systolic blood pressure at the heart.

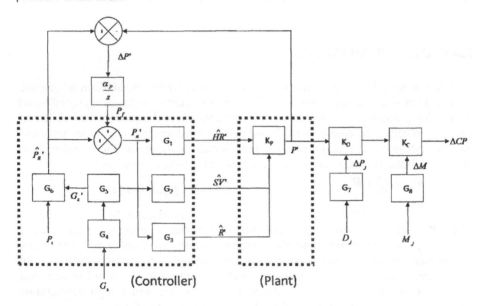

FIGURE 1.1 Human Information Processing Model Block Diagram

$$P_i = \rho \cdot g(1G_z) \cdot h_{eye} \qquad \textit{(Equation 4)}$$

$$P(t) = \rho \cdot g(t) \cdot h_{eye} \qquad \textit{(Equation 5)}$$

$$\Delta P(t) = P(t) - P_i \qquad \textit{(Equation 6)}$$

$$P_R(t) = P_{Hi} + \Delta P(t) \qquad \textit{(Equation 7)}$$

Although the above equations provide a method for calculating the approximate required blood pressure, current technology and U.S. Air Force pilot training methods provide countermeasures that assist the heart in generating compensating pressure. The first aid is the standard G-suit that squeezes blood, pooling in the legs, back up toward the heart. Accordingly, venous blood return to the heart is improved (decreases attenuated) which translates to higher stroke volumes and increased systolic blood pressure when compared to subjects without G-suit inflation, but only for a short duration (~6-12 seconds) (Tripp, et al, 1994).

The second method of reducing the physiological costs of high acceleration is a technique known as the anti-G straining maneuver, which incorporates a series of Valsalva maneuvers (Jennings, et al., 1990). Although effective in short durations, this constant contraction is physically demanding and eventually succumbs to muscle fatigue and loses its potency.

Due to the transitory nature of acceleration countermeasures, the influence of the G-suit and anti-G straining maneuver was implemented using an equation describing the "effective G_z" on the human body that asymptotically declines over time. For example, if the standard G-suit provides +1 G_z of protection, and the G_z at time "t" is +5 G_z, the "effective G" on the human operator is only +4 G_z. This effect was modeled with element G_5 and defined by equation 8.

$$G_z' = \left(0.32 + 0.68 \left(1 - \sin\left(\pi \cdot \frac{t'}{2} \right) \right) \right)$$ (Equation 8)

Substituting G_z' into equations 4 yields the following:

$$\hat{P}(t) = \rho \cdot G_z'(t) \cdot h_{eye}$$ (Equation 9)

This augmented pressure gradient value is then fed into equation 6 and 7 to yield the required pressure now denoted as $\hat{P}_R'$.

Analogous to the isometric control model developed by Phillips (2000), element G_4 represents a model component that generates a normalized time (t') in relation to the total time to fatigue (T_F) for the human operator and the time to failure of the G-suit (T_F). The equation for this element was developed by Phillips (2000) and can be found in equation 10. For this modeling effort, T_F was set to 55 seconds.

$$t' = \frac{\Delta t}{T_F} \sum_{j=1}^{m} j$$ (Equation 10)

Element G_1 provides the proportional control equation that is driven by the signal error between the required heart-level systolic pressure to maintain the 1 G_z eye level blood pressure and the current systolic heart-level blood pressure (P_e). Governing heart rate control function is displayed in equation 11 where HR_i denotes the initial heart rate. Here, the value was set to 70 beats/min.

$$HR = 5(P_e) + HR_i$$ (Equation 11)

The second major contributing factor to blood pressure was the stroke volume (SV) defined in element G_2. The error signal again drives the SV element defined by equation 12, where SV_i is the initial value for the stroke volume set at a value of 82.6 ml.

$$SV = SV_i - 4(P_e)$$ (Equation 12)

The final controller block is element G_3 that generates the peripheral vascular resistance (R) for the system model. Equation 13 defines the relationship between the pressure error signal (P_e) and the peripheral resistance where R_i denotes the initial value at +1 G_z (2.0 Pa*min/ml).

$$R = R_i + 0.2(P_e)$$ *(Equation 13)*

Control blocks G_1, G_2, and G_3 provide direct input into element K_p (plant) which defines the systolic blood pressure at the heart level. This is simply the product of the three variables HR, SV, and R (equation 14). For the feedback loop, the pressure is normalized by dividing the value at time "t" by the maximum blood pressure value (equation 15).

$$P = HR \times SV \times R$$ *(Equation 14)*

$$P' = \frac{P}{P_{max}}$$ *(Equation 15)*

The pressure at the eye level is then calculated using equation 16.

$$P_{eye} = P' - \hat{P}(t)$$ *(Equation 16)*

Beginning with the feedback loop from P' to the summation block, the remaining mathematical relationships were defined using equations defined by Phillips (2000). First, the difference between the normalized systolic blood pressure from the system output and the normalized required blood pressure to maintain adequate eye-level blood pressure was defined by equation 17,

$$\Delta P = P_R' - P'$$ *(Equation 17)*

where P_R' was quantified by equation 18:

$$P_R' = \frac{P_R}{P_{R\,max}}$$ *(Equation 18)*

The feedback pressure (P_f) is then written as:

$$P_f = \frac{1}{T_f} \sum_{i=1}^{n} \Delta P_i \cdot dt$$ *(Equation 19)*

$$P_f = \alpha_f \sum_{i=1}^{n} \Delta P_i \cdot dt$$ *(Equation 20)*

Where α_P is the reciprocal time constant of the pressure feedback via baroreceptor and mechanoreceptor output (see equation 21).

$$\alpha_P = \frac{1}{T_f}$$ *(Equation 21)*

HUMAN INFORMATION PROCESSING MODEL

With the cardiovascular parameters defined, it was possible to derive the relative regional oxygen saturation in the major brain areas of interest. New model elements were added to describe this relationship mathematically.

The first additional block is element G_7, which requires the relative distance measurement (D_j) from the eye point to brain structure j. G_7 uses this distance to then calculate the pressure difference between the eye level and the brain structure of interest using the following equation, where the letter j references the brain structure.

$$\Delta P_j(t) = \rho \cdot g(t) \cdot h_j \qquad \textit{(Equation 23)}$$

Therefore, the pressure at the jth brain structure was defined as equation 24.

$$P_j = P_{eye} - \Delta P_j(t) \qquad \textit{(Equation 24)}$$

With the blood pressure defined in each relative brain area, it was possible to describe its relationship with relative oxygen saturation (element K_O). By investigating cerebral oxygen saturation data (rSO_2) from subject 1 in the "peripheral information processing" experiment described by McKinley, et al. (2008), the relative oxygen saturation equation was derived. A corresponding data set from the "rapid decision making" experiment also described by McKinley et al. (2008) was used to validate the equation (shown below).

$$rSO_2(j) = 0.0023(P_j) + 71.439 \qquad \textit{(Equation 25)}$$

To accurately predict the relative change in neural activity in any given brain area, it was important to also consider the basal metabolic rate. Regional metabolic rates of glucose were obtained from Volcow et al., (2001) and Bassant, Jazat-Poindessous, and Lamour, (1996). Because the brain uses aerobic metabolic processes almost exclusively to generate necessary energy, equation 26 was used to calculate the relative change in neural metabolism based on oxygen content (element G_8). Here, M_{ji} refers to the initial basal metabolic rate of glucose of brain area j.

$$M_j = (M_{ji}) \left(1 - \frac{100 - rSO_2(j)}{100} \right) \qquad \textit{(Equation 26)}$$

Lastly, the change in cognitive performance (ΔCP) of the task (element (K_c) of interest was derived by creating a directly proportional relationship with the change in regional metabolism (for brain areas involved in execution of the task) and the predicted task performance. This relationship was defined using objective task performance data collected during the "motion inference" experiment. A complete description of the task can be found in McKinley, et al., (2008).

The purpose of the analysis performed on the motion inference task data in the HIPDE report was to determine whether the subjects were significantly early or late in their estimation of time. Therefore, each data point describing the amount of error between the location of the target hash mark and the location the subject actually stopped the circular target were given a positive or negative sign: positive indicated an early estimation, while late indicated a late response. However, the purpose of this effort is to describe the human performance changes during $+G_z$ acceleration. Therefore, the absolute value of each error data point was calculated and then compared to the average error value calculated from the baseline ($1\ G_z$) data. The percent change from baseline was then calculated for each data point and averaged across subjects and days.

The averaged data from the 3G and 7G 15-sec plateau profiles were then used to build the proportional equation used in element K_C. Equation 27 provided the change in metabolic rate for brain region j based on the change in regional oxygen saturation and region specific metabolic rate. The resulting cognitive performance (CP) can then be calculated using equation 28.

$$\Delta M_j = \left(M_{ji}\right)\left(\frac{100 - rSO_2(j)}{100}\right) \qquad \textit{(Equation 27)}$$

$$CP_j = (-18.483)\left(\Delta M_j\right) + 101.11 \qquad \textit{(Equation 28)}$$

Equation 28 was then verified using data from the 5G plateau and 7G SACM profiles. Validation of the model was completed using data from a completely different HIPDE experiment entitled "precision timing." This task is also described by McKinley, et al., (2008). As in the motion inference experiment, the positional error metric was assigned a positive sign (late response) or negative sign (early response). To determine the percent change from baseline performance irrespective of direction, the absolute value was computed for each data point. The data were then averaged across subjects and days for comparison with the model predictions.

MODEL VERIFICATION AND ANALYSIS

Model predictions were verified and validated using data sets from the Human Information Processing in Dynamic Environment (HIPDE) program sponsored by the United States Air Force. This series of 12 experiments provided both rSO2 and objective task performance data, but did not offer cardiovascular parameters such as heart rate or blood pressure to verify the accuracy of the cardiovascular model. However, the cardiac values were compared to those from other studies (Grygoryan, 1999; Jennings, et al., 1990; Tripp, et al., 1994) to verify the range of values was reasonable and accurate for the Gz levels applied.

The cognitive performance algorithm was developed using part of the data set from the "motion inference" HIPDE experiment and then compared to data from the "precision timing" experiment. Model accuracy was quantified using the "agreement" approach described by Griffin, (2001). This methodology is founded on the principle that high correlation (values approaching 1) between two data sets is insufficient to prove a model's accuracy when validating with measured data. The author states that two data sets can have perfect correlation when the values of one set are exactly half the values of the other set. As a result, additional metrics are necessary to ensure predicted values are representative of measured values. First, Griffin (2001) described the necessity of plotting all the values with the measured values on the ordinate axis and the predicted values on the abscissa. In this way, one can plot the linear best-fit trend and record the slope. A slope close to one indicates high agreement between the two data sets. Lastly, Griffin states that the mean percent error between measured and predicted values should be relatively low. All three metrics (correlation coefficient, linear best-fit slope, and mean percent error) were used to quantify agreement.

RESULTS

REGIONAL CEREBRAL OXYGEN SATURATION

Only two experiments from the HIPDE program utilized the INVOS 4100 cerebral oximeter to collect regional cerebral oxygen saturation (rSO_2) data (McKinley, et al., 2008). As a result, the rSO_2 prediction algorithm defined in the previous section was developed using data from one subject in the "peripheral information processing" experiment and then validated using the data set from a separate HIPDE study that tested "rapid decision making." Plots of the rSO_2 predictions from the model (output of element KO) and the measured data averaged across subjects for each G_z profile can be found in Figure 1.2.

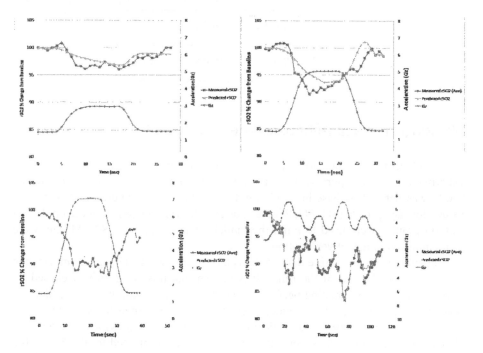

FIGURE 1.2 Measured and Predicted rSO2 during 3 G_z Plateau (top left), 5 G_z Plateau (top right), 7 G_z Plateau (bottom left), and 7 G_z SACM (bottom right)

The correlation coefficients, linear best-fit slope (on a plot of measured versus predicted values), and mean percent error between measured and predicted rSO_2 values are provided in table 1. Correlation coefficients for all data sets were determined using equation 29, where X and Y are the predicted and measured data sets, respectively.

$$Corr(X,Y) = \frac{\sum(x - \bar{x})(y - \bar{y})}{\sqrt{\sum(x - \bar{x})^2 \sum(y - \bar{y})^2}} \qquad \textit{(Equation 29)}$$

Table 1.1 rSO$_2$ Model Agreement Metrics

Gz Profile	Corr. Coeff.	Linear Best Fit Slope	Mean % Error
3G	0.8687	0.5760	0.75
5G	0.8803	0.7099	1.42
7G	0.7483	0.8191	3.33
7G SACM	0.8637	0.9484	2.73

MOTION INFERENCE PERFORMANCE

All objective task performance data collected during the "motion inference" experiment indicate a percentage change from baseline (1 G$_z$) performance where "100" was defined as baseline. Figure 1.3 displays the predicted and measured values for each profile (3G, 5G, 7G plateaus and the 7G SACM). Data from the 3G and 7G profiles were used to build the model. The 5G and 7G SACM data were then employed in model verification.

The plot of the 5G profile data contains an extremely high change from baseline performance at the end of the run. It is known that this is an artifact caused by subjects prematurely arresting performance of the task believing that data collection was over once they felt the induced acceleration had returned to baseline levels. McKinley, et al., (2008) and McKinley, et al., (2005b), previously noted this phenomenon. As a result, agreement calculations were completed for the 5G profile without this data point. The results of all four analyses between the measured and predicted values can be found in table 1.2.

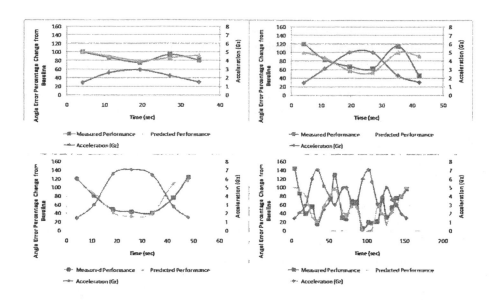

FIGURE 1.3 Motion Inference Predicted vs. Measured Performance during 3 G$_z$ Plateau (top left), 5 G$_z$ Plateau (top right), 7 G$_z$ Plateau (bottom left), and 7 G$_z$ SACM (bottom right)

Table 1.2 Motion Inference Model Agreement Metrics

Gz Profile	Corr. Coeff.	Linear Best Fit Slope	Mean % Error
3G	0.7103	0.5494	6.35
5G	0.9451	0.8206	12.88
7G	0.8827	0.9144	17.11
7G SACM	0.8253	0.7416	38.21

PRECISION TIMING PERFORMANCE

Data from the precision timing task collected during the HIPDE program were used to validate the model predictions. All data indicate a percentage change from baseline (1 G$_z$) performance where "100" was defined as baseline. Figure 1.4 displays the predicted and measured percentage change from baseline performance values for each profile (3G, 5G, 7G plateaus and the 7G SACM).

As in the motion inference data set, the plot of the 5G profile data contains an unusually high decrease from baseline performance at the final data point of the run. Again, subjects began to stop performing the task when the G_z was unloaded because they felt the run was over. Consequently, the agreement calculations were again completed for the 5G profile without this final data point. The results of all four analyses between the measured and predicted values can be found in table 1.3.

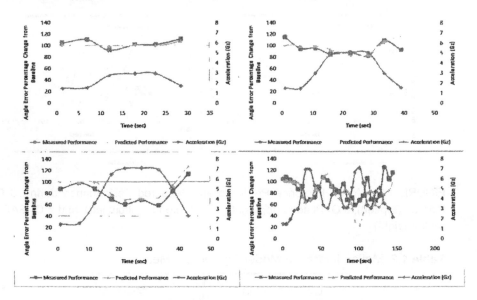

FIGURE 1.4 Precision Timing Predicted vs. Measured Performance during 3 G_z Plateau (top left), 5 G_z Plateau (top right), 7 G_z Plateau (bottom left), and 7 G_z SACM (bottom right)

Table 1.3 Precision Timing Model Agreement Metrics

Gz Profile	Corr. Coeff.	Linear Best Fit Slope	Mean % Error
3G	0.5321	0.2225	4.99
5G	0.7676	0.5795	6.30
7G	0.9726	1.027	12.07
7G SACM	0.6856	0.8513	17.28

DISCUSSION

Selection of the motion inference and precision timing task for this modeling effort provided both a complex and simplistic task for comparison to ensure the validity of the basic underlying theory. Although inherently similar in structure, they probe two separate cognitive functions and are vastly different in complexity and level of difficulty. This difference is explained by analyzing the brain areas active in execution of each task. As described previously, the precision timing task requires a short amount of time to complete each trial. Consequently, the cerebellum (the body's internal clock) is able to process the perception of the time and enable the human operator to respond at the appropriate moment. Given that the motion inference task removes the target from view, provides a distracter task, and requires more than 5 seconds to complete, it is evident that the human operator will require engagement of working memory processes (hence the DLPFC) to correctly retain the initial velocity of the target and subsequently infer the movement of the target across the arc. This is an important distinction because the cerebellum lies a full 4.6 cm below the level of the DLPFC. Because the drop in arterial blood pressure is a function of height above the level of the heart, the cerebellum retains higher local arterial blood pressure (hence oxygenated blood profusion) when compared to the DLPFC. Additionally, the cerebellum's basal metabolic rate is approximately 30% lower than that of the DLPFC indicating that its metabolic need is less.

With the knowledge concerning cognitive decrements elucidated from the HIPDE program, the objective of this effort was to develop and validate a computational model capable of predicting the cognitive performance fluctuations across different tasks using acceleration data as the only model input. Thus, the focus of the analysis was evaluation of the accuracy and precision of the predicted cognitive task performance when compared to collected measured data. Human research performance models with correlation coefficients in the range of 5-6 are considered very good, and coefficients around 8 as excellent (Keppel and Wickens, 2004).

REGIONAL CEREBRAL OXYGEN SATURATION MODEL

The predicted eye-level arterial blood pressure data directly fed the algorithm that calculated expected regional cerebral oxygen saturation (rSO_2). This equation was developed using only one subject's rSO_2 data from the "peripheral information processing" experiment performed within in the HIPDE program as McKinley et al, (2005a) had previously shown this data to be highly repeatable between subjects. The predicted rSO_2 values were then objectively validated using data collected during the "rapid decision making" HIPDE experiment. Referring to table 1, agreement between the measured and predicted rSO_2 values was extremely high across acceleration profiles given that all maintained error averages below 5% (0.75-3.33), high correlation (0.7483-0.8687), and high overall linear best-fit slopes (0.5760-0.9484). Because the 3G profile produced much smaller reductions in

rSO_2, the measured values were more sensitive to signal noise caused both by error in the cerebral oximeter and small variations in average subject rSO_2. As expected, the rSO_2 agreement calculations during the 3G profile are lower. However, a close inspection of the plot (Figure 1.2) provides evidence that the predictions closely match the magnitude of the rSO_2 decline and are temporally precise.

It is noteworthy that following the decline of acceleration in each of the four G_z profiles, the predicted rSO_2 values recovered to baseline values prematurely. Likewise, predicted rSO_2 recovery following the first 7G peak within the SACM profile was much greater than the measured values. Given that rSO_2 is directly proportional to eye-level blood pressure, this may indicate that the cardiovascular model is predicting eye-level blood pressure recoveries that fall outside the normal average. Nevertheless, the impact of this phenomenon on overall model accuracy was minimal and closeness of fit between the predicted and measured rSO_2 data remained high.

COGNITIVE PERFORMANCE MODEL: MOTION INFERENCE

Minor perturbations (noise) in the measured task performance data can influence agreement results to a greater extent when the change from baseline is itself minor (low signal-to-noise ratio). Hence, as in the rSO_2 model, the predicted motion inference performance during the 3G acceleration trial resulted in the lowest agreement with the measured data of the four acceleration profiles (Correlation: 0.7103, Linear Best-fit Slope: 0.5494, Mean Percent Error: 6.35). Regardless of this fact, the predicted magnitude of motion inference performance decline was accurate as illustrated in Figure 1.3. This coupled with the low mean percent error between the measured and predicted values and high correlation indicates the model produced an accurate prediction of human motion inference performance during the 3G plateau profile.

Assuming the validity of removing the final data point from the 5G profile (as described in the methods section), predictions for motion inference performance during the 5G (Correlation: 0.9451, Best-Fit Slope: 0.8206, Mean Percent Error: 12.88%) achieved the highest overall agreement between the measured and predicted values of the four acceleration profiles.

Because motion inference performance data collected during the 7G plateau profile were used to develop the coefficients in the linear equation relating regional cerebral oxygen saturation to predicted change in performance, it was intuitively pleasing that performance predictions during this profile produced excellent overall agreement (Correlation: 0.8827, Best-Fit Slope: 0.9144, Mean Percent Error: 17.11%). In addition, motion inference performance declines were quite pronounced relative to signal noise during this acceleration plateau creating a high signal-to-noise ratio and minimizing the effect of noise on agreement. Overall agreement across all four profiles yielded high agreement with the collected motion inference performance data evidenced by correlation coefficients close to one (0.7103-0.9451), relatively low mean percent error (6.35%-38.21%), and overall high linear best fit slopes (0.7416-0.9144) (with the exception of the 3G profile).

Cognitive Performance Model: Precision Timing

To ensure the accuracy of the model predictions across tasks, data from the precision timing experiment were used in final model validation. Because these data did not serve in the development of any of the model algorithms, they provided a pure and unbiased assessment of the model's ability to predict human cognitive performance under various levels of acceleration stress. Prior to evaluating the model, measured precision timing data revealed that performance on the task did not significantly change from baseline (1G) levels under the 3G profile. Consequently, the measured signal was comprised exclusively of noise. As noise was deliberately not modeled, predictions of precision timing performance during the 3G profile were constant. Any attempt to correlate a constant value signal with another that oscillates around the constant value will inevitably lead to poor agreement results. Consulting table 1.3, it is clear that the model agreement results for the 3G profile confirmed this expectation (Correlation: 0.5321, Best-Fit Slope: 0.2225, Mean Percent Error: 4.99%). In this case, even though the agreement calculations as defined by Griffin (2001) do not indicate high agreement between the measured and predicted precision timing performance, it does not signify that the model predictions are inaccurate. The model accurately predicted that performance remained at baseline levels throughout the profile and even calculated the moderate improvement in performance over baseline levels after the acceleration stress had ceased (evidenced by the low mean percent error). Combined, these parameters provide evidence the model accurately predicts precision timing performance during single peak, 3G acceleration.

As in the motion inference analysis, the final data point of the 5G plateau was removed based on evidence from McKinley, et al., (2008) that suggested subjects prematurely stopped performing the task once the G_z stress approached baseline levels. With the data point removed, the predicted precision timing performance during the 5G plateau yielded an extremely close fit with the average measured data (Correlation: 0.7676, Best-Fit Slope: 0.5795, Mean Percent Error: 6.30%).

Measured and predicted precision timing performance during both the 7G plateau (Correlation: 0.9726, Best-Fit Slope: 1.027, Mean Percent Error: 12.07%) and 7G SACM (Correlation: 0.6856, Best-Fit Slope: 0.8513, Mean Percent Error: 17.28%) demonstrated exceptional agreement. Nevertheless, predictive accuracy did not remain constant throughout the 7G SACM profile. Evaluation of Figure 1.4 indicated the prediction was most accurate during the first half of the profile. The large dip in precision timing performance predicted by the model during the second 7G peak did not appear in the collected data. The reason for this inaccuracy is not immediately known, but could indicate that there was a higher overall G_z tolerance in the subject group that participated in the precision timing experiment when compared to those in the motion inference study.

REFERENCES

Bassant, M.H., Jazat-Poindessous, F., Lamour, Y. (1996). Effects of Metrifonate, a Cholinesterase Inhibitor, on Local Cerebral Glucose Utilization in Young and Aged Rats. *Journal of Cerebral Blood Flow & Metabolism, 16*:1014–1025.

Carlson, Neil R. (2007). Physiology of Behavior. (9th ed.). Boston: Allyn and Bacon.

Ernsting, J., Nicholson, A.N., and Rainford, D.J. (eds.). Aviation Medicine 3rd edition. (1999) pp. 43-58. Butterworth Heinemann, Oxford, England.

Fraisse, P. (1984). Perception and estimation of time. *Annual Review of Psychology, 35*:1-36.

Griffin M.J. (2001). The Validation of Biodynamic Models. Clinical Biomechanics. 16 (1, Suppl.): 81S-92S.

Grygoryan, R. (1999). Development of a Hemodynamics Computer Model of Human Tolerance to High Sustained Acceleration Exposures. Air Force Research Laboratory Technical Report, AFRL-HE-WP-TR-2006-0143.

Jennings, T., Tripp, L.D., Jr., Howell, L., Seaworth, J., Ratino, D., Goodyear, C. (1990). The Effect of Various Straining Maneuvers on Cardiac Volumes at 1G and During +Gz Acceleration. *SAFE, 20*(3):22-28.

Johnson, K.A., Becker, J.A. "The Whole Brain Atlas." 1999. Harvard Medical School. 25Feb2009. <http://www.med.harvard.edu/AANLIB/home.html>.

Keppel, G. & Wickens, T.D. (2004). Design and Analysis: A Researcher's Handbook, 4th edition. Upper Saddle River, NJ: Prentice Hall.

Krnjevic, K. (1999). Early Effects of Hypoxia on Brain Cell Function. *Basic Sciences, 40*(3):375-380.

Lee, J.M., Grabb, M.C., Zipfel, G.J., Choi, D.W. (2000). Brain tissue responses to ischemia. *The Journal of Clinical Investigation, 106*(6):723-731.

Mangels, A. J., et al. (1998). Dissociable contributions of the prefrontal and neocerebellar cortex to time perception. *Cognitive Brain Research, 7*:15-39.

McKinley, R.A., Tripp L.D. Jr., Bolia S.D., Roark M.R. (2005a). Computer Modeling of Acceleration Effects on Cerebral Oxygen Saturation. *Aviat Space Environ Med, 76*: 733-738.

McKinley R.A., Fullerton K.L., Tripp L.D. Jr., Esken R.L., Goodyear C. (2005b). A Model of the Effects of Acceleration on a Pursuit Tracking Task. Air Force Research Laboratory Technical Report, AFRL-HE-WP-TR-2005-0008.

McKinley R.A., Tripp L.D. Jr., Loeffelholz, J., Esken R.L., Fullerton K.L., Goodyear C. (2008). Human Information Processing In the Dynamic Environment. Air Force Research Laboratory Technical Report, AFRL-HE-WP-TR-2008-0008.

Naval Aerospace Medical Institute. G-Induced Loss of Consciousness (G-LOC). Chapter 7: Neurology. In: United States Naval Flight Surgeon's Manual: Third Edition; 1991.

Newman D.G., White S.W., Callister R. (1998). Evidence of baroreflex adaptation to repetitive +Gz in fighter pilot. *Aviat Space Environ Med, 69*:446-451.

Nichelli, P et al. (1996). Perceptual timing in cerebellar degeneration. *Neuropsychologia, 34*:863-871.

Phillips, C.A. (2000). Neuromuscular Control Systems. In: Human Factors Engineering. Wiley, New York, (Chapter 9).

Rubia., K., Smith, A. (2004). The neural correlates of cognitive time management: a review. *Acta Neurobiologica, 64*:329-340.

Salzmann, E.W., Leverett, S.D. (1956). Orthostatic venoconstriction studies by miniature balloon technique. *Fed Proc, 15*:160-161.

Tripp, L.D., Jr., Jennings, T.J., Seaworth, J.F., Howell, L.L., Goodyear, C. (1994). Long-Duration $+G_z$ Acceleration on Cardiac Volumes Determined by Two-Dimensional Echocardiography. *J Clin Pharmacol 34*:484-488.

Tripp, L.D., Chelette, T, Savul, S.A., and Widman, B.S. (1998). Female exposure to high-G: Effects of simulated combat sorties on cerebral and arterial O2 Saturation. *Aviat Space Environ Med 69*(9):869-874.

Volcow, N.D., Chang, L., Wang, G., Fowler, J.S., Franceschi, D., Sedler, M.J., Gatley, S.J., Hitzemann, R., Ding, Y., Wong, C., Logan, J. (2001). Higher Cortical and Lower Subcortical Metabolism in Detoxified Methamphetamine Abusers. *Am J Psychiatry, 158*:383-389.

Chapter 8

Static and Dynamic Discriminations in Vigilance: Effects on Cerebral Hemodynamics and Workload

*Matthew E. Funke[1], Joel S. Warm[1],Gerald Matthews[2], Victor Finomore, Jr.[1],
Michael A. Vidulich[1], Benjamin A. Knott[1], William S. Helton[3],
Tyler H. Shaw[4], and Raja Parasuraman[4]*

[1]Air Force Research Laboratory
WPAFB, USA

[2]University of Cincinnati, USA

[3]University of Canterbury, NZ

[4]George Mason University, USA

ABSTRACT

In this study, a distinction was drawn between dynamic and static type vigilance tasks. The former are tasks in which signal/noise discriminations are difficult or moderately difficult and the focus of performance evaluation is upon the accuracy of signal detection. In the latter, signal/noise discriminations are relatively easy or are not required; the mere appearance of a predetermined stimulus event is a critical signal for detection. In this case, the focus of performance evaluation is upon response time to signal detection since detection accuracy is at a ceiling. At the heart of the dynamic/static issue, is a long-standing belief that accuracy and speed are interchangeable indices of the same underlying process in vigilance (Buck, 1966). The present study shows that there are major differences between these two types of tasks in terms of cerebral hemodynamics and perceived mental workload. Clearly, the view that per cent correct and response time are equivalent measures of a common underlying process in vigilance is incorrect and vigilance researchers need to consider the dynamic/static task distinction in their models of vigilance performance.

Keywords: Cerebral Blood Flow Velocity, Cerebral Hemodynamics, Response Measure Equivalence, Vigilance, Workload

INTRODUCTION

Vigilance or sustained attention tasks require observers to maintain their focus of attention and to detect infrequent and unpredictable targets over prolonged periods of time (Davies & Parasuraman, 1982). The ability of observers to sustain attention is of considerable concern to human factors/ergonomic specialists because of the vital role that vigilance plays in many automated human-machine systems including military surveillance, air-traffic control, cockpit monitoring, airport and border security, industrial process control, and medical functions such as cytological screening and the inspection of aesthesia gauges during surgery (Warm, Parasuraman, & Matthews, 2008). Several studies have shown that accidents ranging in scale from minor to major are the result of vigilance failures on the part of human operators in automated systems (Molloy & Parasuraman, 1996). Hence, understanding the factors that influence performance and their underlying mechanisms is a critical human factors concern for system reliability and public safety and health (Nickerson, 1992).

Traditionally, vigilance tasks have been considered as tedious but benign assignments that place little demand upon operators and the decrement function, the decline in efficiency over time that typifies performance in vigilance tasks (Davies & Parasuraman, 1982), has been viewed as resulting from task underload and consequent under arousal (Nachreiner & Hanecke, 1992). However, as summarized by Warm and his associates (Warm et al., 2008), recent studies have indicated that while they are tedious, vigilance tasks impose a substantial demand upon the information-processing resources of observers and are highly stressful. This more recent view has emerged from studies examining neural measures of resource demand and perceived mental workload. Neurological evidence of task demand comes from studies using a non-invasive brain imaging system known as Transcranial Doppler Sonography (TCD) that employs ultrasound signals to monitor changes in cerebral blood flow velocity (CBFV) and thereby to gauge changes in metabolic activity during task performance (Tripp & Warm, 2007). These studies have shown that the decrement function is accompanied by a corresponding temporal decline in CBFV when observers are actively engaged in the vigilance task but not when they simply view the task for an equal amount of time without a work imperative, an assignment that should lead to a maximum loss of arousal (Warm & Parasuraman, 2007; Warm, Matthews & Parasuraman, 2009). Evidence of high mental workload comes from studies using the NASA Task Load Index (NASA-TLX), a widely employed subjective workload measure (Wickens & Hollands, 2000).These studies indicate that global workload scores in vigilance tasks typically fall within the upper level of the NASA-TLX scale and that mental demand and frustration are the major components of the workload associated with vigilance tasks (Warm, Matthews, & Finomore, 2008). The finding of high information-processing demand in vigilance challenges arousal theory and supports an attentional resource view proposed by Parasuraman and Davies (Davies & Parasuraman, 1982; Parasuraman & Davies, 1977) that the workload imposed by vigilance tasks reflects the impact of focused mental effort and a drain on information-processing resources over time (Johnson & Proctor, 2004; Warm, Parasuraman, et al., 2008).

All of the studies described above utilized what might be termed *dynamic type* vigilance tasks involving moderate to difficult signal/noise discriminations in which the vigilance decrement was observed in terms of a decline in the percentage of correct signal detections over the course of the vigil. However, vigilance tasks can be more *static* in character in which signal /noise discriminations are either quite easy or not required; critical signals for detection are defined simply by the appearance of a predefined stimulus event. In these cases, critical signals are almost always detected and the vigilance decrement appears in terms of an increase in response time to signal detections over the course of the watch (Davies & Parasuraman, 1982; Dinges & Powell, 1985).

The dynamic static distinction brings to mind a long-standing debate over the equivalence of accuracy and speed of signal detection measures in psychology in general and vigilance in particular (Kantowitz, 1985; Kemp, 1984; McGrath, 1963; Santee & Egeth, 1982; Teichner, 1974). With regard to vigilance, a careful review of the literature led Buck (1966) to conclude that these measures are interchangeable indices of the same underlying process. He maintained that when a high degree of signal detection accuracy over time is accompanied by a temporal rise in reaction time, the level of sustained attention is declining yet remaining above the level at which missed signals enter the picture. A position of this sort would suggest that dynamic-type vigilance tasks in which the focus of performance evaluation is upon the accuracy of signal detection and static-type tasks in which the focus of performance evaluation is upon the speed of signal detection are moderated by the same attentional processes. Buck's (1966) conclusion was reached many years ago before the findings regarding CBFV and perceived mental workload were available. Accordingly, the goal for the present study was to assess the equivalence of dynamic and static type vigilance tasks in terms of CBFV and workload and the ability of a resource model to account for performance with these types of tasks. In this way, the study would provide a further evaluation of the validity of Buck's (1966) claim that accuracy and speed are interchangeable indices of the same underlying process in vigilance.

METHOD

Observers assumed the role of air traffic controllers monitoring the flight pattern of a squadron of four jet fighters on a circular display divided into four quadrants as illustrated in Figure 1. Each quadrant contained a triangular jet icon. In a dynamic task condition, the squadron flew in either a clockwise or a counterclockwise direction (defined by the "noses" of the planes) throughout the vigil, so that on any given display exposure, observers needed to differentiate between two possible flight directions in order to detect critical signals. Critical signals for detection in the dynamic task were cases where one of the planes was flying in an inappropriate direction relative to the direction of the others so that a collision could occur. Given the multidirectional flight paths employed, a plane that was at fault when the flight was in one direction would not be at fault when the flight was in the other direction. The display was updated 30 times/min with a dwell time of 1000 msec. Twelve critical signals appeared at random intervals during each consecutive 10-min period of a 40-min watch. An equal number of critical signals appeared in each quadrant of the monitored display. Observers indicated their detection of critical signals by pressing the space bar on a computer keyboard. Responses occurring within 1000 msec after the appearance of critical signals were recorded automatically as correct detections. All other responses were recorded as errors of commission or false alarms.

In a static task condition, signal/noise discriminations were not required; the mere exposure of the display constituted a critical signal for detection to which observers responded by the key press response described above. To control for critical signal frequency, the flight display in this condition appeared 12 times at random intervals within each 10-min period of the 40-min vigil. As in the dynamic task, the dwell time of the display was 1000 msec. The rules for determining correct detections and false alarms were identical to those in the dynamic task. To control for disparities in the display images presented to the observers in both task conditions, the displays viewed by observers in the static condition were the critical signal displays presented in the dynamic condition. Stimulus presentations in all experimental conditions were orchestrated by a Dell computer using SuperLab software. The computer also recorded response accuracy and response time (to the nearest msec) in all conditions.

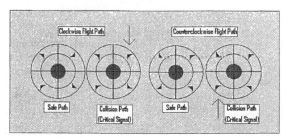

Figure 1. Examples of neutral events and critical signals in the flight path display employed in the dynamic task condition. In the static task condition, appearances of the display were the critical signals for detection and neutral event/critical signal discriminations were not required.

Twelve observers (sex equated) were assigned at random to each task condition. An additional 12 observers served as passive controls to assure that time-based changes in CBFV were task-determined. These observers viewed the flight display without an information-processing imperative, and were instructed to simply gaze at the display for 40 min until the session ended. Six of the control observers experienced the signal/ noise display format employed in the dynamic task and the remainder experienced the signal-only display format employed in the static task. All of the observers in this study were right handed and had normal or corrected-to-normal vision.

Within the two active flight conditions and the passive control condition, bilateral hemovelocity measurements were taken from the left and right medial cerebral arteries of all observers using a Nicolet Companion III TCD unit. Changes in CBFV during task performance were expressed as a proportion of a resting baseline. Subjective workload was measured by the NASA-TLX which was administered immediately upon the conclusion of the vigil.

RESULTS

CORRECT DETECTIONS AND RESPONSE TIME

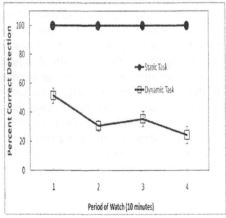
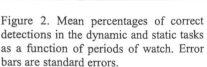
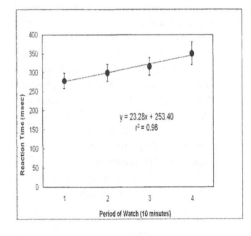

Figure 2. Mean percentages of correct detections in the dynamic and static tasks as a function of periods of watch. Error bars are standard errors.

Figure 3. Mean response times (ms) to critical signals as a function periods of watch in the static task. Error bars are standard errors.

Mean percentages of correct detections in the two task conditions are plotted as a function of periods of watch in Figure 2 above. It is evident in the figure that detection probability in the static task started and remained at 100% across all periods of watch while the mean detection scores in the dynamic task were much lower and declined considerably over the course of the vigil from approximately

54% in period 1 to approximately 24% in period 4. These Impressions were confirmed by a 2 (tasks) × 4 (periods) mixed-analysis of variance (ANOVA) of the arcsines of the percentage scores which revealed significant main effects for task $F(1, 22) = 333.01, p < .001, \eta_p^2 = .51$ and periods of watch, $F(2.11, 46.45) = 11.23, p < .001, \eta_p^2 = .34$, and a significant Task × Periods interaction, $F(2.11, 46.45) = 11.23, p < .001, \eta_p^2 = .34$. A supplementary simple effects ANOVA indicated that the decline in the accuracy of signal detection over time in the dynamic task was statistically significant, $F(2.11, 23.23) = 11.23, p < .001, \eta_p^2 = .51$. In these and all subsequent ANOVAs, the Box correction (Maxwell & Delaney, 2004) was used when appropriate to correct for violations of the sphericity assumption. There were no false alarms in the static task and the false alarm rate in the dynamic was less than 1.5 %. Consequently, false alarms were not analyzed further.

Means of median response times in the static task are plotted as a function of periods of watch in Figure 3 above. It is evident in the figure that response time to signal detection in this condition increased over the course of the vigil. A one-way repeated measures ANOVA revealed that the increase was statistically significant, $F(1.67, 18.33) = 8.08, p < .001, \eta_p^2 = .42$ and that it was linear in form, $F(1, 11) = 11.51, p < .01$, so that response time increased by 23.28 msec per period of watch. Thus, while performance efficiency in the static task remained stable over time when viewed in terms of detection probability, this task was not immune to the vigilance decrement; the decrement function appeared in terms of a temporal elevation of response times to signal detection.

CEREBRAL HEMODYNAMICS

Inspection of the *resting baseline* data indicated that the CBFV scores in the dynamic and static vigilance task conditions and in the passive control conditions were similar in both the left and right cerebral hemispheres. Thus, subsequent CBFV effects in the two cerebral hemispheres associated with active and passive observing cannot be attributed to sampling artifacts in the original resting baselines. A 2 (hemispheres) × 4 (periods of watch) repeated-measures ANOVA of the experiment-based CBFV scores among the *passive control* observers revealed no significant sources of variance, $p > .05$ for all components of the analysis, indicating that time-related changes in the experiment-based CBFV scores among the active observers were task-related.

A 2 (tasks) × 2 (hemispheres) × 4 (periods of watch) mixed-model ANOVA of the experiment-based CBFV data for the active observers revealed that the overall level of CBFV was significantly greater in the right ($M = .99$) than in the left ($M = .95$) hemisphere, $F(1, 22) = 22.18, p < .001, \eta_p^2 = .50$. In addition, the Task x Hemisphere, $F(1, 22) = 22.62, p < .001, \eta_p^2 = .50$, and the Periods x Hemisphere, $F(1.43, 31.48) = 5.54, p < .05, \eta_p^2 = .20$, interactions were statistically significant and the Task x Periods interaction closely approached significance, $F(1.48, 32.61) = 3.36, p = .06, \eta_p^2 = .13$. All other sources of variance in the analysis were not significant, $p > .05$ in all cases

The Task x Hemisphere and Task x Periods interactions are displayed in Figures 4 and 5, respectively. It is clear in Figure 4 that the overall hemispheric difference in CBFV was task dependent. CBFV was greater in the right than the left hemisphere only in the context of the dynamic task. The CBFV scores were approximately the same across hemispheres in the context of the static-task. Figure 5 reveals that there was a temporal decline in the level of CBFV and that this effect was also was also task dependent; it was limited to the dynamic task. Supplementary tests of simple effects indicated that the decline over time in the dynamic task was statistically significant, F (2.06, 22.68) = 8.30, $p < .01$, $\eta_p^2 = .43$, There was no significant periods effect in the static task, $p > .05$.

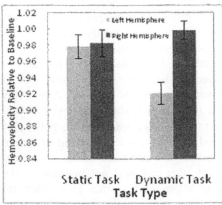

Figure 4. Mean cerebral blood flow velocity scores in the left and right hemispheres for the static and dynamic tasks. Error bars are standard errors.

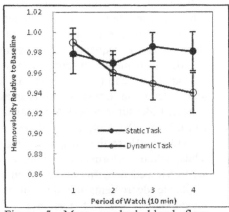

Figure 5. Mean cerebral blood flow velocity in the static and dynamic tasks as a function of periods of watch. Error bars are standard errors.

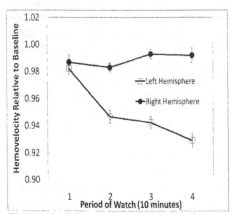

Figure 6. Mean cerebral blood flow velocity scores in the left and right hemispheres as a function of periods of watch. Error bars are standard errors.

The Hemisphere × Periods of Watch interaction is illustrated in Figure 6 wherein the CBFV scores in the right and left cerebral hemispheres are plotted as a function of periods of watch. The figure shows that the temporal decline in CBFV was also hemispheric dependent; it was limited to the left cerebral hemisphere. Tests of simple effects indicated a significant time on task effect for the left hemisphere, $F(1.83, 42.08) = 11.06$, $p < .001$, $\eta_p^2 = .33$, but not the right, $F(1.27, 29.24) = 0.72, p > .05$.

WORKLOAD

Following a procedure recommended by Nygren (1991), observers in the two task-type conditions rated their workload on each of the six subscales of the NASA-TLX. The ratings could range from 0 to 100. The overall mean of an observer's ratings provides a composite index of global workload while the scores for each subscale provide an index of the relative contribution of the individual subscale dimensions to the workload profile. Mean workload scores for the static and dynamic tasks are presented for each subscale in Table 3. A 2 (tasks) x 6 (subscales) mixed ANOVA of the workload data revealed that observers performing the dynamic task rated their global workload to be significantly greater than those performing the static task, $F(1, 22) = 16.05$, $p < .001$, $\eta_p^2 = .42$. Indeed, as can be seen in the table, the overall mean for the dynamic task fell above the midpoint of the scale while that for the static task fell substantially below the midpoint. In addition to the main effect for task, there was a significant main effect for subscales, $F(4.27, 93.93) = 12.89$, $p < .001$, $\eta_p^2 = .37$, and a significant Task × Subscales interaction, $F(4.27, 93.93) = 2.62$, $p = .037$, $\eta_p^2 = .11$. Bonferroni-corrected t-tests with alpha set at .05 indicated that observers in the dynamic task condition rated their vigilance assignment to be more mentally and temporally demanding than those in the static task condition and that observers in the dynamic task felt that they were less successful in accomplishing what they were asked to do than their static task cohorts (higher scores on the Performance subscale indicate poorer self-evaluations). Perusal of Table 1 will reveal that the mean Mental Demand and Temporal Demand ratings for observers in the dynamic task were well above the midpoints of the scales while those for observers in the static task were well below the midpoints of the scales.

Table 1. Mean NASA-TLX Scale Scores for the Static and Dynamic Tasks. Standard Errors are in Parentheses. Scores for Each Subscale and for the Composite Mean Across Subscales can Range from 0 to 100)

Condition		MD	PD	TD	P	E	F	Mean
				Subscale				
Static Task		39.58	15.00	36.25	39.58	52.50	42.50	37.57
		(9.16)	(4.56)	(7.69)	(3.87)	(7.63)	(6.92)	(4.41)
Dynamic Task		70.42	20.83	79.58	50.83	65.83	62.92	58.40
		(4.41)	(6.96)	(4.50)	(7.88)	(7.15)	(7.11)	(2.76)
	Mean	55.00	17.92	57.92	45.21	59.17	52.71	
		(6.79)	(5.76)	(6.10)	(5.87)	(7.39)	(7.02)	

DISCUSSION

This study assessed the equivalence of dynamic-type vigilance tasks in which the accuracy of signal detection is of primary concern and static-type vigilance tasks in which response time to signal detection is of interest since signal detection accuracy is at a ceiling . Comparisons between the tasks were made in terms of CBFV and perceived mental workload and consequently, the ability of a resource model to account for performance with these types of tasks. In this way, the study provided an evaluation of the validity of Buck's (1966) claim that accuracy and speed are interchangeable indices of the same underlying process in vigilance.

The decrement function, the decline in performance efficiency over time that is typical in vigilance tasks, was the principal envelope in which task comparisons were made. Such a decline was observable with both the dynamic and static tasks. In the case of the former, it appeared as a temporal decline in the accuracy of signal detection, in the case of the latter, it took the form of a temporal elevation in response time over the course of the vigil. While the decrement function occurred in both types of tasks, results with respect to CBFV and perceived mental workload indicated that the mechanisms that underlie the decrement function are quite different in the two types of tasks.

With the dynamic task, the overall level of CBFV was maximal in the right cerebral hemisphere and the temporal decline in performance efficiency was accompanied by a temporal decline in CBFV that was maximal in the left hemisphere, indicating that cerebral metabolic activity was decreasing as time on task progressed predominately in the left hemisphere. These effects were not present in the context of the static task. With that task, there was no laterality in CBFV and CBFV did not decline over time. In addition, perceived mental workload fell at the upper and lower levels of the NASA-TLX for observers in the dynamic and static tasks, respectively, and observers in the dynamic task reported higher levels of mental and temporal demand than those in the static task.

With regard to CBFV and workload, the results with the dynamic task are consistent with previous findings supporting the view that the workload imposed by vigilance tasks reflects the impact of focused mental effort and a drain on information-processing resources over time (Davies & Parasuraman, 1982; Johnson & Proctor, 2004; Parasuraman & Davies, 1977; Warm, Parasuraman, et al., 2008). This is clearly not the case with the static task. Given the low workload induced by that task, one might argue the arousal model might account for the decrement function in the context of the static task, but this seems unlikely given the CBFV results. If arousal is declining over time, one might anticipate that brain metabolic activity as reflected in CBFV would also decline over time, a result that did not occur in the static task. Rather than modifications in attentional processes, Davies and Parasuraman (1982) have suggested that response time changes over the course of a vigil may reflect criterion shifts in decisions to respond. In addition, the temporal elevation in response time may be due to increased motor fatigue. Whatever the explanation for the appearance of the decrement function in the static task, two important points are evident from the results of this study: (1) vigilance researchers need to consider the dynamic/static task distinction in their models of vigilance performance and (2) Buck's (1966) claim that accuracy and speed are interchangeable indices of the same underlying process in vigilance is not correct.

REFERENCES

Buck, L. (1966). Reaction time as a measure of perceptual vigilance. *Psychological Bulletin, 65*, 291-304.

Davies, D.R., & Parasuraman, R. (1982). *The psychology of vigilance.* London: Academic Press.

Dinges D.F, & Powell J.W. (1985). Microcomputer analyses of performance on a portable, simple visual RT task during sustained operations. *Behavior, Research Methods, Instruments and Computers 17*,652-5.

Johnson, A., & Proctor, R. W. (2004). *Attention: Theory and practice.* Thousand Oaks, CA: Sage.

Kantowitz, B.H. (1985). Channels and stages in human information processing: A limited analysis of theory methodology. *Journal of Mathematical Psychology, 29*, 135-174.

Kemp, S. (1984). Reaction time to a tone in noise as a function of the signal-to-noise ratio and tone level. *Perception & Psychophysics, 36*, 473-476.

Maxwell, S. E., & Delaney, H. D. (2004). *Designing experiments and analyzing data: A model comparison perspective* (2nd ed.). Mahwah, NJ: Erlbaum.

McGrath, J.J. (1963). Some problems of definitions and criteria in the study of vigilance performance. In D.N. Buckner & J.J. McGrath (Eds.), *Vigilance: A symposium* (pp. 227-237). New York: McGraw-Hill.

Molloy, R., & Parasuraman, R. (1996). Monitoring an automated system for a single failure: Vigilance and task complexity effects. *Human Factors, 38*, 311-322.

Nachreiner, F., & Hanecke, K. (1992). *Vigilance.* In A. P. Smith & D. M. Jones (Eds.), *Handbook of human performance* (pp 262-288). San Diego, CA: Academic Press, Inc.

Nickerson, R.S. (1992). *Looking ahead: Human factors challenges in a changing world.* Mahwah, NJ: Erlbaum.

Nygren, T. E. (1991). Psychometric properties of subjective workload measurement techniques: Implications for their use in the assessment of perceived mental workload. *Human Factors, 33*, 17-33.

Parasuraman, R., & Davies, D.R. (1977). A Taxonomic analysis of vigilance. In R.R. Mackie (Ed.), *Vigilance: Theory, operational performance, and physiological correlates* (pp. 559-574). New York: Plenum.

Santee, J.L., & Egeth, H.E. (1982). Do reaction time and accuracy measure the same aspects of letter recognition? *Journal of Experimental Psychology: Human Perception and Performance, 8*, 489-501.

Teichner, W.H. (1974). The detection of a simple visual signal as a function of time on watch. *Human Factors, 16*, 339-353.

Tripp, L.D., & Warm, J.S. (2007). Transcranial Doppler sonography. In R. Parasuraman & M. Rizzo (Eds.), *Neuroergonomics: The brain at work* (pp. 82-94). New York: Oxford University Press.

Warm, J.S., Matthews, G. & Parasuraman, R. (2009) Cerebral hemodynamics and vigilance performance. *Military Psychology, 21*, S75-S100.

Warm, J. S, Matthews, G., & Finomore, V. S. (2008). Workload, stress, and vigilance. In P.A. Hancock & J.L Szalma (Eds.), *Performance under stress*. Brookfield, VT: Ashgate.

Warm, J.S., & Parasuraman, R. (2007). Cerebral hemodynamics and vigilance. In R. Parasuraman & M. Rizzo (Eds.), *Neuroergonomics: The brain at work* (pp. 146-158). New York: Oxford University Press.

Warm, J.S., Parasuraman, R., & Matthews, G. (2008). Vigilance requires hard mental work and is stressful. *Human Factors, 50*, 433-441.

Wickens, C. D., & Hollands, J. G. (2000). *Engineering psychology and human performance* (3rd ed.). Upper Saddle River, NJ: Prentice-Hall.

CHAPTER 9

4th Dimensional Interactive Design For Dynamic Environments

Manuela Galli*, Frazer McKimm+
Hiromichi Yanagihara°, Veronica Cimolin*

+ DHS, Dublin
*Bioeng. Dept., Politecnico di Milano, Milan, Italy
° Toyota Motors Europe

ABSTRACT

OBJECTIVE: This study examined the effect of the use of three different interfaces of navigator system on driving of 20 young subjects.

BACKGROUND: Various studies have indicated that the non immediate comprehension of traditional Satellite Navigation (Sat Nav) map data may lead to increased driver distraction. At the moment the information presented in a Sat-Nav Telematic system is designed using a static view visual model, which does not take sufficiently into consideration the dynamic time space nature of the car environment. DHS set about designing a different kind of interface which takes into consideration specific time space stimuli restrucutring the data feed so that the mind is better tuned in the dynamic car environement.

For this reason two new navigator system interfaces, where the indications are based on specific landmarks, are proposed. These navigator system interfaces were tested and compared to a Standard Sat Nav system in order to evaluate the effects produced on driver' 3D kinematics.

METHOD: the drivers followed an urban route in a driving simulator following the voice commands given to them by three different navigation systems: the first Satellite navigation traditional system (Sat-Nav), the second one based on the use of landmarks able to provide detailed information about the map (2-Me system) and the third one with the same landmarks but with a different map representation (C-Me).

The systems were evaluated in terms of the effects on driver distraction. In particular an optoelectronic system for kinematic analysis of head driver's movements was used. The number of head movements to the navigator display and the time spent in watching the navigator were key evaluation parameters. In this way the distraction occurring in the different navigator modalities was evaluated.

RESULTS: the 2-Me system was characterized by low number of the head movements to the navigator display and by less time spent by the drivers to read and to understand the navigator displayed.

CONCLUSION: The evaluated navigation interface 2-Me showed promising elements for a new navigator interface planning.

APPLICATION: The messages provided by the 2-Me interface are not currently supplied by existing navigation systems and they could possibly be added to them in the next future.

Keywords: Navigator systems, kinematics, driver distraction,

INTRODUCTION

In understanding the interpretation of visual information (like Sat Nav maps in a car) we need to understand how the human brain responds to visual stimuli in a dynamic environment. The brain has evolved to understand a reality generally based upon a near static or slowly changing environment. For modern men and women, running or rapid movement (outside the context of sport) is a relatively unnatural state for our brains. Decisions and precise calculations are made in this dynamic situation; however in such a state we are fully engaged with our sensory abilities and sensitive to a multitude of surface sounds and external stimuli. From an anthropological historical perspective the running state was a normal one for primitive man, and the sedentary state of the modern mind is a recent one. Our minds are a summation of a very long evolutionary past as so less influenced by recent "modern" history as Ward notes "Because the human mind and behavior are products of evolution, we must reconstruct the selective pressures that shaped our lineage in order to understand ourselves today" (Ward, 2003). The driving experience in terms of the brains interpretative abilities can thus be considered an even more abnormal state for us, as our body is static and sensory detection is visually centric. From inside the car, the driver's view or perspective appears as a movement down a "tunnel" like space with a precise time relation. The visually centric nature of the information in the dynamic car state presents an interpretative dilemma for the brain. On one hand we have evolved to understand dynamic running based state and engage all our senses to interpret the environment in real time in order to make and /or anticipate events affecting our safety.

On the other hand we are physically static in a dynamic state with a truncated sensory capacity (visual and limited sound). This has the effect of dulling our decision making capacity and increasing our likely misinterpretation of the data.

Currently the information presented in a Sat-Nav Telematic system is designed using a static view visual model. This does not take sufficiently into consideration the dynamic time space nature of the car environment. In general the dynamic nature of this information is essential and has to be considered because the driver moves with the car and feels instantaneous events sequences.

From the drivers perspective the information presented by the Sat-Nav (Telemetric) system is often unclear and can influence driving behavior negatively with respect to road traffic safety. Also by our estimation it does not improve driver orientation. Current navigation systems represent detailed information as a plan view or perspective map data, but in general we believe the human brain can better perceive mono sensory (visual) Sat Nav map information if it is carefully structured. This is especially the case when it is presented along with everything else as a kind of data flood typical of a driving situation. Our brains have a limited capacity to manage information in the short time frame of a dynamic space time environment. We also know that through training or experience, we may get improvements in information perception. However a new kind of navigation system based on a simplified and more meaningful information structure is welcome. Observing the above considerations, the characteristics of the information presentation should be both dynamic and compressed. Following this hypothesis we hope to improve current navigation system in both quality and quantity of data.

In published literature on driver safety there have been numerous studies documenting the effects of distracting tasks or of engaging in cell phone conversation. In general these studies demonstrated that distracted drivers have slowed responses to critical traffic events and are more likely to miss external events such as changing traffic light, among other effects (Alm, 1993; McKnight, 1993; Strayer and Johnson 2001; Hancock, 2003; Lees, 2007; Horrey, 2008). Many studies are also dedicated on the evaluation of the distraction effects related to the use of different kind of navigation systems. Dingus et al. (1994) reported some general performance results for six navigation configurations. In particular the results showed that the condition without voice required a higher demand on visual attention. The addition of voice substantially reduced the visual attention requirements. These results are reached considering mainly the driver eye glance behavior as well as variance of lateral acceleration, longitudinal acceleration and the number of large steering wheel corrected for trip time (Dingus, 1997).

Starting from the literature in this topic and from the limits previously described actually present in the current navigation systems, the DHS study was designed which proposed a different kind of interface, that took into consideration specific time space stimuli.

The main aim of this work is to study how a dynamic "Just in time" restructured, dynamic and compressed data feed could improve driver's interface comprehension and performance reducing the distraction and consequently increasing the safety. In particular the effect produced by three different navigator interfaces on the human behavior in a simulation car driving was set.

MATERIAL AND METHODS

The effects produced on driving by 3 interface navigation solutions (Sat-Nav, 2-Me interface and C – Me interface) were tested in this work. The Sat-Nav interface is the

Satellite navigator system proposed actually by the market (Figure 1). The 2-Me and C-Me interfaces, developed by DHS Company, are characterized by the use of specific landmark to indicate the way that the driver has to follow. Instead of the command proposed by the Sat-Nav system "after 140 m turn right", in the 2-Me and C-Me modality the command is substituted by "at the church turn right". In this way specific landmarks related to the landscape are used in order to better identify the street changes required by the driver. The selection of specific landmarks, related to the urban landscape perceptible as the easiest way by driver, were chosen in order to answer the compressed information requirement.

The two modalities (2-Me and C-Me) differ one to each other as described here following. 2-Me essentially presents information in segments that arrives at the driver as he or she drives down a given road (Figure 2). C-Me essentially gives the driver an over view of the trip in one "flash" (Figure 3). It connects the memory icons (drawings, photos of places of people) in a continuous strip. We structured our 2-Me and C-Me to allow for the 4-Dimensional Awareness model of the driver being in the car, with data elements arriving in sequence at him/her in a time space "tunnel" situation. We also used pattern recognition and abstraction scale to construct our 2-Me and C-Me solution.

20 subjects (11 males; 9 females, mean age 34 +/- 12 years) used to a navigation system driving were selected for the tests. The subjects were asked to drive in a simulated car environment, when in front of them a video taken from a real driving experience was presented. In particular the subjects were tested in these tasks:

- Task 1- go to the destination following Sat -Nav system.
- Task 2- go to the destination following 2-Me interface system.
- Task 3- go to the destination following C-Me interface system.

In all the navigation interfaces the given command was both visual and vocal. The aim of this experimental set-up was to evidence how many times and how long the drivers looked at the navigator device in order to better understand the map.

The subjects were analyzed during the simulated driving, analyzing their biomechanical behavior, with the 3D quantitative kinematic evaluation, and their opinion about the different interfaces, with a questionnaire.

In particular the effect produced by the 3 different interface was evaluated analyzing: 1) the number of the head movement corresponding to the number of time in which the driver look to the navigator display; 2) the time spent to maintain the position of the head versus the navigation display corresponding to the time spent by the driver looking to the navigator display instead of the road.

An optoelectronic system (Elite2002; BTS, IT) able to measure the 3D coordinates (x, y, z) of passive marker was used for the test. The markers were placed in specific points (head, upper limb, on the navigator) and in this way the 3D kinematics versus time was reconstructed. Three markers were placed on the head (in order to monitor its 3D behavior), and bilaterally, one on acromion, one on elbow, one on wrist and one on second metacarpal finger head. In order to define the position of the trunk, one marker was placed on the back, on C7, and one marker in the front side, on sternum. In order to identify the position of the navigator in the simulated driving environment, 4 markers were placed on the monitor's corners of the navigator system (Figure 4).

In order to know the head movement number to the navigator during the three different tasks, the angle α, in the horizontal plane, indentified by the central markers placed on the head and the x axis (corresponding to the anterior direction) was computed (Figure 5). When α is $=0°$ it means that the subject is looking to the road in front of him, when α shows a maximum as shown in Figure 6, it means that the subject is looking to the navigator display.

It is possible to observe (Figure 6) that when the face of the subject is directed to the navigator system, and there is a plateau, the duration of the plateau corresponds to the time spent by the driver reading the information inside the navigator display.

During this period the attention of the driver is far from the street and for this reason it represents a risk for safety and a risk for distraction.

This behavior is very important because the more time spent in the evaluation of the screen is, the more distraction produced by the navigator interface is.

Starting for the angle α versus time these following parameters were considered:

- Number of the head movements: i.e. how many time the subject looked to the navigator systems in order to better understand the way that he/she is following.
- Duration of movements, i.e. the time spent looking to the navigator in order to read the map and to understand the information

These parameters were computed for the three different tasks and a statistical analysis was conducted. All the previously defined parameters were computed for each subject and then the mean values and standard deviation relating to all indices were calculated for the group in the three tasks execution (interface 1, interface 2 and interface 3).

A one-way between groups analysis of variance (ANOVA) was applied to compare interface1, interface 2 and interface 3; the assumptions of the ANOVA model were tested by evaluating the fit of the observed data to normal distribution (Kolmogorov-Smirnov test) and homogeneity of variances (Levene's test). Specific effects were evaluated by means of the post-hoc comparisons of means (Bonferroni test). Null hypotheses were rejected when probabilities were below 0.05.

Two kinds of questionnaire were proposed to the drivers:

The first one (based on 19 questions) was used to test if the drivers are or not experienced with navigator system driving and with electronic equipments in general. These data are useful in order to establish if the tested subjects could be considered as a homogeneous group.

The second one (summarized in Table 1) was proposed in order to know the opinion of the drivers about the use of the three different interfaces. For each question of this questionnaire a score from 1 to 5 (1 not satisfied- 5 fully satisfied) was asked in order to evaluate each kind of Nav system performance.

Figure 1: Sat-Nav interface

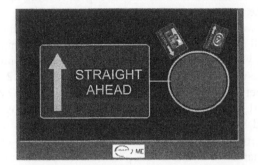

Figure 2: 2-Me interface

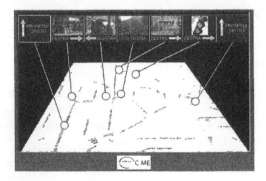

Figure 3: C-Me interface

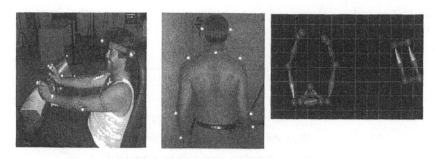

Figure 4: Marker position (lateral view, back view) and 3D reconstruction (top view during driving)

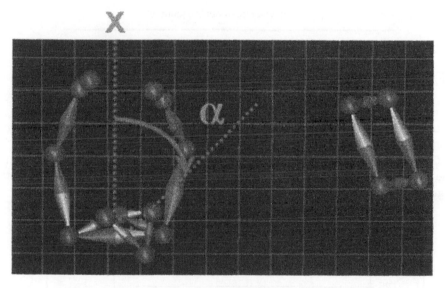

Figure 5: Identification of angle α for the head movement analysis.

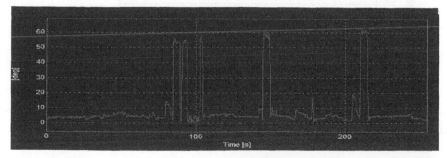

Figure 6: angles α versus time. Peak values correspond to the position of the driver face to the navigator. The plateau duration indicates the time spent by the driver analyzing the navigator screen.

Table 1: Proposed questionnaire for the evaluation of the opinion of the drivers in using the three different navigation interfaces (Note All questions are translated from Italian language original)

1	Has it been easy to follow the route?
2	Were the indications were clear?
3	Were they given early enough?
4	Were the turning points easy to recognize?
5	Is the terminology used easy to understand?
6	Had followed the street instead of the path on the screen?
7	Was the information enough?
8	Did you remember the journey?

RESULTS AND MAIN CONSIDERATIONS

According the results of the first questionnaire all the subjects resulted well trained in the use of navigation system during driving and of electronic interfaces (internet, cell phone,). From this point of view the subjects represented a homogenous group.

Concerning the kinematics, the number of the head movements (Figure 7a) and the time spent to look to the navigator system, i.e. the movement duration parameter (Figure 7b) are represented in the three different tasks.

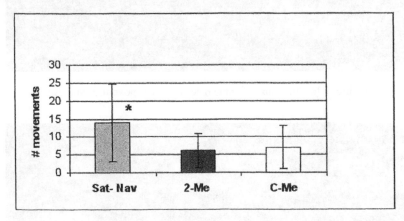

Figure 7a: Number of movement of the head to the navigator display (mean and s.d.). *= p<0.05, Sat-Nav versus 2-Me and C-Me.

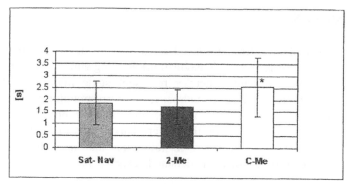

Figure 7b: Time spent to look to the navigator display (mean and s.d.). *=p<0.05, C-Me versus Sat-Nav and 2-Me.

It is important to highlight that the 2-Me and C-Me systems produced a lower number of movement to the navigator, i.e. the trip is better understood in these two modalities with respect to the Sat-Nav. These results are supported by the statistical analysis. In this way the subjects do not need to look to the navigator as often (many times) and it produces a safer driving (Reduced Distraction).

On the other hand the C-Me system is the modality which required more time, statistically significant, to be understood (duration of head movement parameter). It means that in 2-Me and in C-Me the subjects looked less the navigator, but when it happened, C-Me navigation required more time to show clearly the meaning of the road presented in this modality. From this analysis it seems that the 2-Me navigator is the interface which produced lower number of head movement to the navigator display and with a lower distraction. It means that the 2-Me navigation was the safer and less distractive modality.

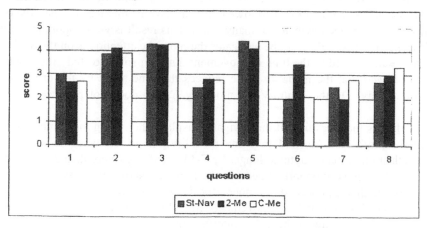

Analyzing the results related to the answer given by the drivers (Figure 8) in the second questionnaire, it is important in particular to highlight the results of the question n. 6. The drivers recognized that the 2-Me navigator interface allowed them to better follow the road confirming the more safety way to drive in this modality.

DISCUSSION

This study evaluated the effect produced on drivers by three different navigation interfaces (Sat- Nav, 2-Me and C-Me navigation) working with voice commands.

Current navigation systems generally represent static map and fully detailed information. They represent a static landscape, not corresponding to the real state of the driver inside a moving car environment. They can show too much details, which are not always easily understood by the driver. This results in confusion in driving orientation and a driver action delay in following the instruction given by the navigator, i.e. by the time the driver understands exactly the direction to be followed, it is too late.

The main characteristics of the new navigation interfaces (2-Me and C-me) consist in the use of commands with a cognitive meaning. The driver is asked to turn right or turn left at a "church" or at a "school" instead of to turn right ore left after "140 m" or other distance command. In this way very simply identifiable landmarks in an urban landscape and in the driving context (lights, banks or commercial buildings, architecture elements etc.) are easily recognized by the drivers, and are supported by vocal commands.

The evaluation of the drivers movement kinematics and in particular the head movement evaluations was done using a new approach based on optoelectronic system, i.e. an equipment able to provide 3D kinematics of body part marked with specific surface passive markers.

The results showed that the drivers understood very well the change direction required by the navigator when the navigator interface is supported by the use of landmarks, present in 2-Me and C-me interfaces, both in the vocal command and in the map. The concept is clearly demonstrated by the statistically significant reduction of the number of head movements to the navigator system. So the introduction of landmarks makes the map to be followed clearer and, for this reason, the drivers do not need to look at the navigator so many times. This result is very important because the less the driver is looking to the navigator the more the attention is on the road.

The results related to the time of movement duration spent to study the navigator display showed that the modality 2-Me reduces the time spent by the driver to the monitor display. On the contrary the modality C-Me, does not produce the same effect.

The drivers look less time at the navigator display when using the commands with landmarks (both in 2-Me and in C-me modality); however, when looking at the display, their attention is captured longer by C-Me modality (more than 2.5 seconds).

This results is very important because the literature shows that any single display glance longer then 2.5 seconds is inherently dangerous (Bhise 1986). As the mean values of the duration of head movements in C-Me interface is very close to this value, our concern is about the C-Me interface modality because it does not allow a safe driving.

The main conclusion of this paper can be summarized as follow:
1) The new interfaces (2-Me and C-Me) reduce the use of the navigation display information thanks to the introduction of landmarks;

2) The safest navigation system seems to be the 2-Me interface as the complexity of the map representation of C-Me interface is full of details that could be at risk for the drivers distractions.

In this way we confirm quantitatively the improvement of navigation system interface with the introduction of a dynamic and compressed information feed system. A term to describe this process would be adjusted multi modal data delivery whereby a dynamic external reality is mirrored by a compressed dynamic data feed which in effect reconstructs in the mind of the user a more appropriate mental model for the delivery of Navigation Data feed. The multimodal support in the above test cases was audio identification of landmarks. However future research will be aimed to expand on the modality of sensory stimuli to further enhance the rapidity of comprehension and further reduce the frequency and duration of glance away time.

REFERENCES

Ward C. (2003) *The Evolution of Human Origins*. American Anthropologist. 105 (1): 77-88

McKnight AJ, Mcknigh AS. (1993) *The effect of cellular phones use upon driver attention*. Accident Analysis and prevention, 25: 259-965.

Alm H, Nilsson L. (1994) *Changes in driver behavior as a function of hands free mobile phones – a simulator study*. Accident Analysis and prevention, 26:441-451.

Strayer DL, Johnston WA. (2001) *Driven to distraction: dual –task studies of simulated driving and conversing on a cellular phone*. Pysic Sci., 12:462-466.

Hancock PA, Lesch MF, Simmons L. (2003) *The distraction effect of phone use during a crucial driving maneuver*. Accident Analysis and prevention, 35:501-514

Lees MN, Lee JD. (2007) *The influence of distraction and driving context on driver response to imperfect collision warning systems*. Ergonomics, 50:1264-1286.

Horrey W.J, Lesh MF, Garabet A. (2008) *Assessing the awareness of performance decrements in distracted drivers*. Accident analysis and prevention, 40:675-682.

Dingus TA, Hulse MC, Mollenhauer MA, Fleischmann RN, McGehee DV, Manakkall N. (1997) *Effects of age, system experience, and navigation technique on driving with an advanced traveler information system*. Human factors , 39:177-199.

Dingus TA, Hulse MC, McGehee DV, Manakkall N, Fleischmann RN. (1994) *Driver performance results from the Trav –Tek IVHS camera car evaluation study*. In Proceeding of the Human factors and Ergonomics society 38th annual meeting, (pp.1118-1122). Santa Monica, CA: Human factors and Ergonomics society.

Bhise VD, Forbess LM, Farber EI. (1986) *Driver behavioral data and considerations in evaluating*, Washington D.C.

CHAPTER 10

Developing Methods for Utilizing Physiological Measures

Lauren Reinerman-Jones, Daniel Barber,
Stephanie Lackey, Denise Nicholson

Institute for Simulation and Training
University of Central Florida

ABSTRACT

Many test environments have been constructed to support experiments with human participants. Some, such as the Multi-Attribute Task Battery (MATB) and Virtual Battlefield System 2 (VBS2), are used widely throughout the research community. However, each is limited, particularly in their integration, or lack thereof, of physiological measures. The Mixed Initiative Experimental (MIX) testbed was constructed to capitalize on the benefits afforded by MATB and VBS2 and address the weaknesses present in those test environments. Theory driven tasks, integrated physiological and logging capabilities, and an easily reconfigurable interface define the MIX testbed. For those reasons, the MIX testbed is proposed as the next generation approach for testing environments.

Keywords: Testing Environment, EEG, ECG, GSR, Eye Tracking, Change Detection, Target Detection, Situation Awareness

INTRODUCTION

Neuroergonomics combines cognitive psychology and human performance with neuroscience and physiological indices (Parasuraman & Rizzo, 2007). As an emerging field, challenges are present. Perhaps the greatest challenge is not new, but an old issue spanning all of research concerning laboratory methodology involving human participants. That problem is building a test environment capable of laboratory research with controls, but simultaneously capable of field testing and assessing real-world impact. Often laboratory batteries are designed from an idea that a researcher would like to test, stimuli are simplified to eliminate the possibility of artifacts, and tasks are unitary to understand a specific phenomenon. However, real-world environments are dynamic and complex and the brain is as intricate and diverse as the cognitive and performance questions that experiments seek to answer. Thus, one might conclude that studies should only be conducted in the application environment to produce the best ecologically valid results. In contrast, without laboratory experiments, field researchers would lack critical insight into problem spaces regarding where to begin training or selecting personnel, improving brain-computer and human-computer interfaces, or triggering automation by physiological response or performance decrement. Additionally, laboratory environments allow for greater control and thus better statistical power and often at a lower cost. A compelling solution to the laboratory versus real-world research dilemma put forth in the current paper is to construct a testbed based on theory and field knowledge. Integrating fundamental methodological principles with an eye towards application might lead to the best data transfer.

EXISTING EXPERIMENTAL ENVIRONMENTS

Several simulation and task battery environments have been developed which provide researchers repeatable experimental environments for testing operator performance. The Multi-Attribute Task Battery (MATB), developed by NASA, incorporates tasks analogous to activities that aircraft crewmembers perform in flight in addition to auditory communications, resource management, and time scheduling (Comstock & Arnegard, 1992). MATB is a useful laboratory tool for multiple experiments testing a variety of hypotheses because of its ability to simulate single or multiple parallel tasks with varying levels of workload. With the explicit control of the tasks provided, experiments lend themselves to Meta-Analysis because of the direct comparisons possible across studies.

Although MATB provides a controlled environment for experimentation, it does not accurately replicate the dynamic high fidelity environments inherit to other task domains such as reconnaissance and surveillance activities required by the operators of unmanned systems. According to the Department of Defense's (DoD) Unmanned System Roadmap (U.S. Department of Defense, 2007), Reconnaissance and Surveillance and Target Identification and Designation are prioritized capability needs for the Warfighter. Within these tasks operators typically monitor visual feeds from camera payloads on Unmanned Air Systems (UAS) and Unmanned Ground Systems (UGS), (Barber, Leontyev, Sun, Davis, Nicholson, & Chen, 2008; Chen & Barnes, 2008). The visual monitoring tasks of MATB are low fidelity and do not include a dynamic live or virtual environment for events to take place.

The requirement of a high fidelity environment for experimentation has led to the use of more modern 3D simulations and commercial video games like Virtual

Battlefield System 2 (VBS2; Bohemia Interactive, 2010). VBS2 provides a fully interactive 3D environment for the development of scenarios and simulations for training or experimentation of unitary tasks, not a multi-tasking environment. Tools like VBS2 provide data collection and scripting capabilities, however they do not lend themselves well to the integration of Physiological sensors which require precise timing and feedback for event related measurements. Integration of these devices is also problematic when multiple computers are used to perform an experiment because systems like MATB and VBS2 record timing information relative to the start of a scenario or task and not to an external/global time (e.g. UTC). Although there has been some success using devices like EEG with MATB (Wilson & Russell, 2003) and VBS2 (Coufal & Martin, 2008), the addition of these sensors occurred as an afterthought to the experimental system, not part original design. During the integration process, modifications or additional modules were required for systems like VBS2 (Coufal & Martin, 2008) to allow for event synchronization with performance data logs.

Based on analysis of MATB, VBS2 and other similar environments, future testbed designs need to explore tasks that are independent and interact within complex dynamic environments while still supporting the controls demanded of a laboratory. Moreover, these systems must have built in methods for inclusion of measures collected from multiple computers, third-party applications, and participants through external time synchronization techniques. Physiological sensors must be accounted for in any task battery as an alternative means of assessing observer performance and state and for predicting and augmenting within a given environment.

THE NEXT GENERATION TESTBED

The Applied Cognition and Training in Immersive Environments (ACTIVE) Laboratory at the Institute for Simulation and Training (IST) has worked with its DoD partners to construct a testbed that capitalizes on the benefits of the MATB concept and VBS2 dynamic environment and addresses the limitations of those test platforms. The Mixed Initiative Experimental (MIX) Testbed incorporates theory driven tasks integrated into a moderately high fidelity simulation system based on DoD employed icons and terrains. Figure 1 shows the MIX testbed operator interface consisting of a mini-map (top left), a streaming video feed (top right), and a situational map (bottom). Additional advantages of the MIX testbed include the opportunity for assessing performance in a multi-tasking environment, the detailed logging capabilities that permit the use of multiple sensors (e.g. eye tracking, EEG, ECG, and GSR) simultaneously synchronizing with performance events, and the easily reconfigurable display allowing for countless investigations. Given the theoretical foundation of tasks, sensor integration, and reconfigurable features provided by the MIX testbed, it represents an experimental setting capable of bridging the gap between traditionally sterile laboratory testbeds and dynamic real-world environments.

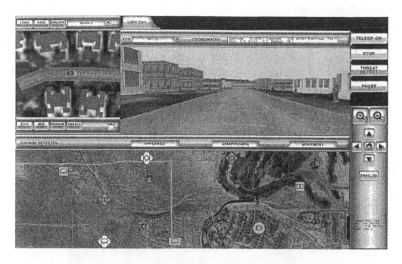

FIGURE 1. MIX testbed operator interface.

THEORY DRIVEN TASKS

Four tasks- change detection, threat detection, situation awareness, and auditory response- have generally been employed in the MIX testbed with particular interest to automation expectations, fitting with the unmanned system roadmap (U.S. DoD, 2007). That said, the MIX testbed is not limited to those tasks, but it was constructed with them in mind due to theory and literature supporting incorporation. Further, the testbed is not restricted to examining automation hypotheses, although manipulated levels of automated control include manual, adaptive, and static automation and the above mentioned tasks have relevance to adaptive conditions particularly because of the potential to trigger automation based on performance and physiological response to the tasks. The implications of that statement are driven primarily by resource theory, which is the idea that as task load increases, mental resources required to complete the assignment are depleted and not replenished (Matthews, Davies, Westerman, & Stammers, 2000). Therefore, the need exists to alleviate some resources for allocation to other tasks (Kahneman, 1973) and in this instance manual control is the relieved task.

Change detection research is founded on principles of change blindness in that certain types of change, the number of changes, the similarity of change to the original environment, the other tasks required, and many other factors influence the difficulty of detecting a change (Rensink, 2002). One step further, change blindness has been linked to the visual pathways of the brain and to eye tracking (Haffenden & Goodale, 2000; Hayhoe, Bensinger, and Ballard, 1998). The ability to manipulate task load of change detection and the physiological relation led to the MIX testbed change detection capability. The change detection task utilizes military icons located throughout a map of the vicinity in which a scenario takes place. Types of changes include appearance, disappearance, movement, color (affiliation), and shape (unit). Manipulations of task load consist of the number of icons present on the display, the amount of time between changes, and the type of change. Increasing or decreasing task load is time-based. Performance logs encompass an operator detecting that a change occurred, detecting the correct type of change that occurred, and reaction time. Figure 2 illustrates a recent study utilizing three types of change (appear, disappear, and movement) with varied timing between events (8 sec, and 15 sec) and task load manipulated by the number of icons present on the display (8 or 24). That task load increased or decreased applied after four and a half minutes on the task out of a 9-min scenario.

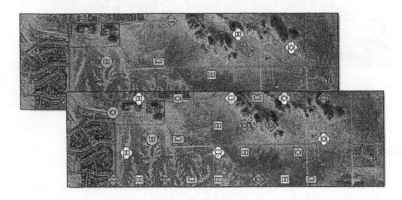

FIGURE 2. Example of change detection task capabilities.

Signal detection theory (SDT) was the driving model for the threat detection task. SDT entails the discrimination of events or signals from other items or noise (Wickens & Hollands, 2000). Detecting the difference between signal and noise results in four general classifications: hits, misses, false alarms, and correct rejections. Signals are said to invoke a stronger neural response than when not present, but it is possible for neural firing to occur even when a signal is not present and that would be classified as a false alarm (Wickens & Hollands, 2000). Shown in figure 3, the threat detection task integrated in the MIX testbed requires participants to discriminate between signals, Armed Civilians and Enemy Military, and noise, Friendly Civilians and Friendly Military. Given the significant impact of the story told by using military based scenarios and stimuli, the sensitivity-response bias trade-off is important. To improve ecological validity, 23 different signal stimuli and 30 different noise stimuli were used. The stimuli are able to be placed anywhere in the video feed screen, but in keeping with SDT they are spaced to maintain the one to three signal to noise ratio. In order for detection to occur, operators were required to first use a computer mouse to depress a threat button on the display and then select the threat identified on the display. That allowed logging of performance for detection of a threat present in the display, correct identification of a threat detected, reaction time for noticing a threat, categorization of the four SDT classifications, and error notation (i.e. multiple identification of the same target and inaccurate mouse clicks near a stimulus).

FIGURE 3. Examples of stimuli present in the threat detection task.

Endsley et al. (2000) provide an applicable model of situation awareness (SA) consisting of three levels. Level one SA addresses perception of the tasking environment. Level two SA encompasses comprehension of the tasking environment. Level three SA advances to projection within the tasking environment. Difficulty increases with each respective level. The MIX testbed implementation of SA assessment is based on SA probe delivery (Endsley, Bolte, & Jones, 2003) in which task-relevant probes or questions determined from subject matter experts are presented either visually or auditorily. In instances of visual presentation, a black screen replaces the display and a typed response is recorded. For auditory presentations, scenarios continue without interruption and the operator responds verbally. Both cases allow for time-based and/or location-based probes. Any other type of auditory response task can be included in the same manner as the SA probes. For example, experiments have implemented call sign response as a task, requiring operators to answer yes when their call sign is heard (Bravo 41) and no when any other call sign is presented (Alpha 36). Tasks such as these enhance ecological validity as they resemble radio communications.

PHYSIOLOGICAL SENSOR INTEGRATION

The present chapter alludes to the integration importance of physiological capabilities. The evidence for physiological links provided in the tasks above and the possibility of invoking automation from physiological state lead to the integration of Electroencephalogram (EEG), Eye Tracking, Electrocardiogram (ECG), and Galvanic Skin Response (GSR). Experiments conducted in the MIX testbed utilize any one or any combination of those sensors. Figure 4 shows all physiological sensors in use. The current sensors consist of Advanced Brain

Monitoring (ABM) 6 and 9-channel EEG units, Arrington EyeFrame Eye Tracking Systems, and Thought Technology ECG and GSR systems.

FIGURE 4. EEG, Eye Tracking, ECG, and GSR connected to a single operator.

To support data analysis and event synchronization, the MIX Testbed employs a standardized logging format that includes a common set of information that is used to correlate data post-hoc or during real-time data collection. Every data record that is added to a log file includes a Universal Coordinated Time (UTC) and Simulation time value. The UTC value is a global timestamp and is stored in two formats: Day:Hour:Minute:Second:Milliseconds, and the time converted to seconds. Simulation time is stored in the same format as UTC, but is relative to when the program or simulation started, similar to the time values from MATB log files. If a program records data on a different computer than the simulation, the data can be synchronized post-hoc using the UTC timestamps because they are global time values referenced externally between computers. To automate the re-tagging of Simulation timestamps in data logs post-hoc, the MIX Testbed also generates a "Simulation Time Log," which records scenario events such as: Start, Pause, Resume, and End. For each event in the "Simulation Time Log," UTC and Local Time Zone (e.g. Eastern Time) timestamps are recorded. By recording multiple time formats, it is possible to integrate events from other third-party applications post hoc as long as they also record this same information.

CONCLUSION

Developing an experimental testbed with flexibility to apply to a variety of problem situations and with built-in capabilities for integration of physiological measures is a long standing challenge. The MIX testbed seems the most advanced environment currently available for going beyond the default capacities of other existing systems like MATB and VBS2 to achieve the best controlled and yet applicable testing environment. Clearly each task employed in the MIX testbed is carefully selected based on theory and literature, but the environment is not limited by those tasks that have previously been employed, thus expanding its application to limitless problem situations. The advanced integration of physiological sensors increases the utility for future system and product development. Stepping beyond the single UGS setting, the MIX testbed has been utilized for investigating Remote Weapon Systems (RWS; Ortiz, Barber, Stevens, & Finkelstein, 2009) and control over multiple unmanned ground vehicles (results in progress). Future research in the MIX testbed might encompass testing reactor environments, intensive care monitoring, or unmanned air vehicles. The potential to assess physiological state in any dynamic environment is innumerable.

ACKNOWLEDGMENTS

This work was funded by the Army Research Laboratory Award # W91CRB08D0015.

REFERENCES

Barber, D., Leontyev, S., Sun, B., Davis, L., Nicholson, D., Chen, Y.C. (2008). The Mixed-Initiative Experimental Testbed for Human Robot Interactions. *Proceedings of the 2008 International Symposium on Collaborative Technologies and Systems (CTS 2008), pp. 483-489.*

Bohemia Interactive (2010). Virtual Battlefield Simulator 2 (VBS2) Homepage. Retrieved from http://virtualbattlespace.vbs2.com/.

Chen, Y.C., Barnes, M. J. (2008). Robotics operator performance in military multi-tasking environment. *Proceedings of the 3rd ACM/IEEE International Conference on Human Robot Interaction*, pp. 279-286.

Comstock, J. R., Arnegard, R. J. (1992). The Multi-Attribute Task Battery for Human Operator Workload and Strategic Behavior Research (NASA Technical Memorandum 104174). Hampton, Virginia: National Aeronautics and Space Administration.

Coufal, J., Martin, M. (2008). Simulation Framework for ERP Experiments. *Proceedings of the 9th International PhD Workshop on Systems and Control: Young Generation Viewpoint*, Izola, Slovenia.

Endsley, M.R., Holder, L.D., Leibrecht, B.C., Garland, D.J., Mattews, M.D., & Graham, S.E. (2000). *Modeling and measuring situation awareness in the infantry operational environment* (Research Rep. No 1753). Alexandria, VA: U.S. Army Research Institute for Behavioral and Social Sciences.

Endsley, M.R., Bolte, B., & Jones, D.G. (2003). *Designing for situation awareness: An approach to human-centered designing.* London: Taylor & Francis.

Haffenden, A. M., & Goodale, M. A. (2000). Independent effects of pictorial displays on perception and action. *Vision Research, 40,* 1597-1607.

Hayhoe, M. M., Bensinger, D. G., & Ballard, D. H. (1998). Task constraints in visual working memory. *Vision Research, 38,* 125-137.

Matthews, G., Davies, D.R., Westerman, S.J., & Stammers, R.B. (2000). *Human performance: Cognition, stress and individual differences.* London: Psychology Press.

Ortiz, E., Barber, D., Stevens, J., & Finkelstein, N. (2009). Simulation to assess an unmanned system's effect on team performance. Proceedings for the annual *Interservice/Industry Training, Simulation, and Education Conference,* Orlando, FL.

Rensink, R. A. (2002). Change detection. *Annual Review of Psychology, 53,* 245-277.

US Department of Defense (2007). Unmanned Systems Roadmap: 2007-2032. Arlington, VA: Directorate of Information Operations and Reports.

Wickens, C.D & Hollands, J.G. (2000). *Engineering psychology and human performance* (3rd ed.). Upper Saddle, NJ: Prentice Hall.

Wilson, G.F., Russell, C.A. (2003). Real-Time Assessment of Mental Workload Using Psychophysiological Measures and Artificial Neural Networks. *Proceedings of Human Factors, Vol. 45, No. 4, pp. 635-643.*

CHAPTER 11

Methods from Neuroscience for Measuring User Experience in Work Environments

Dieter Spath, Matthias Peissner, Sandra Sproll

Fraunhofer Institute for Industrial Engineering IAO

ABSTRACT

User Experience (UX) is identified to be a key factor for the success of interactive products. Human-oriented work environments support positive emotions and thus boost motivation and productivity. This paper gives an overview of methods from neuroscience with potentials in measuring UX. Chances and constraints of relevant methods in real work environment are discussed. Especially, their potential to identify crucial moments of human-computer interaction in real time can contribute to a deeper understanding of UX.

Keywords: User Experience, Neuroscience, Human-Computer Interaction, Emotion, Work Environments, Psycho-Physiological Methods, Imaging Methods

INTRODUCTION

Positive emotions contribute a lot to psychological health and play a prominent role for productivity, creativity and coping with pressure and stress. The measurement of "User Experience" (UX) gains in importance against the background of the wish for high quality of life and human-oriented, supporting work environments. The understanding of UX as a consequence of fulfilling basic human needs and the effects of a positive UX show clearly, that the insights from UX research are also meaningful for work science. Methods from neuroscience provide the opportunity to measure emotions during Human-Computer Interaction. Thus, factors for boosting positive emotions can be detected and further studied.

The aim of this paper is to give an overview of relevant methods for measuring emotions in relation with UX. Some of the presented psycho-physiological methods are suitable for measuring arousal; others capture the valence of an emotion. Neuroimaging methods detect brain indices for emotion. The potentials, but also the constraints of methods for measuring UX in realistic work environments will be discussed.

USER EXPERIENCE IN WORK ENVIRONMENTS

Emotion, motivation, joy-of-use, aesthetics and flow-experience are often discussed aspects of UX. These aspects reveal the subjective and positive nature of UX which is central for our UX model described below. Today UX research is mainly focused on consumer products. However, people spend most of their time in work environments and a lot of interaction with technology takes place at work.

USER EXPERIENCE MODEL

Our understanding of UX bases on a project of Fraunhofer IAO, Folkwang University, Stuttgart Media University (HDM), and Deutsche Telekom Laboratories, where we developed a UX model for telecommunication products.

An experience is an ongoing reflection on events (Hassenzahl, 2008) or an ongoing stream of self-talk (Forlizzi and Battarbee, 2004). We are interested in the experience referring to events as instances for human-computer interaction. Part of the self-talk is always the momentary feeling of pleasure or pain (Kahnemann, 1999) which regulates our behavior. UX is thereby seen as a momentary, evaluative (good-bad) and subjective feeling while interacting with technology. The origin of the positive or negative feeling is seen in the fulfillment of human needs. Thereby, a positive UX is the consequence of fulfilling human needs like relatedness, popularity, competence, stimulation, security or autonomy. The salience of these needs depends more on the situation (task or state of a person) than on personality. In work environments, e.g. while preparing a presentation, the need for competence might come to the fore, and a positive UX derives from a software enhancing experiences of competence.

BENEFITS OF POSITIVE USER EXPERIENCE

A key concern in Human Factors was adapting technology to the human nature. According to Hassenzahl (2008), "nature" referred primarily to perceptual and cognitive processes ensuring efficiency in task performance. This was especially valid in work environments where technology is seen as a tool to attain certain goals efficiently. However, recent discussions indicate that this view is too limited. Technology itself can be a source of pleasure (Hassenzahl, 2008), competence or autonomy experiences. According to the Self-Determination Theory, satisfaction of basic needs motivates the self to initiate behavior for its own sake because it is interesting and satisfying in itself (Deci and Ryan, 2002). If intrinsic motivation is enhanced at work through positive UX, both the organization as well as the individual benefits. For example, "flow" – the experience caused by an optimal balance of challenges and skills in a goal-oriented environment (Csikszentmihalyi, 1975) is the positive UX derived from fulfilling the need for competence (Hassenzahl, 2008). Moreover, the attachment of employees to their organization might be strengthened by emotional bonds with products belonging to their work environment (cf. Mugge, 2008). Positive UX with work equipment conveys appreciation of the employee and contributes thereby to a positive motivation.

MEASURING USER EXPERIENCE

Our definition of UX reveals two fields for assessing UX. Firstly, the response to needs in an interaction can be measured. Questionnaires or interviews are an efficient way to do this. Secondly, the experience itself – the positive or negative feeling – is of interest. Methods from neuroscience are considered to be especially promising for measuring emotions.

NATURE OF EMOTION

Human emotions are complex phenomena affected by different subsystems with a multi-component character (Mahlke and Minge, 2008). The *emotion triad* comprises the components subjective feelings, motor expresses and physical activation (Izard, 1977). The most prominent categorical approach by Ekman (1972) suggests six basic universal and cultural-independent emotions: Facial expressions are associated with happiness, fear, sadness, surprise, disgust and anger (Ekman and Friesen, 1978). However, Herbon et al. (2005) showed for HCI that basic emotions could not be easily mapped to physiological data because of their variations. They recommend dimensional approaches to structure emotion. A dimensional approach is Russell's *circumplex-model* with the dimensions valence (unpleasant – pleasant) and arousal (calm – excitement) (Russel and Pratt, 1980). Emotions can be arranged on these dimensions. Dimensional approaches enable to quantify and characterize emotions according to parameters and manifest changes within the *emotion triad*.

OVERVIEW ON METHODS

According to the *emotion triad* there are mainly three fields for measuring UX: Self-Reports based on subjective feelings and cognition, observation of behavior/expression and measuring biological correlates of internal processes.

At first, individuals´ ability for introspection in self-reports is constricted. Questionnaires also lack insight into the user's changing experience (Hirshfeld et al., 2009). Thus, questionnaires are an instrument for an overall assessment, but fail in identifying critical events responsible for a good or bad experience. Continuous measurement is the aim of observation methods. Using the Facial Action Coding System (FACS) based on Ekman, a trained person categorizes the observed pattern of activity for basic emotions. FACS serves as a basis for optical emotion recognition (e.g. FaceReader™ or SHORE™). A major limitation of these methods is their failure in detection of slightly shown emotions. An extensive facial expression is probably not usual in HCI. Neuroscience methods are a more immediate and sensitive way of measurement. Problems related to self-reports like biases due to attributions errors, social desirability and memory capacity are eliminated (Mandryk, et al., 2006)

METHODS FROM NEUROSCIENCE

The following paragraph gives an overview of relevant psycho-physiological and neuroimaging methods and discusses their suitability for measuring UX in work environments.

Psycho-Physiological Methods

Psycho-physiological methods are used to measure changes in the activity of the body correlating with psychological events (Grings and Dawson, 1978). These methods refer to the Autonomous Nervous System (e.g. EDA, ECG, EMG) that plays a prominent role for investigating emotional behavior. They also measure activity in the Central Nervous System (e.g. EEG) which consists of the brain and the spinal cord and is enclosed within the meninges.

Electro-Dermal Activity (EDA)

Cognitive activity and emotions increase the activity of the sympathetic nervous system whereby the perspiratory glands are activated. The skin conductance level is reduced, even if the perspiration has not reached skin surface (Stern et al., 2001). Usually skin conductance is recorded using two electrodes placed on the palms of the hands. In work environments where the person uses the hands to control a technical device, the feet can be used. Skin conductance reaction occurs 1-3 seconds after a specific stimulus (Dawson et al., 2007), but also without an identifiable external event as an unspecific spontaneous reaction (Schleicher, 2009). This implies that EDA is not specific to a single type of stimulus and can change due to multiple causes. Moreover, the changes can be caused by cognitive activity and various emotions. Merely arousal and thus the intensity of an experience can be measured, while the valence of an emotion cannot be determined definitely. However, previous research suggests that higher decreases in EDA are correlated with more negatively valenced situations (Herbon et al. 2005; Ward and Marsden, 2003). Summarizing, EDA can provide additional information about the experience, but should be combined with other methods (e.g. self-reports).

Electrocardiography (ECG)

A variety of parameters from the raw signal recorded by the ECG can be analyzed: Time-related parameters are heart rate, inter-beat-interval and heart rate variability (HRV) (Fahrenberg, 2001). Studies using the heart rate show inconsistent results for predicting emotional valence in HCI. Ward and Marsden (2003) found a decrease of heart rate in negative valenced situations, while other studies show a positive correlation between heart rate and valence (Bradley et al., 1993). Ekman, Levenson and Friesen (1983) reported a strongly increasing heart rate for anger, fear and sadness, a slight increase for pleasure and surprise and a decrease for disgust. The decrease for disgust was confirmed by following research, while for the other emotions inconsistent patterns occur (Schleicher, 2009). It is assumed that heart rate is a more reliable indicator for arousal or mental workload than for emotional valence (Fahrenberg, 2001). However, HRV seems to be a promising parameter for measuring UX. Some studies show a correlation between negative affect, mental workload, psychological stress and an increased Low/High-Frequency-Ratio (Hjortskov et al., 2004). Vice versa, a contrary effect can be found with positive affect (Bhattacharyya et al, 2008). Moreover, flow-experiences might be identified and quantified by HRV measures (Mueck-Weymann, 2002). Limitations of HRV measurement are related to big individual differences. For measuring UX, ECG should be combined with other methods (e.g. video-based protocols). ECG can easily be used in work environments by wearing it under the clothes. However, influences of physical activity have to be considered.

Electromyography (EMG)

Sensors placed on muscle regions detect minimal potential differences caused by muscle activity. For emotions, spontaneous reactions in the face are especially relevant. Facial EMG studies have shown that the activity above the eyebrow (Corrugator supercilii, muscle for frowning) varies inversely with the emotional valence. EMG-activity above the cheek (Zygomaticus major, muscle for smiling) is positively associated with positive affect (Cacioppo et al., 2000). Regarding the measurement of UX, the potential to discriminate between positive, neutral and negative emotions is valuable. Muscle activity provides more immediate information on the valence of an experience than self-reports. Thus, crucial events of an experience can be detected. In contrast to optical recognition of emotions, EMG is a more sensitive method. Emotions can be detected, even if they are not shown extensively. The limitations are adulteration through talking and laughing. This occurs in interactive situations - typical for work environments-, or could arise by other methods (e.g. "Think-Aloud"). However, the interference can be reduced by using needles instead of electrodes (Mandryk and Atkins, 2006). This could be distracting in real work environments. Summarizing, EMG can indicate the potentially unconscious valence of an experience, if the context of use is adequate.

Electroencephalography (EEG)

Traditionally, attention or states-of-consciousness as well as information processing for external and internal stimuli are explored with the EEG (Schandry, 1998). The working brain creates electrical signals in active regions. Potential differences between two electrodes on the scalp are measured in varied combinations, amplified and recorded in waves. Thereby, temporal resolution is very good. Facing the critics of only moderate spatial resolution the improvement of electrodes, more powerful computers for measurement with many electrodes aim to improve the localization of activity (Hüsing et al., 2006). For measuring UX localization of activity is important in order to identify emotions. EEG is highly susceptible to biological artifacts, i.e. bio-signals not originated by the brain that certainly appear while measuring UX (e.g. eye-movement). Instructions (e.g. to relax or to blink in advance) can reduce the artifacts, but this might be difficult in real work environments. Moreover, methodic adaptations (e.g. using low-pass filter) or additional measurements (e.g. EOG to control eye-movement) are helpful (Schandry, 1998). A big problem in work environments might be the susceptibility to electrical interference. In contrast to other methods, the set-up for EEG allowing for localizing activity is complex (Lee and Tan, 2006). Summarizing, the EEG can be used besides measuring cognitive workload and arousal to localize brain activity during product interaction. The susceptibility to artifacts should be reduced, so that the method can be used in real work environments (Minnery and Fine, 2009).

Imaging Methods

Like a "window to the brain" imaging techniques enable to explore the structure and functional activity. The principle is to create an image out of a huge amount of data gathered from the measurement of physical variables (Hüsing et al., 2006). The spatial resolution is illustrated by luminance and color-coding. For measuring UX those techniques are relevant that can provide images of the working brain.

Functional Magnetic Resonance Imaging (fMRI)

In recent years the fMRI has become one of the most important methods in cognitive neuroscience, because active structures can be illustrated with relatively high spatial resolution. A key component of the blood, hemoglobin, is responsible for transporting oxygen from the lungs to active cells. The response to this oxygen utilization is to increase blood flow to regions of increased neural activity. Through a process called hemodynamic response, the ratio between oxygenated and deoxygenated blood increases. fMRI utilizes the different magnetic characteristics of oxygenated and deoxygenated hemoglobin and measures the distribution and thereby the activity (BOLD-effect) (Hüsing et al., 2006). An issue of the fMRI is the indirect measurement. Not the activity itself is measured, but the hemodynamics as a result of activity (Hüsing et al., 2006). As a consequence, also the relatively low temporal resolution can be criticized. Besides exploring brain activity regarding perception, motion and memory, fMRI has also been used for measuring emotions and their functional neuroanatomy: In comparison to neutral visual stimuli, processing of emotional stimuli leads to an increased activation of various cortical areas, including the amygdale, the medial prefrontal cortex and sensory areas (Phan et al., 2002). Moreover, fMRI experiments show that unexpected rewards elicit activation in brain regions that utilize dopamine (D'Ardenne et al., 2008). Minnery and Fine (2009) take this as evidence of the potential to identify aspects of the interaction a user finds pleasing. An issue of fMRI for measuring UX is besides costs and complexity the fact that subjects have to remain relatively motionless in a scanner with a high noise-level. The interaction with e.g. computers is not possible, because metal objects cannot be used in the scanner. Moreover, it is doubtful that there is a realistic work experience at all. The interaction could be replaced by an alternative representation (e.g. movies of the interaction), but the stimuli might be too weak to cause activation. Summarizing, fMRI proved to be a method enabling to "see" brain activity, but its drawbacks diminish its ability to measure UX in real work environments.

Functional Near-Infrared Spectroscopy (fNIRS)

A less capital-intensive method, relatively new technology for non-invasive real-time brain activity measurement has been introduced to HCI. Like fMRI, fNIRS allows reflecting the dynamics of the cerebral blood flow. fNIRS uses light sources in the near infrared wavelength range (650nm-850nm) and optical detectors to probe brain activity (Izzetoglu et al., 2004). The changes in the absorbed light can be ascribed to changes in oxygenated and deoxygenated hemoglobin and provide relevant markers of neural activity in the brain. The light attenuation allow for tissue imaging at depths up to 2-3 centimeters. Many higher cognitive functions of interest in HCI like planning, problem solving, memory retrieval, attention, visual-spatial processing are localized in the outmost layer of brain tissue and are accessible to fNIRS (Girourd et al., 2008). For detection of emotions also deeper structures of the brain are essential like the limbic system. In studies fNIRS measured changes in brain activation of the occipital cortex modulated by visual emotional stimuli, at least for deoxygenated hemoglobin (Herrmann, et al., 2006). Moreover, the insula or the orbito-frontal cortex relevant for a cognitive representation of emotion can be explored (Hofman, 2010). The current improvement of the fNIRS technique itself will probably allow a deeper exploration of emotions in the future. Helpful for measuring UX are experiments of Solovey et al. (2009) conducted to develop guidelines for using fNIRS in real environments. Their goal was to examine whether physical behavior typical for computer usage interferes with fNIRS. Limitations like clicking, typing, minor head and eye movements or ambient light inherent in other brain sensing techniques are not factors when using fNIRS or could be controlled. Altogether, the characteristics of fNIRS make it suitable to be used in real work environments. fNIRS is easy to use, safe, portable, less invasive than other imaging techniques, and could be used wirelessly.

Conclusion and Outlook

Psycho-physiological methods measure correlates of psychological processes and enable an immediate measurement of an experience. As some methods focus on arousal, others on valence, chances are seen in method combinations. Combined with self-reports they provide markers for critical events as basis for interviews. Imaging methods shed light on the experience by localizing and characterizing emotional processes. Limitations are most notably the complex set-up, the interference with external artifacts not avoidable in real work environments, and the disturbance of the experience by the method itself. Nevertheless, the technical progress of these techniques improves the ability to measure UX efficiently, especially in real environments. The fNIRS seems to have great potential for both evaluation of interaction as well as for real time input to systems in order to improve UX (Girouard et al., 2008). Systems could adapt their behavior to current information measured from the brain not only in order to reduce mental workload, but also to enhance positive emotions.

Summarizing, methods from neuroscience will not replace classical self-reports and observation. The multi-component character of emotions should always be addressed by different methods (Mahlke and Minge, 2008). Especially the chance to identify critical moments of the experience in real time contributes a lot to a deeper understanding of UX.

REFERENCES

Bradley, M.M., Greenwald, M.K., and Hamm, A.O. (1993), Affective Picture Processing and the semantic differential, *Journal of Behavioural Therapy and Experimental Psychiatry, 25,* 204-215.

Bhattacharyya, M.R., Whitehead, D.L, Rakhit, R., and Steptoe, A. (2008), Depressed Mood, Positive Affect, and Heart Rate Variability in Patients With Suspected Coronary Artery Disease, *Psychosom Med*, 70(9), 1020-1027.

Cacioppo, J.T., Berntson, G.G., Larsen, J.T., Poehlmann, K.M., and Ito, T.A. (2000), The psychophysiology of emotion, in: Haviland-Jones, J.M. (Ed.), *The Handbook of Emotions*. New York: The Guilford Press.

Csikszentmihalyi, M. (1975), *Beyond Boredom and Anxiety*, SF: Jossey-Bass.

D'Ardenne, K, McClure, S.M., Nystrom, L.E., and Cohen, J.D. (2008), BOLD Responses Reflecting Dopaminergic Signals in the Human Ventral Tegmental Area. *Science*, 319, 1264-1267.

Dawson, M. E., Schell, A. M., and Filion, D.L. (2007), The Electrodermal System, in: Cacioppo and Tassinary (Eds.), *Principles of Psychophysiology: Physical, Social and Inferential Elements* (pp.159-181). Cambridge: Cambridge University Press.

Deci, E. and Ryan, R. (2002), *Handbook of self-determination research*, Rochester, NY: University of Rochester Press.

Ekman, P. and Friesen, W.V. (1978), *Facial Action Coding System: Investigator's Guide*, Palo Alto: Consulting Psychologists Press.

Ekman, P., Levenson, R. W., and Friesen, W. V. (1983), Autonomic Nervous System Activity Distinguishes among Emotions, *Science*, 221, 1208-1210.

Fahrenberg, J. (2001), Physiological fundamentals and measuring methods of cardiovascular activity, in: Rösler, F. (ed.) *Grundlagen und Methoden der Psychophysiologie* (pp. 317–454), Gottingen: Hogrefe.

Forlizzi, J. and Battarbee, K. (2004), Understanding experience in interactive systems, in: *Proceedings DIS 04: processes, practices, methods, and techniques*, New York: ACM.

Girouard, A., Hirshfield, L.M., Solovey, E., and Jacob, R.J.K. (2008), Using functional Near-Infrared Spectroscopy in HCI: Toward evaluation methods and adaptive interfaces, *Proceedings CHI 2008, Florence: ACM*.

Grings, W.W., and Dawson, M.E. (1978), *Emotions and bodily responses: A psychophysical approach*, New York: Academic Press.

Hassenzahl, M. (2008), User Experience (UX): Towards an experiential perspective on product quality, in: *IHM '08: Proceedings of the 20th French-speaking conference on Human-computer interaction*.

Herbon, A., Peter, C., Markert, L., van der Meer, E., and Voskamp, J. (2005), Emotion Studies in HCI, *Proceedings of the HCI*, Mahwah/NJ: Erlbaum.

Herrman, M.J., Huter, T., Plichta, M.M., Ehlis, A.C., Alpers, G.W., Muehlberger, A., and Fallgatter, A.J. (2008), Enhancement of Activity of the Primary Visual Cortex During Processing of Emotional Stimuli as Measured With Event-Related fNIRS and ERP, *Human Brain Mapping*, (29), 28-35.

Hirshfield, L.M., Solovey, E.T., Girouard, A., Kebinger, J., Jacob, R.J.K., Sassaroli, A., and Fantini, S. (2009), Brain Measurement for Usability Testing and Adaptive Interfaces. An Example of Uncovering Syntactic Workload with Functional Near Infrared Spectroscopy, in: Greenberg, S., Hudson, S.E., Hinckley, K., Morris, M.R., and Olsen, D.R. (Eds.), *Proceedings CHI 09*, Boston: ACM.

Hofmann, M.J. (2010), retrieved Feb 22nd, 2010, from http://www.languages-of-

120

emotion.de/de/dine/nirs.html.

Hjortskov, N., Rissén, D., Blangsted, A. K., Fallentin, N., Lundberg, U., and Søgaard, K. (2004), The effect of mental stress on heart rate variability and blood pressure during computer work, *European Journal of Applied Physiology*, 92 (1-2), 84-89.

Hüsing, B., Jäncke, L., and Tag, B. (2006), *Einblick ins Gehirn – Hirnuntersuchungen mit bildgebenden Verfahren.* TA-SWISS Studie, Bern: vdf Hochschulverlag AG.

Izard, C.E. (1977), *Human Emotions*, NY: Plenum Press.

Izzetoglu, K., Bunce, S., Onaral, B., Pourrezaei, K., and Chance, B. (2004), Functional Optical Brain Imaging Using Near-Infrared During Cognitive Tasks, *International Journal of Human-Computer Interaction,* 17 (2), 211-231.

Kahneman, D. (1999), Objective happiness, in: Kahneman, D., E. Diener, E., and Schwarz, N. (Eds.), *Well-being: The foundations of hedonic quality* (pp.3-25), New York: Sage.

Lee, J.C. and Tan, D.S. (2006), Using a Low-Cost Electroencephalograph for Task Classification in HCI Research, *Proceedings, UIST 2006*, Montreux: ACM.

Mandryk, R.L. and Atkins, M.S. (2006), A fuzzy physiological approach for continuously modelling emotion during interaction with play technologies, *International Journal of Human-Computer Studies,* 65, 329-347.

Mandryk, R.L., Inkpen, K., and Calvert, T.W. (2006), Using Psychophysiological Techniques to Measure User Experience with Entertainment Technologies, *Behaviour and Information Technology,* 25 (2), 141-158.

Mahlke, S. and Minge, M. (2008), Consideration of Multiple Components of Emotions in Human-Technology Interaction, in: Peter, C. and Beale, R. (Eds.), *Affect and Emotion in HCI,* (pp.51-65), Berlin: Springer.

Minnery, B.S. and Fine, M.S. (2009), Neuroscience and the Future of Human-Computer Interaction, *Interactions*, 16 (2), 70-75.

Mugge, R. (2008), *Emotional Bonding with Products*, Saarbrücken: VDM.

Mueck-Weymann, M. (2002), retrieved Feb 22nd, from www.hrv24.de.

Phan K.L., Wager T., Taylor S.F., and Liberzon, I. (2002), Functional neuroanatomy of emotion: A meta-analysis of emotion activation studies in PET and fMRI, *Neuroimage,* 16, 331–348.

Russel, J.A. and Pratt, G.A. (1980), A Description of the Affective Quality Attributed to Environments, *Journal of Personality and Social Psychology,* 38, 311-322.

Schandry, R. (1998), *Lehrbuch Psychophysiologie – Körperliche Indikatoren psychischen Geschehens*, Weinheim: Beltz Verlag.

Schleicher, R. (2009), *Emotionen und Peripherphysiologie*, Lengerich: Pabst Science Publishers.

Solovey, E.T., Girouard, A., Chauncey, K., Hirshfield, L.M., Sassaroli, A., Zheng, F., Fantini, S., and Jacob, R.J.K. (2009), Using fNIRS Brain Sensing in Realistic HCI Settings: Experiments and Guidelines, *Proceedings UIST*, Canada: ACM

Stern, R.M., Ray, W.J., and Quigley, K.S. (2001), *Psychophysical Recording*, New York: Oxford University Press.

Ward, R. D. and Marsden, P. H. (2003), Physiological Responses to Different Web

Page Designs, *International Journal of Human-Computer Studies*, 59, 199-212.

CHAPTER 12

Subjective and Objective Measures of Operator State in Automated Systems

Lauren Reinerman-Jones, Keryl Cosenzo, Denise Nicholson

University of Central Florida, Army Research Laboratory
and University of Central Florida

ABSTRACT

Automation is the future alternative for alleviating the performance decrement that occurs when an operator in a military or civilian job becomes highly taxed with increased task load. Assessing the state at which automation is needed is pertinent to effectively creating adaptive systems. Self-report questionnaires, such as the NASA-TLX, are means for quantifying workload. However, questionnaires do not continuously evaluate workload state throughout the completion of a task. Instead, administration of such measures requires interruption to the task and thus would not be a useful invocation tool for adaptive automation. An alternative is the use of physiological measures such as Eye Tracking and Electrocardiogram. The present study sought to compare workload assessment instruments including the NASA-TLX, Nearest Neighbor Index, Heart Rate Variability, and Interbeat Interval. Physiological response results show promise for the future of adaptive systems.

Keywords: Automation, ECG, Eye Tracking, Nearest Neighbor Index, Workload

INTRODUCTION

In both civilian and military applications individuals are often tasked with more responsibilities than they are able to handle. To ameliorate this task load, automations are often provided. Automation can range from a simple alert to a decision aid. An issue with automation, however, is that if misappropriately applied it can yield performance decrements and not the gains intended. One approach to offset the challenges of static is adaptive automation. Adaptive automation solutions may moderate operator workload and preserve situation awareness (SA) by applying automation when the task demands and the operator's cognitive resources reach a critical point. There are several ways to invoke automation, based on critical event, a model of operator performance and events, actual operator performance or physiological measurement in real-time, or a hybrid approach that combines the invocations mechanisms. Adaptive automation uses mitigation criteria that drive an invocation mechanism to maintain an effective mixture of operator engagement and automation for a dynamic multi-task environment.

If an adaptive automation solution is developed that is triggered by operator performance and or physiological measurement we need to be able to quantify the operator's state. Workload is a significant contributor to performance and is often associated with changes in physiological activity. Workload is a transactional process with regard to the human's evaluation of a given task and more specifically is the cost accrued by engaging in task requirements (Hart & Wickens, 1990). Typically high workload is associated with performance deterioration (Wickens, 2002), which is best accounted for by resource theory. Resource theory is the dominant approach to the assessment of perceived workload (Wickens, 2002). Workload from a resource theory viewpoint is described by the relation between an individual's cognitive resources and the situational demands (Norman & Bobrow, 1975). Succinctly, workload is the operator's resource demand associated with task performance (Wickens & Hollands, 2000).

A common goal expressed throughout the history of workload research addresses capturing the complexity of workload into a quantifiable measure that allows investigators to control, predict, and interpret workload levels on a given task. One approach to quantify workload is by self-report questionnaires. The NASA-Task Load Index (NASA-TLX; Hart & Staveland, 1988), a well regarded self-report measure of perceived workload or the resource demands associated with task performance (Wickens & Hollands, 2000), evaluates perceived workload on six sub-scales: mental demand, physical demand, task demand, performance, effort, and frustration. Participants rate their perception of each of these sub-scales for a given task by providing ratings of 0-100 with 100 representing the highest perceived level of workload (Hart & Staveland, 1988).

As noted previously, workload often relates to changes in physiological activity, thus another strategy for quantifying workload is through physiological measures. Physiological correlates of workload include changes in EEG patterns, heart rate, and eye scanning. Previous research reported limited success with measures of heart rate variability (HRV), galvanic skin response (GSR), pupil dilation, and event-related potentials (ERP) (Kramer, 1991; Parasuraman, 1990). More recently, investigations into workload measured via eye-movements using sophisticated eye tracking technology occurred. McCarley and Kramer (2007) provided a comprehensive review of the role of eye tracking in predicting and describing workload on various tasks such as driving, baggage inspection, and cockpit instrumentation. The user-friendly, affordable eye tracking systems have resulted in improved comprehension, explanations, and predictions of a variety of tasks and processes, such as cognitive state or mental workload level, task load imposed by a secondary task, and interface design (McCarley & Kramer, 2007). To that end, the present study aims to strengthen the foundation of eye tracking as a valuable measure of workload and to provide clarity to the conflicting results surrounding IBI and HRV.

NEAREST NEIGHBOR INDEX (NNI)

The idea of objectively measuring workload through eye tracking led to the innovative application of nearest neighbor index (NNI), a concept of randomness, implemented to quantify eye tracking for differentiating workload levels on a task. NNI is a distance statistic that expresses the proximity of each point (e.g. object or instrument) on a plane of space relative to all other surrounding points (Clark and Evans, 1954). It measures the randomness of fixation patterns. When applied to eye movements, NNI takes into consideration all fixations. NNI is a ratio of the average of the observed minimum distances between points or fixations and the mean distance that one would expect if the distribution were random. According to Di Nocera, Camilli, Terenzi, and Nacchia (2007, p. 6), "This ratio is equal to 1 for a distribution that is random. Values lower than 1 suggest grouping, whereas values higher than 1 suggest regularity (i.e. the point pattern is dispersed in a non-random way). Theoretically, the NNI lies between 0 (maximum clustering) and 2.1491 (strictly regular hexagonal pattern)." In other words, as the NNI values move away from 1 in either direction, the distribution becomes less random and becomes either clustered or regular depending on the respective direction. Clustering occurs when fixation patterns group together in areas of the display. Fixation patterns are regular when they take on a strictly hexagonal pattern (six-sided geometric figure). NNI has been found sensitive to workload variations and Figure 1 depicts that relation.

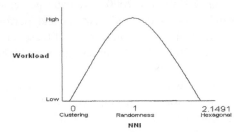

FIGURE 1. Workload as a function of NNI (left) and entropy (right).

Di Nocera and his colleagues (Camilli, Terenzi, & Di Nocera, 2007; Di Nocera, Camilli, Terenzi, in press; Di Nocera, Camilli, Terenzi, & Nacchia, 2007; Di Nocera, Terenzi, & Camilli, 2006) have consistently found randomness in high workload conditions and non-randomness in low workload conditions. These advocates of NNI have argued that this direction is geometrically logical because a regular check of interface space would be more reasonable when workload is low and time is not an issue (Di Nocera, Terenzi, & Camilli, 2006). Stated differently, it seems more likely that an observer in a high workload condition would have more random scan patterns, patterns that jump around only to the instruments or areas on a display needed at that moment. An observer in a low workload condition would have less random scan patterns because the condition would allow more time for the observer to make regular checks of all instruments or areas on a display.

NNI studies have most often employed dynamic and continuous situations to evaluate fixation pattern randomness (Di Nocera et al., 2006). The idea that NNI could provide a continuous or real-time measure of workload is exciting because many complex tasks induce varying levels of workload throughout their execution. Camilli et al. (2007) confirmed concurrent validity of NNI as a measure of workload by conducting a study that elicited results of the same caliber from the NNI, NASA-TLX, and P-300 amplitude. The present study took this validation approach by including the NNI, NASA-TLX, and HRV. The purpose of validating NNI is to have an accurate, objective, continuous, and non-intrusive measure of workload. This is important in the effort to eliminate response bias, and because many real-world tasks are unremitting, meaning that interruptions are not feasible. The latter factor is even more significant if the ultimate goal is to step past simply describing workload to predicting workload, which would be useful for invoking automation or temporarily relieving a person of his or her duties.

METHOD

PARTICIPANTS

Forty-one participants (28 males and 13 females) ranging in age from 18-40 years were recruited from the Central Florida area and were compensated for their time. Each participant completed two phases, one training and one testing, totaling five hours. The testing session was completed within seven days of training.

THE MIXED INITIATIVE EXPERIMENTAL TESTBED (MIX)

The Mixed Initiative Experimental Testbed (MIX) is a simulated environment designed to test human-robot interactions and operator performance in various automated conditions (Barber, Davis, Nicholson, Finkelstein, and Chen, 2008). The present experiment utilized the MIX testbed to create a multi-tasking environment upon which to assess operator's physiological responses throughout six 9-minute scenarios. That multi-tasking environment consisted of a three condition control task in which the participant was required to navigate a ground vehicle through a route. In the manual or teleoperation control condition, participants steered the vehicle with a joystick through pre-designated routes. In the automated or static condition, the vehicle was directed through routes via waypoints. The adaptive automation condition consisted of beginning the scenario in manual and ending the scenario in static or vice versa. Figure 2 illustrates the route map (top left) used for direction reference when in manual control or waypoint reference when in static control and the streaming video feed of vehicle view (top right) used for traveling through the terrain. In the latter region, participants were also required to identify threats portrayed as armed civilians or enemy soldiers.

FIGURE 2. Display of the MIX testbed.

The bottom half of figure 2 employed a change detection task in which operators were asked to detect changes in icons. Three types of changes–appearance, disappearance, and location movement–occurred an equal number of times at an equal number of two rates, 10 or 15 sec randomized, to limit the possibility of participants anticipating a pattern. Change detection task load was manipulated by increasing or decreasing the number of icons present in the display. Therefore, low task load conditions contained roughly 8 icons and high task load conditions altered about 24 icons. The approximation is due to the continuous, non-discrete progression of the task. An increase or decrease in task load happened in each of the six scenarios with that change implemented at 4min and 30 sec out of the 9-min scenario. As previously mentioned, automation is able to be invoked by a number of triggers and in this instance automation in the two adaptive scenarios was turned on when task load was high. See figure 3 for the design.

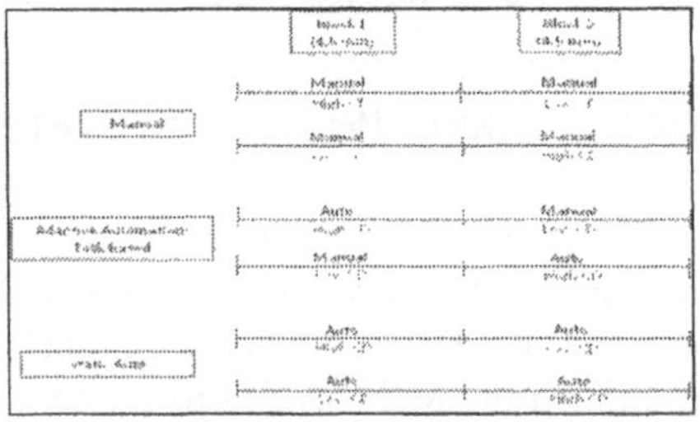

FIGURE 3. Design for control (auto = automation), block, and task load (CD = Change Detection)

MEASURES: NASA-TLX, ELECTROENCEPHALOGRAM (ECG), AND EYE TRACKING EQUIPMENT

The NASA-TLX was administered following each scenario. A Thought Technology ECG unit was employed for the collection of interbeat interval and heart rate variability. Three electrodes, two placed on the collar bone and one below the sternum, collected data that was recorded on a connected Dell desktop computer. An Arrington EyeFrame Head-Mounted Eye Tracker measuring infrared corneal reflectance was worn by the observer to record fixations onto the same Dell desktop computer. Fixations were eye positions with maximum change of 1° of visual angle for a minimum of 100 ms. Performance on the tasks described above is outside the scope of this paper.

PROCEDURE

During the training phase, an informed consent form was collected from participants before acquainting them with the tasks of the MIX testbed. Training consisted of a self-paced PowerPoint instructional guide with the ability to reference the MIX testbed on a separate computer monitor positioned next to the training screen. Manual control, automated control, threat detection, and change detection were practiced individually for 2 min each. Once participants were comfortable with each task separately, they completed one entirely manual 4-min scenario and one entirely static 4 min-scenario with all tasks integrated.

The testing phase began with a brief PowerPoint review of the tasks and completion of two 4-min adaptive scenarios involving all tasks. Physiological measures were then connected and calibrated to the participants. Participants were instructed to place the three ECG electrodes at the appropriate sites and a five minute resting baseline was taken. The eye tracking unit was dawned and calibrated by participants following a series of 16 squares on the screen with their eyes. Slip correction was used between each scenario. ECG and eye tracking measures were continuous throughout all six scenarios. Scenario order was counter-balanced and the NASA-TLX was administered following each scenario.

RESULTS

All ANOVAs utilized SPSS calculations of the Bonferroni correction for alpha and Greenhouse-Geisser to account for violations of sphericity.

NASA-TLX

A 3 (control: manual, adaptive, and static) x 2 (task load: low and high) repeated measures ANOVA was conducted for physical demand, temporal demand, mental demand, performance, effort, and frustration sub-scales. A significant main effect was found for control on the physical demand sub-scale, $F(1.877, 75.070) = 13.196$, $p < .001$, such that perceived physical demand was greater for adaptive ($M = 33.293$) than static ($M = 27.378$) control, but manual ($M = 39.329$) was greater than both static and adaptive with all post-hoc comparisons significant at $p < .05$.

A significant main effect was found for control on the frustration sub-scale, $F(1.883, 75.327) = 3.624$, $p < .031$, such that perceived frustration was greater for adaptive ($M = 42.866$) than static ($M = 38.841$) control, but manual ($M = 43.171$) was greater than both static and adaptive. Post-hoc comparisons showed no significant difference using Bonferroni correction, however, without the correction, the difference between adaptive and static was significant ($p = .038$) and between manual and static was significant ($p = .030$). No significant effects were found for task load.

ECG

Interbeat Interval(IBI)

A 3 (control: manual, adaptive, and static) x 2 (block: 1 and 2) x 2 (task load: low and high) repeated measures ANOVA was conducted for IBI of each condition compared to baseline IBI. A main effect for task demand was found, $F(1,40) = 8.325$, $p = .006$, such that low demand elicited slower IBI ($M = .945$) than high demand ($M = 1.698$). A main effect for block, $F(1,40) = 13.618$, $p = .001$, showed that IBI was faster for block one ($M = 1.822$) than block two ($M = .821$). A significant control by demand interaction was found, $F(2,80) = 6.505$, $p = .002$, such that in adaptive control slower IBI occurred for low task load ($M = .112$) and faster IBI was invoked for high task load ($M = 2.229$). That difference was significant ($p = .002$). A control by block interaction was also revealed, $F(2,80) = 4.578$, $p = .013$, such that when in autonomous control IBI was faster for block one ($M = 2.482$) than block two ($M = 1.589$) and when in manual control IBI was faster for block one ($M = 1.610$) than block two ($M = -.094$). Both differences were significant ($p = .030$ and $p < .001$, respectively).

Heart Rate Variability (HRV)

A 3 (control: manual, adaptive, and static) x 2 (block: 1 and 2) x 2 (task load: low and high) repeated measures ANOVA revealed no significant effects for HRV.

NNI

Two methods for calculating NNI have been used throughout the literature. Smallest rectangle defines the smallest possible rectangle surrounding all points being examined and convex hull locates three points to create a temporary hull and then defines triangles from each point. A 3 (control: manual, adaptive, and static) x 2 (block: 1 and 2) x 2 (task load: low and high) repeated measures ANOVA revealed no significant effects for either smallest rectangle or convex hull methods with the defined area being the computer display. Figure 4 illustrates the convex hull NNI calculation of a participant that is representative of the participant pool in this study.

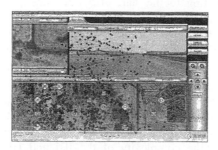

FIGURE 4. Convex hull NNI calculation overlaid on the MIX display.

Since NNI is based on fixations, particularly the number of fixations, an additional analysis was conducted to eliminate that influence. Based off of the fixation locations found in Figure 4 and other screen captures, a 3 (area of interest: route map, streaming video feed, and change detection) x 3 (control: manual, adaptive, and static) x 2 (block: 1 and 2) x 2 (task load: low and high) ANOVA for median fixation duration was conducted to examine the possibility of one area demanding more visual resources than another. A significant main effect for area of interest was found, $F(1.414, 56.560) = 12.279$, $p < .001$, such that median duration time was the longest for the streaming video feed ($M = 185.507$). Median fixation duration was longer for change detection ($M = 176.979$) than the route map ($M = 156.470$). All were significantly different, $p = .013$. A significant main effect for block, $F(1, 40) = 7.746$, $p = .008$, showed longer duration for block 1 ($M = 175.974$) than block 2 ($M = 171.209$; $p = .008$). A significant main effect for task load, $F(1, 40) = 20.704$, $p < .001$, revealed longer duration for high task load ($M =177.667$) than low task load ($M = 169.516$; $p < .001$).

DISCUSSION

The NASA-TLX conveyed differences present in control conditions, indicating perceived task load to be more demanding in the manual condition than either adaptive or static. That finding provides support for the utility of automation in reducing attentional resources allowing allocation for other tasking. It is logical for the adaptive automation condition to elicit greater significant physical demand than the fully autonomous condition because some element of motor ability is required, but perhaps even that result would be mitigated depending on automation invocation method.

The goal was to identify a physiological response indexing a heightened workload state that could be the mitigating solution and a measure that is continuous, requiring no interruption in the tasks at hand. IBI results for task load were consistent with the literature in that low task load should have a slower IBI than high task load conditions (Wilson, 1992). The interaction of faster heart rate in block one than in block two in both autonomous and manual control is likely due to the participants acclimating to the start of a new, unexperienced scenario. It is interesting, however, that the interaction between control and task load was due to the adaptive control condition with slower heart rate in low task demand than in high task demand. This is an indication that IBI might be a means of invocation for automation as it seems to be sensitive to workload state that occurs in a control changing environment. HRV, on the other hand, which is often argued to be more accurate at measuring state than IBI, proved to not be an effective assessment of workload state. This is not an unusual outcome as conflicting results are shown throughout literature addressing the utility of HRV as an index of workload state (Kramer, 1991). Given the complexity of the IBI and HRV indices, additional research should be conducted. IBI and HRV responses to each component task included in the MIX testbed should be investigated to determine if it is a single task eliciting the responses or the entire multi-tasking environment.

Initially, eye tracking also seemed an unviable option given the non-significant results of NNI. However, the time spent in the three task regions revealed important differences. Although the change detection task is employed for increasing or decreasing task load, it seems that participants deemed the streaming video feed to be the primary task given the amount of time fixated in that region compared to the other two tasking areas. Even though participants might have perceived the primary task to be located in the streaming video feed, it should not go unnoticed that the high task load condition required more time spent viewing the icons than did the low task load condition, indicating that more resources were still used for high task load, the condition expected to induce a higher workload state. As was the case with IBI, the effect found with block 1 having longer duration than block 2 is not unexpected and is most likely due to a task familiarization period. These findings also suggest the need to individually test the elemental tasks of the MIX testbed to determine the utility of using physiologically recorded workload state or performance on a given task as a trigger for turning automation on or off in a system. The questions to answer concern the emphasis placed on a particular task, as well as task location on the display.

HRV and NNI did not produce the expected results for assessing workload, but deviates of those measures did show promise. The NASA-TLX findings are also relatively limited. Thus, further research is needed before implementing any invocation method into a real-world system.

ACKNOWLEDGMENTS

This work was funded by the Army Research Laboratory Award # W91CRB08D0015.

REFERENCES

Clark, P. J., & Evans, F. C. (1954). Distance to nearest neighbor as a measure of spatial relationships in populations. *Ecology, 35,* 445-453.

Camilli, M., Nacchia, R., Terenzi, M., & Di Nocera, F. (2007). ASTEF: A simple tool for examining fixations. Retrieved May 21, 2007 from www.astef.info.

Camilli, M., Terenzi, M., & Di Nocera, F. (2007). Concurrent validity of an ocular measure of mental workload. In D. De Waard, G. R. J. Hockey, P. Nickel, & K. A. Brookhuis (Eds.), *Human factors issues in complex system performance* (pp. 1-13). Maastricht, NL: Shaker Publishing.

Di Nocera, F., Camilli, M., Terenzi, M., & Nacchia, R. (2007). *Cognitive aspects and behavioral effects of transitions between levels of automation.* EOARD Technical Report FA8655-05-1-3021.

Di Nocera, F., Terenzi, M., & Camilli, M. (2006). Another look at scanpath: Distance to nearest neighbour as a measure of mental workload. In D. De Ward, K. A. Brookhuis, & A. Toeffetti (Eds.), *Development in Human Factors in Transportation, Design, and Evaluation* (pp. 1-9). Maastricht, Netherlands: Shaker Publishing.

Hart, S.G., & Staveland, L.E. (1988). Development of NASA-TLX (Task Load Index): Results of empirical and theoretical research. In P.A. Hancock & N. Meshkati (Eds.), *Human mental workload* (pp. 139-183). Amsterdam: North-Holland.

Hart, S. G., & Wickens, C. D. (1990). Workload assessment and prediction. In H. R. Booher (Eds.), *MANPRINT: An approach to systems integration.* New York: Van Nostrand Reinhold.

Kramer, A. (1991). Physiological measures of workload: A review of recent progress. In D. Damos (Ed.), *Multiple task performance* (pp. 279-328). London: Taylor and Francis.

McCarley, J. S., & Kramer, A. F. (2007). Cerebral hemodynamics and vigilance performance. In R. Parasuraman & A. M. Rizzo (Eds.), *Neuroergonomics: The brain at work* (pp. 95-112). Cambridge, MA: MIT Press.

Parasuraman, R. (1990). Event-related brain potentials and human factors research. In J. W. Rohrbaugh, R. Parasuraman, & R. Johnson, (Eds.), *Event-related brain potentials: Basic and applied issues* (pp. 279-300). New York: Oxford University Press.

Norman D. A., & Bobrow, D. G. (1975). On data-limited and resource-limited processes. *Cognitive Psychology, 7,* 44-64.

Wickens, C. D. (2002). Multiple resources and performance prediction. *Theoretical Issues in Ergonomics Science, 3,* 159-177.

Wickens, C. D, & Hollands, J. G. (2000). *Engineering psychology and human performance* (3rd ed.). Upper Saddle River, NJ: Prentice Hall.

Wilson, G. F. (1992). Applied use of cardiac and respiration measures: Practical considerations and precautions. *Biological Psychology, 34,* 163-178.

CHAPTER 13

Brain Power: Implementing Powerful Neurally-Inspired Mechanisms in Computational Models of Complex Tasks

Amy Santamaria

Alion Science and Technology
Boulder, CO, USA

ABSTRACT

Neuroergonomics requires an understanding of both the system and the human in real work environments. Computational models can help us to make concrete predictions about complex and dynamic behaviors. This chapter describes MS/RPD, a neurally-inspired, integrated modeling approach that represents the human and the system. MS stands for Micro Saint, a task network modeling tool. Task network modeling is a powerful and accessible way to represent systems but has difficulty representing the subtleties of the human in the system. RPD stands for Recognition Primed Decision (see Klein, 1998), the inspiration for an underlying decision/cognitive model to augment task networks with simple and powerful learning and memory mechanisms. MS/RPD was used to model the Three Block Challenge, a complex environment with pressured multi-attribute judgments. Because MS/RPD is computational, bottom-up, and neurally-inspired, it marries computational power with efficiency, allowing models to be scaled up in size and complexity to represent problems of interest to the field of neuroergonomics.

Keywords: Computational modeling, neurally-inspired, behavior representation

INTRODUCTION

Neuroergonomics is the study of cognitive and brain function in complex work environments (Parasuraman & Rizzo, 2007). In a neuroergonomic approach, we need to understand both the system and the human, and computational models are a critical tool in this process. To understand and make concrete predictions about the complex behaviors and interactions that occur in real work environments, neuroergonomic theory and conceptual models need to be translated into computational models and simulations. This chapter describes MS/RPD, a neurally-inspired integrated modeling approach that represents both the human and the system at useful levels of abstraction, allowing for efficient development of models that can be scaled up in size and complexity. First, a conceptual organization of different ways of modeling human behavior and cognition is presented to provide a context for this approach. Next, the functional and implementational details of the approach are described. The next section describes an example of a model of a complex military task, resource allocation in the Three Block Challenge. The chapter concludes with a discussion of insights garnered from this modeling approach and benefits that neurally-inspired computational models can offer the field of neuroergonomics.

MODELING SYSTEMS

There are many ways to represent a system, including concept maps, documentation, checklists, written requirements, websites, and management software and tools. All of these approaches can be somewhat cumbersome and opaque, however, as a system grows larger. Few systems that involve humans are simple and straightforward. Many interacting parts (e.g., tasks, steps, components, communications) lead to behaviors and effects that emerge from a system. Without a functional way to represent interactions and emergent effects, it is easy to get lost in the details of a system. One successful way to functionally represent a complex system is with task network models.

Task network models are a powerful and accessible computational modeling tool. They are a standard tool in human performance modeling and can generate useful predictions about the behavior of a system. However, task network models have difficulty representing complexities of the human in the system (see Kieras, 2003). When a network branches into two or more possible paths or behaviors, a path is typically selected using either probability distributions or conditional rules. While this provides a description of behavior, it is not predictive and cannot capture underlying mechanisms and dynamic effects of cognition.

There are many ways to improve the sophistication of human behavior representation within a system model. The next section discussions a conceptual organization of approaches to modeling the human in the system.

MODELING HUMANS

Researchers and developers have taken a variety of approaches to modeling human behavior and cognition. These can be organized with a series of bifurcations, as illustrated in Figure 1. On the left side of the figure are system representations, as discussed above. The right side of the figure shows approaches to representing the human.

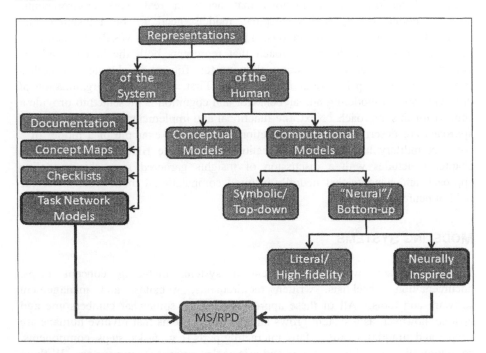

FIGURE 1. A conceptual organization of approaches to modeling the human and the system. MS/RPD combines a neurally-inspired, bottom-up computational model of the human with a task network model of the system.

Conceptual and Computational Models

Within human representations, the first bifurcation is into conceptual and computational models. While conceptual models are certainly important tools to guide neuroergonomic research, it is critical to go beyond high-level descriptive models to build mechanistic models and simulations.

There are several reasons that computational modeling is a useful tool over and above conceptual models and theories. First, computational models provide predictive power and a mechanism for new hypothesis generation. Second, computational models allow researchers to discover constraints that are not obvious in a descriptive theory. Because they must actually be built, computational models avoid "black box" and testability problems to which descriptive theories can fall prey. Third, computational models lead naturally to technology applications. Decision and training aids, intelligent agents, and large-scale simulations often fall short of realistic representations of human cognition. Neuroergonomic theories can inform and improve these applications, but to do so, descriptive theories must be translated into computational terms.

Symbolic and "Neural" Models

Within computational models, there are symbolic, or top-down models, and "neural", or bottom-up models. "Neural" is in quotation marks because bottom-up models may not strictly follow neural constraints but they are based on brain processing in some way and share important characteristics. In symbolic/top-down models, problems are solved with built-in knowledge or rules. Often they require hand-coding by experts. Many early AI approaches to well-defined problems like chess fall into this category, but symbolic models are still used, especially for complex domains. "Neural", or bottom-up, approaches build on simple, low-level mechanisms to create emergent behavior in a model. Information about the environment is captured implicitly, without explicit instruction, and learning occurs gradually over time.

Many of cognitive tasks that occur in real-world, complex work environments are not easily characterized in terms of rules. An advantage of "neural"/bottom-up models is that they can capture essential aspects of these tasks in a general way, without building in assumptions about them ahead of time. In real life, people learn in a noisy, changing environment, and they can pick up on the statistical regularities of this environment in a way that seems effortless and automatic. Bottom-up approaches capture this "effortless" learning by leveraging powerful, general-purpose mechanisms.

Literal and Inspired Models

Within "neural" models, there are both literal models, which strive for biological fidelity at one or more levels, and inspired models, which leverage aspects of brain-based learning for a functional purpose. Literal models attempt to incorporate as much neural biology as possible, so that they may accurately represent processing in as it occurs in the brain. These are ambitious and worthy aims, but the large scale makes it difficult to interface with models of systems, especially as tasks grow more complex. Neurally-inspired models take a looser interpretation and only implement those functional aspects of brain processing that are most useful.

Characteristics of neurally-inspired models may include:

- Parallel processing
- Memory and experience-based behavior
- Learning

- The ability to represent errors (rather than just optimal behavior)
- General purpose mechanisms
- Flexibility
- Emergent behaviors

The modeling approach described in the next section is an example of a computational, bottom-up, neurally-inspired approach. In an attempt to improve the fidelity and flexibility of representations of cognition within task network models of systems, we have leveraged simple but powerful learning and memory mechanisms. This allows us to capture important characteristics of human behavior and cognition without a lot of knowledge engineering. The goal is a modeling approach that can account for many different cognitive phenomena with low implementational overhead, so that it can scale to more complex, realistic tasks.

THE MS/RPD MODELING APPROACH

MS/RPD is a modeling approach inspired by work on naturalistic decision making (Klein, 1998). It combines a model of the system, the task network model, with a model of the human. MS stands for Micro Saint, a task network modeling tool, and RPD stands for Klein's (1998) Recognition Primed Decision, the inspiration for our underlying decision model. MS/RPD is a framework for defining experience-based, recognitional decision making in task network modeling environments, and it has been validated with human performance data for a variety of tasks, including decision making in a resource allocation task, probability matching behavior in a binary choice task, monitoring behavior, category learning, adversarial behavior, and probabilistic category learning (Santamaria & Warwick, 2007, 2008, 2009; Warwick & Fleetwood, 2006; Warwick & Hutchins, 2004; Warwick & Santamaria, 2006). The approach uses a modified version of Hintzman's multiple trace model of long-term memory (1984, 1986a, 1986b). Each decision making episode is recorded as a single trace in long-term memory, and a similarity-based recognition routine is used to compare the current decision making situation to all past episodes. A composite echo of past situations is returned, and a single course of action is chosen and implemented. The entire experience, including outcome, is then stored as a new trace in long-term memory. This happens each time a decision is made (for more details, see Warwick et al., 2001.)

To specify a decision in this framework, the modeler needs to do three things: 1) Identify courses of action, which are branching points in a task network model. 2) Map simulation variables to the cues that prompt recognition. These can be enumerated or a fixed set of binned values. 3) Define successful outcomes for each course of action. These are written as Boolean-valued assertions given a simulation state, and they are not necessarily exhaustive or exclusive. Figure 2 shows how branching paths and variables in the task network model are mapped to courses of action and cues in the decision model.

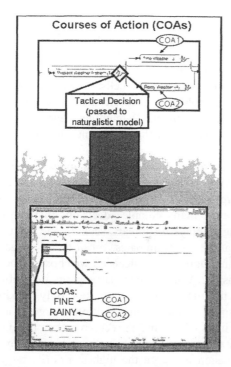
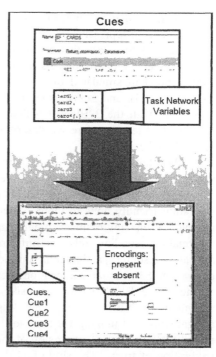

FIGURE 2. The connection between the task network model (top) and the decision model (bottom). Left box: courses of action are defined and then specified in the RPD model; Right box: variables in the task network environment are mapped to cues in the RPD model. (Figure published in Santamaria & Warwick, 2008.)

To illustrate how MS/RPD can be used to model a real-world, complex task, the next section describes an example model of the Three Block Challenge.

MODELING THE THREE BLOCK CHALLENGE

Decision making by Command and Control (C2) teams in complex urban terrain encompasses what the military calls the three block challenge. The "three block challenge" refers to balancing three contexts - humanitarian, peacemaking, and combat missions - in the same place at the same time. This is a challenging environment that requires experienced decision making, information sharing, and shared situation awareness, all under time pressure.

A SIMULATED MULTI-CONTEXT COMMAND AND CONTROL TASK

The concept of the three block challenge has been captured in a task developed by Yen and colleagues (Fan et al., 2006). The Three Block Challenge task is a human-in-the-loop simulation environment used to study C2 performance in an urban environment. It represents the trade-offs of the real-life Three Block Challenge but simplifies it into a task that resembles a video game, with operators under time pressure interacting with objects in an environment, earning a score based on how they respond to simulated threats. There are three C2 cells: Intelligence (S2), Operations (S3), and Logistics (S4). These roles can be performed by 1) humans interacting with the 3 Block Challenge environment, 2) intelligent agents, or 3) human-agent teams. Here we describe a simulation that represents the human (a particular take on 2).

In the Three Block Challenge, there are several threats that can appear: crowds, insurgents, and improvised explosive devices (IEDs). The job of the command and control cells (human or agent) is to respond appropriately to these threats by allocating friendly units to them. There are many complex decisions that can be modeled in this environment: monitoring displays for threats, seeking and filtering information, sending relevant information to teammates, prioritizing targets, allocating resources, and tracking mission progress. The decision we chose to focus on was the S3's resource allocation decision. When human participants interacting in the Three Block Challenge make this decision, they are not only making assignments, but searching the display, monitoring multiple targets and assignments, and judging the trustworthiness of automated recommendations. Our model makes a simplified version of the allocation decision, without monitoring or calibration of trust. The simplified decision is: given the highest priority target, which units do I assign? This decision was deemed to be at a level of abstraction that was non-trivial but tractable.

The Three Block Challenge is a complex environment with asymmetric conflict and time-pressured, multi-attribute judgments. The decision of resource allocation requires cue gathering, implicit learning, and adapting to a dynamic environment. Threats appear, disappear, and move, the available resources are constantly changing, and mission progress and success are dynamic variables.

IMPLEMENTING THE THREE BLOCK CHALLENGE TASK IN MS/RPD

The simulation environment for the Three Block Challenge was implemented in Micro Saint (MS). The MS environment can 1) generate threats (crowds, insurgents, and IEDS), 2) track unit assignments and combat readiness of units, and 3) support rich interactions between decisions and outcomes.

The decision making model was implemented within the MSS model. In constructing our model of the resource allocation decision, we made several modeling choices. The specific decision we modeled was: for a given target, for a given unit, should the unit be assigned to the target? We opted to loop through this simple decision rather than model the higher-level decision of: for a given target, which units should be assigned; we did this because of the online nature of the learning in our decision model. A related representation choice we made was defining courses of action as "assign" and "don't assign", rather than selecting from all the permutations of possible assignments. Another modeling decision was the cues we selected, which were 1) distance to target, 2) number of units already assigned to the current target, and 3) whether the unit was already assigned to a target. We also decided how to define success: the model was successful when it assigned a unit and 1) the distance from the unit to the target was

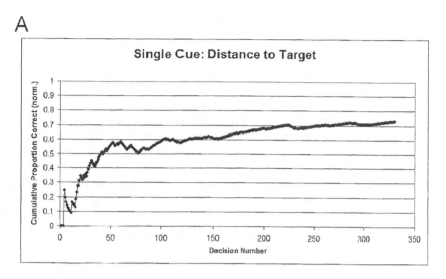

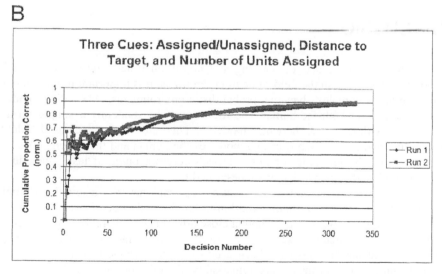

FIGURE 3. Three Block Challenge model performance. A. Cumulative success as a function of time for a single cue: distance to target. B. Cumulative success as a function of time for a combination of three cues: distance to target, number of units assigned to current target, and whether current unit is assigned to a target.

less than half the maximum screen distance, 2) fewer than two units had been assigned to the target, and 3) the unit was not already assigned. Conversely, the model was successful if it did not assign the unit and these three conditions were not met. This was a simplified measure of success to evaluate whether the model was exhibiting meaningful behavior and appropriately picking up on the cues in its environment.

The model showed evidence of meaningful behavior and learning. It picked up on the predictive value of individual cues and combinations of cues. Figure 3a shows cumulative success as a function of trial, over 330 decisions, for a representative single run of the model, where the model was given one cue: distance to target. For this test, we used a limited definition of success, including only the distance to target criterion described above, to determine whether the model could pick up on that particular cue. The graph shows that the model showed increased success over time, demonstrating that it was responsive to the individual cue. Figure 3b shows cumulative success as a function of trial, over 330 decisions, for two representative runs of the model, where the model was given three cues: distance to target, number of units assigned, and whether the unit was already assigned. The measure of success was our default definition, as described in the previous paragraph. In both runs, the model increased success over time. Because success was defined in terms of all three cues, this demonstrates that the model was responsive to the combination of three cues.

DISCUSSION

Computational models of learning and cognition usually represent simplified laboratory tasks, but in the field of neuroergonomics, it is important to build models that operate in complex, dynamic, probabilistic environments. MS/RPD provides a way to represent both the human and the system, by modeling a simple form of cognition within a task network model. This allows for increased fidelity, flexibility, and predictive power in a model of the human and the system. We have demonstrated the feasibility of this approach across a range of tasks, including a more complex, dynamic task that we described here.

Our approach is computational, bottom-up, and neurally-inspired. Complex behaviors can emerge without hand-built, hand-tuned, rules, strategies, or rich semantic representations, and without a close adherence to literal biological fidelity, which can be difficult to scale. The same multiple-trace memory and simple reinforcement mechanisms have been used to model a range of behaviors. These neurally-inspired mechanisms are remarkably flexible; because of their ability to pick up on the statistics of the environment, they have been able to model a variety of learning tasks. In addition, their simplicity allows them to scale up to more and more complex and realistic contexts, making them a critical component of the in neuroergonomic toolkit.

REFERENCES

Fan, X., Sun, B., Sun, S., McNeese, M., and Yen, J. (2006). RPD-enabled agents teaming with humans for multi-context decision making. In *AAMAS*. Hakodate, Hokkaido, Japan.

Hintzman, D. L. (1984). MINERVA 2: A simulation model of human memory. *Behavior Research Methods, Instruments & Computers, 16,* 96-101.

Hintzman, D. L. (1986a). Judgments of Frequency and Recognition Memory in a Multiple-Trace Memory Model. Eugene, OR: Institute of Cognitive and Decision Sciences.

Hintzman, D. L. (1986b). "Schema Abstraction" in a Multiple-Trace Memory Model. *Psychological Review, 93,* 411-428.

Kieras, D. (2003). Model-based evaluation. In Jacko, J. A. & Sears, A. (Eds.), *The Human-Computer Interaction Handbook* (pp. 1139-1151). Mahwah, NJ: Lawrence Earlbaum Associates.

Klein, G. (1998). *Sources of Power: How People Make Decisions.* Cambridge, MA: The MIT Press.

Parasuraman, R. & Rizzo, M. (2007). *Neuroergonomics: The Brain at Work.* New York: Oxford University Press.

Santamaria, A. & Warwick, W. (2007). A naturalistic approach to adversarial behavior: Modeling the prisoner's dilemma. *Proceedings of the 16th Conference on Behavioral Representations in Modeling and Simulation.*

Santamaria, A. & Warwick, W. (2008). Modeling probabilistic category learning in a task network model. *Proceedings of the 17th Conference on Behavioral Representations in Modeling and Simulation.*

Santamaria, A. & Warwick, W. (2009). Using a "Naturalistic" Mechanism to Capture Cognition and Dynamic Behaviors in Task Network Models: An Overview. *Proceedings of the 18th Conference on Behavioral Representations in Modeling and Simulation.*

Warwick, W. & Fleetwood, M. (2006). A bad Hempel day: The decoupling of explanation and prediction in computational cognitive modeling. *Proceedings of the Fall 2006 Simulation Interoperability Workshop. Orlando, FL. SISO.*

Warwick, W. & Hutchins, S. (2004). Initial comparisons between a "naturalistic" model of decision making and human performance data. *Proceedings of the 13th Conference on Behavior Representation in Modeling and Simulation.*

Warwick, W., McIlwaine, S., Hutton, R. J. B., & McDermott, P. (2001). Developing computational models of recognition-primed decision making. *Proceedings of the 10th Conference on Computer Generated Forces.*

Warwick, W. & Santamaria, A. (2006). Giving up vindication in favor of application: Developing cognitively-inspired widgets for human performance modeling tools. *Proceedings of the 7th International Conference on Cognitive Modeling.*

Chapter 14

ESP^2: A Platform for Experimental Design in Cognitive Ergonomics

Roberto Sottile, Luigi Yuri Di Marco, Lorenzo Chiari

Department of Electronics, Computer Science and Systems
Alma Mater Studiorum – Università di Bologna
Bologna, Viale Risorgimento 2 – 40136 ITALY

ABSTRACT

ESP^2 (an Experimental Stimuli Presentation Platform) is a prototypical implementation of a thin software layer on top of the Java™ environment for coding experiments in scientific disciplines like cognitive psychology, cognitive ergonomics and behavioral sciences. The platform is composed of a set of libraries providing a basic application programming interface (API) for the implementation of cognitive experiments, and a lightweight application framework for stimuli definition, presentation, and subsequent processing of user response.

Experiments can either be executed in stand-alone mode, or driven by an external unit which remotely controls experiment execution on an arbitrary number of target machines concurrently and transparently, thus allowing to implement parallel experimental sessions, when applicable, with no additional costs in terms of experimenter's time or effort. The internal libraries allow simultaneous streaming of information regarding presented stimuli and user response timings to an arbitrary number of external devices (e.g. EEG monitors, ambulatory polygraphs, external computing units), by using a simple custom network protocol that may also be used to synchronize or annotate recorded data.

In the present chapter the ESP^2 platform is presented, and its main characteristics – cross-portability, software extensibility, network integration, timing accuracy, performance – introduced and discussed.

Keywords: experiment design, stimuli presentation, response synchronization

INTRODUCTION

Designing experiments in behavioral sciences often involves the assessment of human performance during the execution of given tasks. Research and analysis of possible relationships between the dependent and independent variables of selected experimental conditions may help to better understand the underlying processes of human cognition, information processing, interaction design, human-computer interaction and related disciplines.

At the design level, a given experiment requires a set of actions and decisions to be taken in order to improve the quality and validity of the collected data (Gawron, 2008). Direct control of basic experimental conditions and experiment parameters plays a central role for a correct validation of the experimental hypotheses. Consequently, experiments are often implemented in software: they can be executed on general purpose computing platforms, like normal personal computers or laptops, and parameters can be programmatically customized depending on the experimenter's needs.

The panorama nowadays offers a rich selection of professional solutions that help creating complex experiments with little or no effort, hiding the underlying complexity of the mechanisms for real-time stimulus generation and presentation; they provide highly accurate timing measures, numerous functionalities for logging data, precise synchronization mechanisms for interfacing with input/output devices, different modalities of stimuli presentation, integrated tools for data analysis. Some of the available products offer hardware-level synchronization between the system and a number of different typologies of acquisition devices – for example EEG (electroencephalography) amplifiers, ambulatory polygraphs, fMRI (functional magnetic resonance imaging) scanners – which is a valuable feature when the experimental sessions are designed for neurocognitive assessment. In Schneider et al. (2005) a brief but comprehensive overview of currently available solutions for experiment authoring is provided, along with characterization of their basic approach to experiment generation: script-based systems, HyperCard, form-based systems, experiment diagram approach, cross-linked lists, graphical interfaces.

ESP^2 is a prototype software tool for experiment generation whose goal is to integrate natively with the Java environment, thus providing a cross-platform set of libraries and simplified API for generation of stimuli of different nature (text, visual, auditory), general purpose routines for user input acquisition from different sources (commonly, keyboard and mouse), exposing accurate timing performance in the majority of platforms and operating systems, being capable of presenting any designed experiment in real-time. Thanks to the solid networking layer offered by the Java environment, ESP^2 is able to take full advantage of a distributed approach to experiment design, allowing for example both remote logging of user input and stimuli onset, and allowing to synchronize the captured events with external devices. Moreover, the same experiments can be run in both stand-alone and remote-controlled modes; the latter case may offer additional advantages over a traditional approach, since the experiments do not need to reside physically on the target machine: programs can be uploaded and run on the target machine without physical intervention by the experimenter – the distributed approach is not new to the subject of behavioral experiments: see Birnbaum (2004) for a review of laboratory and Web-based research methods and approaches to distributed data collection.

DESIGNING THE TOOL

Generally speaking, software aimed at experiment authoring must provide a set of functionalities for a correct execution of the experiment and validity of collected data. In the next paragraphs an overview of the principal characteristics that software should offer is presented.

TIMING ACCURACY

One of the most widely used methods for measuring human performance is *time*: if the duration of a task involving user interaction can be measured, the relation between human response time and the task duration may be investigated as potential source of evaluation for measuring human performance in the specific task. If the simulated experiment is implemented in software, accurate timing measures are mandatory. A tool aiming at addressing such applications should therefore provide the most precise resolution and accuracy on measuring time intervals – accuracy should be guaranteed not to fall below one millisecond. In Plant et al. (2004), in the context of measuring the timing accuracies of general purpose hardware, the reader is asked to pay attention to the possible drawbacks resulting from the use of computers as hardware platforms for running experiments in behavioral sciences, when time is the studied dimension.

Nowadays, code generated by C/C++ and Java compilers can execute at comparable speed on modern machines, and able to reach millisecond accuracy in the most common operating systems. The pitfalls on measuring time delays, however, often reside in how and at which point of execution the system clock is being queried. For example, some implementations retrieve the user response in the form of a software event which is passed to the application by the graphical subsystem, which is responsible for rendering the experiment state to the output display and handling events from the input devices. Although input events can be acquired by the underlying hardware at high sampling rates, generally output devices are much slower: often, liquid crystal displays operate at average refresh rates around 60 Hz (especially on laptops), which means that the effective moment in time when the stimulus is presented to the user could suffer from poor timing accuracy: on 60 Hz displays, the average time resolution for stimuli presentation could be as low as 16.67 milliseconds.

SYNCHRONIZATION

Actual consequences to the validity of experimental data caused by timing accuracy issues depend strongly on the nature of the experiment itself; however it is possible to take into account for the effects of the phenomenon if, for example, the time interval (delay) between stimulus generation and subsequent presentation is somehow measurable. Anyway, the problem becomes more delicate when a form of synchronization between stimulus and some external device is needed: in neuroergonomics research, for example, it is not uncommon to have the experimental setup involve external acquisition equipment such as electroencephalographs, ambulatory polygraphs, fMRI scanners, etc. In such cases, accurate alignment between experiment events and recorded signals is fundamental – a comprehensive reading about the involved methods and the targeted range of applications can be found in Parasuraman and Rizzo (2007).

In some circumstances, however, hardware synchronization may not be achievable for various reasons relying either at software or hardware levels, or even both: the software not supporting the feature, or the hardware providing limited or no GPIO capabilities. In these cases, synchronization is still possible, and one opportunity may be provided by network connectivity.

NETWORKING

The communication opportunities provided by the network layer are countless, and the applications are many. Often, computers residing in the same LAN (local area network) share software and hardware resources, and this feature can be exploited in many ways (one example is described in Lahl and Pietrowsky, 2008). For example, one single advantage of having two or more hosts interconnected could be a distributed approach to computation. Restricting the focus to behavioral research, a number of solutions have been provided in order to take advantage from the network topology. One application area targets Web-based assessment and data collection (Birnbaum, 2004). Anyway, driving experiments through the Internet might not be the right option if accurate time measures are required: web-browsers are far from being accurate time-measuring platforms, the underlying communication protocol among hosts (TCP/IP most of the times) could suffer from high latencies due to the internal protocol mechanisms responsible for data reliability and delivery guarantees; moreover the actual network topology and routing protocols between source and destination hosts are out of the control of the communicating parties.

In this case, a local area network topology may be a suitable solution. The simplest network topology, shown in Figure 1, depicts two possible scenarios in which experiments may take advantage of the intercommunication channel provided by the network.

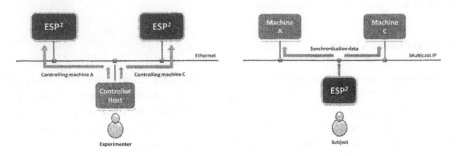

FIGURE 1 – Advantages of the network architecture: a) remote control of target machines by a single controller host; b) synchronization of experiment events and data among multiple hosts through multicast communication

In Figure 1, the same network architecture is used for two different purposes: Figure 1a) outlines a simple scenario for execution of experiments on target machines, remotely controlled by a single machine on the same network. This architecture may show advantages over the single-machine stand-alone setup: experimental sessions can be driven in parallel by a single experimenter, thus optimizing time, logistical and hardware resources; in addition, laboratory conditions, as well as environmental factors that may affect experiments and collected data, are shared among participants.

Figure 1b) shows how networking can be used for synchronization purposes. A target machine running the experiment streams information about presented stimuli, user reaction and other annotation events in real time over a selected network channel. Depending on the network protocol adopted and the network adaptors capabilities, the receiver target can be a single host or a group of different host. The latter solution is often implemented involving IP multicasting (Deering, 1989), which is the technology often adopted for streaming multimedia contents over the Internet due to the low latencies that, in laboratory setups, can be controlled and maintained under one millisecond, thus becoming transparent to the network peers.

SOFTWARE EXTENSIBILITY

Another fundamental feature for experiment authoring software is the ability given to the experimenter of designing his/her own experiments by using the features and tools made available by the software. The available paradigms for generating the sequence of actions composing the experiment, rendered by the software in real-time in the form of stimuli of different nature (textual, auditory, visual the most common) presented to the subjects, are many. Regardless of the paradigm, the most important factor considered by the experimenter is the possibility of modifying the programs at will. Anyway, like any other piece of software, the tradeoff between expressive power and ease of use has to be considered: most tools offer graphical user interfaces (GUI), other rely on custom scripting languages, some other mix both of the functionalities, covering the whole range of possible skill levels of their users.

Another approach is to provide the experimenter with a set of software libraries (e.g. a collection of software algorithms under a unified, simplified API) with bindings for different programming languages: in these cases, the targeted users are requested to actually code their experiments by implementing them using a real programming language. The disadvantages of this approach, which requires the experimenter to have computer programming skills, may be counterbalanced by the improved expressive power gained. The result is close to optimal, direct control on the execution of the stimulation, plus the ability to extend the software functionalities provided by the authoring tool with custom code, data structures, and analysis routines.

THE ESP^2 PLATFORM

In this section, each of the above mentioned characteristics is considered with particular focus on the ESP^2 platform and the underlying implementation philosophy.

FOREWORD

The Experimental Stimuli Presentation Platform, ESP^2, is a prototypical implementation of a thin software layer on top of the Java environment. It is being developed at the Biomedical Engineering Group, Department of Electronics, Computer Science and Systems, Faculty of Engineering, Alma Mater Studiorum – Università di Bologna, Italy.

Aim of this project is to create a tool for experimental design in cognitive ergonomics, in order to develop a software platform where to implement known experimental paradigms useful to investigate human behavior and performance when complex tasks are executed. The purpose is to analyze those processes involved in human-machine interaction both at the performance/behavioral and physiological levels in many different task paradigms, which may vary from simple tasks to complex multiple task batteries. Moreover, the design should be kept as open as possible to future extension and usable both at academic, didactic level and for actual research.

TIMING ACCURACY

To date, available implementations of the Java Virtual Machine (JVM) offer the necessary accuracy over the system time clock of the underlying machine architecture. Time intervals can be measured with nanosecond precision (although

148

some hardware could be able only to expose micro- or millisecond accuracy), which is enough to achieve the required time resolution of one millisecond.

The graphical subsystem is accessible natively with the extensive API provided by the standard development kit, and it can benefit from hardware acceleration if provided by the graphic hardware – however the consideration outlined above regarding actual display refresh rates, and consequently display time resolution for presented stimuli, still apply, since those are hardware related issues.

SYNCHRONIZATION AND NETWORKING CAPABILITIES

The experimental setup adopted in a set of current studies within our group involves the acquisition, processing and analysis of several electrophysiological signals: ECG (electrocardiogram), PPG (pulse-plethysmogram), Respiration, EEG, GSR (galvanic skin response), acquired by two distinct pieces of equipment, and recorded during the task execution. Therefore, a robust synchronization technique is required to align the acquired signals to the source of cognitive stimulation. A general diagram of the experimental setup involved in our studies is depicted in Figure 2.

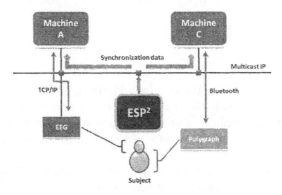

FIGURE 2 – Experimental setup for neuroergonomics research using ESP^2

The EEG amplifier (Braintronics BV Brainbox EEG-1166), which acquires the electroencephalographic signals of the tested subject, is connected to the same network of machine A through an Ethernet adapter (Braintronics BV Ethernet-102); machine A acts as a controller to the amplifier. In this configuration, Ethernet is the only technology for intercommunication between machine A and the EEG.

Machine B, connected to the same network, runs the ESP^2 software, which in turn executes the selected cognitive experiment. Machine C, connected to the same network of the other elements, drives the acquisition and preprocessing software of the ambulatory polygraph (g$^®$.Tec g.MOBIlab+), responsible of acquiring the subject's ECG, PPG, Respiratory and GSR signals, via Bluetooth® connection.

Synchronization among these devices is realized through the definition of an application protocol on a per-experiment basis. The EEG amplifier features an 8-bit digital I/O hardware channel; this initially suggested the design of the synchronization protocol data to have size of 1 byte, allowing 256 different levels for annotations in its simplest form.

Each time a stimulus is presented to the subject, a visual cue may be shown on the subject's screen, if accurate synchronization is needed (via opto-coupling sensors). As an alternative, the active devices may be informed of events occurring within the stimulation software environment by connecting to a multicast address, where the ESP^2 platform (machine B) acts as the sender, and machine A and C as the receivers. The advantage of multicast transmission is that each packet is sent only once, independently of the number of listening parties, and packets are delivered to all the receiving parties at the same time. The protocol overhead, being based on top of UDP, is very low thanks also to the small header size for each packet. However, UDP does not guarantee packet delivery or sequence integrity, ordering or uniqueness of the streamed data; thus simple redundancy and basic control of order of arrival on the client side are involved. The overall performance of the mechanism is such that, when the LAN is dedicated only to the experiment setup and no additional actor is present, the transmission latencies can be kept below one millisecond.

SOFTWARE EXTENSIBILITY AND PORTABILITY

The software architecture is composed of a set of libraries for the definition of the main classes for experiment design, and a lightweight application framework to actually run the experiment. These components are put together to realize a skeleton to run the stimulation code in a modular fashion: the design allows external experiments to be uploaded to a target machine by a controller unit, and be remotely executed without physical intervention by the experimenter.

Following this approach, multiple parallel experiments can be conducted, often useful for training sessions with new subjects or novel experiments. Moreover, the implementation of each experiment adheres to a common structure, hence it is decoupled from the platform itself, allowing the platform be easily extended and improved with reduced probability of recompiling experiments already produced.

Lastly, the use of Java technology as both the programming language and experiment definition language, allows the ESP^2 users to augment the functionalities of the platform with any additional features, algorithms, data structures or stimulation paradigms with all the freedom provided by such an approach.

Concerning portability, the platform is implemented in native Java without the use of additional external libraries. Being the JVM available for a wide selection of target platforms, cross-platform portability (at least, among compatible architecture families) is inherently satisfied. Indeed, the platform itself is being implemented and maintained interchangeably under the Linux and Windows® environments.

CONCLUSIONS

In this paper a general overview of the main characteristic of the ESP^2 – an Experimental Stimuli Presentation Platform – has been presented, along with preliminary considerations about its use for cognitive ergonomics experiments.

Although in its early development stage, ESP^2 is designed around a set of key points playing a key role in the investigation of factors related to human performance and cognitive assessment. Details on current experiments driven in our laboratories will be described separately, along with accurate reports about system performance, stability and synchronization protocol robustness.

Once completed the internal development, the ESP^2 platform will be proposed to large-scale beta testing by releasing it to the open source community.

REFERENCES

Schneider, W., Bolger, D. J., Eschman, A., Neff, C., and Zuccolotto, A. P. (2005), *Psychology experiment authoring kit (PEAK): Formal usability testing of an easy-to-use method for creating computerized experiments*. Behavior Research Methods, 37, 312-323.

Gawron, V. J. (2008), *Human performance, workload and situation awareness measures handbook*, 2nd ed. Boca Raton, FL: CRC Press.

Suzuki, R., Niki, K., Fujimaki, N., Masaki, S., Ichikawa, K., and Usui, S. (2008), *Neuro-Imaging Platform for Neuroinformatics*. In *Neural information Processing: 14th international Conference, ICONIP 2007, Kitakyushu, Japan, November 13-16, 2007, Revised Selected Papers, Part II*, M. Ishikawa, K. Doya, H. Miyamoto, and T. Yamakawa, Eds. Lecture Notes In Computer Science, vol. 4985. Springer-Verlag, Berlin, Heidelberg, 895-904.

Birnbaum, M. H. (2004), Human research and data collection via the Internet, Annual Review of Psychology, 2004. Vol. 55:803-832.

Plant R. R., Hammond N. and Turner G. (2004), *Self-validating presentation and response timing in cognitive paradigms: How and why?*, Behavior Research Methods, Instruments and Computers, 2004, 36 (2), 291-303.

Parasuraman, R. and Rizzo, M. (2007), *Neuroergonomics: the brain at work*, Series in Human-Technology Interaction, New York: Oxford University Press

Deering, S. (1989), *RFC 1112 – Host extensions for IP multicasting*, Network Working Group, Request for Comments #1112, http://www.faqs.org/rfcs/rfc1112.html

Lahl, O. and Pietrowsky, R. (2008), *Tracer: a general-purpose software library for logging events in computerized experiments*, Behavior Research Methods, 2008, 40 (4), 1163-1169.

CHAPTER 15

Modeling Sleep-Related Activities from Experimental Observations - Initial Computational Frameworks for Understanding Sleep Function(s)

Mary Ann Greco, John Murray, Carolyn Talcott

SRI International, Menlo Park, California, USA

ABSTRACT

Sleep disorders affect ~70 million Americans, an estimate that is likely an underestimate, based on recent data. Though sleep is considered an essential behavior, the biological function of sleep is not known, which has limited the development of safe and effective therapies. We hypothesize that sleep serves some sort of restorative function that affects the entire organism. We propose to model activities associated with sleep using a "bottom-up" approach that utilizes computational modeling to link sleep-related protein activities. In this paper, we show how data from our investigation of proteins correlate to sleep-wake in the cortex of rats. We believe that understanding the underpinnings of sleep at all levels of the body's organizational hierarchy holds great promise for the future of neuroergonomics research and practice.

Keywords: Sleep-wake behavior, biological timing, cellular underpinnings of sleep, interactions across brain regions, organ-organ interactions, protein expression, sleep deprivation, neuroergonomics, health maintenance.

INTRODUCTION

A comprehensive and ever-growing literature on the evolutionary conservation of sleep and sleep patterns [Zeppelin 2000] and the debilitating effects of sleep deprivation [Bonnet 2000] in all animals studied to date [Cirelli, Huber, Gopalakrishnan, Southard & Tononi 2005] indicates that sleep is an essential behavior, similar to eating, drinking, and mating. How and why the brain orchestrates shifts in vigilance states are fundamental questions in sleep research. That a balance between waking and sleep is important to maintain health and productivity in humans is supported by statistics associated with sleep disturbances. Sleep disorders affect approximately 70 million Americans, with associated costs estimated at billions of dollars per year. Similar to other physical anomalies, sleep disorders generally result from extremes in behavior- i.e., too much or too little sleep. While it is recognized that these extremes in behavior are determined by a combination of genetics and environment, sleep deprivation is the more prevalent of these extremes in our 24 h, 21st century society and individuals that consistently do not get enough sleep, exhibit cumulative decrements in cognitive and psychomotor performance [Owens 2001]. For example, cognitive performance was decreased by 30-40% in military personnel deprived of sleep for one night, with further declines in performance (60-70%) after a second night of sleep deprivation [Westcott 2005]. In addition, as people age, alterations in sleep (the inability to fall asleep, and/or stay asleep) are also correlated with cognitive and/or psychomotor performance [Ohayon, Carskadon, Guilleminault & Vitiello 2004]. Moreover, recent findings in adolescent and young adults indicate there are positive correlations between inadequate sleep and the development of obesity, cardiovascular disease and type-2 diabetes [Knutson & Van Cauter 2008; Mullington, Haack, Toth, Serrador & Meier-Ewert 2009]. Thus chronic sleep deprivation may negatively impact the function of several different organs in the body. Hence, from the point of view of neuroergonomics research, understanding the underpinnings of sleep at all levels of the body's organizational hierarchy holds great promise for improving and sustaining physical and mental capabilities and performance. As discussed below, the biological function (s) of sleep (i.e., what is restored during sleep) is not known, though an overall restorative function for sleep is widely accepted. To appreciate sleep function however, requires an understanding of the cellular correlates and mechanisms that underlie sleep, an area of sleep research that remains poorly characterized. The following sections provide an overview of the complexities that underlie sleep and our efforts to identify biological function (s) using systems biology approaches.

CHARACTERISTICS OF SLEEP

The evolution of sleep research began with classical "top-down" experimental approaches that revealed a complex hierarchy underlies this relatively simple behavior. Sleep-wake behavior is orchestrated by a number of brain regions interacting with one another via the differential release of a variety of neurotransmitters and peptides. In addition, the lack of sensitive, high-throughput biotechnologies to characterize the cellular correlates of sleep in these regions did not become available until relatively recently. As a result, the specific biological function(s) of sleep remain unclear. Proposed functions include the maintenance of body temperature [McGinty & Szymusiak 1990; Wehr 1992], energy homeostasis [Benington & Heller 1995; Adam 1980; Walker & Berger 1980], immune function [Majde & Krueger 2005; Opp 2005], synaptic plasticity [Tononi & Cirelli 2006] and memory consolidation/reconsolidation [Born, Rasch & Gais 2006; Stickgold & Walker 2005; Stickgold & Walker 2007]. One clinical consequence of these gaps is that drugs designed to treat sleep disorders may have a number of undesirable side effects, including addiction. Considering the number of people with sleep disorders and the potential that insufficient sleep can lead to other pathologies, there is an urgent need to develop safer and more effective drugs to treat sleep disorders.

The identification of sleep-wake states is based on a combination of observational and electrophysiological recordings, parameters that are similar in all vertebrates studied. Characteristics include a specific sleeping site, typical body posture, physical quiescence, an elevated arousal threshold, rapid state reversibility, and regulatory capacity, i.e., compensation for sleep after sleep loss. Waking is characterized by irregular, low voltage fast waves and high muscle activity. Non-rapid eye movement sleep or slow wave sleep (SWS) is characterized by high voltage slow waves and decreased muscle activity. In humans, this state is divided into four stages, with Stages 1 and 2 considered "light" sleep, and Stages 3 and 4 regarded as "deep" sleep or SWS. Rapid eye movement (REM) sleep, the "other" sleep state, is characterized by low voltage fast waves in the EEG, and virtual atonia of the neck muscles, as well as rapid eye movements. Together, SWS and REM states provide the basis of sleep architecture. Though sleep architecture and sleep stage classifications differ between rats and humans, the process of falling asleep is similar and can be characterized by a progressive decrease in wakefulness that is followed by SWS. Under normal conditions, SWS precedes REM sleep.

Sleep-wake behavior is regulated by a combination of circadian and homeostatic factors [Borbely 1982; Borbely & Achermann 1999]. The circadian control of sleep-wakefulness is associated with the regulation of sleep timing and emanates from the master pacemaker in the suprachiasmatic nucleus (SCN) [Edgar 1995]. Homeostatic factor(s) are associated with the regulation of sleep drive, sometimes expressed as sleep need. The homeostatic influence on sleep is suggested by the observations that longer periods of wakefulness result in an increased need for sleep, cumulative bouts of SWS and REM sleep are necessary to dissipate the sleep drive [Levine, Roehrs, Stepanski, Zorick & Roth 1987; Stepanski, Lamphere, Roehrs, Zorick & Roth 1987], and sleep-wake transitions persist even after lesion of the suprachiasmatic nucleus, the master circadian clock [Tobler, Borbely & Groos 1983]. The nature of the homeostatic drive is unknown; however, an increase in an endogenous factor(s) that accumulates during wakefulness and dissipates during sleep has been postulated to regulate homeostatic drive. At the electrophysiological level, this buildup is reflected by a gradual increase in slow wave activity (SWA; 0.1–4 Hz) during waking that dissipates with sleep. In rats, an increase in SWA accompanies sleep deprivation, reaching an asymptote after ~12 h of continuous waking [Tobler & Borbely 1986]. SWA declines exponentially during the recovery sleep that follows SD. As a result of these considerations, sleep research has primarily used total and selective SD paradigms as tools to investigate the homeostatic component of sleep and to distinguish SWS and REM effects on sleep. However, recent demonstrations that the targeted disruption of core circadian clock genes affects sleep duration, sleep structure and EEG delta power and core circadian gene expression in the cortex appears dependent on prior sleep-wake history [Franken & Dijk 2009] underscore the complexities involved in studies designed to distinguish circadian from homeostatic effects at the cellular level. Our data examining Per-2 levels, a circadian transcription factor, following SD/RS are consistent with these findings [Greco, unpublished]. Thus while homeostatic and circadian processes have different origins, at the cellular level, they may interact directly with one another to control behavior.

INITIAL STEPS

To address the question of sleep function(s), a "bottom-up" analysis was undertaken to identify putative cellular correlates of sleep within the brain. Using high through-put mRNA and protein technologies, changes in the expression of proteins [Vazquez, Hall & Greco 2009; Vazquez, Hall, Witkowska & Greco 2008; Basheer, Brown, Ramesh, Begum & McCarley 2005] and/or mRNAs [Cirelli, Gutierrez & Tononi 2004] associated with energy metabolism, re-dox state, and synaptic plasticity were shown to underlie sleep-wake bouts that occur during the latter portion of the lights-on period. In addition, comparison of protein profiles across spontaneous sleep in young and old rats indicate that processes like synaptic plasticity that are controlled by phosphorylation [Ramakers 2002] may be compromised in old animals from damage to intracellular organelles and macromolecules (i.e., DNA, proteins, and lipids) caused by the accumulation of reactive oxygen and nitrogen species, ROS and RNS, respectively [Balaban, Nemoto, Finkel 2005; Calabrese, Giuffrida Stella, Calvani, Butterfield 2006; Joseph, Shukitt-Hale, Casadesus, Fisher 2005]. ROS/RNS species may also affect other post translational modifications (i.e., acetylation) involved in the regulation of other systems integral to the maintenance of circadian effects on biological timing [Borrelli, Nestler, Allis, Sassone-Corsi 2008]. SD also generates ROS and RNS species; thus mitigation of the responses described herein may also provide insights into aging.

Our experimental approach has several unique features. In particular, the specific mix of sleep-wake behavior is controlled with high precision at the time of sacrifice [Vazquez, Hall, Witkowska, Greco 2008]. Once sleep- and/or wake-related proteins and their related activities are determined in the brain, putative interactions between brain regions and with other organs across these states can be mapped and compared using systems biology approaches. With this strategy in mind, we have embarked upon some initial protein analysis studies and we are starting to use the results to develop a predictive computational systems biology model of the dynamics of sleep.

As sleep deprivation studies are traditionally used to differentiate sleep timing (circadian) from sleep need (homeostatic regulation), we believe that using this paradigm will provide insight into mechanisms associated with both time of day and sleep-wake components of biological timing.

INTEGRATED BIOLOGICAL SYSTEM MODELING

Our approach to developing a predictive model of the biological systems underlying the responses to sleep deprivation combines statistical analysis, logical representation of subsystems, processes (also known as executable or symbolic systems biology) and interactions, model abstraction and hybrid systems methods that accomodates different levels of detail, types of measurments, and missing information. Subsystem models can be combined using module calculi that are part of the logical framework and linking rules. This approach will be applied to integrating models of multiple subsystems and pathways, with behavioral, realtime, and homeostatic measurements. We are not aware of any existing work that attempt to model such diverse subsystems or such a the range of data types.

Traditionally, biological models have been represented using simplified drawings (cartoons) capturing relations between key components (A activates/regulates B) and tables relating model components and experimental observations. Such models are invaluable to build initial insights and give overall structure, and can be used to guide our initial model development. However these models don't scale, it is difficult to understand the consequences when two models are combined, and they are not directly suited to computational processing.

Computational systems biology is an approach to overcome limitations of informal models, and to bring the power of computation to understanding the results of biological data. Such models have generally been based either on systems of differential equations or statistical analysis of high throughput data. The differential equations approach can be used to answer questions about dynamic (kinetic) aspects of a system, such as change of concentration or expression of one or more molecules over time, under different conditions. One problem this approach faces is the lack of experimental data for rate parameters, thus techniques for inferring or fitting parameters must be used. Simple curve fitting is often used to measure responses with circadian components. Recent work has shown that adding biochemical components to the equations can improve model fidelity. The statistical approach can be used to infer correlations between changes in different components and is frequently applied to understand transcriptomic data. The resulting interaction graph can be used, for example, to identify highly connected components, to determine subgraphs that correspond to biological function or processes. Such models give useful high level insights, but are not adequate to explain underlying mechanisms or to predict effects of change.

SYSTEM-WIDE MODELING APPROACH

Integration of proteomics and gene expression data from non-human animal studies into a computational model of baseline behaviors is a rational and innovative means to examine key links between sleep deprivation , the resulting behavioral responses (adaptive or dysfunctional), and other effects of inadequate sleep. We envision that these models may eventually be used to support the neuroergonomics community, in the study of human sleep characteristics and effects. In addition, they may offer a means to assay the efficacy of existing sleep-wake drugs, to aid the discovery /development of novel drugs to modulate sleep-wakefulness, and to facilitate studies designed to establish safe and effective drug treatment regimens *in silico*.

To examine putative functional interactions between proteins identified by mass spectrometry and sleep-wake behavior within the frontal cortex, we have used Pathway Logic, a symbolic computational modeling system and signaling knowledge base, to look for common functional path(s). One resulting "hit" linked proteins associated with cellular transport/cytoskeletal support and signal transduction to synaptic plasticity, a property of neurons which underlies higher executive behaviors like memory, cognition and learning [Guzman-Marin, Ying, Suntsova, Methippara, Bashir, Szymusiak, Gomez-Pinilla, McGinty 2006; Stickgold, Walker 2007]. We are currently testing the signal transduction pathway identified by Pathway Logic. Our plan is to develop a symbolic systems biology model to support the interpretation of the data resulting from sleep studies to identify biomarkers and mechanisms underlying adaptive and maladaptive responses to sleep deprivation and to predict the effects of modulating target markers on the response.

The symbolic computational system model provides a framework for integrating the diverse types of measurements. As data is incorporated into the model, we will be able to look for underlying mechanisms that correlate the different observations, across data types, time and sleep conditions. The intent is to identify significant correlations and dependencies between model features, and by using the underlying computational system model to identify processes/pathways affected by components with significant changes. These analyses will be used to explain and validate potential markers that are linked to adaptive and maladaptive responses to sleep disorders.

Our underlying hypothesis is that the restorative function(s) fulfilled during sleep affect the entire body. This lends itself to using a broad systems approach to sleep exploration, which draws us to examine the signaling processes between the brain and the rest of the body – specifically across the blood-brain barrier. We therefore plan to look for markers of sleep deprivation in blood, as well as unique protein expression in liver, which may be signals controlling the metabolic processes that correlate with effective sleep states. By adopting this system-wide approach, we anticipate that a connected whole-body model will emerge, which is supported by the data and adds confidence to the theory. For example, perhaps the blood is carrying sleep-related signals from the brain, that control what the liver is doing.

The modeling effort involves the construction of a baseline system level model, which can then be used to predict the effects of modulation and to interpret experimental results. The proposed multi-level baseline model will incorporate and integrate models from several subsystems, and will include both non-human and human sub-models (based on data availability). It will incorporate biological similarities (functional, structural) to assist in the creation of a broad human model of the intracellular underpinnings of adaptive/maladaptive responses to sleep disorders. Figure 1 shows the architecture of the system model.

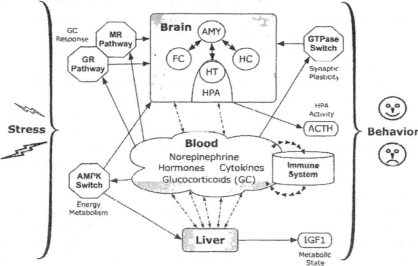

FIGURE 1. System level model of response to stress (sleep deprivation), reflecting interaction with its environment via stress inputs (on left, bold represents high stress, light represents low stress) and behavioral response on the right (managing stress, smile; negative stress effects, frown).

The figure shows three main system model components - brain, blood, and liver - along with signals flowing between components that serve as modulators of behavior. Brain submodels correspond to key regions: Frontal Cortex (FC), Amygdala (AMY), Hippocampus (HC), and Hypothalmus (HT). HPA (Hypothalamic-Pituitary-Adrenal axis) modulates cognition by propagating signals from the HT. ACTH (Adrenocorticotropic hormone) is an indicator of HTA activity. Blood serves as a transport system for signaling molecules such as norephinaphrin, hormones, glucocorticoids (GC), and cytokines, including a providing a connection with the immune system. Liver cell behavior is modulated by signals from the blood system and from the AMP activated protein kinase (AMPK) switch. Insulin growth factor, IGF1, is a key indicator of the metabolic state. Octagons represent pathways through which received signals (arrows into octagons) are transduced into signals modulating behaviors of target components (arrows out of octagons). Thus, the GR and MR pathways are glucocorticoid-like receptor pathways that control energy metabolism depending on glucocorticoid level. The GTPase switch integrates signals from molecules including ephrins, and neurotrophins to modulate features such as synaptic plasticity.

Information for the baseline system model will be curated from existing knowledge about interactions, behaviors, and phenotypes available in the literature and databases. As noted earlier, we expect model construction to initially focus on the major subsystems - brain, liver, and blood. Beyond the high-level systems view, information about specific processes, switches and pathways involving selected proteins and metabolites will be curated and linked into the system level model.

We plan to use existing systems biology knowledge bases (KBs), including Pathway Logic (pl.csl.sri.com), PANTHER (pantherpathway.org), and BioCyc (biocyc.org), as a starting point and collect additional information as needed from external pathway databases, such as KEGG and the NCI Nature Protein Interaction databases, and published literature. The collected information supporting the computational model will be stored using a formal KB representation with a well-defined schema/ontology. A system-level signaling KB helps to organize data to enable extraction of specific models and search for relevant patterns and pathways. Model elements will be extracted from the resulting KB by query. The KB will also be a source of facts for initial model validation and consistency tests.

Computational model elements will be represented using a combination of logical constraints and a rule-based formalism integrated using an algebraic signature. They will be combined using logical module operations and analyzed using formal reasoning tools such as model checking, constraint solving, and model abstractions. A neural model will be developed from data obtained from regions of the brain linked to the modulation and control of adaptive/maladaptive cognitive response(s) to stress. Sleep-related inputs and outputs for each region will be modeled and will include regional interconnections.

CONCLUSION

Our long-term vision for this work is to develop a comprehensive, system-level computational modeling and analysis framework for exploring sleep-related processes and responses in the brain, liver, and other organs. For example, a neural model of the effects of sleep across brain regions will help characterize the modulation and control of the cognitive impacts of SD/RS. A model that maps sleep-related processes in the liver will help describe the relevant effects of energy metabolism and other expression markers. A third model, using the levels of signaling molecules in blood, provides a means to represent the blood as a transport system for organ-organ intercommunications. The result will be a high-level, holistic system view of the dynamics of sleep throughout the body.

This type of computational model can be used to understand interactions between subsystems, as well as to predict possible consequences of perturbing the system. Genetic variation in primary protein sequence represents one example of such a perturbance, where the altered genes may produce non-functional proteins, proteins that are always active, or do something else that deviates from the norm. Similarly, altered neural interconnections can occur, hence sending signals to unexpected places. We anticipate that our computational modeling framework will prove to be a valuable tool for neuroergonomic analysis of these important issues.

ACKNOWLEDGEMENTS

This work has been supported by NIH grants HL69706, NS045791 (MAG) and NSF grant IIS-0513857 (CT). The authors much appreciate the assistance of Jasmin Joseph with the preparation of this paper.

REFERENCES

Adam, K. (1980), "Sleep as a restorative process and a theory to explain why." *Progress in Brain Research*, 53, 289-305.

Balaban, R.S., Nemoto, S., & Finkel, T. (2005), "Mitochondria, oxidants, and aging." *Cell*, 120, 483-495.

Basheer, R., Brown, R., Ramesh, V., Begum, S., McCarley, R.W. (2005), "Sleep deprivation-induced protein changes in basal forebrain: Implications for synaptic plasticity." *Journal of Neuroscience Research*, 82, 650-658.

Benington, J.H., & Heller, H.C. (1995), "Restoration of brain energy metabolism as the function of sleep." *Progress in Neurobiology*, 45, 347-360.

Bonnet, M.H. (2000), "Sleep Deprivation." in *Principles and Practice of Sleep Medicine*, Kryger, M.H., Roth, T., & Dement, W.C. (eds), W.B. Saunders Co., Philadelphia, 53-71.

Borbely, A.A. (1982), "A two process model of sleep regulation." *Human Neurobiology*, 1, 195-204.

Borbely, A.A., & Achermann, P. (1999), "Sleep homeostasis and models of sleep regulation." *Journal of Biological Rhythms*, 14, 557-568.

Born, J., Rasch, B., & Gais, S. (2006), "Sleep to remember." *Neuroscientist*, 12, 410-424.

Borrelli, E., Nestler, E.J., Allis, C.D., & Sassone-Corsi, P. (2008), "Decoding the epigenetic language of neuronal plasticity." *Neuron*, 60, 961-974.

Calabrese, V., Giuffrida Stella, A.M., Calvani, M., & Butterfield, D.A. (2006), "Acetylcarnitine and cellular stress response: roles in nutritional redox homeostasis and regulation of longetivity genes." *Journal of Nutritional Biochemistry*, 17, 73-88.

Cirelli, C., Huber, R., Gopalakrishnan, A., Southard, T.L., & Tononi, G. (2005), "Locus cerulus control of slow-wave homeostasis." *Journal of Neuroscience*, 25, 4503-4511.

Cirelli, C., Gutierrez, C.M., & Tononi, G. (2004), "Extensive and divergent effects of sleep and wakefulness on brain gene expression." *Neuron*, 41, 35-41.

Edgar, D.M. (1995), "Control of sleep-wake cycles by the mammalian suprachiasmatic pacemaker." *Brain Research Bulletin*, 1, 2-7.

Franken, P., & Dijk, D.J. (2009), "Circadian clock genes and sleep homeostasis." *European Journal of Neuroscience*, 29, 1820-1829.

Guzman-Marin, R., Ying, Z., Suntsova, N., Methippara, M., Bashir, T., Szymusaik, R., Gomez-Pinilla, F., & McGinty, D. (2006), "Supression of hippocampal plasticity-related gene expression by sleep deprivation in rats." *Journal of Physiology*, 575, 807-819.

Joseph, J.A., Shukitt-Hale, B., Casadesus, G., & Fisher, D. (2005), "Oxidative stress and inflammation in brain aging: nutritional considerations." *Neurochemical Research*, 30, 927-935.

Knutson, K.L., & Van Cauter, E. (2008), "Associations between sleep loss and increased risk of obesity and diabetes." *Annals of the New York Academy of Sciences*, 1129, 287-304.

Levine, B., Roehrs, T., Stepanski, E., Zorick, F., & Roth, T. (1987), "Fragmenting sleep diminishes its recuperative value." *Sleep*, 10, 590-599.

Majde, M.A., & Krueger, J.M. (2005), "Links between the innate immune system and sleep." *The Journal of Allergy and Clinical Immunology*, 116, 1188-1198.

McGinty, D., & Szymusiak, R. (1990), "Keeping cool: a hypothesis about the mechanisms and functions of slow-wave sleep." *Trends in Neuroscience*, 13, 480-487.

Mullington, J.M., Haack, M., Toth, M., Serrador, J.M., & Meier-Ewert, H.K. (2009), "Cardiovascular, inflammatory, and metabolic consequences of sleep deprivation." *Progress in Cardiovascular Disease*, 51, 294-302.

Ohayon, M.M., Carskadon, M.A., Guilleminault, C., & Vitiello, M.V. (2004), "Meta-analysis of quantitative sleep parameters from childhood to old age in healthy individuals: developing normative sleep values across the human lifespan." *Sleep*, 27, 1255-1273.

Opp, M.R. (2005), "Cytokines and sleep." *Sleep Medicine Reviews*, 9, 355-364.

Owens, J.A. (2001), "Sleep loss and fatigue in medical training." *Current Opinion in Pulmonary Medicine*, 7, 411-418.

Ramakers, G.J., (2002), "Rho proteins, mental retardation and the cellular basis for cognition." *Trends in Neuroscience*, 25, 191-199.

Stepanski, E., Lamphere, J., Roehrs, T., Zorick, F., & Roth, T. (1987), "Experimental sleep fragmentation in normal subjects." *International Journal of Neuroscience*, 33, 207-214.

Stickgold, R., & Walker, M.P. (2005), "Memory consolidation and reconsolidation: what is the role of sleep?" *Trends in Neuroscience*, 28, 408-415.

Stickgold, R., & Walker, M.P. (2007), "Sleep-dependent memory consolidation and reconsolidation." *Sleep Medicine*, 8, 331-343.

Tobler, I., & Borbely, A.A. (1986), "Sleep EEG in the rat as a function of prior waking." *Electroencephalography and Clinical Neurophysiology*, 64, 74-76.

Tobler, I., Borbely, A.A., & Groos, G. (1983), "The effect of sleep deprivation on sleep in rats with suprachiasmatic lesions." *Neuroscience Letters*, 42, 49-54.

Tononi, G., & Cirelli C. (2006), "Sleep function and synaptic homeostasis." *Sleep Medicine Reviews*, 10, 49-62.

Vazquez, J., Hall, S.C., & Greco, M.A. (2009), "Protein expression is altered during spontaneous sleep in aged Sprague-Dawley rats." *Brain Research*, 1298, 37-45.

Vazquez, J., Hall, S.C., Witkowska, H.E., & Greco, M.A. (2008), "Rapid alteration in cortical protein profiles underlie spontaneous sleep and wake bouts." *Journal of Cellular Biochemistry*, 105, 1472-1484.

Walker, J.M., & Berger, R.J. (1980), "Sleep as an adaptation for energy conservation functionally related to hibernation and shallow torpor." *Progress in Brain Research*, 53, 255-278.

Wehr, T.A. (1992), "A brain-warming function for REM sleep." *Neuroscience and Biobehavioral Reviews*, 16, 379-397.

Westcott, K.J. (2005), "Modafinil, sleep deprivation, and cognitive function in military and medical settings." *Military Medicine*, 170, 333-335.

Zepelin, H. (2000), "Mammalian Sleep." in *Principles and Practice of Sleep Medicine*, Kryger, M.H., Roth, T., & Dement, W.C. (eds), W.B. Saunders Co., Philadelphia, 82-92.

Chapter 16

Physiological Day-to-Day Variability Effects on Workload Estimation for Adaptive Aiding

G. F. Wilson[1], C. A. Russell[2], J. Monnin[1], J. Estepp[1], J. C. Christensen[1]

[1]US Air Force Research Laboratory, WPAFB, OH

[2]Archenoetics, Inc., Honolulu, HI

ABSTRACT

Psychophysiological data has been used to determine the mental workload state of operators. This has been successfully made use of to drive adaptive aiding. In order to be employed in workstations the psychophysiological data must consistently provide accurate information on the state of the operator. Using a complex task and an artificial neural network classifier highly accurate classification of operator state was found within one day. However, the accuracy of the estimates decreased when data from different days were used to test the classifier. Adjustments to operator state classification procedures may have to be made to accommodate the day-to-day variability of the psychophysiological measures.

Keywords: Psychophysiology, operator functional state, artificial neural network, EEG, day-to-day variability

INTRODUCTION

The ability to correctly classify operator functional state (OFS) using the operator's physiological signals while performing complex tasks has been demonstrated (Berka, et al., 2004; Freeman, Mikulka, Prinzel & Scerbo, 1999; Gevins, et al., 1998; Wilson & Fisher, 1991; Wilson & Russell, 2003a; 2003b). Further, this information has been used to modify the task via adaptive aiding to enhance system performance in high cognitive workload situations (Freeman, Mikulka, Prinzel & Scerbo, 1999; Wilson & Russell, 2007). This provides the ground work for systems that are capable of monitoring the functional state of operators and that have the ability to modify the task demands to assist the operator in times of mental overload. These systems should produce improved overall system effectiveness and reduce catastrophic errors. However, the effects of day-to-day fluctuations in the operator's physiology have not been assessed using complex tasks. The stability of EEG signals has been investigated using eyes open/eyes closed conditions or simple laboratory procedures. In general, they have been found to be fairly stable over time within each individual (Burgress & Gruzelier, 1993; McEvoy, Smith & Gevins, 2000; Pollock, Schneider, & Lyness, 1991; Salinsky, Oken, & Morehead, 1991). With regard to complex task performance in applied settings where adaptive aiding may be implemented it is necessary to determine the stability of the psychophysiological signals during complex task performance. Because the adaptive aiding will be implemented to assist operators it must function correctly every day in order to be effective as well as accepted by system designers and system operators. Day-to-day variations in the operator's internal milieu may make it necessary to adjust the physiologically driven OFS classifiers in order to achieve optimal performance day after day. The purpose of this study was to assess the changes in operator's physiology while performing a complex task over a four week period to determine the stability of the physiological responses that are related to the performance of the task.

METHODS

The Multi Attribute Task Battery (MATB) was used to provide three levels of task difficulty (Comstock & Arnegard, 1992). The lights and dials, radio frequencies, and fuel management tasks were presented in unison. The difficulty of each task was adjusted so that three levels of overall difficulty were presented. Reaction times were collected from the lights, dials and radio frequencies, and error scores from the fuel management task. Eight adult subjects (5 female), with a mean age of 21.1 years, were trained on the MATB until a consistent level of performance was achieved. Stable performance was declared when the performance parameters attained asymptote with minimal errors. This reduced learning effects and allowed subjects to become familiar with the laboratory setting. This took approximately 3 hours over one to two days.

Psychophysiological data were recorded from the subjects while they performed the MATB. EEG from nineteen channels was recorded at sites positioned according to the International 10-20 electrode system (Jasper, 1958).

Reference and ground electrodes were positioned on the mastoids, with impedances below 5K ohms. Horizontal and vertical electrooculogram, heart, respiration and blink rate were also recorded. Corrections for eye movement and blinks were made and the data were stored at 256 samples per second. The EEG data were filtered using elliptical IIR filter banks. The passbands were consistent with the five traditional bands of EEG; delta (0.5-3 Hz), theta (4-7 Hz), alpha (8-12 Hz), beta (13-30 Hz) and gamma (31-42 Hz). The data were segmented into forty-second windows with a 35-second overlap. Log power of five bands from the 21 sites was used, resulting in 105 EEG features as inputs to the classifiers. The interval between eye blinks, interbeat intervals, and the interval between breaths, were also used as input features, resulting in 108 total features.

Three trials consisting of five minutes each of randomly ordered low, medium and high cognitive demand were collected over five days distributed over one month. The five sessions were separated by one day, one week, three weeks and four weeks. The data from the low and medium conditions were pooled and became a nominal workload condition (easy). In order to determine the functional state of the operators an artificial neural network (ANN) was used. The preprocessed psychophysiological data were used as inputs to the ANN.

Equal numbers of samples from the easy and difficult conditions were used to train the ANN. Data representing twenty five percent of the data from the training sets were randomly removed and used to test the trained classifier's ability to identify the easy from the difficult condition. Data from each of the five days were separately used to train the classifier and the data from the other days were used as test sets to examine the classifier's ability to discriminate between easy and difficult task data.

A feedforward backpropagation neural network was used (Widrow and Lehr, 1990; Lippmann, 1987). The network learned the input-output classification from the set of training vectors. Then, after training, the ANN acted as a classifier for new vectors. During training, the ANN adjusted the weights and biases based on the training data set. A separate validation set representing 25% of the data were randomly selected and were used to stop the training in order to prevent overlearning which prevented the classifiers from generalizing to new input data (Wilson & Russell, 2003a; Bishop 2006). Once trained, network weights were fixed and the ANN acted as a pattern classifier. As a classifier, the ANN examined input data it had never seen and predicted the class of the input data as either easy or difficult.

RESULTS

Analysis of the performance data revealed that the differences between the easy and difficult conditions produced significantly different mean responses (see Table 1). For the communication, dials and lights tasks the difficult condition produced significantly longer reaction times. The difficult condition also resulted in significantly greater error scores in the resource management task. The two levels of task difficulty produced significantly different performance from the operators while the main effect for days was not significant in all cases.

In order to assess the reliability of the classifier to correctly discriminate between the easy and difficult task conditions, using the psychophysiological data, the mean correct classifications for the five days were examined. Figure 1 shows the means when each of the five day's data were separately used to train the classifiers and then the held out data from that day (same) and the data from the other four days (different) were classified as representing the easy or difficult task condition. The classification accuracies for the same day were quite high. The mean classification accuracy for the same day's data from the five days was 99.7% for the easy condition and 99.9% for the difficult condition. However, the ANN was not nearly as able to distinguish between the easy and difficult conditions on the data from the non-training days. The mean accuracies were 76.2% for the easy condition and 47.4% for the difficult condition from the different day's data. This represents an overall 23.5% decline for the easy condition and a 52.5% decline for the difficult condition.

Table 1 Mean and standard deviations (SD) of reaction times, in seconds, for communication, dials and light tasks and mean error for the resource management tasks. The F values and probabilities for the comparison between the easy and difficult task levels are presented in the bottom row.

	DIALS	LIGHTS	RESOURCE MANAGEMENT	COMMUNICATION
EASY	2.52 (1.20)	2.82 (1.68)	1.69 (0.57)	495.25 (132.42)
DIFFICULT	3.11 (1.19)	3.78 (1.67)	2.32(0.76)	747.65 (270.64)

There were significant main effects for difficulty ($F(1, 7) = 11.36$, $p < 0.01$) and day effects ($F(1, 7) = 329$, $p < 0.01$. There was a significant interaction between difficulty and day conditions ($F(1,7) = 117.6$, $p < 0.01$). It appears that the psychophysiological features change more for the difficult task than for the easy task because the classifier accuracies fell to chance levels for the difficult task data.

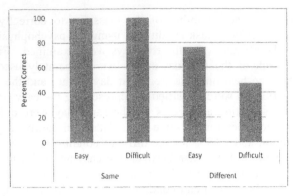

Figure 1. Mean classifier accuracies for the ANN on the single day data with the test data from the training day (same) or from the other four days (different). The easy and difficult task results are shown.

DISCUSSION

The results of the present study replicate earlier reports by demonstrating that ANNs trained and tested on psychophysiological data from the same day can very accurately determine which of two levels of task difficulty produced the data (Berka, et al., 2004; Freeman, Mikulka, Prinzel & Scerbo, 1999; Gevins, et al., 1998; Wilson & Fisher, 1991; Wilson & Russell, 2003a; 2003b). The classifier accuracies from each of the five days, separated by up to one month, was 99% for the ANN when the test data were recorded on the same day as the classifier training data. This is important for applied situations with complex tasks in which OFS estimation is required or adaptive aiding is to be implemented using a classifier trained with psychophysiological data. However, the results also show that the critical psychophysiological features change over time. When tested on data from days that were not included in the training set the accuracies declined. Interestingly, the decline was much more precipitous for the difficult task condition. The mean easy task accuracy for the data collected on different days was 80.7%. The difficult task data were correctly classified at only the chance level of 49.3%. The classification of the same day's data, regardless of which day or which combination of days was used was over 99%. The psychophysiological features are stable within a given day and provide sufficient information to correctly discriminate between the easy and difficult task performance with a very high level of precision. However, the critical features for each day change. Apparently the critical features from the psychophysiological data change more during the difficult task performance than during the easy task performance. It is possible that, despite the validation procedure used during training, the ANN was over trained to recognize the data that were in the training data set. When presented with new data the ANN was not able to accurately recognize the psychophysiological features from the difficult data set as well as it could correctly recognize the data from the easy set which had not changed as much. Alternatively, it may be that the psychophysiological measures used to train the classifiers vary enough from day-today to render the classifier model invalid for generalizing across days. If the measures change enough in the model input space the classifier cannot accurately differentiate the classes and the result is lower classification accuracies.

The low level of correct classification of the difficult task by the classifier is puzzling. If one assumes that the EEG features are produced by the same brain structures from day-to-day then much less variance would be expected. This is especially true in the situation where the operators have been trained to stable levels of task performance as evidenced in the performance data. Once a task has been learned it would seem that the same cortical structures and thereby the same pattern of EEG changes would be found. It is also possible that the same brain structures are utilized but produce different EEG patterns. It may be that different cortical structures are involved in solving the problem on different days. The data showed that from day-to-day different features were selected as the most salient by the two classifiers. It is possible that other factors may strongly influence the EEG that was recorded in the present study. Various artifacts may have been present in the data. Ocular artifacts were removed and visual inspection showed that the corrections were effective. However, it is possible that small amounts of the contaminants were still present that may have systematically influenced the classifiers. The influence of high frequency electromyographic (EMG) artifacts may have been present. It is well known that EMG frequencies can extend into the alpha band region. The overlap of EMG frequencies with these bands, especially the beta and gamma bands, make it very difficult to separate them. Examination of the EEG did not show contamination by muscle bursts that would be associated with body movement. However, it is very possible that low levels of tonic EMG could have contaminated the EEG. Whitham et al (2007) and others have reported that EEG activity above 20 Hz increases with increased cognitive activity. They extended this result by showing, via paralysis of the muscles of the body, that the correlation between increased beta and gamma band activity and higher cognitive demands was removed when the muscles did not contribute to the EEG. We have earlier suggested that the beta and gamma band increases during higher levels of task demand might be due to the inclusion of muscle related activity (Wilson & Fisher, 1995).

Taken together these results demonstrate that psychophysiological measures can be used to very accurately determine if an operator is in a cognitively demanding state while performing a complex task. Further, day-to-day variability in these features may make it necessary to determine each day which features will yield the optimal classifier. In applied settings this may require some form of classifier adjustment or retraining each day. The optimization procedures to meet the requirement of highly accurate classification and the time and facility constrains of real world operational settings remains to be determined. It may be possible to make adjustments to a trained classifier based on daily examination of EEG band power. Collecting data from short laboratory type tasks might provide sufficient information to correctly derive accurate classifiers. On the other hand, it may be necessary for operators to perform typical task scenarios to provide data that can be used to adjust or train the classifiers.

REFERENCES

Burgess, A. & Gruzelier, J. (1993), "Individual reliability of amplitude distribution in topographical mapping of EEG." *Electroencephalography and clinical Neurophysiology*, 86, 210-223.

Berka, C., Levendowski, D. J., Cvetinovic, M. M., Petrovic, M. M., Davis, G., Lumicao, M. N., Zivkovic, V. T., Popovic, M. V., & Olmstead, R. (2004), "Real-Time Analysis of EEG Indexes of Alertness, Cognition, and Memory Acquired With a Wireless EEG Headset." *International Journal of Human-Computer Interaction*, 17, 151-170.

Bishop, C. M. (2006). *Pattern Recognition and Machine Learning* (p. 32). Springer

Comstock, J. R. and Arnegard, R. J. (1992). "The multi-attribute task battery for human operator workload and strategic behavior research". *NASA Technical Memorandum* No. 104174.

Freeman, F. G., Mikulka, P. J., Prinzel, L. J. and Scerbo, M. W. (1999), " Evaluation of an adaptive automation system using three EEG indices with a visual tracking task". *Biological Psychology*, 50, 61-76.

Gevins, A., Smith, M.E., Leong, H., McEvoy, L, Whitfield, S., Du, R. and Rush, G. (1998), "Monitoring working memory load during computer-based tasks with EEG pattern recognition methods". *Human Factors*, 40, 79-91.

Jasper H. H. (1958), "Report of the Committee on Methods of Clinical Examination." *Electroencephalography and clinical Neurophysiology,* 10, 370-375.

McEvoy, L. K., Smith, M. E. & Gevins, A. (2000), "Test-retest reliability of cognitive EEG." *Clinical Neurophysiology*, 111, 457-463.

Pollock, V. E., Schneider, L. S. & Lyness, S. A. (1991) "Reliability of topographic quantitative EEG amplitude in health late-middle-aged and elderly subjects." *Electroencephalography and clinical Neurophysiology*, 79, 20 -26.

Ruck, D. W., Rogers, S. K., Kabrisky, M. Oxley, M. E. & Suter, B. W. (1990), "The multilayer perceptron as an approximation to a Bayes optimal discriminant function." *IEEE Transactions on Neural Networks*, 1(4), 296-298.

Salinsky, M. C., Oken, B. S. & Morehead, L. (1991), "Test-retest reliability in EEG frequency analysis." *Electroencephalography and clinical Neurophysiology*, 79, 383-392.

Whitham, E. M, Pope, K.J., Fitzgibbon, S. P., Lewis, T., Clark, C. R., Loveless, S., Broberg, M., Wallace, A., DeLosAngeles, D., Lillie, P., Hardy, A., Fronsko, R., Pullbrook, A., & Willoughby, J. O. (2007). "Scalp electrical recording during paralysis: Quantitative evidence that EEG frequencies above 20 Hz are contaminated by EMG." *Clinical Neurophysiology,* 118, 1877-1888.

Wilson, G. F., & Fisher, F. (1991), "The use of cardiac and eye blink measures to determine flight segment in F4 crews." *Aviation, Space and Environmental Medicine,* 62, 959-961.

Wilson, G. F. & Fisher, F. (1995), "Cognitive task classification based upon topographic EEG data." *Biological Psychology*, 40, 239-250.

Wilson, G. F. and Russell, C. A. (2003a), "Operator functional state classification using psychophysiological features in an air traffic control task." *Human Factors*, 45, 635-643.

Wilson, G. F. & Russell, C. A. (2003b), "Real-Time Assessment of Mental Workload Using Psychophysiological measures and artificial neural networks." *Human Factors*, 45, 635-643.

Chapter 17

Challenges of Using Physiological Measures for Augmenting Human Performance

Martha E. Crosby, Curtis S. Ikehara

Department of Information and Computer Sciences
University of Hawaii at Manoa
Honolulu, Hawaii 96822, USA

ABSTRACT

Discussed are the challenges of using physiological measures for augmenting human performance. The main focus is on sensor issues including problems with eye-tracking and passive physiological sensors including a pressure sensitive computer mouse. Also discussed is how a human performance model relates to the implementation of a human performance enhancing augmented cognition system.

Keywords: Physiological sensors, biosensors, eye-tracking, pressure mouse, augmented cognition.

INTRODUCTION

Although manufacturers of physiological sensor systems emphasize the ease of use, accuracy and the vast quantities of data that can be collected, there are several factors to consider: before starting augmented cognition research, when collecting data, when analyzing data and when implementing an augmented cognition system to improve human performance based on physiological measures. Assuming cognitive load (i.e., load on working memory) is being acquired from physiological measures, the model of cognitive load being used needs to describe the relationship between cognitive load and physiological changes. Assuming cognitive load can be accurately obtained via physiological measures, then how will that information be used to guide changing the task based on a human performance model.

The cognitive load model also needs to address several issues such as individual differences and task type. Individual differences includes: gender, age, aptitude, self-efficacy, and many more characteristics relevant to the task being performed. Differences in the task type, in relation to individual differences, can significantly change an individual's cognitive load on a task. For example, when doing math problems, a person poor in math may have a higher cognitive load than a person good at math. Task type can also impact the analysis of sensor data. For example, for eye-tracking, the duration of fixation for reading can be significantly different than for searching a picture.

A human performance model will need to guide task variations that are related to the measured cognitive load so that performance can be optimized. Ikehara, Chin and Crosby (2003, 2004) discuss how an adaptive information filtering model could reduce cognitive load. The human performance model used will need to account for many issues including: motivation, learning, stress, and fatigue. Motivation of an individual, whether the source is internal or external can have a significant impact on performance. Learning can change the aptitude of the individual with the rate of learning varying depending on the individual and the situation. Stress, either too high or too low, can have a detrimental effect on performance. Too little stress may result in boredom that is associated with inattention and fatigue. Too much stress may result in poor performance, reduced motivation and lower self-efficacy. Fatigue negatively impacts human performance and can occur from several sources (e.g., lack of sleep, boredom, excessive physical or mental activity, jet lag, etc.).

It becomes clear that measuring more than cognitive load is necessary to meet the information requirements of both cognitive load and human performance models. Individual data can come from demographic and task related factor questionnaires while performance data can come from rating scales, physiological measures and performance measures (Charleton, 2002). Questionnaires that can assess gender, aptitude, self-efficacy, motivation, ability to handle stress and potential sources of fatigue (e.g., staying up all night). The questionnaire may have questions that require minimum updating, while having others that require frequent updating (e.g., How many hours did you sleep last night). The rating scales can be used to assess the user's perception of the task. Physiological measures can provide real-time information on cognitive load, stress and fatigue. Real-time performance measures such as task response time and accuracy can be used to also assess fatigue, but more importantly assess if changes to the task based on physiological data input to a human performance model is achieving its desired effect of improved performance.

DATA COLLECTION

Table 1 shows the different types of sensors, physiological measures and secondary measures related to mental states collected from an individual by our laboratory. Calibration is an important part of data collection. We also collect demographic and task related data previously discussed and real-time performance data when available.

Table 1. Sensors, physiological measures and secondary measures

Sensors	Physiological Measures	Secondary Measures
Eye Position Tracker	Gaze Position, Fixation Number, Fixation Duration, Repeat Fixations, Search Patterns	Difficulty, Attention, Stress, Relaxation Problem Solving, Successful Learner, Higher Level of Reading Skill (Andreassi 1995), (Sheldon 2001)
	Pupil Size, Blink Rate, Blink Duration	Fatigue, Difficulty, Strong Emotion, Interest, Mental Activity - Effort, Familiar Recall, Positive / Negative Attitudes, Information Processing Speed (Andreassi 1995)
Mouse Pressure	Pressures Applied to the Mouse Case and Buttons.	Stress, Certainty of Response, Cognitive Load (Ikehara, Crosby & Chin, 2005; Ikehara & Crosby, 2005)
Skin Conductivity	Tonic and Phasic Changes	Arousal (Andreassi 1995)
Temperature	Finger, Wrist and Ambient Temperature	Negative Affect (Decrease), Relaxation (Increase) (Andreassi 1995)
Relative Blood Flow	Heart Rate and Beat to Beat Heart Flow Change	Stress, Emotion Intensity (Andreassi 1995)

Although some sensor systems provide an abundance of data, every sensor system has its problems as shown in Table 2. Our laboratory's approach has been to use a variety of sensors as a strategy to mitigate the problems of a single sensor system.

Different data sources can have different data output rates. It is essential when collecting data from several sources that the data be synchronized. One of our former PhD students developed open architecture software to synchronize sensor data from several sources into human readable XML that also conforms to statistical packages. EventStream (Aschwanden and Stelovsky, 2003) is an open source program designed to collect data from multiple sensor sources (see http://sourceforge.net/projects/eventstream/). The software also can graphically display the data.

Table 2. Sensor positives and negatives

Sensors	Positive	Negatives
Eye-Tracker	Able to determine several cognitive states during task.	Does not provide information when there are no visual targets. Rest periods go unmonitored. Pupil size is affected by sudden changes of image intensity.
Mouse Pressure	Provides information primarily during movement and clicking.	Does not provide information during rest periods.
Skin Conductivity	Provides continuous information.	Affected by user movements and slow rise and decline in rate relative to the trigger event.
Temperature	Provides continuous information.	Very slow rise and decline in temperature change relative to the trigger event.
Blood Flow	Provides continuous information.	Slow rise and decline in heart rate relative to the trigger event.

EYE TRACKING

From Table 1, it is apparent that eye-tracking is a rich source of information about the individual. The ASL eye-trackers in our lab use the relationship between the center of the pupil and corneal reflection to estimate the point-of-gaze.

A desktop eye-tracker is placed just outside of the viewing monitor and has the advantage of not requiring the user to wear anything on their head, though in some cases head tracking sensors have been placed on the user's head to improve eye-tracking. Figure 1 shows the desktop eye-tracker located just below the computer screen. For calibration, targets are placed on the screen and the user is asked to look at each target.

Figure 1. Desktop eye-tracker with numbers on the screen for calibration.

Determining the accuracy of eye-tracking over the entire computer screen can be difficult since after a few minutes of asking the user to focus on different locations on a screen, it becomes apparent from the increased eye-tracking measurement error that fatigue is setting in. Even with a fast calibration system, the calibration of the eye-tracker is depends on the initial calibration of the user.

Our lab has developed a glass eye that has a simulated pupil and real corneal reflection that changes when the glass eye is gazing at different locations on the screen. The glass eye also has a method of pointing the glass eye to any position on the viewing screen (see Figures 2, 3 & 4). The glass eye makes it easier and provides greater accuracy when evaluating the gaze position measurements from different areas of the viewed screen.

Figure 2. Glass eye used for eye-tracker calibration.

Figure 3. Glass eye with simulated pupil in the center and corneal reflection that changes depending on where the eye is pointed.

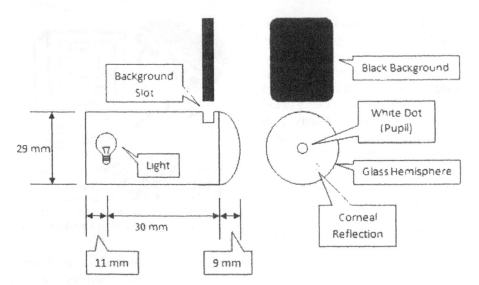

Figure 4. Glass eye internal specifications.

The glass eye has two modes of operation called aiming and eye-tracking. Knowing where the glass eye is aimed at is done by turning on the light in the back of the glass eye, which is also a lens, and removing the black background tab. The glass eye projects a spot of light that is related to where it is aimed at (see Figure 5). When using it for eye-tracking the light is turned off and the black background is inserted (see Figure 6).

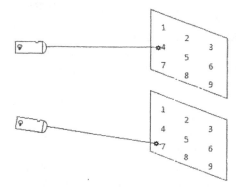

Figure 5. Glass eye pointing at different target locations where the spot of light illuminates the location.

The accuracy of eye-tracking results can depend on where on the screen the user is looking. Calibration over the entire screen can yield locations with significant eye-tracking error. Reduction of the errors can be done by readjustment of the system or screen placement of viewed targets. The glass eye can also be used to determine if there is drift in the eye position measurements and the drift rate since it can be left in a fixed position for extended periods of time, unlike the user's eye.

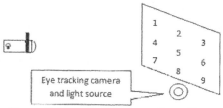

Figure 6. Glass eye in operation with the black background inserted.

Table 3 lists some of the problems and solutions with the table mounted or desktop eye-tracker. Many of the problems are common sense, but all of the solutions must be followed to optimize the eye-tracking session.

Table 3. Desktop eye-tracker problems and solutions

Problems	Solutions
User Fatigue - Usually, tasks have a large amount of eye movement or long periods of staring at a location which causes eye fatigue. This can result in long blinks that can cause eye-tracking to be lost and reduce task performance.	Avoid long eye-tracking sessions since fatigue usually occurs near the end of a session. Limit the total task time including calibration to less than an hour.
Cosmetics with Glitter – The reflection of cosmetics with sparkling material (e.g., glitter) can cause the eye-tracker to mistake the reflection from the glitter as a corneal reflection.	Request ahead of using the eye-tracker that any cosmetics with sparkling material should not be worn.

Table 3. (Continued) Desktop eye-tracker problems and solutions

Problems	Solutions
Moving Contact Lens – Contact lens slippage can create errors in tracking and blinking exacerbates the problem by changing the position of the lens.	Request the user bring computer glasses.
Glossy Rim Eye Glasses – The reflection from eye glasses with glossy rims can be mistaken by the eye-tracker for a corneal reflection causing data errors.	Have a thin piece of flat black tape ready to place on the rim of the eye glass.
Turning During Instructions – Loss of eye-tracking when a user turns to face the experimenter when instructions are given.	Before starting, instruct the user to always look at the screen.
Looking Off the Screen – Some users look off the screen when thinking causing the loss of eye-tracking.	Before starting, instruct the user to always look at the screen.
User is Too Tall or Too Short for Optimal Eye-Tracking - There is an optimal head position of the user for the eye-tracker.	Provide an adjustable chair and calibrate for an average height person when sitting.
Swaying and Moving Forward and Backwards.	Before starting, instruct the user to try not to move around too much.

The head-mounted eye-tracker is worn on the head of the user like a cap. It has a cable extending behind and can be connected to a backpack. See Table 4 for problems and possible solutions.

Table 4. Head-mounted eye-tracker problems and solutions

Problems	Solutions
Slippage on head from cable.	Reduce cable strain and tighten head-mount.
Head-mount on too tight causing pain or headache.	Loosen head-mount. This will increase the slippage potential.
Lack of calibrated reference due to head movement.	This is solved with some eye-trackers that have position references in the field of view available or with head tracking sensors.
The head-mounted system will be used by a variety of users which could be the source of head borne health conditions (e.g., skin disease, head lice, etc.).	Have a disposable surgical cap worn by the user.

CUSTOM PHYSIOLOGICAL SENSOR SYSTEM

A custom designed electrically isolated physiological sensor system is used to obtain galvanic skin conductivity (GSR), peripheral temperature, relative blood flow and the pressures applied to a computer mouse (see Figure 7). The computer mouse equipped with pressure sensors inside the body and the buttons of the mouse is used to detect pressures applied during task performance (see Figure 8). Pressure applied by the user to both buttons and the case are collected for analysis. These pressures have been linked to the cognitive load of a task (Ikehara & Crosby, 2005).

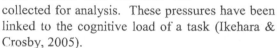

Figure 7. Sensors place on the hand with electronics in the background.

When analyzing data a norm for the individuals click signature is necessary to determine what a distorted click signature is. Distortion in click signature can be related to cognitive load.

Figure 8. Internal view of a computer mouse with pressure sensors on the buttons.

Another factor when analyzing a mouse click is that under certain conditions, the first part of the click signature provides different information than the latter part. When clicking a mouse button in normal use, a feature is activated when the button is released, but with games and other fast paced activities, the downward press causes an activation of a feature that changes the display. In the case of a game, the click signature should be partitioned at the point of display change. The initial click press being indicative of the decision process, while the release that occurs after the change of display indicating the reaction to the displayed result (Ikehara & Crosby, 2005).

Skin conductivity, temperature and relative blood flow is taken from the wrist and fingers of the user, but some tasks require that both hands be free. Our lab has successfully taken skin conductivity, temperature and relative blood flow from the feet and toes of users. Gloves should be used for cleaning of the user's foot and placement of the sensors. Also, disinfecting of the sensors should be done after each use.

IMPLEMENTATION

Real-time data collection and analysis poses several problems. The shear amount of data being collected taxes computing resources needed to handle: the computational requirements of the task being done; input and synchronization of sensor data sources; data storage to hard disk or memory if possible; computational requirements of the data analysis algorithms; output to change the task to increase performance; and evaluation of real-time task data to determine the efficacy of the change of task. The computation can be allocated to different computer servers, but this adds another level of complexity requiring communication and synchronization.

Changing the task in response to physiological changes to improve performance is the goal, but as mentioned before, issues such as motivation, learning, stress and fatigue needs to be accounted for. Augmented cognition could be best used where there is clear personal gain for being more productive using the system (e.g., more money, better grades, praise from boss, etc.). The entry of parameters such as maximum stress and maximum fatigue desired could be entered by the individual. A user need to know her or himself since there is a risk of over use, where the user's motivation is so great that the user employs the performance enhancing system until excessive stress or fatigue occurs.

A supervisor could enter the stress and fatigue parameters for the user at the risk of the user spoofing the system. A user could spoof the performance enhancing system by pretending to act fatigued knowing the system will detect the signs of fatigue and reduced the workload. For example, a user could purposely blink slowly. The eye-tracker would detect the slow blinking interpreting this signal as increased fatigue. The system would then change the task to something easier or more fun. Once the user learns to manipulate the performance enhancing system to reduce performance and is motivated to take it easy, the system value is lost.

The discussion so far has concentrated on measuring an individual, but with the cost of sensors dropping, this increases the possibility of improving human performance of groups based on measures from physiological sensors. Ikehara and Crosby (2003) discuss the possibility of enhancing distance education by using passive physiological sensors to monitor all students and passing that information to the instructor so that the instructor can modify the instruction to optimize class learning.

CONCLUSION

Using real-time physiological measures to guide the improvement of human performance has great potential, but there are several issues that require significant research before successful implementation can be achieved. This paper outlines some of those issues including: the human performance model; assessing individual differences; physiological sensor use; data acquisition; and implementation. Fortunately, advances in computers, sensors, and human performance modeling makes real-time physiological measures to guide the improvement of human performance a realizable goal.

REFERENCES

Andreassi, J.L. (1995). Psychophysiology: Human Behavior and Physiological Response (3rd Ed.). New Jersey: Lawrence Erlbaum.

Aschwanden, C. and Stelovsky, J. (2003) "Measuring Cognitive Load with EventStream Software Framework", 36th Annual Hawaii International Conference on System Sciences.

Charleton, S. G., (2002) "Measurement of Cognitive States in Testing and Evaluation," Handbook of Human Factors and Evaluation, S. G. Charleton and T. C. O'Brien, Eds., 97-126.

Ikehara, C., Chin D. N. and Crosby, M. E. (2003). "A Model for Integrating an Adaptive Information Filter Utilizing Biosensor Data to Assess Cognitive Load," Proceedings of the 9th International Conference on User Modeling, Pittsburgh, PA.

Ikehara, C., Chin, D. N. and Crosby, M. E. (2004) "A Modeling and Implementing an Adaptive Human-Computer Interface Using Passive Biosensors," Proceedings of the Hawaii International Conference on System Sciences, Kona, Hawaii.

Ikehara, C. and Crosby, M. E. (2003) "Real-Time Cognitive Load in Educational Multimedia," Proceedings of the 2003 World Conference on Educational Multimedia, Hypermedia & Telecommunications, Honolulu, HI.

Ikehara, C. S. and Crosby, M. E. (2005) "Assessing Cognitive Load with Physiological Sensors," 38th Annual Hawaii International Conference on System Sciences.

Ikehara, C. S., Crosby M. E. and Chin, D. N. (2005) "A Suite of Physiological Sensors for Assessing Cognitive States" 11th International Conference on Human-Computer Interaction.

Sheldon, E. (2001). Virtual agent interactions, PhD Thesis, Orlando: University of Central Florida.

Chapter 18

Infrasonic Analysis of Human Speech: An Index of Workload

Christopher K. McClernon[1], Mathew A. Middendorf[2], Benjamin A. Knott[1], Gregory J. Funke[1], Michael J. Harter[1]

[1]Air Force Research Laboratory
Wright-Patterson AFB, OH, USA

[2]Middendorf Scientific Services
Medway, OH, USA

ABSTRACT

This study investigated whether sub-auditory, infrasonic measures of vocal cord microtremors are sensitive to workload manipulations in a collaborative command and control (C2) task. Research participants performed 16 hours of simulated military C2 scenarios during which time task demand and access to collaborative tools were manipulated. NASA-TLX scores were used to assess subjective workload while infrasonic measures of vocal cord undulations were used as an objective measure of workload. The results of this study suggest that during high task demand trials infrasonic measures increased. In addition, during trials with access to collaborative tools that have previously been found to decrease subjective workload, infrasonic measures also decreased. Subjective workload scores and infrasonic measures were also found to correlate during trials. These results provide promising empirical evidence for the efficacy of this non-intrusive, non-invasive measure of operator workload.

Keywords: voice stress analysis, workload assessment, infrasonic measures, command and control

INTRODUCTION

The notion that military human performance systems can be understood by considering individual operators interacting with dedicated interfaces is quickly becoming outdated. Military operations increasingly rely on the effective interaction between individuals in distributed teams with multiple system interfaces. With this shift in focus, the assessment and measurement of various aspects of team process and performance is receiving greater consideration. For example, Cooke and Gorman (2009) emphasize the need for metrics of cognition and behavior that are distributed across people and their environment. Moreover, they suggest these measures should be predictive, diagnostic, unobtrusive, and capable of near-real time assessment (Cooke & Gorman, 2009).

Verbal communication is one overt and measureable team behavior that might be exploited for understanding factors influencing team performance. Indeed, Brenner, Doherty, and Shipp (1994) suggested that measures such as speech rate, fundamental frequency and speech loudness could serve as valuable indices of workload due to their unobtrusive nature. However, speech-based measures could be problematic if a continuous real-time assessment is desired during non-speaking portions of a task.

One alternative to audible speech measures of workload may be derived from sub-auditory, or *infrasonic*, components of speech (Inbar & Eden, 1976; Steeneken & Hansen, 1999). Infrasonic signals result from naturally occurring microtremors in the vocal cords and surrounding muscles. These microtremors are typically in the 8-12 Hz frequency range and spectral features (amplitude and frequency) of the signals can change as stress and workload fluctuate (op. cit.). Infrasonic measures are particularly appealing because they are present whether or not an individual is actually speaking providing continuity in measurement during speaking and non-speaking phases of a mission. In addition to the persistence of infrasonic measures, they are also non-intrusive and non-invasive, potentially allowing assessment of operator states without the subject being aware of the assessment.

There is much debate regarding the effects that stress and/or workload have on infrasonic measures, and inconsistent results can be attributed to three barriers to infrasonic measurement. First, infrasonic measures may only be sensitive to large fluctuations in workload and stress. When conducting two separate studies testing the efficacy of infrasonic measures for detecting lying, Barland (2002) only found significant results when questioning criminal suspects (high stress) as opposed to questioning college students (low stress). In an Airborne Command and Control (C2) environment, Congleton, Jones, Shiflett, Mcsweeney, and Huchinson (1997) found infrasonic measures sensitive when stress was either absent (preflight) or present (simulated combat operations), but not during different levels of workload and stress during simulated combat operations.

Second, infrasonic measures may be greatly influenced by individual differences. Early clinical work determined that microtremors of various muscle groups may not only vary between individuals but also within the same individual over time (Lippold, 1971). In a flight simulator context, Schneider and Alpert (1989) observed changes in an individual's infrasonic measures following manipulations in workload, but due to individual differences, the effect of overall mean workload across subjects did not reach statistical significance.

Third, infrasonic research has suffered from studies using sterile laboratory environments and tasks, thus making the generalization of results outside of the laboratory environment problematic. Many studies on infrasonic measurement address the effect of workload and stress on portions of speech such as phonation, repetitive words, etc. (e.g., Horii, 1979). Researchers have also focused extensively on the utility of infrasonic measures in a truth telling, polygraph setting. However, these results lend very little to the assessment of general workload and stress in human factors research as this application is heavily reliant on the subjects' ability to moderate their own physiology (see Ruiz et al., 1990).

In attempt to overcome the barriers to infrasonic measurement, the purpose of the current study was to explore the effects of task demand on infrasonic measures as an index of workload in an operationally realistic synthetic environment. To this end, following manipulations of task demand and the introduction of collaborative tools, infrasonic measures were evaluated for corresponding changes. In addition, potential correlations between infrasonic and subjective measures of workload were examined.

METHODOLOGY

One hundred five participants (70 men and 35 women) completed the experiment in five-person teams, yielding a total of 21 experimental teams. Participant ages ranged from 18 to 30 ($M = 21.9$, $SD = 3.2$). Participants were monetarily compensated for their participation and randomly assigned to one of three roles (weapons director, strike operator, tanker operator) within each team.

A $3 \times 2 \times 2 \times 2$ mixed design was employed in this experiment. Team position was a between-participants factor with three levels (weapons director, strike operator, tanker operator). Within-participants factors included two levels of task demand (low, high), two levels of data-display (tabular, graphical), and two levels of team communication (standard, enhanced). Each team completed 2 trials in each experimental condition, for a total of 16 trials in each experimental session.

This experiment employed Aptima, Inc.'s Distributed Dynamic Decision-making (DDD) software (version 3.0; MacMillan, Entin, Hess, & Paley, 2004). DDD provides a scriptable, low-to-moderate fidelity, team-in-the-loop simulated environment. DDD has successfully been used to simulate team C2 tasks and to study realistic and complex team processes in a variety of military and civilian research projects (MacMillan et al., 2004). The simulation required nine networked PCs and five participant workstations each equipped with a keyboard, mouse, and a 17-inch Dell 1707FPv LCD monitor. Each workstation was also equipped with a push-to-talk foot pedal, and participants were instructed to depress the pedal to communicate with team members. Activation of the foot pedal initiated audio recording and release terminated it.

The DDD scenarios employed teams of five to collaboratively complete a simulated C2 air defense task. This task has previously demonstrated to be sensitive to experimental manipulations of task demand and collaborative technologies (e.g., Finomore, et al., 2007; Funke et al., 2009). Participant teams were composed of two weapons directors, two strike operators, and one tanker operator. Within the simulation, the weapons directors' responsibilities were to allocate friendly fighters to appropriate enemy targets, schedule fighters for refueling and resupply, and communicate their plans with other team members. Strike and tanker operators' responsibilities included maneuvering team assets as instructed, performing requested activities (e.g., engage enemy targets, refuel friendly aircraft), and communicating pertinent information to teammates concerning asset resources.

The air defense task was presented on a tactical display which included information about friendly and enemy assets (see Figure 1). The simulation comprised three battle management zones: a "preferred attack zone" where enemy assets entered the scenario (right portion of display), a sensitive "friendly territory" (middle portion of display), and a highly sensitive "home territory" (left portion of display) that contained an air base and ground units. The objective of the simulation was to defend friendly assets against attacks from enemy aircraft.

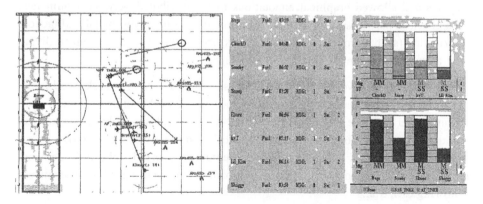

Figure 1. The DDD simulation. The tactical display (left) showing whiteboard marks (blue and black lines) a participant created to indicate asset and target route information. The tabular (center) and graphical (right) data displays. Both displays included information concerning remaining fuel and weapons of team assets.

The task demand condition of each trial determined the number of enemy targets present in each scenario. To ensure a relatively constant level of demand throughout each trial, as participants successfully prosecuted enemy aircraft, new enemy targets would enter the scenario from the right side of the screen. The low demand condition featured 4 enemy targets throughout each scenario and the high task demand condition featured 6.

Asset resource information was available to team members in a data display that contained either tabular or graphical information (see Figure 1) depending on the trial's data display condition. In the tabular display condition only strike and tanker operators had direct access to asset weapon and fuel status, presented in a digital format. Weapons directors, therefore, had to rely on teammates for judicious resource updates. In the graphical display condition, asset resources were displayed in an analog format, and this display was available to all team members. The graphical display also conveyed supplemental fuel information to team members – fuel gauges changed to an amber color when fuel reserves were low, and it featured a black bar which indicated the minimum reserve fuel required to rendezvous with a tanker asset.

Communication between teammates in this experiment was manipulated through the team communication factor. In the standard communication condition, participants could communicate orally over a shared radio channel using a radio headset. In the enhanced communication condition, participants could communicate using the radio or using two collaborative tools: chat messaging and a virtual whiteboard. The virtual whiteboard was integrated with participants' tactical displays and allowed graphical annotations to be distributed between teammates. This provided participants an alternative means to communicate spatial and tactical information (such as routes, enemy locations, etc.) without forcing them to divide their attention across multiple displays (see Figure 1).

Operator subjective workload was assessed using the NASA Task Load Index (TLX; Hart & Staveland, 1988), a standard measure of workload that is widely used in human performance research. The NASA-TLX provides a global index of task workload on a scale of 0 to 100 and identifies the relative contributions of six sources of workload: mental demand, temporal demand, physical demand, performance, effort, and frustration. Participants completed the TLX immediately following each trial.

The duration of the experiment was approximately 16 hours, conducted across two 8-hour sessions. The first session was devoted to training, and the subsequent session to experimental data collection. Upon arrival at the laboratory, participants completed an informed consent document, were introduced to their teammates, and given a brief overview of the experiment. They were then randomly assigned to a team position and received written and oral instructions detailing the DDD simulation, the team's goals, the control of team assets, and the roles and responsibilities of each team position. The instructional period concluded with a quiz to ensure that participants understood the DDD task and the requirements of all three team roles. Next, teams completed 13 practice trials designed to give them an opportunity to collaborate and strategize under conditions identical to those they would experience during experimental trials. These trials, and all subsequent experimental trials, were 10 minutes in duration. During the experimental session, teams completed 16 experimental trials; the experimental schedule of conditions was randomized and counterbalanced across teams to control order effects.

DATA ANALYSIS AND RESULTS

The focus of analyses for this study was on the effects of the experimentally manipulated factors on subjective and infrasonic measures of workload. Three of the 21 groups had incomplete audio recordings and were not considered for subsequent analysis. Six of the remaining 18 groups were removed from the analysis due to inadequate audio recording quality. Specifically, these audio recordings included audio signal clipping to the extent that power spectral analyses were not possible. Twelve groups ($N = 60$) remained for analysis.

The audio data was sampled at 8000 Hz and processed by a fourth order band pass filter. The low and high cutoff frequencies of the filter were five and 15 Hz, respectively. A hanning window was applied to the filtered data to improve the accuracy of the spectral analysis, and a power spectrum was computed (see Figure 2).

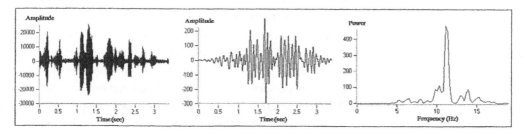

Figure 2. A sample of the data used for analysis. Raw data (left) is filtered and windowed (center) prior to power spectrum (right).

The average weighted power was computed for two six-hertz wide triangular windows, with a low power (LP) window centered at 8.5 HZ and a high power (HP) window centered at 11.5 Hz. Window widths were selected to allow for individual differences in the natural infrasonic frequency. As is depicted in Figure 2, the power spectrum clearly shows an infrasonic spike at approximately 11 Hz indicating a relatively high HP infrasonic response.

All audio recordings of the 12 groups (40 males, 20 females) were processed to generate the LP and HP infrasonic measures. Due to the HP and LP measures being highly skewed, they were submitted to a log transformation to retain the data's normal distribution properties for subsequent Analyses of Variance (ANOVA) testing. For each participant in each trial, separate LP and HP PSDs were calculated for each recorded utterance. From these values, mean LP and HP scores were calculated for each participant in each trial. As mentioned previously, infrasonic measures are likely to exhibit large individual differences across speakers. Therefore, each participant's data was standardized using the formula:

$$Z = (TP - MP) / SDMP \tag{1}$$

where TP = participant's mean power (LP or HP) for each trial, MP = participant's mean power calculated across all trials, and $SDMP$ = participant's standard deviation in power calculated across all trials. This approach resulted in infrasonic responses measured in standard deviations from each participant's mean.

Mean infrasonic measures were tested for statistically significant differences between conditions using 3 (team position) x 2 (task demand) x 2 (data display) x 2 (team communication) mixed-model ANOVAs. An alpha level of 0.05 was used to identify significant effect of team position, task demand, data display, and team communication on low and high infrasonic power.

Analyses indicated significantly higher LP and HP response when participants used the tabular data display compared to the graphical display, $F(1, 57) = 16.01$, $p < .05$ and $F(1, 57) = 11.65$, $p < .05$, respectively (see Figure 3). When measuring workload using LP, a statistically significant interaction between team position and task demand was also observed, $F(2, 57) = 3.51$, $p < .05$. Follow-up post hoc paired sample t-tests for the team position × task demand interaction revealed higher weapons director LP response during high task demand trials compared to the same operators during low demand trials, $t(59) = 2.80$, $p < .05$ (see Figure 3). No other sources of variance in the analysis were statistically significant (all $p > .05$). As can be seen in Figure 3, there was a trend for higher HP response during high task demand trials for the weapons directors.

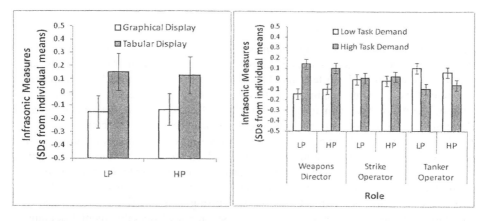

Figure 3. Infrasonic workload metrics of participants using the tabular and graphical data displays (left) and of each role during low and high task demand (right).

Mean NASA-TLX, LP, and HP values were compared using bivariate correlations. All TLX subscales were included to further define the infrasonic measures of workload. As can be seen in Table 1, the majority of TLX subscales displayed a modest correlation with the infrasonic measures. LP and HP values were also correlated with a global measure of TLX workload (calculated as the mean TLX rating across all subscales), $r = 0.30$ and $r = 0.29$ respectively, $p < .05$.

Table 1 Pearson correlations between infrasonic measures and subjective workload.

	Correlations						
Measure	*1*	*2*	*3*	*4*	*5*	*6*	*7*
1. LP							
2. HP	0.94*						
3. Mental	0.30*	0.28*					
4. Physical	-0.11	-0.08	-0.16				
5. Temporal	0.37*	0.37*	0.75*	-0.02			
6. Performance	0.07	0.06	0.27*	0.07	0.28*		
7. Effort	0.30*	0.28*	0.81*	-0.09	0.81*	0.26	
8. Frustration	0.32*	0.29*	0.32*	0.33*	0.51*	0.52*	0.38*

Note: N = 60. * $p < .05$.

DISCUSSION

The purpose of this research was to explore the utility of infrasonic measures for assessment of individual workload within the context of team collaboration. Results indicated that participants' infrasonic scores (LP and HP) were higher while using the tabular display. This finding is consistent with Funke et al. (2009), who found higher subjective workload scores when participants used a tabular resource display in the same C2 task. This effect may be due to increased cognitive demands placed on operators in that condition. Infrasonic measures were also sensitive to manipulations of task demands, with higher observed LP values for weapons directors in high task demand conditions. These results are inconsistent with Congleton et al. (1997) who, using a different voice stress analysis technique, did not find any effects of operational task demand on weapons directors' infrasonic measures in a C2 task, showing further promise for the techniques used in the current study. Finally, the positive correlations observed in this study between infrasonic measures and subjective workload ratings support the use of infrasonic measures as a means for online, real-time assessment of operator mental demand, temporal demand, effort, and frustration.

This study helped overcome some of the barriers to infrasonic measurement. First, the manipulation of task demand in this study was relatively small (4 or 6 enemy targets appearing during a 10 minute scenario) and infrasonic measures were sensitive to this manipulation. Second, the infrasonic calculations and the standardization method used in this study's analyses (see Equation 1) allowed effective comparison across research participants in the presence of individual differences. Finally, the results of this study and the operational task used (i.e., an airborne C2 task) show promise for the implementation of infrasonic measures outside the laboratory environment.

Other potential applications include the implementation of infrasonic measurement to facilitate distribution of workload across team members in a collaborative environment or distribute workload between man-machine interfaces (e.g., dynamic function allocation). These measures may also be used to assess adversary workload in order to identify combatants (e.g., suicide bomber) and assist in military interrogations. Although the techniques used in this infrasonic analysis calculated measures during speaking portions of the C2 task, these same measures can theoretically be collected during non-speaking portions of a task. The current ongoing research endeavor is investigating infrasonic analyses in this manner.

The results of this study make a clear connection between infrasonic measures and workload, but the neurological pathology of infrasonic measures is still unclear. Some have linked human microtremors to the body's autonomic nervous system activity (Barland, 2002), the spinal cord, adrenalin circulation (Lippold, 1971), and, more generally, natural physiological trembling of the muscles surrounding the vocal cords (Ruiz et al., 1990). These factors may help define the psychophysiological underpinning for infrasonic measures, but more human factors research in this area is needed.

Overall, the results of this experiment are encouraging, indicating that infrasonic measures may provide a sensitive, non-invasive measure of operator workload. Given the potential operational utility of this type of workload assessment, infrasonic measurement can be seen as a critical tool for future research in neuroergonomics.

REFERENCES

Barland, G. H. (2002). Use of voice changes in the detection of deception. *Polygraph, 31*, 145-53

Brenner, M., Doherty, E. T., & Shipp, T. (1994). Speech measures indicating workload. *Aviation,Space, and Environmental Medicine, 65*, 21-26.

Congleton, J.J., Jones, W.A., Shiflett, S.G., Mcsweeny, K.P., Huchingson, R.D. 1997). An evaluation of voice stress analysis techniques in a simulated AWACS environment. *International Journal of Speech Technology, 2*, 61-69. doi:10.1007/BF02539823

Cooke, N. J. & Gorman, J. C. (2009). Interaction-Based Measures of Cognitive Systems. *Journal of Cognitive Engineering and Decision Making, 3*, 27-46. doi:10.1518/155534309X433302

Finomore, V.S., Knott, B.A., Nelson, W.T., Galster, S.M., & Bolia, R.S. (2007). The effects of multimodal collaboration technology on subjective workload profiles of tactical air battle management teams. *Proceedings of the International Symposium on Aviation Psychology, 14*, 190-196.

Funke, G.J., Russell, S.M., Knott, B.A., & Miller, B.T. (2009). What can a multidimensional measure of stress tell us about team collaborative tools? *Proceedings of the International Symposium on Aviation Psychology, 15*, 480-485.

Hart, S.G., Staveland, L.E. (1988). Development of NASA-TLX (Task Load Index): Results of empirical and theoretical research. In P.A. Hancock & N. Meshkati (Eds.), *Human mental workload* (pp. 139-183). Amsterdam: Elsevier Science. doi:10.1016/S0166-4115(08)62386-9

Horii, Y. (1979). Fundamental frequency perturbation observed in sustained phonation. *Journal of Speech and Hearing Research, 22*, 5-19.

Inbar, G.F., & Eden, G. (1976). Physiological stress evaluators: EMG correlation with voice tremor. *Biological Cybernetics, 24*, 165-167. doi:10.1007/BF00364119

Lippold, O. C (1971). Physiological Tremor. *Scientific American, 224(3)*, 65-73. doi:10.1038/scientificamerican0371-65

MacMillan, J., Entin, E.B., Hess, K.P., & Paley, M.J. (2004). Measuring performance in a scaled world: Lessons learned from the Distributed Dynamic Decisionmaking (DDD) synthetic team task. In S.G. Schiflett, L.R. Elliott, E. Salas, & M.D. Coovert (Eds.), *Scaled worlds: Development, validation, and applications* (pp. 154-180). Brookfield, VT: Ashgate.

Ruiz, R., Legros, C., Guell, A. (1990). Voice analysis to predict the psychological or physical state of a speaker. *Aviation, Space, and Environmental Medicine, 61*, 266-71.

Steeneken, H.J.M., & Hansen, J.H.L. (1999). Speech under stress conditions: Overview of the effect on speech production and of system performance. *Proceedings of the IEEE International Conference on Acoustics, Speech, and Signal Processing, 4*, 2079–2082.

Schneider, S. J. & Alpert, M. (1989). Voice measures of workload in the advanced flight deck: Additional studies. *Technical Report No. NASA-CR-4258, NASA.*

CHAPTER 19

Real-Time Classification of Neural Signals Corresponding to the Detection of Targets in Video Imagery

Jonathan Touryan, Laurie Gibson, James H. Horne, Paul Weber

Science Applications International Corporation
Louisville, CO 80027, USA

ABSTRACT

Recent advances in neuroscience offer unprecedented insight into human perception and cognition. Powerful neuroimaging methods, such as functional magnetic resonance imaging (fMRI) and electroencephalography (EEG) allow scientists to directly measure dynamic and subtle brain states. It is easy to imagine how this work could lead to far-reaching improvements in medical diagnosis and treatment. The implications for revolutionary advances in the way that humans and machines interact in the real world are less obvious. In this paper, we describe an application of neurotechnology that combines the complementary power of the brain and the computer to process complex natural imagery. The goal of this work is a system that can rapidly scan a very wide field-of-view for threats and alert the user to their presence before they can harm him. An automated algorithm is able to quickly process images over a wide area to identify potential threat objects. However, these algorithms also detect many uninteresting objects (*i.e.*, a bush moving in the wind or rock that looks like a truck). If there are too many of these false alarms, the operator is perpetually alerted and the system is useless. By incorporating the human visual system to sort through the objects that the algorithm identifies, the number of algorithm-generated false alarms can be reduced to an acceptable level.

Our paradigm uses a target cuer to find potential threats within the larger scene. In this case, threat objects included principally people and vehicles, both moving and stationary. The images containing the possible threats are presented to the human operator in a mode called rapid serial visual presentation (RSVP). The operator passively views this pre-screened imagery while the system captures and analyzes his brain activity via EEG. Using only the recorded brain signals we can determine, with remarkable accuracy, whether or not the operator saw a threat object in any of the imagery presented. The acquisition and classification of brain signals is accomplished seconds after imagery is presented allowing for real-time adaptation and feedback. We show reliable and accurate classification of the EEG signal, even for long-duration experiments (two hours) both in and out of the laboratory. These results are incorporated into a custom, lightweight, semi-portable system for the real-time classification of EEG signals (including artifact detection and mitigation). This system was tested with live video and demonstrated a level of reliability sufficient for integration into commercial technology.

INTRODUCTION

In this paper, we describe a real-time neuroimaging system that is based on established scientific observations describing how the human visual system processes and classifies complex imagery. For over a decade, it has been known that the visual cortex can perform complex categorization of visual stimuli within a few hundred milliseconds of the stimulus presentation (Hillyard & Anllo-Vento, 1998, Thorpe, Fize & Marlot, 1996). In the original laboratory experiment, subjects were asked to determine if a photograph, presented for only a few milliseconds (ms), contained an animal. Using electroencephalography EEG, researchers were able detect an animal/no-animal categorization signal emanating from frontal and parietal cortex far earlier (150ms) than any behavioral response could be initiated. Since the initial finding, various groups have expanded this work by having subjects perform more complex categorizations (Scott, Tanaka, Sheinberg & Curran, 2008, Tanaka & Curran, 2001, Tanaka, Curran & Sheinberg, 2005). This idea has also been extended to incorporate sequentially presented stimuli, known as the rapid serial visual presentation (RSVP) paradigm, for both still and motion imagery (Luo & Sajda, 2009).

Laboratory studies that use EEG to measure and classify perceptual states often use repeated trials to improve the signal-to-noise ratio and subsequent results. The underlying neural signatures associated with a perceptual state can often be clearly shown by averaging over repeated trials. However, for many applications of this technology, including the one described here, it is critical to minimize the number of required stimulus presentations and to rapidly process the EEG signals. Here we used a real-time system that processes the perceptual state of the subject on a trial-by-trial basis.

METHODS

The study had two components: a series of experiments under controlled conditions and a field test outside of the laboratory. The experiments were conducted in our EEG laboratory while the field test was held at Yuma Proving Grounds, Ariz. (YPG). For the laboratory experiments, imagery was carefully prepared from high-resolution video that had been previously collected in the field. The field test made use of live video from sensors positioned at the test site, recording ongoing events.

Subjects

Thirty-four naïve subjects participated in the laboratory experiment. The subjects (16 female and 18 male) ranged in age from 20 to 59 with a mean age of 38. Subjects were both right-handed and left-handed (30 right-handed, four left-handed). Eleven of the 34 subjects participated in the initial phase of the experiment where data was only collected for the purpose of building a general EEG classification model. Twenty-three of the 34 subjects participated in the second phase of the experiment where individual models were developed for each subject. These 23 subjects participated in a total of two experimental sessions. The data collected from the first session was used to build the individual classifier model employed in the second session. For this group, the classification of neural signals occurred in real time, as the signals were collected during the experiment. Five trained subjects participated in the field test at YPG. Trained subjects participated in at least four laboratory sessions before the field test.

Stimuli

Stimuli consisted of small regions of interest (ROIs) within a large natural landscape. Each ROI was made up of five sequential images (captured with a 2Hz video camera) of a square (300 by 300 pixel) region. All images were 8 bit grayscale. Each ROI belonged to one of two main categories based on the content of the imagery. If ROI images contained objects of interest (defined as vehicles, people, or objects perceived to be possible threats), they were classified as targets. Targets that were not vehicles or people were identified as possible threats by consensus review (with four subjects) of all stimuli. If the ROI did not contain an object of interest, it was classified as clutter. In the laboratory experiments, all target ROIs were selected manually, while clutter ROIs were selected both manually and through various algorithms that identified areas of high saliency (that could potentially be confused with targets). In the field tests, all ROIs were identified with a neuromorphic target cuer (Schachter, 2008). Both target and clutter ROIs could contain objects that were either moving or stationary throughout the sequence of images. There were a total of 282 target ROIs and 1,683 clutter ROIs used in the laboratory experiments. Example images shown in Figure 1 are representative of the type of stimuli used in the study.

Procedure

The subjects were seated in front of a computer monitor at a distance of approximately 100 centimeters. All images were displayed at the center of the monitor and subtended a visual angle of about 8.25° horizontally and 8.25° vertically. The laboratory experiment consisted of 12 to 16 blocks containing roughly 240 ROIs presented in rapid sequence (*i.e.*, the rapid serial visual presentation paradigm). Each frame of the ROI was presented for 100ms (500 ms per ROI) and there were no breaks between ROIs. The effective frame rate of the ROI sequence was 2Hz (*i.e.*, two complete ROI presentations per second). Subjects were instructed to fixate at the center of the monitor and respond, via button press, when they saw a target ROI. There was a pause at the end of each 240 ROI block and the subject started the next block, via a button press, at their discretion. Thus, the subjects experienced approximately two minutes of RSVP, followed by a self-paced rest period.

The field test procedure was the same, with the exception that the number of blocks varied from session to session, depending on the output of the target cuer, which in turn depended on the events taking place at the site. The subjects were not required to respond with a button press when they saw a target ROI: however, they were given the option to press an unattached button as an aid to maintain engagement in the task.

EEG recording

Scalp EEG was collected with a 128-channel HydroCel Geodesic Sensor Net™ (Electrical Geodesics Inc., Eugene, Ore.) connected to an AC-coupled 128-channel, high-input impedance amplifier (200 MΩ, Net Amps™, Electrical Geodesics Inc., Eugene, Ore.). Individual sensors were adjusted until impedances were less than 50 kΩ. Amplified analog voltages (0.1- to 100-Hz bandpass) were digitized at 250 Hz. Recorded voltages were initially referenced to a vertex channel. EEG was excluded from analysis if it contained eye movements (vertical electro-oculogram channel differences greater than 70 µV) or more than five bad channels (changing more than 100 µV between samples, or reaching amplitudes over 200 µV). Data from individual bad channels were replaced using a spherical spline interpolation algorithm. Event-related potentials (ERPs) were time-locked to the first frame of the ROI. ERPs were baseline-corrected for the 200 ms interval prior to the first frame of the ROI.

Real-time classification

The real-time classification of the EEG signal was accomplished through analysis of the ERP following the presentation of each ROI. A particular component of the ERP, the P300 (or positive deflection beginning 300 ms after stimulus onset), has been shown to be associated with the detection of targets in visual stimuli (Handy, 2005, Squires, Wickens, Squires & Donchin, 1976, Thorpe et al., 1996). The P300 increases when the evoking stimulus is rare, recognized, or meaningful. In this paradigm, target objects, by definition, are meaningful to the subject and thus evoke a larger P300 response.

The subtle nature of the signal (relative to ongoing EEG activity) and significant variation across individuals required that single-trial classifier parameters be customized for each subject. The goal was to find a linear combination of the components of the ERP that most reliably discriminated between the responses to target versus clutter stimuli. While the classification method was the same for all subjects, the classifier parameters were derived individually. For each ERP, a large number of features were generated by first linearly transforming the raw EEG signal using principal components analysis (PCA), and then calculating windowed fast Fourier transforms (FFT) for a variety of window sizes and starting times relative to the stimulus onset. The approach described in (Perkins, Lacker & Theiler, 2003) was used to automatically find the most significant and stable features in the EEG signal that correlated with target detection.

Once the features were identified and the appropriate parameters were selected, the resulting linear classifier was applied to each ERP to generate a single, unbounded score. Positive scores indicated likely target ROIs, while negative scores indicated likely clutter ROIs. Trials containing eye movements or more than five bad channels were not scored and the corresponding ROI was re-shown to the subject. Additionally, if the score for a specific ROI did not clearly indicate whether the ROI was target or clutter, the ROI was re-shown to the subject, with scores averaged across multiple (up to four) showings of a single ROI.

LABORATORY EXPERIMENT

Subjects were asked to indicate when they saw targets in RSVP of natural imagery (see Methods; Figure 1A). All subjects were able to perform the task with sufficient accuracy. The average target detection accuracy for naïve subjects was 92.4% with a reaction-time (RT) distribution of 498.5 ± 54.0ms (mean ± std). The trained subjects had an average target detection accuracy of 95.1% with an RT distribution of 464.4 ± 44.3 ms.

A subset of subjects (N = 23) participated in the laboratory experiments where the real-time classification system analyzed the neural response associated with each ROI after it was presented to the subject (see Methods). For valid trials (free of eye movements or bad channels), the classifier calculated a single score for the corresponding ROI. The score indicated how likely it was that the stimulus contained a target. While these scores were unbounded, they were typically centered on zero, with positive values indicating likely target ROIs and negative values indicating likely clutter ROIs. To quantitatively assess the performance of the classifier, we calculated the area under the receiver operating characteristic (ROC) curve (Green & Swets, 1966) for each subject (Figure 1C). The average area under the ROC curve for these naïve subjects was 0.847 (max = 0.971, min = 0.584). This was in contrast to the trained subjects (N = 5), who participated in at least four experimental sessions. The average area for these subjects was 0.952 (max = 0.965, min = 0.942). While the trained subjects were selected based on their initial classifier performance, multiple experimental sessions improved behavioral accuracy (ANOVA, F = 7.64, p < .01) and classifier scores (ANOVA, F = 72.27, p < .01). However, the behavioral accuracy was only significantly better in one of the five subjects (ANOVA, F = 14.33, p < .01).

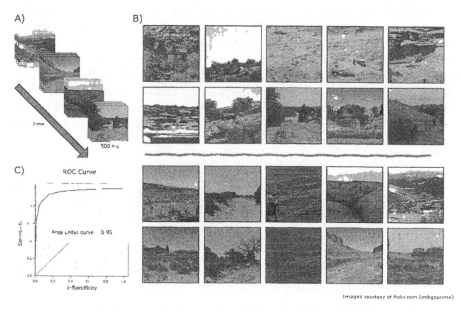

Images courtesy of flickr.com (indigoprime)

Figure 1. Experimental paradigm and classifier results. Panel A: Stimuli consisted of an RSVP sequence containing short (500ms) video clips of potential targets and background clutter. Panel B: Video clips ranked by the neural response with colored border indicating ground-truth (red = target, blue = clutter); only the 10 highest and lowest scoring clips are shown. Panel C: Classifier performance for one subject quantified in an ROC curve (area under curve = 0.95). Note: the actual images used as stimuli have not been approved for public release; images above are representative.

The classifier scores were based on a linear weighting of the ERP elicited by each ROI. While the target detection ERPs for each subject varied, they had a characteristic time course. Figure 2A shows the average ERP elicited for all target ROIs for an example subject, typical of the population. The first significant activation starts before the P300 in the occipital scalp electrodes, around 100 ms after stimulus onset. The P300, or target-detection signal, starts with a frontal activation (beginning around 300 ms) which spreads to a frontal-parietal activation at the later stages of the ERP (between 400 and 600 ms after stimulus onset). The early activation over the frontal scalp is speculated to be an orienting response, possibly from the anterior cingulated (Menon, Ford, Lim, Glover & Pfefferbaum, 1997). The later frontal-parietal activation is likely the neural correlate of the post-sensory processing of stimuli and decision making (Philiastides & Sajda, 2006, Ratcliff, Philiastides & Sajda, 2009).

The average ERP associated with target and clutter ROIs is significantly different, and thus distinguishable, across many channels. However, due to the ongoing EEG activity and environmental noise, single trial classification of ERPs presents a significant challenge. Through a machine learning approach, we developed a method to identify the most reliable features in the EEG signal that correlate with target detection. These features are typically contained within a smaller number of channels and over a more restricted time course than the average ERP. Figure 2B shows the maximally discriminant features of the ERP. To isolate these features and score each ERP, linear weighting functions were derived for each subject (see Methods). For the real-time classification system, the individualized weight file was applied to the ERP to generate the classification score.

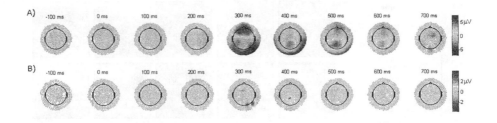

Figure 2. Event Related Potential (ERP). Panel A: Average ERP associated with target ROIs for the same subjects as Figure 1C. ROI presentation begins at 0 ms and continues through 500 ms. Panel B: Linear weighting function applied to ERP in panel A. The classification score for each ROI is generated by the sum of the weighted ERP corresponding to that ROI.

FIELD TEST

A field test using the real-time classification system was conducted at YPG in February 2009. Although the field system utilized roughly the same hardware and software components as the laboratory system, there were several significant differences between the laboratory and field experiments. First, the field experiments incorporated live imagery acquired from the surrounding environment. Targets (*i.e.*, people and vehicles) were introduced into this environment at random time intervals while a high-resolution digital camera continuously acquired imagery of the scene. This imagery was then immediately processed by the target cuer algorithm (Schachter, 2008) which transmitted ROIs to the subject. Second, an additional task was added. After viewing a sequence of ROIs from the target cuer, the subject manually confirmed or rejected ROIs identified by the neural classifier as containing a target. Lastly, the test environment was noisier and had more distractions than the laboratory.

The results of the EEG classifier for the five subjects over 17 sessions at the field test are shown in Figure 3. Sessions varied significantly in terms of length (17 minutes to 28 minutes), number of ROIs presented (964 to 2,200), number and type of target, and in the difficulty in discriminating the targets from the background. Performance differed significantly among the subjects (ANOVA, F = 3.5, p < .05).

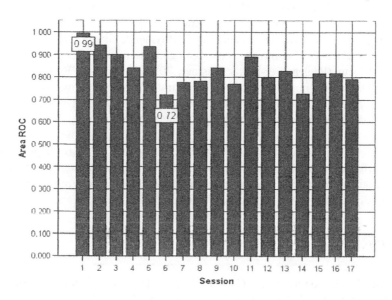

Figure 3. Results of the 17 sessions of the field test expressed as area under the ROC curve.

CONCLUSIONS

This paper demonstrates a real-time neural classification system for the detection of targets in video imagery. Using classifier models, customized to the individual, we were able to achieve classification performance of greater than 0.95 area under the ROC curve. Importantly, the classification system performed well in a test outside the laboratory in a more operational environment. While the field test suggests that the distraction and stress inherent in these environments affects the neural classification performance, and likely the underlying ERP activation, the degradation was not severe or universal.

Previous studies have identified the neural correlates of target detection in video imagery using a similar paradigm (Luo & Sajda, 2009); however, to our knowledge, this is the first report of real-time classification of the target detection signal at this level of accuracy. Any system that seeks to augment the performance of the human, via neural signal classification, must be both rapid and robust. This has been an ongoing challenge for the burgeoning areas of neuroergonomics and brain-computer interface (Parasuraman & Rizzo, 2007). Here, we demonstrate the feasibility of using such a brain-in-the-loop technology to augment the performance of the human.

One open question that relates to the usefulness of this technology as a neuroergonomic tool is the attentional resources required for the target detection task. While it has been shown that attentional blink (Kranczioch, Debener & Engel, 2003, Vogel, Luck & Shapiro, 1998) can dramatically affect the detection of targets in RSVP paradigms (Chun & Potter, 1995, Raymond, Shapiro & Arnell, 1992), there is evidence that complex scene categorization occurs outside of direct attentional focus (Peelen, Fei-Fei & Kastner, 2009, VanRullen, Reddy & Fei-Fei, 2005). Additionally, image analysts with years of training show complex scene categorization early in the evolution of the ERP (at the P1 and N1 stage), before the circuits of decision making are engaged (Curran, Gibson, Horne, Young & Bozell, 2009). Thus, it may be possible to incorporate this paradigm within a dual-task setting, even though traditional studies indicate that the behavioral performance should suffer from interference (Pashler, 1994). This approach may be well suited for situations where trained experts are required to engage in ongoing, critical tasks while rapidly scanning their environments for potential threats.

REFERENCES

Chun, M.M., & Potter, M.C. (1995). A two-stage model for multiple target detection in rapid serial visual presentation. *J Exp Psychol Hum Percept Perform, 21* (1), 109-127.

Curran, T., Gibson, L., Horne, J., Young, B., & Bozell, A. (2009). Expert image analysts show enhanced visual processing in change detection. *Psychonomic Bulletin & Review, 16* (2), 390-397.

Green, D.B., & Swets, J.A. (1966). Signal detection theory and psychophysics. (p. 479). New York, New York: Wiley.

Handy, T.C. (2005). Event-related potentials: a methods handbook. (p. 404). Cambridge, MA: The MIT Press.

Hillyard, S.A., & Anllo-Vento, L. (1998). Event-related brain potentials in the study of visual selective attention. *Proc Natl Acad Sci U S A, 95* (3), 781-787.

Kranczioch, C., Debener, S., & Engel, A.K. (2003). Event-related potential correlates of the attentional blink phenomenon. *Brain Res Cogn Brain Res, 17* (1), 177-187.

Luo, A., & Sajda, P. (2009). Comparing neural correlates of visual target detection in serial visual presentations having different temporal correlations. *Front Hum Neurosci, 3*, 5.

Menon, V., Ford, J.M., Lim, K.O., Glover, G.H., & Pfefferbaum, A. (1997).

Combined event-related fMRI and EEG evidence for temporal-parietal cortex activation during target detection. *Neuroreport, 8* (14), 3029-3037.

Parasuraman, R., & Rizzo, M. (2007). Neuroergonomics: The Brain at Work (p. 430). New York, N.Y.: Oxford University Press

Pashler, H. (1994). Dual-task interference in simple tasks: Data and theory. *Psychological Bulletin, 116*, 220-244.

Peelen, M.V., Fei-Fei, L., & Kastner, S. (2009). Neural mechanisms of rapid natural scene categorization in human visual cortex. *Nature, 460* (7251), 94-97.

Perkins, S., Lacker, K., & Theiler, J. (2003). Grafting: Fast, incremental feature selection by gradient descent in function space. *Journal of Machine Learning Research, 3*, 1333-1356.

Philiastides, M.G., & Sajda, P. (2006). Temporal characterization of the neural correlates of perceptual decision making in the human brain. *Cereb Cortex, 16* (4), 509-518.

Ratcliff, R., Philiastides, M.G., & Sajda, P. (2009). Quality of evidence for perceptual decision making is indexed by trial-to-trial variability of the EEG. *Proc Natl Acad Sci U S A, 106* (16), 6539-6544.

Raymond, J.E., Shapiro, K.L., & Arnell, K.M. (1992). Temporary suppression of visual processing in an RSVP task: An attentional blink? *Journal of Experimental Psychology: Human Perception and Performance, 18*, 849-860.

Schachter, B. (2008). Biological models for automatic target detection. 6967 (p. 69670Y): SPIE.

Scott, L.S., Tanaka, J.W., Sheinberg, D.L., & Curran, T. (2008). The role of category learning in the acquisition and retention of perceptual expertise: a behavioral and neurophysiological study. *Brain Res, 1210*, 204-215.

Squires, K.C., Wickens, C., Squires, N.K., & Donchin, E. (1976). The effect of stimulus sequence on the waveform of the cortical event-related potential. *Science, 193* (4258), 1142-1146.

Tanaka, J.W., & Curran, T. (2001). A neural basis for expert object recognition. *Psychol Sci, 12* (1), 43-47.

Tanaka, J.W., Curran, T., & Sheinberg, D.L. (2005). The training and transfer of real-world perceptual expertise. *Psychol Sci, 16* (2), 145-151.

Thorpe, S., Fize, D., & Marlot, C. (1996). Speed of processing in the human visual system. *Nature, 381* (6582), 520-522.

VanRullen, R., Reddy, L., & Fei-Fei, L. (2005). Binding is a local problem for natural objects and scenes. *Vision Res, 45* (25-26), 3133-3144.

Vogel, E.K., Luck, S.J., & Shapiro, K.L. (1998). Electrophysiological evidence for a postperceptual locus of suppression during the attentional blink. *J Exp Psychol Hum Percept Perform, 24* (6), 1656-1674.

Acknowledgements

This work was supported by The Defense Advanced Research Projects Agency (Government Contract No. W31P4Q-08-C-0262).

CHAPTER 20

Understanding Brain Arousal and Sleep Quality Using a Neuroergonomic Smart Phone Application

Daniel Gartenberg, Raja Parasuraman

George Mason University
Fairfax, VI 22031, USA

ABSTRACT

The increasing prevalence of mobile computer technology in everyday life allows for the collection of data that were previously inaccessible or expensive to obtain. Such data provide unique information different to that collected in the lab or through the use of questionnaires because they are obtained under naturalistic conditions—a requirement for neuroergonomics practice. Additionally, if users perceive the technology as either beneficial or enjoyable, longitudinal data become easier to collect. With these factors in mind, an iPhone application was developed to provide personalized feedback on sleep, sleep arousal, and variables that co-vary with sleep. The device included a vigilance task to measure arousal, an alarm clock to track sleep duration and time of sleep, and a diary that displayed sleep statistics to the user. The effect of sleep duration on vigilance task performance was investigated in a case study utilizing this device to collect naturalistic data. Three variations of the Psychomotor Vigilance Task (PVT) were administered, consisting of a 2 second window of response (for a total-time-on-task of 1 minute), 5-second window of response (for a total-time-on-task of 2.5 minutes), and 10-second window of response (for a total-time-on-task of 5 minutes). Performance efficiency on all three tasks was related to sleep deprivation. The results suggest that shorter duration tasks that require spatial discrimination and have higher motor demands than the conventional PVT are sensitive to the homeostatic component of sleep. Potential uses for this and similar applications to the field of neuroergonomics are also discussed.

Keywords: Neuroergonomics, Brain Arousal, Vigilance, Sleep, Health Care

INTRODUCTION

Neuroergonomics is an emerging field that combines neuroscience, human factors psychology, and engineering with the aim to optimize mental functioning during cognitive and physical work (Parasuraman, 2003). To accomplish this goal, work and brain function must be evaluated in naturalistic environments because there may be differences between behavior assessed in the laboratory and in real-world settings (Rizzo, Robinson, & Vicki, 2007).

The human brain arousal system has a major impact on work and health in general. There are many laboratory-based tests and questionnaires that have been used to assess arousal and human performance (Matthews, Davies, Westerman, et al., 2000). However, assessment techniques are needed that can be used with minimal training in naturalistic settings. Moreover, one goal of neuroergonomics is to use assessment of brain functioning to optimize mental functioning and performance in work or everyday settings. Towards that end, we developed a smart phone application that evaluates sleep in a naturalistic setting, analyzes the results, and then displays feedback designed to improve mental and physical functioning.

Insufficient sleep is prevalent. In 2007, the United States Department of Health and Human Services estimated that 64 million Americans regularly suffer from insomnia each year. Additionally, many more Americans intentionally go without sleep. The health and behavioral risk associated with sleep deprivation is severe, making this statistic particularly staggering.

Both simple and complex tasks are affected by sleep deprivation since performance decreases as sleep deprivation increases (Balkin, Rupp, Picchioni, et al., 2008). For example, epidemiological studies found an increased incidence of sleep-related crashes in drivers reporting six or fewer hours of sleep per night on average (Slutts, Wilkins, Osberg, et al., 2003). Studies also suggest that sleep is essential to higher-level cognitive processes such as processing sense of humor and learning new information (Thomas, Sing, Belenky, et al., 2000; Maquet, 2001).

Sleep deprivation also impacts health, where short-term effects include increased blood pressure, snacking on fatty foods, insulin resistance, and a weaker immune system. Furthermore, long-term effects of continual sleep deprivation are linked to higher rates of morbidity (Kripke, Garfinkel, Windegard, et al., 2002).

COMPONENTS OF SLEEP

The arousal system is regulated by three major components: (1) the homeostatic component, which is controlled by regions in the brainstem, such as the reticular formation, and serves to innervate the cortex in order to regulate sleep need, (2) the circadian component, which is instantiated by the interaction between hypothalamic oscillators in the suprachiasmatic nucleus of the hypothalamus and visual input, i.e.

light, and (3) sleep inertia, which is affected by sleep stage when awakened and is characterized by decreased cerebral blood flow after awakening from sleep. Understanding these components of sleep may promote healthier sleep decisions.

[1] Homeostatic Component

Sleep amount has a linear relationship with arousal. As the length of time awake increases, measures of arousal decrease. The rate of change in this relationship is dependent on various factors. For example, sleep need is a trait-like characteristic (Van Dongen, Baynard, Maislin, et al., 2004). Additionally, if sleep debt is high due to habitually sleeping for shorter periods, sensitivity to the effects of sleep loss increases (Balkin et al., 2008). However, there are general guidelines to sleep need where when sleep is routinely restricted to 7 hours or less, the majority of motivated healthy adults develop cognitive performance impairments; yet when nightly sleep periods are between 8-9 hours, no cognitive deficits are typically found (Mallis, Banks, & Dinges, 2008). Furthermore, at approximately 24 hours of sleep deprivation, the effects of sleep deprivation tend to plateau (Bonnet, 1991).

[2] Circadian Component

The circadian component interacts with the homeostatic component and has a sinusoidal relationship with arousal that peaks at around 3-5 p.m. (highest arousal) and troughs at around 2-4 a.m. (lowest arousal) in a 24 hour period (Dijk & Czeisler, 1995). Similar to the homeostatic component, this component is trait-like where "evening types" have a later shift in the circadian component and "morning types" have an earlier shift (Baehr, Revelle, Eastman, 2000). Additionally, environmental cues, called zeitgeber can influence this component of sleep. The strongest zeitgeber is exposure to light (Mallis, et al., 2008).

[3] Sleep Inertia

Sleep inertia is a third component of sleep that affects arousal after awakening from sleep and dissipates in an asymptotic manner that can take up to 4 hours (Jewett, Wyatt, Ritz-De Cecco, et al., 2008). However, the time course of sleep inertia is usually about a half hour (for a review see, Tassi & Muet, 2000). The severity of sleep inertia is largely affected by prior sleep deprivation and awakening from sleep near the circadian nadir (Naitoh, Kelly, & Bobkoff, 1993).

VIGILANCE

Typical vigilance tasks involve the detection of signals over a long period of time. The signals are intermittent, unpredictable, and infrequent. As the task progresses, performance steadily declines, and at about 10 minutes, task performance steeply declines (Davies & Parasuraman, 1982; Parasuraman, 1986; Boksem, Meijman,

Lorist, 2005). The steep decline is known as the vigilance decrement.

Wilkinson (1970) was the first to demonstrate that an auditory vigilance task is sensitive to sleep loss. Dinges and Powell (1985) then modified this task, and developed a 10-minute visual vigilance task called the psychomotor vigilance task (PVT). This task was found to be sensitive to all the components of sleep (Dinges, Orne, Whitehouse, et al., 1987; Van Dongen, et al., 2001). Vigilance tasks are more sensitive to the components of sleep than subjective measures of sleepiness and fatigue because people frequently underestimate the cognitive impact of sleep deprivation (Van Dongen, Maislin, Millington, et al., 2003).

The traditional explanation of this decrement is that it is due to boredom or motivation decline (Mackworth, 1968); however, another explanation of the vigilance decrement is that the workload associated with the vigilance task is high and that this depletes mental resources in a time-on-task driven manner (Warm, Parasuraman, Mathews, 2008). The latter hypothesis was recently supported by an fMRI study on the neural basis of the vigilance decrement, which showed that the vigilance decrement activated a right fronto-parietal attentional network that lateralized to the basal ganglia and sensorimotor cortices (Lim, Wi, Wang, et al., 2010). This activation was found after a time-on-task of about 20 minutes, which is longer than the typical PVT. Additionally, pre-task levels of CBF in the thalamus and right middle frontal gyrus were predictors of the vigilance decrement.

The 20 minute time-on-task required to detect the neural basis of the vigilance decrement suggests that the traditional 10 minute PVT may be activating cognitive processes other than those associated with the vigilance decrement. Indeed, performance on brief cognitive tasks that require speed of cognitive throughput, working memory, and other aspects of attention have been found to be sensitive to sleep deprivation (Mallis, et al., 2008) and these do not necessarily involve the vigilance decrement. However, like the PVT, tasks that measure sleep will have to be repeatedly administered without a learning effect.

Neuroergonomic Smart Phone Application

Technology to detect drowsiness in real-world environments must be unobtrusive to the user and able to calculate drowsiness in real-time (Mallis, et al., 2008). This requirement was met and exceeded in the development of the neuroergonomic smart phone application described here. A real time measure of drowsiness was evaluated using vigilance tasks and the application tracked sleep quality by integrating the vigilance task with an alarm clock. Furthermore, sleep quality and arousal is affected by more than just the components of sleep. Medication, drug use, food and caffeine intake, daily energy expenditure, and mood can also affect these tasks. This application can be used to measure these various factors that can affect sleep quality and arousal. User compliance was promoted by displaying these behaviors to the user in a manner that allows easy self-evaluation of behaviors. This can be used to adjust behaviors to improve cognitive performance. Detailed depictions of the application can be found at www.proactivesleep.com.

[1] Tracking Sleep

In order to assess sleep quality, an alarm clock was integrated in the application where sleep onset was estimated based on setting the alarm. Pressing either <wake naturally> or <set alarm> brings up another screen where a button can be pressed to indicate if the user is permanently awakened. This enabled for the tracking of sleep without needing the alarm to go off. Additionally <wake naturally> enabled users to use the application without setting an alarm. When the alarm goes off or <permanently awakened> is pressed, the vigilance task is triggered. Awakening from sleep is determined by when the user completes the task.

[2] Vigilance Task

Figure 1. Vigilance Task

In this task a stimulus, in this case, a sun, randomly appeared at a location on the screen for 1.5 seconds (see Figure 1). When the sun appeared, users were instructed to tap on it as quickly as possible. There were three variations of this tasks included. These tasks differed in the time duration of the stimulus onset window or in other words, in the time interval that the stimulus may appear. This time interval was either short (2 seconds), medium (five seconds), or long (ten seconds). Thus, in the short condition there was a higher frequency of motor activity, but in the long condition the onset of the stimulus was more unpredictable. This means that the long condition more in line with typical vigilance tasks. All tasks had 30 trials, resulting in the short task lasting 1 minute, the medium task lasting 2.5 minutes, and the hard task lasting 5 minutes.

[3] Usability

The sleep diary is the feedback mechanism that provided users with information on there behaviors (see Figure 2). In this version, the behaviors were sleep related, but other versions may include information on other behaviors,

Figure 2. Sleep Diary

like mood, exercise, and diet. Gestalt based color principles were used in this interface where green indicates healthy behavior, yellow indicates warning, and red indicates danger. Additionally, both daily behavior (each individual entry) and average behavior (see the bottom of Figure 2) were displayed to the user. The colored bars for each entry display sleep duration and time of sleep, which can be edited. The color represents sleep amount, where users set a sleep goal and the colors indicated if they were meeting, almost meeting, or not meeting the goal. The clock icons corresponded to the result of a question about how long it took to fall asleep. Again, how long it took was represented by the color of the response, and additionally, reinforced by more time on the clock. Performance on the vigilance task was also displayed to users (see upper right of Figure 2) and users could write additional notes about sleep or other behaviors that were displayed below entries.

METHOD

[1] Participants

Twenty-five George Mason University undergraduate students participated for course credit. All participation was voluntary. Participants were required to own either an iPhone or an iPod-touch running the 3.0 operating system or higher.

[2] Design and Procedure

The sleep diary was disabled in order to promote experiment control and each participant was randomly assigned to one of three vigilance task conditions: short window of response, medium window of response, and long window of response.

Email was the only medium used to communicate with participants. After signing up for the study, instructions were provided on downloading the application to an iPhone or iPod-touch device and how to use the application. Participants were instructed to go to the settings tab and select which vigilance task would be administered by pressing one of three buttons. After completing this task, participants were told not to enter the settings screen again.

Participants were given an overview of how to interact with the application. It was emphasized that participants should not deviate from their normal sleeping patterns. After the overview, participants completed a practice session using the device. In this practice session, they were instructed to set the alarm, press the <set alarm> button, and then press <permanently awakened>. This prompted the idle screen of the vigilance task and participants were told the objective of the task, which was to tap the image of the sun as quickly as possible. Then participants were told to press the <start game> button, and instructed to complete a session of the game.

The actual task consisted of setting the application at night and playing the vigilance task upon awakening from sleep. The task lasted for three weeks. One day using the application entailed setting the alarm by pressing <set alarm> or

<wake naturally>, sleeping, and playing the vigilance task when the alarm is triggered, which when completed, shuts off the alarm music and phone vibrator. Another option available to the participant when the alarm is set is to press the <snooze> button, which results in a ten-minute delay before the alarm goes off again. All data was collected using the Google App Engine. This system acts as a server for smart-phone applications. Thus, any interaction that users have with the application can be exported to the server provided that the Google App Engine script is embedded in the application and the device has Internet connection.

[3] Measures

Sleep onset was determined based on when <set alarm> or <wake naturally> was pressed and awakening was determined when the vigilance task was completed. From this, sleep duration was estimated. Consistent with how PVT performance is typically assessed, measures were taken on the fastest 10% of reaction times, the slowest 10% of reaction times, and misses (Dinges et al., 1985).

RESULTS AND DISCUSSION

Four subjects were eliminated because they had less than seven entries of sleep data, which resulted in twenty-one total participants. There were seven participants in each condition, resulting in 69 entries for the short window condition, 85 entries for the medium window condition, and 76 entries for the long window condition. The number of entries per subject ranged from seven to twenty-one. Only sleep durations of more than two hours were included in the analyses because the aim of this study was to investigate the effects of at least a night of sleep.

A between subjects ANOVA was run to check that sleep durations did not systemically vary across the three conditions. Results suggested sleep duration was not different between the short window condition ($M = 7.63$ hrs, $SD = 1.60$ hrs), medium window condition ($M = 7.50$ hrs, $SD = 1.94$ hrs), and long window condition ($M = 7.10$ hrs, $SD = 2.10$ hrs), $F(2, 229) = 1.60, p = .20$.

In order to understand the relationship between sleep duration and task performance, sleep duration was mean centered around the participant's average duration of sleep. The reasons for this were that sleep need shows considerable inter-individual variability and because it was necessary to compare sleep duration between participants. Pearson correlations were then run between sleep duration and the three performance measures (10% fastest RTs, 10% slowest RTs, and misses) across all three conditions (see Table 1).

Table 1. Pearson Correlation Table Between Sleep Duration and Vigilance Performance.

Variables	Fastest 10% RT	Slowest 10% RT	Misses
Sleep Dur. Short Window (N = 69)	-.25**	-.24*	-.22†
Sleep Dur. Medium Window (N = 85)	-.26*	-.06	.27*
Sleep Dur. Long Window (N = 76)	-.37**	-.24*	-.42**

† $p < .10$. * $p < .05$. ** $p < .01$.

There was a consistent trend where as sleep duration increased, performance on all three vigilance tasks improved. This improvement in task performance involved decreased 10% of fastest and slowest reaction times to detect the stimulus and decreased misses when detecting the stimulus. Based on Cohen's guidelines, all of the correlations were in the medium effect size range, with the exception of slowest 10% of reaction times in the medium window task.

GENERAL DISCUSSION

The results of this study show that effects of sleep quality on performance can be detected under naturalistic conditions using smart phone technology. As such the design application described here represents an example of a neuroergonomic assessment technique that can be used with minimal training by users in their everyday lives. The smart phone application we developed evaluates sleep in a naturalistic setting, analyzes the results, and then displays feedback to the user. Such feedback could be used by users themselves to monitor and if necessary improve their mental and physical functioning.

A neuroergonomic application centered on the human arousal system may be an important step in addressing the widespread and severe problem of inadequate sleep. This application provides unique data related to the arousal system because it can be administered in naturalistic setting. The data can predict task performance, which may be especially useful for designing job schedules that have erratic hours (i.e., doctors and pilots). Additionally, since behavioral treatments play a role in alleviating many sleep and mood disorders (Mori, Bootzin, Buysse, et al., 2006), this application may be a vehicle for monitoring the efficacy of clinical treatments.

The evaluation study found that all three versions of the vigilance task were sensitive to sleep duration, although the degree of sensitivity varied with task type. In particular, the finding that even the 1-minute task was sensitive to sleep duration is notable. The longer the duration of a vigilance task, the less likely that participants and users will be compliant and finish the task when testing themselves in their everyday routine. Thus, a short-duration task will promote more repeated use of the application.

We did not obtain evidence of vigilance decrement in any of the three task conditions. While under certain conditions the vigilance decrement can occur after as little as 5 minutes of continuous work (Nuechterlein, Parasuraman, Jiang, 1983), the necessary task conditions (low target salience or high working memory load) were not features of the task used in the present application. The findings suggest that the vigilance decrement is not a requirement to detecting the homeostatic component of sleep

A future research direction is to determine if other tasks may be more sensitive to the detection of arousal. Tasks that involve the prefrontal cortex, which implicates divergent thinking and working memory, are also adversely affected by sleep loss. (Mallis, et al., 2008). Furthermore, a vigilance task developed by Cagglano and Parasuraman (2004) involved the dual processes of signal detection of a spatial location and working memory. Such a task may be a stronger predictor of the components of sleep.

References

Baehr, E., Revelle, W., & Eastman, C. (2000). Individual differences in the phase and amplitude of the human circadian temperature rhythm: with an emphasis on morningness-eveningness. *Journal of Sleep Research, 9,* 117-127.

Balkin, T., Rupp, T., Picchioni, D., Wesensten, N. (2008). Sleep Loss and Sleepiness. *Chest, 134,* 653-660.

Boksem, M. A., Meijman, T. F., Lorist, M.M. (2005). Effects of mental fatigue on attention: an ERP study. *Brain Res. Cogn. Brain Res. 25,* 107-116.

Bonnet, M. H. (1991). The Effect of Varying Prophylactic Naps on Performance, Alertness and Mood throughout a 52-Hour Continuous Operation. *Sleep, 14,* 4, 307-315.

Caggiano, D., & Parasuraman, R. (2004). The role of memory representation in the vigilance decrement. *Psychonomic Bulletin and Review, 11,* 5, 932-937.

Davies, D. R. & Parasuraman, R. (1982). *The Psychology of Vigilance.* London: Academic Press.

Dijk, D. J., & Czeisler, C. A. (1995). Contribution of the circadian pacemaker and the sleep homeostat to sleep propensity, sleep structure, electroencephalographic slow waves, and sleep spindle activity in humans. *Journal of Neuroscience, 15,* 3526-3538.

Dinges, D. F., Orne, M. T., Whitehouse, W. G., Orne E. C. (1987). Temporal placement of a nap for alertness: contributions of circadian phase and prior wakefulness. *Sleep, 10,* 313–329.

Dinges, D. F., & Powell, J.W. (1985). Microcomputer analyses of performance on a portable, simple visual RT task during sustained operations. *Behav Res Meth Instr Comp, 17,* 652–655.

Jewett, M. E., Wyatt, J. K., Ritz-De Cecco, A., Khalsa, S. B., Dijk, D., & Czeisler, C. A. (2008). Time course of sleep inertia dissipation in human performance and alertness. *Journal of Sleep Research, 8,* 1, 1-8.

Kripke, D., Garfinkel, L., Winegard, D., Klauber, M., & Marler, M. (2002).

Mortality associated with sleep duration and insomnia. *Arch Gen Psychiatry, 69,* 131 – 136.

Lim, J., Wi, W., Wang, J., Detre, J. A., Dinges, D. F., & Rao, H. (2010). Imaging brain fatigue from sustained mental workload: An ASL perfusion study of the time-on-task effect. *Neuroimaging, 49,* 3426-3435.

Naitoh, P., Kelly, T., & Babkoff, H. (1993). Sleep inertia: Best time not to wake up? *Chronobiology International, 10I,* 2, 108-118.

Nuechterlein, K., Parasuraman, R., & Jiang, Q. (1983). Visual sustained attention: Image degradation produces rapid sensitivity decrement over time. *Science, 220,* 327-329.

Mackworth, J. F. (1968). Vigilance, arousal, and habituation. *Psychol. Rev., 75,* 308-322.

Mallis, M., Banks, S., & Dinges, D. Sleep and circadian control of neurobehavioral functions. Ed. Parasuraman, R., & Rizzo, M. *Neuroergonomics: The Brain at Work.* New York: Oxford Univsersity Press, 2007.

Matthews, G., Davies, D. R., Westerman, S. J., & Stammers, R. B. (2000). *Human performance: cognition, stress, and individual differences.* Philadelphia: Taylor and Francis.

Maquet, P. (2001). The role of sleep in learning and memory. *Science, 294,* 1048–1052.

Mori, C., Bootzin, R., Buysse, D., Edinger, J., Espie, C., Lichstein, K. (2006). Psychological and Behavioral Treatment of Insomnia: Update of the Recent Evidence (1998-2004). *Sleep, 29,* 11.

Rizzo, M., Robinson, S., & Neale, V. The Brain in the Wild: Tracking Human Behavior in Naturalistic Settings. Ed. Parasuraman, R., & Rizzo, M. *Neuroergonomics: The Brain at Work.* New York: Oxford University Press, 2007.

Parasuraman, R., (2003). Neuroergonomics: research and practice. *Theor. Issues. Ergon. Sci. 4,* 5-20.

Parasuraman, R. (1986). Vigilance, monitoring, and search. In K. Boff, L. Kaufman, & J. Thomas (Eds.), *Handbook of perception and human performance. Vol. 2: Cognitive processes and performance* (pp. 43.1-43.39). New York: Wiley.

Stutts, J. C., Wilkins, J. W., Osberg, S. J., & Vaughn, B. V. (2003). Driver risk factors for sleep-related crashes. *Accident Analysis and Prevention, 35,* 321-331.

Tassi, P. & Muet, A. (2000). Sleep Inertia. *Sleep Medicine Reviews, 4,* 4, 341-353.

Thomas, M., Sing, H., Belenky, G., Holcomb, H., Mayberg, H., Dannals, R., Wagner, Jr., H., Thorne, D., Popp, K., Rowland, L., Welsh, A., Balwinski, S., and Redmond, D. (2000). Neural basis of alertness and cognitive performance impairments during sleepiness: Effects of 24 h of sleep deprivation on waking human regional brain activity. *Journal of Sleep Research, 9,* 335-352.

Van Dongen, H. P., Maislin, G., Mullington, J. M., & Dinges, D. F. (2003). The cumulative cost of additional wakefulness: Dose-response effects on neurobehavioral functions and sleep physiology from chronic sleep restriction and total sleep deprivation. *Sleep, 26,* 117-126.

Van Dongen H.P., Baynard, M.D., & Maislin G. (2004). Systematic inter-individual differences in neurobehavioral impairment from sleep loss: evidence of trait-like differential vulnerability. *Sleep, 27,* 423-433.

Warm, J.S., Parasuraman, R., Mathews, G. (2008). Vigilance requires hard mental work and is stressful. *Hum. Factors, 50,* 433-441.

Wilkinson, RT. (1970). Methods for research on sleep deprivation and sleep function. In: E Harmann(Ed), *Sleep and dreaming*, Little Brown, Boston.

CHAPTER 21

A Neuroergonomic Perspective on Human-Automation Etiquette and Trust

Ewart de Visser, & Raja Parasuraman
George Mason University

ABSTRACT

Different facets of politeness and etiquette can markedly influence collaboration between humans and automation, in a similar way to human-human relationships. We identify theories relevant to understanding the nature of human-automation etiquette and explore several ways to evaluate how etiquette affects performance in complex human-machine systems, including behavioral and computational methods. We complement these approaches with a *neuroergonomic* perspective (Parasuraman, 2003; Parasuraman & Rizzo, 2007). Recent research has shown that examining theories and empirical findings on human brain function can bolster our understanding of such aspects of human performance at work as mental workload, vigilance, adaptive automation, and individual differences (Parasuraman & Wilson, 2008). Accordingly, we examine human-automation etiquette from the perspective of both human behavioral and neural mechanisms.

Keywords: human-automation etiquette, trust, automation, neuroergonomics

INTRODUCTION

With the advent of intelligent technology, there has been a call for research that not only examines how automation should work, but how it should *behave* in conjunction with human users (Miller, 2004). To that end, it has become increasingly necessary to design automation that can conform to social structures. Products such as navigation devices and automated customer phone services attempt to interact with users in a natural, conversational way. However, interactions between automated technology and humans often lead to frustration, distrust, and even anger. Some have argued that one of the reasons for such frustration is that the behavior of an automated device does not conform to human-human etiquette (Brown & Levinson, 1987; Miller, 2004). From this perspective, designers should make automation conform to the norms of human-human protocol and social expectations.

Nass and colleagues have shown that the way people behave towards computers shows striking similarities to their interaction with other people . For example, Nass et al. (1994) examined how people responded to a computer rating its own performance on a task. When the computer rated its own performance, participants rated the computer as more competent and friendly when they filled out the ratings on the same computer they had used for the test. These ratings were significantly lower when participants responded with pencil and paper or on another computer. Nass et al. proposed that this finding supports the social norm that people are polite to others, including computers. Remarkably, participants did not think they would rate the computer higher when they were told about the experimental manipulation. The authors concluded that computers are often seen as social actors if they behave similarly to humans.

Recent technological advances have led to computer systems possessing even greater agent-like qualities than the simple computers used in the studies by Nass and colleagues. Whereas at one time automation primarily involved sensing and mechanical actions, modern automation often entails information-processing or decision making functions. The trend began in aviation but has extended to most domains at work, the home, transportation, and many forms of entertainment. Furthermore, older automation, while able to sense the world and control action, was immobile. But modern robots and unmanned vehicles possess eyes, ears, hands, feet, as well as rudimentary thinking brains.

These developments inevitably invite greater examination of issues concerning social aspects of communication with automated systems. As automation capabilities have become more human like, it is not surprising that studies of human interaction with automation will need to consider issues that were once thought of as unique to human social intercourse—such as trust and reliance. Trust has been an important factor in understanding human-automation interaction—even before automation possessed a face, could move, or exhibit a "personality." Two factors were important in motivating interest in studies of human trust of automation. First, despite the best intentions of designers, automated systems have typically not been used in ways that they were intended to, with instances of misuse and disuse being prevalent (Parasuraman & Riley, 1997). Second, most automated systems do not exhibit 100% reliability: variable human trust is the consequence of imperfect automation.

There is now a large literature of trust and its role in human-automation interaction (Lee & See, 2004). Much of the theoretical development has stemmed from extensions of studies into human-human trust (Muir, 1987; Madhavan & Wiegmann, 2007). There is also a growing empirical literature on the neural basis of human trust (Krueger et al., 2007). We propose that trust is an important construct that must be evaluated when considering issues of politeness and etiquette in human-computer interaction. We are entering an era where humans may be required to form "relationships" with automated systems. If the trend continues, automation technologies will need to be designed to be consistent with expected social norms in the work place as well as in personal settings.

TRUST AND HUMAN-AUTOMATION ETIQUETTTE

Trust significantly impacts the effectiveness of human-automation collaboration, particularly in the willingness to share and allocate tasks as well as to exchange information and create an impetus for supportive behavior . For the purposes of this paper we will adopt the definition of trust given by Lee & See (2004): "the attitude that an agent will help achieve an individual's goals in a situation characterized by uncertainty and vulnerability". This definition fits the idea that social actors are rational and that individuals maintain some sense of face (i.e. vulnerability).

Empirical evidence showing the impact of polite strategies on trust have been demonstrated in recent studies (Parasuraman & Miller, 2004; de Visser et al., 2009). In one study, participants were required to perform a multiple task battery simulating different flight tasks. One of the tasks involved diagnosing and correcting engine malfunctions with the help of automation advisories. The other two tasks involved tracking and fuel management. When providing advice, the automation either displayed good etiquette (defined as being patient and non-interruptive) or poor etiquette (defined as being impatient and interruptive). In addition, the advisories provided by the automation were either correct most of the time (high reliability) or only some of the time (low reliability). When participants interacted with this automation, their trust was lower in the poor etiquette condition compared to the good etiquette condition. Interestingly, in the low reliability condition good etiquette increased participants' rating of trust close to the trust levels in the highly reliable, but poor etiquette condition. This effect for trust was also found in a recent similar etiquette study using simulated unmanned aerial vehicles (de Visser et al., 2009). This suggests that having automation exhibit good etiquette has an impact on collaboration behaviors. Important to note here is that interruption per se, which can be annoying, could not explain this effect. Participants experiencing neutral interruptions had lower trust compared to the good etiquette conditions.

A social actor evaluates observed information to develop trust in another agent. This can be accomplished in three ways via performance, process, and purpose (Lee & See, 2004). Performance refers to set of observable actions about the other agent. If the trustee performs well on a certain task, for example, he or she can be trusted to a greater extend. Process refers to the consistency and manner of those actions over a period of time. Finally, purpose refers to the overall goal and intentions of the trustee. Based on the observable information of performance, process, and purpose that form the basis of trust, the information is then evaluated according to either of three distinct processes that lead to the trust attitude: analytical, analog and emotional processes. These processes vary in the speed with which the information is evaluated. Analytical processes are slow, analog a little faster and emotional are the fastest.

NEURAL MEASURES OF TRUST

As stated above, trust also plays a critical role in human-automation interaction (for a review see Lee & See, 2004). A robot such as Asimo will have to gain the trust of, for example, the elderly population in performing assistive tasks in their households. In this section, we review some of the recent advances in the understanding of the neural basis of social and reward learning and expectation processes associated with trust. Moreover, the potential application of the insights arising from current and future findings of this social issue to the design of human-machine interfaces will be discussed.

Various methodologies and conceptual approaches have been used to examine the neural mechanisms of trust. We will succinctly discuss a few of these studies that we believe adequately illustrate significant advances in three fundamental aspects: a theoretical framework, hormonal orchestration and cerebral neural networks associated with the processing of this psychological construct.

As indicated by the psychological research, there are several stages that can be identified in the trust process. Similar stages have been uncovered when using neuroscience to identify regions and systems related to the trust mechanism. These regions one of the following information processing stages: 1) initial perception 2) learning over time and 3) the evaluation. Each of these steps can be unified with the psychological steps previously mentioned.

Research that illustrates these steps was shown by researchers investigating the effects of trustworthiness of faces. A variety of virtually emotionless faces were rated both explicitly (rating based on trustworthiness) or implicitly (rating based on age) (Winston, Strange, O'Doherty, & Dolan, 2002). They found that when participants rated faces explicitly for trust the superior temporal sulcus (STS) showed activation. This area of the brain has been known to be involved with the perceptual properties of stimuli (Adolphs, 2003). Furthermore, they found amygdale and right insula activation when implicit ratings were made to untrustworthy faces. Finally, the orbitofrontal cortex was activated for explicitly rated trustworthy faces and implicitly rated untrustworthy faces. It is therefore possible that the faces are first rated for their perceptual properties in the STS and then evaluated unconsciously by the amygdala and consciously by the orbitofrontal cortex (Adolphs, 2003).

A popular paradigm for investigating trust in neuroscience research is the trust-game (Camerer & Weigelt, 1988). In this game, an investor makes a monetary investment that can be tripled in value when it is sent back by the trustee. There is a risk by the investor to trust the trustee because this person my not return any or less of the initial investment. In these games, trust is operationalized as the initial money sent from the investor to the trustee. One such study used a multi-round version of the game to investigate trust and reciprocity in the brain for two people simultaneously (King-Casas et al., 2005). They postulated an 'intention to trust' within the trustee's brain depending on either benevolent or malevolent behavior of the investor's decisions. They found an area called the caudate nucleus that was activated in early games *reactively* just after the investor had made the investment decision. However, in later rounds of the game this signal had moved 14 seconds earlier making it a *predictive* signal. This observed signal matches reinforcement learning models (Montague, King-Casas, & Cohen, 2006). As described previously, learning is a central element in the ability to build trust. Furthermore, the fMRI BOLD response in the caudate nucleus was higher for benevolent than for malevolent reciprocity, suggesting that this brain structure computes information associated with fairness of a partner's decision.

Delgado and colleagues (2005), also investigating the role of the striatum in adjusting future behavior on the basis of a reward feedback, have successfully shown that previous moral and social information about an individual can affect the effective processing of the neural circuitry of his/her interacting partner, in some cases annulling these predictive correlations. In this study, participants were given previous information about their hypothetical partners, vividly depicting them as 'good', 'bad' or 'neutral'. Their choices during a trust game were recorded both behaviorally and neurophysiologically (fMRI). As expected, participants were more cooperative, displaying more trust, with the morally 'good' partners and taking less risky choices with the 'bad' profiles. Moreover, even when tested with equivalent reinforcement rates for all three moral categories, participants would persistently display more trust in the previously depicted 'good' partners. Interestingly, the caudate nucleus showed robust differential activity between positive and negative feedback, consistent with the hypothesis of its involvement in the processing of learning and adaptive choices (O'Doherty et al., 2004; Tricomi et al., 2004) but only for the 'neutral' partner. There was no significant differential activity for the 'good' partner, and it was very weak for the 'bad' partner. This result shows that when an individual has an a priori moral expectation of his/her interacting partner, this belief not only influences the behavioral choices but also disrupts the regular encoding of evidence, diminishing reliance on feedback mechanisms and its associated neural circuitry (Delgado et al., 2005). It seems that the existence of moral information prior to the interaction reduces the differential activity in the caudate nucleus leading the individual to discount feedback information and failing to adapt its choices to the learned outcomes. This is a striking example of the flexibility and potential unreliability of these systems as 'strong' predictors.

Another brain area consistently associated with cognitive control, in the sense of conflict monitoring and resolution is the cingulate cortex (Carter *et al.*, 1998; Ridderinkhof *et al.*, 2004). Even in the Delgado and colleagues study (2005), unlike the caudate, the differential activation of the cingulate cortex was consistent with previous studies, indicating that although prior moral information affects the neural activity of reward and feedback learning systems, it does not influence the neural activity of areas associated with conflict monitoring. Differential activity of the cingulate cortex has also been associated with 'intention to trust'. For example, in the previously described study from King-Casas and colleagues a significant correlation was shown between activity in the investor's middle cingulate cortex (MCC) and the trustee's anterior cingulate cortex (ACC). The MCC of the investor displayed strong activity when he/she lodged a decision and the ACC of the trustee when the investor's decision was revealed.

A popular way to examine trust in real-world situations has been to use the neurotransmitter oxytocin. Oxytocin is a small peptide that acts as a neurotransmitter, with a dual role as a hormone and a neuromodulator. It is produced in the hypothalamus and targets a variety of brain areas associated with both emotional and social aspects of human behavior (i.e. amygdala, nucleus accumbens, etc.). Its presence and influence has also been studied in animals, where oxytocin has been found to influence male-female and mother-infant bonding and to modulate approach behavior. By diminishing the natural reluctance to avoid the close approach of another individual, thereby increasing trust in that person, oxytocin also facilitates the development of social cohesion and cooperation (Carter, 1998; Insel & Shapiro, 1992). Oxytocin is produced in the hypothalamus and impacts the amygdala and the nucleus accumbens. Hypothesized to be involved in reducing natural resistance facilitating approach behavior. Approach may be a critical measure of trusting behavior because people will approach more if they trust more. Oxytocin is said to facilitate overcoming aversion to betrayal which is then combined with the effects of reward that make up approach behavior. Excessive amounts of oxytocin may be released that may lead to indiscriminatory trusting of people. Studies examining oxytocin typically use adjusted economic games. There are two games that are typically played. One is the trust game and one is the risk game. In the trust game an investor can choose to send money to a trustee. This money is tripled for the trustee who can then decide to send money back to the investor. The risk game is identical to the trust game with the difference that the payback selection is conducted randomly by a computer.

In a recent study, Kosfeld and colleagues (2005) explored the interaction between individuals playing the roles of either 'investor' or 'trustee' during a monetary exchange. Their results show that investors injected with oxytocin show higher trust levels in the trustees than a placebo group. On the other hand the behavior of the trustees is not altered by oxytocin. Based on these results, it was concluded that this neurotransmitter can elicit an increase of trust without affecting the objective reliability of the agent. It was hypothesized that during real human interactions oxytocin is released in specific brain regions when a person perceives a certain social configuration. This in turn leads to modulation of neural networks to produce an increase of trusting behavior related to another individual or group. This mechanism would act as a facilitator of interpersonal interactions that could lead to both the individual and group's success.

Another study examined the effects of oxytocin on different brain regions (Baumgarter et al., 2008). After administrating oxytocin to a group of people, they found that they maintained their trust even though it was breached several times compared to a placebo group. For those in the oxytocin group there was a specific reduction of activation in the amygdala, the midbrain regions and the dorsal striatum were. The amygdala and midbrain are associated with fear which seems to be reduced when oxytocin is present. This is also consistent with results of lesion studies in the amygdala that show an increase in trust for lesions compared to healthy individuals (Adolphs et al., 1998). Furthermore, the dorsal striatum is associated with learning from feedback responses which seems to be less activated when oxytocin is present. When the striatum is active, responses can be processed for reward feedback and serve as a predictor for future interactions (King-Cases et al. 2005).

Furthermore, behavioral results showed that in the risk game there were no significant differences whereas in the trust game the oxytocin group maintained their trust even though they experienced defection while the placebo group decreased their trust. Neurological results showed stronger activation for the amygdala, caudatus, midbrain regions, and the insula and postcentral gyrus. Higher ACC actication was found for the placebo group. This fits with the idea that the ACC is involved in cognitive control. The authors suggest that participants in the oxytocin group are less afraid of betrayal. Lack of reliance on feedback information is critical. This may be influenced by several factors including oxytocin and vivid character descriptions.

In an alternative research paradigm researchers studied the actual oxytocin levels found in the blood of participants (Keri et al., 2009). In this study, the researchers were interested in measuring oxytocin in both healthy subjects and patients with schizophrenia during neutral and trust-related interactions. The neutral interaction involved writing down a neutral message to the experimenter who in turn responded with a neutral message. In the trust condition, participants were asked to write down an important secret in their life to which the experimenter also wrote down an important secret of their life. They found that in the trust condition, oxytocin levels increased for healthy subjects but remained low in patients with schizophrenia. The oxytocin levels predicted the negative symptoms of schizophrenia.

Recently, Krueger and colleagues (2007) have extended previous work by parsing trust into two strategies: *conditional trust* and *unconditional trust*. Conditional trust is characterized by the assumption that one's partner is self-interested and therefore the risks of interaction with this person in the attainment of reward must be carefully calculated. In contrast, unconditional trust begins with the assumption that one's partner is trustworthy, resulting in a quickly developed cooperative relationship. These trust strategies were tested in a reciprocal trust game while both players were simultaneously scanned using fMRI. Conditional trust was based on responses of first movers that defected. This group showed less activation in the paracingulate cortex than the non-defector group as well as higher activations in the ventral tagmental area, a region associated with the evaluation of expected and realized reward. Unconditional trust was linked to activation in the septal area. The septal area plays a role in the control of anterior hypothalamic functions, including the release of oxytocin that modulates social attachment.

CONCLUSION

The neuroergonomic perspective provides the necessary methods and theories to explore the different facets of human-automation etiquette. One important topic within this area is the construct of trust. Trust has been studied in the context of human-automation etiquette and found to be an important predictor of reliance. In addition, studies in neuroscience have revealed different types of trust that could further help the study of human-automation etiquette.

REFERENCES

Adolphs, R. (2003, March). Cognitive neuroscience of human social behavior. *Nature Reviews Neuroscience, 4*, 165-178.

Baumgartner, T., Heinrichs, M., Vonlanthen, A., Fishbacher, U., & Fehr, E. (2008). Oxytocin Shapes the Neural Circuitry of Trust and Trust Adaptation in Humans *Neuron, 58*(4), 639-650.

Camerer, C. F., & Weigelt, K. (1988). Experimental tests of a sequential equilibrium reputation model. *Econometrica, 56*, 1-36.

Carter, C. S., Braver, T. S., Barch, D. M., Botvinick, M. M., Noll, D., & Cohen, J. D. (1998). Anterior cingulate cortex, error detection, and the online monitoring of performance. *Science, 280*(5364), 747.

de Visser, E.J., Shaw, T., Rovira, R., & Parasuraman, P. (2009). Could you be a little nicer? Pushing the right buttons with automation etiquette. In *Proceedings of the International Ergonomics Association conference*, Beijing, China.

Delgado, M. R., Frank, R. H., & Phelps, E. A. (2005). Perceptions of moral character modulate the neural systems of reward during the trust game. *Nature Neuroscience, 8 (11)*, 1611-1618.

Insel, T. R. & Shapiro, L. E. (1992). Oxytocin receptor distribution reflects social organization in monogamous and polygamous voles. *Proceedings of the National Academy of Sciences, 89(13)*, 5981-5985).

Kéri S., Kiss I., & Kelemen O. (2009). Sharing secrets: oxytocin and trust in schizophrenia. *Social Neuroscience, 4*(4), 287-93.

King-Casas, B., Tomlin, D., Anen, C., Camerer, C. F., Quartz, S. R., & Montague, P. R. (2005, April). Getting to know you: reputation and trust in a two-person economic exchange. *Science, 308*, 78-83.

Kosfeld, M., Heinrichs, M., Zak, P. J., Fischbacher, U., Fehr, E. (2005). Oxytocin increases trust in humans. *Nature, 435*, 673-676.

Krueger, F., McCabe, K., Moll, J., Kriegeskorte, N., Zahn, R., Strenziok, M., Heinecke, A., & Grafman, J. (2007). Neural correlates of trust. *Proceedings of the National Academy of Sciences USA*, epub ahead of print.

Lee, J., & See, K. (2004). Trust in automation: Designing for appropriate reliance. *Human Factors, 46*(1), 50-80.

Madhavan, P., & Wiegmann, D. (2007). Similarities and differences between human-human and human-automation trust: an integrative review. *Theoretical Issues in Ergonomics Science, 8*(4), 270-301.

Muir, B. M. (1987). Trust between humans and machines, and the design of decision aids. *International Journal of Man-Machine Studies* 27, no. 5-6: 527-39.

Montague, P. R., King-Casas, B., & Cohen, J. D. (2006). Imaging valuation models in human choice. *Annual Reviews of Neuroscience, 29*, 417-448.

Nass, C., Steuer, J., & Tauber, E. R. (1994). *Computers are social actors*. Paper presented at the Proceedings of the SIGCHI conference on Human factors in computing systems: celebrating interdependence.

O'Doherty, J. P., Dayan, P., Schultz, J., Deichmann, R., Friston, K., & Dolan, R. J. (2004). Dissociable roles of ventral and dorsal striatum in instrumental conditioning. *Science, 304 (5669)*, 452-454.

Parasuraman, R. (2003). Neuroergonomics: Research and practice. *Theoretical Issues in Ergonomics Science, 4*, 5-20.

Parasuraman, R., & Miller, C. A. (2004). Trust and etiquette in high-criticality automated systems. *Communications of the ACM, 47*(4), 51-55.

Parasuraman, R., & Riley, V. (1997). Humans and automation: Use, misuse, disuse, abuse. *Human Factors, 39*(2), 230-253.

Parasuraman, R., & Rizzo, M. (2007). Introduction to neuroergonomics. In R. Parasuraman & M. Rizzo (Eds.), *Neuroergonomics: The brain at work*. (pp. 3-12). New York: Oxford University Press.

Parasuraman, R., & Wilson, G. F. (2008). Putting the brain to work: Neuroergonomics past, present, and future. *Human Factors*, in press.

Ridderinkhof, K. R., Ullsperger, M., Crone, E. A., & Nieuwenhuis, S. (2004). The role of the medial frontal cortex in cognitive control. *American Association for the Advancement of Science, 306*, 443-447.

Winston, J. S., B. A. Strange, J. O'Doherty, & Dolan, R. J. (2002). Automatic and intentional brain responses during evaluation of trustworthiness of faces. *Nature Neuroscience* 5, no. 3: 277-83.

CHAPTER 22

Neuroethics: Protecting the Private Brain

Joseph R. Keebler[1], Scott Ososky[1], Grant S. Taylor[2]
Lee W. Sciarini[2], Florian Jentsch[1]

[1]Team Performance Laboratory

[2]Applied Cognition & Training in Immersive Virtual Environments
(ACTIVE) Laboratory

ABSTRACT

This paper serves as a call to establish guidelines and considerations necessary for proper ethical treatment of those who will be measured and categorized based on neurotechnological methodologies. Included in this discussion is the notion of "brain privacy" with respect to the protection of identity and free-will. An examination of current technologies will be presented, along with a look towards potential future technologies, and the ethical implications that both may hold. The ethical dilemma of brain privacy and measurement will be grounded in classical and modern philosophies to establish a basis for ongoing discussion in the ever developing realm of Neuroscience.

Keywords: Neuroscience, Neuroethics, Ethics, Brain privacy, Free will

INTRODUCTION

This chapter intends to help educate practitioners with respect to the possible implications for a future embedded with Neurotechnology by highlighting issues that exist now or may become a problem in the future. Current and potential future technologies will be reviewed. We encourage the reader to consider the dimensions of brain privacy, brain information discrimination, and brain/mind control within each of these examples. The measurement, manipulation and use of the brain and its data is something we must be concerned with as both individuals and as a species, and we must be certain to not place the responsibility of making educated decisions solely on the individual, but also on the governing bodies of the world. To this end, an investigation into past and modern philosophy will also be conducted to establish a basis for Neuroethics.

Neuroethics can be described as the ethical issues arising from neuroscience (Farah, 2005). In the context of Neuroergonomics, which is concerned with the measurement of brain and behavior at work (Parasuraman, 2003), it is important to remember that the data gathered from such techniques is undeniably private, and possibly the most important data to the individuals from which it is taken. The development and application of ethical principles guiding Neuroergonomics is relatively new, with the name "Neuroethics" first appearing in scientific literature in 2002 (Farah, 2005). Currently, few guidelines exist regarding the ethical use of Neurotechnology. It is the authors' hope that this chapter will add to the discussion.

CURRENT CAPABILITIES

Although concepts such as "mind reading" and brain manipulation have traditionally been the exclusive providence of science fiction fantasies, advances in neuroscience are bringing us ever closer to the kinds of technologies once considered impossible. The rudimentary beginnings of these capabilities already exist to varying degrees of success, and the one prediction that can be made with certainty is that these fields will only continue to develop.

MEASURING MENTAL ACTIVITY

The average human brain contains an estimated 100 billion neurons, with each neuron connecting to 1000 other neurons, resulting in roughly 100 trillion individual synaptic connections. Given that this is all contained within a space the size of a large grapefruit, and surrounded by a quarter-inch of solid bone, saying that measuring mental activity is a daunting task is a gross understatement. However, neuroscience has led to the development of some remarkable capabilities in a short period of time. It wasn't until 1920, with the development of the EEG (electroencephalogram), that scientists were capable of making even rudimentary recordings of brain activity, and yet in under a century's time we are able to non-invasively measure at the millimeter level. However, even our current best (non-invasive) methods measure at a resolution that averages the activity of roughly 100 thousand neurons to form a single point, obviously making the process of measuring mental activity a difficult one.

Current efforts to measure mental activity focus on decoding very specific neurological processes, such as aspects of vision, emotion, language, etc. Recently, scientists studying the physiological underpinnings of vision were capable of reproducing simple images viewed by research participants based entirely on recordings of neurological activity (Miyawaki et al., 2008). The researchers began by presenting a set of known images to participants while measuring brain activity with an fMRI. These images were very simple, 10x10 grids of consisting of a random pattern of grey or checkered squares. Neurological correlates of each stimulus type (grey or checkered) were found for each grid position, allowing the researchers to reconstruct new images shown to the participants solely by monitoring their brain activity. While this process still falls a great deal short of what most people think when they hear "mind reading", as it is limited to simple 10x10 grids and requires the participant to be in a very un-portable fMRI machine, it is still simultaneously both an exciting and frightening first step toward our collective future.

CONTROLLING MENTAL ACTIVITY

The concept of controlling someone else's mental actions is unquestionably the most potentially risky aspect of future neuroscience applications. It is also arguably the area most prevalent in modern society. Twenty-six percent of the American adult population suffers from diagnosable mental disorders (Kessler, Chiu, Demler, & Walters, 2005). Even if only half of these patients are prescribed a psycho-pharmaceutical, this still results in a significant portion of the population who allow their minds to be manipulated by an outside source every day. Psycho-pharmaceuticals are simply the crude ancestors of the more precise emerging and future methods of manipulating brain activity.

Recent developments in both direct and indirect brain stimulation show promise of more precise manipulation of brain activity than that currently allowed by traditional drug therapies. Direct brain stimulation involves, as the name suggests, the application of electrical current directly to the brain. This technique has proven successful in alleviating symptoms of Parkinson's disease (Benabid, 2003), among other neurological disorders. However, this process has the considerable downside of being highly invasive, requiring a major surgery to implant electrodes directly into the brain. This has been overcome with the development of transcranial magnetic stimulation, which uses the electromagnetic field generated by a current passing through a coil to induce a current within the cortex. The process is entirely non-invasive, and has shown promise in alleviating symptoms of numerous psychological disorders, including schizophrenia (Hoffman & Stanford, 2008). However, transcranial magnetic stimulation is restricted by both its limited range and spatial resolution, making it incapable of manipulating subcortical brain structures or highly localized areas. Nevertheless, both direct brain stimulation and transcranial magnetic stimulation are undoubtedly just the beginning, given the many promising benefits of brain manipulation.

LOOKING TOWARD THE FUTURE

One intention of this chapter is to stimulate the discussion on brain privacy as we look toward an inevitable future of increasingly sophisticated brain measurement technology. Although we cannot predict, with any certainty, the manner in which this technology will develop or the ways in which it will be installed within our society, we can present a number of possible futures for further examination into proactive solutions. For this, we need look no further than science fiction to imagine how the concept of brain privacy might factor into our brave new world.

NATURAL SELECTION THROUGH UNNATURAL MEANS

The film, *Gattaca*, imagines a world in which genetic engineering is common, although not fully embraced by all members of society. Humans are then further classified as valid or invalid, which, in turn, determines their social class and eligibility for employment. The film follows a genetically pure, "invalid" man who forges his identity by purchasing the identity of a paraplegic "valid" individual, in an effort to become an astronaut in the country's space program. The story highlights the irony of his triumph through shear will and determination, versus those who fail despite given every genetic advantage to succeed.

What then, if anything can genetic or brain state information tell us about the actual outcomes an individual will ultimately produce? Although the Genetic Information Nondiscrimination Act (2008) prohibits employment and insurance discrimination on the basis of genetic information, there are no measures in place that would prohibit similar discrimination on the basis of brain state information. Can a brain state indicative of productivity really ensure that an individual will be productive on the job? Is it acceptable to continue to monitor employees to maintain their productivity? If so, would it then be acceptable for an individual to manipulate their brain to subvert these screening tests? More importantly, would it be acceptable for someone other than the individual to manipulate their brain state to achieve certain ends?

Certain arguments, perhaps, can be made in support of such an approach given the premise of a complex, high-profile task or in the interest of shareholder wealth in the private sector. However, given a future in which brain state monitoring is possible, we also face a reality similar to that portrayed in the film, *Minority Report,* where individuals are arrested and prosecuted based upon the prediction of future actions. Now consider a similar line of questioning: Can a brain state indicative of criminal activity really ensure that an individual will commit a crime? Can these brain states be manipulated to subvert the screening tests? Furthermore, how does one differentiate between brain states that are based in reality and those that are not (e.g. committing virtual "crimes" in a video game)?

FOR ENTERTAINMENT PURPOSES ONLY

Video games, social media, and other entertainment outlets, present a unique challenge to policy makers to educate the public on the ramifications of their actions. How do you protect the brain privacy of individuals when the information is freely given? For example, a participant may be unwilling to provide the name of their wireless carrier or favorite soft drink to a surveyor at the local mall, but will freely offer the information in a questionnaire masquerading as entertainment on a social networking website. It is not so unreasonable to assume that someday, brain state information could be required to play a video game, earn some sort of artificial prestige within an online social network, or even uploaded to some *YouTube*-eqsue sharing website. It is equally, if not more reasonable, to assume that this information can also be collected and sold by the organizations that provide those services.

THE PRICE OF PIECE-OF-MIND

If, ultimately, our brain states can be captured and recorded, this information might also be bought and sold for a variety of purposes. Such is the case in the film, *Strange Days*, where recorded memories and emotions are sold like illegal drugs. An apparatus similar in appearance to the fNIR is used to relive the memories and emotions of others as if they were their own. The relevance is: what price, if any, can be associated with our thoughts and emotions? Perhaps we would be willing to part with our thoughts in the same way we sell plasma, or surrender our demographic information in order to enter sweepstakes.

A next logical step in this process is to construct and insert artificial thoughts and experiences into our brains. Given that so many individuals are willing to take prescription drugs to alter their brain states, it follows directly altering or introducing new thoughts may someday become a requested treatment method. The film, *Total Recall,* shows us a future in which virtual vacations can be taken from the comfort of an office laboratory. How much would these vacations cost, and how can we ensure that our physical bodies and minds will be protected while we are "away"? And, at what physical and mental cost will these activities take place after the monetary price is paid? Perhaps an ethical discussion of brain privacy must also be supplemented by a conversation surrounding brain state manipulation addiction.

BRAIN BACKUPS: SCIENCE OR FICTION?

Perhaps, in the distant future, we will someday find a way to back up the contents of our brains, much in the same way we backup our computers, or create copies of discs and other media. Clearly, many would find such relief in knowing that a "back up" copy of their mind exists, should something happen to the version currently in use. Moreover, this same technology might be used to preserve the minds of our society's greatest scientists, artists, or even criminals. What governing body, if any, will decide where these backups are stored, if they can be subpoenaed in an investigation, or even downloaded into the mind of another individual?

Again, science fiction provides a sandbox for this future in the television series, *Dollhouse*. Scientists at the dollhouse are able to erase the minds of individuals, and implant them with new personalities, talents, and abilities (see also, *The Matrix*). These "dolls" are requested and customized by wealthy clients for engagements, similar to an escort service. The original personalities of the "dolls" are stored on backup hard drives, and restored to their bodies once their contracted service to the dollhouse is complete.

It is situations like these that strike at the core of our private brains, both in their preservation and protection. Simple guidelines will not apply to the shades of grey that currently exist within our society: mental illness, brain trauma, free will, etc. By presenting a number of possible futures, it is our hope that the discussion surrounding brain privacy will be a proactive one.

ETHICS OF BRAIN PRIVACY

To better understand the Neuroethics of brain privacy, we must have a firm grasp of ethics in general. We aim to use these ideas to provoke thinking, and to establish that when it comes to ethics, there are no easy answers. We will begin with a brief discussion of free will and the mind-body problem, and then progress through the philosophies of eliminative materialism, deontology, and existentialism. This section will conclude with an interesting thought problem based around conscious experience controlled by machinery.

FREE WILL AND THE MIND BODY PROBLEM

The idea of body and mind has been a topic of discourse for thousands of years and continues to this day. The main issues concerning brain privacy in relation to the mind/body problem are as follows: if the brain and mind are not separate, then brain states should be a precursor to action; if instead they are fundamentally separate, then brain data is not a reliable source for predictive analysis. Obviously, in the case of the brain and mind being one indistinguishable unit, we must be very careful with what happens to our brain data. In the second case, it doesn't really matter if our brain data is private or not, because it in no way is any more than a correlation and does not explain anything about our intentions, desires or inner world of thought. This leads to a problem concerning free will. As Moreno (2003) suggests, the mental, if reducible to the physical, can implicate that there is no free will.

The idea of free will is undeniably tied to brain privacy. Simply put, we must accept that humans do have free will when it comes to ethical decisions concerning the raw data of our minds. This leaves it to the individual to do what he or she wishes with one's own brain data. It is as private as, let's say, our health data. Some may argue it should be even more private, considering that it is literally a direct measure of our consciousness. Governments must also be careful to respect this. Although the majority of law in the U.S., for example, supports respectful treatment of human data, when an issue such as national security (Canli et al., 2007) is brought into the fray, privacy takes a back seat. Our very thoughts and intentions now become something extremely valuable to governing bodies, for instance, due to the predictive analyses that may stop a terrorist attack.

Free will must be considered in the area of brain enhancement as well. Imagine a world where you can be made more intelligent by implanting an integrated circuit array into your cerebral cortex. Imagine a world where this becomes necessary for certain job applicants, or where this becomes a standard in the military. Simply look at the use of prescription brain enhancing drugs such as Adderall and Ritalin. These amphetamine-like drugs are prescribed to children to enhance their minds in educational environments. Is it so hard to believe that we would instead implant a microchip in infants if such a device could guarantee success in multiple aspects of life?

Although it is still unclear as to whether a 1:1 ratio in brain measurement will ever be possible, there are technologies that seem to be stepping in the direction of meaningful brain measurement. If we can measure brain activity and derive what the individual being measured is feeling, thinking, or experiencing, then brain privacy is an absolutely crucial and important human right.

RELEVANT ETHICAL PHILOSOPHIES

Ethics, as it relates to Neurophilosophy and brain privacy, can be broken into a few key views. These views go from the extremes of questioning science and our own ability to correctly perceive our world, to practical and pragmatic views that may be useful for establishing guidelines for current and future technologies.

Eliminative Materialism

Perhaps one of the most extreme of the modern day Neurophilosohies is the idea of eliminative materialism. Eliminative materialism proposes that our everyday perception is unreliable, and that our common sense perspective may be profoundly mistaken (Ramsey, 2007). Eliminative materialism is further broken into two schools of thought. The more extreme supposes that beliefs are simply not real, while the less extreme school instead believes that all of our mental constructs are just neurological brain states (Ramsey, 2007). The mind exists solely due to brain states, and is therefore subject to the physical laws of the universe. If this is true, the idea of free will becomes debatable as well, as described above. How can an individual make free willed decisions if their entire mental process is pre-determined by the physical laws governing their brain and body?

The problem with eliminative materialism is that it stands on unstable footing. By acknowledging that no thought system is a reliable analysis of the world, this idea risks calling itself false. What is meant by this is that it is a philosophy that argues that we cannot use our minds as accurate instruments to gauge the world. Therefore, even this philosophy is not an appropriate mechanism, because it refutes itself. In order for us to get a better grasp on the ethics of brain privacy, we will need to look beyond eliminative materialism.

Kantian Deontology

One of the most respected philosophical ethicists of the enlightenment was Immanuel Kant (1724-1804). Kant's ethical philosophy, better known as deontology, simply states to "act as to treat humanity, whether your own person or any other, in every case as an end and never merely as a means only" (Kant, 2000, p. 298). Kant believes that we not only have a duty to each other, but to ourselves as well. Kant's philosophy is based on a moral *categorical imperative*, which is a maxim that states if it is not right for everyone to take a certain action, it is not right for anyone to take the same action.

Modern deontologists take Kant's ideas a step further and make the case that being servile is immoral (Hill, 1973). What we can say in relation to brain privacy is that it is the duty of each individual, to themselves, to want brain privacy, and it is the duty of the makers of brain technology, as well as the governing bodies of the world, to keep the public informed and educated about all risks associated with measuring and/or controlling the brain and mind. Kant's ideas, considered revolutionary at the time, are actually still profoundly useful and pertinent in the current world. Overall, we must have respect for ourselves and our fellow man in all branches of science, and the use of neurotechnology and measurement needs to adhere to this as well. Later we will address the idea of immersing ourselves in an experience machine, and Kant's ideas will be fundamental to why this action goes against the very nature of being human.

Existential Humanism

Existentialism, according to Jean-Paul Sartre (1905-1980), is an atheistic philosophy that defines the human condition in a unique and modern way. Sartre argues five maxims of existentialism: we are not determined by heredity or environment; we must define ourselves; we are completely responsible for our actions, therefore we create our own morality; we must exist in anguish and despair; we should celebrate the fact that we are creators of our own values (Sartre, 2000).

Existential humanism is closely tied to brain privacy, due to a world view based on subjectivity. Like Kant's deontology, it is up to each individual to prescribe values and morals to themselves and others. Unlike Kant, there is no categorical imperative on which to compare values. Instead, existentialism places responsibility solely on the individual. In the realm of brain privacy, this means that we, as individuals, are responsible for protecting our most vital and private data. What is important to note about existentialism is that it describes the human reality as existence preceding essence (Sartre, 2000). We are brought into existence in the world long before we ever have a defined purpose. This gives us the ultimate responsibility as both an individual and a representative of the human race to make decisions, and these decisions are what create the reality of being a human. According to existentialism, it would be wrong to willingly alter one's brain chemistry or to install hardware into one's neural circuitry if doing so didn't contribute positively to our experience as a human being. As we will discuss in the next section, Sartre (2000) describes the idea of giving one's mind over to a machine, if not for the benefit of living a more fulfilling life, is a pointless endeavor from an existential perspective. Sartre asserts that nothing about our existence is real or tangible except for our actions. Humanity is nothing more than the summation of its actions.

Existentialism, although a profound and exotic thought system, leaves us unsatisfied due to its ambiguous moral tenants. We will now proceed to investigate one of the last frontiers of neurotechnology, the idea of plugging oneself into an experience machine, where action no longer is a reality. Within this theoretical machinery is contained the philosophical standpoint that simply giving one's consciousness over to a machine is on par to ending one's life.

PLUGGING INTO AN EXPERIENCE MACHINE

Robert Nozick (1938-2002) poses an interesting argument: is it a good idea to plug yourself into an experience machine to replace your actual life? Will a human being throw away the actual reality of life for the possibility of any life they want (in this argument, a more pleasurable life)? According to Nozick, most people would not want to be plugged into such a machine. He argues that there are three reasons why having even the most perfect experiences alone are not enough: we want to do things, and not just have the experience of doing them; we want to be a certain way or certain kind of person, and not just be an inanimate husk inside a machine; and thirdly, we want to experience the sublimity of nature, and not the artificial constructs of a man made reality (Nozick, 2000).

Nozick argues that plugging into an experience machine is akin to suicide. Much like the movie *Total Recall*, where the protagonist gets to immerse himself in a virtual reality that takes him out of his boring life into the life of an exciting covert agent, Nozick argues that most individuals would not want this, but rather would have the experience of living in a possibly more mundane, yet infinitely more real, reality. This has important implications for brain technology. Given that Nozick is correct in his prediction, even as technology advances and the brain interface becomes ever more complex, allowing for deeper peering into the private world of our mind, most human beings will simply not want to be a part of the experience machine. According to Nozick, humans want to identify with their own accomplishments and self without the aid and support of self-enhancing technology.

CONCLUSION

As researchers we tend to focus on the slow march of advancing science, but we must also be mindful of the direction we are going, and the potential impact our work can have on the future. While scientific progress will likely never be halted based on its *potential* to lead to ethically questionable future technologies, open philosophical discussion of these risks can lead to the implementation of preventative measures to avoid such dangers without interfering with progress.

We must be vigilant, as scientists and practitioners, to make sure that brain technology is not abused or misused. Individuals must be held responsible as well, to ensure that they treat themselves and their brains with respect. It is in the best interests of science to always keep the public trustworthy, and there is no doubt that with advancements in the future, brain technology will test the bounds of this trust. We must ultimately take care to always treat individuals as ends, and never as means, and to further science without sacrificing the rights of any single individual.

REFERENCES

Benabid, A. L. (2003). Deep brain stimulation for Parkinson's disease. *Current Opinion in Neurobiology, 13*(6), 696-706.

Canli, T., Brandon, S., Caseebeer, W., Crowley, P.J., DuRousseau, D., Greely, H.T., Pascual-Leone, A. (2007). Neuroethics and national security. *American Journal of Bioethics,* 1-15.

Farah, M.J. (2005). Neuroethics: The practical and the philosophical. *TRENDS in Cognitive Sciences, 9,* (1).

Genetic Information Nondiscrimination Act of 2008, Pub. L. No. 110-233, § 122, Stat. 881 (2008).

Hoffman, R. E., & Stanford, A. D. (2008). TMS clinical trials involving patients with schizophrenia. In E. Wassermann, E. M. Wassermann, C. M. Epstein, & U. Ziemann (Eds.), *The Oxford Handbook of Transcranial Stimulation* (pp. 671-684). New York, NY: Oxford University Press.

Hill, T.E. (1973). *The Monist.* La Salle, Il.

Kant, I. (2000). The Moral Law. In L.P. Pojman (Ed.), *The Moral Life* (pp. 297-317). New York, NY: Oxford. (Reprinted from *The Foundations of the Metaphysics of Morality,* by T.K. Abbott, 1873).

Kessler, R. C., Chiu, W. T., Demler, O., & Walters, E. E. (2005). Prevalence, severity, and comorbidity of twelve-month DSM-IV disorders in the national comorbidity survey replication. *Archives of General Psychiatry, 62*(6), 617-627.

Miyawaki, Y., Uchida, H., Yamashita, O., Sato, M., Morito, Y., Tanabe, H. C. Kamitani, Y. (2008). Visual image reconstruction from human brain activity using a combination of multiscale local image decoders. *Neuron, 60*(5), 915-929.

Moreno, J.D. (2003). Neuroethics: An agenda for neuroscience and society. *Nature Reviews Neuroscience, 4*(2), 149-153.

230

Nozick, R. (2000). The experience machine. In L.P. Pojman (Ed.), *The Moral Life* (pp. 615-618). New York, NY: Oxford. (Reprinted from *Anarchy, State, and Utopia*, by Robert Nozick, 1974, HarperCollins).

Parasuraman, R. (2003). "Neuroergonomics: Research and practice." *Theoretical Issues in Ergonomics Science*, 4, 5-20.

Ramsey, W. (2007). Eliminative materialism. *Stanford Encyclopedia of Philosophy* (October 2007 ed.). Retrieved from http://plato.stanford.edu/entries/materialism-eliminative/

Sartre, J.P. (2000). Existentialism is a humanism. In L.P. Pojman (Ed.), *The Moral Life* (pp. 641-650). New York, NY: Oxford. (Reprinted from *Existentialism*, by Jean Paul Sartre, 1947, New York, NY: Philosophical Library).

CHAPTER 23

Activity of Alerting, Orienting and Executive Neuronal Network Due to Sustained Attention Task – Diurnal fMRI Study

Magdalena Fafrowicz [1], Tadeusz Marek [1], Krystyna Golonka [1], Justyna Mojsa-Kaja [1], Halszka Oginska [2], Kinga Tucholska [1], Ewa Beldzik [1,3], Aleksandra Domagalik [1], Andrzej Urbanik [4]

1 Department of Neuroergonomics, Institute of Applied Psychology
Jagiellonian University, Krakow, Poland

2 Department of Ergonomics and Exercise Physiology
Collegium Medicum, Jagiellonian University, Krakow, Poland

3Department of Biophysics, Faculty of Biochemistry, Biophysics and
Biotechnology, Jagiellonian University, Krakow, Poland

4 Chair of Radiology, Collegium Medicum, Jagiellonian University,
Krakow, Poland

ABSTRACT

The study was aimed at analyzing the diurnal variability of neuronal networks activity due to sustained attention task.

Participants and Methods: Fifteen healthy male volunteers (mean age 27.4 years) meeting MR inclusion criteria took part in the study. Subjects were performing a sustained attention task in a MR scanner five times during the day: at 6 am, 10 am, 2 pm, 6 pm, and 10 pm. Each session was combined with 9 blocks (lasted 30 sec each): 5 blocks of fixation point presentations (control conditions) and 4 blocks of targets presentations, each containing 18 stimuli (task conditions).

Subjects were instructed to react, by pressing a button, when two identical words appeared sequentially one after another. They were asked to identify the word's semantic meaning and ignore the color of ink in which the word was printed. MR imaging was performed on the 1.5 T General Electric Signa scanner. Images were analyzed with AFNI software. Interaction effect between times of day was detected using ANOVA.

Results: The analyses revealed significant diurnal differences in activations in brain regions linked with alerting (parietal lobe BA 40), orienting (frontal eye fields – FEF), and executive attention subsystem (fronto-insular cortex – FIC, presupplementary motor area – preSMA, basal ganglia, dorsolateral prefrontal cortex – DLPFC and ventrolateral prefrontal cortex – VLPFC).

Diurnal patterns of neuronal activity of three attention subsystems showed similar profiles with significantly higher activations observed at 6 am in comparison to the other times of day. In case of right BA 40, right and left FIC, activities at 6 am were significantly higher than activities measured at 10 am, 2 pm, and 10 pm.

Conclusion: Observed diurnal profiles appear to present significantly higher levels at 6 am. There is also a visible, but not statistically significant increase of activation at 6 pm in almost all neuronal structures showing diurnal variability. Therefore, both 6 am and 6 pm seem to be related with higher effort invested in performing cognitively complex vigilance task.

Keywords: Attention neuronal network, sustained attention task, diurnal variability, fMRI

INTRODUCTION

Sustained attention also termed vigilance represents a basic attentional function that determines the efficacy of the 'higher' aspects of attention and cognitive capacities of operators. A typical sustained attention task may involve only sensory discriminations or cognitive discriminations (Stollery, 2006). This study was focused on the latter one. Imaging studies revealed anterior and parietal cortical activation associated with sustained attention task performance. It requires activity of the neuronal circuit of attention consisting of alerting, orienting and executive subsystems.

Functions of alerting subsystem includes awareness of action intention and awareness of sensory consequences of movements/action. The former is related with activity of anterior cingulate cortex (ACC) and presupplementary motor area (preSMA), while the latter is linked with activity of inferior parietal cortex (IPC)

(Portas et al., 2004). The orienting subsystem relates to selection of specific information from sensory input and one of the frequently analyzed structure is the area of frontal eye fields (FEF), that exhibits higher activity in response to presented cues referring to where (space), when (time), and to what (object) individuals should direct their attention (Fox et al., 2006). Finally, executive functions involving mechanisms for conflict monitoring and resolving discrepancies among stimulus and responses, relate to the activity of anterior cingulate and lateral prefrontal cortex such as dorso-lateral prefrontal cortex (DLPFC) and ventrolateral prefrontal cortex (VLPFC) as well as parts of basal ganglia (Posner, Rothbart, 2007).

DLPFC seems to be crucial for maintenance and processing of information in the working memory, as well as decision making in goal-oriented behavior.

Is has to be emphasized that with respect to real-life task complexity those differentiations may be difficult. Monitoring a particular source of information requires, at the same time, the selection of such a source and the inhibition of competing sources. Complex sustained attention tasks cannot be performed without additional executive functions.

The aim of present study was to analyze the patterns of neuronal activity of attention networks engaged in sustained attention task from the perspective of diurnal variability.

METHODS

Participants

The participants were 15 healthy adult paid volunteers (mean age 27.4 years). All of them were right-handed males meeting the following inclusion criteria: normal color vision, no history of head injury or neurological disorders, no sleep-related disorders, non-smokers and drug-free.

Each subject was informed about the procedure and goals of the study and provided his written consent. Each also possessed a valid driving license to ensure task competence. The study was approved by the Bioethics Commission of the Jagiellonian University (Krakow, Poland).

Task

In the study subjects were performing a sustained attention task based on Stroop Color-Word task. Participants were instructed to react, by pressing a button, if two identical words (names of colors) appeared sequentially one after another. They were asked to identify the word's semantic meaning and ignore the color of ink in which the word was printed. The task was constructed to activate complex aspects of attentional processes: alerting, orienting and executive one. The focus was placed on alerting network responsible for achieving and maintaining a state of preparedness (readiness for motor reaction in case of occurrence of two identical words sequence). Additionally, executive aspects of attention involved conflict monitoring (attending to one dimension of a stimulus and ignore another one). Presented task was designed to reflect real-life task complexity.

Experimental design

Participants performed a sustained attention task (see Fig. 1) in the MR scanner five times during the day (at 6 am, 10 am, 2 pm, 6 pm, and 10 pm). One session lasted 5 minutes and was combined with nine blocks: five blocks of fixation point presentations and four blocks of target presentations. Stimuli were presented for 1317 msec each, with an interstimulus interval of 350 msec, in a block design with each block containing 18 words, and a total of 30 sec per block. Two identical words that appeared sequentially were presented randomly from 3 to 6 times per each block of targets Subjects practiced the task before scanning in order to eliminate the effect of novelty .

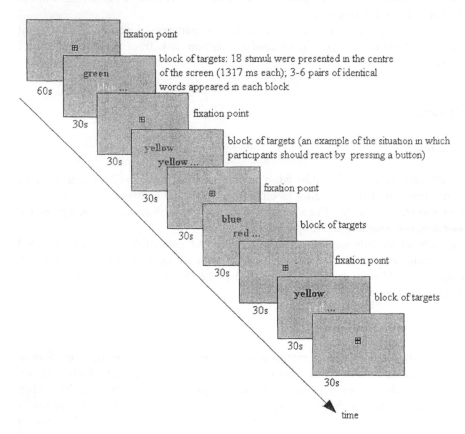

Figure 1. Schematic description of a sustained attention task.

Between MR sessions, participants were asked to perform four sessions of driving tasks, each lasting about two and a half hours. Driving sessions were conducted in a simulator, where subjects performed assignments from some-car games. In sum, the participants spent approximately 18 hours in the controlled conditions of a laboratory (stable temperature, similar physical and cognitive activity, and diet).

Data acquisition

The imaging was performed using a 1.5 T General Electric Signa scanner (GE Medical Systems, Milwaukee, WI). Hearing was protected with earplugs and motion was minimized with soft pads fitted over the ears. T1-weighted, high-resolution whole-brain anatomical images were acquired using 3D spoiled GRASS (SPGR) sequence for co-registration with the fMRI data. A total of 60 axial slices were obtained (voxel dimension = $0.4 \times 0.4 \times 3$ mm^3; matrix size = 512×512, TR = 25.0 s, TE = 6.0 ms, FOV = 22×22 cm^2, flip angle = $45°$. Functional T2*-weighted images were acquired using a whole-brain echo planar pulse sequence (EPI). Each functional session was composed of 50 images for each of 20 axial slices, taken at an interleaved fashion with a TR of 3 s (TE = 60 ms, matrix size = 128×128, FOV = 22×22 cm^2, spatial resolution of $1.7 \times 1.7 \times 5$ mm^3 and flip angle = $90°$). To ensure stability of the magnetic signal, the first three images of each session were excluded from further analysis.

Data analysis

The collected brain images were analyzed with Analysis of Functional NeuroImage (AFNI) software (Cox, 1996). Each 3D image was first time-shifted so that the slices were aligned temporally. Time series were realigned to correct for head motion, zero-padded to match the spatial extent of the anatomic scans, and then co-registered. Anatomical and functional images were transformed into a coordinate system of Talairach space (Talairach, Tournoux, 1988). The data were spatially smoothed using an isotropic Gaussian kernel of 8mm full-width at half maximum. During scaling procedure the low signal intensity voxels corresponding to voxels located outside the brain were excluded from functional images by a clipping function.

Data were subjected to analysis of General Linear Model (GLM), including stimulus, baseline and movement parameters as regressors. The beta coefficient of block stimuli was calculated for each voxel, representing the estimate of the fMRI activity. Firstly, all five scans at each time of day were combined and the general map of activation was created. Statistical maps were averaged across subjects with corresponding T-test (p <0.01 corrected, cluster size >15). Secondly, beta coefficients were calculated for each time of day separately. A 2-way mixed effects ANOVA was used to achieve F-map (p <0.05 uncorrected, cluster size >15) with time of a day as a fix factor. Then, conjunction analysis of the F-map and general T-map was applied with FDR correction (p <0.05 corrected, cluster size >15). Finally, the mean for each cluster beta parameter was extracted in the conjunction map. Diurnal pattern of activity underwent ANOVA test for significance.

RESULTS

As a first step of analyses the clusters obtained with a general activation map were examined. Results (see Fig. 2) revealed significant activations in brain regions linked with the alerting attentional subsystem (parietal lobe – BA 40 and thalamus), orienting attentional subsystem (bilaterally in frontal eye fields – FEF) and the executive subsystem (bilaterally in the fronto-insular cortex – FIC, presupplementary motor area – preSMA, supplementary motor area – SMA, VLPFC and right DLPFC). Activations were also observed in the area of medial frontal (MF), superior frontal, posterior cingulate cortex (PCC), insula, hippocampus and visual area.

The conjunction analysis revealed several clusters significantly differentiating between times of day (Fig. 2). These were mainly located in the regions of the alerting (parietal lobe BA 40), orienting attentional subsystem (frontal eye fields – FEF) and the executive subsystem (fronto-insular cortex – FIC, presupplementary motor area – preSMA, basal ganglia, dorsolateral prefrontal cortex – DLPFC). Additionally, the brain region forming executive system included ventrolateral prefrontal cortex – VLPFC, which was mainly deactivated during the day.

Diurnal patterns of neuronal activity of alerting (left BA40), orienting (bilaterally FEF) as well as executive system (right basal ganglia, right DLPFC, and preSMA) show similar profiles with significantly higher activations at 6 am in comparison with activities measured during the rest of the day. Similarly in case of left VLPFC, activation observed at 6 am was higher than activations at remaining times of day. Finally, right BA 40, right and left FIC activities at 6 am were significantly higher than activities measured at 10 am, 2 pm, and 10 pm.

It is worth mentioning that there is a noticeable, but not statistically significant increase of activation at 6 pm in almost all structures displaying diurnal variability (Fig. 2).

The findings from the study support a model that describes sustained attention as a "top-down" process mediated via anterio-parietal regions. As it is seen from figure 2, the anterio-parietal regions which maintain subject's readiness to detect and discriminate information critical for a given sustained attention task are activated.

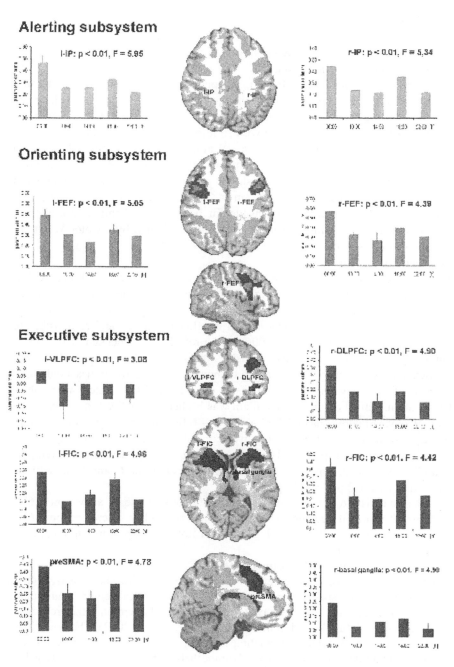

Figure 2. Brain regions showing activations in a sustained attention task (fMRI, block design). Three neuronal subsystems are highlighted, including the alerting, orienting and executive one. The bar graphs indicate mean parameter estimates of clusters which survived conjunction analysis (distinguished by different shades of color). Legend: r-right; l-left hemisphere.

DISCUSSION

All of observed attention subsystems showed diurnal variability.

The alerting and orienting systems

The left inferior parietal lobe (BA40), responsible for spatial shifts of attention (Corbetta et al., 1995) showed diurnal changes of activity level (at 6 am there was a significantly higher activation than at the other times of day). The right parietal lobe (BA 40) showed similar profile with significantly higher activation at 6 am in comparison with activations measured at 10 am, 2 pm, and 10 pm. Likewise, the FEF (in both the left and right brain hemisphere), engaged in the participation of overt eye movements in attentional shifts, proved to be sensitive to temporal changes with a higher activation at 6 am in comparison with further time points. Therefore, neuronal structures responsible for both visual attention controlled by a goal-oriented mechanisms as well as alert state, activated by a Color-Word task were shown to be sensitive to diurnal changes.

The executive system

Diurnal differences in activation level were also observed in brain regions combining executive system. The right basal ganglia – associated with language processing, such as deriving nouns from verbs (Marangolo, Piras, 2008), right DLPFC – thought to support heteromodal conflict resolution and responsible for selecting responses (Raz, Buhle, 2006) and preSMA – associated with conflict monitoring at the stimulus level (Bush, 2004) showed similar profiles with significantly higher activation at 6 am in comparison with activity measured during the rest of the day. Similarly, in case of left VLPFC linked with object identification, activation observed at 6 am was higher than activations at remaining times of day. Finally, bilateral FIC activity – reflecting the degree of subjective salience and involved in various cognitive control mechanisms, including conflict and error monitoring (Crottaz-Herbette, Menon, 2006) also proved temporal variability with significantly higher activity measured at 6 am than activities observed at 10 am, 2 pm, and 10 pm.

There is also a visible, but not statistically significant increase of activation at 6 pm in almost all neuronal structures showing diurnal variability. Therefore, both 6 am and 6 pm seem to be related with higher effort invested in performing cognitively complex vigilance task.

Cholinergic projection and sustained attention

Evidence from functional imaging as well as lesion studies shows activation of anterior cingulate, dorsolateral prefrontal and parietal cortical regions, particularly in the right hemisphere, in subjects performing sustained attention task, irrespective of the modality of stimuli used in these task (Sarter, Bruno, 2002). The crucial role of cortical cholinergic inputs, originating in basal forebrain, in sustained attention was shown in animal experiments (e.g. Himmelheber et al., 2000; 2001). Levels of cortical acetylcholine raised dramatically during performance of tasks requiring sustained attention for detection of a stimulus (Himmelheber et al., 2000; Arnold et al., 2002).

According to Sarter and Bruno (2002) sustained attention is "mediated via the activation of the cortex by cholinergic inputs, specifically anterior-parietal circuits, and the interactions between the modulation of sensory processing by the downstream projections of anterior attention system, including the recruitment of interhemispheric circuits, the direct cholinergic stimulation of sensory areas and thalamic input to these regions".

Strong cholinergic input to the anterior-parietal circuits enhance attention to the environment, making the circuits more responsive to specific features of sensory stimuli. The input to the anterior-parietal regions plays a crucial role in activating "top-down" control mechanisms of sustained attention (Sarter, Bruno, 2000; Sarter et al., 2001). The levels of acetylcholine in the hippocampus and anterior-parietal regions change during different stages of waking and sleep. High levels of acetylcholine during waking set appropriate dynamics for attention to sensory input and encoding of new information.

Higher activation of anterior-parietal circuits at 6 am and at 6 pm suggest stronger cholinergic input. Similar activation effect is observed in both cases, nevertheless evoked by different mechanisms of morning adaptation, i.e. the adaptive effort (6 am) and afternoon/evening compensation, i.e. the compensation effort (6 pm). In both cases, level of mental effort seems to be a crucial factor determining the higher level of activation which reflects the stronger cholinergic input.

The study showed the occurrence of time-of-day-related changes in neural activity of brain regions linked with attentional network. It would seem to be the case that alerting attention operates in an interactive way with executive functions and its diurnal profile seems to be synchronized.

The efficiency of attention system plays important role in every moment of operator's work and provides arguments for the need to explore diurnal profiles of attention subsystems activity. This knowledge may be of importance from the point of the risk prevention programs.

ACKNOWLEDGMENTS

The authors would like to thank Barbara Sobiecka, Justyna Kozub, Adam Swierczyna, and Izabela Gatkowska for their valuable assistance in experimental data acquisition. This research project was supported by a grant from the Polish Ministry of Science and Higher Education (N106 034 31/3110) (2006–2009).

REFERENCES

Arnold, H.M., Burk, J.A., Hodgson, E.M., Sarter, M. and Bruno, J.P. (2002), "Differential cortical acetylcholine release in rats performing a sustained attention task versus behavioral control tasks that do not explicitly tax attention." *Neuroscience*, 114 (2), 451-460.

Bush, G. (2004), Multimodal studies of cingulated cortex. In Posner MI (ed).

Cognitive Neuroscience of Attention. New York : The Guilford Press, pp. 207-218

Corbetta, M., Shulman, G.L., Miezin, F.M., and Petersen, S.E. (1995), "Superior parietal cortex activation during spatial attention shifts and visual feature conjunction." *Science*, 270 (5237), 802–804.

Cox, R. (1996), "AFNI: Software for analysis and visualization of functional magnetic resonance neuroimages." *Computers and Biomedical Research*, 29, 162–173.

Crottaz-Herbette, S., and Menon, V. (2006), "Where and When the Anterior Cingulate Cortex Modulates Attentional Response: Combined fMRI and ERP Evidence." *Journal of Cognitive Neuroscience*, 18(5),766–80.

Fox, M.D., Corbetta, M., Snyder, A.Z., Vincent, J.L., and Raichle, M.E. (2006), "Spontaneus neuronal activity distinguishes human dorsal and ventral attention systems." *PNAS*, 103 (26), 10046–10051.

Himmelheber, A., Sarter, M., and Bruno, J. (2000), "Increases in cortical acetylcholine release during sustained attention performance in rats." *Cognitive Brain Research,* 9 (3), 313-325.

Himmelheber, A.M., Sarter, M., and Bruno, J.P. (2001), "The effects of manipulations of attentional demand on cortical acetylcholine release". *Cognitive Brain Research,* 12 (3), 353-370.

Marangolo, P., and Piras, F. (2008), "Dissociations in processing derivational morphology: The right basal ganglia involvement." *Neuropsychologia*, 46 (1), 196-205.

Portas, C., Maqet, P., Rees, G., Blakemore, S., and Frith, C. (2004), The Neural Correlates of Consciousness. In Frackowiak SJ, Friston KJ, Frith CD, Dolan RJ, Price CJ, Zeki S, Ashburner J, Penny W (eds.). Human Brain Function. New York: Elsevier Science, pp. 269-302.

Posner, M.I., and Rothbart, M.K. (2007), "Research on Attention Networks as a Model for the Integration of Psychological Science". *Annual Review of Psychology*, 28,1–23.

Raz, A., and Buhle, J.(2006),"Typologies of attentional networks." *Nature Reviews Neuroscience*, 7, 367-379.

Sarter, M., and Bruno, J.P. (2000), "Cortical cholinergic inputs mediating arousal, attentional processing and dreaming: differential afferent regulation of the basal forebrain by telecephalic and brainstem afferents." *Neuroscience*, 95, 933-952.

Sarter, M., Givens, B., and Bruno, J.P. (2001)," The cognitive neuroscience of sustained attention: where top-down meets bottom-up." *Brain Research Reviews*, 35(2), 146-160.

Sarter, M., and Bruno, J.P. (2002), Vigilance. In V.S. Ramachandran (ed.) Encyclopedia of the Human Brain. Elsevier Science, USA, pp. 687-699.

Stollery, B.T. (2006), Vigilance. In W Karwowski (ed.) International Encyclopedia of Ergonomics and Human Factors. CRC Taylor & Francis, Boca Raton, London, New York, pp. 965-968.

Talairach, J., and Tournoux, P. (1988), Co-planar Stereotaxic Atlas of the Human Brain. Thieme Medical , New York.

CHAPTER 24

Characteristic Changes in Oxygenated Hemoglobin Levels Measured by Near-Infrared Spectroscopy (NIRS) During "Aha" Experiences

Keisuke Teranishi*, Hiroshi Hagiwara**

*Graduate School of Science and Engineering

**College of Information Science and Engineering
Ritsumeikan University, Shiga, Japan

ABSTRACT

Neuroscience has made rapid progress in recent years, and various workings of the brain continue to be elucidated. In addition to advancing research, these findings are also being used in commercial areas. However, progress has been slow in quantification of the workings of brain activity and subsequent use for commercial purposes. In this study, we focused on "Aha" experiences (inspiration) among the various workings of the brain to quantify brain activity and apply it to Neuromarketing. As a first step, we identified the characteristic changes in hemoglobin concentration in the brain during artificial "Aha" experiences using near-infrared spectroscopy (NIRS). In the control experiment, no characteristic changes were observed. In the main experiment, characteristic changes were observed only during the 30 seconds from the moment when the button was pressed. Consequently, the waveform at this time was considered to indicate characteristics of the time of Momentary Learning. In the main experiment, we found that the latency until the peak (MAX) was smallest in ch1 (center of right prefrontal area) and ch22 (motor areas), and that the response was earlier than in other channels. This may be because the right brain, which is responsible for intuition, was more active than other brain regions. This finding is consistent with results from previous studies using MRI. Thus, it may be possible to conduct experimental measurements more easily using NIRS, which has fewer restrictions and lower costs than MRI.

Keywords: Brain Science, NIRStation, Neuromarketing, "Aha" experience, Momentary Learning

INTRODUCTION

Techniques to measure brain activity without damaging the subject's brain include functional magnetic resonance imaging (fMRI), near-infrared spectroscopy (NIRS), and magnetoencephalogram (MEG). Various workings of the brain continue to be elucidated rapidly using these systems. As a result, in addition to research applications, these findings are being applied to commercial areas. In neuroscience, the response of a subject's brain is measured to analyze the subject's mental and action changes. Conventional questionnaires cannot measure unconscious responses. Neuromarketing is the goal of business to apply marketing research. Methods to quantify the various workings of brain activity and their subsequent use in Neuromarketing have not been developed. In this study, we focused on "Aha" experiences (inspiration) among the various workings of the brain to quantify brain activity for use in Neuromarketing. MRI has been used in neuroscience in recent years, but it requires a large equipment to measure brain activity with magnetism and requires a high degree of subject restraint. Using this approach, we cannot easily perform experiments because of the restricted experimental conditions and the large expense.

However, the advantage of NIRS is that it has almost no restrictions about experimental conditions, and the experimental equipment is inexpensive. It uses near-infrared to measure the subject, and the only equipment needed is a probe touching the subject's head. fMRI cannot use metal items because of the powerful magnetism, and the noise is very loud. Furthermore fMRI can only be used when the subject is lying down. In contrast, NIRS uses another equipment simultaneously as near-infrared. The subject can exercise moderately in the range of the optical fiber cable of NIRS. Because NIRS requires a lower cost of operating and maintenance, we can perform experiments easily and frequently. Moreover, for "Aha" experiences (inspiration), there is precedence established by fMRI, but there is almost no data from NIRS. Therefore, in our study, we first identified the characteristic changes in hemoglobin concentration in the brain during "Aha" experiences. However, it is extremely difficult to reproduce "Aha" experiences in daily life. In this study, we produced artificial "Aha" experiences by extracting the characteristics of these experiences and defining them. We used the three characteristics of inspiration: momentary understanding, emotion, and release from stress. "Aha" experiences (inspiration) were defined as Momentary Learning, and we were able to reproduce "Aha" experiences in our experiments. Subjects were also asked to press a button at the time of Momentary Learning during the performance test. We also conducted an additional control experiment in which only the button input motion was performed to investigate the effects of the button input motion on the brain. For this experiment, subjects were asked to watch animation in which some items disappeared from the screen or gradually changed color. The moment when they noticed the change on the screen was considered to be an "Aha" moment (Momentary Learning), and changes in hemoglobin levels in the brain at that moment were measured. In the frontal lobe (ch1, 10, 19), which is involved in cognitive judgment, we examined 46 fields (ch25) associated with attention control and execution. Motor areas (ch22) were also investigated. In addition, relationships between each part of the brain were examined. During the experiment, the Rokan Arousal Scale (RAS), Alpha Attenuation Coefficient (AAC), electroencephalogram (EEG), and electrocardiogram (ECG) were used to obtain data.

METHOD

SUBJECTS

Twenty healthy, non-medicated subjects (9 males, 11 females; 21-23 years old) participated in the experiment. All subjects obtained adequate sleep and performed the experiment without excessive eating or drinking the previous night. In addition, subjects refrained from alcohol and caffeine ingestion and avoided napping or engaging in prolonged or strenuous exercise before the experiment. Informed, written consent was obtained from each subject prior to participation in the study. The room temperature during the experiment was 24°C.

EXPERIMENT

During the experiment, RAS, AAC, EEG, ECG, and oxygenated hemoglobin levels were measured. We used an EEG1100 equipment (Nihonkoden) to measure EEG and ECG. We used near-infrared spectroscopy (NIRS-SHIMAZDU) to measure oxygenated hemoglobin levels in the brain. We obtained EEG using the international 10-20 system. Four EEG channels were applied (O1/O2). EEG and ECG were recorded at a sampling frequency of 500 Hz. We attached the probe to channel points to measure oxygenated hemoglobin levels with the NIRS's holder. The probes were spaced in the holder every 3 centimeters. We analyzed ch1, 10, 19, 22, and ch25 (Figure 1).

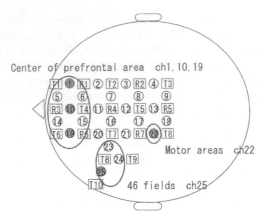

Figure 1. Attachment of the probes

ALPHA ATTENUATION COEFFICIENT (AAC)

When the subjects were quiet, had their eyes closed, were in a relaxed state, and were awake and concentrating, the amplitude of the alpha waves decreased immediately and appeared at the onset of quiet. Alpha waves are known to appear from the back of the head when one's eyes are closed and the subject is relaxed. Alpha waves decline when the eyes are opened and the subject is sleepy. Moreover, Alpha waves increase when the brain is active and concentrating. In contrast, alpha waves during sleeping have small amplitudes. Thus, we analyzed the alpha waves and calculated the ratio of the average power of closed eye (30 seconds ×3) alpha waves and the average power of opened eye (30 seconds ×3) alpha waves and thus defined the Alpha Attenuation Coefficient (AAC) (Figure 2). This calculation was used as the quantitative evaluation index of the degree of awakening.

$$AAC(\text{Alpha Attenuation Coefficient}) = \frac{Averagepowereyeclosed}{Averagepowereye\ opened} \quad \text{Eq. 1}$$

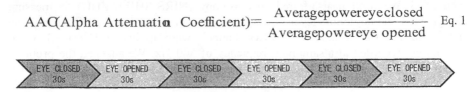

Figure 2. Experimental design of the Alpha Attenuation Test (AAT)

ROKEN AROUSAL SCALE (RAS)

RAS is a quantitative index that evaluates six states (Sleepiness, Activation, Relaxation, Strain, Difficulty of Attention and Concentration, and Loss of Volition) and uses twelve questions to show the load of mental work. The six states were analyzed with two similar questions. We defined the average of the two similar questions as the value of the states.

MORPHING ANIMATION TEST (PERFORMANCE TEST)

We made animations in which some items disappeared from the screen or changed color (rate of change is 0.05-0.1%) by morphing, to induce Momentary Learning in the subject (Figure 3).

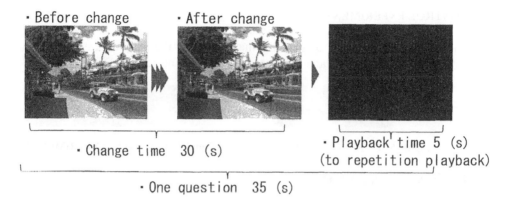

Figure 3. Morphing Animation Test (Performance Test)

EXPERIMENTAL PROCEDURE

MAIN EXPERIMENT

The subjects were equipped with EEG electrodes and NIRS probes. Rest was recorded for 30 seconds before beginning RAS1. Eyes open and eyes closed were performed for a total of 3 minutes (AAT1) for 30 seconds at a time. After AAT1, the subjects did the performance test. To prevent the order effect from influencing the result, ten performance tests were experimented on at random. Subjects rested for 30 seconds before beginning RAS2 (Figure 4).

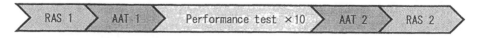

Figure 4. MAIN EXPERIMENTAL PROCEDURE

CONTROL EXPERIMENTS

The subjects were equipped with EEG electrodes and NIRS probes. Subjects rested for 30 seconds before they did the performance test. In control experiment 1, pattern 1 (Figure 6) was shown. In control experiment 2, pattern 2 (Figure 7) was shown. Subjects rested for 30 seconds before the end of the experiment (Figure 5).

> Performance test ×5 (2 patterns) >

Figure 5. CONTROL EXPERIMENTAL PROCEDURE

CONTROL EXPERIMENT 1

To consider the influence of screen changing, the subjects pushed the A-button as soon as the screen changed (Figure 6).

CONTROL EXPERIMENT 2

To consider the influence of pushing the A-button, the subjects pushed the A-button about 15 seconds later after having changed the screen (Figure 7).

Figure 6. CONTROL EXPERIMENT 1   Figure 7. CONTROL EXPERIMENT 2

DATA ANALYSIS

For analysis, the waves that appeared 30 seconds before and after the Momentary Learning were averaged for each subject, with a total of 60 seconds being recorded. Normalization of the 30 seconds before and after the Momentary Learning was performed so that relative changes from the standard could be obtained. In addition, in this normalized experiment, the first maximum value of the waveform from the button input time was defined as the peak (MAX). Control experiments were analyzed in the same manner. The AAC and RAS values of each subject were individually averaged and compared before and after the experiment. In addition, in this normalized experiment, the first maximum value of the waveform from the button input time was defined as the peak (MAX). Control experiments were analyzed in the same manner.

RESULTS

The value of AAC2 decreased significantly ($P < 0.05$) compared with AAC1 (Figure 8a; AAC). In RAS, the value of sleepiness increased significantly ($p<0.1$) after the experiment. The value of relax decreased significantly ($p<0.05$) after the experiment. As for the other states, significant differences were not observed (Figure 8b; RAS).

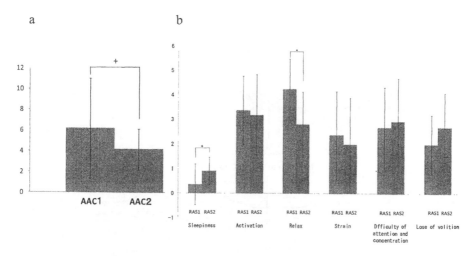

Figure 8. Results of AAC (a) and RAS (b). Bars indicate S.D, n − 8, +: p < 0.1, *: p < 0.05

In the control experiment, no characteristic changes were observed. Before and after the moment the button was pressed for the button input motion or the change on the screen, no significant differences were noted. Characteristic changes were observed only during the 30 seconds from the moment when the button was pressed during the main experiment, and significant differences were noted (Figure 9 a, b, c). During the main experiment, the waves before and after the Momentary Learning characteristic wave showed a peak 6-7 seconds after the Momentary Learning. Every channel after 50-60 seconds returned the same level before 0-10 seconds (Figure 9a). In control experiments 1 and 2, the wave fluctuated slightly after the button was pressed. The fluctuation, however, was not significantly different before vs. after the button was pressed (Figure 9 b, c).

We found that the latency until the peak (MAX) was smallest in ch1 and ch22. Except for ch22 without a direct relation to thought, the order of the latency is ch1 (the right brain), ch19 (left brain), ch25 (Cautious-control/ execution), and ch10 (center of the prefrontal area).

DISCUSSION

In the main experiment, the obvious wave change with a peak (MAX) during Momentary Learning was observed in 6-7 seconds after A-button pushed. The value of Momentary Learning was compared with the peak (MAX) in ch19 and showed that the peak (MAX) was higher with Momentary Learning (p < 0.05). In the other channels, a high tendency was observed (p < 0.1). The peak (MAX) of all channels was different. We thought that there was a difference in the brain activity time. All channels reached the peak (MAX) in 6-7 seconds. In previous studies using MRI, a time lag of 5-10 seconds was observed for the increase in the amount of blood flow, reflecting reaction of the nerve cells of the brain. Thus, we thought

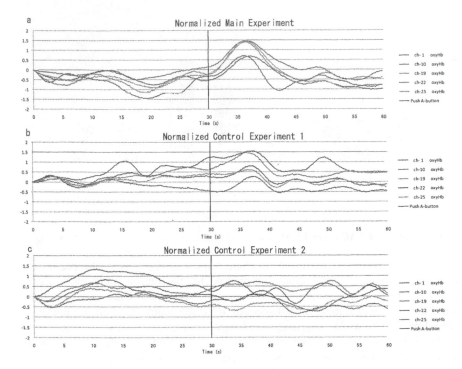

Figure 9. Wave patterns of each channel in the main experiment (a), normalized control experiment 1 (b), and normalized control experiment 2 (c). The button was pressed at 30 seconds.

Table 1: The peak (MAX) coordinates

	Oxygenated Hemoglobin levels in brain	Seconds
ch1	1.468	36.010
ch10	0.691	37.310
ch19	1.451	36.530
ch22	0.711	36.010
ch25	1.396	36.660
	The latency until the peak (MAX) order	Seconds
	ch1	6.110
	ch22	6.110
	ch19	6.630
	ch25	6.760
	ch10	7.410

Table 1. Oxygenated hemoglobin levels and seconds of peak (MAX), as well as the seconds and ranking from Momentary Learning to peak (MAX) are shown.

that the brain reacted until it reached the peak MAX, and that it was necessary to consider the relation between each part of the brain. The results of control experiment 1 didn't show changes in waves with changes in the screen. An influence of pushing the button was not observed. In all channels, the value of Momentary Learning was compared with the peak (MAX) in the main experiment, and significant differences were not observed. We observed the same trend in control experiment 2 and control experiment 1. Every channel that changed during Momentary Learning to the peak (MAX) returned to the same level 0 - 10 seconds later. Thus, we defined this reaction that was early in the order of the peak (MAX) as Momentary Learning. The rises during fatigue and sleepiness were observed similarly as the results of AAC and RAS. Thus, we obtained the opposite result in which sleepiness goes up and concentration is maintained. We thought that this sleepiness reflected tiredness rather than boredom with the experiment.

CONCLUSION

We found that the latency until the occurrence of peak (MAX) was the shortest in ch1 and ch22, and was longest in ch10. Momentary Learning was judged by ch10 in connection with recognition and judgment rather than ch1 in connection with a hunch and an image. ch19, which is connected with logical thinking and ch25, which is involved in cautious control/execution showed nearly simultaneous peaks (MAX). Thus, the brain was working simultaneously with respect to logical thinking and cautious control and execution. ch10, which is connected with recognition and judgment, was later in every state. We thought that ch10 generalized all channels. The value of AAC2 fell, rather than the value of AAC1 to a level of 10%. Thus, we thought that the degree of activity fell during the experiment. Moreover, the result of RAS was also the same. In RAS, the value of sleepiness increased significantly ($p<0.1$) after the experiment. The value of relax decreased significantly ($p<0.05$) after the experiment. As for the other states, significant differences were not observed. However, the values of activation and strain and loss of volition did change substantially. Thus, we thought that the subjects maintained their concentration during the main experiment. However, we thought that fatigue and wearing of the experimental equipment and the time required for the experiment influenced the body and thus influenced the results of AAC and RAS. During the experiment, the subjects had to concentrate. Results shown in Figure 9 a, b, c and Table 1 illustrate that the prefrontal area worked complexly during Momentary Learning. The right brain, which is involved with hunches and images, worked predominantly. We thought that a reaction similar to a reflex was carried out rather than logical thinking. Our RAS results are in agreement with results from previous studies using MRI. Thus, it may be possible to obtain experimental measurements more easily using NIRS, which has fewer restrictions and lower costs than MRI.

REFERENCES

Shirley M Coyle, Tomas E Ward, Charles M Markham (2007) "Brain-computer interface using a simplified functional near-infrared spectroscopy system", J. Neural Eng. 4, 219-226

Yumi Shibagaki, Kozue Ogawa, Hiroshi Hagiwara (2010) "Evaluation of Physiological Indices to Indicate Sleepy or Relaxed States Using Illuminate Stimulation", Transactions of the Society of Instrument and Control Engineers. 46(1), 65-71

CHAPTER 25

Performance Under Pressure: A Cognitive Neuroscience Approach to Emotion Regulation, Psychomotor Performance and Stress

Amy J. Haufler, Bradley D. Hatfield

Cognitive and Motor Neuroscience Lab
Department of Kinesiology
University of Maryland
College Park, MD, 20742, USA

ABSTRACT

Self-regulation of the psychological state when military personnel are inserted in combat operations is of critical importance for effective decision-making, prevention of fatigue and effective control of motor behavior. The association between mental state and an individual's ability to cope with stress has been well established. Specifically, the response of the central nervous system and major body systems to stress and its subsequent effect on an individual's performance indicates that there is a high level of coupling in which one may influence the other to result in either positive or negative performance outcome. Investigations of the brain processes during skilled performance have revealed that superior visuo-motor performance is marked by efficient cortical dynamics which likely reflects refined

252

task-relevant networks that are associated with decreased variability in movement trajectory and enhanced quality of movement output. In addition, examination of peripheral physiology in experts shows that they are able to manage their stress or level of arousal in the moments prior to task execution, so as to not compromise their performance in the face of challenge. In this manner, level of skill is important when examining the brain-performance-arousal relationship. A regression to lower levels of performance is expected under conditions of challenge, particularly when the performer can not manage the stress response or level of arousal. A neuro-cognitive model of psychomotor performance is illustrated (Figure 1) in which the relationship between cortical processes and the motor loop are postulated to impact the quality of motor output via muscle activation. As such, "neuro-motor noise" in higher cortical regions would negatively impact the quality of motor performance. Performance under pressure requires emotion regulation to preserve the "quiet" mind for superior performance output. Figure 1: Input from the association areas of the cerebral cortex in the form of "neuro-motor noise" contributes to nonessential input to the motor loop to result in disregulated corticospinal outflow and degraded motor performance.

Figure 1

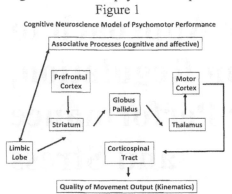

Cognitive Neuroscience Model of Psychomotor Performance

Keywords: Psychomotor performance, stress, neuroergonomics, emotion regulation, military, expertise, brain, EEG, marksmanship

SUPERIOR PERFORMANCE IS CHARACTERIZED BY PSYCHOMOTOR EFFICIENCY

Neuroscientific investigations of marksmanship performance have collectively revealed a robust relationship between cerebral cortical activation and quality of precision target shooting (Bird, 1987; Haufler et al., 2000; Hatfield et al., 1984; 1987; Kerrick et al., 2004). Specifically, lower activation was typically noted in the left temporal region of the brain (T3), as indicated by greater EEG alpha power, relative to that observed in the right homologous region (T4) during the aiming period leading up to the trigger pull (Hatfield et al., 1984; 1987). These findings are interpreted as suppression of verbal-analytic and explicit memory processes, both associated with left hemispheric functioning, while the maintenance of right temporal activity (i.e., indicated by EEG alpha stability or suppression) appeared consistent with the task-specific visual-spatial processes associated with the right hemisphere (Left Brain – Right Brain).

Haufler et al. (2000) compared cerebral cortical activation between expert marksmen and novice shooters during the aiming period and observed less activation in the frontal, central, temporal, parietal, and occipital regions in the experts. The group difference was of greatest magnitude in the left temporal region and indicates that the experts accomplished the task in a more efficient or cortically refined manner as indicated by the EEG. Brain electrical activity maps (Figure 2) reveal higher cortical activity (i.e., EEG gamma power illustrated in pink) in novice shooters (higher in the left-temporal and -frontal, right-temporal and occipital areas) as compared to expert marksmen and associated exemplar aiming point trajectory. Furthermore, higher levels of cortico-cortical communication (EEG coherence computed for association and motor planning regions) were associated with greater aiming point variability in the expert group while no such relationship was revealed in the novices (Deeny et al., 2009). These results confirm that skilled performance is associated with cortical refinement yoked to superior, controlled psychomotor performance. Modeling of these kinematic data revealed that experts and novices employed a similar strategy to the visuo-motor aiming task, but were differentiated by the level and management of neuro-motor noise relative to task execution (Goodman et al., 2009).

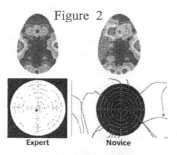

Figure 2

Expert   Novice

STRESS AND PERFORMANCE: A CASE FOR EMOTION REGULATION

Conditions of stress or challenge affects not only neural systems involved in learning and retention of emotionally arousing information but also disrupts psychomotor performance such that a regression from higher levels of performance is observed under conditions of challenge. The psychological state of an individual can be highly influenced by environmental factors, yet there is great variation in the abilities of each individual to cope with and subsequently perform within various environments. While some individuals can cope adaptively to stress with relative ease and maintain performance levels, approximately 60% of the general population is characterized by a specific genetic variation that has recently been shown to alter the brain's ability to react to stressful conditions (Hariri et al., 2002). Specifically, carriers of either one or two copies of the short allele (i.e., S/S or S/L genotypes) of the serotonin transporter gene (5-HTT) demonstrates brain hyper-reactivity to stressful stimuli and increased fear and anxiety-related behaviors. In essence, S allele carriers may be considered "stress-prone" while LL carriers may be considered "stress-regulators." Therefore, "stress-prone" individuals may exhibit disrupted or noisy cortical dynamics which translates to greater variability in their performance relative to individuals who may be "stress-regulators".

One population that experiences challenge with mental stress are military personnel. Soldiers face increasing and unprecedented information processing and psychomotor challenges critical to mission success and survival. Technological advancements place a premium on the human operator's (i.e., soldier's) attention capacity, decision-making processes, and motor control thereby creating a situation in which the war fighter is challenged to realize the advantages and exploit the limits of these technologies. Further, the elicitation of intense emotional states and uncontrolled arousal under battle conditions can consume and degrade the critical mental resources and adaptive decision-making processes that future force warriors will need to execute their responsibilities. Although the basic fight-flight response initially advanced by Cannon (1932) is certainly adaptive for survival in many situations (i.e., those well served by gross motor and sympathetic excitation), it is not consistent with the fine-tuned self-regulation of emotional control required for complex decision-making while operating in threatening and physically demanding environments. A critical component of high-level cognitive-motor performance is emotion or affective regulation.

The psychological control involved during self-initiated emotion regulation starts with the executive processes (i.e., inhibitory) that are largely housed in the dorsal lateral prefrontal cortex (DLPFC) that, in turn, impact the activity of the emotional processes that are largely housed in the inferior regions of the frontal lobe. The dampening of such emotion-related processes is enabled by the anatomic connection between the inferior frontal lobe and the amygdalae. This pathway can be strategically controlled by the DLPFC which would serve as a central and pivotal brain region that would exert enormous influence in the self-initiated control of stress/fear and attendant arousal-related sequelae (Ochsner & Gross, 2005). Emotion regulation can preserve the cortical dynamics necessary for superior psychomotor performance. See Figure 3 for a schematic of multiple factors involved in the brain-mind-body interaction for superior performance.

Figure 3 **Stress/Fear Circuit Model**

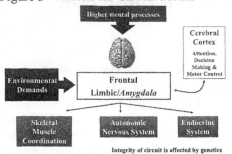

CONCLUSIONS

Studies of brain activity during psychomotor performance have shown that a fundamental marker of expertise is efficiency of neural processes characterized by refinement of attention processing, relative absence of effortful cognition, and the emergence of efficient limb movements. Stress or heightened challenge can negatively moderator the neural processes involved in learning, memory and performance. While individual differences exist regarding self-regulation during challenge, superior performance under pressure necessitates arousal regulation. The relevance of this work is positioned with the American Warfighter who simultaneously processes information, makes decisions and performs highest levels while experiencing extreme challenge.

REFERENCES

Cannon, Walter B. (1932). The wisdom of the body, 2nd Edition, 1939, Norton Pubs, New York.

Deeny, S., Haufler, A. J., Saffer, M. & Hatfield, B. D. (2009). Electroencephalographic coherence during visuomotor performance: a comparison of cortico-cortical communication in expert and novice marksmen. Journal of Motor Behavior, 41(2), 106-116.

Goodman, S., Haufler, A., Shim, J., & Hatfield, B. (2009). Regular and random components in aiming-point trajectory during rifle aiming and shooting. Journal of Motor Behavior, 41(4), 367-384.

Hariri, A. R., Mattay, V. S., Tessitore, A., Kolachana, B., Fera, F., Goldman, D., Egan, M. F., Weinberger, D. R. (2002). Serotonin transporter genetic variation and the response of the human amygdala. Science, 297, 400-403.

Hatfield, B. D., Landers, D. M., & Ray, W. J. (1984). Cognitive processes during self-paced motor performance: An electroencephalographic profile of skilled marksmen. Journal of Sport Psychology, 6, 42-59.

Hatfield, B.D., Landers, D.M., & Ray, W.J. (1987). Cardiovascular-CNS interactions during a self-paced, intentional attentive state: elite marksmanship performance. Psychophysiology, 24 (5), 542-549.

Haufler, A. J., Spalding, T. W., Santa Maria, D. L., & Hatfield, B. D. (2000). Neuro-cognitive activity during a self-paced visuospatial task: comparative EEG profiles in marksmen and novice shooters. Biological Psychology, 53, 131-160.

Ochsner, K. N., & Gross, J. J. (2005). The cognitive control of emotion. TRENDS in Cognitive Science, 9(5), 242-249.

CHAPTER 26

A Framework for Improving Situation Awareness of the UAS Operator through Integration of Tactile Cues

Matthew Johnston[1], Kay Stanney[1], Kelly Hale[1], Robert S. Kennedy[2]

[1]Design Interactive, Inc.
Oviedo, FL 32765, USA

[2]RSK Assessments
Orlando, FL 32803, USA

ABSTRACT

Unmanned aerial systems (UAS) have high mishap rates as compared to general aviation. Among other factors, UAS operator error has been associated with instrumentation/sensory feedback systems and channelized attention. Multimodal displays, and in particular exteroceptive tactile cueing, may be a useful means of compensating for UAS operator's sensory isolation, alleviating high cognitive demands, and directing attention. Exteroceptive tactile cueing could be used to supplement auditory and visual displays in improving the situation awareness (SA) of UAS operators. Given past use of tactile cues as effective alerting and feedback mechanisms, a framework is presented which represents how tactile cues can be used to enhance object recognition awareness, spatial awareness, and temporal awareness. The framework aims to improve UAS operator training effectiveness, throughput, and operational performance and has applicability to the military and medical domains. Future research opportunities will be suggested.

Keywords: unmanned aerial systems, tactile cues, haptic interaction

INTRODUCTION

Unmanned aerial systems (UAS) have high mishap rates as compared to general aviation (UAS: 100s of mishaps per 100,000 flight hours as compared to a general aviation rate of 1 mishap per 100,000 flight hours; Hing & Oh, 2008; Tvaryanas, Thompson, & Constable, 2005). Among other factors, UAS operator error has been associated with instrumentation/sensory feedback systems and channelized attention. With regards to instrumentation/sensory feedback, UAS separate the aircraft from the operator, which deprives the operator of a range of multi-sensory cues that are inherent to actual flight (McCarley & Wickens, 2005). Specifically, UAS operators generally receive remote vehicle-camera imaging covering a limited field-of-view; missing cues include kinesthetic and vestibular input, ambient visual information, and spatialized audio. This restricted, unimodal view places high cognitive demands on operators. However, the effect this sensory deprivation and high workload has on UAS operator performance is not well understood. Multimodal displays may be a useful means of compensating for the UAS operator's sensory isolation, and in turn help alleviate high cognitive demands (McCarley & Wickens, 2005). In particular, the inclusion of exteroceptive tactile cueing could be used to supplement auditory and visual displays in improving the situation awareness (SA) of UAS operators. Tactile cues have successfully been used to alert operators to system failures (Calhoun et al, 2002; Dixon et al, 2003), support navigation in flight (van Erp et al., 2004), and represent feedback such as turbulence one would only normally receive in the cockpit (Ruff et al, 2000). Tactile cues have also been successfully used to abstractly communicate with and alert operators through the use of tactile symbols displayed on a wearable display (Fuchs et al, 2008). Given these successes, the current effort presents a framework for improving the SA of UAS operators through the use of tactile cues and haptic feedback. This framework represents how tactile cues can be used to enhance object recognition awareness, spatial awareness, and temporal awareness. These have been previously suggested as constructs critical to maintaining SA (Hale, Stanney, & Malone, 2009; Hale et al., 2009). This paper will discuss how tactile cueing and/or haptic feedback could be used to maintain or improve each component of awareness for the UAS operator and future research opportunities will be suggested.

UAS SA FRAMEWORK

Human operators of UAS are generally involved in supervisory control of one to multiple systems, potentially while on the move and under enemy fire. This presents high cognitive workload and attention demands, which could contribute to UAS mishaps. When reviewing mishaps, Tvaryanas et al. (2005) found that workload and attention factors associated with mishaps included, among other factors, issues with workstation design and UAV operator attentional focus. Interventions to address these factors could focus on multimodal display design to reduce high workload, direct attention, and enhance SA. Specifically, the UAS SA framework shown in Figure 1 posits that tactile cues can play a significant role in achieving optimal SA for a UAS operator. This framework suggests that tactile cues should be integrated into a multimodal environment such that exteroceptive changes in the UAS environment are conveyed to the UAS operator in a manner that supports multiple components of their SA (i.e., object recognition, spatial awareness, and temporal awareness). The enhanced SA will then translate into the selection of a course of action with given levels of accuracy and efficiency, which in turn, changes the active state of the UAS and is experienced as a change in the environment, and so on.

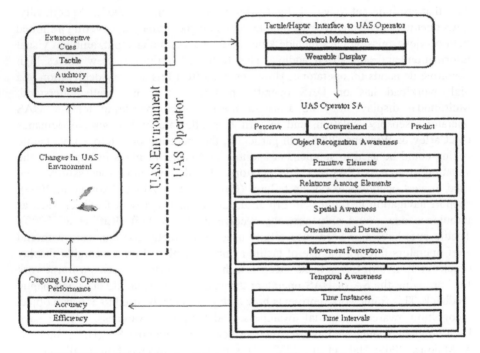

FIGURE 1. UAS operator SA development through exteroceptive cues.

While pilots generally receive exteroceptive cues during flight to include visual, auditory, and tactile cues, UAS operators are somewhat removed from these cues, as they relate to changes in the UAS environment and the operator is not collocated with the UAS as a traditional pilot would be[1]. The lack of exteroceptive cueing

---

[1] The exact situation of the UAS operator can vary. Specifically, Air Force UAV operators fly from a vehicle-centric perspective (e.g., from within the UAV via a nose camera image), while Army and Navy/Marine UAV operators fly from an

limits the ability of the UAS operator to achieve high levels of SA. The framework shown in Figure 1 and described in the following sections suggests that tactile cueing and haptic feedback can be successfully used in UAS training and operational environments to assist the UAS operator in recognizing objects in their environment and improving spatial awareness of the UAS with respect to safe flight and mission parameters, such as route navigation. The framework also suggests that tactile cues could be used to improve the temporal awareness of the UAS operator through attention cueing to critical faults / events or provision of tactile pacing solutions. Haptic feedback could also be used to simulate the typical tactile exteroceptive cues a traditional pilot would receive during flight, such as turbulence, by providing feedback related to the UAS whether this is through contact with instrumentation or possibly wearable tactile displays.

OBJECT RECOGNITION AWARENESS

The detection and recognition of objects in real-time from a UAS is the basis of many mission scenarios (e.g., detection and recognition of ships at sea, vehicles, buildings). For example, a UAS operator may be required, during an ongoing mission, to monitor multiple displays for the intrusion of a hostile aircraft in the UAS vicinity (Gunn et al., 2002). This is generally designed as a visual task that is using resources from an already taxed visual system that is required to monitor UAS status and perform route navigation. Such object recognition information could be communicated to an individual via a tactile display such that it reduces the complexity of information traditionally portrayed visually or audibly, consequently reducing cognitive workload. With such displays, a UAS operator could perceive the presence of a hostile or critical object through presentation of an intuitive tactile symbol. Such tactile symbols can be learned quickly and the bandwidth of the human sensory system to learn such symbols is quite broad (e.g., Fuchs et al [2008] demonstrated that participants were able to learn a 56 symbol vocabulary consisting of military hand and arm signals in less than 3.5 hours, on average. The language was subsequently improved and the same vocabulary was learned to 90% criterion in 90 minutes, on average, and participants retained 89%, on average, of the vocabulary 2 days later with no further training [Johnston et al., in press]).

Future research is necessary to determine whether tactile representations of objects relevant to the UAS operator mission is an effective means of presentation and what information, if any, can be effectively presented to the operator such that cognitive workload is reduced and information presentation is optimized. Future research must also consider the appropriate modality for presentation of this type of information and consider when a certain modality is optimal for presentation.

---

exocentric perspective (e.g., observing the UAV from a position aside the runway) (Tvaryanas et al., 2005).

SPATIAL AWARENESS

Although the UAS operator is not collocated with the vehicle, spatial information is still critical with respect to maintaining control of the aircraft and understanding its relative position with respect to the intended route and mission parameters. Tactile displays have been used to maintain spatial awareness, prevent spatial disorientation, and reduce cognitive workload in operational fixed wing aircraft (Raj et al., 1996; van Erp et al., 2006; Rupert et al, 1996), rotary wing aircraft (van Erp et al., 2003a; van Erp et al., 2003b; McGrath et al., 1998), and UAV operation during landing (Aretz et al., 2006). Van Erp et al (2006) demonstrated that a tactile display could be used to successfully counter spatial disorientation; participants, disoriented by a spinning chair, were able to successfully reorient themselves. The Tactile Situation Awareness System (TSAS, Raj et al, 1996; Rupert et al, 1996) has been successfully used to help pilots maintain fixed wing aircraft orientation. Pilots were asked to perform maneuvers, such as straight and level, altitude recovery, and loops, while receiving tactile orientation information. The tactile cues enabled pilots to successfully maintain control of the aircraft even with no visual cues during straight and complex maneuvers. Van Erp et al (2003a) tested the effectiveness of a tactile display on maintaining altitude in a rotary wing aircraft. Two conditions of tactile information were displayed to the pilot – simple information in which the direction of the desired altitude was presented and complex information in which current aircraft motion was added to the simple information. The results indicated that the presence of tactile information reduced altitude error significantly without significantly increased mental workload. These results were also seen under night vision conditions. In a separate evaluation, van Erp et al (2003b) demonstrated that a tactile display could be used to maintain helicopter hover during usage of night vision goggles in an operational rotary wing aircraft without increasing attentional or cognitive demands on the pilot, regardless of salient visual cues. The TSAS system (Raj et al, 1996; Rupert et al, 1996) has also been used to provide aircraft information to pilots of rotary winged aircraft in an effort to reduce visual workload (McGrath et al, 1998). Roll and pitch cues were presented to pilots using the TSAS system, as well as airspeed and heading error, while pilots performed a variety of maneuvers. Pilots were able to solely rely on tactile cues for all attitude information while successfully performing the maneuvers, however, heading error and airspeed presentation proved problematic during simultaneous presentation. A challenge in using tactile cues in an operational environment is representing the appropriate information to optimize SA while not overloading the haptic channel.

Taken together, these studies indicate that tactile displays have shown utility in improving spatial awareness of pilots of fixed and rotary wing aircraft, successfully providing information relevant to aircraft attitude and preventing or recovering from errors in aircraft altitude and attitude. The use of tactile displays in UAS environments is less substantial. One area where it may be particularly helpful is in addressing landing errors, which account for a majority of human factors related UAS mishaps (Tvaryanas et al., 2005). Using a tactile vest, Aretz et al (2006) provided Predator UAS pilots with cues regarding altitude deviation. A Predator control station generally provides operators with a visual map display and a monitor that combines a forward camera with a head up display. The Predator operator generally receives no auditory, kinesthetic, of vestibular feedback typically available to a traditional pilot. The experimenters chose to simulate kinesthetic and vestibular sensations by providing tactile sensations related to altitude deviations and demonstrated that novice UAS operators achieved a passing score for landings of Predator UAVs in significantly less trials when assisted with a tactile display compared to a no tactile display condition. Conversely, Ruff et al (2000) improved SA of UAS operators to the onset of turbulence by the addition of tactile cues through a control stick. In both cases kinesthetic and/or vestibular feedback was provide in a different manner. Future research must investigate the appropriate way to convey this type of information.

The work described above focuses on the use of tactile cues for providing information largely relevant to the status of the aircraft, such as pitch, roll, altitude, heading, and airspeed, and the onset of turbulence. Tactile displays also have utility in providing spatial information regarding mission parameters, such as in waypoint navigation. In military applications, success has been achieved in using tactile displays for the purpose of land, sea, and air navigation, as well as pilot orientation (Chiasson, McGrath, & Rupert, 2003; Dobbins & Samways, 2002; van Erp & van Veen, 2003, van Erp et al, 2004). For example, van Erp et al (2004) describe the use of horizontal tactile vibrations around a pilot's waist in a Gazelle helicopter. Pilots were successfully able to navigate between waypoints when only provided a tactile cue representing the direction of the next waypoint and a different tactile cue when the waypoint had been passed. Similar applications could prove beneficial to UAS operations.

Future research must examine what information is necessary to optimize the spatial awareness of the UAS operator. In particular, is it important to replicate or simulate traditional sensations, such as turbulence, or is it more effective to focus on presentation of related types of information or can both be done effectively. Hing and Oh (2008) describe the development of a motion cueing platform for UAS operators that can be used to provide the traditional sensations of the cockpit. Future efforts should determine if the operational or training environment need go this far in representing traditional sensations or whether tactile cueing on a smaller scale is effective. A motion platform, while providing vestibular cues, would be competing with other tactile cues and the simultaneous presentation of tactile information could compete for the limited attentional resources of the UAS operator.

It is also important to understand the tradeoffs not only between different forms of tactile or haptic presentation but potential interference with other modalities. Multimodal tradeoffs regarding how tactile presentation of spatial information can integrate with visual and auditory cues must be considered. Gunn et al (2002) describe a study in which visual cueing, spatialized audio and haptic force feedback controls were used to assist participants in a target acquisition task. In this case the haptic feedback was provided through a joystick controller. The results indicated no significant differences between the methods for target acquisition and suggest that they could be used interchangeably. Future research must identify when modalities can be interchanged, what type of information can be effectively presented via tactile cues, how tactile cues relevant to target cueing/acquisition and route guidance should be presented such that it doesn't interfere with UAS performance. Future attempts must take a decidedly multimodal approach such that any tactile cues provided to the UAS operator are optimized with respect to information provided by visual and or auditory displays.

TEMPORAL AWARENESS

Motor skill/coordination or timing deficiency have been found to be preconditions of UAS mishaps (Tvaryanas et al., 2005). This could be attributed to a lack of temporal awareness due to the disassociation between the UAS operator and the vehicle. Such temporal awareness could be enhanced through tactile cueing. Information conveyed via tactile cues could include pacing information regarding flight segments or attention to critical task sub-components (Tan, et al., 2003). The occurrence of events, whether it be intrusion of a hostile aircraft in their airspace or alert to critical faults (Calhoun et al., 2002; Dixon et al, 2003) with respect to system state or interval presentation, could also be presented via tactile cues. The previous section discussed tactile representation of where an object is, but the ability to compare spatial position relative to self requires knowledge of position over time. In this case, tactile cues could be used to track relative position by providing continuous presentation of position of relevant entities. Tactile systems can effectively display temporal information in isolation (ETSI, 2002); however, combined with vision, the haptic sense can provide more accurate timing information compared to vision alone. The combination could enable a UAS operator to better formulate a plan and predict future states. A particular future state of interest to the UAS operator is position of self relative to a target. Van Veen, Spape, and van Erp (2004) investigated the effectiveness of different temporal patterns on conveying distance through a tactile display during a route navigation task. They discovered no significant differences between coding distance via temporal cues versus other coding methods. Prediction of future position is typically inferred based on current position and speed information that is typically presented on a visual display. Given the success of tactile information as a navigation tool, one could surmise that its use may be successful in providing UAS operators a tool to inform them of current position relative to an upcoming waypoint and a predicted arrival time based on the rate of tactile presentation. A UAS operator could also be provided with a pacing cue to dictate the optimal speed of travel to make prediction of future state easier.

CONCLUSIONS

Over and above the savings in pilots lost, the economical and tactical advantages of UAS cannot be denied. However, accident rates comprise a face evident measure of systems (person-in-the-loop) performance and recent studies have shown that flight control of uninhabited air systems exhibit 100 times the accident rate of general aviation (Hing & Oh, 2008; Tvaryanas et al, 2005). Clearly, research into improvements in such systems performances of UAS should be conducted. It is our view that flight control accidents are a reasonable proxy for the other in-flight task elements of UAS performance and argue that holistic studies of the three underlying predictors of situation awareness (viz, object recognition, temporal, and spatial awareness) are likely to offer a research framework to serve as a model for research in this area.

For reasons stated above, we propose that a coordinated research program should be undertaken. First there should be creation of a task taxonomy and analysis of the UAS tasks; then the relatively unexplored areas of the contributor of gravi - receptor sensory input pathways (i.e., haptic, proprioceptive, kinaesthetic, vestibular) to mission performance should be explored. Furthermore, because large amounts of predictive variance are likely to be attributable to reliable individual capabilities (e.g., 50 – 70% is not unexpected) and practice / training effects (e.g., 15 – 40% is not unusual), we propose that this research program should follow a common overarching research strategy (e.g., economical multi factor fractional factorials; Simon, 1970; Jones, Kennedy, & Stanney, 2004) so that meta analyses will be possible at various stages throughout development of the scientific progress.

ACKNOWLEDGEMENTS

This material is based upon work supported in part by the Office of Naval Research (ONR) under SBIR contract N00014-08-C-0203. Any opinions, findings and conclusions or recommendations expressed in this material are those of the authors and do not necessarily reflect the views or the endorsement of ONR.

REFERENCES

Aretz, D., Andre, T., Self, B., and Brenaman, C. (2006), "Effect of tactile feedback on unmanned aerial vehicle landings." *Proceedings of the Interservice/Industry Training, Simulation, and Education Conférence (I/ITSEC)*, Orlando FL.

Calhoun, G.L., Draper, M.H., Ruff, H.A., and Fontejon, J.V. (2002), "Utility of a tactile display for cueing faults." *Proceedings of the Human Factors and Ergonomics Society 46th Annual Meeting*, 2144-2148.

Chiasson, J., McGrath, B., and Rupert, A. (2003), *Enhanced situation awareness in sea, air, and land environments*. DTIC Technical Report; Naval Aerospace Medical Research Laboratory Pensacola, FL.

Dobbins, T., and Samway, S. (2002), "The use of tactile navigation cues in high-speed craft operations." *Proceedings of the RINA conference on high speed craft: technology and operation"* (pp. 13-20). London: The Royal Institution of Naval Architects.

Dixon, S.R., Wickens, C.D., and Chang, D. (2003), "Comparing quantitative model predictions to experimental data in multiple-UAV flight control." *Proceedings of the Human Factors and Ergonomics Society 47th Annual Meeting*, 104-108.

European Telecommunications Standards Institute [ETSI] (2002), Human factors: guidelines on the multimodality of icons, symbols, and pictograms (Rep. No. ETSI EG 202 048 v 1.1.1). Sophia Antipolis Cedex, France.

Fuchs, S., Johnston, M. Hale, K.S., and Axelsson, P. (2008), "Results from pilot testing a system for tactile reception of advanced patterns (STRAP)." In *Proceedings of the Human Factors and Ergonomics Society 52nd Annual Meeting*, 1302-1306.

Gunn, D.V., Nelson, W.T., Bolia, R.S., Warm, J.S., Schumsky, D.A., and Corcoran, K.J. (2002), "Target acquisition with UAVs: Vigilance displays and advanced cueing interfaces." *Proceedings of the Human Factors and Ergonomics Society 46th Annual Meeting*, 1541-1545.

Hale, K.S., Stanney, K.M., and Malone, L. (2009), "Enhancing virtual environment spatial awareness training and transfer through tactile and vestibular cues." *Ergonomics*, 52(2), 187-203.

Hale, K., Stanney, K., Milham, L., Bell, M.A., and Jones, D. (2009), "Multimodal sensory information requirements for enhancing situation awareness and training effectiveness." *Theoretical Issues in Ergonomics Science (TIES)*, 10(3), 245-266.

Hing, J., and Oh, P.Y. (2008), "Integrating motion platforms with unmanned aerial vehicles to improve control, train pilots and minimize accidents." *Proceedings of the ASME 2008 International Design Engineering Technical Conferences & Computers and Information in Engineering Conference (IDETC/CIE 2008)*. August 3-6, 2008, Brooklyn, New York.

Johnston, M.R., Axelsson, P., and Hale, K.S. (2009, in preparation), "*Learning and retention of a tactile language and vocabulary*." Design Interactive, Oviedo, FL.

Jones, M. B., Kennedy, R. S., and Stanney, K. M. (2004), "Toward systematic control of cybersickness." *Presence,* 13(5), 589-600.

McCarley, J.S., and Wickens, C.D. (2005), *Human factors concerns in UAV flight*. Technical Report AHFD-05-05/FAA-05-01. Institute of Aviation, Aviation Human Factors Division. University of Illinois at Urbana-Champaign.

McGrath, B.J., Suri, N., Carff, R., Raj, A.K., and Rupert, A.H. (1998), "The role of intelligent software in spatial awareness displays." In *3rd Annual Symposium and Exhibition on Situational Awareness in the Tactical Air Environment*, 2-3 June 1998. Piney Point, MD.

Raj, A.K., et al. (1996), "From benchtop to cockpit: Lessons learned from the vibrotactile orientation system prototype development and flight test program." *Aviation, Space, and Environmental Medicine*. 67(A9): p. 48.

Ruff, H.A., Draper, M.H., Poole, M.R., and Repperger, D.W. (2000), "Haptic feedback as a supplemental method of altering UAV operators to the onset of turbulence." In *Proceedings of the IEA 2000/ HFES 2000 Congress*, 2000, pp. 3.14-3.44.

Ruff, H.A., Narayanan, S., and Draper, M.H. (2002), "Human interaction with levels of automation and decision-aid fidelity in the supervisory control of multiple simulated unmanned aerial vehicles." *Presence, 11*, 335-351.

Rupert, A.H., et al. (1996), "Tactile interface to improve situation awareness." *Aviation, Space, and Environmental Medicine*, 67(A10): p. 53.

Tan, H.Z., Gray, R. Young, J.J., and Traylor, R. (2003), "A haptic back display for attentional and directional cueing." *Haptics-e*, 3(1), Available from: http://www.haptics-e.org.

Simon, C. W. (1970), *Reducing irrelevant variance through the use of blocked experimental designs* (Technical Report No. AFOSR-70-5). Culver City, CA: Hughes Aircraft Company.

Tvaryanas, A.P., Thompson, W.T., and Constable, S.H. (2005), "The U.S. military unmanned aerial vehicle (UAV) experience: Evidence-based human systems integration lessons learned." In *Strategies to Maintain Combat Readiness during Extended Deployments – A Human Systems Approach* (pp. 5-1 – 5-24). Proceedings of the RTO-MP-HFM-124, Paper 5. Neuilly-sur-Seine, France: RTO. Available from: http://www.rto.nato.int/abstracts.asp.

van Erp, J.B.F., and van Veen, H.A.H.C. (2003), "A Multi-purpose tactile vest for astronauts in the international space station." *Proceedings of Eurohaptics 2003*. Dublin Ireland: Trinity College, pp. 405-408, 2003.

van Erp, J.B.F., Veltman, J.A., and van Veen, H.A.H.C. (2003a), "A tactile cockpit instrument to support altitude control." In *Proceedings of the Human Factors and Ergonomics Society 47th Annual Meeting*. Human Factors and Ergonomics Society, Santa Monica, CA.

van Erp, J.B.F., Veltman, J.A., van Veen, H.A.H.C., and Oving, A.B. (2003b), "Tactile torso display as countermeasure to reduce night vision goggles induced drift." In *Spatial Disorientation in Military Vehicles: Causes, Consequences and Cures. RTO Meeting Proceedings* 86, pp. 49-1 - 49-8. Neuilly-sur-Seine Cedex, France: NATO RTO.

van Erp, J.B.F., Jansen, C., Dobbins, T., and van Veen, H.A.H.C. (2004), "Vibrotactile waypoint navigation at sea and in the air: two case studies." In *Eurohaptics 2004 conference*. Munchen, Germany.

van Veen, H.A.H.C., Spapé, M., and van Erp, J.B.F. (2004), "Waypoint navigation on land: different ways of coding distance to the next waypoint." *Proceedings of Eurohaptics 2004*, pp 160- 165. München, Germany: Technische Universität, 2004.

van Erp, J.B.F., Groen, E.L., Bos, J.E., and van Veen, H.A.H.C. (2006), "A tactile cockpit instrument supports the control of self-motion during spatial disorientation." *Human Factors*, 48(2), pp. 219-228.

CHAPTER 27

Predicting Perceptual Performance From Neural Activity

Koel Das[2], Sheng Li[3], Barry Giesbrecht[1, 2], Zoe Kourtzi[4],
Miguel P. Eckstein[1, 2]

[1]Institute for Collaborative Biotechnologies

[2]Department of Psychology
University of California, Santa Barbara

[3]Department of Psychology
Peking University

[4]School of Psychology, University of Birmingham
Edgbaston, Birmingham

ABSTRACT

A key aim of neuroergonomics is to gain an understanding of human neural function in relation to cognitive and behavioral performance in real world tasks. Here we investigated the relationship between neural activity and human performance in a rapid perceptual categorization task. We compared two different modalities to indirectly measure neural activity, functional magnetic resonance imaging (fMRI) and electroencephalography (EEG). We applied a multivariate pattern classifier to predict the individual variability in perceptual performance during a difficult visual categorization task using single trial EEG and fMRI separately. Twenty observers perceptually categorized images of cars and faces embedded in filtered noise while EEG activity from 64 electrodes was concurrently recorded. Another twenty observers performed the same task while their fMRI-

BOLD responses were recorded. Our results showed significant correlations between the neural measures and perceptual performance (p <0.05; r = 0.69 for EEG; r =0.66 for fMRI). We were able to reliably identify from their neural activity the best performing individual from two randomly sampled observers (84% for EEG; 75% for fMRI; chance = 50 %). Finally, we demonstrated that EEG activity predicting the performance across individuals was distributed through time starting at 120 ms and sustained for more than 400 ms post-stimulus presentation, indicating that both early and late components contain information correlated with observers' behavioral performance. Together our results highlight the potential to relate individual's neural activity to performance in difficult perceptual tasks and show a convergence in predictive ability across two different methods to measure neural activity.

Keywords: neural correlates, visual perception, pattern classifier, single trial EEG and fMRI

INTRODUCTION

Perceptual tasks remain important in many life-critical professions including airport traffic control, satellite imagery surveillance, airport security screening, and medical imaging. Yet, human errors in these visual tasks are not uncommon. There is also large variability in perceptual performance across individuals (e.g., medical images (Beam et al. 2002)). The traditional approach in human factors and imaging science has been to optimize displays and task performance through direct measurement of human performance or to use metrics of image quality. The rising field of neuroergonomics provides an interesting new approach. Its goal is to utilize emerging knowledge of brain function to design technologies that are well-adapted to neural coding and that lead to optimized human operations. To achieve such a goal, an initial critical step is to develop computational methods that identify neural activity related to an observer's cognitive state, knowledge of the state of the world, or impending decision.

In the last decades, the field of cognitive neuroscience has used different techniques such as electroencephalography (EEG) (Mangun & Hillyard 1990; Vogel & Machizawa 2004), functional magnetic resonance imaging (fMRI) (Grill-Spector et al. 2004; N. Kanwisher et al. 1997; Kamitani & Tong 2005; Harley et al. 2009a), magnetoencephalography (MEG) (Lu et al. 1992), positron emission tomography (PET) (Chugani et al. 1987) to establish a link between brain activity and behavior. For example, a recent study (Harley et al. 2009b) found a significant correlation between behavioral performance of radiologists detecting abnormal growths in chest radiographs and activations in FFA and lateral occipital cortex (LOC). Another study (Ben-Shachar et al. 2007) reported a significant correlation between a widely used measure of reading ability (phonological awareness) and motion responsivity in children's motion selective cortex (MT).

With the advance of computing technologies, analysis techniques have evolved from trial averaging and single electrode/voxel analyses to single trial, multi-variate pattern analysis. Machine learning techniques have been particularly effective for single-trial EEG/fMRI analysis. Use of multivariate pattern classifiers allows to integrate neural activity into a single decision variable to predict either an observer's behavioral response or the stimulus presented (Philiastides &. Sajda 2006; Kamitani & Tong 2005). Arguably, these multivariate techniques provide more powerful tools to relate neural activity to behavioral performance.

In this context, the primary objectives of the present study were to: 1) investigate the neural correlates of perceptual performance in a challenging perceptual task using EEG and fMRI separately by using pattern classifiers to predict the individual differences across observers performing a visual categorization task and 2) to use pattern classifiers and the high temporal resolution of EEG to illustrate the time-epochs that encode neural information predictive of observers' perceptual performance. Improving our ability to relate neural activity to behavioral performance could potentially allow for neural based measures of perceptual performance and image quality that can complement or even replace traditional behavioral measures.

MATERIALS AND METHODS

The observers' task was to identify the correct category (face/car) of the images presented in the screen while their neural signals were acquired concurrently. The details of the experiment for both EEG and fMRI are given in the following section.

STIMULATION AND DISPLAY

The stimuli consisted of 290 x 290 pixel 256-level grayscale images, 12 faces (six frontal views, six 45° profile) and 12 cars (six frontal views, six 45° profile). All images were filtered to achieve a common frequency power spectrum (the average of all images). Twelve images of each class, face and car, (six frontal view, six 45°rotated) were used as stimuli. Gaussian white noise was then added to these 24 base images to build a stimuli set of 720 images (360 face, 360 car). Noise was generated by filtering independent white Gaussian noise fields (standard deviation of 3.53 cd/m^2) by the average power spectrum of the car/face stimuli. The noise fields were added to the original car/face images.

EXPERIMENTAL SET UP

For EEG recording, observers sat 125 cm from a monitor with each image subtending 4.57° of visual angle and for fMRI, observers were situated 65 cm from the image surface with each image spanning 5.13° of visual angle. Contrast energy (CE) of all face and car stimuli were matched to be 0.3367^{o2} ,where CE is defined as the sum of the squared contrast values of the stimuli multiplied by the spatial extent of a pixel.

OBSERVERS AND PROCEDURES:

EEG: Twenty naive observers (ages: 18–26) participated in the study. Observers were initially presented with 1000 stimulus-familiarization trials on the first day and 100 more practice trials immediately preceding the current experiment on the second day. The actual study consisted of 1000 trials split into 5 successive blocks, each having 200 trials. The observers fixated on a central cross and pressed a mouse button to indicate the beginning of the trial (Fig 1). The stimulus appeared for 40 ms. after a variable delay of 0.5-1.5 seconds. The stimulus was followed by a blank screen presented for 0.5-1.5 seconds followed by the response window. Observers were asked to rate how confident they were that they saw either a face or a car, with a rating of 1 indicating complete confidence that a face was presented and a rating of 10 indicating complete confidence that a car was presented. Confidence responses were recorded by mouse clicks on the rating buttons of the response window.

a) EEG timeline

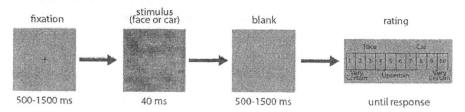

b) fMRI timeline

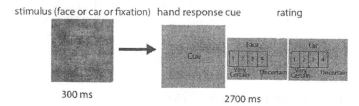

FIGURE 1 Illustration of psychophysical procedure for EEG and fMRI study

fMRI: Twenty different naïve observers with normal vision from the University of Birmingham participated in the experiment. 720 images (360 cars, 360 faces) divided into 8 blocks were used as stimuli with each trial matched for history (2-back). Each blocks consisted of 127 test trials (45 instances cars, faces, and fixations, along with 2 trials at the beginning of the block to equate the history for trials 3 and 4). Each block had a 9 second fixation period at the beginning and end.

Stimulus was presented for 300 ms followed by an interval of 2700 ms for response. Observers responded using button box and both hands and were informed beforehand by means of response cues which category corresponds to which hand. The response cues were alternated with half the time the right hand used for a car response while the left hand was used for a face response, and vice versa (Figure1). Each button box had four buttons corresponding to observers' fingers (thumb excluded) and represented confidence ratings with index always being the highest rating and little finger the lowest. Observers completed 7-8 blocks.

DATA ACQUISITION AND PREPROCESSING

EEG: Each subject's electroencephalogram was recorded from 64 Ag/AgCl sintered electrodes mounted in an elastic cap and placed according to the International 10/20 System. The horizontal and vertical electrooculograms (EOG) were recorded from electrodes placed 1 cm lateral to the external canthi (left and right) and above and below each eye, respectively. The data were sampled at 512 Hz, re-referenced offline to the signal recorded from the central midline electrode (Cz), and then band-pass filtered (0.01-100Hz). Trials containing ocular artifacts (blinks and eye movements) detected by EOG amplitudes exceeding ±100 mV or by visual inspection were excluded from the analysis. The average ERP waveforms in all conditions were computed time-locked to stimulus onset and included a 200 ms pre-stimulus baseline and 500 ms post-stimulus interval.

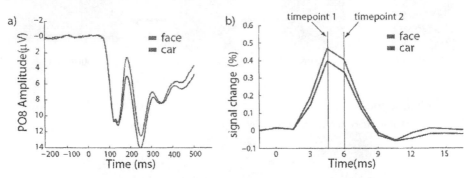

FIGURE 2a) Event-related potential (ERP) for face and car trials for PO8 electrode averaged across all observers. b) Time course for face and car stimuli from the FFA region from one of the observers in the fMRI sessions showing increased response to faces compared to cars.

fMRI: A 3T Achieva scanner (Philips, Eindhoven, and The Netherlands) was used for the experiments conducted at the Birmingham University Imaging Centre. EPI and T1-weighted anatomical (1 x 1 x 1 mm) data was collected with an eight channel SENSE head coil. 24 slices (whole brain coverage, TR: 1500 ms, TE: 35

ms, flip-angle: 73 degrees, 2.5 x 2.5 x 4 mm resolution) were acquired for EPI data (Gradient echo-pulse sequences) for the face/car categorization task. Localizer scans used 32 slices (whole brain coverage, TR: 2000 ms, TE: 35 ms, flip-angle: 80 degrees, 2.5 x 2.5 x 3 mm resolution). Data preprocessing was done using Brain Voyager QX (Brain Innovations, Maastricht, The Netherlands) and included slice-scan time correction, motion correction, temporal high-pass filtering (3 cycles) and linear trend removal. No spatial smoothing was applied on the functional data. The fMRI images for each observer were aligned to a T1-weighted high resolution (1 x 1 x 1 mm) anatomical data and finally all observers' data were transformed into Talairach space at the standard resolution of 3 x 3 x 3 mm.

DATA ANALYSIS USING PATTERN CLASSIFIERS

Recently, multivariate pattern classifiers has been successfully used in research related to functional magnetic resonance (Haynes and Rees, 2005; Norman et al., 2006; Kamitani and Tong, 2005) and/or EEG (Philiastides and Sajda, 2006b,a). In this study, we used a regularized linear discriminant analysis (LDA) (Fisher 1936) for classification analysis on both modalities' data sets. LDA is perhaps the most widely used feature extraction technique. The objective of LDA is to perform dimensionality reduction while enhancing class separability, normally by maximizing an objective function. Details for both modalities are given below.

EEG: Input data to the classifier was taken for the time epoch beginning at stimulus presentation through 512 ms post-stimulus, yielding 256 time points. Each trial thus provided 16,128 independent inputs (256 time points for 63 electrodes) rendering traditional LDA unfeasible. We have used Principal Component Analysis (PCA) to reduce dimension of the EEG signals thereby avoiding the small sample size problem before applying LDA. We used a 10-fold stratified cross validation where the dataset was randomly divided into 10 non overlapping folds of equal size, each having 100 trials. One of the folds was designated as test data while the remaining 9 folds constituted the training set.

fMRI: Various regions of interest (ROI) were identified for each observer including : 1) Retinotopic areas , 2) lateral occipital complex (LOC), and 3) fusiform face area (FFA) using standard ROI mapping procedures (Engel et al. 1994; DeYoe et al. 1996; Sereno et al. 1995; Kourtzi & N Kanwisher 2000; N. Kanwisher et al. 1997). For each region of interest, voxels were sorted based on a t-statistic computed by comparing responses to all stimulus conditions versus a control fixation on a blank screen condition. The time course for each voxel was normalized separately for each session to account for baseline differences across runs. Input data to the classifier consisted of the raw blood-oxygen-level dependent (BOLD) signal sampled at 2 time points (3 and 4.5 seconds post-stimulus in order to account for the hemodynamic response lag; see Figure 2b). We implemented a leave-one-session-out validation wherein the pattern classifier was tested on one session (90 test trials: 45 faces, 45 cars) and trained on the remaining sessions. The pattern classifier performance was evaluated by using a non-parametric measure of area under the Receiver Operating Curve (ROC), AUC, which quantifies the relationship between hit rate and false alarm rate for all possible decision criteria. To reduce the variance of estimated area under the curve (AUC), the overall cross validation procedure was repeated 8 times, each time selecting a new testing session and the overall performance is given by the mean AUC.

FIGURES OF MERIT TO EVALUATE NEURAL METRICS

The ability of various metrics to predict behavioral perceptual performance was evaluated by using a Pearson linear correlation and rank ordering of individuals' performance based on neural metrics. Rank ordering was performed by selecting the best performing individual out of two randomly sampled observers from their neural activity and validating based on the behavioral performance of the chosen observers. The rank ordering measure is more similar to a Spearman Rank Correlation which does not penalize departures from linearity and is complementary to the Pearson correlation.

RESULTS

We used LDA to predict the perceptual performance of observers performing the visual task. The pattern classifier was evaluated using AUC. Figure 3a shows the statistically significant positive correlation (r=0.69; p = 0.0005) between pattern classifier performance (AUC) using EEG activity and behavioral performance (AUC). Similar analyses using fMRI response yielded consistent results. Figure 4 shows the correlation between behavioral performance (AUC) and classifier performance (AUC) using fMRI (see also Table 1). Consistent with the EEG results, there was a statistically significant (r=0.66; p = 0.0005) correlation between classifier performance identifying car/face stimulus and subjects' behavioral performance.

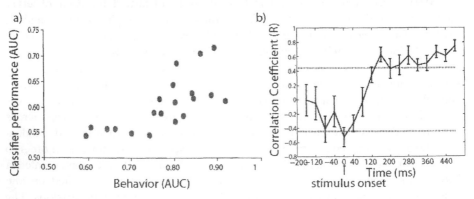

Fig3:a) EEG measures: AUC of pattern classifiers trained on only EEG trials vs. the

behavioral performance (AUC) of all 20 observers. b) Correlation between AUC of pattern classifiers of 20 observers taken over time intervals and their behavioral performance(AUC).(|r| >= 0.444 for statistically significant R using 95% confidence interval is marked with dotted line).

To further evaluate the ability of the EEG/fMRI metric to predict the variation in perceptual performance across individuals we performed a simple simulation calculating the accuracy of the different metrics in rank ordering the behavioral performance of two randomly sampled observers based on their neural activity (190 simulation trials). Table 1 shows the mean percent correct rank ordering of two random observers using either EEG or fMRI. Rank ordering of behavioral performance produced statistically significant results (chance=0.5) consistent with the correlation for both EEG (proportion correct=0.84) and fMRI (proportion correct=0.75).

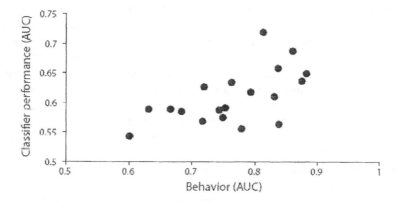

Fig4: fMRI measures using pattern classifiers. Scatter plot showing AUC of pattern classifiers vs. the behavioral performance (AUC) of all 20 observers

Finally, we analyzed the EEG signals in the temporal domain by dividing the signals into 40 ms intervals starting from 200 ms before the stimulus was presented to 480 ms post-stimulus. Each of the 40 ms time epochs was separately fed to the pattern classifier. The correlation coefficient between the pattern classifier performance and the observers' behavioral performance for each 40 ms temporal window starting at 200 ms pre-stimulus onset is shown in Figure 3b. The correlation was negligible pre-stimulus onset and it increased monotonically, showing a rapid increase until about 120 ms, followed by a more gradual increase continuing beyond 400 ms. There was a statistically significant correlation between observers' behavioral performance and the EEG measure obtained from pattern classifier from 170 ms onwards demonstrating that significant information discriminating face and car was coded during that time epoch.

Table 1 The correlation coefficient between behavioral data and EEG measures and rank ordering using pattern classifiers under for 20 observers

Modality	Correlation Coefficient	Rank Order (proportion correct)
EEG	0.69 (±0.15)	0.84(±0.01)
fMRI	0.66 (±0.17)	0.75(±0.09)

DISCUSSION AND CONCLUSION

Recent EEG and fMRI studies have shown that pattern classifiers can be used to infer from neural activity the identity of visual stimuli presented to humans for objects, faces (Philiastides & Sajda 2006) and orientation of simple patterns (Kamitani & Tong 2005). The current study extends previous work by applying pattern classifiers to EEG and fMRI data to predict variations in perceptual performance across a large number of individuals in a categorization task (faces vs. cars). We show comparable ability to predict observers' performance using fMRI and EEG signals in conjunction with pattern classifiers, demonstrating the convergence in predictive power across neuroimaging and electrophysiology modalities. We also demonstrated that the efficacy of EEG activity in predicting the performance across individuals increased 120 ms post stimulus presentation and continued for more than 400 ms. Our finding is consistent with the existing studies (Philiastides & Sajda 2006) using similar stimuli and demonstrates the presence of both early and late components which contain significant information that correlates with observers' behavioral performance.

Together, our results show the potential of neural measures to infer an observer's knowledge about a briefly presented perceptual stimulus. These neural metrics could potentially be used to rapidly evaluate the quality of a visual display. For example, a recent study (Luo & Sajda 2006) has shown that during a very rapid sequence of image presentation, EEG in conjunction with the mean response time, obtained from behavioral response of observers, can be used to estimate the target onset time more accurately compared to using only behavioral response. In addition, neural activity could allow for a measure of an individual's knowledge of the presence of an object for tasks in which an observer is occupied with a different primary task (e.g. piloting a vehicle) and is not explicitly providing a report about the presence/absence of the object of interest.

ACKNOWLEDGEMENTS

Funding for this study was graciously the Institute for Collaborative Biotechnologies through contract no. W911NF-09-D-0001 from the U.S. Army Research .

REFERENCES

Beam, C.A., Conant, E.F. & Sickles, E.A., 2002, Factors affecting radiologist inconsistency in screening mammography. *Acad Radiol.* 9(5):531-40.

Ben-Shachar, M. et al., 2007. Contrast responsivity in MT+ correlates with phonological awareness and reading measures in children. *NeuroImage*, 37(4), 1396-1406.

Chugani, H.T., Phelps, M.E. & Mazziotta, J.C., 1987. Positron emission tomography study of human brain functional development. *Annals of Neurology*, 22(4), 487-497.

DeYoe, E.A. et al., 1996. Mapping striate and extrastriate visual areas in human cerebral cortex. *PNAS,* 93(6), 2382-2386.

Engel, S.A. et al., 1994. fMRI of human visual cortex. *Nature*, 369(6481), 525.

Fisher, R.A., 1936. The use of multiple measurements in taxonomic problems. *Annals of Eugenics*, 7, 179–188.

Grill-Spector, K., Knouf, N. & Kanwisher, N., 2004. The fusiform face area subserves face perception, not generic within-category identification. *Nat Neurosci.*, 7, 555-562.

Harley, E.M. et al., 2009a. Engagement of fusiform cortex and disengagement of lateral occipital cortex in the acquisition of radiological expertise. *Cerebral Cortex*, 19(11), 2746-2754.

Kamitani, Y. & Tong, F., 2005. Decoding the visual and subjective contents of the human brain. *Nat Neurosci.*, 8, 679-685.

Kanwisher, N., McDermott, J. & Chun, M.M., 1997. The Fusiform Face Area: A Module in Human Extrastriate Cortex Specialized for Face Perception. *J. Neurosci.*, 17(11), 4302-4311.

Kourtzi, Z. & Kanwisher, N., 2000. Cortical regions involved in perceiving object shape. *The Journal of Neuroscience*, 20(9), 3310-3318.

Lu, Z.L., Williamson, S.J. & Kaufman, L., 1992. Behavioral lifetime of human auditory sensory memory predicted by physiological measures. *Science,* 258(5088), 1668-1670.

Luo, A. & Sajda, P., 2006. Using single-trial EEG to estimate the timing of target onset during rapid serial visual presentation. *IEEE Engineering in Medicine and Biology Society. Conference*, 1, 79-82.

Mangun, G.R. & Hillyard, S.A., 1990. Allocation of visual attention to spatial locations: Tradeoff functions for even-related brain potentials and detection performance. *Perception & Psychophysics*, 47, 532-550.

Philiastides, M.G. & Sajda, P., 2006. Temporal Characterization of the Neural Correlates of Perceptual Decision Making in the Human Brain. *Cereb. Cortex*, 16(4), 509-518.

Sereno, M. et al., 1995. Borders of multiple visual areas in humans revealed by functional magnetic resonance imaging. *Science*, 268(5212), 889-893.

Vogel, E. & Machizawa, M.G., 2004. Neural activity predicts individual differences in visual working memory capacity. *Nature*, 428, 748-751.

Chapter 28

The Motion of Emotion: Affective Influences on Movement Planning and Execution

Christopher M. Janelle[1], Stephen A. Coombes[2], Kelly M. Gamble[1]

[1]Performance Psychology Laboratory
University of Florida
Gainesville, FL 32653, USA

[2]Department of Psychiatry
Department of Kinesiology and Nutrition
University of Illinois at Chicago
Chicago, IL 60612, USA

ABSTRACT

Critical errors in movement planning and control compromise the efficiency and effectiveness of human motor behavior, leading to fatigue, frustration, injury, and performance failure. Our objective is to summarize how the arousal and valence components of emotional reactivity influence the motor parameters that underlie simple and complex functional behaviors. We have determined that (1) exposure to unpleasant emotional conditions leads to faster, more forceful ballistic upper extremity movements compared to pleasant and neutral conditions, (2) greater emotional arousal increases force production while attenuating force decay of movements that are sustained at moderate force levels, and (3) emotion-modulated force production is qualified by individual differences in anxiety and depression. In addition to modulation of upper extremity movements, we have also recently demonstrated that (4) emotions influence whole body movements, particularly gait initiation. Intense unpleasant emotional states accelerate initial motor responses, but pleasant emotions facilitate the initiation and maintenance of forward gait as evidenced by step length, stride length, and step velocity. Implications are discussed concerning how the compatibility of emotional reactions and directional movements can be coupled to facilitate execution of goal directed actions. The potential for manipulating affective input to enhance human motor performance and risk mitigation will be highlighted by considering how a computational neuroergonomics approach might be used to model the impact of emotional reactivity on motor function.

Keywords: Emotional arousal, valence, motor function, motor control, performance, approach and avoidance, motivational direction

INTRODUCTION

Empirical and theoretical efforts to understand the psychological attributes that enable proficient motor performance have largely neglected advancements in the study of emotion. More critically, *how emotional reactions* are translated to *emotional actions* has been scarcely examined (deGelder, 2004). In many domains, high level performance mandates the coordination and execution of skilled movements. Herein, we briefly highlight recent efforts to wed mainstream affective science with traditional and contemporary questions that have been confronted by performance psychologists. These novel perspectives can help inform unanswered questions concerning how the fundamental movement parameters that are necessary for completion of virtually all functional voluntary activities are altered under different affective contexts In reviewing this work, we aim to provide considerations for alternative approaches and design options that will advance the collective understanding of human performance under inherently fluctuating emotional conditions, while also providing some implications for how extant issues could be addressed through computational neuroergonomics.

EMOTION AND MOTOR FUNCTION: AN INTEGRATED APPROACH

Human emotion and motor action are largely intertwined and reciprocally interrelated. A key issue among contemporary affective scientists concerns the interaction between motivational priming and the direction of an intended movement. In general, unpleasant emotions (excluding anger, cf., Gable & Harmon Jones, 2008) activate defensive brain circuitry, which prime avoidance behaviors and facilitate movements away (i.e., lever pushing / arm extension) from the body. Pleasant emotions, in contrast, activate appetitive brain circuits that prime approach behaviors and facilitate movements toward (i.e., lever pulling / arm flexion) the body (e.g., Cacioppo, Priester, & Berntson, 1993; Chen & Bargh, 1999; Centerbar & Clore, 2006). Researchers have also manipulated movements themselves and examined the influence of motor perturbations on emotional experience. Such manipulations have yielded strong evidence that preferences and attitudes are modulated by inhibition, alteration, and exaggeration of motor action (Niedenthal, 2007). Seminal work in our lab and elsewhere has demonstrated that activation of appetitive and defensive brain circuitry modulates the speed (Coombes, Cauraugh, & Janelle, 2007a, 2007b; Coombes, Janelle, & Duley, 2005), and accuracy (Coombes et al., 2008), of upper extremity movements and whole body actions, which are briefly overviewed next.

EMOTION AND UPPER EXTREMITY MOVEMENT EXECUTION

Movements shorter in duration than the typical reach to grasp (i.e., ballistic movements) primarily rely on the planning system for successful execution (Glover, 2004). Emotional states alter the manner by which short duration ballistic movements are executed (Coombes et al., 2007a). Movements made after presentation of unpleasant stimuli (visual images) exhibit a greater rate of change of force production and faster premotor RT's compared to pleasant conditions. Moreover, discrete emotional reactions influence movement quality above and beyond the broad dimensions of pleasant and unpleasant affect. Specifically, Coombes et al. (2007b) demonstrated that threat inducing images would prime extension movements more so than disgust inducing images. Exposure to threat images led to speeded RT, driven by accelerated premotor time. Thus, perceived threat enhances the rapidity by which extension movements are executed when speed is the primary goal of the movement, suggesting that motivational direction and affective valence differentially modulate movement. Such effects seem to be qualified by dispositional differences in emotional susceptibility. Recent work (Coombes, Higgins, Gamble, Cauraugh, & Janelle, 2008) has demonstrated that individuals who are comparatively more anxious are susceptible to compromised movement efficiency under highly emotionally arousing conditions when performing low level target force contractions. Emotions clearly impact the motor planning processes that precede execution, but emotional state has also been shown to affect control processes. As movements progress over time, execution is increasingly regulated by the control system. Adapting a popular protocol in the motor control literature, we (Coombes et al., 2008) found that subjects who were required to sustain low level forces produced less force decay over time when executing a movement during intense emotional states. Variability of movement execution at these low level forces was not altered by emotional priming. Our recent work has shown that the ability to sustain such forces under different emotional conditions is reliably altered by dispositional depression (Gamble, Coombes, & Janelle, in review).

EMOTION AND WHOLE BODY MOVEMENT

In addition to studying the upper extremity, we (Gamble, Joyner, Hass, & Janelle, in review) have also sought to determine how pleasant and unpleasant emotional states impact lower body kinematics and whole body movements. Such tasks as forward gait initiation (GI) and walking provide unambiguous tasks with which to investigate compatibility between emotional responses and directional movements. With regard to the initiation of forward gait, three notable findings have emerged: (1) highly arousing *unpleasant* conditions accelerate reaction time during GI (see Figure1), (2) as the time course of movement unfolds, *pleasantly* valenced emotional conditions (relative to unpleasant) clearly facilitate the anticipatory postural adjustments and center of pressure transfer that permit larger and higher velocity stepping movements during forward gait (see Figure 2), and (3) encountering stimuli that motivate disgust emotional responses significantly shorten stride length and step velocity. These findings demonstrate that highly arousing unpleasant emotional states accelerate the initial motor response, but pleasant emotional states facilitate the initiation of forward gait due to the approach-oriented directional salience of the movement. That high arousing unpleasant pictures (attack) speeded reaction times on the GI task may seem counterintuitive given that forward gait is an unambiguous approach oriented movement. From an evolutionary perspective, however, speeded initial responses are likely rooted in abrupt feedforward preparations to support immediate fight or flight defensive reactions upon detecting threat cues. Moreover, this apparently paradoxical result supports the notion that faster movements, regardless of approach or withdrawal orientation, are

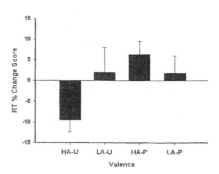

Figure 1. Mean and SE percent change scores for the reaction time data across valence conditions. Emotional conditions are indicated by slide type (HA-U = High arousal unpleasant; LA-U = Low arousal unpleasant; HA-P = High arousal pleasant; LA-P = Low arousal pleasant).

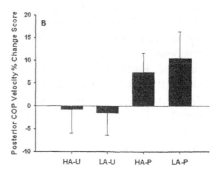

Figure 2. Mean and SE percent change scores across the valence conditions for the *velocity* of the COP data in the posterior direction during the S1 time period. Emotional conditions are indicated by slide type (HA-U = High arousal unpleasant; LA-U = Low arousal unpleasant; HA-P = High arousal pleasant; LA-P = Low arousal pleasant).

primed in threatening situations. Such findings are also consistent with our prior work investigating comparatively simple nondirectional and extension oriented upper extremity movements (e.g., Coombes, Janelle, & Duley, 2005). These results, however, also highlight a critical limitation of prior work; namely, that reaction time provides a rudimentary, narrow, and potentially misleading measure of the impact of emotional experience on approach / withdrawal motor function. Process oriented measures that capture the neural mechanisms that underlie performance alterations over extended durations of time are necessary. Neuroergonomics offers other avenues to evaluate such processes, and to do so at a deeper level than can be achieved through behavioral analysis.

IMPLICATIONS FOR COMPUTATIONAL NEUROERGONOMICS

According to Parasuraman (2003), the primary goals of neuroergonomics are to generate knowledge of the neural basis of mental function to design safer and more efficient technologies and work environments, while also developing a finer understanding of the brain mechanisms that underlie real world performance. The increased sophistication afforded by modern measurement tools has permitted researchers to peer into the inner workings of the brain with unparalleled vividness in pursuit of answers consistent with each of these objectives. Such efforts have led to unique and timely discoveries concerning the attentional mechanisms that enable performance proficiency, with a robust corpus of work indicating reliable psychophysiological differences in among relative experts and non experts (see Janelle & Hatfield, 2008, for review). Seminal findings inferred from spectral EEG and ERP recordings have been recently supported by fMRI work (e.g., Milton et al., 2004, Krakauer et al., 2009) that has revealed the likely cortical seats that afford the expert advantage. Unquestionably, emotions impact attentional allocation. Appropriate attention allocation cannot be overstated in determining performance outcome. Studies of the neurological basis of attentional concepts can be readily assessed using current technology, and are routinely employed in the assessment of expertise in sport and other performance skills (Janelle & Hatfield, 2008). Assessment of the neurological basis of how emotion affects *movement*, however, is inherently difficult. We contend, however, that those who study high level performance / productivity and risk mitigation must consider another imperative seat of performance variability; namely how emotions affect the parameters that yield overt motor action.

The behavioral level of analysis has clearly dominated our recent work and we have deliberately taken a behavioral approach given the vast experiential and physiological database that exists to account for emotional reactivity. Upon considering the two primary missions of neuroergonomics, however, it is clear that studying the neurological basis of emotion modulated movement to inform these aims is particularly challenging. Current assessment tools that are used to assess constructs such as *attention* and *emotion*, for example, often cannot be readily implemented to understand *movement* due to artifacts induced by movement itself. As such, scientists must be careful to evaluate movements that have practical relevance and implications for real world performance, but can be studied in a manner that is unmaligned by movement artifact.   Favored tasks include movements such as pinch and power grips, wrist extension and flexion, lever pushing and pulling, and most recently, lower extremity movements and postural adjustments. While seemingly simple, such tasks embody real world significance, particularly with regard to workplace tasks. Deficits in the planning and control of movement result in a high rate of accidental injury and death in workplaces. Close examination of epidemiological statistics reveals that an alarmingly high rate of accidental injuries and deaths occur in occupations necessitating precise movement accuracy and control [USDOL, 2006], particularly transportation incidents, contact with objects and equipment, and falls. Understanding the neurological basis of errors in motor planning and execution will undoubtedly aid in curbing this accident rate.

Recognizing the implications of understanding interactions of emotion and motor function, scientists have begun to investigate these interactions while considering their neurological underpinnings. For instance, recent evidence has shown that emotion alters excitability of the corticospinal motor tract (Coombes et al., 2009; Hajcak et al., 2007). Related tracing studies in animals and functional magnetic resonance imaging in humans extend these findings by demonstrating that the prefrontal cortex and basal ganglia (which each project to motor cortex)

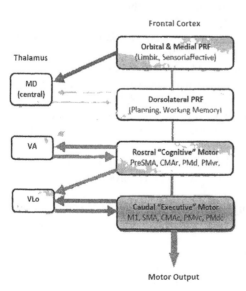

Figure 3.   Proposed schema of information flow between thalamic relay nuclei and frontal cortical areas. Thalamic areas central MD, VA, and VLo are depicted on the left and the corresponding prefrontal, premotor and motor cortical areas on the right. Black lines between cortical regions demonstrate the diverse corticocortical interconnections between adjacent frontal cortical areas. Colored gradients in boxes indicate the functional association between particular thalamic and frontal cortical areas (from most limbic, red, to motor, blue). Arrows illustrate the major thalamocortical and corticothalamic connections between areas. Information is transmitted in a feedforward manner through strong reciprocal thalamocortical-thalamic connections and prominent nonreciprocal corticothalamic inputs from more rostral, cognitive, or limbic association areas. Adapted from McFarland & Haber, (2002). Thalamic relay nuclei of the basal ganglia form both reciprocal and nonreciprocal cortical connections, linking multiple frontal cortical areas. *Journal of Neuroscience, 22,* 8117-8132.

integrate emotion and motor processes (Doron & Goelman, 2010; McFarland & Haber, 2002, see Figure 3).

Emerging evidence strongly suggests that the specific basal ganglia nuclei that inhibit movement (Aron & Poldrack, 2006) or control executed movement (Vaillancourt et al., 2004) are also the same areas that integrate emotion and motor processes. Brain imaging work in humans is underway to examine this very issue and computational neuroergonomics can play a prominent role in achieving this objective by modeling the cortical and subcortical networks that are responsible for integration of emotion and motor behavior.

REFERENCES

Aron A.R., & Poldrack, R.A. (2006) Cortical and subcortical contributions to Stop signal response inhibition: role of the subthalamic nucleus. *Journal of Neuroscience, 26,* 2424-2433.

Bureau of Labor Statistics, U.S. Department of Labor, *Nonfatal occupational injuries and illnesses.* 2006, Washington, D.C.

Bureau of Labor Statistics, U.S. Department of Labor, *National census of fatal occupational injuries in 2005.* 2006, Washington, D.C.

Cacioppo, J., Priester, J.R., & Berntson, G.G. (1993). Rudimentary determinants of attitudes. II: Arm flexion and extension have differential affects on attitudes. *Journal of Personality and Social Psychology, 65,* 5-17.

Chen, M., & Bargh, J. A. (1999). Consequences of automatic evaluation: Immediate behavioral predispositions to approach or avoid the stimulus. *Personality and Social Psychology Bulletin, 25,* 215-224.

Centerbar, D.B. & Clore, G.L. (*2006).* Do Approach-avoidance actions create attitudes? *Psychological Science, 17,* 22-29.

Coombes, S. A., Cauraugh, J. H., & Janelle, C. M. (2007a). Dissociating motivational direction and affective valence: specific emotions alter central motor processes. *Psychological Science, 18,* 938-942.

Coombes, S. A., Cauraugh, J. H., & Janelle, C. M. (2007b). Emotional state and initiating cue alter central and peripheral motor processes. *Emotion, 7*(2), 275-284.

Coombes, S.A., Janelle, C.M., & Duley, A.R. (2005). Emotion and motor control: Movement attributes following affective picture processing. *Journal of Motor Behavior, 37,* 425-436.

Coombes, S. A., Gamble, K. M., Cauraugh, J. H., & Janelle, C. M. (2008). Emotional states alter force control during a feedback occluded motor task. *Emotion, 8*(1), 104-113.

Coombes, S. A., Tandonnet, C., Fujiyama, H., Janelle, C. M., Cauraugh, J. M., & Summers, J.J. (2009). Emotion and motor preparation: A transcranial magnetic stimulation study of corticospinal motor tract excitability. *Cognitive, Affective, and Behavioral Neuroscience.*

de Gelder, B., Snyder, J., Greve, D., Gerard, G., & Hadjikhani, N. (2004). Fear fosters flight: a mechanism for fear contagion when perceiving emotion expressed by a whole body. *Proc Natl Acad Sci U S A, 101*(47), 16701-16706.

Doron O, Goelman G (2010) Evidence for asymmetric intra substantia nigra functional connectivity—application to basal ganglia processing. *NeuroImage, 49,* 2940-2946.

Gable, P.A., & Harmon-Jones, E. (2008). Approach-motivated positive affect reduces breadth of attention. *Psychological Science, 19,* 476-482.

Gamble, K.M., Coombes, S.A., & Janelle, C.M. (2010).Individual differences in subclinical depression modulate the impact of emotion on force control at low and moderate target force levels. Manuscript in review.

Gamble, K.M., Joyner, J.A., Coombes, S.A., Hass, C.J., & Janelle, C.M. (2010). Emotional state affects the initiation of forward gait. Manuscript in review.

Glover, S. (2004). Separate visual representations in the planning and control of action. *Behav Brain Sci, 27*(1), 3-24; discussion 24-78.

Hajcak, G., Molnar, C., George, M. S., Bolger, K., Koola, J., & Nahas, Z. (2007). Emotion facilitates action: a transcranial magnetic stimulation study of motor cortex excitability during picture viewing. *Psychophysiology, 44*(1), 91-97.

Janelle, C. M., & Hatfield, B. (2008). Visual attention and brain processes that underlie expert performance: Implications for sport and military psychology. *Military Psychology, 20,* 117-134.

McFarland, N.R., & Haber, S.N. (2002). Thalamic relay nuclei of the basal ganglia form both reciprocal and nonreciprocal cortical connections, linking multiple frontal cortical areas. *Journal of Neuroscience, 22,* 8117-8132.

Milton, J.G., Small, S.S., & Solodkin, A. (2004). On the road to automatic: Dynamic aspects in the development of expertise. *Journal of Clinical Neurophysiology, 21,* 134-143.

Niedenthal, P.M., *Embodying Emotion.* Science, 2007, *316,* 1002-1005.

Vaillancourt DE, Mayka, M.A., Thulborn, K.R., Corcos, D.M. (2004). Subthalamic nucleus and internal globus pallidus scale with the rate of change of force production in humans. *Neuroimage, 23,* 175-186.

Wolpert D.M. (2007). Probabilistic models in human sensorimotor control. *Human Movement Science, 26,* 511-24.

Yarrow, K., Brown, P., Krakauer, J.W. (2009). Inside the brain of an elite athlete: the neural processes that support high achievement in sports. *Nature Reviews Neuroscience, 10,* 585-596.

Chapter 29

South Side of the Sky: Evidences for Two Different Scanning Strategies in the Upper and Lower Visual Fields

Francesco Di Nocera[1], Marco Camilli[1], Amelia D'Arco[1], Sabrina Fagioli[2]

[1]Department of Psychology
Sapienza University of Rome, Italy

[2]CeNCA - Applied Cognitive Neuroscience Center
Rome, Italy

ABSTRACT

The aim of this study was to investigate the hypothesis that the relative position of the stimuli within the lower and the upper visual fields influences the ocular scanning of a dynamic real-time scene. According to Previc, distinct brain systems code information coming from different spatial regions. In particular, Previc describes a Peripersonal space and an Extrapersonal space, suggesting that they are involved in vertical visual field asymmetry effects. Here we verified that functional segregation of the space affects the ocular behavior, by testing the hypothesis that fixation scattering in the upper and the lower visual hemi-field reflects the functional asymmetry of neural processes. The NNI provided an index of the total scattering of the fixations as a function of the visual hemi-field. This suggest that when cognitive requirements are high, subjects adopt the strategy of a global exploration of the space; in contrast, in a low cognitive requirement situation, subjects adopt a more punctual strategy consisting of exploring larger portion of space in the lower hemi-field than the ones in the upper hemi-field.

Keywords: Eye-Tracking, Visual Field Asymmetries, Mental Workload, Fixations Distribution

INTRODUCTION

When we explore the visual world, we get the impression of a coherent, uniform space. Nevertheless, several behavioral and neurophysiological studies demonstrated that the human performance changes depending on the position of the objects in the space. For instance, it has been showed (Danckert & Goodale, 2001; Khan & Lawrence, 2005) that the performance to visual search and recognition tasks improves when the stimuli are presented in the upper visual field (UVF), whereas the performance to visuo-motor coordination tasks takes an advantage when the stimuli occur in the lower visual field (LVF). Moreover, Genzano ct al. (2001) showed that object relocation is performed better when objects are presented in the lower than in the upper visual field. The evidence of vertical visual field asymmetries strongly supports the hypothesis that the human brain codes the space in a modular, rather than uniform fashion.

Previc (1990) accounted for these effects identifying distinct brain systems that deal with information coming from peripersonal and extrapersonal space. Specifically, the peripersonal space would be involved in reaching and manipulative behaviors, being controlled by the dorsal visual system. Conversely, the extrapersonal space would control search and scanning tasks, as well as navigation and orientation to the environment, and would use input from the ventral visual system. Previc described a vertical bias in these systems due to the ecological relationship of the lower and upper portion of the visual field to the near and far space, respectively. Indeed, most manipulative behaviors in the near space occur below eye-level, whereas most information coming from the far space is located in the upper hemi-field.

Despite much empirical research has been provided showing an impact of the functional segregation of the space on the cognitive processing, it is still unknown how the vertical visual field asymmetries affect the gaze behavior.

Here, we addressed the hypothesis that the relative position of the stimuli within the LVF and the UVF influences the scanning strategy of a dynamic real-time scene. We investigated the cognitive workload associated to the visual scene in order to reveal specific scan patterns in relation to vertical visual field asymmetry.

A simple visuo-motor task inspired to the PC-based game known as "Asteroids" was used as experimental task. Participants were required either to avoid collision with the asteroids without shooting them (high task load condition: hard) or to avoid collision with the asteroids firing them as they preferred (lower task load condition: easy). Details about this task and its relation to eye scanning patterns can be found in Di Nocera et al. (2006). The type of spatial distribution produced by the fixations' pattern was computed with a spatial statistics algorithm called the Nearest

Neighbor Index (NNI; Clark & Evans, 1954). Recent studies, most of them carried out in our lab (e.g. Di Nocera et al., 2007; Camilli, et al. 2007; Fidopiastis, 2009), showed that the NNI is sensitive to mental workload variations. Di Nocera and Bolia (2007) proposed that two processes might respectively contribute to dispersion and grouping in visual scanning behavior: temporal demand and visuo-spatial demand. This hypothesis was corroborated in a specific study where the temporal and the visuo-spatial mental workload components were isolated in two different task conditions (see Camilli et al., 2008). Results showed that the spatial distribution of eye fixation was more dispersed (i.e. higher NNI values) when the temporal demand was the most loading workload component rather than when workload was moderate. Differently, when the most loading workload component was the visuo-spatial demand, fixations spatial distribution was more grouped (i.e. less random, lower NNI values). Briefly, the NNI showed to be a suitable mental workload indicator for discriminating both different levels and different types of task demands. Furthermore, the index was also used over fixations recorded during small epochs of time showing its suitableness for detecting rapid changes (about 1 min) in the level of mental load experienced by the individuals. Finally, "A Simple Tool for Examining Fixations" (ASTEF: Camilli et al., 2008) was developed in our lab with the aim of providing companion software for encouraging the use of this novel method by the Human Factors / Ergonomics community.

STUDY

METHODS

Participants

Twenty individuals (10 males) aged 21-26 years (mean age 21.85) volunteered in this experiment. All participants were right-handed with normal or corrected-to-normal vision.

Apparatus and Stimuli

Participants were asked to play the "Asteroid" computer game (see figure 1). Two different difficulty levels were produced by varying the cognitive workload: in the low-level difficulty condition (easy) participants were instructed to avoid the asteroids by piloting the space ship using the arrow keys of the computer keyboard, and by using the blank key for shooting at the asteroids; in the high-level difficulty condition (hard) participants avoided the collision with the asteroids only by piloting the space ship. In each experimental condition, twenty-four asteroids were presented on the screen. In order to add dynamism, four of them were programmed to chase the space ship, whereas the remaining asteroids randomly changed their direction. The shooting tool had inter-shot rest period of 2 seconds.

Eye movements' were recorded throughout the game sessions by an infrared-based eye-tracking system and analyzed using ASTEF. The Nearest Neighbor Index (NNI) was considered as a dependent measure.

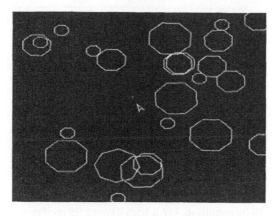

FIGURE 1. The Asteroids game scenario.

Procedure

Participants performed a 10-minute long training session, and underwent a calibration procedure for the eye-tracking. Successively, they were asked to play the two game sessions (easy and hard). Each session lasted 10 minutes with a few minutes break between the two. At the end of each session, participants were required to fill the NASA-TLX. The Easy-Hard conditions order was balanced across participants.

RESULTS

The ANOVA design Task Load x Hemi-field (UVF vs. LVF) carried out using NNI scores as dependent variables showed a significant main effect of Task Load ($F_{1,19}$=5.2 p<.01) due to the fixations being more dispersed in Hard rather than in the Easy condition (see figure 2).

The NNI scores used in the previous analysis were computed using all the available fixations and dividing them according to the hemi-field in which they were located. Nevertheless, it should be considered that a meaningful pattern of fixations is one that includes only successive fixations. For this reason, another analysis was carried out using only the patterns composed of consecutive fixations within the UVF and the LVF. Given that these sequences always included less than 50 fixations, computing the NNI was strongly discouraged. In this case, the convex hull (the area of the polygon defined by the outermost fixations) was used instead.

Results showed a significant Task Load x Hemi-Field interaction ($F_{1,19}$=8.8 p<.01; figure 3). Duncan post-hoc testing showed that this interaction was due to a difference in the Easy condition in which the dispersion of fixations was larger in the Lower rather than in the Upper Visual Field.

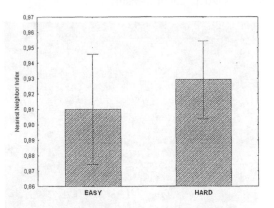

FIGURE 2. NNI by condition. Error bars denote .95 confidence intervals.

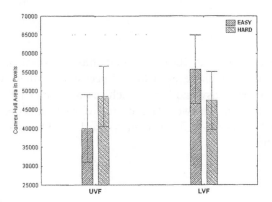

FIGURE 3. Extension of the area scanned by condition and visual field. Error bars denote .95 confidence intervals.

The space ship movements in the two hemi-fields were recorded during the task and they were analyzed in terms of behavioral strategies performed by participants. The number of the ship transitions between the two hemi-fields was used as dependent variable in a repeated measures ANOVA design using the Task Load (Easy vs. Hard) as repeated factor. Results showed a larger number of transitions performed in the more loading task load condition ($F_{1,19}$=8.6, p<.01). This result was indeed expected, because without firing (i.e. in the harder condition), participants had to maximize the avoidance strategy by continuously moving the spaceship.

The proportion of time spent by the spaceship in the two hemi-fields was used as dependent variable in a Hemi-Field (UVF vs. LVF) x Task Load (Easy vs. Hard) ANOVA design. Results showed a significant interaction between the hemi-field and the task load condition ($F_{1,19}$=9.4, p<.01; figure 4). During the harder task condition, the ship spent more time in the UVF rather than in the LVF. Conversely, the space ship spent more time in the LVF during the easier task condition. Overall, the ship spent more time in the lower visual hemi-field ($F_{1,19}$=140.3, p<.01).

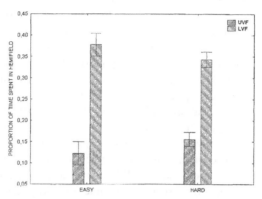

Figure 4. Time spent by the ship separately for hemi-field and for task load. Error bars denote .95 confidence intervals.

DISCUSSION AND CONCLUSIONS

The aim of this study was to investigate the hypothesis that the relative position of the stimuli within the LVF and the UVF influences the scanning strategy of a dynamic real-time scene. According to Previc (1990), distinct brain systems code information coming from different spatial regions. In particular, Previc describes a Peripersonal space and an Extrapersonal space, suggesting that these spaces are involved in vertical visual field asymmetry effects.

Here we verified that the functional segregation of space affects the ocular behavior, by testing the hypothesis that fixation scattering in the upper and the lower visual hemi-field reflects the functional asymmetry of neural processes.

The cognitive workload had an impact on the Nearest Neighbor Index (NNI) that showed higher values in the Hard condition, confirming that that a temporal demand increment was associated to more dispersed distributions of fixations. The distribution of fixations also varies as a function of the hemi-field. Particularly, when cognitive requirements are high, subjects adopt the same ocular strategy independently from the hemi-field; in contrast, in a less engaging situation, subjects show a concentration of fixations in the UVF (suggesting lower mental workload) and a spread of fixations in the LVF (suggesting higher mental workload).

Considering that the ocular strategy used by an individual reflects the amount of resources invested in a task, these results suggest that when the temporal demand increases (i.e. during the harder task condition) participants maximize their visual search by enlarging the region of visual exploration, thus eliminating the asymmetries between hemi-fields. The analyses on gaming strategies, instead, showed that participants preferentially located the spaceship in the lower rather than in the upper hemi-field. This finding is reasonable because, in this type of task, the workload could be moderated by using active controlling strategies (i.e. firing the asteroids) that should be functionally allocated in the lower hemi-field (because of the peripersonal space advantages in handling).

With that in mind we may conclude that these data confirm a functional specialization of the lower visual hemi-field for visuo-motor-search tasks that also affect eye scanning. This is consistent with the theory of Previc concerning a functional segregation of the 3D space and provides hints for design. For example, placing visual indicators and controls (e.g. gauges) that require minimum cognitive effort for the operator in the upper part of the visual field may minimize the risk of procedural errors, due to the specialization of this hemifield for visual scanning tasks. Conversely, arranging controls that afford visuo-motor responses (e.g. button and switches) in the lower visual hemi-field may be a viable strategy for reducing operators' errors, because of the functional specialization of the lower hemi-field for manipulative tasks. Future studies will be aimed at exploring the advantages of particular layouts designed to account for the functional segregation of space.

REFERENCES

Camilli, M., Nacchia, R., Terenzi, M., & Di Nocera, F. (2008). ASTEF: A simple tool for examining fixations. *Behavior Research Methods*, 40(2), 373-382.

Camilli, M., Terenzi, M., & Di Nocera, F. (2007). Concurrent validity of an ocular measure of mental workload. In D. de Waard, G.R.J. Hockey, P. Nickel, and K.A. Brookhuis (Eds.), *Human Factors Issues in Complex System Performance* (pp. 117-129). Maastricht, the Netherlands: Shaker Publishing.

Camilli, M., Terenzi, M., & Di Nocera, F. (2008). Effects of Temporal and Spatial Demands on the Distribution of Eye Fixations. *Proceedings of the Human Factors and Ergonomics Society Annual Meeting*, 52(18), 1248-1251.

Clark, P. J., & Evans, F. C. (1954). Distance to nearest neighbor as a measure of spatial relationships in populations. *Ecology*, 35, 445-453.

Danckert, J., & Goodale, M.A. (2001). Superior performance for visually guided pointing in the lower visual field. *Experimental Brain Research*, 137, 303-308.

Di Nocera, F., & Bolia, R.S. (2007). PERT networks as a method for analyzing the visual scanning strategies of aircraft pilots. *Proceedings of the 14th International Symposium on Aviation Psychology* (pp. 165-169). Dayton, OH: Wright State University.

Di Nocera, F., Terenzi, M., & Camilli, M. (2006). Another look at scanpath:

distance to nearest neighbour as a measure of mental workload. In D. de Waard, K.A. Brookhuis, A. Toffetti (Eds.), *Developments in Human Factors in Transportation, Design, and Evaluation* (pp. 295-303). Maastricht, the Netherlands: Shaker Publishing.

Di Nocera, F., Camilli, M., & Terenzi, M. (2007). A random glance to the flight deck: pilot's scanning strategies and the real-time assessment of mental workload. *Journal of Cognitive Engineering and Decision Making*, 1(3), 271-285.

Fidopiastis, C.M., Drexler, J., Barber, D., Cosenzo, K., Barnes, M., Chen, J.Y.C., & Nicholson, D. (2009). Impact of Automation and Task Load on Unmanned System Operator's Eye Movement Patterns. In D.D. Schmorrow et al. (Eds.): *Augmented Cognition, HCII 2009, LNAI 5638* (pp. 229–238), Berlin Heidelberg: Springer-Verlag.

Genzano, V.R., Di Nocera, F., & Ferlazzo, F. (2001). Upper/lower visual field asymmetry on a spatial relocation memory task. *NeuroReport*, 12(6), 1227-1230.

Khan, M.A., & Lawrence, G.P. (2005). Differences in visuomotor control between the upper and lower visual fields. *Experimental Brain Research*, 164, 395–398.

Previc, F.H. (1990). Functional specialization in the lower and upper fields in humans: Its ecological origins and neurophysiological implications. *Behavioral Brain Science*, 13, 519-575.

Chapter 30

Visual Attention in a Team C2 Change Blindness Task

*Benjamin A. Knott[1], Gregory J. Funke[1], Camilla C. Knott[2],
Brent T. Miller[1], April Rose Panganiban[1]*

[1]Air Force Research Laboratory
Wright-Patterson Air Force Base, OH 45433, USA

[2]Aptima, Inc., Fairborn, OH 45324, USA

ABSTRACT

To date, little research has been conducted on change blindness (CB) in complex military environments even though substantive differences exist in the context, meaning, and consequences of changes in those settings compared to other research environments. The current experiment represents an attempt to characterize the relationship between CB and several military-relevant task factors. Single-task participants were required to perform a change detection task which occurred in the context of an ongoing air defense task. Dual-task participants completed the change detection task while also managing the air defense task. The results suggest that CB may occur at a relatively high rate in military settings, and that speed of detection may be influenced by practice and context. Results of eye tracking analyses suggest that such tracking may be useful triggers for automated CB mitigation. Overall, the results of the experiment suggest that CB in military environments is relatively complex and may be difficult to eliminate.

Keywords: Change blindness, visual attention, teams, command and control

INTRODUCTION

Change detection involves both localization and identification of change. Change blindness (CB) is the lack of awareness of considerable changes in a visual scene when the change occurs simultaneous to visual transients or distractions (e.g., eye movements, Rayner & Pollatsek, 1983; screen flashes, Rensink, O'Regan, & Clark, 1997). Visual transients that are naturally produced by change in the environment, such as motion, signal the visual system that a change occurred by drawing attention to the source of the change, thereby allowing the change to be incorporated into observers' mental representations of the environment. CB is generally attributed to the absence of attention from the source of a change (Rensink, et al., 1997; Simons & Levin, 1997). Comparisons of a scene before attention is diverted from, and after its return to, the source of a change may be limited by capacities of memory and attention (Rensink et al., 1997; Simons & Levin, 1997).

Recent research suggests that operators in military command and control (C2) environments may be particularly susceptible to change blindness due to the attentional demands and complex nature of their duties (e.g., DiVita, Obermayer, Nugent, & Linville, 2004; Durlach, 2004b). For instance, air battle management (ABM) teams rely on information-rich tactical displays in developing and maintaining common awareness of a battlespace. These visual displays include operational boundaries, terrain data, and dynamic information updates concerning the locations of friendly and enemy assets. In addition to situation assessment, operators engage in voice and chat communications and manipulate separate radio consoles, requiring operators to shift attention across multiple displays. The multi-task nature of ABM is likely to increase the risk of CB occurrence. Indeed, previous research suggests that CB is exacerbated under dual- or multi-task conditions (e.g., McCarley et al., 2004; Muthard & Wickens, 2003).

Effective ABM relies on teams of experienced operators to detect critical changes in a battlespace and communicate those changes to other operators in the field. The occurrence of CB in this and similar environments (e.g., air traffic control, emergency response management, etc.) could be catastrophic. The sudden disappearance of a friendly aircraft symbol from a tactical display may signal a crash. Operator failure to immediately detect this change may affect the success of subsequent search and rescue missions. Additionally, CB by one or more members of a team may affect common awareness of the tactical battlespace and lead to potentially devastating outcomes caused by delayed consensus and decision making. While there is a considerable literature on the effects of various stimuli on detection, few studies have investigated CB in tasks in which failure to detect changes has important consequences to the observer(s). This is particularly relevant to ABM where consequences are likely to drive behaviors and strategy that may influence CB.

Moreover, certain CB paradigms such as 'flicker' or 'mudsplash' tasks ignore top-down influences on detection such as the relevance and meaning of changing stimuli. Durlach, (2004a) noted that in addition to the perceptual characteristics of a stimulus, the meaningfulness of the stimuli may affect the likelihood of detection. Hollingworth & Henderson (2000) also noted that changes affecting meaning are detected faster than changes that do not. Together, these findings suggest that meaningful tasks with specified outcomes, e.g., domain-specific tasks, may reduce CB (DiVita et al., 2004), and domain experts are more likely than novices to know when and where task-relevant changes may occur.

Experience may also affect CB performance in ways that are of interest to military operations. Several researchers have shown that prior knowledge and experience may indeed influence CB performance. For example, the mere knowledge that changes are likely to occur during a task reduces CB (Simons & Mitroff, 2000). Werner and Thies (2000) provided evidence that football experts were better than novices at detecting meaning-altering scene changes, suggesting that domain expertise reduces susceptibility to CB.

CB in teams of observers is an area of the research literature that has not been deeply investigated. However, given that CB occurs due to an observer's inattention to the source of a change, is it reasonable to predict that if more than one observer monitors a scene, then the likelihood increases that at least one observer will be attending to the source of the change when it occurs. In one of the few published studies on CB in teams, Knott et al. (2007) found that increasing the number of observers reduced CB, but only if there is a mechanism such as verbal communication to allow observers to coordinating their activities.

THE PRESENT STUDY

The present study employed a novel CB paradigm using realistic ABM simulations. These introduced changes and transients that naturally occur during a mission (e.g., appearance of enemy fighters; interruptions due to communications, etc.) and must be detected to efficiently meet mission demands. Moreover, change detection will have important consequences in terms of team performance. Therefore we endeavored to measure and demonstrate CB effects with a meaningful and operationally relevant task, in which change detection is part of, and relevant to task performance. Analysis of eye-gaze behavior during task performance was also conducted to provide insight into observers' allocation of attention during task performance.

The experiment manipulated team type (single-task, dual-task) experimental session (1-3), task demand (low, high), detection responsibility (individual, team), and amity designation (neutral, hostile).

Based on findings that domain expertise reduces CB (Werner & Thies, 2000; DiVita et al. 2004), and the notion that experience can affect resource allocation, and the way information in a particular domain is interpreted, we hypothesized that change detection performance would improve with practice. Based on findings that CB is exacerbated under dual- or multi-task conditions (e.g., McCarley et al., 2004; Muthard & Wickens, 2003), CB was hypothesized to be worse under dual-task conditions which required operators to manage an air defense operation concurrent with a change detection task. Further, based on the findings of Knott et al. (2007) that increasing the number of observers reduced CB, it was hypothesized that CB would be reduced for team detection events. Finally, based on findings that high workload exacerbates CB (Muthard & Wickens, 2003), CB was hypothesized to be relatively poor under high compared to low task demand.

METHOD

PARTICIPANTS AND EXPERIMENTAL DESIGN

Twenty-four people (14 men, 10 women) between the ages of 18 and 30 ($M = 21.17$, $SD = 3.54$) served as participants in this experiment. Participants were students recruited from local universities and were compensated for their participation. The experiment also included three confederates. Confederates were compensated at the same rate as participants. In total, the experimental sample included twelve teams; six dual-task teams, each consisting of two participants and three confederates, and six single-task teams, each consisting two participants.

This experiment utilized a $2 \times 3 \times 2 \times 2 \times 2$ mixed design. The between-subjects factor was team type (single-task, dual-task). Within-subjects factors included experimental session (1-3), task demand (low, high), detection responsibility (individual, team), and amity designation (neutral, hostile). These factors were combined factorially, yielding 48 experimental conditions.

APPARATUS

This experiment employed Aptima, Inc.'s Distributed Dynamic Decision-making (DDD) software (version 3.0; MacMillan, Entin, Hess, & Paley, 2004). DDD provides a scriptable, low-to-moderate fidelity, team-in-the-loop simulated environment. The simulation required eleven networked PCs communicating under TCP/IP protocol. Each workstation was equipped with a keyboard, mouse, and monitor. The DDD tactical display was projected to participants on a 17-inch Dell 1707FPv LCD monitor.

The DDD scenario employed was composed of two interconnected tasks – a simulated air defense task and a change detection task. Participants assigned to dual-task teams were required to complete both DDD tasks simultaneously. Participants on single-task teams only completed the change detection task.

During the experiment, eye tracking data was collected from participants in the dual-task condition during trials 4-8 of each experimental session. This period was selected to provide experimenters adequate eye tracking information with minimal discomfort to participants. Due to temporal constraints, eye tracking data could not be collected from participants in the single-task condition. Eye tracking was accomplished using two ASL model 501 head-mounted eye trackers. The systems measure infrared corneal reflectance and pupil and head position to determine point of gaze. Fixations were points of gaze with maximum change of 2° visual angle for a minimum of 100 ms. Eye movements were recorded at 120 Hz for all participants.

Air Defense and Change Detection Tasks

The air defense task required teams of five to collaboratively complete a simulated C2 mission. Dual-task team composition included two weapons directors (WDs), two strike operators (SOs), and one tanker operator (TO). WDs were always participants, SOs and TOs were always confederates. Single-task teams were not assigned team roles.

Within the simulation, the WDs' responsibilities were to match friendly fighters with appropriate enemy targets, schedule fighters for refueling and resupply, and communicate their plans with other team members. Strike and tanker operators' responsibilities included maneuvering team assets as instructed, engagement of enemy targets, and communication of pertinent information to teammates concerning asset resources. Further information about the air defense task can be found in McClernon et al. (this publication).

The task demand condition of each trial determined the number of enemy targets present in each scenario. To ensure a relatively constant level of demand throughout each trial, new enemy targets would enter the scenario as participants prosecuted enemy aircraft. The low demand condition featured 4 enemy targets and the high task demand condition featured 6.

The change detection task employed in this experiment was modeled on the real world duties of ABM, which include monitoring air traffic as it relates to military tactics. Within a battlespace, unidentified aircraft may be relatively prevalent (given that it is not uncommon for 50 to 100 contacts of interest to be present on tactical situation displays). These aircraft must be monitored for safety reasons, but also for amity information. If these aircraft are identified as enemy targets, WDs must then engage the situation appropriately.

The change detection task employed in this experiment required participants to monitor unidentified aircraft within the simulated battlespace. Unidentified aircraft would enter the simulation from the right side of the screen and could thereafter change into neutral or hostile aircraft (representing an automated amity assessment). Participants were instructed to monitor all unidentified aircraft and upon detection of an amity change, click on the change target's on-screen icon, thereby indicating that they had detected the change. All target icons employed in this experiment were adaptations of Military Standard 2525B symbols (Department of Defense, 1999). In addition, following an amity designation, participants in the dual-task condition were instructed to engage the target appropriately (i.e., engage hostile targets, avoid neutral targets).

Each trial featured a total of 12 change events. These events were evenly divided by detection responsiblity – 6 team events and 6 individual events. Team events could be detected and responded to by either participant. Individual events were configured so they would only appear on one participant's tactical display, and could therefore only be detected and responded to by that participant. This allowed a comparison of individual change detection performance with performance of a team. The amity designations of the team and indivual change events were also evenly divided between neutral and hostile targets, resulting in a total of 3 neutral team targets, 3 hostile team targets, 3 neutral individual targets, and 3 hostile individual targets per trial (the amity of individual targets was counterbalanced across trials to ensure that both participants experienced the same number of neutral and hostile change events).

Aspects of the change detection task were also dependent on team type. As mentioned previously, dual-task teams performed the air defense and change detection tasks simultaneously. However, because of the interrelation of these two tasks, it was not possible to effectively control aspects of the change detection task which could influence likelihood of detections (such as number or density of screen icons, etc.). To overcome these limitations, single-task teams served as yoked-controls to the dual-task teams. Single-task teams were randomly yoked to a dual-task team and then observed recordings of that team's DDD trials. Their task was to watch the recording for the same change events that dual-task teams had experienced and to indicate their detections by clicking the on-screen icons that had changed. The simulation recordings included all DDD events (asset movements, enemy engagements, change events, etc.) minus those associated with dual-task participants' performance of the change detection task (i.e., on-screen mouse clicks).

PROCEDURE

The duration of the experiment was approximately 10 hours, conducted across three experimental sessions. Participants completed an informed consent document, and were given a brief overview of the experiment. They then received written and oral instructions detailing the DDD simulation, the air defense and change detection tasks, and the roles and responsibilities of each team member. Following the instructional period, participants were required to complete a proficiency demonstration, which required them to display adequate team role knowledge and ability. Next, teams completed 3 practice trials to give them an opportunity to collaborate and strategize under conditions identical to those they would experience during experimental trials. These trials, and all subsequent experimental trials, were 12 minutes in duration.

Next, participants completed eight experimental trials. The experimental schedule of conditions was randomized and counterbalanced across teams to control order effects. After completing all experimental trials, participants were dismissed for the day. They were required to return for the two subsequent experimental sessions in which participants began with one practice trial, followed by eight experimental trials.

RESULTS

A full and detailed accounting of the results of this experiment is beyond the scope of this manuscript. As such, the focus of analysis here is on performance of the change detection task and eye fixation data.

During each experimental trial, software recorded the number of correctly detected change events and participants' reaction times to those events. To normalize the distribution of the reaction time data, raw reaction times were submitted to a log-transformation before further analysis. From these data, mean percent correct detections and mean log reaction times were calculated for each experimental condition. Unless otherwise noted, data were tested for statistically significant differences between conditions by means of a $2 \times 3 \times 2 \times 2 \times 2$ mixed analyses of variance (ANOVAs). In all subsequently reported analyses including a repeated measures factor with three or more levels, the Box/Geisser-Greenhouse epsilon correction was applied to adjust the degrees of freedom for violations of the sphericity assumption. In addition, because of the relatively large number of factors manipulated in this experiment, and the potential need for subsequent post hoc tests to explicate the effects of these factors, all reported omnibus tests were conducted using an alpha level of .05, and all follow-up post hoc tests were conducted with an alpha level of .01.

CORRECT CHANGE DETECTIONS AND REACTION TIMES TO CHANGE EVENTS

For the analysis of the mean percent correct detections, the results indicated a statistically significant main effect of team type, $F(1, 10) = 10.77$, $p < .05$. Single-task teams correctly detected a higher percentage of change events (M = 98.96%, SE = .29%) than did the dual-task teams (M = 79.51%, SE = 2.40%). No other sources of variance in the analysis were statistically significant (all $p > .05$).

The analysis of the mean reaction time data indicated statistically significant main effects of team type, $F(1, 10) = 133.63$, experimental session, $F(1.85, 18.46) = 9.55$, detection responsibility, $F(1, 10) = 43.02$, and amity designation, $F(1, 10) = 13.41$, and statistically significant interactions between team type and experimental session, $F(1.85, 18.46) = 6.39$, and between team type and amity designation, $F(1, 10) = 11.90$ (all $p < .05$). There were no other statistically significant sources of variance in the analysis (all $p > .05$).

Further examination of the detection responsibility main effect revealed that participants detected team targets more quickly ($M = 4.54$ s, $SE = .79$ s) than individual targets ($M = 5.85$ s, $SE = 1.05$ s).

To further explore the team type × experimental session interaction, separate repeated measures ANOVAs were calculated for each team type condition examining changes in reaction time across experimental sessions. The results of these analyses indicated that single-task teams showed no change in reaction time across sessions ($p > .01$), but dual-task teams did, $F(1.81, 9.06) = 11.67$, $p < .01$. As can be seen in Figure 1, reaction times in the dual-task condition steadily decreased across experimental sessions. In exploring the team type × amity designation interaction, separate paired-samples t-tests were computed for each team type condition to examine them for differences in reaction times to hostile and neutral targets. The results of these analyses indicated that single-task teams showed no differences in reaction times across amity designations ($p > .01$), but, as can be seen in Figure 1, reaction times in the dual-task condition were faster for hostile than for neutral targets, $t(5) = 3.68$, $p < .01$.

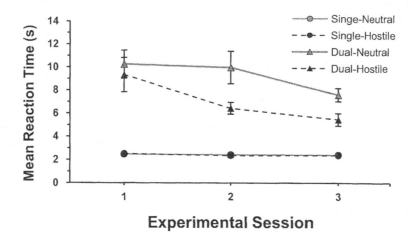

FIGURE 1. Mean reaction times to change events as a function of team type, experimental session, and amity designation. Error bars are standard errors.

FREQUENCY AND DURATION OF FIXATIONS AND FIXATIONS OF MISSED EVENTS

Based on the results of Nelson and Loftus (1980), all fixations occurring within two degrees visual angle of a change event and the duration of those fixations were calculated for each dual-task participant for a) the period of initial entry of an unidentified target into the simulation until it changed, and b) the period of time from the change event to the time it was identified or 45 seconds elapsed (whichever came first). From this data, the mean number of fixations per minute and the mean duration of fixations were calculated for the pre- and post-change periods of both missed and detected events. These data were then submitted to separate 2 (period) × 2 (detection success) repeated measures ANOVAs.

For the mean number of fixations, the analysis indicated a statistically significant main effect of period, F (1, 11) = 21.46, p < .05. Prior to a change, participants fixated in the vicinity of unchanged (i.e., unidentified) targets less frequently (M = 3.89 fixations/min, SE = .68) than after they had changed (M = 20.41 fixations/min, SE − 4.09). No other sources of variance in the analysis were significant (all p > .05).

Analysis of the mean duration of fixations indicated statistically significant main effects of period, F (1, 11) = 32.48, and detection success, F (1, 11) = 24.00, and a statistically significant interaction between period and detection success, F (1, 11) = 18.37 (all p < .05). Follow-up post hoc paired sample t-tests for each period indicated that, pre-change, participants did not fixate for different durations on missed or detected targets, but post-change, participants fixated for longer durations on detected targets ($M = 398.45$ ms, $SE = 24.01$ ms) than on missed targets($M = 223.18$ ms, $SE = 29.69$ ms).

The fixation data were also examined for information regarding missed events. Of specific interest was the frequency of missed events without fixations in their vicinity either before or after a change event (or both). Misses in these cases may be attributable to poor or incomplete information, unlike change events with fixations before *and* after that may be described as more interesting change blindness events. In classifying these events, the data indicated that 7.5% of the missed events included no fixations by either participant in their vicinity, 23.75% of missed events included at least one fixation by a participant in the vicinity either before *or* after a change (not both), and 68.75% of changes included at least one fixation before *and* after a change by at least one participant.

DISCUSSION

The goal of the current experiment was to determine the influence of several military-relevant factors on the prevalence of CB in a complex, team-based C2 environment.

Consistent with previous research (e.g., McCarley et al., 2004) and initial hypotheses, CB rates were higher under dual-task conditions. Speed of detection was also impaired under dual-task conditions, but this effect was somewhat mitigated by practice and target amity. This suggests that some "costs" associated with multi-tasking in military environments may offset by practice and training, but that additional countermeasures, such as the intelligent change alert system proposed by Durlach and Meliza (2004), are likely to be necessary to reduce CB occurrences.

Also consistent with initial hypotheses, speed of detection improved with practice, but only in the dual-task condition. This may suggest that dual-task teams were learning to more efficiently allocate attentional resources across both the air defense and detection tasks.

Contrary to initial predictions, trial task demand (i.e., number of enemy targets) did not substantively influence performance of the change detection task. This may indicate that the relatively small change in the number of on-screen targets in each demand condition was not sufficient to alter detection performance.

The results of the current experiment support initial hypotheses and research by Knott et al. (2007) in that team change events were detected more quickly than individual events. As this was consistent across both single- and dual-task teams, but did not change with practice, it is likely that the observed benefit in this experiment is due to increases in scanning capacity associated with multiple observers rather than more complex team search strategies.

As initially predicted, change relevance was found to influence change detection in dual-task teams. Since single-task teams did not complete the air defense task, the amity designation of targets lacked meaning for them, resulting in similar detection speeds for both neutral and hostile targets. For dual-task teams, on the other hand, the hostile amity designation had direct implications for execution of the air defense task. Given the observed difference in detection times for neutral and hostile targets by dual-task teams, it is likely that they favored an "enemy-focused" scanning strategy, resulting in faster detections of hostile targets.

Examination of the fixation data suggested that participants fixated more frequently on change targets after they changed amity, and analysis of the duration data suggested they fixated for longer periods on changes they successfully detected. These results suggest that fixation frequency and duration may be useful triggers for CB mitigation. While this will not eliminate the problem, its likelihood may be reduced by monitoring operators' fixations and automatically directing attention to changes that operator did not fixate before or after a change (or both). In addition, the duration of fixation may be useful in distinguishing changes the operator is likely to have observed from those they were not. These mitigations could be deployed separately or in conjunction with the change alert system mentioned previously (Durlach & Meliza, 2004).

Overall, the results of this experiment suggest that change blindness in military environments is quite complex and likely to require sophisticated countermeasures to overcome.

REFERENCES

Department of Defense. (1999, January 30). *Department of Defense interface standard, common*
    warfighting symbology (MIL-STD-2525B). Reston, VA: Defense Information Systems Agency, Center for Standards.

DiVita, J., Obermeyer, R., Nugent, T.E., & Linville, J.M. (2004). Verification of the change blindness phenomena while managing critical events on a combat information display. *Human Factors*, 46, 205-218.

Durlach, P.J. (2004a). Change blindness and its implications for complex monitoring and control systems
    design and operator training. *Human-Computer Interaction, 19*, 423-451.

Durlach, P.J. (2004b). *Army digital systems and vulnerability to change blindness* (Accession No.
    ADA433072).

Durlach, P.J., & Meliza, L.L. (2004). The need for intelligent change detection in complex monitoring and control systems (Tech. Rep. SS-04-03-
018). In J. Gunderson & C. Martin (Chairs), *Interaction between humans and autonomous systems over extended operation*. Paper presented at Association for the Advancement of Artificial Intelligence (AAAI) Spring Meeting, Menlo Park, CA.

Hollingworth & Henderson, (2000). Semantic informativeness medieates the detection of changes. *Visual Cognition, 7*

Knott, C.C., Nelson, W.T., McCroskey, M.K., & Miller, B.T. (2007). Do you see what I see? Effects of communication on scanning strategies in change detection by individuals and teams of observers. *Proceedings of the Human Factors and Ergonomics Society*, 51, 1186-1190.

McCarley, J.S., Vais, M., Pringle, H., Kramer, A.F., Irwin, D.E., & Strayer, D.L. (2004). Conversation disrupts scanning and change detection in complex visual scenes. *Human Factors*, 46, 424-436.

Muthard, E. K., & Wickens, C. D. (2002). *Factors that mediate flight plan monitoring and errors in plan revision: An examination of planning under automated conditions* (AHFD-02-11/NASA-02-8). Savoy, IL: University of Illinois, Aviation Human Factors Division.

Rayner, K., & Pollatsek, A. (1989). *The psychology of reading*. Englewood Cliffs, NJ: Prentice Hall.

Rensink, R.A., O'Regan, J.K., & Clark, J.J. (1997). To see or not to see: The need for attention to perceive changes in scenes. *Psychological Science*, 8, 368-373.

Simons, D. J., & Levin, D. T. (1997). Change blindness. *Trends in Cognitive Science*, 1, 261-267.

Simons, D. J., & Mitroff, S. R. (2001). The role of expectations in change detection and attentional capture. In L.R. Harris & M. Jenkin (Eds.), *Vision and Attention* (pp. 189-207). Springer Verlag.

Werner, S. & Thies, B. (2000). Is "change blindness" attenuated by domain-specific expertise? An expert-novices comparison of change detection in football images. *Visual Cognition*, 7, 163-173.

Chapter 31

Training Needs for Units with Small Unmanned Aerial Systems

Paula J. Durlach

U.S. Army Research Institute
12423 Research Parkway
Orlando, FL 32826

ABSTRACT

Unmanned aerial systems are considered small (SUAS) if they are man-portable and their employment does not require an established infrastructure (such as a runway or airport). In the Army, instruction concerning SUAS here-to-fore has focused almost entirely on training the operator how to fly the vehicle. The job of operator, however, is not designated as an occupational specialty, and trainees do not receive supplemental training in valuable enabling skills such as tactics, terrain analysis, imagery interpretation, or communication. Moreover, leaders in the operator's unit are not required to participate in instruction on system capabilities, man-power and training requirements, maintenance requirements, air space coordination, or tactics, techniques, and procedures related to the system. Co-ordination issues between the operator and commander arise not only because of these training gaps, but also because there is no dedicated communications channel between them. This paper describes methods to improve leader training, enhance leader-operator coordination, and provide simulation-based collective training for SUAS operations.

Keywords: Air space, commander, communication, coordination, decision making, scenario, remediation, team, training, simulation, SUAS, unmanned aerial system

INTRODUCTION

Unmanned aerial systems are considered small (SUAS) if they are man-portable and their employment does not require an established infrastructure (such as a runway or airport). Several types of SUASs are already in use by the military, and fielding is expected to increase. An initial Army vision for SUAS was that they would be dispensable assets fielded at the lowest echelons – to platoons and squads—in order to enhance their situation awareness by providing information about what was over the next hill, around the next corner, or on the rooftop down the street (Office of the Secretary of Defense, 2002). Operation was meant to be so easy that anyone could do it, and certainly no dedicated personnel would be required. At the current time, however, a Soldier cannot just open his rucksack, take out the SUAS and launch. Many of the reasons for this are technical – current SUASs are simply not yet small enough or simple enough to allow this type of functionality.

For current SUAS, forethought, planning, and coordination are required. There is time needed for assembly and satellite acquisition (for GPS). The availability of an unused radio frequency is required to avoid interference from or to other systems. And very importantly for safety, coordination with other vehicles in the air space is necessary, because current systems are large enough to pose a danger to manned vehicles. The need to obtain air space clearance means that most SUAS missions are pre-planned. While there are provisions for "immediate requests," these must be approved at brigade level and are therefore unlikely to be immediate in the true sense of the word.

In addition, use of current SUAS may require multiple personnel. Operation requires training; it is not self-evident. It also requires the complete attention of the operator; and some systems even recommend a two-person team, with one person piloting the vehicle and the other interpreting sensor imagery, making tactical decisions, and/or relaying information to remote team members. Because these operators are preoccupied with operation, additional personnel may be required to provide them with security, depending on the nature of the launch point.

Perhaps because of all these issues, SUASs today, are typically controlled by a battalion or company commander, not a platoon or a squad. At the present time, a maneuver platoon would be challenged to handle an organic SUAS, because of manpower and communications limitations (Durlach, 2007). A battalion commander typically chooses how he wishes to assign his SUASs. He may put them under the control of a dedicated company or platoon, while maintaining control of its missions at battalion level; or he may delegate systems to company commanders to employ at their own discretion.

This paper is about the training required for efficient and effective use of today's SUASs. Essentially, available training has not yet come to terms with the actual SUAS capabilities (or lack there-of). The training currently in place is more suitable to the original vision of the small eye-in-the-sky that can be launched any time on-demand. Current mandated instruction focuses entirely on the operator. That instruction focuses on system capabilities, maintenance, and operation (all required material); but it neglects enabling skills, such as tactical considerations for use, terrain analysis, imagery interpretation, and communication skills. Glaringly missing is training for the officer or sergeant in charge of the mission the SUAS will be used to support, and team training in which the SUAS is employed in the context of a realistic mission, where the information obtained via the system is used to make real-time decisions. Battalions preparing for deployment may not have had any collective training using their SUASs prior to their final pre-deployment live training exercise at the National Training Center or the Joint Readiness Training Center.

ADAPTIVE TRAINING FOR LEADERS

Under the auspices of a research project on intelligent computer-based training design, the U.S. Army Research Institute (ARI) has developed an adaptive computer-based training prototype (with Boeing Research and Technology), which we have called SUAS-COMPETE (SUAS Company Employment Training Exercise). The training aims to educate leaders about the factors they ought to consider in the use of SUAS, with a focus on air space considerations. The prototype provides both didactic instruction and decision making practice in the context of realistic branching scenarios, in accord with current theory on scenario-based training (e.g., Salas, Priest, Wilson, & Burke, 2006). Scenarios cover SUAS mission planning, preparation, and execution.

One of the research goals was to design a system that could respond to a heterogeneous training audience. We anticipated two types of heterogeneity: rank and prior knowledge. There is no standard chain of command by which SUASs are commanded. A decision made by a company commander in one battalion might be made by a platoon sergeant in another battalion. We therefore tried to focus more on the types of decisions that need to be made during mission planning, preparation, and execution, and less on who was making the decisions. Thus, we tried to make the training "rank-agnostic;" during the course of a scenario, the trainee might change rank such that at one point he is told he is the company commander, at another point he is told he is the platoon sergeant.

With respect to prior knowledge, we anticipated that some trainees might have no prior knowledge about SUAS operations, others might have some knowledge (some leader training is available, though not mandatory), and still others might have extensive knowledge gained during actual prior deployment history. The prototype therefore includes a pre-test intended to assess depth of in-coming knowledge, and to adapt the instruction to the needs of the student. The pretest determines which learning objectives (if any) require instruction for each individual, prior to their entering the scenario context. On the basis of the test results, relevant instructional materials will be selected by the system for the student to review. A student showing proficiency can be routed directly to a scenario. A student requiring instruction on a limited set of learning objectives will receive that instruction and then go on to the scenario. A student showing little to no proficiency will received instruction and scenarios by phase (e.g., the planning phase instruction and then planning phase scenario). The intention here is to avoid a lengthy period of instruction without opportunity to apply the knowledge.

Nine terminal learning objectives were selected based on the scope of our project and analysis of the knowledge leaders need to bring to bear on SUASs operations. These are listed in Table 1. Each terminal learning objective has multiple associated enabling learning objectives, for a total of 48 enabling objectives. For example for Airspace mission requests, there were three types of requests to be covered: planned, immediate, and dynamic. Didactic materials to teach knowledge of the enabling objectives were constructed by selecting information from relevant doctrine and other Army publications. Subject matter experts then constructed scenarios which would allow students to demonstrate knowledge of the learning objectives in the context of tactical missions.

Scenario-based decision making and remediation

Each scenario consists of three phases: planning, preparation, and execution. Each phase consists of three to four decisions points, and each decision point is linked to one or more learning objectives. At each decision point, the student is offered an update of the situation and must decide among one of four possible choices. A challenge of this domain is that the correctness of decisions may be debated. "Doctrine," as written in a field manual, is not necessarily the best practice compared to that which has evolved as a result of real-world experience. We therefore decided to include at each choice point: the doctrinal solution, two non-doctrinal, but viable solutions, and one poor (unacceptable) solution. Each decision is followed by feedback, and then, depending on which decision is selected, the scenario branches to one of three alternate updated situations. In the case of an unacceptable choice, after receiving explanatory feedback, the student is directed to choose again.

During decision-making, a hint button is available, and hints are provided at three levels of specificity. The first level is fairly general. Asking for the first hint does not affect the student model. Asking for the second level hint degrades the learning objective score by half a point, and asking for the third level hint degrades the learning objective score by a full point. Acceptable decisions lead the student to the next node in the decision tree. Unacceptable decisions affect the student model on a terminal learning objective basis. If the total "strikes" for a terminal learning objective reaches three, the scenario is aborted and the student is given relevant remedial instruction. Following remediation, the student restarts the scenario phase they were in before (planning, preparation, or execution). Scenarios end with a review of performance.

Presently, two scenarios have been completed; but, we have not had the opportunity to test the prototype with actual students yet. It has been reviewed by instructors of SUAS operators; they were enthusiastic about the prototype and definitely thought the training would be beneficial; but, they saw institutional barriers to use of it at their school, because their mandate is to train operators.

Table 1 SUAS-COMPETE Learning Objectives

SUAS in Company-level operations
➢ Attack, Defend, Security, Civil Operations
➢ Recon & Surveillance
➢ Target Acquisition & Battle Damage Assessment
➢ Security in SUAS mission (Focus Screen and Force Protection)
➢ Battle Damage Assessment
Company-level SUAS Control Measures (Essential)
➢ Named Area of Interest
➢ Axis of Advance
➢ Engagement Area
➢ Attack by Fire Position
➢ Observation Post
Procedural Controls – Separation
➢ Lateral Separation
➢ Vertical Separation
➢ Time Separation
Airspace Mission Requests/Types of Missions
➢ Planned
➢ Immediate
➢ Dynamic
Necessary Coordination & Situation Awareness
➢ Higher Headquarters coord ination & Situation Awareness
➢ Observation and Reporting
➢ Battle Damage Assessment Reporting
Airspace Environment & Situation Awareness
➢ Power Lines & Communications Interference
➢ Man-made structure (e.g. towers)
➢ Low altitude civilian aircraft
➢ Artillery projectiles and effects
➢ Intervening Terrain and Line of Sight
➢ Positive Control
➢ Close Combat Attack (CCA) Aviation
➢ Enemy Air Defense and Friendly Air Defense Artillery
➢ Close Air Support (CAS)
➢ Weather (e.g. winds, temperature, precipitation, etc.)
Procedure Control
➢ Formal Airspace Coordination Area
➢ Informal Airspace Coordination Area
➢ Restrictive Fire Line, Restrictive Fire Area, No Fire Area
➢ Common Reference Systems (CRS)
➢ Air Corridors
➢ Coordinating Altitudes
➢ Restricted Operating Zones and Areas
➢ Minimum Risk Corridor
Brigade AC2 Control
➢ Air Defense Airspace Management/Brigade Aviation Element (ADAM/BAE)
➢ Battalion S3 and B3 Air
➢ Company Command Post
➢ Destruction/Abort/Emergency Procedures

REASERCH TO DEVELOP TECHNOLOGY AND TRAINING FOR COLLECTIVE USE OF SUAS

In addition to developing prototype individual training for leaders, ARI has supported technology development to support collective employment of SUAS. Our focus has been on communication and coordination between a commander and a remote SUAS operator. Typically, a SUAS has little ability to share information in real time with anyone besides the operator. In some instances, a remote video terminal (RVT) might be available for a non-co-located person to see the sensor imagery as it is being captured; this can be accomplished by setting up a spare operator control unit (OCU) at a remote location that has line of sight with the aerial vehicle. In this case, the RVT is a passive receiver; any on-line communication with the operator has to be conducted by others means (typically radio). More recently, tactical operations centers are being equipped with the One-source remote video terminal (OSRVT), which, similar to the RVT, allows for the passive remote viewing of sensor imagery, potentially from multiple types of aerial platforms; but pays for this with a heavier hardware footprint (Washburn & McIndoe, 2009). The OSRVT also has no provision for two-way communication. Limited technical support for two-way communication between a SUAS operator and a remote mission commander (likely, the consumer of the information provided by the SUAS) is a potential source of misunderstandings, missed opportunities, and frustration. We therefore embarked on two parallel lines of development aimed at ameliorating this situation. The first was intended to provide rapid technology development to enable a dedicated two-way communication channel between a mission commander and a remote operator. This research was supported by a Small Business Innovative Technology Research award (SBIR) by the Office of the Secretary of Defense to Perceptronics Solutions, Inc. The second was to develop a simulation testbed, which could be used to investigate communication and coordination issues, explore how different (simulated) technology solutions might affect them, and develop potential training interventions to avoid them.

Problem Assessment

In order to better understand the communication and coordination issues arising during the use of SUAS, we observed training exercises at multiple locations, involving the employment of the Raven SUAS. These observations were then subjected to task analysis (De Visser, et al., 2008). Briefly, the Raven is a hand-launched fixed-wing SUAS, which can be controlled manually or semi-autonomously using a preprogrammed waypoint-based route. The commander for whom a Raven is being flown is typically remote from the operator and might monitor, or have a designee monitor, an RVT. Communication between the command team and the operator is typically by voice over a standard radio net, often through an intermediary, such as an RTO (radio transmission operator).

Observation of Raven missions revealed several communication bottlenecks. The most dominant of these was establishing and/or maintaining radio communication with the operator. This was variously due to radio net congestion, reliance on intermediaries, and the lack of response at the operator's end. Another impediment to smooth operation was the commander's sketchy understanding of the capabilities of the Raven. This led to requests that the operator was unable to carry out, and the need for multiple communications. A third issue was delayed dissemination of information gained from the SUAS to the team(s) who could make best use of it.

A Technology Solution to Communications Bottlenecks

The technological solution to the observed commander-operator communications bottlenecks, which is being developed by the SBIR project, and is called TECRA (Technology Enhanced Command and Control for Robotic Assets) provides a dedicated communications channel between the commander and operator, using a wireless technology called Cursor on Target (Robbins, 2007). Assuming the commander has the RVT equipment to enable a downlink from the SUAS, also required is a laptop with TECRA-Falconview connected to the downlink equipment. Falconview is software resident on the operator's mission planning laptop; TECRA-Falconview is a version of the same software, but upgraded to facilitate commander understanding of an ongoing SUAS mission and commander-operator two-way communication. No change is required to the operator's setup except for a software plug-in to Falconview on the mission planning laptop. See Figure 1 for a schematic of the standard Raven setup vs. the TECRA setup.

The commander's TECRA interface provides more information about an ongoing Raven mission, and in a much simpler format than the standard RVT display, and early assessments suggest it supports better situation awareness and reduced perceived workload (De Visser, et al, 2010). A major component of the display is an area map, which is synchronized with the Falconview map on the operator's mission planning laptop. Commanders can annotate this map and send the annotations so that they appear on the operator's map. This can include suggested waypoints, routes, or areas or targets of interest. A message window allows two- way text messaging between the commander and operator. Streaming video from the SUAS is also displayed along with selected flight data (e.g., camera heading). The commander can decide to record video or take still photos. These can be saved in a library, with a grid location tag, so that when it is accessed, the location where it was taken appears on the map. Also provided is the ability to markup and export these images to other applications, such as powerpoint. Interviews with Soldiers indicated this would be a very valuable feature allowing support for after action review and/or intelligence dissemination. See De Visser, et al. (2010) for additional detail about development and features.

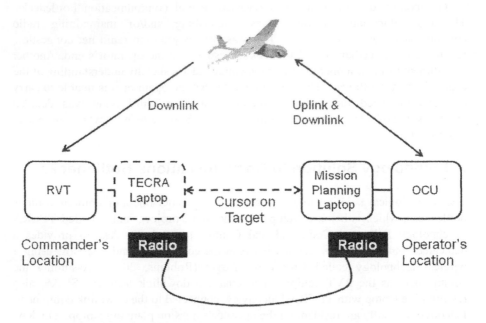

Figure 1 Standard Raven equipment and connections, represented by solid lines and outlines. The TECRA modification is represented by dashed lines and outlines.

The design of TECRA was intended to alleviate the observed communications bottlenecks inherent in the current Raven and other SUAS systems and enhance team performance (Salas, Cannon-Bowers, & Johnston, 1997). In particular, it supports shared situation awareness between the commander and the operator, makes some of the mission constraints easier for the commander to understand, and provides a dedicated channel for communication with the operator. While preliminary testing indicated TECRA improves the commander's situation awareness and workload (De Visser, 2010), these results were not obtained in the context of a realistic mission. Study participants were only required to monitor a simulated video feed, identify targets, and answer questions about the SUAS mission. There was no explicit test of whether communication between a commander and an operator improved; nor was the "commander" assigned any concurrent tasks to attend to, as would be the case in more realistic settings. Therefore, a significant future achievement will be to test the effectiveness of TECRA in more realistic contexts.

Bringing in the Troops

The final section of this paper describes a simulation testbed designed to enable investigation of communication and coordination issues in a full-fledged mission context, explore how different (simulated) technology solutions might affect them, and develop potential training interventions to avoid them. What is required is an environment in which all facets of a SUAS mission can be represented, including the operator's control equipment, the commander's means of coordinating with the operator, and the troops under the commander's authority. ARI developed such an environment, in collaboration with the Institute for Simulation and Training (IST) at the University of Central Florida, and Research Networks, Inc. For this simulation, the SUAS was modeled after a vertical-lift and land ducted-fan prototype, and we initially conducted several experiments examining operator training (Billings & Durlach, submitted for publication; Billings & Durlach, 2008). To the operator training simulation, we added a nominal command and control station (C2node) for the commander, envisioned to be a platoon leader or company commander. Because our aim was to investigate how various technological supports would assist coordination with the SUAS operator, the C2node was designed so that we could configure its capabilities as we wished. For example, we could enable the C2node with blue-force tracking (or not), or we could enable the C2node with streaming video from the SUAS (or not).

In order to bring subordinate Soldiers into the simulation, both the OCU and C2node were networked to an immersive environment provided by modified off-the shelf game-engine technology. The game-engine technology provides for multi-player, distributed first-person interaction in a modeled small-village environment. The SUAS flies in the game environment and can both "see" and be seen in that environment. The commander, SUAS operator, or both, can control avatars in the immersive environment if desired, by providing them with two computers each (one for their control units and one for the game environment). While this is technically not a challenge, we were not entirely sure that a potential research participant would be able to cope with operating both stations at a time, so our initial research was aimed at establishing this (Durlach, Priest, Saffold, & Martin, 2009).

With all the simulation elements in place, another remaining obstacle to initiating research was the need for a human role player for every character in the simulation, including all the forces controlled by the commander, enemies, and bystanders. To overcome this manpower requirement we embarked on the process of improving the artificial intelligence (AI) of software-driven characters (known in the game world as NPCs, or in the simulation world as SAF). Some of these AI capabilities are described in Durlach, Priest, Saffold, and Martin, 2009. Our aim was to be able to run at least a two-squad dismounted infantry mission, with enemy and bystanders with one actual research participant (the commander), a human confederate playing the role of the operator, and only one or two other experimenters orchestrating the rest of the action. We are currently developing scenarios in this context.

Despite the fact that we have yet to conduct a realistic simulated collective training experiment, we believe this testbed illustrates how simulation-based collective training involving SUAS missions could be conducted. It also provides an opportunity to test how different types of technology support for commander-operator coordination could improve the effectiveness of SUAS employment, especially concepts aligned with the original vision of providing SUAS to small infantry units such as platoons.

REFERENCES

Billings, D. R. & Durlach (submitted for publication). "How input device characteristics contribute to performance during training to operate a simulated micro-unmanned aerial vehicle," *Journal of Cognitive Engineering and Decision Making*.

Billings, D.R., & Durlach, P.J. (2008). *Effects of input device and latency on performance while training to pilot a simulated micro-unmanned aerial vehicle*. ARI Technical Report 1234, U.S. Army Research Institute for the Behavioral & Social Sciences, Arlington, VA.

De Visser, E. J., LeGoullon, M., Freedy, A., Freedy, E., Weltman, G., and Parasuraman, R (2008), *"Designing an Adaptive Automation System for Human Supervision of Unmanned Vehicles: A Bridge from Theory to Practice,"* Proceedings of the 52nd Annual Meeting of the Human Factors and Ergonomics Society, New York.

De Visser, E. J., LeGoullon, M., Hovarth, D., Weltmand, G., Freedy, A., Durlach, P., and Parasuraman, R (2010). "TECRA: C2 application of adaptive automation theory", IEEE Aerospace Conference, Big Sky, Montana.

Durlach, P. J. (2007). *PACERS: Platoon Aid for Collective Employment of Robotic Systems*. ARI Research Report 1876, U.S. Army Research Institute for the Behavioral & Social Sciences, Arlington, VA.

Durlach, Priest, Saffold, and Martin (2009), *Developing Collective Training for Small Unmanned Aerial Systems Employment*, MODSIM World, Virginia Beach, VA.

Office of the Secretary of Defense (2002). Unmanned Aerial Vehicles Roadmap, 2002- 2027. Washington, D.C.

Robbins, D. (2007), Unammed aircraft operational integration using MITRE's Cursor on Target, *The Edge, 10*, 7. Downloaded February 18, 2010 from http://www.mitre.org/news/the_edge/summer_07/robbins.html.

Salas, E., Cannon-Bowers, J. A., & Johnston, J. H. (1997), " How can you turn a team of experts into an expert team?: Emerging training strategies". In C. E. Zsambok & G. Klein (Eds.), *Naturalistic decision making (pp. 359- 370)*. Mahwah, NJ: Lawrence Erlbaum Associates.

Salas, E., Priest, H. A., Wilson, K. A., & Burke, C. S. (2006), "Scenario-based training: Improving military mission performance and adaptability". In C. A. C. A.B. Adler, and T.W. Britt (Eds.), *Military life: The psychology of serving in peace and combat (Vol. 2: Operational Stress, pp. 32-53)*. Westport, CT: Praeger Security International.

Washburn, R. and McIndoe, J. (2009), *Ergonomically and Safely Integrating Communication Equipment*. Downloaded February 18, 2010 from http://tardec.army.mil/Documents/AM0709_Ergonomically_and_Safely_In tegrating_Communication_Equip.pdf.

CHAPTER **32**

NeuroGaming: Merging Cognitive Neuroscience & Virtual Simulation in an Interactive Training Platform

Chris Berka[1], Nicholas Pojman1, Jonathan Trejo[1], Joseph Coyne[3],
Anna Cole[4], Cali Fidopiastis[5], Denise Nicholson[2]

[1] Advanced Brain Monitoring, Inc.
Carlsbad, CA, 92008, USA

[2] ACTIVE Laboratory
Institute for Simulation and Training
Orlando, FL 32816, USA

[3] Human-Systems Integration Laboratory
U.S. Naval Research Laboratory
Washington, DC, 20375, USA

[4] Strategic Analysis, Inc.
Arlington, VA, 22203, USA

[5] Department of Physical Therapy
University of Alabama at Birmingham
Birmingham, AL, USA

ABSTRACT

Serious gaming platforms are increasing in popularity for training in both industry and the military. Current training packages do not generally support adaptive training: the ability of a simulator to modify an ongoing training scenario based on real-time assessments of the trainee's peformance and/or psychophysiological state. Creating a platform that can induce modifications to training environments may enhance instruction by optimizing the trainee's pyschophysiological state during training. This paper reports preliminary data on the neurophysiological correlates of threat detection in a training simulation environment using Electroencelpholographic (EEG) metrics. Event-Related potential (ERP) and power analysis (ERD/ERS) distinguished threat type, threat difficulty, and predicted errors in performance. These neural signatures in a virtual environment serve as a foundation for developing a neurophysiologically driven adaptive simulation engine, the NeuroGaming Platform.

Keywords: EEG, Closed-Loop, Adaptive Learning, Serious Gaming, Video Gaming, Event Related Potential (ERP)

INTRODUCTION

Serious gaming applies sophisticated and immersive video gaming technology to enhance training in industry, academia, and the military. Although this new generation of training technology is engaging and popular with trainees, it represents only an initial step towards a true revolution in creating successful instructional delivery systems. Conventional methods for evaluating instructional design include subjective report, performance metrics and expert observations; all are removed from the actual trainee's experience in both time and space. Neuroergonomics theory suggests that knowledge of brain-behavior relationships can be applied to optimize the design of environments to accelerate learning (Kramer, 2007; Parasuraman, 2005). The convergence of recent advances in ultra-low power consumer electronics, ubiquitous computing and wearable sensor technologies enables real-time monitoring of cognitive and emotional states providing *objective, timely, and ecologically valid* assessments. These technologies allow access to the psychophysiological states associated with learning. Increasing evidence suggests that physiological correlates of attention, alertness, cognitive workload, arousal, and other fundamental constructs essential to training can be identified. In addition to accessing the underlying cognitive and emotional states of the trainee in real-time, neurosensing offers the ability to adapt an ongoing tactical scenario based on near-to-real-time assessments of the trainee's performance and/or physiology. Introducing modifications in real-time should enhance training experiences by optimizing the trainee's cognitive state and by tailoring the information delivery to an individual's or team's evolving skill level (Schmorrow, Stanney, Wilson, & Young, 2005; Stripling et al., 2007). An early demonstration of adaptive automation using an EEG-driven system improved performance on a Multiple Attribute Task (Pope, Bogart, & Bartolome, 1995; Prinzel, Freeman, Scerbo, Mikulka, & Pope, 2000). The Adaptive Intelligent Training Environment (AITE) program sponsored by the Office of Naval Research has defined a functional architecture for a closed loop training platform (Kemper, 2007; Nicholson, 2007). Other successful closed-loop systems include EEG-based drowsiness alarms in a driving simulator and EEG-workload in complex tasks such as the Aegis radar and Tactical Tomahawk Weapons simulations (Berka et al., 2004; Berka, Levendowski, Westbrook, Davis, Lumicao, Ramsey, et al., 2005; Poythress et al., 2006). In these examples, the physiological thresholds and mitigations employed in the closed-loop adaptive systems assessed within specific training application and were not designed to extend to alternative simulation-based training environments. Thus, there is a need for a closed loop implementation that can generalize across simulation-based training platforms providing an environment for adaptive training research, development, and applications.

This paper presents such a platform which integrates neuroscience-based evaluation technologies and simulation training with the goal of creating an interactive learning environment adaptive to the skill levels and needs of trainees. Electroencephalographic (EEG) and other physiological measurements such as heart rate and skin conductance are easily monitored and aligned in combination with performance metrics such as accuracy, speed, and efficiency. Interactive adaptive computer-based training systems can then be designed to facilitate mitigations within the simulation environment to accelerate skill acquisition and provide quantitative evidence of successful training in a variety of task domains.

NEUROPHYSIOLOGY OF THREAT PERCEPTION IN A VIRTUAL ENVIRONMENT

Creating a closed loop training platform requires an understanding of the physiological signatures that characterize perceptual encoding of key elements in the training environment, as well as the cognitive states associated with successful learning. The physiological signatures of success or failure in training provide the framework for selection of mitigation strategies to be implemented. Multiple intervention strategies including slowing or speeding the pace of the scenario; introducing alarms or prompts to direct attention; or changing the level or complexity of the task can be implemented and tested to determine impact on accelerating learning. The Virtual Battles Space 2 (VBS2) platform is a complex and realistic simulation designed for military training that can be used to train observation skills, specifically by identifying threats. A study was conducted to characterize the neural signatures associated with this skill using commonly encountered threats including People, Vehicles and Improvised Explosive Devices (IEDs).

MATERIALS AND METHODS

Thirteen healthy participants recruited from the UCF experimental recruitment website were studied at the IST ACTIVE Laboratory and five healthy participants at Advanced Brain Monitoring's (ABM) Laboratory.   EEG was acquired with the 9-channel B-Alert wireless EEG sensor headset developed by ABM for applications in operational environments (Berka, et al., 2004; Berka, Levendowski, Ramsey, et al., 2005; Mathan S., 2007; Poythress, et al., 2006).   Referential recordings from F3, Fz, F4, C3, Cz, C4, P3, POz, and P4 referenced to linked mastoids were acquired at 256Hz with 24-bit resolution. The headset is easily applied with each sensor dispensing a small amount of conductive cream (Synapse Cream, Med-Tek, Joliet, IL).   Automated impedance measurements are employed and impedance is maintained below 30kΩ.   EEG is amplified, digitized, and then transmitted over Bluetooth to a PC. A PC was used to display the experimental testbed using a flat screen LCD monitor running E-Prime software (Psychology Software Tools, Inc.) to record and compute performance measures. EEG was synchronized to testbed events via parallel port signaling coupled with ABM's External Sync Unit to provide the millisecond level timing precision required for event locked physiological analysis. All stimuli consisted of static virtual scenes from VBS2 containing either threatening or nonthreatening Vehicles, People or IEDs (10 threat, 10 non-threat per category, see **Figure 1** for examples).   Sixty images (30 threat, 30 non-threat) were each displayed twice in randomized order, for a total of 120 trials per session. Participants viewed one example stimulus per category prior to testing and were instructed to respond as quickly and accurately as possible.   Images were displayed for a fixed two-second duration to allow users to scan for potential threats before responding. A two-second fixation cross was presented between stimuli.

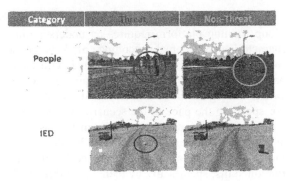

Figure 1. Sample stimuli. The image on the right shows and an example of a people non-threat and IED non-threat stimulus. The image to the right shows an example of people-threat and IED-Threat stimulus.

ANALYSIS AND RESULTS

A total of 14 subjects were used for all data analysis. Four subjects were excluded from analysis due to equipment malfunction or failure to meet minimum data quality standards. EEG epochs with excessive artifact were not included in analysis. Event related synchronization (ERS) and desynchronization (ERD) analysis were computed using PSD. ERPs, ERDs and ERSs were averaged across all valid trials by subject for each stimulus type, and averaged across all subjects to compute grand means. If appropriate, data was randomly down sampled to maintain sample sizes for inter-group comparisons. **Figure 2** summarizes performance by category and across the entire session. ANOVA analysis of reaction time and accuracy (percent correct) revealed IED stimuli most difficult as indicated by significantly lower accuracy [$F(2,15)=19.11$; $p<.0001$] and longest reaction times [$F(2,15)=47.89$; $p<.0001$] compared to People and Vehicle stimuli. Longer reaction times for IEDs likely reflect a greater intrinsic cognitive load for IED discrimination relative to the other threat categories. The easiest discrimination was for the People stimuli, as indicated by highest accuracy.

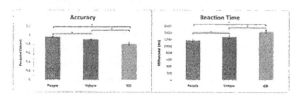

Figure 2. Performance -results showed the IED category was significantly more difficult to identify as indicated by lowest accuracy and longest reaction times when compared to both Vehicle and People stimulus types.

Figure 3 shows that the positive identification of an IED threat resulted in the largest and most consistent late positive component (LPC) across all scalp sites and across all participants for correctly identified stimulus. Mean IED amplitude differences were not only distinct compared to both Vehicle and People categories, but also had the smallest threat vs. non-threat differences. The distinct and small mean amplitude differences combined with an observed larger LPC amplitude may reflect increased decision making and cognitive processing demands in assessing IED threat presence as compared to the other two categories (Polich et al., 1997; Polich & Kok, 1995) or may be characteristic of the perception of an imminent, potentially life-threatening event.

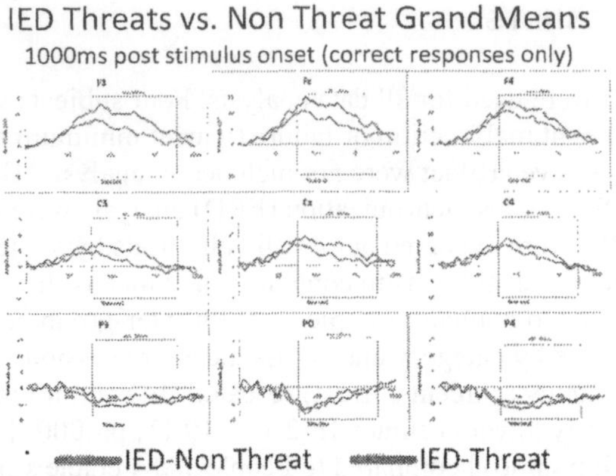

318

IED Threats vs. Non Threat Grand Means
1000ms post stimulus onset (correct responses only)

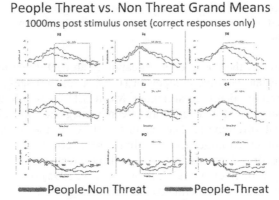

IED-Non Threat   IED-Threat

Figure 3. ERPs 1000ms post stimulus onset for IED-threat and IED non-threats. IED ERPs (both threat and non-threat) showed largest LPCs across Stimulus categories.

Distinctive laterality of the LPC was observed between People-threat and People-Non-Threat. In contrast to the other stimulus classes (Vehicles and IEDs), People threat stimulus elicited greater LPC in the left frontal and central regions with a right hemisphere dominance for the identification of the non-threat People (illustrated in **Figure 4**)(Botzel & O.J., 1989). These results are similar to the reported laterality of late positive components during facial recognition tasks (Cuthbert, 2000; Laurian, 1991; Schupp, 2004)

People Threat vs. Non Threat Grand Means
1000ms post stimulus onset (correct responses only)

People-Non Threat   People-Threat

Figure 4. Grand Mean ERPs for People-NT and People-T for 1 second following stimulus onset. People-T had greater late negativity lateralized to right hemisphere.

Both threats and non-threats across all categories elicited an observed LPC. Visual inspection of individual participant's mean ERPs were used to define a measurement window of 400-900 ms post-stimulus, corresponding to the required decision-related response for each stimulus (Polich & Kok, 1995). ERP differences were computed within category (Non Threat-Threat) and averaged to compute the mean amplitude difference for each category from 400-900ms following stimulus onset for correct responses only. ANOVA was performed using 3 category by 3 laterality (Left F3,C3,P3), Central (Fz,Cz,Pz) Right (F4,C4,P4). Analysis revealed a main mean amplitude difference affect by category type, $F_{(2,26)} = 4.25$; $p < .05$, and affect by laterality, $F_{(2,26)} = 14.35$; $p < .05$ as shown in **Figure 5.**

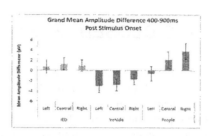

Figure 5. Shows mean amplitudes differences (Non-Threat minus Threat) from 400-900ms post stimulus onset for correct responses. ANOVA analysis revealed a main mean amplitude difference affect by category type, $F_{(2,26)} = 4.25$; $p < .05$ and affect by laterality $F_{(2,26)} = 14.35$; $p < .05$. IED Non-Threat minus Threat differences were closest to 0, reflecting greater difficulty in threat and non-threat distinction for this category.

Previous research on cognitive workload and attention suggests alpha suppression is correlated with the processing of complex tasks involving perceptual judgment, memory and motor demands (Gevins, Smith, McEvoy, & Yu, 1997; W. Klimesch, 1999). Upper alpha band (10-12 Hz) suppression (indicated by decreased alpha power) is correlated with better performance during stimulus related cognitive tasks (W. Klimesch, 1999; W. Klimesch, Schimke, H. & Pfurtscheller, G. , 1993).

There was a trend towards incorrect responses being preceded by increased alpha (8-12) power in comparison to correct responses in the 1000ms prior to stimulus onset. The results support previous findings that higher alpha power preceding stimulus presentation is *predictive of errors*. Alpha disparity was highest in occipitoparietal sites (P3, PO, P4). Local changes in alpha power are easily computed in real time and thus could prove valuable in the closed loop training environment. This alpha metric can be used to determine whether or not a subject is cognitively prepared for an upcoming task. A Mitigation strategy (i.e. increase inter- stimulus intervals, decrease task complexity or task-related messaging) could be implemented to enhance training. Perceptual training may also be enhanced using neurofeedback training (NFT) techniques to decrease alpha power during cognitive tasks. Prinzel, (2002) reported NFT improved performance and self-reported workload in a simulated adaptive environment. NFT completed independently or in parallel with adaptive training may enhance performance during training.

The results from this study showed that unique neural signatures of threat detection can be identified in association with VBS2 imagery. Importantly, EEG predictors of error in detection were also characterized. These neural signatures form the foundation for mapping a highly detailed description of the brain's capability to identify and respond to both threatening and non-threatening stimuli. Results suggest that unique neural signatures exist which distinguish threat type, threat difficulty, and performance in a threat perception task. These metrics can be used as inputs for altering a training environment in real time and provide objective evidence of learning in a simulation environment.

THE NEUROGAMING CLOSED-LOOP

One major limitation in previous closed loop systems is the requirement for adaptive elements to be embedded within each unique training platform with no transferability to alternative simulation environments. To address the issue of cross compatibility, a modular, flexible, and adaptive platform was developed to support rapid implementation across simulations and for development of a comprehensive closed-loop system. The NeuroGaming platform (**Figure 6**) allows for automated adaptive training based on both physiological (e.g. EEG, EKG, GSR,) and non-physiological (performance, subjective rating, expert observation) metrics.

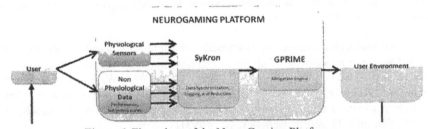

Figure 6. Flow chart of the NeuroGaming Platform.

The B-Alert Wireless Sensor headset was designed to record high quality physiological signals including EEG, HR, EOG, and/or EMG in operational environments with real-time analysis of the EEG including the computation of indices of alertness, engagement, arousal, and mental workload (Berka, Levendowski, Ramsey, et al., 2005; Poythress, et al., 2006). B-Alert has been used to characterize alertness and memory in patients with sleep apnea; identify individual differences in susceptibility to the effects of sleep deprivation (Berka, Levendowski, Westbrook, Davis, Lumicao, Olmstead, et al., 2005); and distinguish between sleep deprivation and sleep-disordered fatigue. In addition, EKG can be acquired to compute HR and heart rate variability (HRV) to monitor parasympathetic and sympathetic nervous system activation (Camm, Malik, Bigger, Breithardt, & Cerutti, 1996; Pojman, 2009). The system has been successfully integrated into real-time, closed-loop automated computing systems to implement dynamic regulation and optimization of performance during a driving simulation task and in the Aegis C2 and Tactical Tomahawk Weapons simulation environments (Berka, Levendowski, Ramsey, et al., 2005; Berka, Levendowski, Westbrook, Davis, Lumicao, Ramsey, et al., 2005; Poythress, et al., 2006).

The Synchronous Operational Psychophysiological Sensor Suite or SyKron is a set of software libraries and tools developed by UCF's ACTIVE Laboratory. SyKron integrates multiple disparate sensor technologies for experimentation and data analysis. SyKron supports all task-relevant data streams including physiological (EEG, HR, GSR, Eye Tracking, etc.) and non physiological inputs (performance, subjective rating, etc). SyKron logs physiological sensor inputs and has data processing tools which allow for real time analysis of physiological signal. SyKron's playback tool enables it to replay previously recorded sensor data. The ability to simulate and replay physiological signals facilitates iterative assessment of adaptive mitigation and threshold development.

The Warfighter Human-Systems Integration Laboratory at the U.S. Naval Research Laboratory has developed the General Purpose Real-Time Mitigation Engine (GPRIME), a software platform that can support streaming data from multiple IP addresses, allowing for mitigations to be triggered by data variables streaming from multiple computers on a local network. In the NeuroGaming platform, GPRIME receives processed real-time (or near real-time) physiological inputs along with subjective and/or performance data from SyKron as variables to create Boolean logic (If, And, Or, >, =, etc.) rules that are saved and evaluated in near real-time to assess when it is appropriate to perform a mitigation. When the streaming data inputs meet the threshold rules, pre-recorded keyboard and mouse click macros are triggered to modify the training scenario. GPRIME is a user friendly GUI designed for flexibility, allowing it to seamlessly interface with new simulators quickly and easily. Adaptive thresholds and mitigations can be easily saved and implemented across simulation trainers, allowing for previous work to be leveraged into novel environments or upgraded platforms. G-PRIME enables the rapid development of a large set of operator-defined manipulations without requiring code changes to the application or simulation control systems.

Before creating specific thresholds and designing appropriate mitigations for an adaptive system, physiological correlates of successful training must first be characterized in a controlled environment. Our next project will assess convoy scenarios developed in the VBS2 environment in collaboration with USMC trainers at Camp Pendleton, CA. The testbed consists of detailed training and train-to-criteria tasks, followed by four test missions, each with unique demands (such as verbal threat identification, spatial recall, change observation, etc.). Each scenario was developed with the goal of both training and assessing perceptual and communication skills needed for active deployed military: spatial and working memory, visuospatial processing, selective attention, verbal communication, and decision making. Preliminary results revealed physiological trends related to task demands as well as performance. Example potential mitigations for these existing scenarios include adding or subtracting key elements (people, objects, etc.) or distracters; changing the speed of the vehicle; and altering mission instructions which can be easily introduced as brief text messages to the participant.

CONCLUSION

The NeuroGaming platform has many applications where automatic real-time detection of cognitive state could be leveraged to improve training effectiveness: imagery analysis, monitoring tasks, interpreting medical images, surveillance system monitoring, critical infrastructure protection, and other tasks where operator error may result in catastrophic outcomes. The scope of the NeuroGaming platform has been designed to meet the requirements of the serious gaming community, but also has potential to be easily implemented into any educational environment and may have applications to the broader base of video gaming enthusiasts.

ACKNOWLEDGEMENTS

This work was funded by the Office of the Secretary of Defense SBIR Award # N00014-09-M-014.

REFERENCES

Berka, C., Levendowski, D., Cvetinovic, M., Petrovic, M., Davis, G., Lumicao, M., et al. (2004). Real-time Analysis of EEG Indices of Alertness, Cognition, and Memory Acquired with a Wireless EEG Headset. *International Journal of Human-Computer Interaction, 17*(2), 151-170.

Berka, C., Levendowski, D., Ramsey, C. K., Davis, G., Lumicao, M., Stanney, K., et al. (2005, May). *Evaluation of an EEG-Workload Model in an Aegis Simulation Environment.* Paper presented at the Proceedings of SPIE Defense and Security Symposium, Biomonitoring for Physiological and Cognitive Performance during Military Operations, Orlando, FL.

Berka, C., Levendowski, D., Westbrook, P., Davis, G., Lumicao, M., Olmstead, R., et al. (2005, May). *EEG Quantification of Alertness: Methods for Early Identification of Individuals Most Susceptible to Sleep Deprivation.* Paper presented at the Proceedings of SPIE Defense and Security Symposium, Biomonitoring for Physiological and Cognitive Performance during Military Operations, Orlando, FL.

Berka, C., Levendowski, D., Westbrook, P., Davis, G., Lumicao, M., Ramsey, C., et al. (2005, July 22-27). *Implementation of a Closed-Loop Real-Time EEG-Based Drowsiness Detection System: Effects of Feedback Alarms on Performance in a Driving Simulator.* Paper presented at the 1st International Conference on Augmented Cognition, Las Vegas, NV.

Botzel, K., & O.J., G. (1989). Electric brain potentials evoked by pictures of faces and non-faces: a search for "face-specific" EEG potentials. *Experimental Brain Research, 77*, 349-360.

Camm, A. J., Malik, M., Bigger, T. J., Breithardt, G., & Cerutti, S. (1996). Heart rate variability: Standards of measurement, physiological interpretation, and clinical use. *European Heart Journal*(17), 354-381.

Cuthbert, B., Schupp, HT., Bradley, MM., Birbaumer, N., Lang, PJ. (2000). Brain potentials in affective picture processing: covariation with autonomic arousal and affective report. *Biological Psychology, 52*(2), 16. doi: 10.1016/S0301-0511(99)00044-7

Gevins, A., Smith, M. E., McEvoy, L., & Yu, D. (1997). High-resolution EEG mapping of cortical activation related to working memory: effects of task difficulty, type of processing, and practice. *Cerebral Cortex, 7*(4), 374-385.

Kemper, D., Davis, L., Fidopiastis, C. M., & Nicholson, D. (2007, July 22-27, 2007). *Foundations for creating a distributed, adaptive user interface.* . Paper presented at the Foundations of Augmented Cognition, Third International Conference, FAC 2007, Held as Part of HCI International 2007, Beijing, China.

Klimesch, W. (1999). EEG alpha and theta oscillations reflect cognitive and memory performance: a review and analysis. *Brain Res Brain Res Rev, 29*(2-3), 169-195.

Klimesch, W., Schimke, H. & Pfurtscheller, G. . (1993). Alpha frequency, cognitive load and memory performance. *Brain Topography, 5*, 1-11.

Kramer, A., Parasuraman, R. (2007). Neuroergonomics: Application of Neuroscience to Human Factors. In J. T. Caccioppo, Tassinary, L.G., Berntson, G.G. (Ed.), *Handbook of Psychophysiology* (3 ed., pp. 704-722). New York: Cambridge University Press.

Laurian, S., Bader, M., Lanares, J., Oros, L. (1991). Topgraphy of event related potentials elicited by visual emotional stimuli. *International Journal of Psychophysiology, 10*(3), 8. doi: 10.1016/0167-8760(91)90033-T

Mathan S., W. S., Dorneich M., Ververs P., Davis G. (2007). Neurophysiological estimation of Interruptibility: Demonstrating Feasibility in a Field Context. In D. Schmorrow, Nicholson D., Drexler J., Reeves L. (Ed.), *Augmented Cognition* (4th ed., pp. 51-58). Arlington, VA: Strategic Analysis, Inc.

Nicholson, D. M., Fidopiastis, CM, Davis, LD, Schomorrow, DD, Stanney, KM. (2007). An Adaptive Instructional Architecture for Training and Education. In D. D. Schmorrow, Reeves, L.M. (Ed.), *Augmented Cognition* (pp. 380-384). Berlin: Springer-Verlad Berlin. (Reprinted from: 2007).

Parasuraman, R. (2005). *Neuroergonomics: The Brain at Work.* New York: Oxford University Press.

Pojman, N., Johnson, R., Kintz, N., Behneman, A., Popovic, D., Davis, G., Berka, C. . (2009). Assessing Fatigue using EEG Classification Metrics during Neurocognitive Testing. *Sleep, 32*, A161.

Polich, J., Alexander, J. E., Bauer, L. O., Kuperman, S., Morzorati, S., O'Connor, S. J., et al. (1997). P300 topography of amplitude/latency correlations. *Brain Topogr, 9*(4), 275-282.

Polich, J., & Kok, A. (1995). Cognitive and biological determinants of P300: an integrative review. *Biol Psychol, 41*(2), 103-146.

Pope, A. T., Bogart, E. H., & Bartolome, D. S. (1995). Biocybernetic system evaluates indices of operator engagement in automated task. *Biol Psychol, 40*(1-2), 187-195.

Poythress, M., Russell, C., Siegel, S., Tremoulet, P., Craven, P. L., Berka, C., et al. (2006). Correlation between Expected Workload and EEG Indices of Cognitive Workload and Task Engagement. In D. Schmorrow, K. Stanney & L. Reeves (Eds.), *Augmented Cognition: Past, Present and Future* (2 ed., pp. 32-44). Arlington, VA: Strategic Analysis, Inc.

Prinzel, L. J., Freeman, F. G., Scerbo, M. W., Mikulka, P. J., & Pope, A. T. (2000). A closed-loop system for examining psychophysiological measures for adaptive task allocation. *International Journal of Aviation Psychology, 10*(4), 393-410.

Schmorrow, D., Stanney, K. M., Wilson, G. F., & Young, P. (2005). Augmented cognition in human-system interaction. In G. Salvendy (Ed.), *Handbook of human factors and ergonomics (3rd edition)*. New York: John Wiley.

Schupp, H., Ohman, A., Junghofer, M., Weike, AI., Stockburger, J., Hamm, AO. (2004). The Facilitated Processing of Threatening Faces: An ERP Analysis. *Emotion, 4*(2), 11.

Stripling, R., Coyne, J. T., Cole, A., Afergan, D., Barnes, R. L., Rossi, K. A., et al. (2007). Automated SAF Adaptation Tool (ASAT). *Augmented Cognition 2007 Conference*, 346-353.

Chapter 33

Adaptive Training in an Unmanned Aerial Vehicle: Examination of Several Candidate Real-time Metrics

Ciara Sibley[2], Anna Cole[2], Gregory Gibson[1], Daniel Roberts[3], Jane Barrow[3], Carryl Baldwin[3], Joseph Coyne[1]

[1]Naval Research Laboratory

[2]Strategic Analysis

[3]George Mason University

ABSTRACT

The present study examined the sensitivity of several candidate metrics of real-time workload within the spatial component of an unmanned aerial vehicle (UAV) task. Advanced Brain Monitoring's (ABM) wireless B-Alert system was used to collect participant's EEG workload and engagement data. Eye tracking data was also collected. The UAV simulation required participants to report heading information of moving vehicles, as seen from the UAV. There were four blocks of difficulty, over which a significant performance decrement was shown. Additionally, participants rated their workload significantly higher and pupil diameter significantly increased across blocks of increasing difficulty, as well as within each block during periods of highest mental demand. ABM's workload and engagement metrics however did not show a significant change over or within blocks. The results showed that pupil diameter shows promise as a correlate of mental workload.

Keywords: Mental Workload, Training and Simulations, Augmented Cognition, Adaptive Training

INTRODUCTION

Augmented Cognition emphasizes the use of a closed-loop system using real-time physiological assessment to improve human performance (Schmorrow & Stanney, 2008). In a training environment closed-loop systems could reduce the time required to train an individual by keeping workload at an optimal level for learning (Coyne, Baldwin, Cole, Sibley, & Roberts, 2009). Several metrics such as pupil diameter and electroencephalographic (EEG) have been shown to vary predictably with increases and decreases in workload. Monitoring these different metrics allows training to be optimized in computer based training (CBT) environments. Ultimately, this research will impact the way CBT is conducted by establishing the foundation for adaptive automation through monitoring neural resources.

EEG and eye tracking metrics have been extensively investigated as a means of assessing cognitive workload. For example, (Berka et al., 2007) developed a mental workload metric based on an individual's EEG signal that tracks task demand in mental arithmetic and digit span tasks. Other researchers have focused on eye tracking metrics and found changes in pupil diameter, fixation duration, and blink frequency to be predictive of various levels of cognitive demand in a task (Tsai, Viirre, Strychacz, Chase, & Jung, 2007; Van Orden, Limbert, Makeig, & Jung, 2001; Veltman & Gaillard, 1996). Additionally, many researchers have had success using artificial neural networks (ANN) to accurately classify different operator states for individuals (Wilson, 2005; Wilson & Russell, 2003) and improve performance with the aid of adaptive automation (Wilson & Russell, 2007).

Recent advances in eye tracking and EEG technologies have made utilizing closed-loop systems based on physiological measures more feasible. For example, accurate and unobtrusive off-the-head eye trackers now allow and account for head movements and can collect and process data in real-time. Furthermore, technologies like wireless EEG caps and dry, no-prep electrodes have recently been developed (Christensen, Estepp, Wilson, & Davis, 2009); both of which reduce the prep time normally required. Both EEG and eye tracking data can also now be collected and run using affordable personal computers that are capable of processing and storing large amounts of data. These and other similar advances have made it viable to utilize this type of technology in a CBT environment.

The ultimate goal of this multi-year effort is to build an automated training environment where objective physiological metrics along with subjective workload ratings and quantifiable performance measures can be used to classify an individual's workload and guide desktop training simulations. The purpose of the current study, reported here, was to examine neurophysiological markers of workload in a simulated UAV task at varying levels of difficulty.

METHOD

PARTICIPANTS

All participants (N= 15) were volunteers recruited from the Naval Research Laboratory. None of the participants had any prior experience with UAV simulators. Two were dropped from the study: one was due to second day attrition and the other because of partial dropped eye tracking data. Therefore, thirteen participant's eye tracking and performance data were analyzed and only the last nine participant's electroencephalographic (EEG) data were analyzed due to a hard drive error that caused four participant's data to be lost.

MATERIALS

Advanced Brain Monitoring's (ABM) wireless B-Alert system was used to collect participant's EEG data. The system uses a wireless six channel head cap that transmits data via Bluetooth to a PC running ABM's B-Alert software. ABM's classification algorithms assessed raw EEG and provided a second by second workload and engagement metric on a scale of 0-1. In addition, the Tobii X120 off-the-head eye tracker was used to collect pupil diameter and gaze position data. The unit was placed in front of the participant and just below the surface of the monitor running the simulation. The system recorded both eyes at 120 samples per second.

Virtual Battlespace 2 (VBS2) by Bohemia Interactive, Australia was used to construct the UAV simulation scenarios. VBS2 is a high-fidelity, 3-D virtual training system used for experimental and military training exercises. One Windows PC ran the UAV scenario, while a second PC recorded the eye tracking data, and a third recorded the EEG data. All computers were time synched using network time protocol in order to ensure accurate post-hoc data analysis.

TASKS AND PROCEDURES

UAV DESKTOP SIMULATION

After receiving a brief PowerPoint training about the task, participants engaged in a UAV desktop simulation created from videos using VBS2 where they were trained to report information on enemy targets as seen from a UAV. A continuous video stream from the UAV was shown on the monitor (Image 1) and participants were asked to report heading information about the target vehicles crossing the screen. Participants were given the heading of the UAV and were required to estimate the heading of the vehicle on the ground. A graphical depiction of a compass facing due north with 30 degree increments was provided to the participant for reference. After entering the target heading estimation, participants were then asked to rate their mental effort in calculating the target heading.

The difficulty of the task progressed over four blocks of trials. Only one vehicle was shown on the screen at a time and a total of sixteen vehicles were shown within each block. Difficulty was manipulated by varying the UAV heading as well as the possible target heading. For example, the easiest level (block one) showed the UAV heading at only 0 degrees and the target's heading could be either 0, 90, 180 or 270 degrees. The most difficult level (block four) showed the UAV heading at various 30 degree increments, which changed after every two targets, and the target heading could be any 30 degree increment.

Since this simulation is ultimately intended to help train a UAV operator, the order of difficulty levels were not randomized. On the first day, participants only completed one block, referred to as the baseline block, which was the equivalent difficulty level of block four. On the second day of the experiment, participants progressed through the task from block one to block four. This was done in order to assess learning, by comparing performance on the baseline block and block four. Each block took approximately eight minutes to complete.

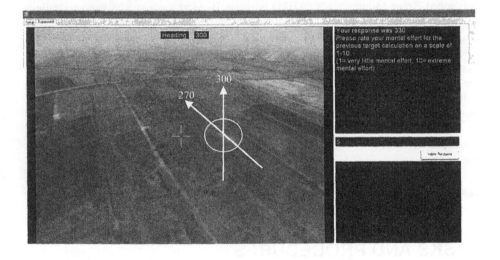

IMAGE1. Screenshot of the UAV simulation. Note the dust trail of an enemy vehicle just to the right of center. Based on the given UAV heading of 300, the participant would correctly report this vehicle heading as approximately 270°.

THE EXPERIMENT

All participants took part in two, one hour sessions over two days. At the beginning of each day, participants were prepped for EEG recording with ABM's six electrode wireless headset. Both EEG and eye tracking data were collected while participants were engaging with the UAV simulation.

On the first day, participants completed ABM's thirty-minute vigilance task. This task was developed by ABM as a means to filter out noise and uniquely fit classification algorithms to a participant in order to assess various levels of cognitive state. The vigilance task and software are part of ABM's real-time EEG classification system. After completing that task, any subsequent EEG data was run through ABM's classification algorithm to provide an individual's workload and engagement in real-time. After this process, participants reviewed a PowerPoint presentation that contained an overview of the tasks and training on how to complete the heading determination task. Participants were given a brief practice on the task and they next completed the experimental baseline block.

On the second day, participants were prepped for EEG and the experimenter briefly reviewed the task instructions. Following the instructions, participants began the UAV simulation while participant performance, EEG, and eye tracking data were collected, along with subjective mental effort ratings. All participants proceeded from blocks one through four with targets appearing at the exact same time, in the same order.

RESULTS

BEHAVIORAL PERFORMANCE

Analysis of performance data for blocks one through four confirmed effective manipulation of difficulty among levels within the UAV simulation. A significant difference existed among blocks one through four in heading error, $F(3, 36) = 16.52$, $p = .000$, $\eta^2 = .75$, subjective workload ratings, $F(3, 36) = 43.47$, $p = .000$, $\eta^2 = .78$, and for errors of omission, $F(3, 36) = 4.50$, $p = .006$, $\eta^2 = .29$. Heading error was computed by dividing the error from correct heading answer by 180 degrees; subjective ratings were on a scale of one to seven; and errors of omission were averaged over the entire block. See Figure 1 for a depiction of these effects.

While a statistically significant difference does not exist between heading error on the baseline block ($M = 0.16$, $SD = 0.08$) and block four ($M = 0.13$, $SD = 0.07$), the average error did decrease slightly and errors of omission decreased from 1.69 on the baseline block to 0.92 on block four.

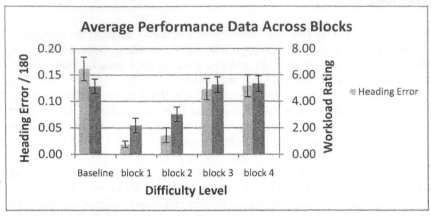

FIGURE 1. Average heading error and subjective workload ratings across all blocks

ABM'S WORKLOAD AND ENGAGEMENT INDICES

Preliminary analysis of the ABM workload and engagement metrics showed almost identical levels of workload and engagement when the metrics were averaged within each block and then compared across block levels. Thus, we further investigated the metrics by averaging each classification over the three seconds preceding participant response for each target heading. This time was chosen because it should correspond with when the participant is calculating the target heading, and thus is most cognitively loaded. Still, results revealed no significant difference in the ABM's workload metric across blocks one through four, $F(3, 24)$ = 1.62, p = .211. Similarly, no significant difference existed in ABM's engagement metric across blocks one through four, $F(3, 24)$ = 1.41, p = .265. See Figure 2.

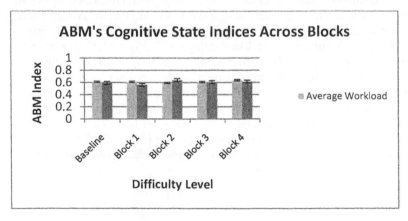

FIGURE 2. ABM's engagement and workload indices averaged three seconds prior to participants providing their heading response

PUPILLOMETRY

Pupil dilation was investigated as a measure of mental workload, and consequently pupil diameter was averaged within each entire block and compared across difficulty levels. Analysis revealed significant differences in pupil diameter among blocks one through four for the left eye ($F(3, 36) = 6.9$, $p = .005$, $\eta^2 = .37$) as well as the right eye ($F(3, 36) = 6.9$, $p = .008$, $\eta^2 = .37$), as shown in Figure 3. Analysis of the pupil diameter between the baseline and block four yielded some interesting results. A significant difference between the baseline ($M= 3.34$, $SD= .46$) and block four ($M= 3.25$, $SD = .39$) did exist for the left eye, $F(1, 12) = 5.18$, $p = .042$, $\eta^2 = .30$. However no significant differences existed between the baseline ($M= 3.34$, $SD= .36$) and block four ($M= 3.31$, $SD= .40$) for the right eye, $F(1, 12) = 0.17$, $p = .688$, $\eta^2 = .02$.

Further investigation of pupil dilation also prompted averaging pupil size over the immediate seconds preceding participant response for each target heading. Increments of one, three, and ten seconds were investigated and all yielded similar results. In particular, pupil diameter across blocks one through four was significantly larger one second preceding heading response when averaged across the whole block for the left eye, $F(1, 12) = 64.96$, $p = .000$, $\eta^2 = .84$, and the right eye, $F(1, 12) = 88.11$, $p = .000$, $\eta^2 = .88$. This suggests that pupil dilation is sensitive to phasic changes in workload over a small amount of time and confirms pupil dilation as a highly promising correlate of workload. See Figure 4 for a comparison of the different average time increments.

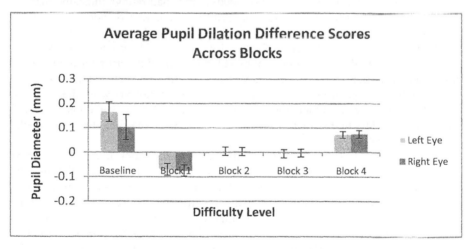

FIGURE 3. Average of all participants' pupil size difference from his or her average pupil size for each block

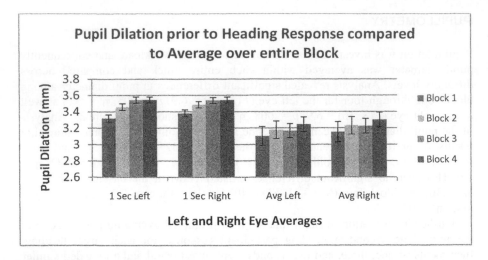

FIGURE 4. Average pupil dilation averaged one second prior to heading response compared to pupil dilation averaged over the entire block

DISCUSSION

Analysis of the performance data and subjective workload ratings indicated that the various levels of difficulty were successfully manipulated across the task. Subjective workload ratings, errors of omission, and heading error all increase in accordance with increasing levels of difficulty (i.e. from block one to block four). While a significant difference does not exist between heading error for the baseline block and block four, about half as many errors of omission occurred on block four (i.e. a failure to respond due to time pressure or simply not knowing the answer). Thus, heading errors on block four could be influenced by fewer omissions, and therefore be slightly higher than if the omission rate were the same between blocks four and the baseline block.

Comparison of performance between the baseline and block four are of interest as a means of assessing the UAV simulation as a potential training simulation. Due to potentially a lack of power and other factors, pre (baseline block) and post (block four) test training effects weren't statistically different. However, the total time allotted to training was only about forty minutes over both days, since each block took about eight minutes to complete. Hence, with more time to train an individual at each level in a real world training simulation, one would expect to see smaller heading errors and errors of omission by the end of training, compared to the baseline trial. In addition, one would expect ratings of workload to be significantly lower on the post test than the pre test.

Neither ABM's workload index nor engagement index were sensitive to changes in this task across difficulty levels. Changes were also not apparent when the index was calculated three seconds prior to heading response, when workload and engagement should have been highest within the block. On account of these findings, future studies will not be using ABM's cognitive state classification algorithms, but instead will investigate the use of artificial neural networks as a means of assessing workload in a UAV training simulation.

The most promising results of this study were systematic changes in pupil dilation as a function of difficulty level. The initial analysis of pupil diameter was performed by averaging an individual's pupil diameter over each eight minute block. Simply comparing average block pupil dilations yielded significant differences in pupil size across blocks (see Figure 3). Further investigation showed that average pupil size one second prior to submitting heading response was significantly higher compared to pupil size during the rest of the block (see Figure 4). This pre-response computation was also calculated at three and ten seconds preceding response, and yielded similar effects; indicating that this effect was likely not due to some kind of response initiation. Therefore, pupil dilation is not only sensitive to changes in workload over large periods of time, but also is sensitive within the demands of a task. These results substantiate the robustness of pupil dilation as a means of assessing cognitive load.

One surprising result was the large difference between average left and right pupil diameter for the baseline block, that actually yielded differing results when comparing dilation between the baseline block and block four. Left eye data is consistent with research that suggests differences in workload across difficulty levels should diminish with practice (Berka et al., 2004). However, data from the right eye would suggest that this is not the case. At present, further investigation is necessary before any firm conclusions can be drawn.

Future studies are planned to investigate how measures of workload change with practice within a difficulty level. In particular, other eye tracking metrics, such as blink frequency/duration, fixation frequency/duration, and divergence will be investigated. Analysis of blink data were not possible for this study, due to the inability to reliably differentiate lost eye tracking data from blinks using the Tobii eye tracking system. Future studies will use EOG to solve this problem. Fixation data and nearest neighbors analyses also were not possible to analyze because of too much error in the Tobii calibration. This problem has been resolved with new software that will be incorporated into future studies.

Another area of interest will be collecting physiological data when a participant is overloaded. We intend to increase the difficulty level of the hardest block in order to purposely overload the participant. Additionally, fewer blocks will be necessary since it is difficult to distinguish four distinct levels in the performance data. Three levels with more trials in each level will be used in follow up studies.

Overall, these findings show promise for using pupil diameter as a means of assessing workload. More data collection is necessary to investigate other eye tracking and EEG correlates. Using spectral analysis of the EEG recordings may prove more sensitive than the ABM engagement index explored in the present study. Ultimately, with the combination of performance, subjective ratings, eye tracking data, and EEG, we are confident that we will be able to successfully predict user workload and eventually perform mitigations within a closed loop system.

ACKNOWLEDGMENTS

This research was supported by the Office of Naval Research's Human Performance and Education Program.

REFRENCES

Berka, C., Levendowski, D. J., Cvetinovic, M. M., Petrovic, M. M., Davis, G., Lumicao, M. N., Zivkovic, V. T., Popovic, M. V., & Olmstead, R. E. (2004). Real-Time Analysis of EEG Indexes of Alertness, Cognition, and Memory Acquired With a Wireless EEG Headset. *International Journal of Human-Computer Interaction, 17*(2), 151 - 170.

Berka, C., Levendowski, D. J., Lumicao, M. N., Yau, A., Davis, G., Zivkovic, V. T., Olmstead, R. E., Tremoulet, P. D., & Craven, P. L. (2007). EEG correlates of task engagement and mental workload in vigilance, learning, and memory tasks. *Aviation Space and Environmental Medicine, 78*(5 II).

Christensen, J. C., Estepp, J. R., Wilson, G. F., & Davis, I. M. (2009). A Demonstration of a Dry/No Preparation Electrode System for EEG. *Human Factors and Ergonomics Society Annual Meeting Proceedings, 53*, 1652-1653.

Coyne, J. T., Baldwin, C., Cole, A., Sibley, C., & Roberts, D. (2009). *Applying Real Time Physiological Measures of Cognitive Load to Improve Training.* Paper presented at the Proceedings of the 5th International Conference on Foundations of Augmented Cognition. Neuroergonomics and Operational Neuroscience: Held as Part of HCI International 2009.

Schmorrow, D., & Stanney, K. M. (2008). *Augmented cognition: A practitioner's guide.* Santa Monica, CA: Human Factors and Ergonomics Society.

Tsai, Y. F., Viirre, E., Strychacz, C., Chase, B., & Jung, T. P. (2007). Task performance and eye activity: Predicting behavior relating to cognitive workload. *Aviation Space and Environmental Medicine, 78*(5 II).

Van Orden, K. F., Limbert, W., Makeig, S., & Jung, T.-P. (2001). Eye Activity Correlates of Workload during a Visuospatial Memory Task. *Human Factors, 43*(1), 111.

Veltman, J. A., & Gaillard, A. W. K. (1996). Physiological indices of workload in a simulated flight task. *Biological Psychology, 42*(3), 323-342.

Wilson, G. F. (2005). *Operator functional state assessment for adaptive automation implementation.* Paper presented at the Proceedings of SPIE - The International Society for Optical Engineering.

Wilson, G. F., & Russell, C. A. (2003). Real-Time Assessment of Mental Workload Using Psychophysiological Measures and Artificial Neural Networks. *Human Factors, 45*(4), 635-643.

Wilson, G. F., & Russell, C. A. (2007). Performance enhancement in an uninhabited air vehicle task using psychophysiologically determined adaptive aiding. *Human Factors, 49*(6), 1005-1018.

Chapter 34

Learning Effects in Physiologically Activated Adaptive Aiding

J.C. Christensen, J.R. Estepp, I.E. Davis, G.F. Wilson

Human Effectiveness Directorate
US Air Force Research Laboratory

ABSTRACT

Psychophysiological monitoring has been demonstrated to be a highly effective means of controlling task allocation and system adaptation in an adaptive aiding paradigm. However, work to date has often used laboratory tasks or simulations wherein the adaptation would be difficult or impossible to apply in a real-world setting. This project sought to advance the application of the adaptive aiding paradigm by using a more realistic multi-UAS control simulation with physiologically activated partial automation and priority cuing. In order to establish the efficacy of the adaptive system, physiological activation was directly compared with manual activation or no activation of the same automation and cuing systems. In the first data collection session, manual activation resulted in slightly better performance than either no activation or physiological activation. However, performance in the physiological activation condition improved over sessions, and by the third session this condition produced the highest performance. Despite the appeal of physiological activation for more closely matching operator needs, the interaction between operator and aiding system under this approach is complex and requires adaptation or learning on the part of the operator as well as the system to achieve maximum effectiveness.

Keywords: Psychophysiology, operator functional state, adaptive aiding, artificial neural network, EEG, training

INTRODUCTION

Adaptive aiding and adaptive automation have been studied and discussed for quite some time. The idea of reallocating tasks between human and machine (Rouse, 1976) to produce maximum overall system performance has an intuitive appeal that has led to great interest over the years. There are many possible methods of determining task allocation in an adaptive system, but one that has shown particular promise is based on psychophysiological monitoring (Pope, Bogart & Bartolome, 1995; Byrne & Parasuraman 1996; Scerbo, 2007) and real-time modification (Wilson & Russell, 2003; Wilson & Russell, 2007).

Wilson and Russell (2007) present a particularly effective demonstration of the potential of such systems. They used a reasonably realistic simulation of multi-UAS control in a ground attack mission, while monitoring electroencephalographic (EEG), electrocardiographic (ECG) and electrooculographic (EOG) data. These data were used to classify operator workload and initiate aiding when high workload was detected. The aiding consisted of reducing the airspeed of those vehicles that were approaching a target by 50%. They demonstrated that this aiding increased performance much more when delivered based on the physiologic workload classification than when delivered randomly. Reducing the airspeed by half and consequently reducing the rate at which events had to be processed by the operator is clearly a highly effective mitigation; however in combat applications this is not a realistic option.

In an effort to advance the application of physiologically-activated adaptive aiding, this study sought to test more realistic workload mitigations in the UAS simulation used by Wilson and Russell (2007). Rather than reducing the airspeed of the vehicles, when high workload was detected the system implemented a suite of mitigations, corresponding to stages two and three in the hierarchy proposed by Fuchs et al. (2007): directing attention via cuing and optimizing salience of critical events via decluttering/fog layer. In addition, partial automation paired vehicles and targets, subject to operator override. This extends the similar study performed by Parasuraman et al (2009) by testing attentional aiding in addition to automation. In order to assess the effectiveness of this suite, physiological activation was compared with manual activation and a no mitigation baseline.

METHODS

Six participants have completed the study to date. These participants were either paid student volunteers recruited locally, or unpaid government employees who volunteered their time for no specific compensation. Five were male and one female, with a mean age of 23 years.

The simulated UAS operation task was the same as used in Wilson and Russell (2007). Participants monitored the progress of eight or more autonomous vehicles, on two abutted 20-inch (~51-cm) diagonal computer screens, as they flew a preplanned bombing mission. When the vehicles reached designated way points, simulated radar images of the target area were available to the operators. The operators gave commands to download and view the images and then performed a visual search of the images. They were required to mark targets for bombing before the vehicle reached the minimum weapons release distance. If the targets were not selected and/or the weapons release command was not given in time, the weapons from that vehicle could not be released, reducing the number of targets successfully engaged. Participants were not allowed to double back to reengage targets.

Psychophysiological data were recorded from the subjects while they performed the task. EEG from five channels was recorded at F7, Fz, Pz, T5, and O2, positioned according to the International 10-20 electrode system (Jasper, 1958). Reference and ground electrodes were positioned on the mastoids, with impedances verified below 5K ohms. Horizontal and vertical electrooculogram, heart, and blink rate were also recorded. Corrections for eye movement and blinks were made and the data were stored at 200 samples per second. The EEG data were filtered using elliptical IIR filter banks. The passbands were consistent with the five traditional bands of EEG; delta (0.5-3 Hz), theta (4-7 Hz), alpha (8-12 Hz), beta (13-30 Hz) and gamma (31-42 Hz). The data were segmented into ten-second windows with a nine-second overlap. Log power of five bands from the five sites was used in addition to ECG and EOG measures, resulting in 37 features as inputs for workload classification. Workload classification was accomplished via a feedforward artificial neural network (ANN) trained via backpropagation with two output nodes corresponding to low and high workload.

Participants were extensively trained prior to commencing the study. Each participant completed a staged series of training scenarios with the UAS simulation until criterion performance of ~80% ground targets successfully engaged in the high workload condition was achieved. This required an average of fifteen days of training, with one two hour session per day spread out over approximately four to eight weeks. During the course of this training, the number of vehicles controlled simultaneously was increased; the low workload condition was defined as eight vehicles while the high was sixteen. In the last two training sessions, participants were introduced to the mitigation suite and allowed to manually activate it via a single keypress. This mitigation suite identified the highest priority targets and vehicles based on time to contact, color coded (red/yellow/green) the top three pairs, and desaturated (fogged) all of the non-priority targets and vehicles. In addition, target-vehicle pairing was performed automatically by pairing each target with the closest vehicle with the correct weapons type. This pairing occasionally produced errors due to lack of awareness of later targets, and thus required operator oversight. Participants did not experience physiological activation as opposed to manual activation prior to the experimental sessions.

The experimental sessions consisted of three identical sessions run on three different days. Days one and two were sequential, while day three was one week after day one. During each session, participants completed a five-minute warmup with a task training scenario, a four-minute eyes-open resting baseline, and then one run of each of the following: low (eight vehicle) and high (sixteen vehicle) workload runs without aiding of approximately 20 minutes each, and three mixed workload runs that alternated from high to low and back lasting approximately 40 minutes. The three mixed runs included one run each of no aiding available, manually controlled aiding, and physiologically activated aiding. Run order was semi-random, with the constraint that the pure low and high difficulty runs had to be completed prior to the physiologically-activated mixed runs, as these runs were used to train the ANN classifier. NASA-TLX data was collected after each run.

To avoid problems with classification accuracy associated with multi-day data collection, the ANN was retrained for each participant and day. 75% of the pure low and high workload runs was used to train the ANN, with the remaining 25% held back to check accuracy and check for overlearning. Mitigation activation required that any two seconds in a three second moving window classify as high workload; once this occurred the mitigation suite was activated for ten seconds. At the end of this period, the last three seconds were examined and immediate, seamless reactivation was possible. If classified workload dropped to low, then the mitigation suite shut off and remained off for a minimum of five seconds. Manual activation mimicked the physiological activation in operation; pressing the key again at any time during activation would trigger another ten-second activation. A digital countdown timer was visible in the upper right of the screen to aid in managing activations.

RESULTS

Subjective workload ratings as assessed with the NASA-TLX confirm that the mitigation suite was effective in reducing workload in the high workload condition, with both manual and physiological activation significantly lower than the no-aiding condition (Fig. 1). Manual and physiological activation were not significantly different from each other.

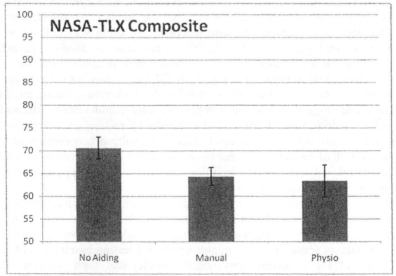

Figure 1. Composite NASA-TLX scores, averaged across participants and days for the mixed high-workload runs. No aiding resulted in significantly higher workload than either of the aiding conditions. Of the subscales, performance, effort, and temporal demand were weighted the highest.

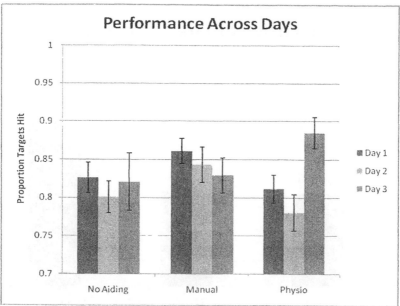

Figure 2. Performance expressed as proportion of ground targets destroyed in the mixed high-difficulty condition. Error bars are standard errors of the mean. Successful training is confirmed by the no aiding condition, which was not significantly different across days and not significantly different from the training criterion value of 80% targets destroyed. "Manual" and "Physio" identify the activation type for the aiding conditions.

Performance reveals an interesting pattern of results (Fig. 2). The no aiding condition provides confirmation that the training regimen has in fact achieved asymptotic performance, with no significant differences observed across days. On the first day of testing, which in many studies would be the only day of testing, there were no significant differences among conditions, though the operator controlled aiding condition resulted in marginally better performance than the other two conditions. All three conditions exhibit a slight drop in performance on day two; this could have been a fatigue effect from the back-to-back days of data collection. On the third day, the only significant difference is the physiologically activated aiding condition, which produced substantially higher performance as compared to the first two days and was the highest performance in the study. Figure 3 is performance just on the third day of the study; the physiologically activated aiding is marginally significantly higher at $t(6)=2.989$, $p \sim .02$, Bonferroni corrected p criterion of .025 due to two planned comparisons.

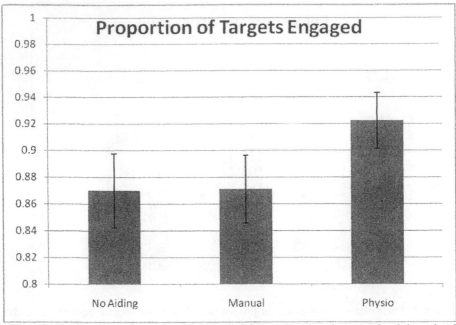

Figure 3. Performance from the third day of data collection from each of the mixed high-difficulty conditions. The physiologically-activated condition resulted in signficantly higher performance than either the no-aiding or manual aiding conditions.

DISCUSSION

This study was designed to advance the application of physiologically-activated adaptive aiding by testing a more realistic suite of mitigations that were designed to both offload a portion of the task to automation and direct attention to time-critical items in order to increase throughput. This set of mitigations was effective at increasing performance and reducing workload, as revealed by comparing either of the mitigated conditions to the no aiding condition. Physiologic activation did eventually result in the best performance observed in the study, but this difference was not apparent until the third day of data collection. As the mitigation suite was identical and participants were trained with manual activation, the most likely explanation for this pattern of results is that the participant must adapt to the nonconsensual physiological activation. This adaptation could be either strategic or physiological; participants could be learning over time to expect the activation and strategize with that expectation, or we could have induced biofeedback training that resulted in participants having some control over activation by the third day. In post test debriefing, no participant identified consciously controlling activation, though one spontaneously stated that "Usually, the physiological activation would trigger just as I was starting to realize that I was in trouble on the task". While that may well have been an ideal example, it illustrates the potential of physiological activation to anticipate conscious realization of an overload state.

It is well worth noting, however, that the performance improvement obtained in this study was much smaller than that observed in Wilson and Russell (2007). This serves to point out a key problem in adaptive aiding systems: the overall system is dependent on the effectiveness of each of the key components. The performance improvement produced by the physiological activation condition is limited by both the accuracy of workload classification as well as the efficacy of the mitigation used. Despite designing the suite of mitigations to address several methods of aiding identified by Fuchs et al. (2007), the performance improvement obtained even in the manual condition is relatively modest, albeit statistically significant. In this and similar studies we have conducted, mitigation effectiveness has been a key issue for applying the paradigm to more complex tasks. While it is relatively easy to identify broad types of mitigation, implementation is inevitably very task and interface specific, and in the case of automation dependent on that automation being intelligent and considerate, after Parasuraman and Riley (1997).

REFERENCES

Byrne, E. A., & Parasuraman, R. (1996). Psychophysiology and adaptive automation. *Biological Psychology*, 42, 249-268.

Fuchs, S., Berka, C., Levendowski, D., & Juhnke, J. (2006). Physiological sensors cannot effectively drive mitigation alone. In D. D. Schmorrow, K. M. Stanney, & L. M. Reeves (Eds.), *Foundations of Augmented Cognition*, 2nd ed. (pp. 193-200). Arlington, Virginia: Strategic Analysis Inc.

Jasper H. H. (1958), "Report of the Committee on Methods of Clinical Examination." *Electroencephalography and clinical Neurophysiology*, 10, 370-375.

Parasuraman, R., Cosenzo, K. A., & De Visser, E. (2009). Adaptive automation for human supervision of multiple uninhabited vehicles: effects on change detection, situation awareness, and mental workload. *Military Psychology* , 21 (2), 270-297.

Parasuraman, R., & Riley, V. (1997). Humans and automation: Use, misuse, disuse, abuse. *Human Factors*, 39, 230-253.

Pope, A. T., Bogart, E. H., & Bartolome, D. S. (1995). Biocybernetic system evaluates indices of operator engagement in automated task. *Biological Psychology*, 40, 187–195.

Rouse, W.B. (1976). Adaptive allocation of decision making responsibility between supervisor and computer. In T.B. Sheridan & G. Johannsen (Eds.), *Monitoring behavior and supervisory control* (pp. 295-306). New York: Plenum Press.

Ruck, D. W., Rogers, S. K., Kabrisky, M. Oxley, M. E. & Suter, B. W. (1990), "The multilayer perceptron as an approximation to a Bayes optimal discriminant function." *IEEE Transactions on Neural Networks*, 1(4), 296-298.

Scerbo, M. W. (1996). Theoretical perspectives on adaptive automation. In R. Parasuraman, & M. Mouloua, *Automation and human performance: theory and applications* (pp. 37-63). Mahwah, New Jersey: Lawrence Erlbaum Associates.

342

Scerbo, M. W. (2007). Adaptive automation. In R. Parasuraman, & M. Rizzo, *Neuroergonomics: the brain at work.* (pp. 239-252). New York, New York: Oxford University Press.

Wilson, G. F. & Russell, C. A. (2003), "Real-Time Assessment of Mental Workload Using Psychophysiological measures and artificial neural networks." *Human Factors*, 45, 635-643.

Wilson, G. F. & Russell, C. A. (2007), "Performance enhancement in an uninhabited air vehicle task using psychophysiologically determined adaptive aiding." *Human Factors*, 49, 1005-1018.

CHAPTER 35

Theoretical Constructs of UAS Training

Erik S Viirre[1], Klaus Gramann[2], Shawn Wing[1]

[1]Naval Health Research Center
San Diego, CA 92106-3521, USA

[2]Swartz Center for Computational Science
University of California, San Diego
San Diego, CA, 92037, USA

ABSTRACT

In the 20th Century, navigation in three dimensions was largely the province of pilots. With ubiquitous deployment of Unmanned Aerial Systems, thousands of people who have not been selected or trained are now operating aircraft, developing intelligence based on the flight of these vehicles and most importantly, depending on that information to be accurate. Ironically, the ease of operation that robotics enables with UAS systems means that operators require little experience to get them in the air. However, the hours and years of flight training bring experience to pilots that remote UAS operators may not have. Further, UAS operators must mentally project themselves into the reference frame afforded by the UAS. Such projection is a difficult task. In this paper we describe the background on mental reference frames in three dimensions that will be critical for UAS operations.

Keywords: Reference Frame, Three Dimensions, Egocentric Allocentric. Operators

INTRODUCTION

Why Study Spatial Navigation?
In the early years of aviation, navigators were selected from the most highly

mathematically skilled volunteers for military service. Then as now, mission success required accurate navigation. However, the low cost and ubiquity of flying platforms with cameras now means that thousands of line personnel are operating *and depending on* information gleaned from aerial vehicles they must navigate. Training personnel to operate flying platforms and to obtain reliable information from the platforms will require nimble minds that understand complex dynamic battle-spaces for situation awareness. What individual differences affect abilities to carry out spatial navigation tasks and affect individual abilities rate of training? Can we detect problems in situation awareness with electrophysiological monitoring? Can we combine theory and technology to optimize training programs to improve performance in spatial navigation and situation awareness?

Behavior and Navigation

There are multiple reference frames that the brain uses to interact with the world. Objects within arms' length are naturally in a body reference frame. Further, studies with animals show that space extended beyond arms' reach is also in a body (or head oriented) reference frame: this is called Egocentric. However, in higher primates and humans the *Allocentric* or external reference frame can be adopted. An allocentric frame is akin to a map, where objects in the world are placed relative to a frame of reference, such as a location in the world and a direction (say an intersection and a road going north). Use of an allocentric frame is important for it allows ready understanding and projection of activities of objects besides the self (the truck is going East on 4th Street to "A" Avenue). An egocentric frame forces constant recalculation of relative location of objects outside the self. However, for self navigation, egocentric and allocentric are relatively similar in computational overhead. Studies of movement in 2 dimensional spaces suggest that individuals are about equally divided as to which frame they use egocentric or allocentric. However, it is relatively unknown which frame people use when experiencing motion in 3 dimensions. The modern warrior must understand 3D battlespaces.

Training and performance for Spatial Orientation

In people required to carry out tasks requiring ongoing spatial navigation there have been two general mechanisms for training: selection and brute force. Self selection has been the beginning. Many people will not even volunteer for a pilot/navigation job because of their recognition of their own limitations or their fear of getting lost. In pilot training programs of the past, navigation tasks were trained by brute force of hundreds of hours of repetition. With the ubiquity of autonomous flying platforms, a more "virtual game"-like environment is created (you don't literally die if the vehicle crashes), but the consequences of poor navigation and target selection are still critical cases of mission failure. The common use of autonomous flying platforms means more people will be required to operate with good 3D understanding. Are there selection procedures and improved training methods that can be implemented to improve performance and reduce training times? In our project we are examining reference frames in 3D. We are determining whether people are ego or allocentric for virtual motion in pitch and yaw (3D).

Understanding reference frames may help predict performance and may predict training necessary for proficiency.

Technology Intervention for Spatial Orientation

In our ongoing project we are also examining neural markers of spatial orientation. We are a determining if we can localize sources of activity in the brain that change whether the subject is oriented or disoriented in a virtual tracking task. Brain activities could be used as feedback sources during training procedures to detect lack of orientation and direct the subject so they can recognize their loss of orientation and return to course.

WHY STUDY SPATIAL NAVIGATION?

Identifying and assessing the mental states of military personnel is of great importance in order to develop technology for improving maintenance of performance on mental tasks. There are a variety of distinct mental states known to adversely affect performance including: fatigue, sleep deprivation, stress, high workload, and motion sickness. During operations using moving platforms (land, sea or air-based) changes in states related to motion sensation occur in many personnel. Such motion related problems include spatial disorientation (SD), impaired situation awareness and even debilitating motion sickness. Even in those people who are not motion sick, disorientation can have severe effects on mission performance. Disorientation and other deleterious mental states can lead to the overall condition of loss of *situation awareness* (SA), where the individual loses perspective of their overall position and direction of action. Loss of situation awareness can lead to failure to complete mission objectives or inability to react to contingencies or anomalies.

To approach the problem of managing loss of situation awareness, we are examining problems that result in spatial disorientation. If a unique neural marker for spatial disorientation could be reliably identified, operations and training could be improved through the use of feedback to operators and training instructors. During simulated or real operations, there are multiple factors that can lead to poor performance. If supervisors or trainers can be given information as to the mental state of personnel, particularly if there are conditions that impair situation awareness, then interventions can be mounted. Thus a brain activity marker that indicates a subject is spatially disoriented could be well used by a trainer. For example, if a subject carrying out a mission in a virtual environment is consistently delivering slow performance, a marker that indicates persistent spatial disorientation would be useful to point to training in that area.

We have already demonstrated data quantitating the often observed/experienced phenomena of one's being confident they know where they are going, when in fact they are lost (high confidence but inaccurate spatial orientation) and we have found neural activities potentially related to this state (Viirre, 2006). The current project will extend our ability to identify such neural markers in three dimensional reference frames.

BEHAVIOR AND NAVIGATION

Background Theory

In his book "The Brain's Sense of Movement", Berthoz (2000, p 99.) gives some background on body frames of reference. There is the space/frame occupied by the body itself, which is extended to the space where one can reach. Notably, the reaching space can be extended by tools and even artifacts like brake peddles where drivers or pilots located far above the ground can "extend" their feet to it. The reference frame that develops coordinates relative to the body is the "egocentric" frame. It can extend to the reaching space and indeed far beyond the body itself. Lower animals use egocentric space. However, higher primates and humans can carry out the mental transformations to "allocentric" or external space. The allocentric space develops coordinates relative to a fixed object and direction, such as a street going north from an intersection. The allocentric space is thus map-like and importantly, its non-moving constituents maintain a constant reference pattern. In contrast, a person who is using an egocentric space and is moving in a room, has constantly changing coordinates relative to all constituents, like doors, windows and furniture. Turns are even more problematic in egocentric space. However, the use of allocentric space requires the subject to carry out the mental transformation of their body position and motion into the map space. The use of allocentric space does readily allow mental simulation of self motion, and also motion of other objects in the environment. The ability to mentally handle allocentric space probably develops in late childhood and reverts back to egocentric in times of stress. Athletes in confrontational running sports such as hockey and football probably take advantage of errant mental projections when "faking out" (or "deking") opposing players when rushing towards them. Animal studies suggest that allocentric mapping activities take place in Occipito-temporal cortex and para-hippocampal areas Galati (2000).

Reference Frames in 3 Dimensions.

Gravity is an external reference frame and provides an "External plumb-line" that can be described as geo-centric. Indeed neurophysiologic studies suggest that the head in mammals (including humans) is stabilized relative to gravity. "It's as if the brain creates a stabilized platform to coordinate movements of the limbs", according to Berthoz (p 101). However, in complex movements, (such as dance) feet rarely touch the ground, thus the ground may be a poor reference frame. Further, there is the gravito-inertial differentiation problem, where linear motion is indistinguishable from gravity. Optic flow is a powerful driver of the sense of the vertical, as illusions in tilted rooms can demonstrate. Thus while it might appear that gravity would provide a solid reference to the vertical, it appears not to be the case.

Reference Frame Descriptions.

Ego vs. Allo in 2D

Map View

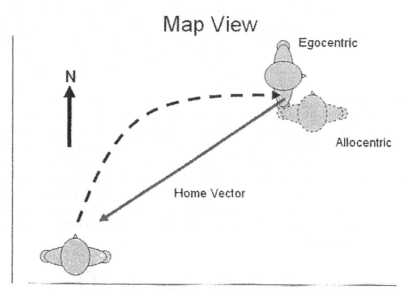

Figure 1. Egocentric vs. Allocentric mental navigation

As described above there are two general possibilities of means of navigation: reference to the self or reference to the environment being navigated. Navigation by reference to the self ("egocentric") means that the movement of the body is monitored by the nervous system with a reference frame which is fixed in relation to the body. This frame usually has "straight ahead" as a vector directed from the face or torso. In contrast, navigation by reference to the environment, ("Allocentric") uses a reference frame related to the local environment. The local environment could be a room, building with references like the door or windows or a geographic area with conventional compass directions. In figure 1 above, we can see a schematization of egocentric vs. allocentric navigation methods.

The graphic is an overhead view of motion of person through an environment. The subject starts moving forward and then turns to the right and stops. In the egocentric mode, the "Straight ahead" position is relative to the nose or chest of the subject, and thus the straight-ahead axis of the body turns 90 degrees. However, in the allocentric mode, the mental reference frame of the subject is fixed to the environment (in this case facing North). At the completion of the motion, the allocentric reference frame is still with the subject facing north. Critically, the vector to point of origination ("home") is very different in the two frames. It is over the right shoulder in the egocentric frame and over the left in the allocentric.

In allocentric navigation, movements are tracked with reference to the environment. Particularly when subjects are navigating *virtually* through an environment, mental imagery of motion and location become critical in representations of motion. Importantly, studies of navigation through 2 dimensional virtual environments show that mental representations by subjects are about equally divided between egocentric and allocentric representations (Gramann, 2006). Further, Gramann has seen that subjects are very fixed in their modes of navigation.

Importantly, egocentric versus allocentric modes of navigation have not been well examined in three dimensions. In the figure below, we demonstrate a motion in the pitch plane and the egocentric versus allocentric frame references.

As with motion in the horizontal plane, we can see the two possibilities of mental orientation after a virtual forward motion with a pitch down. In the Egocentric case, the subject has pitched forward relative to the reference frame, whereas in the allocentric case the body orientation is still oriented erect to the reference frame. As with movement in the horizontal plane, the vector pointing to the original position is different. The egocentric subject has a vector pointing behind and below the head and the allocentric homing vector is pointing back and up. Understanding the divisions of the population that are egocentric and allocentric in the pitch plane is important. Incredibly (including one of the authors and to his surprise), motion in the horizontal plane may be egocentric *and may be allocentric in the pitch plane.* The relative incidence of egocentric and allocentric modes in pitch and yaw is not known.

Ego vs. Allo in 3D

Side View

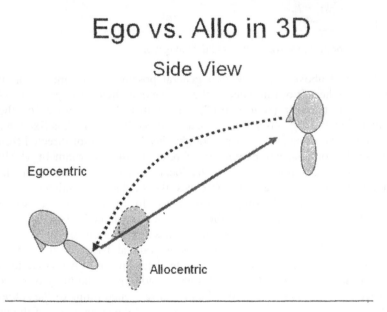

Figure 2. Reference frames in pitch motion.

TRAINING AND PERFORMANCE FOR SPATIAL ORIENTATION

Situation awareness and spatial orientation have historic and current military importance in manned aviation. All three US military branches have carried out research on recovery from unusual attitudes, displays for orientation and post-accident reviews in spatial disorientation incidents. Training programs for avoidance and recovery from SD are part of the traditional military and civilian flight syllabus. That said, there have been no programs for determining *cognitive components* of flight orientation and recovery. Aviation programs have depended on self selection (only people who want to be pilots apply), selection based on basic cognitive skills and training, such as recovery from unusual attitudes. Finally, hundreds of hours of classroom, simulator and supervised flight training are the basis for operational flight safety. Unmanned aerial vehicles take away the risk of loss of pilots, but maintain the risk of spatial disorientation. The risk however, is loss of situation awareness. "Where is my target and what is it doing (relative to me)" is the problem (Navathe, 1994). Training programs for inexpensive (i.e. expendable) UAV platforms are less than a week in duration. Do cognitive capabilities make a difference in the operation and use of intelligence of capabilities of UAV platforms? Are there means to select individuals as good operators (model airplane pilots?) and are there means to optimally train most people to be operators, depending on their individual differences?

It would appear that the egocentric vs. allocentric orientation for subjects would have an influence on their performance on orientation tasks, or at least on their orientation training. Tests, such as those developed by Gramman (2006), where an individual watches optic flow displays on a computer screen and then gives estimates of starting positions, are easy to implement. Other tests of spatial abilities, such as mental rotation tasks can be used to assess cognitive abilities of spatial orientation.

Virtual versus Real Navigation

If *virtual navigation* is mapped onto activity in the *real world*, the mapping modes used by subjects become critical. For example, for piloting an aircraft by instruments and not having visual reference to the ground, we can use egocentric and allocentric reference displays. Indeed, western navigational instruments and Russian navigational instruments are completely different: Western is Egocentric and Russian is allocentric. Tragically, the difference in convention appears to have lead to a fatal commercial jet accident in 2008 near Perm, Russia. A Russian pilot, with long experience with Russian instrumentation was flying an American jet during instrument conditions. Through analysis of the black box recordings, it is believed that the pilot reverted to the interpretation of the flight instrument as he learning in Russian aircraft and inadvertently tipped the aircraft into an unrecoverable attitude Interstate Aviation Committee (2009). Clearly, extensive training does not preclude disorientation if the instrumentation is in a different reference frame.

Common Virtual Navigation Tasks

Fortunately, transition accidents between different primary flight instruments are very rare as flying by instruments is completed safely in tens of thousands of flights every day. However, there are common circumstances where mental or virtual navigation occur and operators must perceive correctly their orientation and motion for critical activities such as targeting and way-finding for remote personnel or systems. Further, training for virtual navigation is a time-intensive and mission critical task. Very commonly, military personnel use 3 dimensional map representations of activity in their daily planning and execution of missions. For example, Falconview^{TM} is software commonly used by the Marines for mission planning. Users must project the motion of themselves and other actors into the virtual mission they create.

Perhaps the most common real-time activity involving complex virtual 3D activity is piloting a UAV. Unlike a piloted aircraft, the UAV operator depends solely on the visual displays for operation of the vehicle and for understanding of their spatial situation: position, orientation and movement. Further, the most common UAV platforms have simple displays and impoverished information. The system used in our study is a simple air vehicle with a fixed camera and a video display with some numeric information (RAVEN B). Training operators to maneuver the vehicles, keep track of the vehicle's position and keep track of objects of interest on the ground can be difficult (Becker, personal communication). UAV operation will be described further below.

TECHNOLOGY INTERVENTION FOR SPATIAL ORIENTATION TASKS

Operation of a remotely piloted Unmanned Aerial Vehicle (UAV) has been selected as a task where situation awareness and spatial orientation are critical to mission success and where intensive, prolonged training is required for successful operation. Operation of an aerial vehicle is inherently more difficult than ground vehicles because of movements in an additional dimension. Our overall research program includes:

1) Simulation of spatially disorientating tasks in UAV training software.

2) Measurement of neural markers of disorientation during simulations.

3.) Planning for engineering implementation of neural marker usage.

The Raven Unmanned Aerial Vehicle (UAV) is a semi-autonomous aircraft. It is used for observation of terrain and "over-the-hill" objects within a few miles of its operator. Raven operations are intended for line personnel, not specially trained pilots. Thus a wide variety of skill levels and experience and abilities will appear in people who are expected to use this platform. Using the simulation of the Raven system from Lockheed Martin Corporation, we are re-creating disorienting vehicle activities and with existing electro-encephalographic (EEG) recording systems, we are recording brain activity during simulated operations where subjects are in control and where subjects show loss of spatial orientation. Using similar techniques as were used in the analysis of the previous experiments, we will examine the brain activity for signature identifiers of loss of situation awareness. Loss of situation awareness is demonstrated in the figure below.

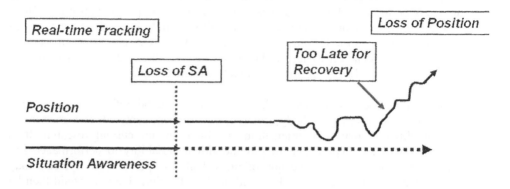

Figure 3. Schematic of loss of situation awareness (SA) during a tracking operation. Undetected deviation from the projected path results in loss of recovery of position information.

The intended program for use of the signature brain markers of loss of situation awareness is shown in figure 4 below. When the system that monitors neural activity determines loss of situation awareness, a signal is sent to the operator warning of the disoriented state, the operator then responds ("checks instruments") and regains control before significant deviation from planned path occurs.

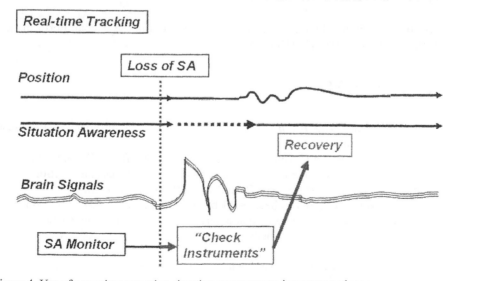

Figure 4. Use of a monitor assessing situation awareness to intervene and recover.

CONCLUSIONS

Unlike many conditions for ground based personnel, UAS platforms offer the unique opportunity and the unique challenges of views of operations from the air in real-time. The operators of such systems need to not only know their location and orientation, but also the location and orientation of the UAS platform and then the location and orientation of targets being viewed via the platform. The mental mathematical transformations for such monitoring are complex.

Compounding the complexity is the variation of mental reference frames that individuals use to navigate: body-centered, environment centered, or a mixture of both in different planes of orientation, as found in our current research. It is unknown what the implications of this individual variability to performance and training requirements are. Selection of individuals to operate UAS systems may require typing as to orientation skills and modes of action. Training could then be individually modified to optimize skill acquisition. For example, subjects that do not typically use an allocentric (map) reference frame could be given extra training on orientation through instruments and simulations. Finally, advanced technologies could go beyond performance monitoring to monitoring of actual neuro-physiologic status. Systems that can detect lack of attention, reduction of situation awareness and even spatial disorientation could be integrated into training systems. Trainees and instructors could be alerted if a student is no maintaining awareness of relevant spatial information.

ACKNOWLEDGEMENTS

This project is funded by the Office of Naval Research in the Human Performance, Training and Education (HPTE) program.

REFERENCES

Berthoz A. (2000) *The Brain's sense of Movement*. Harvard University Press, Cambridge Mass.

Galati G, Lobel E, Vallar G, Berthoz A, Pizzamiglio L, Le Bihan D. The neural basis of egocentric and allocentric coding of space in humans: a functional magnetic resonance study. Exp Brain Res. 2000 Jul;133(2):156-64.

Gramann K, Müller HJ, Schönebeck B, Debus G. The neural basis of ego- and allocentric reference frames in spatial navigation: evidence from spatio-temporal coupled current density reconstruction. Brain Res. 2006 Nov 6;1118(1):116-29.

Interstate Aviation Committee, Air Accident Investigation Commission Final Report. 737-505 VP-BKO.

Navathe PD, Singh B. An operational definition for spatial disorientation. Aviat Space Environ Med. 1994 Dec;65(12):1153-5.

Viirre E.S., Wing S., Huang R.S, Strychacz C., Koo C., Stripling R., Cohn J., Chase B. and Jung T.P. EEG Markers of Spatial Disorientation. (2006) Foundations of Augmented Cognition 2nd Ed. Eds. Schmorrow D.D, Stanney K.M. and Reeves L. Falcon Books, San Ramon, California. pp 75-84.

Chapter 36

Neuroergonomics, From Theory to Practice

Cali M. Fidopiastis[1], Denise M. Nicholson[2]

[1]School of Health Professions
University of Alabama at Birmingham
Birmingham, AL, 35294 USA

[2]Institute for Simulation and Training
University of Central Florida
Orlando, FL 32816, USA

ABSTRACT

Neuroergonomics is a field of research that combines neuroscience and ergonomics to better understand brain function in real world settings. Adapting existing methods and techniques from a basic research paradigm to one that is applied comes with many challenges. We discuss the tenets of the neuroergonomics field and give examples of how the tools and methods of neuroscience are applied within the neuroergonomics framework.

Keywords: Neuroergonomics, applied cognition, adaptive systems, neurosensing

INTRODUCTION

Tenets of the field of neuroergonomics include: 1) cognition and action arise from the brain, 2) the environment shapes brain networks, and 3) brain and behavior together are necessary to understand action in real world spaces (Parasuraman and Hancock, 2004). The research agenda for this emerging applied science includes the use of neuroscience techniques and technologies such as electroencephalography (EEG) and event related potential (ERP) analysis. While these methods have a long history in traditional neuroscientific research, applying them within operational environments is not straightforward (Sarter and Sarter, 2003). In this paper, we discuss current theories of neuroergonomics and provide examples of using neurosensing technology such as Near Infrared Spectroscopy (NIRS) to study brain activity during task performance. We also discuss potential problems with the approach such as translating neuroscience metrics into real world assessments. These potentials and pitfalls are discussed within the framework of the underlying tenets that define the neuroergonomics field.

The first two tenets of neuroergonomics have their foundation in neuropsychology literature, which in turn laid the groundwork for neuroscience to emerge as a field of research in the late 1960's. The second tenet implies that neural circuitry changes structurally and functionally when an operator interacts with the task environment and that these neural changes can be detected using neurosensing technologies. While this tenet is fundamentally true, it also implies learning. Learning does not always require the presence of environmental stimuli. For example, mental practice does not require that the learner be physically present in the environment to experience measurable brain network changes. The third tenet aligns more with the field of cognitive psychology, which places emphasis on ecologically valid or contextualized experimental designs (Neisser, 1967). While cognitive psychology and cognitive neuroscience remain tied to experimental paradigms of behaviorism, applied cognition fields such as virtual rehabilitation and augmented cognition necessitate a shift in experimental design strategies that include near-to-real-time cognitive brain state assessment (Fidopiastis and Wiederhold, 2008). Fidopiastis and Wiederhold (2008) suggest that simulation-based rehabilitation and training environments can achieve the level of control necessary to explore brain activity of operators performing real world tasks with the assistance of portable neurosensing devices. However, these empirical endeavors are not without their challenges. These issues are further discussed in the context of applied experiments.

We present several studies that demonstrate the use of neurosensing technologies within military relevant tasks. These case studies provide a means of exploring methods that directly transfer from neuroscience and those that need to be adapted for this field of use. These examples address the successes and failures of the burgeoning field of neuroergonomics. More importantly, they provide guidance for understanding the way forward.

ADAPTIVE INSTRUCTIONAL TRAINING SYSTEM DESIGN USING PSYCHOPHYSIOLOGICAL MEASURES

Nicholson et al. (2007) proposed an adaptive training system that dynamically monitored both learner-state and trainer-state to optimize the human-machine training experience by triggering the selection of multiple mitigation strategies (e.g., dynamic task reallocation). The driver of such an approach is the instructional design strategy, not the understanding of psychophysiological measures as they relate to learning or training. There is no question that psychophysiological measures can successfully trigger adaptive automation in training simulators (Wilson and Russell, 2007); however, before these training outcomes are meaningful, the psychophysiological measures must be valid, reliable, and generalizable (Reeves et al., 2007).

Two main gap areas when using psychophysiological measures in applied domains include evaluating baseline measures used to calculate the cognitive state changes during task performance and separating task-independent cognitive states and functions (Kruse, 2007). The importance of determining how baseline measures and state-independent cognitive functions affect adaptive learning outcomes is crucial to generalizing the successful outcomes of instructional design across military training. Significant work in developing applied cognition methods such as integrating multiple psychophysiological data streams (Vartak et al., 2008) and metrics (Fidopiastis et al., 2009) for use in adaptive training systems has provided a new means to study instructional design strategies that may accelerate and optimize learning within augmented cognition training systems.

Berka et al. (2010) discussed the use of traditional EEG signatures such as event related potentials (ERP), along with event related synchronization (ERD) and desynchronization (ERD), to identify what type of threat a person has perceived (i.e., vehicle, person, or improvised explosive device, IED). Significant mean amplitude differences in an observed late positive ERP were found that may reflect decision making or cognitive processing demands associated specifically with IED threat presence or threat in general. Finding such brain signatures is a required step to advance adaptive training systems of this type and can be determined using classical neuroscientific approaches. Discerning whether the EEG signature represents a general or specific brain process is a matter of replicating the experiment under different conditions, which is achievable in a simulated military training exercise. These types of replication studies are rarely performed, as are transfer of training studies. Neuroergonomics, as it applies to adaptive system trainers, should include efficacy measures that determine the reliability of the neurobiological signature and the transferability of the combined training approach to real world operations.

OPTICAL IMAGING AND COGNITIVE MAPPING FOR THE REAL WORLD

Neuroergonomics requires a cognitive map of processes (e.g., attention and memory) that is reliably represented by patterns of neuronal firing across brain regions (Sarter and Sarter, 2003). The technology needed to trace and monitor these neuronal pathways has only recently existed as medical imaging devices such as magnetic resonance imaging (MRI). Medical technologies that could monitor both central (MRI and EEG) and peripheral (ECG-electrocardiography) nervous system responses were adapted for use in neuroergonomic experiments (Parasuraman and Wilson, 2008). Algorithms for use within real-time dynamical human-machine systems are still under development.

Combining methods such as MRI and EEG allows for monitoring of the same cognitive construct over time and brain space. These studies provide insight into the mechanisms of brain processing, such as processes supported by neurogenesis (growth of new neuronal cells) versus those that reflect functional changes (changes in neural firing patterns). Both types of brain plasticity, neurogenesis and pattern change, may result in similar behavioral outcomes such as quicker response times. For this reason it is important to understand which type of brain change is necessary for optimizing or understanding performance in an operational setting (Fidopiastis and Wiederhold, 2008).

Functional near-infrared spectroscopy (fNIR) is a brain imaging technique that monitors hemoglobin (Hb) signals thought to be markers for brain activation (e.g., Blood Oxygenation Level-Dependent of the MRI) using near-infrared light (Villringer and Chance, 1997). NIR has the potential to revolutionize the understanding of brain mechanisms and interactions during real task performance (Izzegtoglu et al., 2007). Izzegtoglu et al. (2007) described an fNIR system that monitors hemodynamic changes within the frontal cortex, which is thought to mediate executive brain processes related to memory and problem solving. Others have formulated new methodologies that differentially separate brain areas involved in task performance using high dense systems (Wylie et al., 2009). Image construction and multiple brain area monitoring are available with the higher dense systems (e.g., NIRx).

Keebler et al. (2009) used a high dense fNIR system to explore expert versus novice differences when learning to identify military vehicles. The parietal lobe activation patterns obtained in the study match those cited in the MRI literature for expert versus novice differences. While this was a pilot study of only two novices and two experts, the methodology employed demonstrates that this type of brain imaging technology is usable in the field. Understanding the coordinating efforts of brain regions spatial and temporally has application in clinical as well as occupational fields of use. As with any technology transitioning to the applied domain more work needs to be done in understanding the signal, as well as developing a tool set to process such signals.

The field of cognitive psychology, forty years later, will finally have tools available to answer the challenges of employing more ecologically valid contextual stimuli to their experimental design toolbox. This point highlights the fact that the foundational fields of neuroergonomics, neuroscience and ergonomics, are themselves changing as new methods and tools become available. These foundational advances are running parallel with the need for applied research. Consequently, applied research tends to be based upon the potential of a theory without full benefit of an evolved research concept. For example, augmented cognition research has re-evaluated gaps in its initial approach to integrating psychophysiological measures into adaptive system trainers (Sciarini, Fidopiastis, and Nicholson, 2009). These iterative advances are important in any emerging field and will be no different for neuroergonomics as it continues to grow as a research area.

CONCLUSION

Understanding how the brain carries out tasks in the real world is the ultimate goal of applied fields such as neuroergonomics. Adapting technologies and method from existing research paradigms can either directly translate or require modification. The field of neuroergonomics is currently modifying neuroscience techniques to better match their problem space. The research overview presented outlines the successes and gaps still facing this field of research. While the EEG as an applied science tool has experienced several iterative progressions within the field, near-infrared imaging technology is just starting its evolution. These technologies along with better theoretical grounding will not only change how we understand human-machine work interactions, but brain function itself.

REFERENCES

Berka, C., Pojman' N., Trejo, J. Coyne, J., Cole, J. Fidopiastis, C., and Nicholson, D. (2010), "Merging Cognitive Neuroscience & Virtual Simulation in an Interactive Training Platform." In the *Proceedings of Applied Human Factors and Ergonomics Conference*. Miami, FL.

Fidopiastis, C.M., Drexler, J., Barber, D., Cosenzo, K., Barnes, M., Chen, J.Y., and Nicholson, D. (2009), "Impact of Automation and Task Load on Unmanned System Operator's Eye Movement Patterns."*Augmented Cognition, HCII 2009, LNAI 5638* (pp. 229–238). Springer-Verlag, Berlin.

Fidopiastis, C.M., and Wiederhold, M. (2008), "Mindscape retuning and brain reorganization with hybrid universes: The future of virtual rehabilitation." *The PSI Handbook of Virtual Environments for Training & Education: Developments for the Military and Beyond.* (Vol 3., pp. 427-434). Praeger Security International, Connecticut.

Izzegtoglu, M., Bunce, S.C., Izzetoglu, K., Onaral, B., and Pourrezaei, K. (2007), "Functional brain imaging using near-infrared technology." *IEEE Engineering in Medicine and Biology Magazine*, 26(4), 38-46.

Keebler, J.R., Sciarini, L.W., Fidopiastis, C.M., Jentsch, F., and Nicholson, D. (2009), "Use of functional near-infrared imaging to investigate neural correlates of expertise in military target identification." In the *Proceedings of the Human Factors and Ergonomics Society 53rd Annual Meeting*. San Antonio, TX.

Kruse, A.A. (2007), "Operational neuroscience: Neurophysiological measures in applied environments." *Aviation, Space, and Environmental Medicine,* 78(5), B191-B194.

Neisser, U. (1967), *Cognitive Psychology*. Appleton-Century-Crofts, New York.

Nicholson, D.M., Davis, L., Fidopiastis, C.M., Schmorrow, D., and Stanney, K.

(2007), "An adaptive instructional architecture for training and education." *Foundations of Augmented Cognition, Third International Conference,* (pp. 380-384). Strategic Analysis, Inc., Virginia.

Parasuraman, R., and Hancock, P.A. (2004), "Neuroergonomics: Harnessing the power of brain science for HF/E." *HFES Bulletin,* 47(12), B1-B5.

Parasuraman, R., and Wilson, G.F. (2008), "Putting the brain to work: Past, present and future." *Human Factors,* 50(3), 468-474.

Reeves, L., Stanney, K., Axelsson, P., Young, P., and Schmorrow, D. (2007), "Near-term, mid-term, and long-term research objectives for augmented cognition: (a) Robust controller technology and (b) mitigation strategies." *4th International Conference on Augmented Cognition,* (pp. 282-289). Strategic Analysis, Inc., Virginia.

Sarter, N., and Sarter M. (2003), "Neuroergonomics: opportunities and challenges of merging cognitive neuroscience with cognitive ergonomics." *Theoretical Issues in Ergonomics Science,* 4(1-2), 142-150.

Sciarini, L.W., Fidopiastis, C.M., and Nicholson, D.M. (2009), "Towards a modular cognitive state gauge: Assessing spatial ability utilization with multiple physiological measures." In the *Proceedings of the Human Factors and Ergonomics Society 53rd Annual Meeting.* San Antonio, TX.

Vartak, A.A., Fidopiastis, C.M., Nicholson, D.M., Mikhael, W.B., and Schmorrow, D. (2008), "Cognitive state estimation for adaptive learning systems using wearable physiological sensors." *Biosignals,* 2, 147-152Courage, K.G., and Levin, M. (1968), *A freeway corridor surveillance information and control system.* Research Report No. 488-8, Texas Transportation Institute, College Station, Texas.

Villranger, A., and Chance, B. (1997), "Non-invasive optical spectroscopy and imaging of human brain function." *Trends in Neuroscience,* 20(10), 435-442.

Wilson, G.F., and Russell, C.A. (2007), "Psychophysiologically determined adaptive aiding in a simulated UCAV task." *Human Factors,* 49, 1005-1018.

Wylie, G.R., Graber, H.L., Voelbel, G.T., Kohl, A.D., DeLuca, J. (2009), "Using co-variations in the Hb signal to detect visual activation: A near infrared spectroscopic imaging study." *NeuroImage,* 47, 473-481.

Chapter 37

The Human Factors of Instructional Design and Accelerated Learning

Anna L. Oskorus, Ryan E. Meyer, Terence S. Andre, Kevin C. Moore

TiER1 Performance Solutions

ABSTRACT

With the fast pace of our world, optimizing the time spent in training is of the utmost importance for organizations. From a Human Factors perspective, this includes designing systems with improved usability to support the efficient delivery of instruction. Foundational design principles can be applied to increase the efficiency of learning events and accelerate learners through training. In this paper, we define accelerated learning in the context of our approach to training design. We also provide an overview of the XL Framework, a framework designed to support accelerated learning. This approach involves accelerating learners through content via system design and accelerating the learning process through instructional design. Engagement as a measure of training effectiveness is described, as well as challenges to designing an accelerated learning system.

Keywords: accelerated learning; adaptive learning; instructional design principles; content filtering; human factors; training system design; learner engagement

INTRODUCTION

Talk to any supervisor in the government, commercial, or ·nonprofit sectors, and they'll probably tell you that getting employees up to speed quickly, and maintaining that knowledge over time is essential to the performance of their organization. They want results, and they want them quickly. Training options available today often are one-size-fits-all, passive learning events that ignore what research tells us is essential for effective learning to take place: sustained learner engagement.

Human Factors, at its core definition, is essentially about designing systems to make them easier and more effective for people to use. While this explanation may not be broad enough to cover everything about Human Factors, it certainly seems to resonate with people who are concerned with human performance. Much of the work in Human Factors has led to a collection of design principles, methods, and techniques to make systems, and human interactions with those systems, better (e.g., Wickens, Lee, Liu, and Gordon-Becker, 2004; Norman, 2002). What is interesting in the literature is that there are so few articles about designing learning environments from a Human Factors perspective. This lack of design principles might be due to the fact that there are often changing teaching strategies, technologies, and measurements in the learning space. With these changes, it may be difficult to develop consistent design principles that apply across many learning environments. However, our experience indicates that there are some foundational concepts of learning that can be applied to support learner engagement and accelerate the learning process, ultimately improving human performance. In this paper, we will describe how these learning principles will be applied to create an accelerated learning system for training in any domain.

WHAT IS ACCELERATED LEARNING?

Accelerated learning is often defined in different ways depending on the context in which it is used. Some definitions of accelerated learning are provided in Table 1.1.

From these definitions we operationally characterized accelerated learning as: *the reduction of learner time required to meet learning objectives in a training event.*

To accomplish this goal, we approached acceleration within our system in two ways:
- Adapting the amount of content the learner needs to cover based on their current level of proficiency
- Increasing the efficiency of the learning process

The sections that follow describe how we arrived at this dual approach to system design and instructional design to drive learning acceleration.

SYSTEM DESIGN TO ACCELERATE THE LEARNING PATHWAY

Some learners who will complete a training event may already have some level of mastery of the content to be covered. To accelerate the learning pathway and minimize the time spent on training for these individuals, we devised a three-step process to adapt and personalize the content to be covered:

1. Identify all competencies and learning objectives for the training

2. Use a pretest to analyze learner proficiency of competencies and learning objectives prior to training
3. Adapt the learning pathway through content filtering based on pretest results

Table 1.1 Definitions of Accelerated Learning.

Definitions of Accelerated Learning	Authors
Paraconscious mental activity that can create conditions to automate and use memory, brain and intellectual reserves of people effectively.	Lozanov (1978)
Providing effective training in a short period of time.	Gill and Meier (1989)
Making a superlink between the right and left brain.	Linksman (1996)
Adapting and learning new skills quickly.	Lawlor and Handley (1996)
The ability to absorb and understand new information quickly and retain information.	Rose and Nicholl (1997)
Changing behavior with increasing speed.	Russell (1999)
Fast learning.	Lynn, Akgun and Keskin (2003)
Learning faster and smarter to keep up with change, call on new knowledge and apply new skills.	Landale (2004)

Before a training event can be designed, you must identify the competencies in which you would like learners to be proficient. We define *competencies* as any observable, measureable pattern of knowledge, skills, abilities, behaviors and other characteristics that an individual needs to perform their work roles or occupational functions successfully. Competencies are typically derived through interviews with individuals who have familiarity with the particular content domain to be trained and what the targeted audience needs to do in relation to that content by the end of the training event. Once competencies are defined, they must be broken down into *learning objectives*, statements in specific and measurable terms that describe what the learner will know or be able to do as a result of participating in a learning activity. After competencies and learning objectives for a training event have been developed, they can be used as a guide for learning acceleration.

When the user first interacts with our system, he/she will complete a pre-assessment activity which measures the learner's mastery of each competency. To increase engagement, we've designed our pre-assessment to be a game-based scenario. When the pre-assessment is complete, the user's learning path is dynamically built based on the scores obtained with respect to the various course objectives. If the user displays a mastery of a particular competency, then the module associated with that competency is skipped, or it becomes optional. Otherwise, the user will be required to learn the content that teaches that competency.

At the end of each module, the user must demonstrate mastery by completing a module test that focuses on the specific competency taught in that module. We are using game-based scenarios for module testing, similar to the format of the pre-assessment in our system. Users that do not successfully complete a module receive remediation and are given another opportunity to successfully complete the test. Expert performance research indicates that a deliberate effort to improve performance is more important than the amount of experience a person might have in a particular domain (van Gog, Ericsson, Rikers, and Paas, 2005). Thus, by providing opportunities for deliberate practice, our design will encourage learners to stretch themselves to a higher level of performance.

After all content modules are completed, the learner proceeds to a post-assessment, similar to the pre-assessment. The post-assessment will be used to determine the learner's mastery of each competency. If the user does not successfully complete this post-assessment, they will receive remediation and are given another opportunity to successfully complete the scenario-based game. Once the learner successfully demonstrates mastery of all competencies in the post-assessment, a course conclusion appears and the learner receives completion credit for the course. Figure 1.1 demonstrates how we envision the process of personalizing the learning pathway through content filtering within our system to support acceleration.

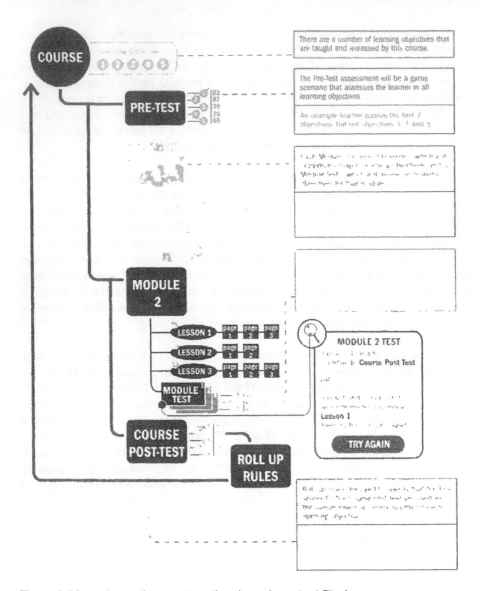

Figure 1.1 Learning pathway adaptation through content filtering.

INSTRUCTIONAL DESIGN TO ACCELERATE THE LEARNING PROCESS

In addition to accelerating the learning pathway, we wanted to make the time the learner did spend with the learning content more effective. We looked for reoccurring themes in the literature to identify the instructional design principles on accelerated learning (Chambers, 2005; Hostetler, 2007; Landale, 2004; Moon, Birchall, and Williams, 2007; Pattison, 2001; Redford, 2006; Tapp, 2007) and neuroscience (Fox, 2008; Rock & Schwartz, 2006; Rock & Schwartz, 2007) that would help a student learn and retain content, thus helping to accelerate the learning process. We also reflected on our own experience as human performance consultants and factored that perspective into our design principles.

The design principles, which were derived from our literature review and our experience, can be summarized into three major categories:

1. Learner-centered focus
2. Emotional engagement
3. Higher-level thinking

Table 1.2 details the three design principle areas we've identified through literature and how these design principles will be applied in our training system.

Table 1.2 Design principles for accelerating the learning process.

Design Principles	Key Points from Literature	Authors	Examples
Learner-Centered Focus			
Gain and keep the learner's attention through: • Meaningful interactivity • Engaging, relevant visual content • Questions • Challenges	Capturing and keeping the learner's attention is essential for learning to take place. Engaging people in more active learning techniques improves retention. Questions and challenges require the brain to work, which promotes learning and remembering.	Moon, Birchall, and Williams (2005) Raybourn (2006) Chambers (2005) Tapp (2007) Rock and Schwartz (2006) Rock and Schwartz (2007) Fox (2008)	• Engage with material that is responsive to learner's actions • Link visual content with audio narration • Use multiple question types throughout content delivery • Introduce "mini-games" to test for knowledge acquisition

Address multiple learning styles	In a training context it is important to use a rich mixture of visual, auditory and kinesthetic activities to address multiple learning styles.	Pattison (2001)
		
		Landale (2004a)
		
		Landale (2004b)
		
		Chambers (2005)
		
		Tapp (2007)
			• Show same content multiple ways:
			  ○ Visual
			  ○ Auditory
			  ○ Kinesthetic
			• Displayed through graphics, text, audio, and animation

Design Principles	Key Points from Literature	Authors	Examples
Exploratory learning/Learner control	Traditional class-room training is not as effective as experiential learning. Encouraging ownership and learner control supports behavior changes. The more in control the learner feels, the more likely they are to take risks and step outside their comfort zone.	Sitzmann, Kraiger, Stewart, and Wisher (2006) Fox (2008)	• Optional additional content
Practice	Practice encourages the strengthening of connections along mental pathways in the brain, ultimately improving performance. Practice boosts confidence, energy and skill level.	Landale (2004a) Chambers (2005) Fox (2008)	• Ability to retry game-based scenarios in the system
Emotional Engagement			
Multiple perspectives/stories	Multiple perspectives and stories encourage the learner to make judgments.	Pattison (2001) Moon, Birchall, and Williams (2005)	• Interviews, case studies
Make emotion part of the experience	Making emotion part of learning enhances our memory.	Pattison (2001) Chambers (2005) Rock and Schwartz (2006)	• Dealing with human characters • Making tough decisions as a supervisor in scenarios
Higher-level thinking			
• Application • Evaluating	It's important for learning to be applied as it would be in real life	Landale (2004a) Moon,	• Applying knowledge and skills learned to authentic scenarios • Analyzing employee

Design Principles	Key Points from Literature	Authors	Examples
• Analysis • Reflection	situations. Providing opportunities for higher-level thinking processes in training helps solidify learning and promote deeper understanding.	Birchall, and Williams (2005) Raybourn (2006) Tapp (2007)	behaviors • Making decisions based on data collected on employees • After-action review

Our findings were synthesized into a comprehensive approach for accelerating learning, the XL Framework.

THE XL FRAMEWORK

We aptly named the resulting framework to support learning acceleration the XL Framework. Figure 1.2 summarizes our dual approach to accelerated learning.

Figure 1.2 The XL Framework

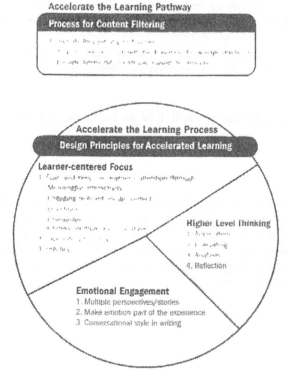

The XL Framework has served as a guide for designing the functionality of our accelerated learning system to create personalized learning pathways, and for the instructional design of the content with which we will populate the system. We believe these design principles could be applied to a variety of learning environments to reduce the time required to achieve learning objectives in a training event. We also can see value in the XL Framework and our resulting system design for compliancy training, as it could potentially reduce time spent in what is typically very passive and repetitive content.

WHY IS ENGAGEMENT IMPORTANT?

We now know that increased interactivity in learning events encourages engagement and supports retention (Fox, 2008). Engagement and high levels of motivation are linked to learner success (Dev, 1997). While motivation remains a complex construct to measure, valid and reliable engagement measures using biosensors are readily available (Berka et al., 2004). We will work with partners to validate the performance measures we have selected for our accelerated learning system and will seek to correlate the learner's psychophysiological responses (i.e., engagement and workload) to aspects of the training that reflect accelerated learning processes. In addition to these variables, other measures such as mental effort and strategies employed by the learner may provide additional insights (van Gog et al., 2005). Through this research, we hope to gain a more holistic view of the learner's cognitive load and engagement level when interacting with the system.

CHALLENGES TO ACCELERATED LEARNING

While we have aligned our approach to what research and our experience says works, there are some challenges we are facing in regards to accelerated learning within our system. Some of these key questions we are trying to answer include:

- What impact does accelerating the learning pathway (i.e., content filtering) have on learner retention?
- Which, if any, of the instructional design principles we've identified have the most impact on learner engagement and improved learner performance, and to what degree?
- How can a learning experience be designed with an engaging and seamless storyline when elements of the training may be filtered from a learner's pathway?

These are just some of the challenges we will face as we build and test our accelerated learning system.

CONCLUSION

We believe that the research supports the development of reliable design principles to accelerate learning. Coupled with a Human Factors perspective on applying these principles, learning environments can be designed to not only engage the learner, but make their learning experience efficient and effective. Through our approach to accelerated learning, we hope to support organizations in the race to get their employees up to speed quickly, and to maintain proficiency in competencies over time.

ACKNOWLEDGMENTS

This material is based upon work supported by the Air Force Research Laboratory (AFRL) under Contract No. FA8650-10-C-6061. Any opinions, findings, conclusions, or recommendations expressed in this publication are those of the authors and do not necessarily reflect the views of AFRL.

REFERENCES

Berka, C., Levendowski, D. J., Cvetinovic, M., Petrovic, M. M., Davis, G. F., Lumicao, M. N., Popovic, M. V., Zivkovic, V. T., Olmstead, R. E., & Westbrook, P. (2004). Real-time analysis of EEG indices of alertness, cognition and memory acquired with a wireless EEG headset. Special Issue of the *International Journal of Human-Computer Interaction on Augmented Cognition, 17*(2), 151-170.

Chambers, P. (2005). Accelerate your learning. *Industrial and Commercial Training, 37*(6), 314-318.

Fox, A. (2008). The brain at work. *HR Magazine, 53*(3), 36-43.

Gill, M. J. and Meier, D. (1989). Accelerated learning takes off. *Training and Development Journal,* 63-65.

Hostetler, E. (2007). Safety at the center: a model that accelerates learning. *Organization Development Journal, 25*(4), 63-66.

Landale, A. (2004b). Mental agility training helps CSA staff to navigate rapid change. *Human Resource Management International Digest, 12*(2), 20-23.

Lawlor, M. and Handley, P. (1996). *The creative trainer: Holistic facilitation skills for accelerated learning.* East Windsor, NJ: McGraw-Hill.

Linksman, R. (1996). *How to learn anything quickly: An accelerated program for rapid learning.* Secaucus, NJ: Citadel Press.

Losanov, G. (1978). *Suggestology and outlines of suggestopedy,* New York, NY: Gordon and Breach.

Lynn, G., Akgun, A.E., & Keskin, H. Accelerated learning in new product teams. *European Journal of Innovation Management, 6*(4), 201-212.

Moon, S., Birchall, D., & Williams, S. (2005). Developing design principles for an e-learning programme for SME managers to support accelerated learning at the workplace. *Journal of Workplace Learning, 17*(5/6), 370-384.

Norman, D. A. (2002). *The Design of Everyday Things.* New York: Basic Books.

Pattison, S. A. (2001). Staff meetings: an opportunity for accelerated training of employees. *Journal of Workplace Learning, 13*(4), 172-178.

Raybourn, E. M. (2006). Applying simulation experience design methods to creating serious game-based adaptive training systems. *Interacting with Computers, 19*, 206-214.

Redford, K. (2006). I think, therefore I can. *Training & Coaching Today,* 6.

Rock, D. & Schwartz, J. (2006). The neuroscience of leadership. Retrieved July 20, 2009 from http://www.strategy-business.com/press/freearticle/06207

Rock, D. & Schwartz, J. (2007). Why neuroscience matters to executives. Retrieved July 20, 2009 from http://www.strategy-business.com/li/leadingideas/li00021

Rose, C. & Nicholl, M. J. (1997). *Accelerated learning for the 21st century*. New York, NY: Dell Publishing Group.

Russell, L. (1999). Fortifying strategic decisions with shadow teams: a glance at product development. *Competitive Intelligence Magazine, 2*, 9-11.

Tapp, L. M. (2007). Better safety training with fun & games. *Professional Safety, 52*(2), 52-55.

Van Gog, T., Ericsson, K. A., Rikers, R. M. J. P., & Paas, F. (2005). Instructional design for advanced learners: Establishing connections between the theoretical frameworks of cognitive load and deliberate practice. *Educational Technology Research and Development, 53*, 73-81.

Wickens, C. D., Lee, J. D., Liu, Y, & Gordon-Becker, S. E. (2004). *An Introduction to Human Factors Engineering* (2nd ed.). Upper Saddle River, NJ: Pearson Prentice Hall.

CHAPTER 38

Serious Games for UAS Training

Michael A. White

Alion Science and Technology
5365 Robin Hood Road
Norfolk, Virginia 23513, USA

ABSTRACT

The proliferation of Unmanned Aircraft Systems (UAS) and their use to support military operations globally has resulted in a significant shortfall in adequate training resources for aircrew flying UAS missions, as well as UAS ground support personnel. Serious Game technologies can be used to provide an engaging and cost-effective training solution. Serious Game technology is proving to be a viable solution for many individual training and learning tasks that rely on well-defined rules and behaviors to support disciplined decision-making.

INTRODUCTION

The proliferation of Unmanned Aircraft Systems (UAS) and their use to support military operations globally has resulted in a significant shortfall in adequate training resources for aircrew flying UAS missions, as well as UAS ground support personnel. The recent Air Force UAS Flight Plan highlights current UAS training shortfalls as an artifact of the UAS acquisition process, the rapid fielding of UAS systems, frequent updates/upgrades to deployed systems, and the availability of the actual UAS to support training functions since they are used 24/7 to support Live missions (USAF Flight Plan, 2009).

Simulation-based training technologies provide an effective means to augment Live UAS training. Simulation systems, however, come with significant development costs and are not the only virtual training technology applicable to the UAS training continuum. Serious Game technologies can be used to provide an engaging and cost-effective training solution. As an emerging training technology, Serious Game technology is proving to be a viable solution for many individual training and learning tasks that rely on well-defined rules and behaviors to support disciplined decision-making. While high-fidelity simulations/emulations may be the best alternative for UAS 'stick-and-rudder' and sensor operator training, Serious Games offer a credible means for training cognitive skills, communications, and/or tactical decision-making related to DoD missions such as Force Protection, Strike, and/or Intelligence, Surveillance, and Reconnaissance (ISR). Serious Games are well-suited for training UAS ground control tasks, spatial awareness and/or air space coordination as well as tactics, techniques and procedures (TTP) related to employment of the systems and their payloads.

The goal of this paper is to introduce the use of Serious Game technology as a viable approach to augmenting existing curricula and methods in UAS training. The intent is to elicit interests of the UAS training community and to advance the state-of-the-art through the use of low-cost, deployable Serious Game technologies. This paper will survey UAS training needs and discuss the benefit of using Serious Game technologies. The paper will also discuss state-of-the-art Serious Game technology, capabilities and limitations, and how its application can potentially be used to improve UAS Training & Readiness (T&R).

UAS Training Needs

Recent research has attempted to address several UAS training needs. Viirre contends that interpreting and understanding sensory cues derived from UAS sensors (a component of spatial awareness) is a critical need for UAS operators (Viirre, et al., 2010), while Johnston acknowledges the need to develop a better understanding of sensory deprivation and the resultant high cognitive demands on UAS crew members (Johnston, et al., 2010). From a slightly different perspective, Durlach's approach to UAS training needs not only includes the psychomotor and cognitive skills an operator needs, but also speaks to enabling skills, such as incorporation of tactical considerations, terrain analysis and communication. She also points out that leader training for those charged with supervising UAS employment is absent (Durlach, 2010). Similar to the Navy recruit training problem discussed in Hussain et al. (2009), UAS training needs include situational awareness, communication and decision-making, at minimum.

At least one segment of the US Air Force attributes UAS training shortfalls to the rapid fielding and frequent updates/upgrades to deployed systems, many of which have bypassed the traditional acquisition and fielding process, and the availability of the actual airframes to support training functions since they are in nearly constant use to support live missions. There are five USAF UAS programs of record, yet only one has a full scale simulator for qualification training. According to the report, initial qualification training has been consistently under-resourced as increased UAS production has led to increased demand for training (USAF Flight Plan, 2009).

A by-product of the proliferation of UAS platforms has been the increase in the number of mishaps. Although the Air Force has not experienced any Class A[1] or Class B[2] mishaps, the risk of these occurrences increases with increased exposure. Citing Manning, et al (2004), Burns, et al (2010) state that "high levels of mishaps reduce operational availability of UAS assets, present potential danger to personnel on the ground and hamper the acceptance of unmanned systems within the US national airspace". One can extrapolate from this statement to include the potential hazards to manned aircraft in a military theater of operations. The authors go on to report on the Nullmeier, et al (2007) finding of significant correlation between UAS mishaps and loss of operator situational awareness.

While the Army has employed enlisted soldiers to crew pilotless aerial vehicles, up to now, the Air Force has depended on experienced pilots to crew uninhabited aircraft. However, the growing demand for operators has influenced Air Force Policy to "increase the use of technology" to meet the burgeoning training need.

---

[1] Class A Mishap: loss of life or severe disability, aircraft destroyed, or total cost is equal to or greater than $1M

[2] Class B Mishap: three or more personnel hospitalized, permanent partial disability or damage between $200K and $1M

SERIOUS GAMES: STATES OF THE ART AND SCIENCE

According to Sorensen and Meyer (2007), "Serious Games are defined as digital games and equipment with an agenda of educational design and beyond entertainment". Serious Games have been adopted for use in defense, education, scientific exploration, health care, emergency management, city planning, engineering, religion, and politics. A serious game may be a simulation which has the look and feel of a game, but corresponds to non-game events or processes, including business operations and military operations (even though many popular entertainment games depict business and military operations). The games are made to provide an engaging, self-reinforcing context in which to motivate, educate and train the players

Games for training and education

Various authors agree that serious games are an effective method for addressing training needs. For example, Engel, et al. (2009) citing the work of Ricci, et al. (1996), conclude that game-based training leads to greater retention than text-based training and the work of Fletcher and Tobias (2007) touting game-based training's improvement of some cognitive capabilities.

As a result of some disagreement between the communities as to the efficacy of using games to educate, Fortugno and Zimmerman (2005) authored an essay designed to bridge the gap between educators and gamers vis-à-vis creation of educational games.

The authors posit that "educational games are first and foremost *games*." Some have logical rules and winning condition... others provide mechanisms for creating and interacting with communities, cities or worlds. Game elements include interactivity (input, output), short-term and long-term goals to shape players' experience, mechanisms for learning the basics of the game incorporated into the design, and structure that facilitates play. Competition and collaboration are both elements of games. Conflict is the basis of the game, but the chance to overcome obstacles is balanced by mutual recognition between players of the game's rules. Games are dynamic, participatory, and process-oriented rather than focused on factual content.

In Fortugno and Zimmerman's opinion, serious games cannot stand alone in the delivery of effective learning, but they can augment other modes of instruction. The use of serious games should be considered in the context of the end-user and the environment within which the game is to be used. They state that the hard problems of developing games that teach include educational standards, developmentally appropriate design and assessment of what the player has learned.

In search of training effectiveness

As the use of games for training and education has become more widespread, investigators have attempted to address how to determine the validity and effectiveness of game-based training. For instance, in their 2006 Spring Simulation Interoperability Workshop presentation, Stoudenmire, et al. (2006) discussed factors affecting performance effectiveness: suspension of disbelief (i.e., immersiveness and presence), stress inducement and environmental fidelity. Their report defined these factors as follows:

- Immersiveness – technical aspects of person's involvement in the game (e.g., use of HMDs); state of captivation (means of increasing presence). The authors supported their contention that more is not always better, citing Morris, et al. (2002); increased (graphical) fidelity is not necessarily better, since "humans naturally enrich and utilize impoverished information"
- Presence – in his Master's thesis at the Naval Postgraduate School, Bernatovich (1999) defines presence as the sense of being there; for instance, the combination of spatial knowledge (such as landmark recognition), procedural knowledge and survey knowledge
- Stress inducement – Stress inducement – "Contextually correct simulations generate higher levels of stress during training, thereby reducing the user's level of stress in actual operational performance" (Morris, et al, 2004)
- Fidelity – physical (looks, sounds, feels like) - functional (acts like) (Alexander, et al., 2005); cognitive fidelity is important because it elicits decision-making and team behaviors that match real-life mental processes (Bell and McNamara, 2005)

A theoretical approach: applying gaming principles to training

In their paper on the effectiveness of using games for training, Engel, et al. (2009) address how gaming principles can be applied to training applications to retain trainee interest. Based on research such as Kraiger, Ford & Salas (1993) in the psychology field and Bloom (1994) in the educational realm, the team identified 14 applicable training outcomes and 11 gaming attributes (TABLE 1).

The objective of their research was to learn more about how gaming attributes, such as feature control, challenge and/or immersion, influence specific training outcomes. In particular, the paper describes the methods, measures and results of experiments investigating the impact of feature control on three training outcomes: declarative knowledge, motivation and application. An unintended by-product of their research; they found that previous experience/training along with greater feature control resulted in a significant increase in training benefit, leading the researchers to conclude that experience level should be one of the factors considered in design of serious games for training.

Training Outcomes	Gaming Attributes
Application	Fantasy
Cognitive strategies	Environment
Declarative knowledge	Conflict
Knowledge organization	Assessment
Adaptation	Action language
Automaticity set	Rules/goals
Compilation	Adaptation
Origination	Challenge/surprise
Psychomotor	Immersion
Attitudinal valuing	Human interaction
Internalizing values	Feature control
Motivation	
Organization	
Receiving/responding phenomena	

TABLE 1 Comparison: Training Outcomes versus Gaming Attributes

A practical approach: use of a conceptual framework

Before embarking on the daunting task of developing a game for UAS training and readiness, the team of users and developers must determine the purpose of the game, what they expect it to do and how it can support the instructional design. FIGURE 1, adapted from Yousef, et al. (2009), illustrates a proposed conceptual framework for games that may be useful (once the purpose of the game is established). Cognitive task analyses and instructional design processes can aid in determining required UAS-related instructional content and in comparing gaming technology training capabilities to assist in establishing intended learning objectives.

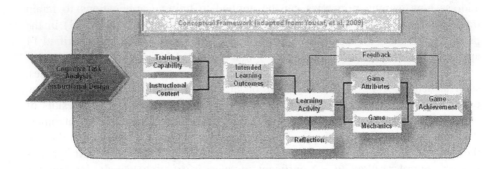

FIGURE 1 Conceptual Framework for Serious Games

From the intended learning objectives, the team can begin an iterative serious game design/development process. Since requirements elicited prior to development are often incomplete, serious game developers prefer to employ an iterative development life-cycle along with agile programming techniques in conjunction with an on-going conceptual framework analyses. These practices enable rapid prototyping and provide greater flexibility than traditional "waterfall" development methodologies.

The game design process includes creating a context for training (story line), determining environmental fidelity (graphics, animation), game features (user interface, progressive game levels, use of AI) and logic providing guidance and feedback tailored to situation and performance outcomes. Other activities include identification and design of desired game attributes that will provide the stimulus and motivation to achieve the desired learning objectives, along with selection of delivery media/mechanism (CD, DVD, web-based or a combination).

SERIOUS GAMES: MILITARY USE

The Army and Marine Corps both have begun to recognize the value of serious games for training. In January 2009, the US Army selected Virtual Battlespace 2 (VBS2), a commercial game-based training platform that blends a first-person shooter immersive environment with scenario editors and after-action review, to help train Company-level and smaller units on multiple tasks and mission rehearsal (US Army, 2010). Since its 2006 inception as a program of record, Marine Corps Deployable Virtual Training Environment (DVTE) has been fielded with VBS as a component of the Infantry Toolkit (ITK). Additionally, every DVTE suite includes the game-based language trainers, Tactical Iraqi, Tactical Pashto and Tactical Farsi. DVTE also employs the Delta3D gaming environment as its visualization engine for the virtual simulators included in the Combined Arms Network (CAN) component.

While the Army and Marines have started to embrace serious games for small unit training and mission rehearsal, the Navy has already made several forays into the realm of serious games for training individuals and crews. Efforts include the Flooding Control Simulator (FCS), Mission Rehearsal Tactical Team Trainer (MRT3), and Virtual Maintenance Performance Aid (VMPA). The FCS focuses on individual training, where the objective is to augment basic training to enhance decision-making, communication procedures, control procedures and situational awareness. Although designed for more senior personnel, VMPA training goals are similar to FCS. The MRT3 hybrid game is targeted at improving or refreshing the cognitive skills of personnel already trained in their jobs.

CAPABILITIES AND LIMITATIONS OF SERIOUS GAMES

As previously alluded, there is no one-size-fits-all approach to using serious games for training. How good is good enough? Because of the innate human quality of naturally enriching and utilizing impoverished information, the gaming environment does not have to identically replicate the representations of physical objects or phenomena to effectively support training. Suspension of disbelief can be achieved without extremely high resolution; "close enough" can provide the physical clues and functional cues required to elicit the appropriate cognitive responses that lead to decision-making and team behaviors that match real-life mental processes.

CAN SERIOUS GAME TECHNOLOGY BE USED TO IMPROVE UAS TRAINING & READINESS (T&R)?

From this author's perspective, the answer to this question is a qualified YES. The three systems described below provide the basis for this hypothesis. Although reports of VMPA and MRT3 success are mainly anecdotal, the Navy has "voted with their checkbook", awarding additional funding for further development of both systems as well as extending the application of the gaming technologies for training crew members of other platforms. On the other hand, in addition to anecdotal evidence, success of the FCS at improving training and readiness was confirmed by a University of Central Florida study of FCS cited in the Hussain paper; i.e., decision making errors were reduced by 50%, communication errors reduced by up to 80% and situational awareness and navigation skills were 50% improved.

Virtual Maintenance Performance Aid

The VMPA familiarizes a Readiness Control Officer (RCO) with the Littoral Combat Ship (LCS), a future addition to the fleet that employs a dramatically reduced personnel complement to crew the ship. The crew members are already experienced sailors who merely need to be introduced to the system specific requirements of the LCS and to maintain proficiency when they are not underway. The training challenges to the austere crew requirements are compounded by the fact that the vessel essentially remains underway constantly, except when changing crews upon completion of a deployment. Unavailability of a live ship on which to train while crewmembers are ashore combined with the limited throughput capability of large-scale shore-based trainers has the potential for atrophy of skills and a concomitant increased ramp-up time for crew members prior to their next deployment.

Mission Rehearsal Tactical Team Trainer

The MRT3 is a hybrid system consisting of simulation and gaming technologies that provides networked crew-level training for crewmembers of aviation platforms engaged in Anti-Submarine Warfare (ASW). Live crew training is expensive, and difficult to plan, since US submarines underway, are loathe to be give away their location by performing as training aids for ASW crews. Similarly, submarine crews of potential adversaries also are reluctant to compromise their secrecy. As might be expected, ASW crewmembers do most of their initial and refresher crew training in large, fixed-facility simulators, which are not located where the majority of ASW units are stationed, thus limiting any other than infrequent live training. The hybrid MRT3 solution enables remote training, wherever the ASW unit members want to use it. The hybrid system does not provide the structure of a pure gaming system (i.e., mechanisms and instructional logic providing guidance and feedback tailored to situation and performance outcome and step-by-step progression through successive levels of difficulty). However, the system does provide repeated exposure to important cue patterns; facility to manipulate variables, view from different perspectives; enables observation of system over time; enables successive tasks that progress towards goals; and enhances motivation by providing a realistic sense of accomplishment, informative feedback, and the same sense of challenge associated with real-world task performance.

Flooding Control Simulator

Using Delta3D, an open source, serious game engine and employing an iterative design process, a team comprised of government, industry and members of academia, developed a game-based Flooding Control Trainer that reinforces skills taught in lectures using a task-based instructional infrastructure that features real-time guidance, feedback, and after action debrief.

In a paper presented at the 2009 Interservice/Industry Training, Simulation and Education Conference (I/ITSEC), Hussain, et al. (2009) discussed their efforts to use serious game technology to address one of the Navy's pressing training needs, transfer of classroom instruction to practical application.

The purpose of FCS was three-fold:
- Supplement classroom instruction,

- Reduce demand on training resources,
- Produce a training platform applicable to multiple training venues: classrooms in technical schools as well as various locations ashore and afloat in the fleet.

Instructional design was a key aspect of their serious game development process, applying a guided discovery strategy, rather than a more traditional didactic approach. The instructional design goal was to balance the presentation of explicit information with provision for students to learn by dynamic free-play.

CONCLUSION

Serious games can be a cost-effective method for addressing training needs. As evidenced by reports on a number of research efforts, when game attributes, capabilities and limitations are matched to desired instructional outcomes, serious games have been shown to enhance individual, team and small unit training.

Based on previous successes cited in this paper, serious games can be used to improve UAS training and readiness, as long as the games are employed within their capabilities and limitations as part of a combined instructional strategy that includes other modes of training. Games are proving to be a viable solution for many individual training and learning tasks that rely on well-defined rules and behaviors to support disciplined decision-making.

Serious Games offer a credible means for training cognitive skills, communications, and tactical decision-making related to DoD missions. Serious Games are well-suited for training UAS ground control tasks, spatial awareness and/or air space coordination and may also serve as a platform for investigating potential solutions to the problems of sensory deprivation and the resultant high cognitive demands on UAS crew members.

Instructional design is a key element of the serious game development process. Along with cognitive task analyses of the proposed instructional content and comparison of capabilities of various training modes, a development team can identify intended learning objectives and pair them with appropriate gaming attributes to form the basis of serious game components, features and mechanics.

Within the context of an overall training strategy, serious games have proven their value to supplement classroom instruction, reduce demand on instructional personnel and increase the effectiveness of limited resources; critical considerations in light of the current economic environment and severe budget constraints.

REFERENCES

Burns, J., Langelier, M., Evans, D., (2010) "Training for Unmanned Aircraft Systems: Operational Challenges, Theoretical Constructs, Training Science, Technologies and Applications", Applied Human Factors and Ergonomics (AHFE) Conference, 2010.

Durlach, P. (2010), "Training Needs for Units with Small Unmanned Aerial Systems," 2010 Applied Human Factors and Ergonomics (AHFE) Conference.

Engel, K., Langkamer, K., Estock, J., Orvis, K., Salas, E., Bedwell, W., Conkey, C., (2009) "Investigating the Effectiveness of Game-Based Approaches to Training," Proceedings of the 2009 Interservice/Industry Training, Simulation and Education Conference (I/ITSEC).

Fortugno, N. and Zimmerman, E. (2005), Soapbox: Learning to Play to Learn – Lessons in Educational Game Design, Gamasutra, April 5, 2005, retrieved from http://www.gamasutra.com/features/20050405/zimmaerman_01.shtml

Hussain, Talib S., et al. (2009), "Designing and Developing Effective Training Games for the U.S. Navy," Proceedings of the 2009 Interservice/Industry Training, Simulation, and Education Conference (I/ITSEC).

Johnston, M., Stanney, K., Hale, K., & Kennedy, R. (2010) "A Framework for Improving Situation Awareness of the UAS Operator through Integration of Tactile Cues", 2010 Applied Human Factors and Ergonomics (AHFE) Conference

Morris, C., Hancock, P., Shirkey, E. (2004) "Motivational Effects of Adding Context Relevant Stress in PC-Based Game Training," Military Psychology, 2004, 16(1), 135-147

Sorensen, B., and Meyer, B, (2007) "Serious Games in Language Learning and Teaching -a theoretical perspective," Proceedings of the Third International Conference of the Digital Games Research Association (DIGRA), retrieved from http://www.digra.org/dl/db/07312.23426.pdf

Serious Games in language learning and teaching – a theoretical perspective," Situated Play, Proceedings of DiGRA 2007 Conference.

Stoudenmire, E., White, M., Roy, K. (2006) "Assessment and Validation of Gaming Technology as Applied to Training: Current State and the Way Ahead," S06-SIW-124, Proceedings of Spring 2006 Simulation Interoperability Workshop.

United States Air Force Unmanned Aircraft Systems Flight Plan (2009), 2009-2047 18 May 2009, http://www.globalsecurity.org/military/library/policy/usaf/usaf-uas-flight-plan_2009-2047.htm.

United States Army Homepage (2010), "Top Army Gamers to Share Best Practices/Ideas at February Conference", Army News, 02/04/2010, http://www.army.mil/-news/2010/02/04/33966-top-army-gamers-to-share-best-practices-ideas-at-february-conference/.

United States Navy Safety Center, Classes of Mishaps (2010), http://www.safetycenter.navy.mil/media/WESS/ClassesofMishaps.htm.

Viirre, E., Grammann, K., Wing, S. (2010) "Theoretical Constructs of Unmanned Aircraft Systems Training", 2010 Applied Human Factors and Ergonomics (AHFE) Conference.

Wikipedia (2010), Serious Game, http://en.wikipedia.org/wiki/Serious_game, retrieved 5 February, 2010.

Yousef, A., Crowder, R., Gilbert, L., and Wills, G. (2009), "A Conceptual Framework for Serious Games," 2009 Interservice/Industry Training, Simulation, and Education Conference

Chapter 39

Auditory Discrimination & Academic Performance

Petra Alfred, Valerie J. Rice, Gary Boykin, Linda Laws

Army Research Laboratory-Army Medical Department Field Element
Ft. Sam Houston, San Antonio, TX, 78234, USA

ABSTRACT

Auditory discrimination, or the ability to attend to and differentiate between similarly sounding words in the presence of background noise, may influence why an individual fails or passes a course. The purpose of this study was to explore the relationship between auditory discrimination, key demographic variables, and academic performance among 68D Operating Room Specialist trainees. Scores on the Woodcock Johnson III Test of Auditory Attention were compared between 68D students considered academically at-risk and students performing well in the course. Volunteers consisted of 48 trainees from the 68D course, 25 high-risk students who failed the course, and 23 low-risk students with an A/B average. On average, high-risk students scored at an 8th-grade level, while low-risk students scored at an 11th-grade level. We also found that Caucasian females scored at a 7th-grade level, while Caucasian males scored at an 11th-grade level. We were able to predict auditory discrimination by using three pieces of information— Caucasian race, female gender, and high risk status. Implications of these findings on classroom design, potential interventions, and future research are discussed.

Keywords: auditory discrimination, academic performance, classroom design

INTRODUCTION

The Army expeditiously trains Soldiers to be proficient, competent medical para-professionals during Advanced Individual Training (AIT). This process takes longer and is more costly if students do not pass their initial training. If trainees do not pass their military occupational specialty (MOS) training on the first attempt, they must repeat the training; this process is known as "recycling." If they fail on their second attempt, they must attend training in a different specialty; this process is known as "reclassification." Recycling and reclassification are costly in terms of funding, time, resources, and morale.

The Army Research Laboratory's Army Medical Department (AMEDD) Field Element, in conjunction with the AMEDD Center and School out of Ft Sam Houston, TX, investigates attrition among medical training specialties. Examination of personal characteristics related to academic performance during combat medic training revealed that visual learners achieved higher grade point averages when compared with verbal (auditory) learners (Rice et al., 2006). A subsequent study noted academically high-risk Operating Room Specialist (MOS 68D) trainees who had failed the course and were awaiting recycle had lower-than-expected grade equivalency scores on the Woodcock Johnson III (WJ III) Auditory Attention Test ($M = 8.47$, $SD = 4.4$) (Alfred & Rice, 2009). The expected grade equivalency on these tests was at least a 10th-grade level, since military coursework is generally written at a 10th-grade level. Students scored at the expected 10th-grade level (or better) on the other 13 WJ III COG evaluations, including visual-auditory learning, auditory working memory, and planning (Ibid.). Since only those at academic risk were evaluated, it was unclear whether this finding was indicative of issues experienced solely by Soldiers encountering academic difficulties or if auditory discrimination difficulties are common to the current generation of new recruits.

The WJ III Auditory Attention Test measures auditory discrimination, or more specifically an individual's ability to attend to and discriminate between similarly sounding words, while "tuning out" increasingly louder background noises. Difficulty with auditory discrimination can negatively impact a person's ability to follow oral directions, listen to and understand a lecture, read, write, and spell. There are different reasons a person can have difficulties processing auditory information, such as health issues (e.g., fluid in the ear canals due to ear infection, damage to the inner ear) or loud environmental background noise (e.g., overhead heating and air conditioning units, other people talking). Niskar et al. (2001) reports one in eight 6 to 19 year olds have hearing loss due to noise exposure. In many cases, the inability to grasp auditory information may be due to central processing problems. Approximately five percent of school children may have a clinical diagnosis of a central auditory processing disorder (CAPD). CAPD, also known as Auditory Processing Disorder (Nemours Children's Health System, 2007), is defined as a difficulty in the efficiency and effectiveness by which the central nervous system (CNS) utilizes auditory information (Paul, 2008). CAPD occurs when the ear and the brain do not coordinate fully, even when hearing tests as normal (Bellis, 2002). A clinical diagnosis of CAPD can be determined only by a professional audiologist. Auditory discrimination is one of the five problem areas that can influence classroom or home performance for a person with auditory processing problems. Others are auditory figure-ground, auditory memory, auditory attention, and auditory cohesion.

Higher signal-to-noise ratios (SNR) are important to assist students in learning, even for those whose hearing is normal (Vause, 2008). The instructor's voice must be louder than the background sounds. Classroom design can contribute to auditory discrimination problems or help prevent lower SNRs. For example, room shape, volume, and surface materials can contribute to reverberation, or the persistence of sound in a room after the source of sound has stopped (Ibid). Higher reverberation creates greater background noise. While all military students have received a basic hearing test prior to active duty entry, more specialized aspects are hearing are not measured, such as auditory discrimination. Knowledge of students' auditory abilities can assist with individualized learning plans. Moreover, it is important to know whether auditory issues are pervasive among students who have academic difficulties so appropriate decisions can be made. For example, if auditory difficulties are identified, measures could be implemented to improve individual performance through training, group performance could be enhanced with environmental design, or screening measures could be implemented should hearing difficulties have career-long implications.

The purpose of this research was to determine the relationship between academic risk category (high versus low risk), key demographic variables (e.g., gender and race), and auditory discrimination (as measured by the WJ III Test of Auditory Attention) among 68D Operating Room Specialist Trainees.

METHODS

PARTICIPANTS

After being briefed on the purpose of the study, 55 students volunteered to participate, passed the inclusion/exclusion criteria, and signed a consent form.

The first 25 Soldiers were recruited from a population of 68D Soldiers who failed either their first or their second Anatomy and Physiology (A&P) exam, and therefore failed the class. They were waiting to recycle into a subsequent class. Because they had failed the course at least one time, this group of students is hereafter referred to as **high risk**. This group participated not only in the WJ III Auditory Attention Test, but also in an intervention program. The results of the intervention program are not yet available, as additional testing and training are being conducted.

The second set of 30 volunteers was recruited from students who had either an A or B average following the A&P exams. Passing the A&P exams is considered a critical juncture for students, as this is the point during training when the greatest numbers of students fail and either recycle or are reclassified into another MOS. Students who have an A or B average after this point are most likely to pass the course in its entirety, and hereafter will be referred to as **low risk**. Due to time constraints, 24 (of the 30) low-risk volunteers completed the WJ III Auditory Attention Test.

Both sets of participants completed a demographic survey as well as the WJ III Auditory Attention Test. The scores for one low-risk volunteer were excluded from the final dataset because English was the Soldier's second language and she had to learn several new words during the Auditory Attention Test training session. She stated she had not previously heard these words or their definitions. Also, her score on the test was considerably lower than her peers' scores. Therefore, it was concluded that her performance was not a true indicator of her auditory discrimination ability, but of her understanding of the English language.

INSTRUMENTS & PROCEDURE

DEMOGRAPHICS SURVEY

The demographics survey contained 24 questions presented in a self-report, Likert scale format. Demographics included age, race, gender, marital status, and highest level of education.

WOODCOCK JOHNSON III

The Woodcock Johnson III (Woodcock, McGrew & Mather, 2001) is a cognitive assessment battery comprised of two batteries: the WJ III Tests of Cognitive Ability (WJ III COG) and the WJ III Tests of Achievement (WJ III ACH). The WJ III COG is used to assess various categories of intellectual ability and the WJ III ACH is used to assess curricular areas of academic achievement. Extensive validity and reliability tests have been conducted on both batteries (Ibid). In addition, scores on the test have been co-normed, meaning all norms were established using the same samples on both tests (McGrew & Woodcock, 2001). The WJ III COG, the battery used in this study, assesses the following seven factors of cognitive ability based on the Cattell-Horn-Carroll (CHC) theory and taxonomy of human cognitive abilities: comprehension-knowledge, long-term retrieval, visual-spatial thinking, auditory processing, fluid reasoning, processing speed, and short-term memory.

The focus of this study is the WJ III Auditory Attention Test (hereafter referred to as AA), part of the auditory processing section of the WJ III COG. The AA Test measures the ability to "overcome the effects of auditory distortion or masking in understanding oral language...requiring selective attention" (Mather and

Woodcock, 2001). Basically, this is a test of auditory discrimination in the presence of background or masking noise. This test is administered using the WJ III flip-chart booklet (Test 14), an audiotape, a tape recorder, and headphones (optional).

Subjects first receive training on the AA task. During training, the researcher reads a word and the subject points to the picture of the word; the training consists of 57 words. If an error is made, the researcher immediately indicates the correct picture, says its name, and if necessary, explains what the picture is. After presenting all 57 items, the researcher retests any missed items.

During AA testing, the researcher shows the subject a series of pictures while the subject listens to an audiotape. The volume of the audiotape is adjusted according to the subject's preference. The test is made up of 50 rows of four items per row. The subject's task is to point to the picture that represents the word they hear on the audiotape. For example, when presented with beans, a bead, bees, and cheese, the subject is instructed to "Point to bees." If the subject does so, he/she earns one point. With each trial, the level of background noise increases, making it more difficult to identify similarly sounding words spoken at a constant volume. Testing is stopped either when the subject completes the final test item, or misses six consecutive items.

Upon completion of the WJ III AA Test, raw scores are put into the WJ III Compuscore and Profiles program, an automated scoring software program that comes with the WJ III Test and uses established norms to compile a report of participants' cognitive ability (Schrank & Woodcock, 2001). The AA Test raw score is entered as the number of correctly identified items and the software program produces a W score (W) and a grade equivalent score (GE), or grade level, for each individual. The W score is useful for statistical procedures as it has equal-interval measurement, while the GE score is useful for interpretation as it represents the individual's performance in terms of the grade level in the normative sample (Mather & Woodcock, 2001).[1]

RESULTS

Participants were mostly young (age range: 18-42 years; $M = 20.7$, $SD = 4.23$), single (83%), Caucasian (44%), female (63%), with a high school diploma (53%), and in the Reserves (73%).

Mean AA GE scores for high- and low-risk Soldiers are provided in Table 1. A two-tailed independent samples t-test was performed to compare the mean AA W scores between risk groups. The t-test failed to find a significant difference ($p =$

[1]All statistical analyses performed within this paper used W scores. GE scores are provided in tables and graphs for interpretation purposes only.

.058) between high ($M = 503.48$, $SD = 6.11$) and low-risk students ($M = 506.35$, $SD = 3.71$).

Table 1 AA GE - Risk Group

	AA GE				
Risk Group	**N**	**Min**	**Max**	**M**	**SD**
High Risk	25	1.4	18.0	8.47	4.37
Low Risk	23	5.8	18.0	10.96	2.84

Mean AA GE scores split by gender are provided in Table 2. A two-tailed independent samples t-test was performed to compare mean W scores between men ($M = 505.72$, $SD = 4.48$) and women ($M = 504.33$, $SD = 5.68$). The t-test failed to find a significant difference between men and women ($p > .05$).

Table 2 AA GE by Gender

	AA GE				
Gender	**N**	**Min**	**Max**	**M**	**SD**
Male	18	3.5	18.0	10.2	3.6
Female	30	1.4	18.0	9.3	4.1

Mean AA GE scores split by risk group and gender are provided in Table 3. Although the mean AA GE scores for men and women in the high risk group appear to be different, a two-tailed independent samples t-test performed on the mean W scores between high risk men ($M = 505.75$, $SD = 6.14$) and high risk women ($M = 502.41$, $SD = 5.98$) failed to find a significant difference ($p > .05$).

Table 3 AA GE by Gender and Risk Group

	AA GE				
	High Risk			**Low Risk**	
Gender	**N**	**M**	**SD**	**N**	**M**	**SD**
Male	8	9.94	4.75	10	10.41	2.52
Female	17	7.78	4.14	13	11.38	3.10

Due to small sample sizes within certain race groups, such as Asian ($N = 5$) and African American ($N = 7$), differences between races could not be examined statistically. Thus, a new variable was constructed that contained only two race categories: Caucasian and Other. The two new categories are meaningful for interpretation and have a larger sample size per group ($N = 21$ and 27, respectively). Subsequent race analyses used this newly constructed variable.

Mean AA GE scores split by gender and the new race variable are presented graphically in Figure 1. Two separate two-tailed independent samples t-tests were conducted on mean W scores by race and gender. Due to the potential for increased Type I error that exists when performing multiple t-tests, a Bonferroni correction was made by dividing the p-value (α) by the total number of comparisons (Abdi, 2007). Using an adjusted p-value ($p = .025$), the first t-test found higher W scores for Caucasian men ($M = 507.44$, $SD = 4.59$) than Caucasian women ($M = 502.33$, $SD = 4.83$) ($t(19) = 2.45$, $p = .024$). The second t-test compared W scores for the Other race category by gender, but failed to find a significant difference ($p > .05$) between men ($M = 504.00$, $SD = 3.87$) and women ($M = 505.67$, $SD = 5.93$).

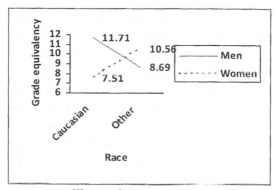

Figure 1 Race x Gender

PREDICTING AUDITORY DISCRIMINATION

To predict auditory discrimination (as measured by AA W scores) from demographic variables, a multiple regression model was constructed using dummy coded variables. According to Hardy (1993), dummy coding is necessary when using categorical independent variables in a regression equation. The following seven variables were entered into the regression equation based on the t-test results and then reduced using a Stepwise approach[2]: High risk, Caucasian, Female, High risk x Caucasian, High risk x Female, Caucasian x Female, and High risk x Caucasian x Female. In addition to a constant, the resulting regression model (Table 4) contained only one variable, the interaction, High risk x Caucasian x Female. This regression model explained a significant proportion of variance in W scores, $R^2 = .13$, $F(1, 46) = 7.01$, $p = .011$.

---

[2] Criteria: Probability-of-F to enter $\leq .05$, Probability-of-F to remove $\geq .10$

Table 4 Summary of Multiple Regression Analysis for AA W Scores (N = 48)

Variable	B	SE(B)	β	t	Sig. (p)
Constant	505.70	.78		646.37	.000
High risk x Cauc. x Female	-5.08	1.06	-.36	-2.65	.011

DISCUSSION

The findings of our study on auditory discrimination are three-fold. First, we found that among 68D trainees, high-risk students performed on average at an 8th-grade level and their low-risk classmates performed at an 11th-grade level. However, the difference in scores between risk groups was not statistically significant. This may have been due to a high degree of variability of scores within the high risk category. Second, we found that Caucasian females had significantly lower auditory discrimination scores than Caucasian male students. Consistent with the gender aspect of this finding, a separate study found in noisy classrooms, male children outperformed female children on complex matrix problems (Christie & Glickman, 2006). Third, we were able to predict to some degree of accuracy auditory discrimination with a linear regression model using three pieces of information-- high risk group, Caucasian race, and female gender. These findings tell us that auditory discrimination difficulty is not simply a characteristic of the current generation of Soldiers, but rather a cognitive ability that is lacking among Soldiers who are having academic difficulties, particularly among Caucasian females. The causes of these difficulties in auditory discrimination among certain groups are unknown, but warrant further investigation.

It is reasonable to assume that auditory discrimination is necessary in academic (learning) environments. Students need to be able to attend to and hear the instructor's lesson, tune out interfering background noises (e.g., classmates talking, overhead HVAC fans), and distinguish among similar-sounding verbal information. Knowing this is a deficiency among certain high-failure-risk Soldiers lends itself to changes aimed at designing the environment, in this case the classroom and the academic program of instruction, according to needs of the student. Specifically, lectures should be held in classrooms with little to no background noise. An experiment with school-aged children with auditory figure-ground difficulties failed to find performance differences in closed versus open-air classrooms (Brown, 1975). However, this may have been due to the fact that noise levels between the open-air and closed classrooms were only significantly different in the morning, and not in the afternoon (Ibid). A similar study on classroom design would need to be conducted with a military population, and with better noise control to see what type of classroom configuration improves performance. In addition, instructors should be advised to speak clearly, and with enough volume (or use a microphone) so all students can hear. Alternately, students with difficulties hearing the lecture should be able to sit to the front of the classroom, closer to the instructor. Lecture materials (e.g., slides and notes), should be provided to all students as a supplement to oral information. In addition, "brain training programs" that incorporate auditory input and claim to strengthen neural networks, such as the Interactive Metronome© (Interactive Metronome, 2009; see Rice et al., 2008) may be helpful for students with auditory discrimination problems. This is another area for future research, to see if auditory discrimination can be improved through training. Lastly, the findings of this research make a strong case for the military to continue to emphasize the importance of hearing protection among service members.

LIMITATIONS

Although this study presents compelling results regarding the relationship between auditory discrimination and academic performance among 68D trainees, additional investigations should be completed to generalize to other military or civilian student populations. In addition, the sample size here was small, limiting the individual race comparisons. Further research is needed to increase statistical power and determine if the findings of this study hold true, and to determine if any other relationships exist between auditory discrimination and academic performance. Lastly, hearing tests were not administered during the study; this information could be used to control for the effects of hearing loss on auditory discrimination.

CONCLUSIONS

This research demonstrated that auditory discrimination is related to academic performance among 68D trainees. Moreover, high-risk Caucasian female students had greater difficulty with auditory discrimination than others. These results warrant additional research into the role of auditory discrimination in academic performance, particularly among different racial/ethnic groups. They also highlight a need for environmental alterations to decrease background noise in classrooms.

REFERENCES

Abdi, H (2007). Bonferroni and Šidák corrections for multiple comparisons. In N.J. Salkind (ed.). Encyclopedia of Measurement and Statistics. Thousand Oaks, CA: Sage. Retrieved Feb 26, 2010 from: http://www.utdallas.edu/~herve/Abdi-Bonferroni2007-pretty.pdf.

Alfred, P. & Rice, V.J. (2009). Cognitive characteristics of academically at-risk Operating Room Specialist trainees. Human Systems Integration Symposium, Annapolis, MD.

Bellis, T.J. (2002). When the brain can't hear: Unraveling the mystery of auditory processing disorder. New York: Atria Books.

Brown, C. (1975). The relationship between auditory figure-ground perception and academic achievement in open area and self-contained classrooms. (Master's thesis).

Christie, D.J. and Glickman, C.D. (2006). The effects of classroom noise on children: Evidence for sex differences. *Psychology in the Schools, 17*, 405-408

Hardy, M. (1993). *Regression with dummy variables*. Newbury Park: Sage Publications.

Interactive metronome (2009). Overview. Retrieved Nov 3, 2009 from: http://www.interactivemetronome.com/ IMPublic/Home.aspx

Mather, N. and Woodcock, R.W. (2001). Examiner's Manual. Woodcock-Johnson III Tests of Cognitive Abilities. Itasca, IL: Riverside Publishing.

McGrew, K.S. and Woodcock, R.W. (2001). Technical Manual. Woodcock-Johnson III. Itasca, IL: Riverside Publishing.

Nemours Children's Health System (2007). Auditory processing disorder. *Kids Health*. Retrieved Feb 24, 2010 from http://kidshealth.org.

Niskar, A., Kieszak, S., Holmes, A., Esteban, E., Rubin, C., & Brody, D. (2001). Estimated prevalence of noise-induced hearing threshold shifts among children 6-19 years of age: The third national health and nutrition examination survey, 1988-94, United States. *Pediatrics, 108* (1), 40-43.

Paul, R. (2008). Auditory Processing Disorder. *Journal of Autism and Developmental Disorders, 38*, 208-209.

Rice, V., Alfred, P., Laws, L., Boykin, G., and Vu, T. (2008). The Interactive Metronome Intervention Study: An Examination of the Interactive Metronome Training Program on 'At Risk' Soldiers' Cognitive, Psychomotor, and Academic Performance; in draft; U.S. Army Research Laboratory: Fort Sam Houston, TX.

Rice, V., Butler, J., Marra, D., DeVilbiss, C., and Bundy, M. (2006). A Prediction Model for the Personal Academic Strategies for Success (PASS) and Academic Class Composite (AC^2T) Tool Supporting Soldiers in AIT at Ft. Sam Houston, Texas; in draft; U.S. Army Research Laboratory: Fort Sam Houston, TX.

Schrank, F.A. and Woodcock, R.W. (2001). WJ III Compuscore and Profiles Program [Computer software]. Woodcock Johnson III. Itasca, IL: Riverside Publishing.

Vause, N.L. (2008). Hearing ergonomics for children: Sound advice. In R. Lueder and V. Rice (Eds). *Ergonomics for Children: Designing products and places for toddlers to teens*. Taylor & Francis: New York, pp. 109-185.

Woodcock, R.W., McGrew, K.S., & Mather, N. (2001). Woodcock-Johnson III. Itasca, IL: Riverside Publishing.

Chapter 40

Self-Reported Sleep & Soldier Performance

Valerie J. Rice, Jenny Butler, Diane Marra

Army Research Laboratory-Army Medical Department Field Element
Ft. Sam Houston, San Antonio, TX

ABSTRACT

Cognitive performance can be impaired by lack of sleep. Healthcare Specialist (68W) Advanced Individual Training (AIT) is rigorous and students must learn and process a great deal of information in a short time. The purpose of this study was to explore the relationships between self-reported sleep (SRS) and the academic and physical performance of students attending 68W AIT, as well as exploring the relationship between SRS and personal variables that may impact performance. Soldiers (n=579) attending 68W AIT completed a questionnaire two weeks after beginning training. The questionnaire addressed hours of sleep, demographics, personal characteristics and coping skills potentially related to performance. Final grades and physical training scores were recorded. SRS was less for those who were older and who currently smoked. Those who were more motivated and willing to complete training, or with higher self-efficacy scores had higher SRS. SRS followed an inverted U-shape, according to stress levels, with SRS being lowest among those with extremely high and extremely low stress. SRS was related to initial and mid-course physical fitness scores ($p < 0.05$), but not to the physical fitness test conducted at the end of the course or to final grade point average or pass/fail status ($p > .05$). The results reveal relationships between SRS and constructs related to Soldier performance. Possible future research and program considerations are provided

Keywords: sleep, academic performance, physical performance, motivation, stress

INTRODUCTION

As our nation is at war, our military services have a greater need for Healthcare Specialists, also known as Combat Medics or 68Ws. Yet, there is a shortage of trained personnel and attrition rates during military occupational specialty training looms at approximately 25% (Whittaker and Parsons, 2009).

Training for 68Ws is intense. The high cognitive demands, coupled with the fast pace required, leaves little time for study and assimilation of new concepts into existing schemes. Anecdotal reports tell of students studying by flashlight at night, even while knowing they must rise for physical training at 0430 the next morning. In a prior study, Soldiers who failed 68W training and those with the equivalent of a B or above, completed a questionnaire and an interview regarding their perceptions of why Soldiers fail training. In addition, their 'battle buddy' or friend also completed a survey and interview. Sleep was a recurrent theme and one of the top issues for those who did not pass training (Rice, Laws, Butler, and Vu, 2008). Over 50% of the failing students reported sleeping between three and four hours per night (DeVilbiss, Rice, Laws, and Butler, in press).

Cognitive performance can be impaired by lack of sleep or even by a person's subjective feelings of fatigue (Druckman and Bjork, 1994). Sleep has also been linked to student grades (Kelly, Kelly, and Clanton, 2001). Shorter sleep times are related to increased sleepiness in class, reduced concentration, and problems focusing on information (Sadeh, Gruber, and Raviv, 2003, cites Epstein et al., 2003). Sleep is also related to learning new information. Students who were deprived of a nights' sleep were less able to recall a new task up to three days after the sleep loss, indicating a need for sleep in order to process, consolidate and retain new memories (Stickgold, 2005, as cited by Winerman, 2006). While the exact role of sleep in memory consolidation is controversial, there is no doubt sleep is important for learning new information and tasks, as well as for memory (Ellenbogen, Hulbert, Stickgold, Dinges, and Thompson-Schill, 2006). Sleep has also been shown to enhance mental performance, such as assisting in the consolidation of motor skill procedural memories (Fenn, Nusbaum, and Margoliash, 2003), preventing memory decay, and fixating declarative memories (Ellenbogen, et. al., 2006). In fact, even the time during which one sleeps is important, as indicated by McCann's research, in which she found the grade point averages of college students who are "night owls" are nearly one letter grade lower than "morning larks" (McCann, 2008).

Not only does sleep impact cognitive performance during education and training, worry about performance can also impact sleep, thus creating a problematic cycle. A study by Tallis, Eysenck, and Mathews (1992, as cited in Kelly, 2004), demonstrated that students who scored higher on academic worry slept for briefer times. Kelly (2004) calculated a simple regression and found worry accounted for six percent of the variance in sleep length. Recently, Ginsberg (2006) replicated the findings, again showing "students who worry more seem to sleep less".

This paper examines whether Soldier's self-reported sleep (SRS), recounted at the start of a training program, differs according to academic (grade point average and pass/fail status) and personal factors that may influence training performance. It also examines the relationship between SRS and physical performance (Army Physical Fitness Test [APFT] scores).

METHODS

Procedures: Soldiers (n = 579) completed a questionnaire within two weeks of arriving at Fort Sam Houston, TX to attend 68W AIT. The Department of Combat Medical Training provided student performance outcome data in terms of final grades, pass/fail status for academic failure, and APFT scores.

Instrumentation

Self-Report Survey. Demographics included age, gender, marital status, ethnicity, military component, and education level. Respondent attributes included self-ratings of study skills, several measures associated with motivation, feelings of tiredness, attitudes toward the training program and faculty, self-efficacy and evaluations of their fear-of-failure, commitment, learning styles, and self esteem.

Self-Reported Sleep (SRS). Soldiers reported their typical number of hours of sleep on week nights and weekend nights. A composite sleep variable was generated by adding the hours of weekday and weekend sleep, with a minimum of four hours required for either variable. For example, if a Soldier reported 5 hours of weekday sleep per day and 9 hours of weekend sleep per night, their score would be 14. As receiving less than four hours of sleep on a regular basis was considered to potentially be an exaggeration (outlier), for either weekdays or weekends, only those who reported sleeping four or more hours were included in the data analysis. This resulted in 558 remaining observations. Other measures were considered, such as weighting weekend or weekday hours or constructing a 'total week' variable. All other constructions were highly correlated with our ultimate sleep variable. SRS for weekend and weekday hours were correlated with our constructed variable (r (557) < .00, p = .83 and r (557) < .00, p = .87).

Performance Measures. This included scores for sit-ups, push-ups and 2-mile run, as well as the total APFT, administered at the beginning, middle, and end of training. Academic measures were grade point average (GPA) and pass/fail status.

Statistics. An Analysis of Variance was used for comparisons, with a Bartlett's test for equal variances. A Pearson's *r* was used for paired comparisons. All analyses used an alpha level of 0.05 and were performed using Stata statistical software (StataCorp, 2005).

RESULTS

Participants were primarily male (male=350, female=228)[1] with an age range of 17-40 years and a mean age of 20.84 years (Male: $M = 21.33 \pm 4.34$; Female: $M = 20.15 \pm 3.79$). The mean of the constructed variable SRS was 12.79 ± 2.15.

On average, for each increase of a year in age, holding other demographics constant, there was a decrease in SRS of three minutes per night ($F (1, 486) = 22.13$, p < .00). Women reported sleeping slightly more than men ($F (1, 541) = 4.34$, p =.04, mean = 13.15 ± 1.86, 12.80 ± 1.94). Non-smokers and past smokers reported similar hours of SRS, while current smokers SRS was 8 and 14 minutes less, respectively ($F (2, 532) = 4.50$, $p = .01$). Reservists SRS was greater than non-reservists' SRS ($F (1, 542) = 7.61$, $p = .01$, mean = 13.44 ± 1.80, 12.84 ± 1.93 respectively). No differences in SRS were found for race, English as a second language, high school grades, education level, or prior medical training ($p > 0.05$).

While prior medical experience was not associated with SRS, prior military service members SRS was less than non-prior service members ($F (1, 538) = 22.51$, $p < .000$, mean = 11.0 ± 1.75, 13.03 ± 1.88, respectively). Higher motivation equated with greater SRS on three motivation measures, and higher levels of satisfaction with the first two weeks of training (Table 1). The highest level of motivation showed SRS as an hour more than the lowest level of motivation (12.08 ± 1.80 vs. 13.09 ± 1.86). Those who were extremely satisfied with the first two weeks of training, reported nearly an hour more sleep than those who were not at all satisfied (13.26 ± 1.72, 12.35 ± 2.06). SRS did not differ for those having 68W training as a personal goal or according to self-ratings of leadership ability (Table 1).

Table 1. Results of Self-Reported Sleep according to Indices of Motivation, Satisfaction with Training, and Leadership.

Motivational Rating Indices (self-report)	Self-Reported Sleep		
	df	F	p-value
Importance of their training to them	2, 538	1.95	0.14
68W training as a personal goal	4, 539	1.62	0.17
Willingness to complete training successfully	2, 540	5.33	0.01
Motivation to complete training	4, 538	3.28	0.01
Interest in being a 68W	4, 537	2.74	0.03
Satisfaction with training received in first 2 wks of class	4, 537	4.11	<0.00
Leadership abilities	4,534	2.13	0.08

---

[1] Total does not equal 579 because one participant did not provide gender information.

Those with higher SRS were satisfied with most program attributes, but there was no relationship between SRS and participants' satisfaction with their own performance (Table 2). No significant relationship was seen between SRS and Rosenberg Self-Esteem scores (F (1, 542) = 2.80, p = .10), yet there was a relationship between SRS and Self-efficacy ratings (F (1, 542) = 6.59, p = .01). SRS was higher for those with higher self-efficacy scores. The relationship between SRS and stress followed a U-shaped distribution with those rating themselves with unusually low (11.44 ±1.74) or unusually high stress (12.36 ± 1.94) reporting sleeping less than those in mild, moderate, or high stress categories (13.41 ± 1.50, 13.20 ± 1.94, 13.0 ± 1.96) (F (4, 535) = 5.80, p < .00). Those who reported feeling overtired more frequently had lower SRS compared with those who rarely or never reported feeling overtired (F (1, 538) = 4.23, p = .04).

Table 2. Results of Self-Reported Sleep according to Self-Ratings of Satisfaction with Program and with Own Performance.

Program Attributes (self-report)	Self-Reported Sleep		
	df	F	p-value
Instructors' helpfulness	4, 536	2.57	0.04
Instructors' availability	4, 536	4.58	<0.00
Unit has strong emphasis on high quality work	4, 529	1.72	0.14
Resource availability (supplies, assistance)	4, 536	2.59	0.04
Commanding Officer supportive of student needs	4, 532	2.66	0.03
Commanding Officer understanding of student issues	4, 533	3.59	0.01
Satisfied with own performance	4, 537	0.35	0.84
Current performance is consistent with their best	4, 536	1.35	0.25

Significant negative correlations were found between SRS and each of three APFT push-up scores, the first two run scores, and the first two total APFT scores ($p \le$ .02), but not between SRS and any of the final APFT scores ($p > .05$). The n for each APFT was 464 for the administration at the start of the course, 416 mid-course, and 336 at the end of the course. SRS did not differ according to GPA or pass/fail status r (557) = .03, p = .55).and F (1, 542) = 2.92, p = .09, respectively).

DISCUSSION

As young men and women enter and progress through adolescence, their total sleep time increases slightly, which some postulate to be related to metabolic changes occurring during growth (Ohayon, Carskadon, Guilleminault, and Vitiello, 2004). Adolescence continues through the teens, some saying into the early twenties, and includes the majority of the participants in this study. Sleep efficiency, the ratio of sleep to the amount of time spent in bed, is reported to remain around 95% at age 20, but to decrease thereafter (Wolfson and Carskadon, 2003). Here, we saw a slight decrease in SRS with an increase in age. It is surmised that SRS is indicative of time spent in actual sleep, rather than time in bed.

Our research revealed current smokers reported sleeping slightly less than past or non-smokers. This supports other research demonstrating cigarette smokers are more likely to report difficulties going to and staying asleep, as well as greater daytime sleepiness (Phillips and Danner, 1995). This could be related to the stimulant property of nicotine interfering with sleep (Zhang, 2008).

No prior research has reported sleep information on reservists vs. active duty while attending AIT. There may be less worry among reservists who know they can return to their homes and jobs following training. Active duty Soldiers typically do not know their next duty station, and should they fail, they will be required to attend training in a different specialty, generally one that is in high demand by the military, but may not be of interest to the individual service member. In addition, those with prior military service tended to sleep less. While the impetus for this is not known, it may be that prior service members understand the implications of doing poorly and push themselves to remain awake, studying longer hours.

Participants who were more interested in being a 68W, and willing and motivated to complete training, reported sleeping more, with those reporting the highest level of motivation having a SRS score of nearly an hour more than those with the lowest motivation. The relationship between motivation and sleep is most likely because those who are feeling rested simply feel better and more enthusiastic, although it may be that those who are more motivated also understand their need for sufficient sleep to optimize their performance. Those who reported sleeping less in our study also reported feeling overtired more frequently. Edens (2006) found college students who experience excessive daytime sleepiness (which he considered as a measure of sleep habits) tend to be motivated by performance goals such as grades, rather than mastery goals. Our questions addressed interest, desire and willingness to work toward a goal, rather than motivation toward performance goals themselves, so a direct comparison with his findings is not possible.

While those who reported sleeping more were more likely to be satisfied with the program, there was no relationship between SRS and self esteem or SRS and satisfaction with participants' own performance. Again, being more satisfied with the program may be due to their feeling more rested and better about the tasks they were about to undertake (training). Even though there was no relationship between SRS and self esteem, those who believe they are capable of passing training (self-efficacy) reported sleeping more. The higher SRS and higher self-efficacy may reflect an analogous finding to those who rated themselves as being more motivated also reporting higher SRS. These findings do not support Edens (2006), who found students with excessive daytime sleepiness had lower scores on self-efficacy.

We did not measure worry, which has been associated with poor academic performance (Ginsberg, 2006). We did examine stress and fear-of-failure. Our results reveal those with extreme stress (either extremely high or extremely low stress) reported sleeping less, but failed to find a relationship between fear-of-failure and SRS. Different constructs exist behind worry, stress and fear-of-failure, although they can be related. For example, worry relates to concern about something that will occur in the future, while stress is a physiological reaction to a current stressor. Fear-of-failure has been linked to feelings of shame and test-anxiety (McGregor and Elliot, 2005), while the same is not necessarily true for worry and stress. Lack of sleep does influence a person's' emotions, increasing their perceived exertion, stress, anxiety, worry, and frustration (Coren, 1996; LeDuc and Caldwell, 1998), and being prone to anxiety can also influence sleep (Vahtera, Kivimaki, Hublin, Korkeila, Suominen, Paunio, and Koskenvuo, 2007).

A clear relationship was seen between sleep and physical performance, but not between sleep and academic performance. However, the relationship between sleep and physical performance was not what we expected. Those who reported more sleep received lower APFT scores on the first two fitness tests. It is unknown if this is because our sleep information was gathered so early during training and therefore is not indicative of actual sleep later during training. It could also be that individuals who require a greater amount of sleep are more negatively impacted by the demands of 68W training than are those individuals who require less sleep. This latter explanation may be more likely, as no individuals reported sleeping exceedingly long hours. Experts recommend that teenagers (high school and college students) sleep between 8 and 9½ hours each night (Maas, 1999) and regular exercise yields an increased need for total sleep time (Kubitz, Landers, Petruzzello, and Han, 1996). Growth and repair occur while there is low metabolic activity and increased secretion of growth hormone during slow-wave sleep (Maas, 1999). While some research indicates cardiovascular performance is not compromised until high levels of sleep deprivation occur (Martin, 1981), others believe a person can accumulate a cumulative sleep debt (Dement and Vaughan, 1999; Walters, 2002). Paraphrasing an example by Walters (2002), if a young soldier needs 8 hours of sleep, but gets only 5½ per night, it would take only 12 days to incur a sleep debt that would negatively impact aerobic performance. Reaction time and detection of visual stimuli are also degraded following moderate sleep deprivation, which could impact maximal efforts for physical performance (Dinges, 1995) during APFTs or during training-related tasks. Thus, it would have been expected that SRS would be higher among those who performed well on the APFT. SRS was not associated with performance on the final APFT. Fewer individuals participate in each APFT as time passes. This is due to the fact that those with musculoskeletal injuries or other illnesses do not participate, and Soldiers become injured or older injuries are revisited during the intense physical training conducted four mornings a week.

The lack of impact of self-reported sleep on academic performance may be because the questionnaire was administered early during training, in the hopes of identifying indicators that would be predictive of academic performance. This early administration may not have been an accurate indicator of sleep patterns later in the training. It is also possible that Soldiers sleep less than the four hours of sleep we considered as necessary for a subjects' data to be included in this study. Other research by this laboratory, in which sleep information was gathered from students at the conclusion of their training, revealed students who failed their training reported sleeping fewer hours, having trouble staying awake during class, and believing lack of sleep was an issue that impacted their academic performance (DeVilbiss and Rice, in press). Cognitive performance declines more rapidly than motor performance as a result of sleep deprivation (Pilcher and Horne, 1996), but SRS was not less for those with lower academic performance in this study.

LIMITATIONS

The limitation of self-report data is the possibility of participants answering to appear competent and capable. This study offered assurance of non-attribution of responses. In addition, supervisors were absent from the room, so they did not know who did or did not volunteer to participate. A second limitation is the subject population, which is all military. It is unknown whether these results can be generalized to other, non-military populations.

CONCLUSION

In general, this study demonstrates the relationship between self-reported sleep and constructs related to Soldier performance during AIT. This study supports earlier study results showing relationships between sleep and age, smoking, and stress.

This study focused on military service members attending AIT. New findings linked past military experience with less reported sleep, while reservists reported more sleep. Self-reported sleep during the first two weeks of training was higher for those who were more satisfied with their training, expressed greater motivation, and believed they were capable of passing their training. More hours of SRS were related to poorer performance on the first two fitness tests. While it would nice to believe Soldiers understand their own need for sleep and the potential impact of lack of sleep on their performance, thus those with higher motivation took care to ensure they received the sleep they needed, the more likely explanation is that those who were receiving more sleep had a more positive outlook on their training and their abilities to meet the training demands. It is also difficult to determine whether sleep patterns reported in the first two weeks of training is indicative of sleep patterns later during training. Further research is warranted, in which Soldiers are asked about their perceived sleep needs (for example, do they typically need eight hours per night or is six hours sufficient), along with repeated, accurate sleep measures throughout the training program, such as wearing an actigraph.

Even though inaccuracies may have resulted from early self-reports, there is utility in the information gained through this study. Those who report getting more sleep are less tired and more motivated and pleased with their training. Coupled with our earlier research on sleep (DeVilbiss and Rice, in press), sleep appears to be an important issue during military AIT. Lack of sleep can impair physical performance, cognitive ability, and emotional stability (Walters, 2002) and has been identified as a factor influencing student performance by the students themselves during interviews (DeVilbiss and Rice, in press). The majority of Soldiers attending AIT enter the military directly after high school and are between 18 and 20 years of age. Adolescents are reported to need more sleep than those of other ages and accommodations such as starting high school classes later in the day have yielded more sleep for students, less daytime sleepiness, less depression, and better grades (Graham, 2000). While civilian higher education programs permit students to schedule their own study and class times, military programs control student times for rising, physical exercise, barracks cleaning, meals, Soldier training in addition to military occupational specialty training (which typically occurs in the evenings after eight hours of daytime classes), and "lights out" time in the evenings. It is suggested that military education programs consider the impact of their programs of instruction and the availability of time for sufficient sleep on student learning, consolidation of memories to enable later recall during times of duress on the battle field, and physical performance (both in terms of physical training and perceptual-motor performance necessary for the Soldiering skills required of a Combat Medic).

REFERENCES

Coren, S. (1996). Sleep thieves: An eye opening exploration into the science and mysteries of sleep. New York: Free Press.

Dement, W. and Vaughn, C. (1999). The promise of sleep. New York: Delacorte, 1999.

DeVilbiss, C. and Rice, V.J. (in press). If you want to know why students fail, just ask them: Self and peer assessments of factors affecting academic performance. Work: A Journal of Prevention, Assessment, and Rehabilitation.

Dinges, D.(1995). An overview of sleepiness and accidents. J Sleep Res, 4(2), 4-14.

Druckman, D. and Bjork, R.A. (Eds). (1994). Learning, rembering, believing: Enhancing human performance. Committee of Techniques for the Enhancement of Human Performance, National Research Council. Washington, D.C.: National Academy Press.

Edens, K.M. (2006). The relationship of university students' sleep habits and academic motivation. NASPA Journal, 43(3), 442-445.

Ellenbogen, J.M., Hulbert, J.C., Stickgold, R., Dinges, D.F., Thompson-Schill, S.L. (2006). Interfering with theories of sleep and memory: Sleep, declarative memory, and associative interference. CurrBiol, 16 (13), 1290-1294.

Fenn, K.M., Nusbaum, H.C. & Margoliash, D. (2003). Consolidation during sleep of perceptual learning of spoken language. Nature, 425, 614-616.

Ginsberg, J. (2006). Academic worry as a predictor of sleep disturbance in college students. The Journal of Young Investigators, 14 (4) retrieved 9 February 2010 from http://www.jyi.org/research/

Graham, M.G. (2000). Sleep needs, patterns, and difficulties of adolescents (4461816 ed). Washington, DC: National Research Council Institute of Medicine.

Kelly, W.E. (2004). Sleep-length and life satisfaction in a college student sample. College Student Journal. Retrieved from http://findarticles.com/p/articles/mi_m0FCR/is_3_38/ai_n6249228/

Kelly, W.E., Kelly, K.E., Clanton, R.C. (2001). The relationship between sleep length and grade-point average among college students. College Student Journal (0146-3934), 35(1).

Kubitz, L.A., Landers, D.M., Petruzzello, S.J.,and Han, M. (1996). The effects of acute and chronic exercise on sleep: A meta-analytic review. Sports Med, 21, 277-291.

LeDuc, P.A. and Caldwell, J.A. (1998). Submaximal exercise increases alertness in sleep-deprived aviators. Aviation, Space, and Environmental Medicine, 69(3), 229.

Maas, J.B. (1999). Power sleep. New York: Harper Perennial.

Martin, B.J. (1981). Effects of sleep deprivation on tolerance of prolonged exercise. Eur J Appl Physiol Occup Physiol, 47, 345-354.

McCann, K. (2008, June 9). Morningness a predictor of better grades in college. Presented at the American Academy of Sleep Medicine Conference, Abstract ID 0728.

McGregor, H.A. and Elliot, A.J. (2005). The shame of failure: Examining the link between fear of failure and shame. Personality and Social Psychology Bulletin, 31(2).

Ohayon, M.M., Carskadon, M.A., Guilleminault, C., Vitiello, M.V. (2004). Meta-analysis of quantitative sleep parameters from childhood to old age in healthy individuals: Developing normative sleep values across the human lifespan. Sleep, 27, 1255-1273.

Phillips, B.A. and Danner, F.J. (1995). Cigarette smoking and sleep disturbance. Arch Intern Med, 155(7), 734-737.

Pilcher, J.J. and Horne, J.A. (1996). Effects of sleep deprivation on performance: A meta-analysis. Sleep, 19, 318-326.

Rice, V.J., Laws, L., Butler, J., and Vu, T.M. (2008). Personal Factors Related to Student Performance & Retention among 68W Health Care Specialists Self & Battle Buddy Reports on Factors Affecting Academic Performance, presented, in part, at the Army Medical Department Attrition Summit, April 2008. Available from AMEDD Field Element, 2421 FSH-Hood Street, Trailer E, San Antonio, Texas 78234.

Sadeh, A, Gruber, R., and Raviv, A. (2003). The effects of sleep restriction and extension on school-age children: What a difference an hour makes. Child Development, 74, 2: 444-455.

StataCorp,(2005). Stata Statistical Software: Release 9. College Station, TX: StataCorp LP.

Vahtera, J., Kivimaki, M., Hublin, C., Korkeila K., Suominen, S., Paunio, T., Koskenvuo, M. (2007). Liability to anxiety and severe life events as predictors of new-onset sleep disturbances. Sleep, 30(11), 1537-1546.

Walters, P.H. (2002). Sleep, the athlete, and performance. Strength & Conditioning Journal, 24(2), 17-24.

Winerman, L. (2006). Let's sleep on it. Retrieved from http://www.apa.org/monitor/jan06/onit.html

Whittaker, D.S. and Parsons, D.L. (2009). Army Medical Department Center and School 68W Attrition Rate. Internal Report.

Wolfson, A.R. and Carskadon, M.A. (2003). Undestanding adolescents' sleep patterns and school performance: A critical appraisal. Sleep Med Rev, 7(6), 491-506.

Zhang, L. (2008). Chest, 133, 427-432. News release, American College of Chest Physicians.

Chapter 41

Videogame Experience and Neurocognitive Timing Performance

Gary L. Boykin, Valerie J. Rice, Petra Alfred, Linda Laws

Army Research Laboratory-Army Medical Department Field Element
Ft. Sam Houston, San Antonio, TX

ABSTRACT

The Interactive Metronome (IM®) is a neurocognitive testing and training device that targets perceptual-motor timing, with visual and auditory cuing and feedback. As such, it might be expected that individuals with prior video gaming experience (VGE) would perform better on an initial IM®. Both IM® tasks and video gaming require information processing, interaction of tactile, kinesthetic, visual, and auditory processes, elicitation of motor responses, and the ability to attend (and respond) over time. The purpose of this research was to examine the relationship between self-reported videogame experience (VGE) and IM® Long Form assessment Task Average (TA) scores. Volunteers (n=25), mean age (20.96) attending Military Occupational Specialty (MOS) 68D Advanced Individual Training (AIT) (surgical technologist program) completed a VGE questionnaire, in which they reported the number of years they have played videogames (SRY), the hours per week they play videogames (SRH), and their level of proficiency on a 1-10 scale (SRP). Pearson product moment correlations between IM® Long Form assessment scores indicated a moderate positive relationship between right sided tasks (r(23)=.402, p=.046) and SRH of video game activity, i.e. Those who played more hours had lower IM® right-side scores. No relationship was found between SRY or SRP and IM® scores. While these results should be interpreted carefully due to the small sample size, they suggest that VGE does not improve timing, as measured by the IM®. Therefore, therapist, coaches, and teachers should not exclude those individuals who have a great deal of VGE from intervention programs aimed at improving timing.

Keywords: Interactive Metronome IM®, timing, rhythm, performance

INTRODUCTION

The Interactive Metronome® was originally developed for use with children who had timing and coordination deficits, as well as difficulty focusing attention, such as Attention Deficit Disorder. Studies have shown improvement in both children and adults who used the IM to improve attention and timing (Kuhlman and Schweinhart, 1999; Kakazu, 2003; Shaffer, et al 2001).

During the IM® assessment an individual listens to, and attempts to match, a reference tone using various movement combinations and tapping a trigger with their hands and/or feet. One possible rationale for use of the IM® to assist in improving timing and integration of physical and visual-spatial tasks is the concept that repetition helps to strengthen neuro-pathways (Green and Bavelier, 2008). In turn, these neural pathways may respond more readily when they are used during subsequent scenarios that require rhythm, timing and sequencing.

Video gaming involves the use of auditory and visual cues and feedback, visual tracking and coordination to perform the task at hand (Schlickum, Hedman, Enochsson, Kjellin, Fellander-Tsai, 2009). The ability to attend over time and timing appear to be key essentials to successful play of video games. Video game simulations are used frequently for training pilots, and health care practitioners to perform complex tasks, and to identify problems and correct errors that could occur in real-time situations (Rosser, et al 2007; Issenberg, et al 1999).

Thus, individuals who have a great deal of videogame experience and expertise might be expected to be better at timing tasks and perhaps not be considered for IM® intervention. More worrisome is the possibility that a parent, therapist or coach could consider that timing and focus could be achieved through videogaming (which is more fun for the child), if, in fact, this was not the case.

The IM® has also been used to improve academic performance among middle school students who do not have a specific diagnosis (Mulder, 2002). Soldiers attending 68D training must perform complex physical tasks that require precise timing in accelerated classrooms, often attaining the knowledge and skills necessary for a job in half the time it typically takes in a civilian educational program. Soldiers' classroom training is typically from 8 AM until nearly 5 PM, with early morning hours (5 AM – 630 AM) dedicated to physical training and barracks duties and evenings spent learning additional Soldiering skills. In addition, the job tasks of an operating room technician require auditory acuity, precise timing, and coordination, as well as focused attention throughout long, involved surgeries during which they assist surgeons. The high level of demands upon Soldiers attending 68W training (and later during their jobs), and the high attrition rate among military allied health care populations, yielded an interest in examining technologies such as the IM as both a screening and an intervention tool (Rice and Butler, 2007; Rice, Butler, and Marra, 2007).

The purpose of this research is to examine the relationships between self-reported VGE, as measured by years of experience (SRY), current hours of videogame play (SRH), and level of proficiency (SRP) with Interactive Metronome performance measures. It is our hypothesis that there is no relationship between VGE and performance results on the Interactive Metronome long form test battery.

METHODS

PROCEDURES

Volunteers for this study were young adults recruited from Military Occupational Specialty (MOS) 68D, Operating Room Specialist training. Individual participants had failed their training on their initial attempt and were considered as candidates to potentially benefit from IM® training. The participants were waiting to either recycle into the next 68D course, or be reassigned into another MOS. Only those who were physically able to perform all tasks without pain were permitted to participate. Each participant received a thorough briefing on the research. Those individuals desiring to participate in the study completed the informed consent document. Individual cubicles were used to isolate each Soldier during Interactive Metronome testing to prevent audible and visual distractions during their assessment. The pre-assessment regimen included verbal instructions and a demonstration of each task, as well as having each volunteer demonstrate correct execution of each task. Table 1 shows the tasks and sequence included in the IM® Long Form while Figure 1 shows the equipment used. Figure 2 shows Task 6, left toe, being executed.

INSTRUMENTATION

Survey. Participants completed a basic demographic questionnaire including age, gender, race, military status, and hand dominance. In addition, they described their personal video gaming experience by reporting the number of years they have been actively involved in video gaming, an estimate of weekly hours devoted to gaming, and provided a self-assessment of their level of proficiency on a scale of 1 to10. For this report, proficiency levels 1 to 4 were categorized as "novice"; levels 5 and 6 as "competent" and levels 7 to 10 as "expert".

Figure 1. Interactive Metronome Hardware.

Figure 2. Left Toe (Task # 6) lower extremity trigger.

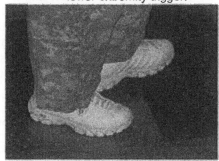

Table 1. IM Long Form Tasks Ordered (1-14)

Task Number	Task Description
Task # 1	Both Hands
Task # 2	Right Hand
Task # 3	Left Hand
Task # 4	Both Toes
Task # 5	Right Toe
Task # 6	Left Toe
Task # 7	Both Heels
Task # 8	Right Heel
Task # 9	Left Heel
Task # 10	Right Hand / Left Toe
Task # 11	Left Hand / Right Toe
Task # 12	Balance Right Foot
Task # 13	11Balance Left Foot
Task # 14	Both Hands with Guide Sounds

Interactive Metronome. Each participant was given a headset to place over their ears, in which an audible reoccurring beat known as reference tone (cowbell) could be heard (Figure 1). A triggering device (sensor) attached to either the palm of one hand or secured to the floor recorded participants' response times in milliseconds in relation to the audible reference tone. That is, the number of milliseconds from the reference tone comprises the individuals' score. The participant's objective was to match the beat of the audible reference tone as closely as possible during the administration of the fourteen exercises by striking the triggering device using their hands, feet or a combination of the two extremities. Participants who wore combat boots to perform lower extremity tasks were asked to change into a shoe that did not have a heel or wide grooves. Flatter sole shoes minimize the likelihood of inaccurate trigger responses.

Participants' individual IM® Pre-Long Form task average scores of 14 tasks were averaged and categorized in accordance with the IM® Certified Provider Training & Resource Binder (IM-CPTRB): Superior, Exceptional, Above Average, Average, Below Average, Severe Deficiency, Extreme Deficiency (IM, 2004). In addition, the (IM-CPTRB) suggested calculations for evaluating the Long Form Assessment results were used for correlations (Box 1).

Box 1. Long Form Assessment Calculations
a.   Hands: Tasks 1, 2, 3, 14
b.   Feet: Tasks 4, 5, 6, 7, 8, 9, 12, 13
c.   Both Hands: Tasks 1, 14
d.   Both Feet: Tasks 4, 7
e.   Left Side: Tasks 3, 6, 9
f.   Right Side: Tasks 2, 5, 8
g.   Bilateral: Tasks 10, 11
h.   Adjusted ms avg: (a+b)/2

STATISTICS

The Statistical Package for the Social Sciences (SPSS), version 15 was used to conduct Pearson product moment correlations ($p>0.05$) between VGE, as measured by SRY, SRH and SRP with IM® performance scores as shown in Box 1. SRP scores were analyzed as interval data, as our anchoring was believed to be sufficient to consider ratings as equal intervals.

RESULTS

Demographic data are shown in Table 2. Participants included 25 Soldiers (17 men and 8 women) who were primarily young (mean age 20.96), male, and Caucasian. The majority of the participants were right hand dominant (76%).

No relationships were found between SRY of videogaming and IM® scores ($p > .05$). Only one significant relationship was found between right side tasks and SRH of videogame activity ($r(23) = .402, p = .046$). As shown in Box 1, right side tasks consist of individual tasks of right hand, right toe, and right heel. In an effort to understand this finding, correlations were done with each subtask and SRH. Results revealed no significant correlations between IM® performance and right hand ($r(23) = .356$, $p=.081$), right toe ($r(23) = .383$, $p = .059$), and right heel ($r(23) = .221$, $p = .289$).

Table 2. Demographic

Demographic Category	Frequency	Percent
Age		
18-20	19	76%
21-30	5	20%
>40	1	4%
Gender		
Male	17	68%
Female	8	32%
Race		
Caucasian	10	40%
Hispanic	7	28%
African American	4	16%
Asian	2	8%
Military Status		
Active	11	44%
Reserve	14	56%
Hand Dominance		
Right Hand	19	76%
Left Hand	6	24%

Self-rated proficiency was correlated with SRY of videogaming ($r(23) = .621$, $p = .001$), but not with SRH of current activity ($r(23) = .213$, $p = .306$). SRP was not correlated with any of the eight IM® scores (p >.05) Figure 3 shows SRP and IM® scores according to the three levels of expertise described in the instrumentation section of this paper. Table 3 shows how SRP compares with the IM® Indicator Chart provided in the IM® training and performance manual.

DISCUSSION

The intent of this research was to discover whether an individual's background and experience in video gaming would impact their initial scores on the IM® long form assessment. These results indicate that neither self-ratings of years of experience or proficiency were related to their IM® performance. However, self-ratings of hours of current videogame activity were related to initial IM tasks involving the right side of the

408

body. Those who reported more hours of current videogaming had lower scores on IM® tasks involving the right side of the body. When the subtasks that made up this larger task category (right hand, toe, and heel) were evaluated, two approached significance (right hand and right toe), while one did not (right heel).

FIGURE 3. Overall Group Performance According to Self-Ratings of Proficiency (lower millisecond average means better performance).

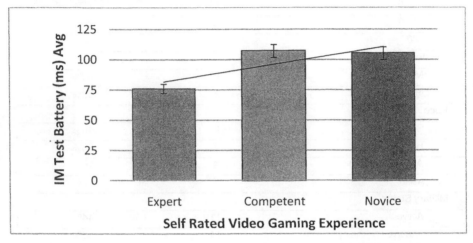

It may be that individuals engaged in more intensive videogaming are more sedentary and they are less able to balance on their left, probably non-dominant, foot, as required during the right toe tapping task. Most participants in this study were right handed, and while being right handed does not necessarily mean the individual is right footed, this is true in most cases (Hannaford, 1992). It could also be that more intensive videogaming, coupled with demanding note taking during training impacts right hand fatigue. In fact, it is not clear why individuals engaged in more hours of videogaming would perform more poorly on a timing task involving the right side of the body. However, the more important finding, in light of our initial question, is that more videogaming experience, hours of play or greater proficiency does not result in better performance in the timing tasks measured by the IM®. It would appear that videogaming and the IM® tapping are two different abilities. However, due to the small sample size, it is highly recommended that further research be conducted to enhance the ability to generalize from these findings.

TABLE 3. Participant Self Reported Proficiency (SRP) in Video Game Activity and IM Placement according to Guidelines for Pre Long Form Assessment of 14 Tasks

IM Performance Categories	Self Reported Level of Proficiency		
	Novice (*n*= 5)	Competent (*n* = 6)	Expert (*n* = 14)
Superior (< 22 ms)			
Exceptional (22-29 ms)			
Above Avg (30-40 ms)	4% (1)		
Average (41-69 ms)	4% (1)	8% (2)	24% (6)
Below Average (70-146 ms)	8% (2)	12% (3)	28% (7)
Sever Deficiency (147-199 ms)		4% (1)	
Extreme Deficiency (> 200 ms)	4% (1)		4% (1)

LIMITATIONS

The primarily limitation for this study is the small sample size. Other potential limitations are the specific population (Soldiers attending Operating Room Technician training) and the self-report data. There is a concern that participants will respond to self-report surveys in a way that makes them appear competent.

CONCLUSIONS

On the basis of our results, it would appear that timing performance, as measured by the IM®, is not improved by greater video game experience, in terms of years of experience, hours currently played, or levels of proficiency. The practical importance of this information is the message it should convey to therapists, teachers, and coaches who are considering such a timing evaluation or intervention for a patient, student, or athlete. That is, there is no reason to preclude someone with a great deal of videogame experience from participating in an IM® evaluation or intervention, as videogaming and the IM® appear to be drawing from different abilities. It is also suggested that similar research be conducted with a larger sample size.

410

ACKNOWLEDGMENTS

We would like to thank the Soldiers who volunteers for this study, as they gained nothing other than the knowledge that their participation might help others in the future. We would also like to thank the U.S. Army Medical Research and Materiel Command for their funding assistance for the overall study, from which this data was extracted. Thanks also to the U.S. Army Medical Department Center and School for encouraging us to continue our examination of methods and interventions to assist Soldiers during training.

DISCLAIMER

The views expressed in this article are those of the authors and do not reflect the official policy or position of the Department of the Army, Department of Defense, or the U.S. Government.

REFERENCES

Green, C.S. and Bavelier, D. (2008). Exercising Your Brain: A Review of Human Brain Plasticity and Training-Induced Learning. *Psychology and Aging. 23*,(4), 692-701.

Hannaford, C. (1992). The Dominance Factor: How Knowing Your Dominant Eye, Ear, Brain, Hand, & Foot Can Improve Your Learning. Virginia: Great Ocean Publishers.

Interactive Metronome, (2004), IM Certified Provider Basic Training Course & Resource Binder.

Issenberg, S., McGaghle, W., Hart, I., Mayer, J., Felner. J., Petrusa, E., Waugh, R., Brown, D., Safford, R., Gessber, I., Gordon, D., Ewy, G. (1999). Simulation Technology for Health Care Professional Skills Training and Assessment. *The Journal of the American Medical Association.*
(282), 861-866.

Kakazu, P. (2003). The Interactive Metronome as an Intervention, Unpublished University of Puget Sound, Retrieved November 6, 2009 from http://www.2ups.edu

Kuhlman, K. and Schweinhart, L.J. 1999. (Retrieved Sept 14, 2009). Timing in Child Development. Retrieved from http://www.highscope.org and available from HighScope Educational Research Foundation, 600 North River Street, Ypsilanti, MI 48198-2898, USAMauro, Daniel G. (2006). The Rhythmic Brain. 5th International Conference of Cognitive Science (pp. 163-164) Institute of Cognitive Science, Carleton University, Ottawa, Ontario.

Mulder, R. (2002). 'Training for the Brain' Technology Yields Gains at St. Thomas Aquinas High School. Technological Horizons in Education, 30 (October).

Rice, V.J. and Butler, J. (2007). Revisiting an Old Question with a New Technology: Gender Differences on a Neuro-Cognitive Temporal Tracking Task. Human Factors & Ergonomics Society Annual Meeting, Baltimore, MD, pp 581-586.

Rice, V.J.B., Butler, J., and Marra, D. (2007). Neuro-Cognitive Assessment, Symptoms of Attention Deficit and Hyperactivity Disorder, and Soldier Performance during 68W Advanced Individual Training. Technical Report No. ARL-TR-4292, US Army Research Laboratory Field Element, Ft. Sam Houston, TX 78234-6125.

Rosser, J., Lynch, P., Cuddihy, L., Gentile, D., Klonsky, J., Merrell, R. (2007). The Impact of Video Games on Training Surgeons in the 21st Century. *Archives of Surgery*, 142,(2) 181-186.

Schlickum, M. K., Hedman, L., Enochsson, L., Kjellin, A., & Fellander-Tsai, L. (2009). Systematic Video Game Training in Surgery Novices Improves Performance in Virtual Reality Endoscopic Surgical Simulators: A Prospective Randomized Study. *World journal of Surgery, 33*(11) 2360-2367.

Shaffer R.J., Jacokes, L.E., Cassily, J.F., Greenspan, S.I., Tuchman, R.F., Stemmer, P.J. (2001). Effect of Interactive Metronome Training on Children with ADHD. *Am Journal of Occup Ther*, *55*(2) 155- 162.

Chapter 42

Inclusive Academic Practices in an Undergraduate Engineering Research Experience

Tonya Smith-Jackson[1], Tamal Bose[2],
Ratchaneekorn Thamvichai[2], Carl Dietrich[2]

[1]Grado Department of Industrial and Systems Engineering

[2]Bradley Department of Electrical and Computer Engineering
Wireless@V.T
Virginia Polytechnic Institute and State University
Blacksburg, VA 24061-0002, USA

ABSTRACT

The Engineering discipline continues to be challenged by recruitment and retention. As early as elementary school, children show little interest in engineering and appear to have limited confidence in their own ability to be successful in the discipline. In response to this ongoing national problem, we present a project that makes use of inclusive practices to increase students' performance and team work in a highly diverse summer research program in cognitive communications. The program helped students work productively on diverse teams and understand their own conflict styles to ensure they contributed to team productivity. The results of the statistical analyses revealed a diverse profile of conflict styles. In addition, students' college academic self-efficacy increased from pre-test to post-test.

Keywords: Inclusive pedagogy, engineering education, self-efficacy, undergraduate experiences

INTRODUCTION

DIVERSITY IN ENGINEERING

Science, Technology, Engineering, and Mathematics (STEM) education in the USA continues to be challenged by recruitment and retention of undergraduate students. Globally, the USA's output of undergraduate degrees in STEM continues to lag behind other countries, which impacts global competitiveness. While representation of US citizens in STEM fields on a global or cross-national level is comparatively lower, within the USA, minority and female under-representation is highly disparate. Yet, diversity has been recognized as one of the most significant factors to enhance the global competitiveness of engineering and science (2000).

The National Science Board (2009) recognized the importance of enhancing the quality of teaching and training of all students while using the best teaching-learning and assessment methods. Researchers and educators alike have explored the factors that contribute to the under-representation of minority groups and women in science and engineering fields. One such factor relates to the design of engineering instruction and academic support (Sunal et al., 2001). Researchers have also identified the importance of minority and majority mentors who are culturally competent (Quaye & Harper, 2007). Of significant impact is the teaching-learning approach used in undergraduate education. Prior research has identified the effectiveness of new learning models that are inclusive of different learning styles, life experiences, and disciplinary knowledge (Artis & Smith-Jackson, 2002; Chepyator, Jeporir, & King, 1996; Marable, 1999; Pruitt-Logan, 1996). The literature on inclusive pedagogy is prolific.

Pavel and Colby (1999) found that American Indian students who began their post-secondary education in community colleges were more likely to acquire baccalaureate degrees. Community colleges are considered more successful at integrating students' real-world experiences with topics associated with STEM, such as using science and math concepts to address problems in their own communities. A similar finding by Marinez and Bernardo (1988) indicated that Chicano and Native American students learned science better through the use of culturally relevant activities and concepts. Likewise, Strutchens (1995) and Wlodkowski and Ginsberg (1995a) identified the importance of relevance and inclusiveness when working with students from different personal and disciplinary backgrounds. Wlodkowsi and Ginsberg (1995b) argued that motivation to learn differs on the basis of how diverse students define and view novelty, gratification, and opportunity. These practices of inclusive pedagogy should be beneficial to all students, including majority male students.

THE WIRELESS@VT INCLUSIVE PEDAGOGY

In light of the ongoing challenges in STEM diversity, we designed and implemented a program to provide a multifaceted educational experience centered on subject and research mastery in cognitive communications. Cognitive communications is a specialty area that focuses on wireless communications using a variety of technologies including cognitive radios, software defined radios, spectrum sensing, wireless networking, and smart antennas. The program was funded by the National Science Foundation and implemented within the Wireless@VT Center at Virginia Tech. The goal of this paper is to describe some of the inclusive features of the program and to focus on our assessment results from the cohort of students who completed the first summer program. While education and training in cognitive communications were the primary goals, the program also targeted a diverse group of students in terms of disciplines, genders, universities, and ethnicities. Inclusive pedagogy served as the main driver of the program activities.

Inclusive pedagogy is the term used to describe pedagogies that will benefit all students, regardless of ethnicity, socioeconomic status, or gender. It includes an integration of multiple perspectives and approaches. Rather than using specific targeted cultural pedagogy, inclusive pedagogy is designed to be effective for all students who are both majority and minority groups within a particular society. In the project described here, our inclusive pedagogy was designed with the mission to produce globally competent and competitive scholars, with an emphasis on ensuring they developed skills to work effectively in environments with cultural and disciplinary diversity.

COGNITIVE COMMUNICATIONS RESEARCH EXPERIENCEES FOR UNDERGRADUATE STUDENTS

The Cognitive Communications (C-COM) Research Experiences for Undergraduates (REU) Program had the following goals:

- Enhance creative and independent thinking;
- Motivate students to pursue graduate studies;
- Help students develop general research skills in an interdisciplinary context;
- Allow students to gain hands-on experience in cognitive radios, wireless networking, and their applications;
- Promote a sense of confidence, team spirit, and an appreciation of the potential of interdisciplinary collaboration in creating new knowledge;
- Expose students to the intellectual excitement involved in research activities; and

- Teach students to effectively assimilate the latest research, assess their own knowledge, present experimental results, effectively prepare reports and publications, and understand the methods for translating research to practice (R2P).

These goals were addressed and ultimately achieved by applying attributes associated with inclusive pedagogy – interpersonal skills development, Paideia practices, and attempts to enhance academic self-efficacy. The features of our inclusive pedagogy are illustrated in Figure 1 and briefly described here.

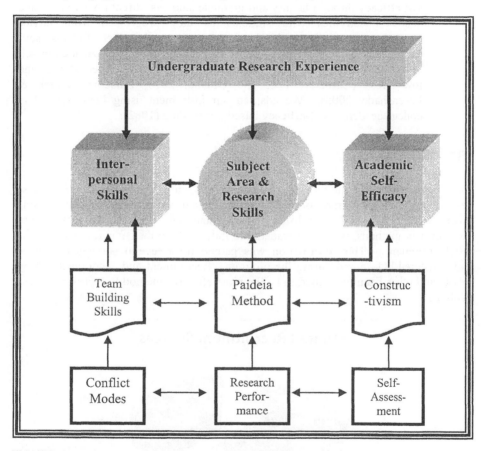

FIGURE 1. Inclusive pedagogy design attributes.

1. **Interpersonal Skills:** Interpersonal intelligence (Gardner, 1983) is one of the strongest predictors of career success. Before beginning the program, students participated in a team-building skills workshop that addressed the phases of team development as well as the importance of working on teams with diverse members and perspectives.

2. **Paideia Method:** Faculty used the Paideia structure in the design and implementation of the curriculum (Adler, 1988). The method that includes the use of students' interdisciplinary backgrounds and experiences to build a common conceptual foundation for knowledge and skill acquisition. The method includes three approaches: didactic teaching, coaching with scaffolding, and use of seminar opportunities for independent mastery. Due to the space limitations, the Paideia attributes of the summer curriculum will not be presented.

3. **Academic Self-Efficacy:** The program was designed to impact academic self-efficacy through faculty and graduate students' direct involvement and hands-on activities with students as they progressed from novice researchers to scholars. Academic self-efficacy describes students' self-confidence in their ability to learn and master concepts. Academic self efficacy has been linked to high motivation, low anxiety, persistence, and low stress (Carberry, Lee, and Ohland, 2010; Zajacova, Lynch & Espenshade, 2005). We adapted our instrument using items related to college academic self-efficacy based on Bandura (1982).

STUDENT DIVERSITY

The effectiveness of an inclusive pedagogy can only be assessed in a diverse environment. The program focused on recruiting a diverse cohort of students using word of mouth, email, and listservs. To determine the effectiveness of our recruitment methods, we asked students to indicate how they found out about the REU program. This information was gained from several surveys that were administered before and during the program. Recruitment and encouragement by REU and other Faculty seemed to be the most effective method to recruit a diverse cohort.

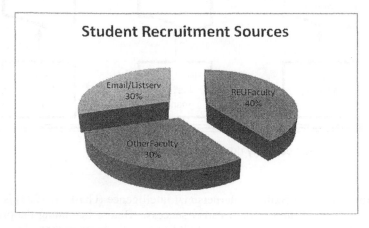

FIGURE 2. Recruitment avenues reported by students.

Ten REU students were selected for the summer 2009 cohort. We set a goal to have 50% of our cohort from underrepresented groups. NSF-defines underrepresented groups in engineering as African-Americans, Latinos, Native Americans/American Indians, and women. The cohort consisted of 5 Caucasian males, 3 Latino males, 1 African-American male, and 1 Latina. Student diversity met the 50% goal and actually exceeded it, since the gender and minority representation combined was 60% (i.e., the Latina was both a gender and ethnic minority). Having only one female in the cohort was an area needing more focus, in spite of having exceeded the overall 50% goal for underrepresented students. Disciplinary diversity was also strong. Of the 10 REU students, 6 were majoring in Electrical Engineering, 4 in Computer Engineering, and 1 in Industrial and Systems Engineering.

INTERPERSONAL AND TEAM-BUILDING SKILLS

In the first week, students participated in a team-building and interpersonal skills workshop. The goal was to demonstrate the importance of working on diverse teams, especially when disciplinary perspectives and world views varied across members. The Thomas-Kilmann Conflict Mode Instrument (TKI; Thomas and Kilmann, 1974) was administered to help students understand how their own conflict styles could be used to facilitate team cohesion and productivity. The TKI was administered on the first day of the summer program and again on the last day of the program. The TKI is a questionnaire formatted as a forced choice list of conflict scenarios. There are five possible styles in which students' responses could be classified – (1) Competitive; (2) Collaborative; (3) Compromising; (4) Accommodating; (5) Avoiding. Frequency values can range from 0 to 12 on any given dimension. Frequency counts are used to classify students into their most dominant conflict style. Each style presents both strengths and weaknesses in different circumstances. Students learned to appreciate all styles, but to recognize how to adapt to a diversity of styles and situations while still maintaining productivity and team cohesion.

Although we did not expect a change in the conflict styles of students, we compared pre-test and post-test measures to better understand the profile of this cohort. There were no changes between pre-test and post-test TKI scores. The pre-test mean values for each style are illustrated in Figure 3.

A key element of the workshop was to bring to surface a number of differences that could lead to conflict, and help students develop ways to use their own conflict styles to ensure they were value-added to the team's productivity. In addition to the discussions related to conflict styles, students were also exposed to Tuckman's (1965) team development stages (Figure 4). Students discussed each of the five stages, with emphasis on the types of issues that could occur on their own teams during the stage with the most conflict – storming. Students walked through different scenarios using a small group format and offered solutions that would maintain cohesion while constructively solving the engineering problem.

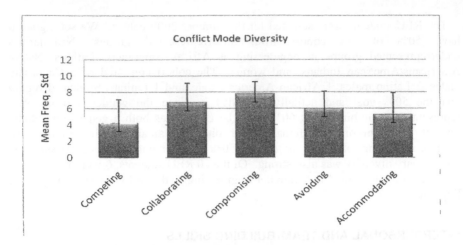

FIGURE 3. Conflict model profile of REU students/cohort.

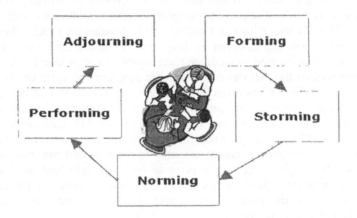

FIGURE 4. Tuckman's (1965) stages of team development.

COLLEGE ACADEMIC SELF-EFFICACY (CASE)

A College Academic Self-Efficacy (CASE) questionnaire was used to determine program effectiveness. We hypothesized a significant increase in CASE from pre- to post-test. Although the one-group pre- and post-test design is not the strongest in terms of internal validity, the use of a control group was not feasible for this project. We used a Likert-type scale with ratings from 1 (not at all) to 5 (quite a lot). The scale reliability was tested using Cronbach's alpha. Alpha reliability was high, $r_{alpha} = .80$. A matched-pairs Wilcoxon Signed Rank test was used to test whether CASE changed from the first day of the program to the last day of the program. We set a significance level of $\alpha = .05$, two-tailed test, and used a difference score as the dependent variable. A significant difference was identified;

participants increased in academic self efficacy from pre- to post test, $W_{(9)} = -22.00$, $p < .05$. Based on these results, the self-efficacy criterion was met. Figure 5 illustrates the means of the total CASE values from pre- to post-test.

Table 1. College Academic Self-Efficacy Scale Items*

1. Understanding cognitive communications.
2. Working on teams in an effective manner.
3. Thinking in a way that integrates more than one subject or discipline.
4. Understanding most of the ideas discussed in class.
5. Translating problems into research questions.
6. Finding journal publications related to the research I have done.
7. Practicing lab skills related to my research.
8. Working with team members.
9. Working with faculty.
10. Producing results that are meaningful to the target group.
11. Applying effective oral communication skills.
12. Applying written communication skills.
13. Understanding complex ideas.
14. Drawing conclusions from the results of my research.
15. Developing skills that will help me in graduate school.
16. Developing self-motivation to pursue graduate study.

Adapted from: Bandura (1982)

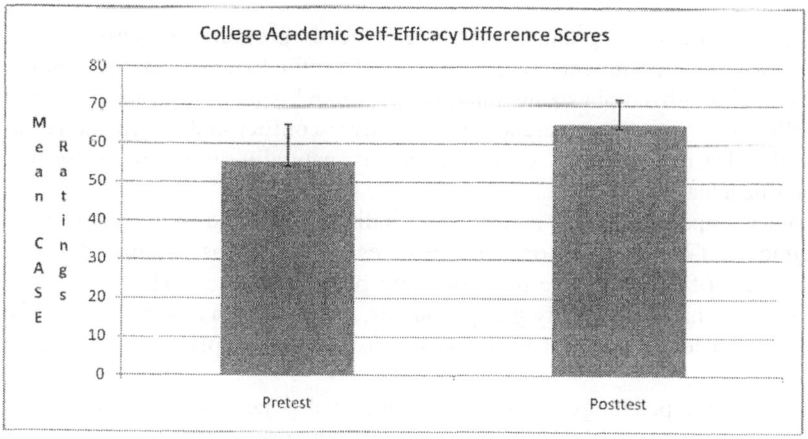

FIGURE 5. Significant differences In self-efficacy scores from pre-test to post-test.

To address specific patterns of change based on culture, we conducted a Friedman Test using the group variable (majority, underrepresented) as a two-level predictor and the two repeated measures as the y-values (pre-test CASE and post-test CASE). A significant interaction (time x group) was identified, $F_{(1, 8)} = 6.41$, $p < .05$. Students in the underrepresented group changed significantly from pre-test to post-test. Although the scores increased as well, the majority group members did not show the same pattern. Figure 6 illustrates a more detailed breakdown by using the actual pre- and post-test CASE ratings as opposed to the difference or change scores, and by using a line graph to better illustrate the interaction.

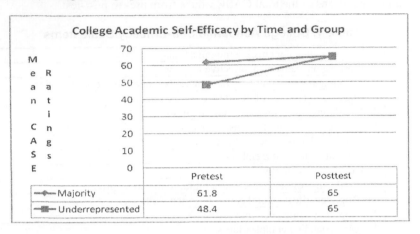

FIGURE 6. Self-efficacy changes by cultural group (interaction).

DISCUSSION AND CONCLUSIONS

Several important program features enhanced program success. Faculty involvement in recruitment contributed significantly to ensuring the cohort was diverse. Through emphasizing interpersonal intelligence, students were able to self-reflect and maintain awareness of their own conflict and interpersonal styles. They also learned what to expect as teams naturally move from forming to completing a team project.

Another important point is that the findings indicated a larger increase in academic self-efficacy among underrepresented groups; demonstrating the effectiveness of our inclusive pedagogy and program design. However, it should also be noted that the majority group students, due to a high academic self-efficacy at the pre-test level, may not have had much room for improvement (e.g., ceiling effect).

Our inclusive pedagogy is a paradigm shift that is necessary due to the need to prepare engineers to be globally competitive with specialty knowledge as well as interpersonal intelligence. We used other features in the program to support specialty knowledge and mastery, but these discussions are beyond the scope of this paper. The primary goal was to communicate the inclusive structure we built into the program and to share some of our assessment results. The project was evaluated by an external evaluator, and overall, was found to have achieved most of the outcomes set forth at the beginning of the project.

ACKNOWLEDGEMENT

This project was supported by National Science Foundation grant numbers 0851400 and 0520418. Any opinions, findings, conclusions, or recommendations expressed herein are those of the authors and do not necessarily reflect the views of the National Science Foundation.

REFERENCES

Adler, M. (1983). *The Paideia Proposal: An Educational Manifesto*. New York: McMillan Publishing Company.

Artis, S. and Smith-Jackson, T. L. (2002). Cultural differences in risk perception. *Proceedings of the 16th Annual International Society for Occupational Ergonomics and Safety Conference.*

Bandura, A. (1982). Self-efficacy mechanism in human agency. *American Psychologist, 37*, 122 – 147.

Carberry, A., Lee, H-S., Ohland, M. (2010). Measuring engineering design self efficacy.*Journal of Engineering Education, 72* – 49.

Chepyator-Thomson, J.R. and King, S.E. (1996). Scholarship reconsidered: Consideration or a more inclusive scholarship in the academy. *Quest, 48*, 165 – 174.

Courage, K.G., and Levin, M. (1968), *A freeway corridor surveillance information and control system.* Research Report No. 488-8, Texas Transportation Institute, College Station, Texas.

Gardner, H. (1983). *Frames of Mind: The Theory of Multiple Intelligences*. New York: Basic.

Marable, T. D. (1999). The role of student mentors in a pre-college engineering program. *Peabody Journal of Education , 74*, 44 – 54.

Marinez, D. I. and Bernardo, R. (1988). *Improving the Science and Mathematic Achievement of Mexican-American Students Through Culturally Relevant Science*. Washington, DC: Office of Educational Research and Improvement.

National Science Board (2009). *NSB STEM Education Recommendations for the President-Elect Obama*. NSF-09-01.

National Science Foundation. (2000). *Land of plenty: Diversity as America's competitive edge in science, engineering and technology*. Arlington, VA.

Pavel, D. M. and Colby, A. Y. (1992). *American Indians in Higher Education: The Community College Experience*. Office of Educational Research and Improvement (ED), Washington, DC.

Pruitt-Logan, A. (1996). Student affairs and inclusiveness. *Journal of College Student Development, 37*, 236 – 238.

Quaye, S. and Harper, S. (2007). Faculty accountability for culturally inclusive pedagogy and curricula. *Liberal Education,* Summer 2007, 32 – 39.

Sunal, D., Sunal, C., Whitaker, W., Freeman, L., Odell, M., Hodges., J., Edwards, L., and Johnston, R. (2001). Teaching science in high education: Faculty professional development and barriers to chance. *School Science and Mathematics, 101*, 232 – 245.

Strutchens, M. (1995). *Multicultural Mathematics: A More Inclusive Mathematics*. Washington, DC: Office of Educational Research and Improvement.

Thomas, K. and Kilman, R. (1974). Thomas Kilman Conflict Modes Instrument. Tuxedo NY: Xicom, 1974).

Tuckman, B. (1965). Developmental sequence in small groups. *Psychological Bulletin, 63*, 384 – 399.

[a]Wlodkowski, R. J, and Ginsberg, M. G. (1995). Diversity and Motivation: Cultural Responsive Teaching. In *Jossey-Bass Higher Education and Adult Education Series, Jossey-Bass Social and Behavioral Science Series*. San Francisco, CA: Jossey-Bass, Inc.

[b]Wlodkowski, R. J. and Ginsberg, M. G. (1995). A framework for cultrally responsive teaching. *Educational Leadership, 53*, 17-21.

Zajacova, A., Lynch, S., and Espenshade, T. (2005). Self-efficacy, stress, and academic success in college. *Research in Higher Education, 46*, 677 – 706.

Chapter 43

Evaluating a Human Factors Training Program for Content Retention and Effectiveness

Conne Mara Bazley

JimConna, Inc.
Carbondale, CO 81623, USA

ABSTRACT

A human factors researcher conducted a study to determine how different learning styles influenced content retention of a mandatory training given to engineers and designers. The training introduced basics human factors concepts and a project-specific Design Guide for their subsequent use. Often, training programs are developed and implemented following a needs assessment. The effectiveness of the training is simply assumed or assessed briefly through an immediate post-training questionnaire. Although the initial development was based on research that should improve retention, it may be that an all-inclusive approach to reach all learning styles is less effective than tailoring training to the learning styles of the attendees. This study identified the learning styles of attendees and examined the relationship between those learning styles and both the ratings of the course and retention of pertinent course information one month post attendance. These results show that the visual learning style was dominate for this group and an all-inclusive presentation style was not as important to information retention. These results are informative for the design of subsequent courses and design guidelines.

Keywords: Training Effectiveness, Content Retention, Learning Styles

INTRODUCTION

A human factors team developed the Human Factors Design Guide (HFDG) for a large production project. The HFDG covers a broad range of human factors topics that pertain to automation, maintenance, human interface, workplace design, documentation, system security, safety, the environment, and anthropometry.

The training program created by the Human Factors Evaluation (HFE) team addressed all learning styles and included four strategies believed helpful for content retention (Brown, Hershock, Finelli, and O'Neal, 2009): a welcoming learning environment, clearly communicated objectives and expected outcomes, interactive presentation styles, and real-world (on-site) examples and narratives. However, there are other factors that can affect the potential success of a training program that include: individual characteristics (of both the instructor and the trainees), such as preferred learning style, motivation, attitude, and basic ability as well as the work environment (Tracey, 1995). Evidence shows that engineers have a significant visual learning-style preference (James-Gordon and Bal, 2001); and there are many learning style indicators (Harris, Sadowski and Birchman, 2008).

The following five learning assessment models most often used are:

1. **Myers Briggs (MBTI)**
   Based on Bloom's Taxonomy, there are a combination of four preferences:
   (1) extraversion versus introversion,
   (2) sensing versus intuition,
   (3) thinking versus feeling, and
   (4) judging versus perceptive.

2. **Kolb Learning Style**
   Based on a four-stage learning theory:
   (1) converger,
   (2) diverger,
   (3) assimilator, and
   (4) accommodator.

3. **Gregorc Style Delineator**
   Based on assessing mediation abilities:
   (1) concrete/sequential,
   (2) abstract/sequential,
   (3) abstract/random, and
   (4) concrete/random.

4. **Herrman Brain Dominance Instrument (HBTI)**
   Based on thinking preferences:
   (1) A-logical,
   (2) B-organized,
   (3) C-interpersonal, and
   (4) D-imaginative.

5. **Gagne Learning Style indicator**

Based on learning taking place through attention, encoding, and retrieval of information:

(1)   Verbal,

(2)   intellectual skills,

(3)   cognitive strategy or logical reasoning,

(4)   attitude, and

(5)   motor skills.

According to Brown et al. (2009), a student's learning style may be defined in large part by the information the student preferentially perceives, which sensory channel is used, how information is organized, how information is processed, and how understanding progresses along.

On the other hand, a teaching style may be defined in terms of the type of information emphasized, the mode of presentation, the presentation organization, the mode of student participation, and the perspective provided on the information. Table 1 provides a side-by-side comparison of learning and teaching styles. The importance of pairing teaching and learning styles for optimum learning and material retention was considered for the HFDG training.

Table 1 Dimensions of Learning and Teaching Styles

Preferred Learning Style		Corresponding Teaching Style	
Sensory /Intuitive	Perception	Concrete/Abstract	Content
Visual/Auditory	Input	Visual/Verbal	Presentation
Inductive/Deductive	Organization	Inductive/Deductive	Organization
Active/Reflective	Processing	Active/Reflective	Student Participation
Sequential/Global	Understanding	Sequential/Global	Perspective

Ideally, retention of presented material is the desired outcome for training or an educational experience. According to http://www.volunteercenter.org (2009), what we know we learn approximately 1% through Taste, 2% through Touch, 4% though Smell, 10% through Hearing, 83% through Sight.

The average learning retention rates from various instructional modes are:

- Lecture – 5%
- Reading – 10%
- Audio Visual – 20%
- Demonstration – 30%
- Discussion by Group – 50%
- Practice by Doing – 75%
- Teaching others – 90%
- Immediate application of learning in a real situation – 90%

Taking into consideration the learning styles, type of training, and limited time to cover a large amount of material, the HFE team was concerned about information usability and retention. Therefore, after the training, the HFE team requested the training participants complete a training evaluation survey and learning-style assessment on a volunteer-only basis. A retention survey was given one month after the training.

METHOD

Management requested that the HFE team provide a mandatory training on the HFDG for Engineers and Designers. The HFE team suggested a short introduction about the importance of human factors be included for those individuals that were not familiar with the profession or the process and reporting of a Human Error Discrepancy (HED). A short learning-style assessment and presentation evaluation was given to participants after the training. A series of retention follow-up questions were asked a month after the training.

RESULTS

The survey participants' age, gender, and job title (Table 2).

Table 2 Table of Characteristics of the HFE Design Guide Training survey

Characteristic	N	%
Age Range		
18-20	0	
21-30	2	8
31-40	1	4
41-50	8	32
51-60	7	28
61-70	7	28
Total	25	100.0
Gender		
Male	16	64
Female	9	36
Total	25	100.0
Job Title		
Engineer	11	44
Designer	14	56
Total	25	100.0

Table 3 presents that most Engineers and Designers believed the HFDG training was adequate and informative and would be useful to their work. They also concluded that it was helpful to know what human factors is and what the HFE team does and can do for them if assistance or information was needed.

Table 3 Highlights of HFE Design Guide Training Survey

	Question	Mean
Q6	Difficulty level about right	4.24
Q8	Applicable to my work	4.20
Q10	Presentation met me needs	4.04
Q12	Trainer is available to me for further help	4.16
Q14	I am confident in using the HFDG	4.08

Engineers and Designers were predominately visual learners (Figure 4). Engineers prefer facts and procedures, Designers not as much. Neither group preferred the step-by-step process. Engineers preferred working alone on projects. On the other hand, Designers preferred working in a group and scored higher overall as visual learners than Engineers.

Figure 4 Results of Learning Style Assessments

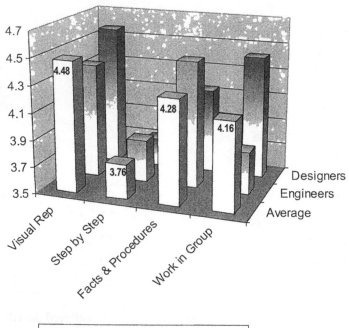

RETENTION QUESTIONS

1. What will you do **differently** in your work and work setting as a result of this training?
- 90% said they would do their work differently because the HFDG had information that would improve the accuracy of their work and be used for reviewing and verification.
- 10% said they would not do anything differently and would not have a use for the HFDG in their line of work. Their type of work did not have a human interface involved or a layout or physical placement of materials or equipment.

2. How can we **improve** this presentation?
- 20 of 25 respondents suggested that the HFE team use more examples that are project-specific and less generalities.
- It was suggested that the presentation be longer because of the large amount of material covered.
- 15 suggestions included using a computer lab so each participant could use the HFDG while examples were being presented.
- Those that found no use for the HFDG in their work suggested that management not make it mandatory training for them.

3. What **additional** training-development education do you require?
- 18 of 25 participants suggested the training be conducted annually.
- 23 suggested that future trainings address job-specific issues, problems and solutions for electrical engineers, electrical designers, procurement specialists, etc.

DISCUSSION

The intended use of the HFDG is to ensure inclusion and considerations of Human System Interfaces in the design. The HFDG provides an easy-to-use source of human factors guidance. The HFDG is not a substitute for in-depth professional human factors engineering practice and expert professional judgment on its application to new systems and equipment.

When designing a training program it is imperative to consider the user or learner. Diversity in jobs, previous experience and education has to be taken into consideration by those teaching the training. In fact, education systems 'train' us to be typically either passive or transaction-based learners. Only when we get to higher education are the more interactive learning types and environments encouraged (Brown, Hershock, Finelli, and O'Neal, 2009).

Corporate learning and trainings often separate work and learning. Poor retention and inappropriate knowledge occur more often than not in a classroom environment. Merging work and learning in a flexible, less-classic state allows for maximum efficacy of the information/learning intervention. Learning environments differ along time and size scales as well as place.

One size does not fit all (McKey & Ellis 2007). The ability to match and align learning environments to organizational needs was an important goal for the HFE Team as well. With further understanding of the critical elements for successful learning environments, the retention rates should be higher. The end goal is to move away from the one-size-fits-all toward a personal learning system. According to McKey & Ellis (2007), ambiguity and 'no-right-answer' are hallmarks of the shift from knowledge-centric to people-centric learning styles and models that the interaction, experiential, and autonomous learning environments provide. The systems must support the learner to manage this perspective. Increased performance can only come from increased skills, responsibility, and improved support systems that allow immediate application of new learning (Holbeche, 2005; and Jonassen, 2003).

CONCLUSIONS

The visual learning style for this group of engineers and designers was the most dominate. Teaching styles often do not match up to learning styles. The HFE team was aware of the learning style of the training group and designed the training to address the visual learning style. The presentation included visual-style preferences such as diagrams, real world examples, and pictures.

Teaching style was not found to be a factor for the material retention. More importantly, it was found that material retention was gained from presentation material that was relevant to specific jobs or tasks at hand and not due to the presentation.

Future training presentations for engineers and designers will focus on job-specific needs and provide more project-specific examples. The HFE team would like to offer the training in the computer lab and conduct a hands-on-learning session rather than present the material in a lecture-style environment.

REFERENCES

Brown, M. K., Hershock, C., Finelli, C. J., and O'Neal, C. (2009). Teaching for retention in science, engineering, and math disciplines: A guide for faculty. Occasional Paper No. 25. Ann Arbor, MI: Center for Research on Learning and Teaching, University of Michigan.

Harris, L., Sadowski, M., Birchman, J. A Comparison of Learning Style Models and Assessment Instruments for University Graphics Educators, Engineering Design Graphics Journal Vol 70 No. 1.

Holbeche, Linda, (2005), "The High Performance Organization", Elsevier, Oxford,UK.

James-Gordon, Yvette, Bal, Jay. (2001). Learning style preferences of engineers in automotive design. Journal of Workplace Learning, 13 (6), 239 – 245.

Jonassen, David, (2003), "Learning to Solve Problems – An instructional design guide", Wiley.

McKey, P., Ellis, A., A Maturity Model for Corporate learning environments, 2007 retrieved from internet Jan, 10, 2010. www.redbean.com.au

Tracey, J.B. (1995) Accounting for Individual Characteristics and the Work Environment Cornell Hotel and Restaurant Administration Quarterly, Vol. 36, No. 6, 36-42.

http://www.engr.ncsu.edu/learningstyles/ilsweb.html

http://www.nwlink.com

"Average Learning Retention Rates" (Google_Search) **2005-07-21** http://www.volunteercenter.org/showyourcharacter/teachers/default.asp?id=30

Diagnosis of Consciousness in the Area of Occupational Hygiene and Safety in Polish Organisations

Katarzyna Boczkowska, Konrad Niziołek

Department of Production Management and Logistics
Technical University of Lodz
90-924 Łódź, ul. Wólczańska 215, Polska

ABSTRACT

This paper presents certain aspects of the researches conducted by the authors in 2000 companies in the framework of the project "The analysis and the assessment of the degree of adjustment of the companies and their OSH services from the Region of Lower Silesia to the changes in the law provisions and the needs of the labor market" financed by the European Union. The project was the diagnosis of the level of awareness of the Polish entrepreneurs in the field of widely understood occupational safety.

Keywords: occupational safety and health, OSH services

INTRODUCTION

The area of occupational safety and health experiences permanent changes, both as far as the law regulations are concerned, in the technical and organizational field and in the field of widely understood human resources management. It is the duty of every employer to adjust the working environment of the subordinate workers to the requirements and standards applicable not only in Poland but also in the European Union. OSH services should help in the fulfillment of the statutory obligations of employers in the fields of occupational safety and health. As the result of the permanent changes, OSH services should constantly increase their professional qualifications and the educational units should dispose of the proper offer of trainings. Meanwhile, as results from numerous observations, both the employers and OSH services are not aware of the upcoming changes and of the consequences that result from them and the educational offer is inadequate. Although nowadays in Poland there are many companies that offer trainings and consultancies in the field of OSH their competencies and the quality of services are neither verified nor monitored. Therefore it is necessary to undertake the activities which will help to prepare the Polish entrepreneurs to the impending changes, keeping in mind the low level of the awareness of employers in this field it makes this problem an important issue. The aim of the project, co-financed by the European Union, in which the authors of this paper participated, is the analysis of the researches on the level of the employers' awareness, OSH services and their preparation for impending and inevitable changes. Thanks to the in-depth analysis it is more likely to achieve the intended soft results of the research.

The results of research can show how: the activities of institutions of business environment can be coordinated, how to diagnose the qualifications of OSH services and the possibilities of their development bearing in mind the incoming changes in law. The project is a diagnosis of the awareness of occupational safety and health in an organization. The project "The analysis and the assessment of the degree of adjustment of the companies and their OSH services from the Region of Lower Silesia to the changes in the law provisions and the needs of the labor market" was realized from 1st December 2008 until 30th November 2009. Its realization consisted of a few stages. During the initial phase the research sample was chosen- 2000 employers. This group reflected the general population of companies in the Lower Silesia- the layered sampling method was used, taking the following criteria into consideration:

- the size of a company in terms of the number of employers,
- the place of business (counties),
- the profile of PKD (Polish Classification of Activities).

BASIC OBLIGATIONS OF AN EMPLOYER TOWARDS THE INSTITUTIONS OF SUPERVISION AND CONTROL

The article 209 item1 of the Labor Code states that "An employer who commences the economical activity, is supposed, within 30 days from the date of commencement of this activity to notify in writing to the relevant regional labor inspector and the proper state sanitary inspector of the location, type and the scope of business." For the non-compliance of that duty, an employer can be penalized by a fine of up to 30.000 PLN (10.000 USD) (Article 283, item2 LC). The analysis of research results shows that approximately 87% of employers (n = 1745) fulfilled the obligation to inform the National Labor Inspectorate and about 62% (n = 1259) to inform the Chief Sanitary Inspection. This means that in many cases the institutions of supervision and control do not know about the existence of these companies until the official complaint by people employed in them is made. The employers are more aware of their obligations towards the National Labor Inspectorate than towards the Chief Sanitary Inspection. It is necessary to mention that the very fact of application to NLI and CSI does not mean that the provisions resulting from the above-mentioned article 209 are fulfilled, that is the notification of an activity within 30 days. One can therefore expect that the assessment of employers' compliance with the above-mentioned requirement would be much worse. The analysis of the size of companies that did not comply with the obligation to inform the National Labor Inspectorate and Chief Sanitary Inspection shows that these are generally small companies employing from 1 to 5 employees.

THE AUDITS OF NATIONAL LABOUR INSPACTORATE

According to the Labor Code article 184 item 1 National Labor Inspectorate is the institution which is mandated to supervise and control "the compliance with the labor code, including the provisions and principles of occupational safety and health" whereas item 2 "Chief Sanitary Inspectorate is responsible for the supervision and monitoring of compliance with rules, regulations, hygiene and conditions of work in the working environment." As results from the analysis of the data, the audits of National Labor Inspectorate were conducted only in 46.5% of companies, which is less than a half of the companies surveyed   (n = 928). More than the half of respondents in the controlled companies was not able to state how frequent such controls are (58%). And if they could provide such information, they stated that they were controlled once in three years (58%) or once every five years (16%). In 6.75 % of the companies the audit was conducted once a year which can mean that both OSH state and the respect of workers' rights are of a very low level.

MACHINERY AND TECHNICAL EQUIPMENT

IDENTIFICATION OF MACHINERY AND EQUIPMENT

In order to monitor the technical state of machinery and equipment each employer should posses an identified technical park. It is difficult to imagine the technical management of machinery and equipment without their inventory. The possession of the list of machinery and equipment, although the law does not require it, shows the level of the technical and organizational culture. It is also a proper technique of management as the lack of such a list of machinery and equipment can lead to a situation, especially in case when the machine park is big, when certain equipment will be omitted in different activities connected with them. This paper deals with the issue of the creation of the list of equipment. The researches show that 25% (n=500) of the companies does not possess the list of equipment.

INSPECTION OF MACHINERY AND EQUIPMENT

According to the Regulation on the minimum requirements for occupational safety and health concerning the usage of equipment by staff members at work item 26 the employer should: "1). Initially control the equipment after its installation and before its exploitation for the first time; 2) control the equipment after its installation in another workstation or elsewhere". The results of researches in this field are presented in table 1.

Table 1. The percentage and number of the documented controls: the initial one and after the installation of a machine (own research).

	The number of companies which conduct the initial documented control of machines	% of companies which conduct the initial documented control of machines	The number of companies which conduct the control of machines after their installation in a different place	% of companies which conduct the control of machines after their installation in a different place
Yes	1327	66.4%	889	44.5%
No	327	16.4%	563	28.2%
Not applicable	346	17.3%	548	27.4%
Total	2000	100%	2000	100%

It results from the Decree of the Ministry of Economy on the minimum requirements for occupational safety and health in the usage of equipment by staff members at work from 30th October 2002 item 28 that: "1 The audit results are recorded and kept at the disposal of the relevant authorities, particularly for the supervision and monitoring of working conditions, for a period of 5 years from the date of completion of these inspections, unless the separate provisions provide otherwise. 2. If machines are used outside the workplace, at the place of their regular usage, a document proving the evidence of the last inspection of a machine should be available" As it is visible the control of the documentation of the technical conditions of machines and equipment is clearly defined by the Polish Law. That is why this issue was also included in the research. It must therefore be underlined that this issue is a very complex one and it is to verify the results of testing the degree of technical inspection of machinery and equipment. The results are presented in table 2.

It results from the presented data (see chart 2) that only 25% of respondents (n=499) defined correctly the number of years to keep the results of control. It must be stated that unfortunately only a half of employers (n=1001, 50.1%) of the companies, which were questioned, stated that they did not know the length of the storage of the control results. Paying more attention to the relations between the issues on control and the documentation of these activities it must be noticed that people who declare that they do not conduct any controls do not know for how long such documentation should be kept. On the other hand, people who conduct such controls provided the researchers with different answers: they pointed at the answer " I do not know", they gave both the correct answers but also gave the different times to keep such documents- from a year, every year until 20, 23 and 24 years but also 27, 29, 35, 36 and even 50 years. It must be noticed that this issue is also to a certain extend the question on the respondent's knowledge as the lack of the realization of control does not mean that an employer is not supposed to posses knowledge on this issue. So in this context the result that was achieved can be considered as highly satisfactory.

Table 2. The percentage and the length of storage of the documents connected with the results of the control of machinery and equipment- the dominant values (own research).

Number of years	Number of units	Percentage
0	1001	50.1%
1	58	2.9%
2	68	3.4%
3	104	5.2%
4	40	2.0%
5	499	25.0%

PROFESSIONAL RISK AND ITS ASSESSMENT

The analysis of the professional risk is the most important element in the accident prevention. The employer is responsible for such an analysis. This duty is imposed by article 226 of the Labor Code and it refers to every workplace. That is why the respondents were asked about the list of posts for which the assessment of professional risk was prepared.

More than a half of people who were questioned openly admits that such a list was not prepared in their companies. It is worrying that in this group, next to small companies employing not more than 10 people which number is the highest (n=1072) also the middle size companies could be found (n=11) and the big companies (n=5). It is difficult to imagine, especially in case of big companies, that the complete assessment of professional risk was prepared without the list of work stands. What is more such a list can only be a tool which identifies the work stands with similar threats for which as the common practice shows the common assessments of professional risks are provided. According to the Article 1 item 2 of the Labor Code "the employer.... informs the employees on the professional risk which is connected with the executed work and on the rules of the protection from the threats...". The Law does not precise in what way employers should be informed about the results of the assessment of risk, however, the first training in a company seems to be the most suitable time. The respondents, when asked about informing their subordinates, gave the answers presented in table 3.

Table 3 The structure of answers that refer to informing the workers on the professional risk in their work stands (own preparation).

Informing the workers on professional risk in work stands	Number	Percentage
Yes	1111	55.55%
No	858	42.90%
I do not know	31	1.55%
Total	2000	100.00%

Although the statutory requirement to inform about the risk has existed in the Polish law for more than 10 years, the results of researches show that only about 56% of the respondents (n = 1111) declared their compliance with that obligation. The form of the presentation of this information was analyzed for this group of respondents. The results are presented in table 4.

Table 4. The ways of informing on professional risk.

Ways of informing the workers on professional risk.	Number	Percentage
Orally	367	33.03%
In writing	682	61.39%
Orally and In writing	51	4.59%
In writing and in the different forms	3	0.27%
I do not know	5	0.45%
Other forms	3	0.27%
Total.	1111	100.00%

The entrepreneurs prefer the written form to inform about the results of the professional risk assessment. The information is provided orally in every third company. There is also a group of respondents (n=51, 4.59%) who, in order to fulfill their duty to inform about the threats in the work environment and the professional risk connected with them, provides this information orally and with the confirmation in the written form. Referring to the article 226 item 1 of the Labor Code "...The employer... prepares the documentation on the risks connected with the work which is executed and takes the necessary preventive steps to decrease the risk" and to the Decree of the Minister of Labor and Social Policy from 26th September 1997 on the general provisions of safety and hygiene of work item 39 a. point 3 "The employer must keep records of the risk assessment ... " It should be noted that, as results from another research questions, only about 41% (n = 812) of respondents declares the documented risk assessment, which means the fulfillment of the requirements of the Polish law. The results obtained in the field of risk assessment are presented in figure 1 in the synthetic way.

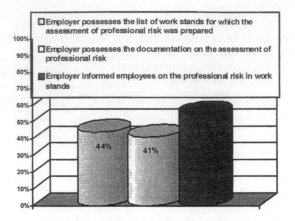

FIGURE 1. The percentage of activities executed by employers participating in the research in the field of the assessment of professional risk (own preparation).

OCCUPATIONAL SAFETY AND HEALTH PROTECTION SERVICES

In the group of companies that were questioned, the different units and people fulfilled the tasks of OSH services. Such possibilities are provided by the Polish law. The answers that were obtained are presented in Figure 2.

The employers constitute the biggest part of this group. They tend to fulfill the tasks of OSH services in small companies employing less than 20 people. In other companies employing more than 20 people specialists from outside of the company fulfill the tasks of such services.

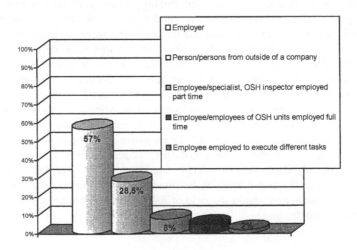

FIGURE 2. Units executing the tasks of OSH services (own preparation).

TASKS AND DUTIES OF OSH SERVICES

The key question in the questionnaire referred to the level of the awareness of employers as to the tasks and duties of OSH services and the fields in which they should operate. The range of activities of OSH services is defined by the Regulation of the Council of Ministers from 2nd September 1997 on occupational safety and health services in item 2. 1. The respondents were provided with a variety of answers. They were constructed in such a way that a part of them was in line with the requirements which are faced by such services in the above-mentioned Regulation. (table 6). The rest of the answers were the proposals contradictory to the Polish Law but the ones that in the opinions of the researchers are common in the employers' opinion. (table 5). All the answers were mixed so as to avoid the suggestion of any proper answers.

As results from the above-mentioned charts (charts 6,7) just a few percent of respondents know the basic range of the activities of such services. The researches supports the statement, which is common although it is not true, that a person employed in a company as an OSH specialist fulfills in a company some supervising functions and is responsible for OSH state in the company. Only few respondents (n=60, 3%) pointed only on control and supervision activities. The answers proposed in the questionnaire are based on the Regulation on OSH services where the legislator defined the aims of such services. Unfortunately 65% of respondents consider that an employee of OSH service is responsible for the state of OSH in the given company.

Table 5. The percentage of answers concerning the tasks of OSH services from the employer's point of view (contradictory to the law) -own preparation.

% of answers	Tasks of OSH services according to an employer
65%	Responsible for OSH situation in the company
55%	Conducts OSH trainings
53%	Supervises the conditions of work
48%	Supervises the compliance with OSH rules and principles
22%	Sends employees for medical examinations
20%	Is legally responsible for irregularities in the OSH field
19.5%	Represents the employer in front of NLI and CSI
17.5%	Measures the parameters of working environment- noise, temperature, light etc.
13.5%	Is financial responsible for the failure to comply with the obligations relating to OSH field (fines imposed by NLI and CSI).
12%	Purchases of the equipment for individual protection
20%	No idea

About the half of respondents stated that this employee was also supposed to supervise the conditions of work (n=1067) and the supervision on the

438

compliance with the OSH rules and principles (n=955). The wrong conviction of the legal responsibility of employee of OSH services was confirmed in the research (n=393, 19.7% of respondents).

Table 6 The percentage of answers in line with the law concerning the tasks of OSH services from the employer point of view.

% of answers	Tasks of OSH services according to the employer
60.5%	Conducts initial OSH trainings
54%	Conducts periodical trainings on OSH
53%	Controls the conditions of work
36%	Deals with and prepares accidents documentation
38%	Conducts post-accidents procedures
33.5%	Prepares periodical reviews on OSH
31%	Participates in the consultations with employees on OSH
30%	Popularizes OSH issues and ergonomics issues
27%	Participates in the works of OSH commission
24.5%	Gives consultancy on the organization of work stands in accordance with OSH rules
13%	Cooperates with a doctor responsible for exercising preventive health care on employees

INCREASE OF THE QUALIFICATIONS OF OSH SERVICES

The assessment of the activities of employers in adjusting the professional qualifications of OSH services to the requirements of the changing law was done on the basis of the question to which the answers are provided in the figure 3.

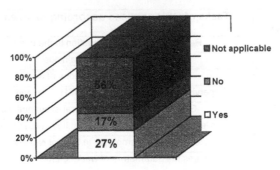

FIGURE 3. Increase of the qualifications of OSH services (own preparation).

The majority of respondents (n=1128) consider that it is not necessary to increase the qualifications of OSH services. It must be added that these are also the employers who fulfill such roles in their companies.

CONCLUSIONS

The research project was conducted on the territory of one of the Polish Voievodships, but as results from the observations and some initial researches in the different parts of Poland, the results achieved can be referred to the whole population of Polish companies. As presented in this paper, the state of OSH in the Polish companies, especially in the micro and medium-sized companies has a very low level. The analysis that is conducted shows that an employer tends to get lost in the big number of provisions (often not clear and not precise) which refer to occupational safety and health. The employer has some awareness as to the level of responsibilities in this field and tries to use the OSH services in their realization.

The creators of this research project hope that the conclusions derived from the research and their dissemination among the bigger group will lead to the increase of the level of employers' and employees of OSH services awareness and the effectiveness of the implementation of modern OSH standards in organizations. It is the aim of the authors of research to expand the horizons that refer to occupational safety and health in order to decrease e.g. the absence from work, the accident rate and the prevention of dismissals of OSH services employees.

REFERENCES

Boczkowska K., Sikora A., "Ocena poziomu profilaktycznej ochrony zdrowia w organizacjach polskich", edited by Ryszard Knosala, Oficyna Wydawnicza Polskiego Towarzystwa Zarządzania Produkcją Opole 2010

Niziołek K., Znajmiecka-Sikora M., "Wypełnianie podstawowych obowiązków prawnych bhp polskiego pracodawcy w świetle badań", edited by Ryszard Knosala, Oficyna Wydawnicza Polskiego Towarzystwa Zarządzania Produkcją Opole 2010

Rozporządzenie Rady Ministrów z dnia 2.09.1997 r. w sprawie służby bezpieczeństwa i higieny pracy

Rozporządzenie Ministra Pracy I Polityki Socjalnej z dnia 26 września 1997 r. w sprawie ogólnych przepisów bezpieczeństwa i higieny pracy

Rozporządzeniem Ministra Gospodarki z dnia 30.10.2002 r. w sprawie minimalnych wymagań dotyczących bezpieczeństwa i higieny pracy w zakresie użytkowania maszyn przez pracowników podczas pracy

Znajmiecka-Sikora M., Boczkowska K., Niziołek K., Sikora A. „Analiza i ocena stopnia dostosowania dolnośląskich przedsiębiorstw i kadr bhp do zmian w przepisach i potrzebach rynku – Raport z badań", Wydawnictwo EGO, Łódź 2009

Ustawa z dnia 26 czerwca 1974r. – Kodeks pracy (tekst jednolity Dz. U. 1998r. Nr 21 poz. 94 ze zmianami)

Chapter 45

Assessment of Soldier Performance: A Comprehensive and Integrated Approach

Leif Hasselquist[1], Edward J. Zambraski[2], Jason S. Augustyn[1],
Louis E. Banderet[2], Laurie A. Blanchard[2], Carolyn K. Bensel[1]

[1]US Army Natick Soldier Research
Development and Engineering Center
Natick, MA 01760, USA

[2]US Army Research Institute of Environmental Medicine
Natick, MA 01760, USA

ABSTRACT

The current battlefields in Afghanistan and Iraq require a highly mobile ground force that faces increasing challenges in diverse environments. The equipment that soldiers use or wear can be an important determinant of their success in completing a mission and their survival. It is the mission of the US Army Natick Soldier Research, Development and Engineering Center (NSRDEC) to maximize the soldier's survivability, sustainability, mobility, combat effectiveness and quality of life by treating the soldier as a system. It is the mission of the US Army Research Institute of Environmental Medicine (USARIEM) to conduct biomedical research to improve and sustain soldiers' health and performance under all conditions. Our combined mission requires us to develop and evaluate equipment and nutritional strategies designed to protect, assist, or improve soldiers' welfare, performance, and completion of required tasks. We at NSRDEC and USARIEM are using a comprehensive, integrated, and collaborative approach to accomplish our mission. This entails working together to assess the biomechanical, physiological, cognitive, and physical performance of the soldier. Our approach is detailed in this paper, using as an example a recently completed study evaluating the effects of load configurations on soldiers' performance. This paper lays out the set of comprehensive evaluation strategies that we have used to address the soldier as a system.

Keywords: Biomechanics, Physiology, Cognition, Physical Performance Evaluation

BACKGROUND

A large amount of research has been completed on evaluating the effects of added load on the body. Much of this work addresses issues related to the energy cost of carrying backpacks, with energy utilization quantified by taking measurements of the rate of oxygen uptake ($\dot{V}O_2$) during performance of physical activities. In studies done on marching with backpack loads, it has been found that energy cost increases with increases in the mass of the load and the speed of walking (Soule et al., 1978).

As is the case with the literature on the oxygen cost of carrying backpack loads, much of the reported research on the biomechanics of load carrying has been focused on the effects of load weight. Research demonstrates that increasing the load weight changes the kinematics, kinetics, and muscle response of the human body during locomotion. In terms of body kinematics, studies have shown that increases in backpack weight result in increases in the forward inclination of the trunk (Harman et al., 2000).

Studies done on the effects of load carriage on agility and performance of soldier-related physical tasks have, like the research pertaining to energy cost and biomechanics, been focused mainly on assessing the impact of load mass. Holewijn and Lotens (1992) reported that backpack mass contributed to performance decrements in maneuvering through obstacles, hand-grenade throwing, running, and jumping. Similarly, other researchers found that performance decreased for an agility run with an increase in load (Martin et al., 1983), and Knapik et al. (1997) found that times to complete a 20-km road march increased as load mass increased. Timed runs of obstacle courses, which require such activities as jumping, crawling, climbing, and balancing, have been used extensively in studies to evaluate different designs of load-carriage equipment (LaFiandra et al., 2003). A repetitive box lift task has been used to assess the efficacy of various physical training programs by comparing the number of lifts accomplished before and after training (Sharp et al., 1989).

According to the Soldier's Manual of Common Tasks (Department of the Army, 2005), a soldier should be able to throw a hand grenade to within 5 m of a selected point 30 m away. Grenade throwing distance and accuracy have been tested in studies of various designs of fighting load equipment (Obusek and Bensel, 1997).

As is the case with handling hand grenades, the 3-5 second rush is a basic activity soldiers are trained to perform (Department of the Army, 2005) that has been adapted as an objective performance test. Soldiers use the rush to move from one covered, protected location to another. The time spent in a standing and running position is limited to 3-5 seconds to avoid enemy fire. Harman et al. (2006) developed a timed test based on the rush. Sharp et al. (2009) analyzed the test-retest reliability of selected measures (30-m grenade throw, box lift, 2-mile march, 80-kg dummy drag, long jump, 30-m rush) of soldier physical performance. These tests were determined to provide a reliable measure of soldier readiness. The highest test-retest results were found for the grenade throw, box lift, long jump, and road march.

The research detailed above demonstrated that arduous load carriage increases energy expenditure, leads to decrements in subsequent physical performance, and increases susceptibility to a variety of musculoskeletal injuries. Much less is known about how physical exertion affects cognitive performance.

In the typical experiment on the relationship between physical exertion and cognitive performance, there is preliminary or baseline cognitive testing (pre-test), followed by exercise of various durations or intensities, which is then followed by cognitive testing (post-test). In this paradigm, there is often a sufficient lapse of time following the bout of strenuous exercise for individuals to recover, marshal their resources, and successfully perform cognitive tasks at a fairly high level. For example, in a study with very high physical demands, soldiers carried loads that ranged between 34 and 61 kg for 20 km. There were no changes in cognitive performance on a complex synthetic task (Knapik et al., 1997). Similarly, in a study of treadmill running to exhaustion, Tomporowski (2003) found no effect on a short-term memory task. In another study of heavy exercise, performance on a signal detection task was improved following a marathon (Gliner et al., 1979). Conversely, several studies that investigated cognition during the exercise period itself seemed more likely to detect changes in cognitive performance, even when the physical requirements of the exercise were relatively mild (Krausman et al., 2002).

It appears that measures of cognitive performance during physical exertion are more likely than those taken immediately after to be sensitive to any effects of exercise. This paradigm offers the advantage of being able to monitor parallel changes in the two domains as physical fatigue begins to develop. From an applied perspective, it is also evident that performance on concurrent cognitive tasks during physical exertion represents a situation that more closely mirrors the demands that confront the dismounted soldier. Furthermore, another difficulty with studies that attempt to relate physical exertion and cognition resides in the failure to specify the level of physical exertion in an objective manner.

APPROACH TO THE EVALUATION OF SOLDIER PERFORMANCE

The evaluation of soldier performance is increasingly complex. It is no longer acceptable to isolate the physical domain from the cognitive domain and we need to address many unique soldier performance issues. In the current environment, loads are not only being added to the backpack, but also to the head, torso, and extremities. The soldier must now be evaluated as a system and a platform for a variety of soldier-borne items (Figure 1).

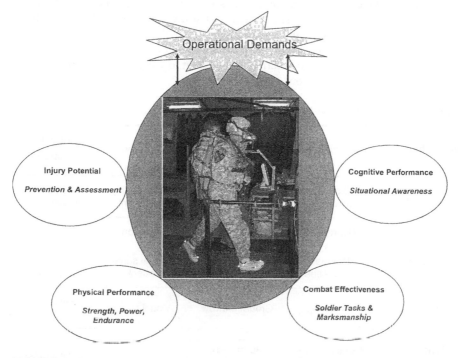

FIGURE 1 The soldier as a system, a comprehensive approach.

In a recently completed study, the goal was to use a comprehensive approach to evaluate the effects of load carriage on biomechanical and physiological measures and on the physical and cognitive performance outcomes of soldiers. The approach described in the following paragraphs and the parameters outlined in Table 1 should be viewed as a comprehensive set of assessment tools used to address these challenges.

In the study of load carriage, the NSRDEC and USARIEM joint Center for Military Biomechanics Research (CMBR) facility was utilized for the capture and analysis of biomechanical measures as soldiers walked and ran at various speeds and grades. These measures included kinematics, kinetics and physiological measures of energy cost. For testing during treadmill walking and running, a force plate treadmill, fabricated by AMTI (Watertown, MA, USA), was used. This treadmill is comprised of two synchronized treadmills on a single platform and is capable of measuring ground reaction force in three planes. A number of variables were derived from these data. The variables included peak vertical, braking, and propulsive forces.

Three-dimensional motion was recorded by ProReflex Motion Capture Unit (MCU) cameras (Qualisys Medical AB, Gothenburg, Sweden) as soldiers walked or ran on the treadmill. These data were used to analyze gait kinematics. The recorded images were processed using dedicated hardware and software (Qualisys Medical AB, Gothenburg, Sweden) to produce files containing time histories of the three-dimensional coordinates of reflective markers located on the body's joint centers of rotation. The Visual 3-D software program (C-motion Inc., Rockport, MD, USA) was used to process the data files to produce histories of numerous kinematic variables describing the soldiers' posture and gait. The kinematic data were analyzed to determine the extent to which gait parameters were affected by the load configurations.

Table 1. Comprehensive Assessment Parameters

Biomechanical	Physiological	Performance	Cognitive	Human Factors
Walking	Oxygen consumption	Road march	Marksmanship-judgment	Basic movements
Running	Heart rate	Obstacle course	Attention	Fit /sizing
Temporal measures	Thermal responses	Rifle marksmanship	Virtual squad attack	Range of motion
Kinematics	Ventilation	Grenade throw	Visual tasks	Perceived exertion
Kinetics	Reaction times	30-m rush		Pain/ discomfort
Muscle activity		Box lift		
Inverse dynamics				

The ParvoMedics TrueMax 2400 metabolic measurement system (Salt Lake City, UT, USA) was used to take $\dot{V}O_2$ measurements during treadmill walking and running. While the soldiers were walking or running on the treadmill, their heart rates were monitored using a Polar Vantage Heart Rate Monitor (Polar USA, Inc., Port Washington, NY, USA).

The NSRDEC Cognitive Performance Lab collaborated with CMBR personnel to collect and analyze dynamic cognitive performance on visual inhibitory control tasks during walking. The Attention Network Test (ANT) was used to assess cognitive performance during bouts of walking with an external load on the body. The ANT is a widely used and reliable test for assessing attentional function (Fan et al., 2002). Brain imaging studies have shown that the pattern of response times on the ANT effectively isolates three functionally and neuroanatomically distinct attentional mechanisms: alerting, orienting, and executive control. Higher order cognitive processes were evaluated in a soldier's virtual environment scenario after the soldiers were physically stressed during a road march. The road march featured a 5-km movement to objective along a course circumnavigating the campus of the Soldier Systems Center. The road march ended at the NSRDEC Cognitive Performance Laboratory, where soldiers engaged in a virtual squad attack (Figure 2). The scenario was designed to test mission-relevant aspects of cognitive performance, including memory, vigilance, and executive function. The primary dependent measures associated with this activity were related to memory performance. Response time and accuracy on the memory tasks were recorded. Additional measurements included marksmanship performance, which was monitored using a motion tracking system.

FIGURE 2 Examples of a soldier conducting the 5-km road march to an objective and the virtual engagement with an opposing force.

Marksmanship was also evaluated in USARIEM's Warfighter Target Engagement and Marksmanship Lab using a modified Engagement Skills Trainer (EST 2000). The EST 2000 is an electronic, laser-based system with digital projector, sound effects, and weapon recoil produced with compressed air (Cubic Simulation Systems, Inc., Orlando, FL, USA.)

The USARIEM EST 2000 has the same capabilities as other Engagement Skill Trainers used world-wide. However, USARIEM's system is unique (Banderet et al., 2009). It was re-engineered to give it capabilities that are essential for research applications. For example, target information, shot coordinates, latency to engage the target, and hit or miss outcome are electronically stored for subsequent analyses. The marksmanship scenario editor was also improved to facilitate construction of equivalent and longer duration scenarios and the timing accuracies and reproducibility of reaction times (latencies) measured by the EST 2000 were certified (Figure 3).

Rifle marksmanship data were acquired in this study as the soldiers fired in prone, kneeling, and standing positions wearing the load configurations under evaluation. Prior to testing, a number of mission scenarios were devised. Tables of fire were created that were appropriate to the scenarios and these determined the pace of target presentations, distance of targets, and time allowed to engage the targets.

Marksmanship was evaluated when the soldiers were rested and after they had performed a unique, exhaustive exercise test. The test, which involves whole-body exercise and torso loading, was developed to measure the impact of fatigue and load configurations on shooting performance. This evaluation utilized personnel

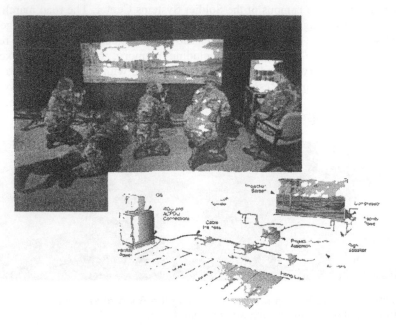

FIGURE 3 Two views of an EST 2000. The upper view shows four trainees and an instructor in a training facility. The lower view is a schematic of the EST 2000 simulator that illustrates the spatial layout of the components.

and facilities of the CMBR and the Warfighter Target Engagement and Marksmanship Lab. The exercise test entailed lifting a box weighing 20.5 kg to a height of 1.55 m while maintaining a pace of 12 lifts per minute. The test ended when the soldier could no longer maintain the pace.

When the exercise test ended, the soldier picked up the box and carried it as quickly as possible to the shooting laboratory to begin the marksmanship evaluation. The variables calculated to quantify marksmanship performance were shooting accuracy (the number of targets hit), trigger pull latency (the time from target presentation until trigger pull), and dispersion of shots.

The CMBR facilities and personnel were utilized for additional testing of the load configurations. The emphasis here was on physical performance and the activities carried out by the soldiers were grenade throws, 30-m rushes, and obstacle course runs. The grenade throwing was conducted outdoors using a training hand grenade, which had the same weight as a real grenade and was thrown at a target 30 m away. The distance thrown and the distance of each grenade point-of-initial-contact from the center of the target (accuracy) were recorded.

The 30-m rush took place indoors at the CMBR. For the rush, the soldiers were required to do five repetitive 30-m sprints carrying an M4 carbine, starting and ending each segment in the prone aiming position. For scoring, the times to complete the individual run segments and the individual transition segments were recorded, along with the total time to complete all five rushes.

The obstacle course was located indoors at the CMBR. The course consisted of hurdles, zigzag run, high crawl, pipe shimmy, wall traversal, straight sprint, mantel ascension, and staircase climb. The time to complete one run of the entire course and times to complete each obstacle or course segment were recorded using electronic timing devices (Brower Timing Devices, Salt Lake City, UT, USA) placed along the course.

The NSRDEC Ergonomics Team evaluated human factors issues associated with the load configurations, including fit, sizing, and mobility considerations. During the human factors portion of the study, the soldiers executed a number of movements using each load configuration. Range of motion, ability to perform certain movements, ease of use, and comfort were assessed for the load configuration conditions.

Subjective measures were also taken throughout the study using several techniques to acquire information from the soldiers regarding the load configurations being tested. The Borg (1970) rating of perceived exertion (RPE) scale was administered to the soldiers in conjunction with their execution of a number of tests that comprised this study. A questionnaire, which is referred to as the rating of pain, soreness, and discomfort (RPSD) questionnaire (Corlett and Bishop, 1976), was also administered to the soldiers at a number of points in the study. The RPE and RSPD were administered at the end of each 10-min bout of treadmill walking and of treadmill running, at the end of torso endurance box lifting, 30-m rushes, 5-km road marching, and upon completion of the obstacle course.

The Biophysics and Biomedical Modeling Division of USARIEM contributed to the study by evaluating the thermal effects of the equipment. Additionally, their physiological monitoring system, the Warfighter Physiological Status Monitor (WPSM), was used throughout the testing to record real time heart rate, skin temperature, and breathing rates of the soldiers. This information will be used to further develop biomedical models for physiological stress prediction. The USARIEM has successfully used ambulatory physiological monitors to evaluate thermal work strain (heat strain) of both soldiers and marines in training scenarios (Buller et al., 2008).

CONCLUSION

Through this collaborative process and utilization of resources at the Natick Soldier Systems Center, we have demonstrated a comprehensive approach that addresses the soldier as a system. By employing the paradigm of evaluating the soldier from the cognitive, physiological, and physical performance perspectives, the processes underlying soldier functioning can be quantified and the interactions among the processes explored. This approach gives a complete picture of what the soldier is challenged with in complex scenarios. Further, the collaboration across the commands within the Army is a true benefit to the soldier.

REFERENCES

Banderet, L.E., Merullo, D.J., Frykman, P.N., Goldman, S.B., Seay, J. (2009) "Use of a Modified Weapon Engagement Skills Trainer to Study Marksmanship Skills and Cognitive Performance in Medical Research," proceedings of Simulation and Alternative Learning Technologies (SALT): New Learning Trends, Orlando, FL.

Borg, G.A.V. (1970) "Perceived exertion as an indicator of somatic stress." Scandinavian Journal of Rehabilitative Medicine, 2, 92-98.

Buller, M.J., Wallis, D.C., Karis, A.J., Hebert, N.J., Cadarette, B.S., Blanchard, L.A., Amin, M.M., DiFilippo, J., Economos, D., Hoyt, R.W., Richter, M.W. (2008) Thermal work strain during Marine rifle squad operations in Iraq (Tech. Rep. T09-01). Natick, MA: U.S. Army Research Institute of Environmental Medicine.

Corlett, E.N. Bishop, R.P. (1976) "A technique for assessing postural discomfort." Ergonomics, 19, 175-182.

Department of the Army (2005) Soldier's manual of common tasks (Training Pub. 21-1-SMCT). Washington, DC: Department of the Army.

Fan, J., McCandliss, B.D., Sommer, T., Raz, A., Posner, M.I. (2002) "Testing the efficiency and independence of attentional networks." Journal of Cognitive Neuroscience, 14(3), 340-347.

Gliner, J.A., Matsen-Twisdale, J.A. Horvath, S.M., Maron, M.B. (1979) "Visual evoked potentials and signal detection following a marathon race." Medicine and Science in Sports and Exercise, 11, 155-159.

Harman, E., Frykman, P., Gutekunst, D., Nindl, B. (2006) "U. S. Army standardized physical training vs. a weightlifting-based program: Effects on soldier physical performance [Abstract]." Medicine and Science in Sports and Exercise, 38(5), S272.

Harman, E., Han, K., Frykman, P., Pandorf, C. (2000) The effects of backpack weight on the biomechanics of load carriage (Tech. Rep. T00-17). Natick, MA: U.S. Army Research Institute of Environmental Medicine.

Holewijn, M., Lotens, W.A. (1992) "The influence of backpack design on physical performance." Ergonomics, 35, 149-157.

Knapik, J., Ang, P., Meiselman, H., Johnson, W., Kirk, J., Bensel, C., Hanlon, W. (1997) "Soldier performance and strenuous road marching: Influence of load mass and load distribution." Military Medicine, 162(1), 62-67.

Krausman, A.S., Crowell, H.P., & Wilson, R.M. (2002) The effects of physical exertion on cognitive performance (Tech. Rep. ARL-TR-28-44). Aberdeen Proving Ground, MD: Army Research Laboratory.

LaFiandra, M., Lynch, S., Frykman, P., Harman, E., Ramos, H., Mello, R. (2003) A comparison of two commercial off the shelf backpacks to the Modular

Lightweight Load Carrying Equipment (MOLLE) in biomechanics, metabolic cost and performance (Tech. Rep. T03-15). Natick, MA: U.S. Army Research Institute of Environmental Medicine.

Martin, P.E., Nelson, R.C., Shin, I. (1983) Effects of backpack frame length, pack load, and participation time on the physical performance of men and women (Tech. Rep. NATICK/TR-82/021). Natick, MA: U.S. Army Natick Research and Development Laboratories.

Obusek, J., Bensel, C.K. (1997) Physiological, biomechanical, and maximal performance of soldiers carrying light, medium, and heavy loads using the Land Warrior and the All-Purpose Lightweight Individual Carrying Equipment (ALICE) systems. Natick, MA: U.S. Army Research Institute of Environmental Medicine. Unpublished manuscript.

Sharp, M.A., Allison, S.C., Walker, L.A., Frykman, P.N., Harman, E.A., Hendrickson, N.R. (2009) "Test-retest reliability of selected measures of Soldier performance [Abstract]." Medicine and Science in Sports and Exercise, 41(5), S271.

Sharp, M., Bovee, M., Boutilier, B., Harman, E., Kraemer, W. (1989) "Effects of weight training on repetitive lifting capacity [Abstract]." Medicine and Science in Sports and Exercise, 21(2), S87.

Soule, R.G., Pandolf, K.B., Goldman, R.F. (1978) "Energy expenditure of heavy load carriage." Ergonomics, 21(5), 373-381.

Tomporowski, P.D. (2003) "Effects of acute bouts of exercise on cognition." Acta Psychologica, 112(3), 297-324.

450

The Effects of Chemical/Biological Protective Patient Wraps on Simulated Physiological Responses of Soldiers

Miyo Yokota[1], Thomas Endrusick[1], Julio Gonzalez[1], Donald MacLeod[2]

[1]US Army Research Institute of Environmental Medicine
Natick MA 01760-5007, USA

[2]US Army Natick Soldier Research Development and Engineering Center
Natick, MA 01760-5007, USA

ABSTRACT

This study used a thermoregulatory model to examine the thermal burden imposed by a new U.S. Army protective patient wrap (PPW) design. The model simulations were conducted for typical desert, jungle, and temperate conditions with and without direct sun. Five PPW configurations (the current baseline, and laminated and non-laminated versions of the PPW with and without fan ventilation) were tested. The results suggested that soldiers would be likely to experience heat illness in < 6 hours when exposed to direct sun light in all simulated environments. Shade is effective in delaying or preventing soldiers from becoming heat casualties.

Keywords: Protective patient wraps, heat strain, modeling, simulation, core temperature, US Army, chemical/biological warfare, thermoregulation

INTRODUCTION

The protective patient wrap (PPW) is an encapsulating sleeping bag like portable, disposable and water-resistant material designed to protect injured soldiers, when necessary, from exposure to harmful chemical and biological materials during triage. It was developed by the US Army in 1980s when the use of chemical and biological weapons became more prominent (US Army Natick Soldier Center, 2007). After 2001, new PPW configurations were developed using more advanced technology. The purpose of this study was to evaluate new PPW designs for their possible thermal impact on soldiers, using a thermoregulatory model. Initial testing was done using a thermal manikin to measure the thermal and water vapor resistance of the PPW. Then the effects of new PPW configurations on patients' physiological responses were simulated for three types of hot climates (i.e., jungle, desert, temperate) with and without direct sun. A thermo-physiological model (Kraning and Gonzalez, 1997) utilized in this study uses first principles of physiology, heat transfer and thermodynamics and represents the human with six compartments (i.e., core, muscle, fat, vascular skin, avascular skin, and central blood). The model predicts physiologic responses over time (e.g., heart rates, core temperature (T_c)) of individuals as a function of metabolic heat production, anthropometry (i.e., height, weight, and percent body fat (%BF)), thermal aspects of the physical environment (i.e., air temperature (T_a), relative humidity (RH), mean radiant temperature (MRT), wind speed (WS)) and clothing characteristics (i.e., thermal and water vapor resistance), and physiological state (e.g., heat acclimatization, hydration).

This evaluation approach provides a convenient means of predicting thermal strain of workers without incurring the risk, cost and time associated with human studies. Predictive modeling is increasingly used as thermal injury prevention and occupational safety assessments.

METHODS

Model simulations to evaluate the different PPW configurations were conducted based on realistic model inputs. These inputs included: subject anthropometric information and resting metabolic rate, the thermal and water vapor resistances of the updated uniform and PPW configurations, and the ambient micro-weather conditions (temperature, humidity, solar load, wind speed). The triage patient was assumed to be heat acclimated with the height, weight, and %BF values (177 cm, 82 kg, 17%) of average active duty US Army male soldiers (Bathalon et al., 2004). The thermal and water vapor resistance characteristics of the current PPW, and the new laminated and non-laminated PPW designs, with and without battery powered fan ventilation, were measured with the USARIEM thermal sweating manikin (Figure 1). The manikin was dressed in T-shirt, Fire Resistant Army Combat Uniform (FR-ACU), and green wool socks, placed inside the PPW, and positioned on a cot elevated two feet above the ground. The metabolic rate of the patient associated with the condition was estimated to be 0.8 MET (~45W/m^2) (ASHRAE, 2001). The simulation of human physiological responses to PPW encapsulation were conducted for typical desert (T$_a$: 49°C, RH: 20%), jungle (T$_a$: 35°C; RH: 70%) and temperate (T$_a$: 35°C; RH: 50%) conditions where deployed soldiers could experience heat related illness or impairment. The MRT for the shade or no-sun condition was assumed to equal the environment's T$_a$. For sunny environments, the MRT was estimated to be 36°F (20°C) greater than T$_a$, using the constant radiant load (175 W•m^{-2}) and radiant heat transfer coefficient, (Matthew et al., 2001). A constant WS of 0.4 m•s^{-1} (0.89 mph) was used for all simulations. Table 1 summarizes the details of three environmental conditions and their radiation levels for full sun (MRTs) and non-sun/shades (MRTn) conditions.

Figure 1. Photographs of the sweating-thermal manikin and protective patient wrap (PPW) test set-up. Left - Manikin placed inside the closed PPW. Right - the filtered ambient air ventilation system attached to the foot of the PPW.

Thirty model simulations were conducted based on the combinations of five PPW configurations and three environmental conditions with each environment in full sun and complete shade (30 = 5 x 3 x 2). Levels of physiological heat strain were assessed based on (1) T$_c$ limit of 38.5°C, representing the point where approximately 25% heat casualty rate is expected to occur (Sawka et al., 2000) and (2) a six hour maximum encapsulation time for PPW in compliance with U.S. Army requirement (Department of the Army, 1985).

Table 1: Meteorological conditions used to simulate desert, jungle, and temperate environments.

Parameters	Environments		
	Desert	Jungle	Temperate
Ta °C(°F)	48.9 (120)	35.0 (95)	35.0 (95)
RH %	20	75	50
DP °C(°F)	20 (68)	30 (86)	23 (73)
V m•s^{-1} (mph)	0.4 (0.89)	0.4 (0.89)	0.4 (0.89)
MRTs °C(°F)	68.9 (156)	55.0 (131)	55.0 (131)
MRTn °C(°F)	48.9 (120)	35.0 (95)	35.0 (95)

Ta: Air temperature, RH: Relative humidity; DP: Dew point; V: Wind speed; MRTs: Mean radiant temperature with full sun; MRTn: Mean radiant temperature with shade

RESULTS

Figure 3a-c summarizes the simulated T_c responses to the different PPW configurations in the three environmental conditions with or without solar radiation. Overall, for the desert condition with and without solar radiation, T_c rises to 38.5 °C more quickly than other conditions. The simulations indicated that soldiers inside PPW would be likely to experience thermal strain (i.e., $T_c > 38.5°C$) during the 6hr exposure in all three climate conditions with the strain developing roughly 25 – 50% faster in the higher levels of T_a and/or RH. In all three environmental conditions, patients in all of the PPWs had consistently lower T_c levels when located in the shade and could safely stay longer than when located in the sun. When the simulated patients were located in a shaded desert or a sunny temperate condition, the fan-powered PPW ventilation system was very effective in helping individuals thermo-regulate, lowering their T_c responses and thus increasing PPW safe stay times by about 15-30% compared to the non-ventilated PPW configurations (Figure 3a, 3c). The differences in T_c responses among the PPWs were small in other conditions.

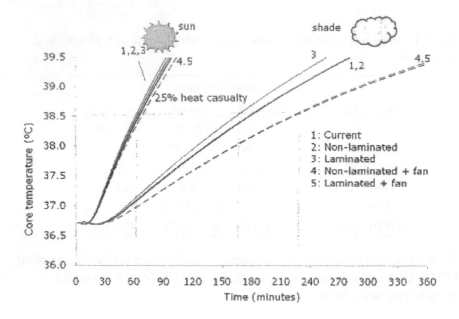

(a) Desert condition (48.9°C/120°F, 20%RH)

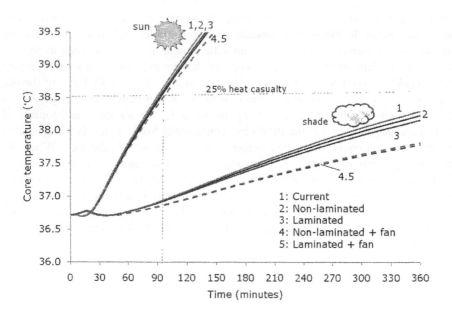

(b) Jungle condition (35°C/95°F, 75%RH)

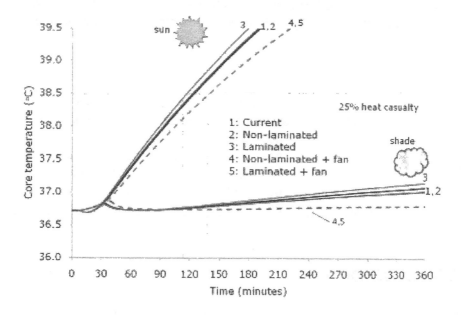

(c) Temperate condition (35°C/95°F, 50%RH)

Figure 3. Core temperature responses over time for five protective patient wrap configurations (current, non-laminated, laminated, non-laminated + fan, laminated + fan) in sunny or shaded (a) desert, (b) jungle and (c) temperate environmental conditions. At a core temperature of 38.5 °C, a 25% heat casualty rate is expected.

Table 2 summarizes tolerance times based on the T_c limit of 38.5 °C by configuration and environmental conditions. Overall, high T_a, high RH, and solar radiation decreased tolerance times. Individuals are more tolerant of encapsulation in PPW in temperate weather than jungle, and least tolerant in the desert environment. Patients in shade are likely able to stay inside a PPW 3-4 times longer than when under direct sun. The model simulations also indicated that individuals would likely become heat casualties in less than 6 hours inside any of the PPW in direct sun. The simulated individuals could tolerate the PPWs for 6 hours only when T_a is equivalent to/less than 35 °C (e.g., temperate, jungle) and MRT is equivalent to/less than T_a. (e.g., shade condition). The use of a fan under sunny temperate and shaded desert conditions increased the estimated time to achieve a T_c limit of 38.5 °C by about 20 min. and 50 min, respectively. The fan had no significant effect under the other conditions.

Table 2: Tolerance time (minutes) to reach core temperature of 38.5°C by protective patient wrap configuration and environmental condition with or without solar load.

Configuration\Solar effect	Desert		Jungle		Temperate	
	Solar	Shade	Solar	Shade	Solar	Shade
Current	66	178	93	>360	124	>360
Non-Laminated + Fan-Off	64	177	91	>360	123	>360
Non-Laminated + Fan-On	68	223	96	>360	140	>360
Laminated + Fan-Off	62	164	90	>360	117	>360
Laminated + Fan-On	68	220	96	>360	139	>360

CONCLUSIONS

This study examined simulated T_c responses of heat acclimated soldiers who were fully encapsulated in a PPW during three different hot-warm environmental conditions. The simulations used to evaluate the different PPW configurations were based on realistic information regarding the subjects' anthropometrics, metabolic rate, PPW biophysical characteristics, and environmental and system operational conditions. The results indicated that patients inside PPW would be likely experience thermal strain faster when T_a and RH increase. Further, the simulations indicated that shading from direct sun is critically important in delaying or preventing individuals from becoming heat casualties. The fan-powered PPW ventilation system was effective only when they were lying under a shaded desert or sunny temperate condition. Otherwise, the differences in T_c responses among the PPWs were minimal. Based on the current U.S. Army criterion for PPW encapsulation targeted time (Department of Army, 1985), soldiers would be likely to experience heat illness in less than six hours when exposed to direct sun light in all simulated environments. Simulated patients, when shaded from solar exposure, could safely endure 6 hours only in the jungle and temperate environments. Thus, shade from direct sun is important in delaying and preventing patients from becoming heat casualties.

The simulation used in this study was assumed to be an "average uninjured" soldier in the US Army. The different somatic forms in a population as well as the patients' medical condition and treatment (Cadarette et al., 1988; Stephenson et al., 1988; Yokota et al., 2008; Bar-Or et al., 1969) could cause variability in physiological responses to the heat stress. For instance, obese individuals in walking in the heat responded to thermal strain with a more rapid rise in T_c than lean individuals (Bar-Or et al., 1969; Shvarts et al., 1973). A multivariate thermal model simulation suggested similar results would be evident in soldiers walking and working on a simple Army task (Yokota et al., 2008). For another example, the usage of atropine as a common treatment to regulate patients' parasympathetic nervous system reduces their sweat rates, and rapidly increases a patient's T_c (Cadarette et al., 1998; Stephenson et al., 1988). Importantly, medical circumstances such as injury, loss of blood, and respiratory problems inside a PPW may also impact the tolerance time of a patient (Cadarette et al., 1998; Stephenson et al., 1988). Thus, thermal responses inside a PPW can vary across individuals.

This study demonstrated that the thermoregulatory model simulations can provide a useful insight into the thermal strain imposed on soldiers/patients encapsulated in PPW. The simulations may be useful not only in understanding the thermal benefits/disadvantage of various PPW prototypes but also assisting in a cost effectiveness analysis of prototype PPWs. The approach taken in this study to assessing the thermal impact on soldiers can be extended to other types of equipment (e.g., micro-climate cooling device, body armor, vehicles) and protective gear/clothing (e.g., body armor, Mission Oriented Protective Posture gear, battle dress uniform). Further, these types of simulations can be applied to various occupational populations other than military (e.g., firefighters, border patrol, police bomb squad) to assess the safety of workers who are exposed to thermal stress during their work.

ACKNOWLEDGEMENTS

The authors would like to thank Drs. Reed Hoyt and Larry Berglund, USARIEM for critical comments on this paper. Opinions, interpretations, conclusions and recommendations contained herein are those of the author and are not necessary endorsed by the US Army.

REFERENCES

ASHARE (2001), *ASHARE handbook fundamentals SI edition*. American Society of Heating, Refrigerating and Air-Conditioning Engineers, Inc., Atlanta.

Bar-Or, H., Lundegren, H., and Buskirk, E. (1969), "Heat tolerance of exercising obese and lean women." *Journal of Applied Physiology*, 26, 403-409.

Bathalon, G., McGraw, S., Friedl, K., et al. (2004), *Rationale and evidence supporting changes to the Army weight control program*. Technical Report T04-08, USARIEM, Natick.

Cadarette, B., Speckman, K., Stephenson, L. (1988), "Physiological assessments of chemical threat protective patient wraps in three environment." *Military Medicine,* 153, 166-179.

Department of the Army (1985), *Letter Requirement (LR) Chemical Warfare Agent Protective Patient Wrap* (USATRADOC CAN 21318). Academy of Health Science. September 18, Fort Sam Houston.

Kraning, K., and Gonzalez, R.(1997), "A mechanistic computer simulation of human work in heat that accounts for physical and physiological effects of clothing, aerobic fitness, and progressive dehydration." *Journal of Thermal Biology,* 22,331-342.

Matthew, W., Santee, W., and Berglund L. (2001), *Solar load inputs for USARIEM thermal strain models and the solar radiation-sensitive components of the WBGT index.* Technical Report T01-13, USARIEM, Natick.

Sawka, M., Latzka, W., Montain, S. et al. (2000), "Physiologic tolerance to uncompensable heat: intermittent exercise, field vs. laboratory." *Medicine & Science in Sports & Exercise,* 33, 422-430.

Shvarts, E., Sarr, E., and Benor, D. (1973), "Physique and heat tolerance in hot-dry and hot-humid environments." *Journal of Applied Physiology* 34,799-803.

Stephenson, L., Kolka, M., Allan, A. et al. (1988), "Heat exchange during encapsulation in a chemical warfare agent protective patient wrap in four hot environments." *Aviation, Space and Environmental Medicine*, 59, 345-351.

US Army Natick Soldier Center (2007) *Chemical protective patient wrap program* [CD]. US Army Natick Soldier Center. Natick, MA.

Yokota, M., Bathalon, G., and Berglund, L. (2008), "Assessment of male anthropometric trends and effects on simulated heat stress responses." *European Journal of Applied Physiology*, 104,297-302.

Chapter 47

The Effects of Encumbering the Soldier With Body Armor

Leif Hasselquist, Karen N. Gregorczyk,
Brian Corner, Carolyn K. Bensel

US Army Natick Soldier Research
Development and Engineering Center
Natick, MA 01760, USA

ABSTRACT

Armor vests are used by military personnel to protect the torso against ballistic challenges. Development of armor to protect the extremities as well was recently undertaken. In the limited testing to date, it was found that wear of extremity armor increases the energy cost of walking and running relative to that with the armor vest alone, a penalty associated with the added weight on the body. The present study was conducted to examine design characteristics of extremity armor that may negatively impact mobility. Three designs of extremity armor were investigated. They were similar in weight, but differed in the area of the body surface covered by ballistic material. Eleven Army enlisted men participated in the test. Measurements were taken of the maximum extent of movement in various planes of the body in the armor vest worn alone and with each of the three extremity armor systems. Compatibility of the extremity armor with military equipment and activities was also examined and participants completed a survey on their opinions regarding the armor. It was found that the three types of extremity armor restricted range of motion to a greater extent that wear of the vest alone. The extremity armor systems also interfered with weapon aiming and use of the standard Army backpack. Further, the three designs of extremity armor differentially affected performance. The poorest mobility and the system least preferred by the participants was the version providing the greatest ballistic coverage of the body. The results of this study indicate that design elements of extremity armor are critical in maximizing warfighter mobility on the battlefield.

Keywords: Extremity Armor, Ballistic Protection, Human Factors

BACKGROUND

Armor vests have been used for decades by dismounted warriors to protect the torso against ballistic challenges. Soldiers and Marines deployed to Iraq and Afghanistan currently wear the Interceptor Body Armor (IBA) tactical vest with groin protector to protect against shrapnel and handgun rounds. Ceramic plates are added to pockets on the front and back of the vest to protect against rifle ammunition. The IBA vest is an effective and highly valued piece of equipment and has saved many lives. The benefits of wearing the IBA appear to outweigh the limitations imposed by increased weight on the body, restricted mobility, and thermoregulation issues (Cadarette et al., 2001; Ricciardi et al., 2008; Woods et al., 1997).

Because of the battlefield threats being encountered by military personnel serving in Iraq and Afghanistan, the Army and the Marine Corps launched initiatives to increase the body area covered by ballistic protective materials to include the arms and the legs. In response, extremity armor was developed in a number of different designs to be worn with the IBA. It can be posited that adding more weight to the body and placing semi-rigid materials on the arms and the legs will increase the energy expended and negatively impact the mobility of military personnel (Martin, 1985; Royer and Martin, 2005; Soule and Goldman, 1969). However, little is known about the effects of extremity armor on the wearer as these systems have only recently been developed.

In what appears to be the first study of extremity armor, Hasselquist et al. (2008) investigated the effects of the IBA vest alone and the vest plus three extremity armor systems on energy costs and movement patterns of soldiers as they walked and ran. Militarily relevant physical activities (box lifting and carrying, 30-m rushes, obstacle course runs) were tested as well to measure maximal effort performance. The extremity armor systems were of similar weight, but they differed in design characteristics, including the materials used to provide ballistic protection. Using rate of oxygen uptake as an index of energy cost, Hasselquist et al. found that oxygen consumed per unit body mass was 22 to 26% higher during walking and 7% higher during running in the extremity armor compared with the vest alone. They ascribed these finding to the weight that the extremity armor added to the body. It was also reported that gait patterns during walking and running differed among extremity armor systems, as did times to complete the maximal effort activities. Hasselquist et al. posited that these findings were due to design differences in the extremity armor systems tested.

The investigation reported here was conducted to further examine soldiers' mobility as affected by wear of extremity armor. Particular emphasis was placed on design characteristics of the systems that may encumber body movements. The maximum extent of motion in various planes of the body was measured to identify movement restrictions. The compatibility of extremity armor with military equipment and activities was also examined, and soldiers participating in the investigation were given a survey devised for the study that included questions on their opinions of the body armor.

METHODS

PARTICIPANTS

Participants were 11 US Army enlisted men (means — Age: 20 yrs.; Stature: 1.8 m; Weight: 79.7 kg) recruited from among the military personnel who serve as human research volunteers assigned to Headquarters Research and Development Detachment, US Army Soldier Systems Center, Natick, MA. Ten of the men (MOS 11B) had just completed infantry advanced training and their mean time in service was five months. One man (MOS 19K, armor crewman) had time in service of 20 months. Informed consent was obtained. The study was approved by the local IRB and conducted in accordance with the Federal Policy for the Protection of Human Subjects. All participants were healthy and without musculoskeletal injuries.

ARMOR

Each participant was tested in four conditions: the Interceptor Body Armor vest, including collar, groin protector, and protective plates (Vest), and three designs of extremity armor (Ext 1, 2, 3), which were worn with the vest. The extremity armor systems were the same ones tested by Hasselquist et al. (2008). The three designs of extremity armor were similar in weight, but varied in the ballistic materials used and in body surface area covered by the materials (Table 1). To determine coverage area, 3-dimensional laser scans of the body surface of each participant were made in the armor conditions under study and surface area covered by armor was calculated. Means calculated over participants are presented in Table 1. The scans in Figure 1 illustrate differences in coverage among the three designs and area coverage provided by the armor vest.

Table 1. Mean Armor Mass and Coverage Area

	Vest	Ext 1	Ext 2	Ext 3
Weight (kg)	8.7	5.6	6.4	5.6
Coverage (m^2)	.411	.717	.775	.926

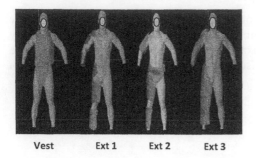

| Vest | Ext 1 | Ext 2 | Ext 3 |

FIGURE 1 Examples of 3-D laser scans of armor vest and extremity armor coverage.

The differences among the three extremity armor systems in the body surface area covered are reflective of differences in the designs of the systems. The upper portion of Ext 3, the system with the greatest body surface area coverage, consisted of two sleeves with shoulder pieces that overlapped the vest. The sleeves completely covered the entire length of the arms, including the elbows. The lower portion of Ext 3, which was a trouser design, extended from the waist to the ankles, covering hips, buttocks, and upper and lower legs. The upper portion of Ext 2, the system with the next greatest area coverage, was a single piece that was secured to the shoulder of the vest and covered the lateral surface of the upper arm. The lower portion was comprised of a piece encircling the waist, a piece that wrapped around the thigh and was secured by straps to the waist piece, and a separate piece that wrapped around the lower leg. The thigh and lower leg pieces had buckles and straps to secure each piece in place. The upper portion of the Ext 1 version was a jacket design made in a mesh fabric. Pads of ballistic material were attached to the mesh and covered the shoulders and upper arms, but not the elbows. A separate piece covering the lower arms was secured to the upper arm piece using straps. The lower portion of the Ext 1 was a similar design approach, with pads of ballistic material attached to a mesh fabric. Ballistic material was placed around the hips and the upper leg. A lower leg piece was attached to the upper with straps. The knee was not covered.

PROCEDURES

To obtain range of motion data, participants executed a series of simple body mobility tasks. They were given three successive trials on each mobility task in each armor condition. The maximum extent of movement possible was obtained by using a gravity goniometer, which measures the angular displacement at a body joint (e.g., elbow, shoulder, knee), or by measurement using a meter stick. The score on a mobility task was the mean over the three trials under a given armor condition. These data were subjected to one-way repeated measures analyses of variance (ANOVAs), with four levels of the body armor variable (Vest, Ext 1-3). Significance level was set at $p < .05$. In those instances in which an ANOVA yielded a significant main effect of body armor, post-hoc analyses in the form of the

Least Significant Difference procedures was performed, with the significance level again set at $p < .05$.

Compatibility of the body armor with military equipment and activities was assessed by observing participants as they assumed weapon firing positions using a simulated M4 carbine, put on and adjusted a military backpack, and used other equipment. At the end of all testing activities, participants completed a survey that was devised to obtain their opinions of the body armor.

RESULTS AND DISCUSSION

RANGE OF MOTION

The means and the results of the statistical analyses applied to the standing trunk flexion movement and the movement that entailed taking five steps forward are shown in Figure 2. On standing trunk flexion, lower scores indicate better performance. For this movement, performance with the extremity armor systems was significantly inferior to performance with the vest alone. There were also differences among the types of extremity armor; Ext 3 scores were significantly better than those for Ext 1. The Ext 1 system differed from the other two extremity armor systems in having a waist belt and pads of ballistic protective material that surrounded the waist and hip areas. It is likely that the bulk around the waist and hips limited flexion at the waist in Ext 1. In the analysis of the stepping data, there were no significant differences among the types of extremity armor. However, the distances that could be traversed in five steps were significantly shorter with the extremity armor than with the vest alone. The more limited motion with the extremity armor could be due to insufficient material allowance in the buttocks area or in the legs.

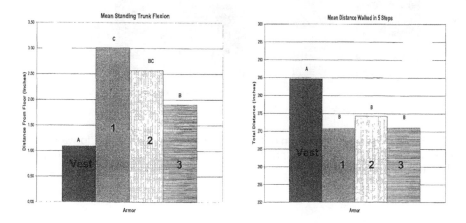

FIGURE 2 Means for each condition on standing trunk flexion and distance walked

464

in five steps forward. Armor conditions that do not share the same letter differed significantly in post-hoc tests ($p < .05$).

The participants performed two arm motions, arm abduction and forward extension. Both required movement at the arm-shoulder joint. The data for the maximum extent of the arm movements in each armor condition are in Figure 3. The results for the two arm movements were similar. There was a significantly greater range of movement with the vest alone than with Ext 2 and Ext 3. The Ext 2 version had the lowest range of movement; it was significantly lower than range of movement for all other conditions (Figure 3). All three extremity armor systems were designed to cover the shoulder and upper arm. The more restricted movement with Ext 2 may be attributable to the stiffness of the ballistic materials incorporated in upper arm piece of this system.

Two leg movements were also executed by the participants, forward extension and flexion. In forward extension, the right leg was moved forward from the hip, with the knee being kept straight. Leg flexion required that the right upper

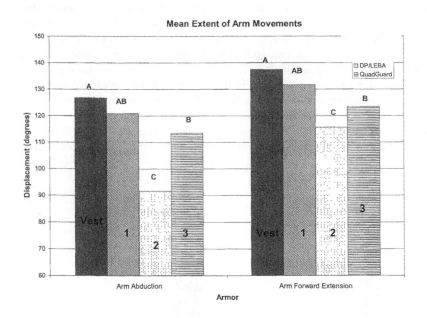

FIGURE 3 Means for each condition of the extent of arm abduction and forward extension. Armor conditions that do not share the same letter differed significantly in post-hoc tests ($p < .05$).

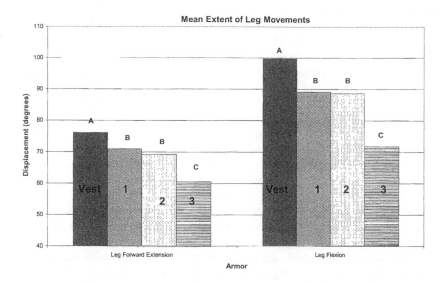

FIGURE 4 Means for each condition of the extent of leg forward extension and leg flexion. Armor conditions that do not share the same letter differed significantly in post-hoc tests ($p < .05$).

leg be lifted up as far as possible while letting the lower leg swing freely from the knee. The analyses of these movements yielded similar results. For both movements, the extent of motion was significantly greater with the vest alone than when the extremity armor was also worn (Figure 4). Among the extremity armor systems, the extent of leg motion was significantly less with Ext 3 than with Ext 1 or Ext 2. The lower portion of the Ext 3 system consisted of trousers made of ballistic protective materials that extended from the waist to the ankles, completely covering the legs, including the knees. The other two extremity armor systems did not cover the knee and their waist portions were separate from the pieces on the leg. It is likely that the materials covering the knee in the Ext 3 version constrained flexion at this joint. Insufficient material allowance through the buttocks may also have limited leg movement in Ext 3.

COMPATIBILITY WITH MILITARY EQUIPMENT AND ACTIVITIES

Participants used the Army's ballistic protective helmet, the Advanced Combat Helmet, along with the body armor during testing. None of the extremity armor systems interfered with wearing the helmet. With regard to weapon use, all participants could assume a kneeling and a prone firing position, regardless of the armor being worn. Participants did have difficulty, however, positioning the butt of the M4 carbine at the junction of the shoulder and the upper arm (i.e., "pocketing" the weapon). In order to aim the weapon, the participants positioned the butt against the shoulder or the upper arm. These methods were not satisfactory for stabilizing the weapon or for aligning the eye properly during aiming of the weapon.

For assessing use of backpacks with body armor, the external frame and the rucksack of the Army's Modular Lightweight Load Carrying Equipment (MOLLE) were used. With all three extremity armor systems, the waist belt of the backpack had to be secured over the waist and hip portions of the extremity armor, with the result that the waist belt could not be tightened securely enough to distribute the backpack load to the participants' hips. Further, when the pack was worn with Ext 2 or Ext 3, the shoulder portions of the extremity armor interfered with passing the pack straps over the tops of the shoulders. The pack straps instead passed over the edge of the shoulders and extended on to the upper arms.

Upon first being exposed to each type of extremity armor, participants required a demonstration of the manner in which the components of each configuration were to be put on and adjusted for a proper fit. After practice, volunteers could don and doff each type of body armor properly and without assistance. Difficulties were encountered securing the shoulder protectors on Ext 2 around the upper arm, a task that had to be done with one hand. Further, after the Ext 3 trousers had been donned, difficulties were encountered with mating the zipper that ran down the lower leg of this armor version. With experience, the volunteers checked to see that the zipper was mated before they put on the trousers.

SURVEY OF PARTICIPANTS' OPINIONS

On the survey administered at the end of testing, participants were asked to rate, on a 5-point scale (from very difficult to very easy), the ease or difficulty of performing activities in the armor conditions tested. The median ratings for each armor condition are in Figure 5. The ratings were generally neutral or on the positive (easy) side of the scale. Further, the vest was rated more positively than the extremity armor systems were. Among the three types of extremity armor, the Ext 3 received the lowest ratings for ease of performing the activities.

On two questions posed on the survey, participants were to select the one type of extremity armor that they most preferred and the one type that they least preferred. Participants were to give reasons for their selections. The Ext 1 was selected as most preferred by 10 of the 11 participants; the remaining man selected Ext 2 as his most-preferred system. With regard to the least-preferred system, 10 of the 11 men selected Ext 3 and one man selected Ext 2. The reasons given for their selections by the participants who favored Ext 2 and by the participants who least preferred Ext 3 are presented in Table 2.

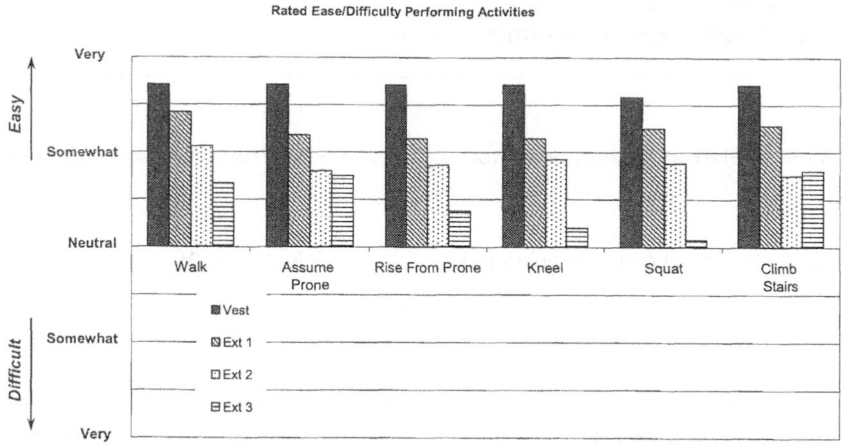

Rated Ease/Difficulty Performing Activities

FIGURE 5 Median rating for each armor condition of ease/difficulty performing movements.

CONCLUSIONS

It was reported previously that the energy cost of walking and running is higher when extremity armor is worn than when it is not, a penalty associated with the weight of the armor. In the present study, quantitative measures of range of motion in various planes of the body revealed that, in addition, wearing ballistic materials on the extremities restricts movement. Further, observations made during the study indicate that extremity armor use introduces incompatibilities between soldiers and their personal equipment. The soldiers participating in this study also judged that executing movements was easier when they were wearing the ballistic protective vest alone than when extremity armor was worn as well.

The data obtained here on three different versions of extremity armor highlight the importance of the design characteristics of the armor in affecting the soldier's mobility. The opinions of the soldier-study participants also indicated awareness of design differences. The extremity armor system associated with the poorest mobility and the system least preferred by the participants was the version providing the greatest ballistic coverage of the body. It was also the only system with ballistic protective materials covering the elbow and knee joints.

Table 2. Reasons Given by Participants for Their Selections of Most- and Least-Preferred Extremity Armor Systems

Most-Preferred Extremity Armor: Ext 1 (Selected by 10 of 11 Participants)	Least-Preferred Extremity Armor: Ext 3 (Selected by 10 of 11 Participants)
Good ballistic coverage of the body without bulk	Bulky
Flexible, good mobility, not restricting	Inflexible, poor mobility, restricts and slows movement
Light weight	Heavy
Cool	Hot
Good fitting	Poor fitting (trousers baggy), too much excess material
	Makes noise when you move in it

Future technological advances in body armor may lead to lighter weight ballistic materials. However, as the results of this investigation show, lighter ballistic materials alone will not be sufficient to reduce the encumbering effects of extremity armor. The design process will still be critical to maximizing the mobility of warfighters while also ensuring their protection against ballistic threats.

REFERENCES

Cadarette, B.S., Blanchard, L., Staab, J.E., Kolka, M.A., Sawka, M.N. (2001) Heat stress when wearing body armor (Tech. Rep. T01-09). Natick, MA: US Army Research Institute of Environmental Medicine.

Hasselquist, L., Bensel, C.K., Corner, B., Gregorczyk, K.N., Schiffman, J.M. (2008) "Understanding the physiological, biomechanical, and performance effects of body armor use," proceedings of the 26th Army Science Conference, Orlando, FL.

Martin, P.E. (1985) "Mechanical and physiological responses to lower extremity loading during running." Medicine and Science in Sports and Exercise, 17, 427-433.

Ricciardi, R., Deuster, P.A., Talbot, L.A. (2008) "Metabolic demands of body armor on physical performance in simulated conditions." Military Medicine, 173(9), 817-824.

Royer, T.D., Martin, P.E. (2005) "Manipulations of leg mass and moment of inertia: effects on energy cost of walking." Medicine and Science in Sports and Exercise, 37(4), 649-656.

Soule, R.G., Goldman, R.F. (1969) "Energy cost of loads carried on the head, hands, or feet." Journal of Applied Physiology, 27, 687-690.

Woods, R.J., Polcyn, A.F., O'Hearn, B.E., Rosenstein, R.A., Bensel, C.K. (1997) Analysis of the effects of body armor and load-carrying equipment on soldiers' movements. Part II. Armor vest and load-carrying equipment assessment (Tech. Rep. NATICK/TR-98/003). Natick, MA: Natick Research, Development and Engineering Center.

Chapter 48

Ergonomic Evaluation of an Exoskeleton Prototype

Jeffrey M. Schiffman[1], Karen N. Gregorczyk[1], Leif Hasselquist[1], Carolyn K. Bensel[1], John P. Obusek[1], David Gutekunst[2], Peter Frykman[3]

[1]US Army Natick Soldier Research
Development and Engineering Center
Natick, MA 01760, USA

[2]Washington University in St. Louis
St. Louis, MO 63130, USA

[3]US Army Research Institute of Environmental Medicine
Natick, MA 01760, USA

ABSTRACT

There is an emerging technology in wearable robots, or exoskeletons, some of which are being developed to assist in carrying loads. These devices are designed to reduce the vertical force of the load on the body and thus reduce the energy expended by the load carrier. This study investigated the effects on metabolic cost and gait biomechanics of walking at 4.8 km·h^{-1} with 20-, 40-, and 55-kg loads while wearing a prototype exoskeleton (EXO). Data were also collected on two maximal performance tests conducted after 8-km marches with and without the EXO and on postural balance as affected by the EXO. Ten U.S. Army enlisted men served as test participants. Using rate of oxygen uptake ($\dot{V}O_2$) as an index of energy cost, mean $\dot{V}O_2$ scaled to body mass and scaled to total mass were found to be significantly higher when the EXO was worn during walking than when it was not. The biomechanical data revealed significant differences in gait patterns between the EXO and the No-EXO conditions, which may have contributed to the higher energy cost with the EXO. The speed maintained by the participants was lower and the distance traversed shorter when the EXO was worn for the 8-km march. Scores on the maximal performance tests were not affected by whether or not the EXO was worn during the march. Postural sway excursions during stationary standing were significantly reduced with the EXO, but there was a greater tendency for body sway to occur without corrective adjustments back to an equilibrium point, a less stable pattern than evidenced without the EXO. This study identified a number of design features of the EXO prototype that contributed to the higher energy cost, changed gait patterns, and reduced balance stability.

Keywords: Biomechanics, Metabolic Cost, Balance, Load Carriage

INTRODUCTION

In industrial and military environments, personnel are often required to carry heavy loads, an activity that can result in physical fatigue and debilitating musculoskeletal disorders (Knapik et al., 1996; Rodrick and Karwowski, 2006). There is an emerging technology in wearable robotic devices, or exoskeletons, that are designed to augment human strength (Guizzo and Goldstein, 2005; Herr, 2009). Lower body exoskeletons to provide assistance in carrying loads have been the focus of recent efforts in wearable robots (Walsh et al., 2007). An assistive device for load bearing that resulted in reduced energy consumption and fatigue of users could have widespread application in industrial and military operations.

There is a growing literature on architectures employed in the designs of lower body exoskeletons, but data on human use of the devices are very limited (Herr, 2009; Kazerooni and Steger, 2006; Walsh et al., 2007). We acquired a prototype of a lower body exoskeleton (EXO), which was developed by a commercial organization to provide load-carriage assistance, and conducted a study on use of the device by soldiers. The EXO consists of: a hip structure with a back plate to which a rucksack is attached; tubular leg struts that incorporate knee and ankle joints; and semi-rigid foot plates, which contain sensors that monitor contact with the ground and are the base for the leg struts (Figure 1). The leg struts are constructed to permit flexion-extension, but not abduction or adduction. The foot plates are 3 cm thick and underlie the length of the user's foot. The total weight of the prototype, including a power unit, batteries, and on-board computer, is 15 kg.

FIGURE 1 Soldier wearing the EXO and a 40-kg load.

The EXO interfaces with the body by use of shoulder straps, a waistbelt, thigh pieces, and foot bindings. The prototype was designed to reduce the vertical force component of the rucksack load exerted on the wearer during locomotion. This is accomplished by application of variable damping to the leg joints, which enables the transfer of the force through the device to the ground. The device does not produce torque to raise the load against gravity or to control the load in the anterior-posterior or the medio-lateral directions. Therefore, the user must control the moment of inertia of the device and the load attached to it, as well as generate the force necessary to propel the device forward. The EXO is designed, however, to store and return some energy during walking via springs at the hip and ankle joints.

In interfacing directly with the user's body and exerting control over aspects of locomotion, exoskeletal devices raise many ergonomic issues. Previously, we carried out initial evaluations using mock-ups of several exoskeleton concepts to assess a user's ability to don and doff the device quickly without assistance and to execute a variety of movements, such as kneeling, squatting, and assuming and rising from a prone position. Fit on a range of body sizes was also evaluated. We acquired information to guide design of aspects of lower body exoskeletons. However, the mock-ups were non-functional and not of representative weights. Therefore, issues related to walking in the devices were not explored.

The prototype we obtained for this study was a functioning exoskeleton, allowing several basic questions to be addressed. The principal question was the extent to which an exoskeleton actually serves an assistive function in load carrying. Physiological research suggests that, if carrying a load on an exoskeleton reduces the vertical force on the body, energy costs will be lower compared with carrying an equivalent load in the traditional manner (Soule et al., 1978). A related question has particular relevance to military operations. Soldiers typically carry loads over great distances and, immediately upon completion of prolonged foot marches, doff a portion of their loads and undertake high-intensity tactical maneuvers that are most critical to mission success. Thus, physical performance in the immediate post-march period as a function of whether or not an exoskeleton was used during the march is an important consideration. A third question pertains to body stability. Research findings indicate that carrying heavy loads compromises postural equilibrium (Schiffman et al., 2006). A concern, therefore, was whether wear of an exoskeleton for load carrying would improve body stability or further compromise it.

This study of the EXO prototype entailed analysis of energy cost through measurement of rate of oxygen uptake during the latter part of 8-min walking trials. Kinematic and kinetic data were also captured to assess gait biomechanics, as changes in walking mechanics can affect the metabolic cost of locomotion (Donelan et al., 2001). Physical performance after a prolonged march of 8 km was evaluated through inclusion of maximal-effort runs of an obstacle course and an upper endurance test. Body stability was assessed using measures of static limits of stability and postural sway during standing.

METHODS

PARTICIPANTS

Ten U.S. Army enlisted men were initiated into the study (Age: 21.1 ± 3.7 yrs.; Stature: 1.76 ± 0.05 m; Weight: 75.3 ± 9.28 kg). All 10 men completed the balance testing. Nine men completed the walking activities and seven completed all post-march activities. The men had recently finished training as infantry soldiers and had carried backpack loads on foot marches as part of the training. They gave informed consent, and the study was approved by the local IRB and conducted in accordance with the Federal Policy for the Protection of Human Subjects.

LOAD CONDITIONS

The 8-min walking trials and the balance assessment were conducted with and without the EXO under three load configuration conditions. Load masses were 20, 40, and 55 kg, exclusive of the 15-kg mass of the EXO. A replica M4 rifle was held in both hands in front of the body, the position in which a rifle is held during foot marches. The 20-kg load consisted of basic clothing (shorts, T-shirt, military field boots), a helmet, and an armor vest. Military items carried during field maneuvers, such as a canteen and ammunition, were placed in pouches attached to the front of the vest. The 40- and the 55-kg loads were comprised of all components of the 20-kg load plus an external-frame rucksack containing soldier items. The rucksack was worn on the back in the usual manner when the EXO was not used. With the EXO, the rucksack was attached to the EXO back plate. The 40-kg load was used for the prolonged marches with and without the EXO. Upon march completion, the rucksack and the rifle were removed, as was the EXO, when applicable, and the obstacle course was completed. Only the basic clothing was worn during the upper body endurance test.

TESTING EQUIPMENT AND PROCEDURES

Over a 2- to 3-day period prior to the start of testing, each participant completed 4 to 6 hours of familiarization with study conditions, including carrying the loads overground and on treadmills with and without the EXO. Most of the pre-test familiarization time was spent walking in the EXO. At the end of the time, participants were able to walk in the EXO for at least 10 min, displaying a repetitive gait pattern and ability to control movements.

8-min Walking Trials

The walking activity was conducted on a dual force plate treadmill (AMTI, Watertown, MA, USA), which is comprised of two separate treadmills, each with its own force plate, arranged fore and aft on a single platform. Voltage outputs of the force plates were sampled at 1200 Hz, filtered with a low-pass Butterworth filter (cut-off frequency of 10 Hz), and converted to physical units (Newtons) using manufacturer-supplied calibration factors. Rate of oxygen uptake ($\dot{V}O_2$) during the walking was measured using a ParvoMedics TrueMax® 2400 metabolic measurement system (Salt Lake City, UT, USA). Three-dimensional motion data were recorded at 120 frames·s^{-1} by eight ProReflex cameras (Qualisys AB, Gothenburg, Sweden). Retro-reflective markers were placed at selected locations on the participant's skin and clothing to expedite motion data capture and processing.

Each participant was tested during two experimental sessions held approximately 7 days apart. A participant carried the three loads at both sessions and wore the EXO at one session. Five participants wore the EXO at the first session; the remaining four were exposed to the device conditions in the opposite order. The order of presentation of the three load configurations was based on a Latin square. At a session, the men walked on the treadmill set at 4.8 km·h^{-1} and 0% grade for 8 min with each load. During the fifth minute, 20 s of kinematic and kinetic data were collected. After 6 min on the treadmill, measurements of $\dot{V}O_2$ were made breath-by-breath for 2 min and the bout of walking was then concluded. Brief rest periods were provided between trials.

The $\dot{V}O_2$ measurements were averaged over 20-s increments for the gas collection period. These data were then ensemble-averaged for the 2-min period to obtain the absolute value of $\dot{V}O_2$ (ml·min^{-1}). This value was scaled to the body mass of the participant ($\dot{V}O_2$ BM; ml·kg^{-1}·min^{-1}) and to total mass ($\dot{V}O_2$ TM; ml·kg^{-1}·min^{-1}), where total mass was calculated as the participant's body mass plus the mass of all clothing and equipment on the body, including the mass of the EXO, when applicable. The images captured with the cameras were processed using dedicated hardware and software (Qualisys AB, Gothenburg, Sweden) to produce files containing time histories of the three-dimensional coordinates of each reflective marker. Kinematic and kinetic variables were calculated using the Visual3D™ software program (C-Motion Inc., Germantown, MD, USA). The data extracted from the walking kinematics and kinetics included various joint angles, joint range of motion (ROM), spatial and temporal gait variables, peak ground reaction forces (GRFs) over a stride scaled to total mass, and times to peak forces expressed as percentages of stride time.

Prolonged March and Post-march Physical Activities

The 8-km march with the 40-kg load was performed on each of two separate days, once with and once without the EXO. A standard treadmill was used and the speed was set at a participant's self-selected speed for carrying the 40-kg load, which was determined during the familiarization sessions. Heart rate was monitored continuously and, if it achieved 85% of age-predicted maximum, the speed of the treadmill was reduced until the heart rate fell below that level.

After completing the march, participants immediately removed the rucksack, the simulated weapon, and the EXO, if applicable, and started the 8-station obstacle course run. They were instructed to complete the course as quickly as possible. Completion time was obtained electronically using an infrared light-beam system with telemetry (Brower Timing Systems, Salt Lake City, UT).

At the end of the obstacle course run, participants immediately doffed all equipment and began the upper body endurance test. The test consisted of lifting a 22.7-kg box from the standing surface, carrying it a distance of 3.05 m, and placing it on a shelf 1.55 m above the standing surface. These activities were repeated as many times as possible in 10 min and the number of cycles completed was recorded.

Balance Testing

Balance data were acquired using a force plate (AMTI, Inc., Watertown, MA, USA) interfaced with a microcomputer that had a data acquisition board and ran LabVIEW 6i. The voltage output from the force plate was sampled at 100 Hz, filtered with a low-pass Butterworth filter (cut-off frequency of 10 Hz), and converted to physical units (N and N·m), eliminating phase shift using forward and backward passes. The limits of stability (LOS) testing preceded postural sway testing and each consisted of a number of trials under each set of conditions. Five participants wore the EXO first while data were collected for the three loads; the remaining five were exposed to the two device conditions in the opposite order. The order in which the participants were exposed to the load configurations was based on a Latin square.

For LOS testing, participants stood on the force plate and leaned as far as possible forward, backward, left, and right while center of pressure (COP) data were recorded. For postural sway testing, the participants stood on the force plate for 30-s intervals while COP data were again recorded. They were instructed to look straight ahead while maintaining a relaxed posture and keeping their weight evenly distributed on both feet.

Static limits of stability were quantified by calculating anterior-posterior and medio-lateral limits (Owings et al., 2000). For postural sway, total excursion lengths for the COP paths in the anterior-posterior and the medio-lateral directions and the resultant planar motion were measured, along with sway area. The maximum ranges of movement in the anterior-posterior and the medio-lateral directions were also calculated (Prieto et al., 1996). Analyses of the postural sway data included application of stabilogram-diffusion analysis to calculate Hurst scaling exponents, a means for assessing the dynamics of postural control mechanisms (Collins and De Luca, 1993).

STATISTICAL METHODS

Statistical analyses were accomplished using SPSS 13.0 (SPSS Inc., Chicago, IL, USA). Data from the 8-min walks were analyzed using a two-factor repeated measures ANOVA with two levels of device condition (EXO, No EXO) and three levels of load configuration (20, 40, and 55 kg). Significant ANOVA findings were followed up with t tests, which were adjusted using a Bonferroni method. The same statistical analysis was conducted on the balance data. For the performance tests, time to complete the obstacle course and number of box lift repetitions were analyzed using a one-factor (EXO, No EXO) repeated measures ANOVA. Alpha was set at .05.

RESULTS AND DISCUSSION

8-MIN WALKING TRIALS

A large number of dependent measures were obtained from the data captured during the walking trials. Means for some variables that yielded a significant effect of device or load condition or both are presented in Table 1. The effects of load when the EXO was not worn were in agreement with past research (Soule et al., 1978). Of particular importance in this study were findings related to energy cost as a function of device condition. The results indicated that use of the EXO did not lower metabolic cost relative to not wearing the EXO, but instead increased it substantially. Mean $\dot{V}O_2$ scaled to body mass and scaled to total mass were significantly higher, by 60% and 41%, respectively, when the EXO was worn. Further, the interaction between device and load conditions was not significant ($p >$ .10) in the analysis of either $\dot{V}O_2$ BM or $\dot{V}O_2$ TM.

The 15-kg mass of the EXO was likely responsible for a portion of the increased $\dot{V}O_2$ (Soule et al., 1978). It is also likely that the distribution of the EXO mass on the user's body was a critical determinant of the magnitude of the increase. The legs and feet comprise 43% of the mass of the EXO, and higher energy costs have been found as loads are placed in more distal locations on the body (Martin, 1985). The kinematic data revealed changes in gait patterns with the EXO that could have contributed as well to the metabolic costs of wearing the device (Table 1). Increased stride widths increase the energy costs of walking (Donelan et al., 2001), and stride width was greater with the EXO than without it. Trunk ROM was greater and knee and ankle ROMs were less when the EXO was worn than when it was not.

It is possible that ROM with the EXO reflected an adaptive response to maintain postural control.

Gait kinetics were also affected by EXO use. Although the data were normalized to total mass, the magnitude of the 1st peak vertical GRF was higher and the time to peak shorter with than without the EXO (Table 1). Thus, damping in the EXO leg joints did not appear to be effective in preventing a high impact, quick onset jolt to the body at heel strike, possibly because of actuator timing. Additionally, peak braking force was higher and time to peak shorter with the EXO, which may be attributable to the design aspect that required the anterior-posterior control of the EXO and the load on it to be accomplished by the user. The high magnitudes of the GRFs with the EXO are likely to contribute to energy cost and the magnitude of the vertical component, in particular, suggests the possibility of increased risk of lower extremity musculoskeletal injury (Knapik et al., 1996). Unlike the vertical and the braking forces early in the stance period, the magnitudes of the 2nd peak vertical and the peak propulsive forces were lower with the EXO than without it. Although the EXO was designed to store and return some energy to assist in propelling the body and the load on it forward into the next step, the user had to raise the load, control the horizontal force components, and generate the forces to move the body forward, all of which are energy-intensive activities.

PROLONGED MARCH AND POST-MARCH PHYSICAL ACTIVITIES

When using the EXO and outfitted with the 40-kg load, participants walked on average only 5.6 km and, at times, their speed was below 4 km·h^{-1}. In some instances in which the march was not completed with the EXO, early termination was due to a heart rate in excess of 85% of age-predicted maximum or muscular discomfort. Without the EXO, participants successfully completed the 8-km march, maintaining a speed of approximately 4.8 km·h^{-1}. Performance on the post-march physical activities was not significantly affected ($p > .05$) by whether or not the EXO was worn on the march. The means for the EXO and the No-EXO conditions were essentially identical on both the obstacle course and the upper body endurance tests.

BALANCE TESTING

The limits of stability analyses indicated that the maximum extent of body lean to the left and right was significantly less with the EXO than without it, whereas forward and backward lean was unaffected by the device condition. The constrained medio-lateral movement with the EXO was likely attributable to the design of the ankle joints, which allowed extension, but not abduction or adduction. The findings for postural sway that were obtained without the EXO showed linear increases in sway measures as load weight increased, results expected based on past research into load effects (Schiffman et al., 2006). When the EXO was used, however, the range of medio-lateral movement and the total sway area changed relatively little with changes in load mass. Further, maximum movement range in the anterior-posterior direction, anterior-posterior excursion length, and planar motion all yielded values for the EXO that were significantly lower, by 20 to 35%, than those obtained when the EXO was not worn. The large base of support provided by the feet of the EXO and the bracing effect of the leg struts and joints are design aspects that may have reduced body sway when the EXO was worn.

Table 1. Means (SDs) of Variables Recorded During 8-min Walking Trials

Variable	Device		Load		
	EXO	No EXO	20 kg	40 kg	55 kg
Oxygen Consumption ($ml \cdot kg^{-1} \cdot min^{-1}$)					
O_2 Uptake-BM	32.7_A	20.4_B	23.5_X	25.9_Y	30.3_Z
($ml \cdot kg^{-1} \cdot min^{-1}$)	(5.3)	(3.2)	(6.5)	(7.1)	(7.8)
O_2 Uptake-TM	19.1_A	13.5_B	16.8_X	15.6_Y	16.3_X
($ml \cdot kg^{-1} \cdot min^{-1}$)	(1.7)	(1.0)	(3.3)	(3.1)	(3.1)
Joint Range of Motion in Sagittal Plane (degrees)					
Trunk ROM	5.5_A	4.1_B	4.5_X	4.9_X	5.0_X
	(1.4)	(1.0)	(1.8)	(1.3)	(0.9)
Hip ROM	48.7_A	48.2_A	45.0_X	48.6_X	51.7_Y
	(5.5)	(6.3)	(5.2)	(5.3)	(5.2)
Knee ROM	62.9_A	69.8_B	68.6_X	66.3_{XY}	64.2_Y
	(6.3)	(4.1)	(6.6)	(5.1)	(6.7)
Ankle ROM	24.0_A	29.4_B	26.5_X	27.0_X	26.6_X
	(2.4)	(2.4)	(4.0)	(3.2)	(3.8)
Peak Ground Reaction Forces ($N \cdot kg^{-1}$) *and Times to Peak* (% stride time)					
1st Peak	12.7_A	11.3_B	11.6_X	12.1_Y	12.3_Y
Vertical	(0.7)	(0.6)	(0.7)	(1.0)	(1.1)
Time to 1st	13.4_A	15.5_B	14.7_X	14.4_X	14.2_X
Peak Vertical	(0.9)	(1.0)	(1.3)	(1.4)	(1.4)
2nd Peak	9.9_A	10.8_B	10.5_X	10.4_X	10.1_Y
Vertical	(0.5)	(0.7)	(0.8)	(0.7)	(0.8)
Time to 2nd	47.5_A	49.4_B	48.9_X	48.1_X	48.3_X
Peak Vertical	(1.8)	(1.1)	(0.9)	(2.2)	(1.9)

Note. Within device and within load, means that do not share the same letter were found to be significantly different ($p < .05$) in ANOVAs or post-hoc tests.

Although sway was reduced with the EXO, analyses of Hurst scaling exponents obtained through application of stabilogram-diffusion analysis suggested that there was a greater tendency with than without the EXO for body sway to drift away from an equilibrium point, unchecked by the postural control system (Collins and De Luca, 1993). These findings indicate less postural stability when the EXO was worn than when it was not. The thick, semi-rigid foot plates of the EXO, along with the leg struts and joints, may have reduced users' ability to sense the position of their base of support, thereby reducing stability while standing.

CONCLUSION

The results of this study revealed that carrying heavy loads with the EXO prototype is not metabolically sustainable for more than brief periods, even by young men who are physically fit. There were a number of factors that contributed to the high metabolic cost of wear of the EXO, including the mass and its distribution on the body, design elements that altered the user's gait patterns, and the requirement that the user control the inertia of the device and the load attached to it. Further, the leg struts and the thick foot plates may have compromised body stability. Use of the EXO negatively affected prolonged marching and did not benefit physical performance after the march. Study findings emphasize the importance of quantifying the effects of emerging designs on human users to determine if exoskeletons can be effective tools to augment human performance in industrial and military environments.

REFERENCES

Collins, J.J., De Luca, C.J. (1993) "Open-loop and closed-loop control of posture. A random-walk analysis of center-of-pressure trajectories." Experimental Brain Research, 95, 308-318.

Donelan, J.M., Kram, R., Kuo, A.D. (2001) "Mechanical and metabolic determinants of the preferred step width in human walking." Proceedings of the Royal Society of London B, 268, 1985-1992.

Guizzo, E., Goldstein, H. (2005) The Rise of the Body Bots. IEEE Spectrum Website: http://www.spectrum.ieee.org/oct05/1901

Herr, H. (2009) "Exoskeletons and orthoses: classification, design challenges and future directions." Journal of NeuroEngineering and Rehabilitation, 6:21. Website: http://www.jneuroengrehab.com/contents/6/1/21

Kazerooni, H., Steger, R. (2006) "The Berkeley lower extremity exoskeleton." Transactions of the ASME, Journal of Dynamic Systems, Measurement, and Control, 128, 14-25.

Knapik, J., Harman, E., Reynolds, K. (1996) "Load carriage using packs: A review of physiological, biomechanical, and medical aspects." Applied Ergonomics, 27, 207-216.

Martin, P.E. (1985) "Mechanical and physiological responses to lower extremity loading during running." Medicine and Science in Sports and Exercise, 17, 427-433.

480

Owings, T.M., Pavol, M.J., Foley, K.T., Grabiner, M.D. (2000) "Measures of postural stability are not predictors of recovery from large postural disturbances in healthy older adults." Journal of the American Geriatric Society, 48, 42-50.

Prieto, T.E., Myklebust, J.B., Hoffman, R.G., Lovett, E.G., Myklebust, B.M. (1996) "Measures of postural steadiness: Differences between healthy young and elderly adults." IEEE Transactions on Biomedical Engineering, 43, 956-966.

Rodrick, D., Karwowski, W. (2006) "Manual materials handling," in: Handbook of human factors and ergonomics (3rd ed), Salvendy, G. (Ed.). Pp. 818-854.

Schiffman, J.M., Bensel, C.K., Hasselquist, L., Gregroczyk, K.N., Piscitelle, L. (2006) "Effects of carried weight on random motion and traditional measures of postural sway." Applied Ergonomics, 37, 607-614.

Soule, R.G., Pandolf, K.B., Goldman, R.F. (1978) "Energy expenditure of heavy load carriage." Ergonomics, 21, 373-381.

Walsh, C.J., Endo, K., Herr, H. (2007) "A quasi-passive leg exoskeleton for load-carrying augmentation." International Journal of Humanoid Robotics, 4, 487-506.

Chapter 49

Using a GPS-Based Tactile Belt to Assist in Robot Navigation

Linda R. Elliott, Elizabeth S. Redden, Rodger A. Pettitt

Army Research Laboratory
Human Research and Engineering Directorate
US Army Infantry Center
Fort Benning, GA, 31905, USA

ABSTRACT

Researchers from the U.S. Army Research Laboratory have conducted a series of experiments aimed at reducing the size of robotic controllers for use by dismounted warfighters. The goal of the research is to reduce the size and weight of the robotic controllers without adversely affecting the human robotic interface. The specific goal of the current experiment was to investigate two alternative robot controller navigation map display configurations with the potential to replace a larger split screen display that presents both a map display and a camera-based driving display side by side on a 6.5 inch screen. The first alternative was a 3.5 inch display that allowed the operator to toggle back and forth between the driving display and the map display. The second alternative added a torso-mounted tactile display to the toggle-based display in order to provide direction information simultaneously with the camera display. Each display option was evaluated based on objective performance data, expert-based observations, and questionnaire items. Findings indicated that operators' navigation performance with the tactile-supported 3.5 inch toggle display was as effective as their performance with a 6.5 inch split screen display. Operator performance was significantly lower with the 3.5 inch toggle display that did not have the tactile display.

Keywords: Tactile display; Multisensory display; Robot controller; Robot navigation; Army robots; Soldier performance

INTRODUCTION

Ground robots have proven to be extremely useful in Army combat operations, to identify potential improvised explosive devices (IEDs), dispose of explosive ordnance, clear mines, and explore territory. When U.S. forces went into Iraq, the original invasion had no robotic systems on the ground. By the end of 2004, there were 150 robots on the ground in Iraq; a year later there were 2,400; by the end of 2008, there were about 12,000 robots of nearly two dozen varieties operating on the ground in Iraq (Singer, 2009). While some of these robots were controlled by operators in vehicles, many were controlled by dismounted warfighters who had to transport the robotic equipment without the aid of motorized vehicles. Dismounted warfighters are already overburdened by the weapons and equipment they must carry on their missions and the additional weight of a large robotic controller increases that burden. Thus, it is important to reduce the size and weight of the controller so that dismounted warfighters can continue to fight and move effectively, even when they are carrying robotic controllers.

This report describes the third in a series of experiments investigating this issue. One experiment compared handheld and weapon-mounted controller options for the dismounted soldier to drive the robot and maneuver the robotic arm (Pettitt, Redden, Carstens, and Elliott, 2008). Another experiment addressed screen size for the dismounted soldier's driving camera display (Redden, Pettitt, Carstens and Elliott, 2008) and showed that a 3.5 inch camera display was as effective as a 10.5 inch camera display. A 3.5 inch screen was found to be effective for a camera display alone; however, operators must also view a map display in order to direct their asset when it is beyond line of sight.

Currently, many robotic controllers show both the camera and map display in a split screen. Since it has been shown that a 3.5 inch screen is sufficient for driving, it would stand to reason that another 3.5 inch screen could be used to show a map display. However, this configuration results in a larger screen size. A smaller option would be to use only one 3.5 inch display and have the operator toggle between the driving camera display and the map display. A question arises concerning whether or not the operator can efficiently drive and navigate using this sequential viewing option. A more intriguing option is to provide map information via another sensory channel (e.g., the tactile channel) so that driving and map information can be accessed simultaneously without requiring an additional visual display. A map display provides location information (i.e., "it" is here); however, it is often used simply to ascertain the correct direction of movement (i.e., "which way"). While a tactile display will not provide location information, it has been shown to effectively provide direction information. Thus, to reduce toggling back and forth from the driving camera view to the map view and the potential loss of situation awareness while switching, a tactile navigation system was added to the toggle display in order to provide direction cues while simultaneously viewing the driving camera display. The operator of this system also retains the ability to toggle to the map display to ascertain location information.

Torso-mounted tactile arrays have been used successfully for personal land navigation (Elliott, van Erp, Redden, and Duistermaat, in press). Displaying navigation information to a different sensory channel allowed simultaneous navigation and cross-country movement without visual overload. In this study condition, torso-mounted tactile cues would provide directional guidance toward the next waypoint in a manner that would enable the soldier to head toward the next waypoint, while using the camera display to drive the robot and negotiate obstacles. Because the information is presented simultaneously, we expected the tactile supported display to be as effective as the split screen and more effective than the toggle screen.

METHOD

This study investigated the effects of three display options on driving, navigating, and performing local surveillance with a small robotic vehicle (TALON) (see figure 1). It took place at Fort Benning, GA and used 33 soldiers from the Officer Candidate School (OCS).

FIGURE 1. TALON robot.

INSTRUMENTS AND APPARATUS

TALON robot.

The TALON is a lightweight robot designed for missions ranging from reconnaissance to weapons delivery. The suitcase-portable robot is controlled through a two-way radio frequency line from a portable operator control unit (OCU). It was developed for the Explosive Ordnance Disposal Technology

Directorate of the U.S. Army's Armament Research, Development, and Engineering Center at Picatinny Arsenal, New Jersey. For this experiment, the TALON was equipped with a video camera that enabled soldiers to maneuver the vehicle and assess enemy activity along the route to the objective. It was also equipped with a global positioning system (GPS) that informed the navigation display of its position and orientation.

TALON OCU.

The tablet hardware used to present the 3.5 in. map and camera display and the split screen display was the AMREL Rocky DR7-M. It is 9.8 in. by 7.4 in. and weighs 1.1 kg, including the battery. It is a 1024 by 768 extended graphics array (XGA). A 6.5-inch display provided a split screen capability. The top portion of the screen presented the driving camera information and the bottom portion presented the map. When in the toggle condition, only the top screen was used to display either the camera view or the map (see figure 2).

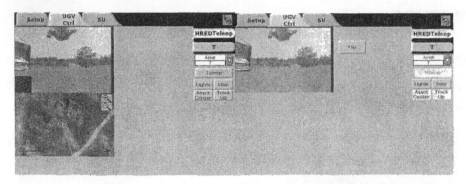

FIGURE 2. OCU split screen and toggle displays

Tactile System.

Figure 3 shows the tactile navigation system used in this study. It was developed for a Defense Advanced Research Project Agency (DARPA) project, using specialized tactors developed to be easily and accurately perceived. Unlike the usual inertial shaker motor found in most cell phones, the C-2 vibrotactile transducer was developed to provide a stronger and more precise signal (Mortimer, Zets, and Cholewiak, 2007). The C-2 tactors are 30 mm diameter, 8 mm deep. The housing serves as a surround that simultaneously contacts the skin and is separated from the moving contactor. The diameter of the moving element is 7 mm, centered in a 9-mm hole in the top surface of the tactor. The tip of the contactor protrudes 0.5 mm above the surface of the surround to ensure firm contact with the skin. Redden, Carstens, Turner, and Elliott (2006) compared the C-2 tactors to standard inertial shaker motor tactors using soldiers in static positions and while performing strenuous movements (obstacle course) in the field. They found that localization was more precise with the C-2 tactors. The tactile belt system was further modified to provide GPS-based direction cues to aid navigation. The display itself consisted of eight tactile drivers that corresponded to the eight cardinal directions.

The control unit, weighing less than 0.2 kilograms, received wireless signals from the GPS onboard the TALON, compared it to the waypoint GPS location, and calculated signals in order to provide the operator the appropriate direction cue in the most intuitive manner. When the TALON was going in the correct direction, the tactor at the front of the soldier's waist activated (200 ms on and 1800 ms off). The object was to drive the robot in the direction that kept the front tactor active. When the TALON was within 5 meters of the waypoint, the tactor's pulse increased to 100 ms on, 200 ms off, 100 ms on, 600 ms off. When it moved to within two meters of the waypoint, the waypoint was considered to be achieved and all the tactors activated for 3000 ms. If a soldier went off course more than two meters, the tactor that corresponded with a "steer to" direction for returning to the correct course route activated at 200 ms on and 1800 ms off.

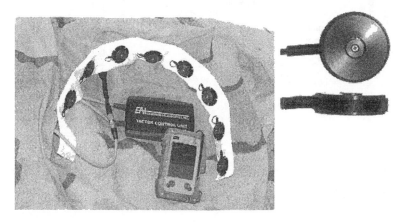

FIGURE 3. Tactile navigation system

Navigation course.

Figure 4 provides the layout of the navigation course. The experiment course consisted of three different lanes, which allowed the soldier to drive on a different lane with each display condition. The total length of each lane was approximately 100 meters. Each experiment navigation lane had three "legs" with different task demands. The first leg of each lane was a marked path approximately 30 meters long. A complex obstacle was placed at the end of the marked path that required the operator to navigate around it. The obstacle did not show up on the map display so the operator was required to drive around the obstacle and continue to the end waypoint (objective). The second leg of the course required the operator to drive as quickly and efficiently as possible using the GPS feedback to the final waypoint. Hand signals were presented by a data collector who was moving with the vehicle to monitor driving errors. The hand signals were presented up to 10 different times (depending on how quickly the operator navigated the course) during the final leg of the course. The data collector ensured that each signal was presented in front of the driving camera so that it could be clearly seen by the TALON operator if he was attending to the driving camera. Soldiers reported when they saw the hand signal and which specific signal they saw to the data collector sitting next to them in the tent who was also timing the course completion. Soldiers tele-operated the TALON from inside a tent during training and actual navigation course negotiation which prevented them from tele-operating the vehicle using line of sight rather than the driving camera display.

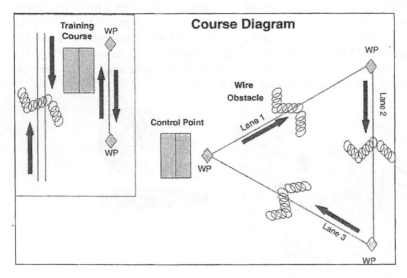

FIGURE 4. Robot navigation course

PROCEDURES

After training on the operation of the TALON control system, each soldier completed operational exercises using each of the three different display concepts. The terrain, targets, and hazards were counterbalanced along with display condition to control for order effects, such as practice, learning, boredom, or fatigue. The display options were evaluated for effects on task performance, workload, and usability, based on objective performance data, data collector observations, and soldier questionnaires. More extensive detail, pictures, and descriptions are provided in a technical report (Redden, Pettitt, Carstens, and Elliott).

RESULTS

PERFORMANCE MEASURES

Table 1 provides mean overall navigation times, driving errors, and visual ID errors for each condition.

Table 1. Means and standard deviations for overall navigation time, number of driving errors, and number of correct visual identifications of hand signals.

	Overall Navigation Time	
	Mean (minutes)	Std. Dev.
A. Split screen	6.10	2.55
B. Toggle screen	7.49	2.59
C. Toggle screen/tactile belt	6.43	2.62

	Driving Errors	
	Mean errors	Std. Dev.
A. Split screen	0.16	0.47
B. Toggle screen	0.12	0.33
C. Toggle screen/tactile belt	0.08	0.28

	Visual Identification	
	Mean correct	Std. Dev.
A. Split screen	96.8	8.5
B. Toggle screen	90.0	13.2
C. Toggle screen/tactile belt	93.2	11.4

Navigation time.

Table 2 indicates the differences between the split screen and the toggle and between the toggle/tactile and the toggle conditions were significant ($p < .01$). The split screen and the toggle/tactile conditions had faster course completion times than the toggle condition. The difference between the split screen and the toggle/tactile course times was not significant.

Table 2. Planned comparisons of course completion times. .

Pair	t	df
Split vs Toggle	3.93*	24
Split vs Toggle/tactile	0.78	24
Toggle vs Toggle/tactile	3.16*	24
*$p < .01$, 1-tailed		

Driving errors.

The mean number of driving errors with each display type is shown in table 1. In each condition, participants averaged less than one error per course completion. Because all conditions were associated with extremely low numbers of errors, differences among the means were not statistically significant.

Visual identification.

Table 1 shows the mean percentage of hand signals correctly identified with each display type. (The data are presented as percentages because the number of hand signals presented varied from 5 to 10). The planned comparisons can be found in Table 3. Table 3 indicates a significant difference between split screen and toggle (split screen had a higher percentage of correct identifications), and no significant difference for the other comparisons.

Table 3. Planned comparisons of visual identification performance

Pair	t	df
Split vs Toggle	3.44*	24
Split vs Toggle/tactile	1.82	24
Toggle vs Toggle/tactile	1.62	24
a: $p < .01$, 1-tailed		

Questionnaire-based ratings.

Questionnaire-based ratings. The soldiers rated each display using a 7 point semantic differential scale (1 = extremely ineffective to 7 = extremely effective). Items regarded how well the soldier performed tasks, such as moving and navigating at different speeds, visual identification (e.g., terrain, hand signals, etc.), interpreting the visual map display, navigating and driving at the same time, navigating around obstacles, and overall awareness of surroundings. Task difficulty ratings for the navigation course were generally worse for the toggle display than for the split screen or tactile displays, such that the toggle display had the lowest rating for 16 out of 17 items. The differences among the displays were not large for most comparisons; however, differences were noteworthy, with toggle being the lowest, for (a) being able to make course corrections, (b) being able to identify if you are on the right course, (c) being able to navigate and drive at the same time, (d) being able to recognize hand signals, and (e) overall ability to perform. Ratings were similar for the split screen and tactile conditions, with little difference in ratings and no trend favoring one over the other.

DISCUSSION

The results from this experiment demonstrate that adding a tactile belt to a robotic control unit for beyond line of sight navigation could be useful as a means of decreasing display size. Soldiers performed equally as well with the smaller display size that included a tactile navigation belt as they did with the larger split screen display.

Sequential operation of a visual driving camera display and a visual map display (i.e., toggle condition) resulted in longer course completion times compared to the split screen and toggle/tactile condition. The longer course completion times for the toggle condition suggest more effort was needed to maintain driving performance. Participants also identified fewer hand signals with the toggle display than with the split screen display. While differences in ratings of workload were not significant, questionnaire-based results indicate that soldiers had more difficulty performing tasks with the toggle display and greatly preferred the toggle/tactile and split screen conditions to the toggle condition.

In summary, this effort was a field-based evaluation of three types of displays, conducted to assess the feasibility of reducing display size of a robot controller used for beyond line of sight navigation. Results suggest that a toggle/tactile display may indeed enable a smaller display size while maintaining performance.

REFERENCES

Elliott, L., van Erp, J., Redden, E., & Duistermaat, M. (In press) Field-based Validation of a Tactile Navigation Device. IEEE Transactions on Haptics.

Mortimer, B., Zets, G., Cholewiak, R. (2007) Vibrotactile transduction and transducers. J Acoust Soc 121(5):2970–2977

Pettitt, R., Redden E., Carstens, C., Elliott, L. (2008) Scalability of robotic controllers: an evaluation of alternatives. ARL-TR- 4457, Army Research Laboratory, Aberdeen Proving Ground.

Redden, E., Pettitt, R., Carstens, C., Elliott, L. (2008) Scalability of robotic displays: display size investigation. ARL-TR-4456, Army Research Laboratory, Aberdeen Proving Ground.

Redden, E., Pettitt, R., Carstens, C., Elliott, L. (2009) Scaling robotic displays: Visual and multimodal options for navigation by dismounted soldiers. ARL technical report, Army Research Laboratory, Aberdeen Proving Ground.

Redden, E., Carstens, C., Turner, D., Elliott, L. (2006) Localization of tactile signals as a function of tactor operating characteristics. ARL-TR-3971, Army Research Laboratory, Aberdeen Proving Ground.

Singer, P. W. (2009). Wired for War: The Robotics Revolution and Conflict in the 21st Century. Transcrpt from a presentation to the Carnegie Council. http://www.cceia.org/resources/transcripts/0114.html

Chapter 50

Development of a Remote Medical Monitoring System to Meet Soldier Needs

William J. Tharion, Mark J. Buller, Anthony J. Karis, Reed W, Hoyt

U.S. Army Research Institute of Environmental Medicine
Natick, MA 01760-5007, USA

ABSTRACT

Medical monitoring systems for military use have unique requirements. The purpose of this evaluation was to determine if changes in system designs improved the fit, comfort, durability, impact on military performance, impact on the body, and overall acceptability of the system. This was accomplished though an iterative process of five studies and four versions of the system. Information from soldiers obtained from these evaluations was provided to the materiel developers to improve the form, fit, and function of this system. The resulting system showed progressive improvements in comfort, durability and acceptability through a reduction in size and improvements in design and materials used.

Keywords: Physiological Status Monitoring, Systems Engineering, Human Factors Design, Wearable, Comfort

INTRODUCTION

The use of physiological monitoring systems may reduce the frequency and severity of injuries to soldiers by providing medical situational awareness during training or actual operational activities. This study examined the development of a medical monitoring system for soldiers; the Hidalgo Equivital VSDS EQ-01 (Hidalgo Ltd., Cambridge, UK), a Food and Drug Administration (FDA) 510k certified device. The VSDS EQ-01 reliably measures heart rate and respiration rate (Tharion et al., 2008), but soldiers found it uncomfortable to wear for extended periods (Tharion et al., 2007).

Physiological monitoring systems can provide useful information, but they must also be comfortable, easy to use, and work reliably in the specific environment for which they are intended to be used (Paradiso et al., 2005). Military environments pose unique demands and certainly differ from home health monitoring or other civilian ambulatory monitoring environments. Sensors embedded within clothing have been shown to be comfortable to the user (Paradiso et al., 2005); however, some of these systems increase the risk of thermal strain because of the insulation factor associated with the added clothing. Additionally, some systems may prove comfortable, but the type or quality of the data obtained is not adequate for medical monitoring of military personnel in harsh environments. The system needs to be small, lightweight, unencumbering, and compatible with other military equipment and clothing worn, easy to clean, capable of functioning over many hours, have low power consumption, ensure privacy of the data, and be of reasonable cost (Pantelopoulos and Bourbakis, 2008).

METHODS

This evaluation used experienced military personnel (n=154) in five studies (Table 1). Soldiers had one or more deployments to Iraq or Afghanistan, had regular chemical biological, radiological, nuclear (CBRN) training, or been engaged in elite small unit operations training as Army Rangers or Special Forces soldiers. Prior to the start of each study, participants were briefed on the purpose of the study and the associated risks and benefits. They were informed of their right to withdraw at any time. Participants gave their written informed consent prior to wearing the system or providing any data. These studies were approved by the Scientific and Human Use Review Committees at the U.S. Army Research Institute of Environmental Medicine. All participants were also briefed on the potential for the VSDS to be used more broadly as a medical monitoring device.

Four versions of the VSDS were evaluated during five military training exercises. The first study at Ft. Polk had infantry soldiers wear the VSDS for approximately 8 hours. Training included wearing Interceptor Body Armor (IBA) and load carriage equipment (i.e., rucksack etc.) during simulated combat scenarios. Volunteers slept during the exercise while wearing the VSDS. The second study was at Aberdeen Proving Grounds and used infantry soldiers participating in military operations in an urban terrain (MOUT). They wore the VSDS for over 90 hours, including one 23-hour sustained operation. Activities included an approach march, room clearing,

Table 1: Study description

VSDS Version #	Personnel and Study Duration	n	Location
1	Infantry Soldiers Duration: 8 Hrs	8	Ft. Polk, LA
2	Infantry Soldiers Duration: 95 Hrs	26	Aberdeen Proving Grounds, MD
3	Civil Support Team – Weapons of Mass Destruction Duration: 4 Hrs	12	North Brookfield, MA
3	Ranger Training Brigade Students Duration: 4 Hrs	77	Ft. Benning, GA
4	Special Forces Students Duration: 90 Hrs	31	Camp McCall, NC

combat in close quarters, and decision-making tactics under various enemy threat levels. The third study was with Civil Support Teams – Weapons of Mass Destruction (CST-WMD) Army National Guard personnel. They wore the VSDS while participating in a search and rescue operation in an enclosed space environment. Activities included obtaining samples of simulated CBRN material while wearing personal protective equipment (PPE). They also rescued simulated casualties. The fourth study used Ranger school students at Ft. Benning, GA who wore the VSDS for approximately 4 hours while participating in a timed road march with a weighted backpack. The last study was with students participating in the Special Forces Small Unit Tactics (SUT) course who wore the VSDS for approximately 90 hours over 10 days including one 24-hour sustained operation at Camp McCall, NC. Volunteers wore a variety of military equipment during their combat training including body armor and load carrying equipment.

The four versions of the VSDS tested are shown in Table 2. The belt has electrocardiograph (ECG) sensors to record heart rate, sensors to detect expansion and contraction of the belt to measure respiration rate, and a skin temperature sensor. The sensor electronics module (SEM) is made of hard plastic and snaps onto the belt in the center of the chest and receives data from the belt sensors. It also has accelerometers to detect body motion and body position. An accompanying health hub worn in uniform pockets, backpacks, etc. was used with the first two VSDS versions. Information from the SEM was transmitted to the health hub using a body area network. The health hub weighed approximately 340 g and measured 12 X 8 X 4 cm. Off-system sensors such as an ingestible thermometer pill that records core body temperature transmitted information directly to the health hub. Software to assess thermal, cognitive capability, hydration, and life sign states was housed in the health hub. The health hub was used to turn the system on, begin data collection and provides the radio network

Table 2: Vital sign detection system (VSDS) description

Version Number	Year Used	System	Change from Previous Version
1	2006	+ Health Hub	
2	2006	+ Health Hub	1. Change belt fabrics 2. Change in stitching 3. Velcro straps used to hold SEM to belt
3	2007-2008		1. Health hub functions incorporated into SEM 2. Belt-SEM connector made more flexible – negating need for side Velcro straps 3. Softer belt fabrics 4. Change in stitching 5. Plastic belt adjuster replaced with small hook adjusters. 6. Securing bungs to SEM
4	2009		1. Added a prototype heat flux sensor

link for transmission of information to remote computers or other devices. The health hub was rendered obsolete in VSDS Versions 3 and 4 where the functions were assumed by the SEM. Once training was completed while wearing the VSDS, a survey was immediately issued to obtain some basic background information (e.g., time in the military, number of deployments etc.) and responses to questions regarding fit, comfort, durability of the system, impact on military performance, physical impact on the body, and overall acceptability. Five or seven-point Likert rating scales were used. For example when asked about how loose or tight the device was, a scale of very tight = 1, neither tight nor loose = 4, and very loose = 7 was used. Yes/no responses (e.g., would you wear the system if it would help save your life?) and open-ended questions (e.g., was the system acceptable to wear, and if not, why not?) were also used. Analyses of variance (ANOVAs) with Tukey's tests were used to determine differences in ratings between system versions with rating scale questions. Frequencies of open-ended and yes/no questions were analyzed with a chi-square test. All data are presented as means ± standard deviations.

RESULTS

The participants from the five studies had 5.7 ± 4.7 yrs of service and were 26.8 ± 5.2 yrs of age. The mean total number of hours the VSDS was worn was 23.8 ± 40.6 hours. Participants in the Aberdeen study (Study 2) wore the system for the longest period of time, a total of 94.6 ± 34.6 hours. The CST-WMD and RTB (studies 3 and 4) participants only wore the system for about 4 hours. A summary the total length of time the VSDS was worn in each study is presented in Table 1.

FIT

When participants were asked to rate the fit of the system, significant differences between system versions existed (p<0.001). Table 3 demonstrates the mean ratings of overall fit of the system using a seven point like-dislike Likert scale. The VSDS felt tight on volunteers but three participants mentioned it became loose over time.

Table 3: Overall fit ratings

VSDS Ver. 1 Ft. Polk	VSDS Ver. 2 Aberdeen	VSDS Ver. 3 North Brookfield	VSDS Ver. 3 Ft. Benning	VSDS Ver. 4 Camp McCall
2.9 ± 1.5^a	3.9 ± 1.6^b	5.9 ± 1.0^c	5.7 ± 1.3^c	5.1 ± 1.5^c

Values across the row with different letter superscripts are significantly different from one another as assessed by Tukey's Test at $p \leq 0.05$. 1 = Dislike Very Much 2= Dislike Moderately, 3= Dislike Slightly, 4 = Neither Like nor Dislike, 5 = Like Slightly, 6 = Like Moderately, 7 = Like Very Much.

COMFORT

Significant differences between system versions existed ($p < 0.001$) in overall comfort. Table 4 shows significant differences among VSDS Versions 1, 2, and 3 or 4, but no difference between versions 3 and 4. Only the first two studies had

Table 4: Overall comfort ratings

VSDS Ver. 1 Ft. Polk	VSDS Ver. 2 Aberdeen	VSDS Ver. 3 North Brookfield	VSDS Ver. 3 Ft. Benning	VSDS Ver. 4 Camp McCall
2.0 ± 0.8^a	3.4 ± 1.2^b	5.5 ± 1.2^c	5.5 ± 1.5^c	4.9 ± 1.7^c

Values across the row with different letter superscripts are significantly different from one another as assessed by Tukey's Test at $p \leq 0.05$. 1 = Very Uncomfortable 2= Moderately Uncomfortable, 3= Slightly Uncomfortable, 4 = Neither Comfortable nor Uncomfortable, 5 = Slightly Comfortable, 6 = Moderately Comfortable, 7 = Very Comfortable.

participants sleep while wearing the system. The VSDS in these studies showed slightly more uncomfortable ratings when trying to sleep (VSDS Version 1 (Study 1): 1.8 ± 0.8, VSDS Version 2 (Study 2): 3.0 ± 1.7) compared to the overall ratings VSDS Version 1 (Study 1): 2.0 ± 0.8, VSDS Version 2 (Study 2): 3.3 ± 1.2).

The comfort ratings of the various belt components which included the cloth electrodes, water proof material that surrounds the electrodes, etc., were also assessed. Only Version 1 had components that received ratings lower than slightly uncomfortable. Those components were the adjuster (3.0 ± 1.5), belt fastener (2.9 ± 1.6), belt stitching (1.9 ± 1.1), and belt elastic (1.9 ± 1.1). No component parts in any other versions received ratings of 3.0 or lower. When volunteers were asked what activities were most uncomfortable while wearing the system, the following activities and number of respondents mentioning that activity across all five studies were: doing prone activities such as shooting or low crawling (n = 7), wearing body armor (n = 6), road marching (n = 6), doing land navigation (n=4), rappelling (n = 3), and when sweating (n = 3). A total of 108 different activities or times when the system was most uncomfortable were cited. It should be noted that not all groups did certain activities. For example only 32 individuals slept in the system, whereas 103 individuals participated in a road march while wearing the system.

DURABILITY OF THE SYSTEM

Volunteers and research staff recorded if the system broke or stopped functioning during testing. There were a significantly higher number of failures for VSDS Versions 1 and 2 compared to Versions 3 and 4 (chi-square: $p < 0.001$) (Table 5). Three system failures in VSDS Version 1 were because the SEM became unsnapped from the belt. Most problems concerning durability were with VSDS Version 2. However that study was the longest and most physically demanding evaluation, resulting in the greatest challenge to the systems. The two system failures in Versions 3 and 4 were due to the SEM becoming unsnapped from the belt.

Table 5: Percent of system failure

VSDS Ver. 1 Ft. Polk	VSDS Ver. 2 Aberdeen	VSDS Ver. 3 North Brookfield	VSDS Ver. 3 Ft. Benning	VSDS Ver. 4 Camp McCall
37.5%	69.2%	0.0%	1.3%	3.4%

For VSDS Version 2 there was a 50% failure rate with the health hubs. The hub was worn either inside a pocket of the Camelbak drinking system (Camelbak, Petaluma, California) or in a small pouch fastened to the soldier's belt. There was a 26% failure rate with the SEMs. Two common failures were that the units could not be turned on, or that the bungs (a small rubber-plastic device that covers some of the electronic pins and acts as the on/off switch) fell out during the exercise. Thirty-nine percent of SEMs became detached from the belt as the snap at the sides of the SEM unsnapped. Three belts had at least one torn metal snap out of the five that are normally present. The other belt had torn foam (used for padding and comfort) near the center of the belt.

IMPACT ON MILITARY PERFORMANCE

Volunteers were asked to rate the impact of wearing the VSDS on military performance. Table 6 shows the ratings by version and study when wearing the VSDS and Advanced Combat Uniform (ACU) alone. Only two studies with the first two versions of the VSDS evaluated impact on military performance when wearing body armor. These ratings show a slightly negative to moderately negative impact on performance: Version 1 (Ft. Polk) rating 3.9 ± 1.5 and Version 2 (Aberdeen) rating 3.4 ± 1.6 on the 1 to 5 point scale (1 = extreme negative impact; 5 = no negative impact). The only other activity that showed ratings below "slightly negative impact" was for military activities performed in the prone position. No significant differences in the performance of activities in the prone position were evident across the various versions of the system.

PHYSICAL IMPACT ON THE BODY

Impact the system had on the body differed across VSDS versions ($p < 0.001$) (Table 7). No differences between Versions 3 and 4 were seen and impact was negligible. The primary impact was with Versions 1 and 2. Discomfort caused by skin irritation was reported in over 90% of Soldiers wearing these versions of the system. The central belt area and the adjustment buckle were the primary areas of complaint. Complaints included skin irritation, redness, sensitivity, abrasion, acne, prickly heat and extreme sweating near the system. Version 3 used an adjustment fastener with small bra-type hooks instead of the plastic buckle.

Table 6: Impact on overall military performance and while performing activities
in the prone position

VSDS Ver. 1 Ft. Polk	VSDS Ver. 2 Aberdeen	VSDS Ver. 3 North Brookfield	VSDS Ver. 3 Ft. Benning	VSDS Ver. 4 Camp McCall
Overall impact on military performance				
4.5 ± 0.6^b	4.1 ± 1.2^a	4.7 ± 0.5^c	4.8 ± 0.6^c	4.8 ± 0.5^c
Impact on military performance while performing activities in the prone position				
3.3 ± 2.6^b	3.2 ± 1.6	3.8 ± 0.5	NA	3.3 ± 2.2

Values across the row with different letter superscripts are significantly different from one another as assessed by Tukey's Test at $p \leq 0.05$. 1 = Extreme Negative Impact, 2= Very Negative Impact, 3= Moderate Negative Impact, 4 = Slight Negative Impact, 5 = No Negative Impact NA = Not Applicable

Table 7: Impact on the body of wearing the system

VSDS Ver. 1 Ft. Polk	VSDS Ver. 2 Aberdeen	VSDS Ver. 3 North Brookfield	VSDS Ver. 3 Ft. Benning	VSDS Ver. 4 Camp McCall
3.0 ± 1.5^a	3.7 ± 1.2^b	4.8 ± 0.4^c	4.9 ± 0.2^c	4.9 ± 0.3^c

Values across the row with different letter superscripts are significantly different from one another as assessed by Tukey's Test at $p \leq 0.05$. 1 = Extreme Negative Impact, 2= Very Negative Impact, 3= Moderate Negative Impact, 4 = Slight Negative Impact, 5 = No Negative Impact

Other changes that occurred between Versions 2 and 3 included changing to a softer belt fabric and changing the stitching material, type, and pattern.

OVERALL ACCEPTABILITY

Volunteers were asked if the system were acceptable to wear for extended periods of time. There was a significant chi-square ($p < 0.001$) difference among versions (Table 8). Only 50% or less of those wearing Versions 1 and 2 found the system acceptable, whereas over 80% of those wearing Versions 3 and 4 found it acceptable. After explaining the possibility that use of the system may potentially save lives, volunteers were asked if they would wear the present system if it was shown to aid in prevention or treatment of injuries that could be life-threatening. A significant chi-square ($p < 0.005$) showed differences by version in frequency of volunteers who would wear the system for life-saving purposes. Less than 50% of those who wore Versions 1 and 2 said they would wear it, while over 80% of those who wore Versions 3 and 4 said they would wear the system. Those who said they would not wear the system were asked why they wouldn't wear it. The leading reasons were that they did not believe the system would actually help save their life,

Table 8: Percent of system acceptability

VSDS Ver. 1 Ft. Polk	VSDS Ver. 2 Aberdeen	VSDS Ver. 3 North Brookfield	VSDS Ver. 3 Ft. Benning	VSDS Ver. 4 Camp McCall
Overall Acceptability of the System – Acceptable to Wear For Extended Periods of Time?				
50.0%	37.0%	91.7%	92.0%	83.9%
System Acceptable if it Saved Your Life?				
62.%%	85.2%	100%	94.7%	74.2%

or it was too uncomfortable to wear even if it could potentially save their life.

DISCUSSION

These results demonstrate improvements in VSDS product quality achieved by providing feedback to the manufacturer after field testing each version. These tests were varied and some of the response differences between studies could have resulted from differences in the military scenarios volunteers perform, or type of volunteers who wore the system. For example, Version 3 of the system generally had the highest ratings, but participants in those studies wore the system for the shortest amount of time. In addition, Version 4 of the system had a prototype heat flux sensor added as a "dongle" (a heat flux disk attached by a wire to the SEM). Any final VSDS product would eliminate any wires and the heat flux sensor would be embedded in the belt. However, this wire and the length of time wearing the system may explain the overall lower ratings of the Version 4 system compared to Version 3.

The overall ratings in all measures – fit, comfort, durability, impact on military performance, impact on the body, and overall acceptability of the system – improved most substantially from Version 2 to Version 3. The most dramatic change to the VSDS also occurred between these two versions, which incorporated all of the functions from the health hub into the SEM. This reduced the added bulk and eliminated a separate piece of the system that was most susceptible to mechanical failure. At this time, some small removable parts to the SEM were secured and no longer made removable. This eliminated the possibility of them falling out while in use. The center chest piece on the belt where the SEM attached was also made slightly more flexible, which reduced the likelihood of the SEM becoming unsnapped from the belt. Another major improvement was the elimination of the plastic adjustment buckle which caused skin chafing. It was replaced that with an adjustment fastener with small hooks like those used on women's brassieres.

After the first two tests it was known that the SEM needed to be reduced in size, a feasible but expensive proposition. Efforts are currently underway to reduce the size of the SEM and implement various firmware upgrades. A prototype, soon to be tested is shown in Figure 6. These changes were initiated in response to direct feedback from soldiers, thereby demonstrating how human factors research can positively influence the design of a medical monitoring system for soldiers.

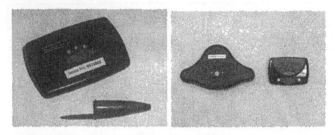

Figure 6. Prototype of the sensor electronics module of the EQ-02.

ACKNOWLEDGEMENT & DISCLAIMER

The authors would like to thank Dr. John Castellani and CPT Lee Margolis who served as co-principal investigators of the Camp McCall study. The views, opinions and/or findings in this paper are those of the authors and should not be construed as an official Department of the Army position, policy or decision. Citation of commercial organizations and trade names in this report does not constitute an official Department of the Army endorsement or approval of the products or services of these organizations.

REFERENCES

Pantelopoulus, A. Bourbakis, N. (2008). "A survey on wearable biosensor systems for health monitoring", proceedings of the 30th Annual IEEE EMBS Conference, Vancouver, Canada, pp. 4887-4890.

Paradiso, R. Loriga, G. Taccini, N. (2005) A wearable health care system based on knitted integrated sensors. IEEE Transactions on Information Technology in Biomedicine, 9: 337-344.

Tharion, W.J., Buller, M.J., Karis, A.J., Mullen, S.J. (2007) "Acceptability of a wearable vital sign detection system", proceeding of the Human Factors and Ergonomics Society 51st Annual Meeting, Baltimore, MD, pp. 1006-1010.

Tharion, W.J., Buller, M.J., Karis, A.J. Mullen, S.J. (2008) "Reliability and validity of a wearable heart rate and respiration rate sensor system" abstracts of the Experimental Biology Meeting, San Diego, CA, no. 1175.7.

Chapter 51

Investigating the Performance of an Autonomous Driving Capability for Military Convoys

James Davis[1], Ed Schoenherr[2]

[1] U.S. Army Research Laboratory (ARL)
Human Research and Engineering Directorate
Aberdeen Proving Ground, MD 21005

[2] U.S. Tank Automotive Research, Development and
Engineering Command (TARDEC),
Warren, MI 48397

ABSTRACT

Force Protection measures impose additional demands on convoy vehicle operators, potentially degrading vehicle control, reducing local area awareness, and impacting overall mission completion. Autonomous driving technologies provide one means of reducing the demands on convoy vehicle operators. The Convoy Active Safety Technologies (CAST) system is one such technology designed to reduce mobility demands on convoy vehicle operators and improve overall convoy performance. Through a series of experiments and demonstrations, the CAST system has demonstrated its usefulness in augmenting convoy driving and allowing the vehicle operator to allocate more attention to local area security tasks. This paper outlines the CAST system's performance for two experiments, highlighting the importance of addressing human factors concerns to achieve optimal performance.

Keywords: Convoy, Autonomous, Semi-autonomous

INTRODUCTION

Sustainment convoys are critical to providing combat commanders with the right support, at the right time and place, and in the right quantities, across the full range of military operations. The ability to conduct sustainment convoys in a variety of hostile environments requires Force Protection measures that address the enemy threat and protect the Soldier. Critical among these measures is maintaining a high degree of awareness in order to detect and identify potential threats as early as possible. These additional security measures impose additional demands on vehicle operators, potentially degrading vehicle control, reducing local area awareness, and impacting overall mission completion.

The military's initiative to reduce manpower has limited the solution space for reducing the task load associated with military driving. Interfaces that are more intuitive for the convoy drivers have afforded some reductions in task load by reducing the mental effort associated with missions tasks. Also, manpower, personnel, and training analyses have provided more effective allocations of tasks among the crewmembers. However, research has shown that task load is still very high even with the optimal task allocation (Mitchell, Samms, Henthorn, & Wojciechowski, 2003). Autorotation provides one of the few remaining solutions for reducing the demands associated with military convoy driving. Considering that a majority of driving during military convoy's requires very basic driving responses, autonomous driving technologies exploit the desirable human qualities (e.g., mission planning and local security) without sacrificing the benefit of automation (e.g., basic path following and collision avoidance).

Goals for incorporating autonomous driving technologies into military convoys are similar to those associated with incorporating adaptive cruise control into the civilian arena: increasing traffic throughput (i.e., improving the number of vehicles moving through an area over a given amount of time) and increasing safety (e.g., reducing the amount of rear-end collisions). Autonomous driving technologies have the potential to improve performance by reducing specific vehicle control demands on the operators (Young & Stanton, 2004), improving vehicle control, preventing collisions, and improving vehicle safety (Ackermann & Bunte, 1997; Kasselmann & Keranen, 1969; Stanton & Young, 1998; Yih & Gerdes, 2005). In addition to reducing mobility demands, autonomous driving technologies can also reduce driver task load, allowing the driver to focus more of his attention on local security.

Autonomous Driving Technology

The Convoy Active Safety Technologies (CAST)[1] system is one such autonomous driving technology designed to increase the safety and efficiency of military convoy and reduce mobility demands on convoy vehicle operators (Figure 1). The CAST system utilizes a suite of commercially available technologies to develop a low-cost, autonomous driving technology for current military convoys. Similar in concept to retrofitable military technologies such as the Standardized Teleoperation System (Omnitech Robotics, 2010), the CAST system is designed to convert most military vehicles into autonomous vehicle following systems. Utilizing both vehicle and path following technologies, the CAST system provides an autonomous driving technology that allows vehicles within a military convoy to navigate operational terrains at a safe designated vehicle-to-vehicle following distance while avoiding salient stationary and mobile obstacles (including non-convoy vehicle interventions). The CAST system is configured such that the lead vehicle is operated manually and the preceding vehicles are controlled using the autonomous driving capability. Every vehicle in the convoy fitted with CAST system, including the lead vehicle, can be reconfigured in any order by resetting the system and changing some of the system parameters.

Figure 1. A 4-vehicle military convoy retrofitted with the CAST system (vehicles left-to-right: one M1078 truck, two M1083 trucks, and M915 Line Haul Tractor).

Vehicle operators interact with the CAST system through the CAST Simple User Interface (SUI) and the Disengage Pedal located to the left of the steering column (Figure 2). The SUI utilizes LED and audio tones to communicate the status of the CAST system to the operator. The LED follows an intuitive color progression, similar to that used for traffic lights (i.e., Green = Fully Operational; Yellow =

---

[1] The CAST project was spearheaded by the Tank-Automotive Research Development & Engineering Center, in partnership with the Combined Arms Support Command and the Army Research Laboratory. Based on the successful implementation, assessment, demonstration, and communication of the CAST system's capabilities in these operational scenarios, the CAST effort received an Army Research and Development Achievement Award for Technical Excellence in 2009.

System Sensors are degraded; Red = System is Inoperable or Sensors are insufficient). The audio tones ('beeps') increase in frequency as the LED's colors moved from yellow to red; there is no tone when the LED is green (i.e., system is fully operational). In cases when the CAST system deviates from the path defined by the vehicle ahead, operators can regain control of the vehicle by pressing the Disengage Pedal, located to the left of the brake pedal, or by pressing the brake pedal (Figure 2). Either pedal disables the CAST system, relinquishing control of the vehicle to the operator. Even when disabled, the CAST systems sensors' remain active so that once the CAST system is re-engaged, it can resume following the vehicle ahead almost immediately.

Figure 2. Current (left) and original (top-right) CAST SUIs. Disengage Pedal (bottom-right)

The effectiveness of the CAST system in military convoys was examined through a series of experiments and demonstrations. Two experiments are outlined below, highlighting several performance and human factors issues associated with incorporating the CAST system into military convoys.

CAST WARFIGHTER EXPERIMENT I:

The first CAST experiment, CAST Warfighter Experiment I (CAST WE I), demonstrated the ability of the CAST system to significantly enhance several aspects of performance for a two-vehicle convoy (Davis, Animashaun, Schoenherr & McDowell, 2008). The two-vehicle convoy completed driving scenarios simulating a tactical re-supply to a Forward Operating Base (FOB) during daylight conditions (at three different convoy speeds: ~20 mph, ~25 mph, ~30 mph). Twelve civilian volunteers (average age of 43.25 ± 9.4[2] years) participated in this experiment, all with experience driving 2.5- to 5-ton military vehicles in convoy-type operations. The experiment utilized a 2x3 within-subjects factorial design, where independent variables were vehicle control mode (Manual and CAST) and speed level (low, medium, and high). Vehicle operators were responsible for: monitoring or controlling the speed and direction of their vehicle, identifying

---

[2] All means are reported in the following format: Mean ± Standard Error

threats (i.e., predefined targets) in their local environment, and responding to unanticipated stopping of the vehicle ahead.

The CAST system enabled operators to improve several aspects of convoy integrity, namely maintenance of designated following distance and responsiveness to stopping of the vehicle ahead. The CAST system allowed for significantly shorter following distances throughout the driving trials ($F_{1, 22} = 10.0$, $p < 0.01$)[3] – see Figure 3. When the CAST system was disengaged, operators tended to increase both their following distance and the variability in their following distance with increasing vehicle speeds. The tendency to increase following distances as speed increases is a critical safety precaution taught in civilian driving based on human response time. The CAST system, indifferent to human limitations and human task load, allowed for shorter and more consistent following distances, even at the higher speed levels.

The CAST system also allowed for significantly shorter braking distances ($F_{1, 22} = 58.2$, $p < 0.01$) and faster reactions times ($F_{1, 22} = 33.5$, $p < 0.01$) when responding to unanticipated stopping of the vehicle ahead (i.e., ~80 ms faster reaction time and ~10 m shorter braking distance) when compared to manual control (Figure 3). Even with specific instructions to stop the vehicle as quickly as possible, operators were unable to respond as quickly without the CAST system. These results demonstrated the system's potential to reduce rear-end collisions in convoy operations and safely achieve shorter following distances, possibly mitigating the "accordion effect" (i.e., large fluctuations in convoy length as a result of variable vehicle speeds) which is prevalent in today's military convoys.

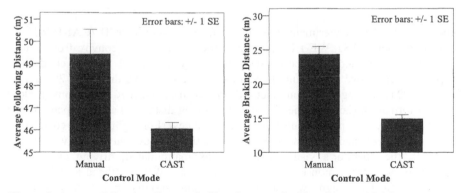

Figure 3. Average following distance (left) and average braking distance (right)

The CAST system was also associated with enhanced performance on the local area security task. While using the CAST system, operators had significantly shorter reaction times ($F_{1, 22} = 10.0$, $p < 0.01$) and significantly greater accuracy ($F_{1, 22} =$

---

[3] Performance data from CAST WE I was analyzed using Linear Mixed Model Analyses with fixed factors of speed level (low, medium, high) and control mode (Manual, CAST), and participant as a random effect.

13.2, p < 0.01) when identifying threats in their local environment. The CAST system enabled a ~1 second (~10%) improvement in operators' reaction times and led to a 6% increase in the quantity of threats identified, revealing the CAST system's potential to augment local security during military convoys. Furthermore, subjective reports of operator workload [via the NASA Task Load Index (Hart & Staveland, 1988)] highlighted the CAST system's ability to significantly reduce the mental demand ($F_{1, 22} = 4.3$, p < 0.05) and effort put forth ($F_{1, 22} = 9.9$, p < 0.01) during the driving conditions.

Even with the evident improvements in threat detection, braking distances and following distances, participants reported their overall performance as being 'better' without the CAST system. It was inferred from exit interviews that the operators' performance assessment was heavily dependent on the system's inability to achieve lateral offsets comparable to their own (lateral offset was defined as the lateral distance between vehicle paths as measured from the centerline of the each vehicle). Though the CAST system enabled significantly greater lateral offset ($F_{1, 22} = 92.8$, p < 0.01), very rarely did this offset (0.81±0.02 m) result in the vehicle exceeding the confines of the path or lane. This brought to question the utility and usability of the system as it was implemented, as operators ultimately have the choice as to whether to use or disengage the technology. Subsequent system modifications and experiments/demonstrations where designed to address and assess these important factors associated not only with vehicle performance, but also with the operator's perception of and willingness to use the technology.

CAST WARFIGHTER EXPERIMENT II

The second CAST experiment, CAST Warfigher Experiment II (CAST WE II), expanded upon CAST WE I by extending the length of the convoy from 2 to 4 military vehicles and by incorporating both day and night driving conditions. Experimenters drove the 1st and 4th vehicle, while operators drove the 2nd and 3rd vehicles. Day driving conditions focused on two basic convoy formations: a close column formation (i.e., low speeds, short following distances) and an open column formation (i.e., fast speeds, long following distances). For night driving conditions, the convoy remained in close column formation throughout the driving trials; however, the operator's visual modality was manipulated, driving either with headlights or Night Vision Goggles (NVGs). CAST WE II utilized a 2x2 within-subjects factorial design, with one independent variable being control mode (Manual, CAST) and the other independent variable being either column formation (for day driving conditions) or visual modality (for night driving conditions). The demographic for the second experiment was similar to the first, with all sixteen active duty volunteers (average age of 36.1 ± 11.1 years) having experience driving 2.5- to 5-ton military vehicles in convoy-type operations.

In addition to changes in the experimental conditions, CAST WE II incorporated several system updates to enhance the performance of the CAST system based on findings from CAST WE I. A critical shortcoming for the system identified during the first experiment was its ability to meet or exceed the operators ability to drive within the path defined by the vehicle ahead (i.e., lateral offset). Engineers reduced the amount of lateral offset by fine-tuning sensory and control algorithms, while also upgrading the CAST system's sensor package to include fiduciary markers that made it easier for the system to determine the position and trajectory of the vehicle ahead. The engineers also implemented human factors suggestions aimed at increasing the usability of the SUI and ride quality of the CAST system. During training exercises from CAST WE II, human factors engineers noticed that that several operators had difficulties interpreting the SUI. During initial system development, the SUI was used by engineers as a means of monitoring specific aspects of the system and diagnosing operational issues. By reducing the amount and complexity of system status indicators, engineers provided a user interface that was more conducive to vehicle operation and not system testing and development (Figure 2). They also incorporated redundant sensory cues when communicating system status updates to the operator, by including a combination of visual and audio cues to indicate the systems status (as described in the section above titled *AUTONOMOUS DRIVING TECHNOLOGY*). To increase the ride-comfort, engineers reduced the severity of the decelerations by using engine braking to slow the vehicle prior to applying the brakes (when applicable). Furthermore, the operational necessity to perform over a wider range of speeds, prompted engineers to provide the CAST system with the ability to achieve dynamic following distances based on convoy speeds as outlined in the Field Manual for Wheeled Vehicles (FM 21-305). To provide a more naturalistic representation of vehicle-to-vehicle distances in military convoys, operators were asked to maintain a 'safe' following distance in CAST WE II. The lead vehicle traveled at a (nearly) constant velocity, so we informed operators that this 'safe' following distance should remain constant for each column formation. So rather than focusing on the ability of the operator to maintain pre-designated following distances, this experiment was designed to examine the natural fluctuations in following distances which can affect the efficiency and throughput during military convoys (e.g., the accordion effect).

It was evident from both reported and realized improvements in convoy integrity and operator performance that the CAST system improved vehicle control and reduced operator task load. For day driving conditions, the operators ability to maintain consistent following distances (as indicated by standard deviation in following distance) was significantly improved by the CAST system for both close and open column formations [($F_{1,15} = 25.9$, $p < 0.01$)[4] and ($F_{1,15} = 8.1$, $p < 0.03$), respectively] -- see Figure 4. Also, improvements to the CAST systems sensing capabilities and control algorithms for CAST WE II reduced the amount of lateral offset seen in CAST WE I by over 0.25 m. No longer was the CAST system associated with greater lateral offset than the operator alone; lateral offset achieved with the CAST system was comparable to lateral offset achieved by the operator alone for both close column (Manual: 0.55 ± 0.12 m, CAST: 0.53 ± 0.12m) and

---

[4] Performance data from CAST WE II was analyzed using Repeated Measures Analyses of Variances with control mode (i.e., Manual or CAST) as a within subject factor.

508

open column (Manual: 0.51 ± 0.04 m, CAST: 0.46 ± 0.06 m) formations. Furthermore, the CAST system afforded significant improvements in the operators ability to accurately identify threats in the local environment in both close and open column formations [($F_{1, 15} = 6.2$, p < 0.03) and ($F_{1, 15} = 15.0$, p < 0.01), respectively]. Due to the limited visibility during night operations, it was more difficult to detect differences between the operators' ability to identify threats with and without the CAST system. However, just as in day driving conditions, driving with the CAST system was associated more consistent following distances than the operator alone, regardless of visual modality [headlights : ($F_{1, 15} = 14.5$, p < 0.02) and NVGs: ($F_{1, 15} = 25.1$, p < 0.01)] – see Figure 4. The CAST system enabled superior vehicle control during night driving conditions by significantly reducing the amount of lateral offset (by nearly a foot) for both visual modalities [headlights: ($F_{1, 15} = 23.2$, p < 0.01) and NVGs: ($F_{1, 15} = 46.4$, p < 0.01)].

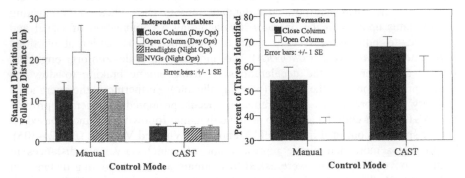

Figure 4. Standard deviation in following distances for close and open column formations (left). Percent of threats identified for close and open column formations (right).

Improvements in the operators' ability to control the vehicle and perform the local security task with the CAST system resulted in more favorable subjective ratings of performance and utility when using the CAST system. Unlike CAST WE I, operator responses from CAST WE II did not purport that they performed better without the CAST system. In fact, about 90% of the operators reported no difference in their ability to maintain a safe following distance, stay directly behind the vehicle ahead, and complete the overall convoy operation with or without the CAST system. This was a critical improvement considering human's affinity for self-enhancement (Fiske, 2003).

Since the convoy was four vehicles long, CAST WE II also provided a basis for examining robot-to-robot following behavior as opposed to simply examining the CAST system's ability to follow a vehicle being driven by a human operator. Results for the first 3 vehicles (all from the Family of Light or Medium Tactical Vehicles) indicated that whether the CAST system was following a human operator or following another autonomously controlled vehicle, there were no practical or statistical differences in vehicle performance (following distance or lateral offset). This is a critical finding, as it makes extrapolating system performance much easier for a homogenous group of vehicles in a military convoy.

FUTURE EFFORTS

Future efforts should examine the effects of the CAST system on convoys under different operational conditions. Operating conditions will often dictate the length, speed and spacing of the convoy, as these factors are critical in avoiding detection and combating enemy tactics to disrupt military operations. The utility of the CAST system in convoy operations will depend heavily on its ability to adapt to a wide variety of operating conditions. Future efforts should also examine the effects of the CAST system on all personnel within and supporting the convoy. Incorporating automation into a task environment not only changes the responsibilities of the operator, but affects the dynamics of the entire crew within and supporting the convoy. An upcoming experiment will begin to address this issue by examining the extent to which the driver can complete additional tasks, commonly delegated to other personnel within the convoy (e.g., responding to radio communications). The effect of the CAST system on operator fatigue is another key issue that could potentially affect its utility in convoy operations. Aside from being a factor that is pervasive in most military operations, fatigue is commonly cited as major contributor in a predominance of vehicle accidents (Lal & Craig, 2001). The system's ability to offload much of the task load associated with driving, allows the operator to put forth more effort into potentially more fatiguing tasks (e.g., local security). Though fatigue was monitored throughout CAST WE I and CAST WE II, neither experiment was designed to accurately quantify the system's effect on operator fatigue (i.e., sleep deprivation wasn't induced and duration for each trial wasn't sufficiently long). Finally, future efforts should examine potential changes in attentional focus related to CAST system operation. For example, if an operator is overly engaged with the autonomous system, it may divert their attention from critical security tasks, consequently reducing overall performance. Eye-tracking data collected during CAST WE II will be helpful in inferring attentional changes, and possibly fatigue, associated with CAST system operation.

SUMMARY

Sustainment convoys are critical to providing combat commanders the right support, at the right time and place, and in the right quantities, across the full range of military operations. Force Protection efforts associated with military operations impose additional demands on the vehicle operators, potentially degrading vehicle control, reducing local area awareness, and impacting overall mission completion. Addressing task load through the use of autonomous driving technologies, such as the CAST system, provides a potential means of mitigating these demands on the driver and also improving overall convoy performance. The CAST system demonstrated its ability to augment convoy driving in a way that not only reduced task loading on the operator, but improved several aspects associated with convoy vehicle control for both day and night driving conditions. Experiments outlined in this paper demonstrated the ability of the CAST system to meet or significantly improve convoy integrity by enabling operators to effectively follow a vehicle at a safe distance, while responding faster to unanticipated stopping of the vehicle ahead than the operator alone. By reducing the task load associated with driving, the CAST system also demonstrated its utility in enhancing operators' ability to perform critical local security tasks. The CAST system demonstrates the potential effectiveness of autonomous driving technologies in military convoys and provides a feasible solution to reduce task loading associated with military convoy driving.

REFERENCES

Ackermann, J. & Bunte, T. (1997). Yaw disturbance attenuation by robust decoupling of car steering. *Control Engineering Practice*, 1131–1136.

Davis, J., Animashaun, A., Schoenherr, E., and McDowell, K. (2008). Evaluation of Semi-Autonomous Convoy Driving. *Journal of Field Robotics*, 25, 880-897.

Fiske, S.T. (2003) *Social Beings: A Core Motives Approach to Social Psychology*. Wiley.

Hart, S. & Staveland, L. (1988). *Development of NASA-TLX (Task Load Index): Results of empirical and theoretical research*. In P. Hancock & N. Meshkati (Eds.), Human mental workload, pp. 139-183. Amsterdam: North Holland B.V.

Kasselmann, J., & Keranen, T. (1969). Adaptive steering. *Bendix Technical Journal*, 2, 26–35.

Lal, S. & Craig, A. (2001). A critical review of the psychophysiology of driver fatigue. *Biological Psychology*, 55, 173 – 194.

Mitchell, D. K., Samms, C. L., Henthorn, T., & Wojciechowski, J. Q. (2003). Trade study: A two versus three-soldier crew for the Mounted Combat System (MCS) and other future combat system platforms (Rep. ARL-TR-3026). Aberdeen Proving Grounds, MD: Army Research Laboratory.

Omnitech Robotics (2010). Standardized Teleoperation System (STS) Applications. Accessed Feb 2010, http://www.omnitech.com/sts.htm.

Stanton, N. A., & Young, M. S. (1998). Vehicle automation and driving performance. *Ergonomics*, 41, 1014–1028.

Department of the Army (1993). Manual for the Wheeled Vehicle Driver (Rep. FM 21-305). Washington, DC: Department of the Army Headquarters.

Yih, P., & Gerdes, J. C. (2005). Modification of vehicle handling characteristics via steer-by-wire. *IEEE Transactions on Control Systems Technology*, 13, 965–976.

Young, M. S., & Stanton, N. A. (2004). Taking the load off: Investigations of how adaptive cruise control affects mental workload. *Ergonomics*, 47(9), 1014–1035.

Chapter 52

Ergonomically Streamlining the Process of Implementing Changes Based on the Behavioral Approach

Zbigniew Wisniewski, Aleksandra Polak – Sopinska, Jerzy Lewandowski

Department of Production Management and Logistics
Technical University of Lodz
Wolczanska 215, 90-924 Lodz, Poland

ABSTRACT

The article contains results of observation and analysis made after working out algorithms of changes process control. The basis for developing the algorithm were identified dynamic properties of organisation subsystems and human behaviour. Reaction models were developed for the research results on the basis of systems dynamic, which enabled working out more effective methods of future implementations of changes with the use of methods based on control theory. The research was carried out through the Ministry of Science and Higher Education funding under grant N115 071 31/3252.

Keywords: Changes management, behavioral dynamics

INTRODUCTION

The principle aim of the research was to identify basic social processes accompanying organization changes and determining dynamic features of these processes. On the basis of collected data concerning attitudes and activities of workers of a few organizations in which various organization changes are implemented, the most important social phenomena occurring in the period of changes and common for different organization contexts, different specificity of activities of particular organizations, their branch, size or functioning characteristics were determined. Through comparison of a situation in a few surveyed organizations and identification of personnel attitudes towards implementations, analytical categories were determined enabling describing progressing of organization change in its social aspect. Application of these analytical categories enables reconstruction of the studied phenomena and formulating them in a model form. It also enables identifying a factor shaping the organization change's process in a social aspect and predicting this process with the use of incidence analyzing, intensity and dynamics of particular phenomena. Thank to the above, persons coordinating organization implementation can follow the personnel attitudes anticipating their influence on the change made and take proper corrective actions depending on situation.

Innovation processes and changes management processes are realised in organisations in the cycles arranged in change clusters. Activities taken by managers to improve implementations or even enable their realization are often reduced to such modification of behaviours and attitudes of organization members, that they start regarding the realised aims as their own or at least indispensable. Usually a danger appears of discarding activities due to not enough spectacular results of actions taken so far and effects not sufficient for sustaining positive approach. The consequence is a risk of choosing easy expedients, which can apparently bring the objective closer or at least satisfy the need of success of the managed persons. Such improvement of situation brings temporary results. If it is a result of incidental activities not resulting from earlier developed rational implementation plan, it does not assure a significant improvement in a global meaning of a change. Innovation and changes implementation processes demand taking an attitude of expecting delayed effects, and makeshift activities should be considered a necessary evil, which can reduce the efficiency of the whole process realization. Therefore it is so important to plan the whole change cycle, including the necessity to take occasional stimulating activities, thorough satisfying the needs for success of people taking part in change implementation. The knowledge of the cycle of influencing of initiating and maintaining stimuli can significantly contribute to improvement of effectiveness of realization changes processes and innovation processes. The use of knowledge of dynamic feature of particular objects taking part in a change or innovation can be the way to improve effectiveness of influence.

Implementation of changes is a dynamic activity because it takes place in a specific interval in which effects are stirred up with often an unexpected character. This results from difficulties with accepting new, imposed conditions of acting. Difficult to predict effects are caused by dynamic character of organisation elements. Dynamism means that reactions to stimuli depend not only on the level of a stimulus but also on the course of events so far and affects so far caused by these stimuli. This is a feature of system dynamics. A lot of managers do not realise that in a transition period, that is before accepting and stabilizing effects after changes implementation, a previously known static dependence between a stimulus and a result does not occur.

In reactions to changes managers root manifestations of resistances resulting from a great number of premises, starting from psychological ones to finish with the technical ones. Not questioning settlements concerning the source of problems appearing during changes implementation, it is worth looking at the process of reacting to stimuli as to the process of influencing dynamic sector having a definite structure: memory, inertia, delay, information, energetic/energy and emotional capacity. The foundations to this approach were built by Marian Mazur, who provided a cybernetic model of human character.

After a period of dynamic influence of single entities (people, teams, workstations, organization units) subjected to input (changes) a stability state is expected. This state occurs in the period of a system "statics", that is when the states of elements outputs depend only on the states of inputs at the same moments. Such state is not interesting from the perspective of changes implementation dynamics, except for the fact that dynamic influence depends on the initial conditions of a given process, i.e. on a static state at the moment of changes initiating.

ANALYSIS OF CHANGES IMPLEMENTATIONS

RESEARCH SAMPLE

18 organisations were included in the research. Among them there were 15 production enterprises and 3 organisations dealing with financial services. The organisations were foreign companies, foreign capital companies as well as companies with Polish capital only.

Collecting data took in total 18 months of research in particular enterprises. During that time processes of team work, preparation and implementation of changes, production and service work were observed. Over 160 of deepened free interviews and informal survey conversations with employees at executive – labour and managements stations were made.

Parameters were registered also with the use of the method of collecting quantitative data about processes state. In part of the cases these methods were based on the system of automatic collection of information about processes (especially in cases of production operations or other operations of a transaction type using computer systems). Other cases demanded registering certain parameters by an employee doing the work or by the process observer.

METHODS

In order to identify the elements of the systems and working of the same system basing on the analysis of responses to input, types of activities possible to take in given conditions by the rule of input, must be determined. In other words: it must be determined if and which activities and situations in organisation must be treated as stimuli inducing reactions and what their features are. These activities play important role in initiating changes because realisation of change can be reduced to the process of exerting an input of a specific character (dynamics) at a given element of organisation. Below most often occurring events identified during observations, which can be described as inputs, are presented. These are examples of organisation activities having the structure close to unit step:

- Single increase of a standard for a performed task
- Change of production order
- Introducing four-team system
- Increasing target value of production capacity indicator
- Improvement of process sigma
- Reducing fraction of faults with a few percentage points
- Dismissal of an employee from a team
- Employing a new worker
- Change of a service or raw material provider
- Installation and introducing into production of a new machining-tool
- Change at post of a superior
- Rising a salary of an employee or a team
- Salary reduction
- Change of income tax rate
- Emergence of a new competitor on market
- Emergence of a substituting good on market
- An employee promotion or demotion
- Passing information about new requirements

From among the situations enumerated above special usability in the practice of dynamics identification was characteristic for those, which were associated with concrete target values of process parameters. In such a case identification is based on comparison of timing of: input and an effect of a given element.

OBSERVATION RESULTS

The observation enabled identifying in an organisation environment (teams of people) a few basic classes of behaviours. They can be described from the dynamics perspective through references to typical dynamic segments in the control theory. The identification was made through evaluation of reaction to appropriate character stimuli.

For particular implementations dynamic parameters were identified. As a result probable dynamic models for every unit taking part in the change implementation were determined. Identification process was realised with the use of AnalyX application developed in Matlab. The result were equations of objects in time function and transfer functions.

For identification an extensive class of models possible to describe with a transfer function has been accepted:

$$G(s) = \frac{ke^{-s\tau_0}}{\displaystyle\prod_{i=1}^{n}(1 + sT_i)} \qquad (1)$$

Due to the character of reactions of objects subjected to input in the studied companies, the model had to be enriched with a possibility of the occurrence of oscillation and the effect of resistance, that is a counter-reaction. This results in taking into consideration of imaginary roots in denominator and in numerator polynomials in degree higher than 1.

As far as the degree of polynomial in denominator is concerned, in practice for the purpose of control of processes with disturbances, maximum third row models are used.

The accepted model of the object, being a subject to identification has the following form:

$$G(s) = \frac{b_2 s^2 + b_1 s + b_0}{a_3 s^3 + a_2 s^2 + a_1 s + a_0} e^{-s\tau_0} \qquad (2)$$

Identification of objects under inputs (implementation of changes) enabled establishing a dynamic model, universal for the analysed categories of objects. Its universal character results from the fact that for all objects transfer functions were obtained that can be reduced to a general transfer having the following features:

- in numerator there is the first degree polynomial, whereas factor of the polynomial b1≤0 (farther on /in the farther part of the deduction it will be marked as T3),
- in denominator there is a polynomial of the maximum second degree in real or imaginary roots,
- there is a constant term of polynomial responsible for a delay,

Now the transfer function of the model has the form:

$$G(s) = k\left(\frac{T_3 s + 1}{T_0^2 s^2 + 2\zeta T_0 s + 1} e^{-s\tau_0}\right) \quad (3)$$

In cases of no over-controlling, oscillations and significant resistance equal still different models were obtained with equally good compliance with empirical data. These competing models are the following:

$$G'(s) = k\left(\frac{1}{T'^2 s^2 + 2\zeta T' s + 1}\right) \quad (4)$$

$$G''(s) = k\left(\frac{1}{T'' s + 1} e^{-s\tau_0}\right) \quad (5)$$

In transmittance (4) denominator has real roots. A difference between the two transfer functions is that the second order inertia element (4) can be substituted by the first order inertia element with a delay element (5). The model (4) seems to be more universal, because with the use of it more configurations of objects can be modelled, under the condition that the occurrence of imaginary roots is accepted. Still this s a necessity in case of the occurrence of oscillations. On the other hand it is postulated to lead the form of models to the lowest degree of the denominator polynomials and substituting higher row inertia with delays.

Yet eliminating inertia for simplification of analytical form is a redundant action, because it is always more useful to use a general model, which gives a possibility of identifying more extensive range of cases. Reduction and excessive simplification of the model may lead to reducing definition of study and failing to notice rare cases (e.g. with oscillations). Nevertheless cases with resistance and oscillations are frequent enough to prevent the authors from resigning from using the most universal model. It has been decided that the most general form (3) will be used, in which the delay element is permanently entangled. Hence there will be no need to differentiate cases du o model class at the identification stage.

THE ENDING

Thanks to man – machine and man with other people relation human dynamics can be better known. In the situation of observing the systems: man – man, man – group, man – environment there is no possibility of objective preparation of the level of signals for testing and identifying dynamic properties of a single man, because the man always enters into relations with the second element of each of the systems, being equally non-deterministic and unpredictable as the observed man. The situation of man - machine coupling gives such a comfort that a machine can be quite precisely identified and its dynamic features can be programmed. Human reactions that can influence the state of machine dynamics are few.

Perception, motor and psychical qualities of a man are not constant but depend on preparation to work, knowledge, habits, health condition, age, motivation for acting, role in a group, the level of needs fulfilment at a given moment and many others. They also depend on relations with the environment, work conditions and characteristics of machines which the person operates. Thus it can be stated that human activity has non-stationary character because human characteristics are changeable. Fortunately the main features determining human dynamic characteristics in relation to machine and the environment (people) do not change in a time shorter than a cycle of a change management.

A man has natural abilities of guiding a system to a stable state. The result of such abilities is among others impressive ability of adaptation in caging conditions. Humans obtain in this field good results thanks to their inherent qualities, but also thanks to proper configuration of control process. Still such a system has its limitations. As in any system they result from limitations of values of parameters with the use of which the process is influenced as well as limits of time, cost, energy etc, to end with natural barriers of human psychology and physiology and social behaviours of a group in which a person is or which a person leads.

REFERENCES

Boonstra, J. J. (2004). *Dynamics of organizational change and learning.* Hoboken, NJ, USA: John Wiley & Sons, Incorporated.
Helfat, C. E. (2007). *Dynamic capabilities: Understanding strategic change in organizations* Wiley-Blackwell.

Hinings, C. R., and Greenwood, R. (1988). *The dynamics of strategic change*. Oxford: Basil Blackwell.

Klein, H. J. (1989). An integrated control theory model of work motivation. *Academy of Management Review, 14*(2), 150-172.

Lewandowski, J., and Wiśniewski, Z. (Eds.). (2007). *Quality and environmental management. continuous change of paradigms*. Łódź: Wydawnictwo Politechniki Łódzkiej.

Mainzer, K. (2007). *Thinking in complexity the complex dynamics of matter, mind and mankind* Springer.

McGarvey, B., and Hannon, B. M. (2004). *Dynamic modeling for business management: An introduction* Springer.

Stacey, R. D. (2007). *Strategic management and organisational dynamics: The challenge of complexity* Prentice Hall.

Wiśniewski, Z. (2009). Dynamics of changes implementation in production systems. In M. Fertsch, K. Grzybowska & A. Stachowiak (Eds.), *Efficiency of production processes* (pp. 109-119). Poznań: Publishing House of Poznan University of Technology.

Wiśniewski, Z. (2007). System approach to the process of changes implementation in the organization. In Grudzewski W. M., Hejduk I.,Trzcielinski S. (Ed.), *Organizations in changing environment. current problems, concepts and methods of management*. (pp. 492-495). Poznań: IEA Press Madison.

CHAPTER 53

Determinants of Implementing Systems of Occupational Hygiene and Safety Management in Polish Enterprises

Konrad Niziołek, Katarzyna Boczkowska

Department of Production Management and Logistics,
Technical University of Lodz
90-924 Łódź, ul. Wólczańska 215, Polska

ABSTRACT

The article is a continuation of the authors research assessing efficiency of implemented management systems in accordance with the standards PN-N 18001 and OHSAS 18001. Efforts made by Polish economic entities in the area of system management of occupational hygiene and safety not always bring expected results. The pursuit for keeping the world standards in broadly understood care for human being should result in improvement of the current state. In the article the experiences of the authors concerning the processes of occupational hygiene and safety implementation processes in economic entities are evaluated. Comparison analysis of their realization in factories and of barriers and problems faced by implementation teams and by entrepreneurs are the subject of the analysis. The objects of the research are enterprises representing various economy sectors, characterized by very different level of organization culture, with different legal statuses. The purposeful choice of units under research including a company with the status of a sheltered work establishment, enabled looking at conditions and the whole process of realizing implementation in an extensive, multi-aspect way. Pointing at the factors decisive for success or efforts efficiency will make it easier for entrepreneurs to take a decision about entering the way of occupational hygiene and safety system management and realization of the objective.

Keywords: Safety management system, safety culture

INTRODUCTION

Problems of occupational safety and hygiene have been for years the object of research and analyses of many researchers. Entering by Poland of the huge market of integrated Europe forces Polish entrepreneurs, especially in the small and medium enterprises sector, to take activities aimed at work costs reduction including costs of occupational hygiene and safety. Accident rates in Poland – in particular the frequency of deadly accidents at work and occupational diseases incidence, despite the observed improvement in the recent years, are still much higher than those noted in the European Union countries. The situation concerns first of all small and medium enterprises, where occupational hygiene and safety problems are treated marginal, and lack of necessity to appoint occupational safety services does not motivate employers to take actions for improvement of occupational safety conditions.

Despite a completed process of adjusting Polish occupational hygiene and safety law to the EU standards, developments of system management rules aimed at proactive activities based on training, motivating for safe work, personnel participation, etc., the level of safety in Polish enterprises is not satisfying. Launching in 2003 of an economic stimulator in the form of a varied accident insurance contribution will not bring expected results without heightening the awareness of the managerial staff, shaping prosafe behaviours and creating proper safety culture. The research carried out in the EU countries indicate strong relationship of culture mechanisms in organization with the level of safety rates. The change in the system of values and attitudes of personnel contributes to improvement of the occupational hygiene and safety positively influencing motivation of workers and their safe behaviours.

RESEARCH METHODOLOGY

In order to evaluate barriers and determinants of implementation of occupational hygiene and safety management in three companies of the Lodz Region a survey was made in the small and medium size enterprises sector employing up to 200 workers. The research was designed and carried out in accordance with the rules of design and realization of statistical research. As the observation technique direct interview was used based on an especially designed questionnaire. The questionnaire included in total 26 fundamental questions and 4 metric questions. The main questions of the questionnaire concerned:

- Engagement of managerial staff,
- communication,
- personnel participation,
- occupational hygiene and safety education,
- analysis of accidents and almost accidents,
- motivating for safe work,

- relationships between employees, their cooperation.

All questions were the closed-ended. The use of dychotomic scale was not justified because to a large extent it could limit the questionnaire real character. Therefore a five-degree–Likert scale was chosen according to the following criteria:

- 1 – absolutely not,
- 2 – rather not,
- 3 – difficult to decide,
- 4 – rather yes,
- 5 – definitely yes.

The variants of replies seem to close the studied problem. The form was developed on the basis of logical and thematic relationships between questions. Metrics took into consideration education and gender of personnel, employment seniority and type of position taken in a company.

The questionnaire was supplemented by polls given to personnel and analysis of accessible documents.

ANALYSIS OF THE SURVEYED GROUP

The following 3 economic entities that introduced occupational safety and hygiene system took part in the research:

Company A – logistics enterprise belonging to an Amerian combine providing logistic services for IT, implementation of OHSAS 18001 system,

Company B – family company, a typical representative of the Polish sector of small and medium sizes enterprises producing foods for the Eastern-European market, implementation of PN-N 18001 system,

Company C – having the status of a sheltered work establishment, electronics, implementation of PN-N 18001 system.

The basic information about enterprises and the studied group is presented In table 1.

Table 1 shows that the groups, despite significant differences in their activities, size and experience on market were surveyed maintaining the criteria concerning the size of the research sample.

RESEARCH RESULTS

In order to compare variety of the levels of functioning of occupational hygiene and safety management systems in the studied enterprises the results obtained for selected questions are presented in the Illustrations 1 – 7.

Managerial Staff of the entities representing foreign capital (companies A, C) approach the problems associated with abiding by the Polish law of occupational hygiene and safety much more seriously. If in case of the company A the high level of managerial staff consciousness is related to the top management staff engagement, their role in the system and building of organizational culture, in case of a the sheltered work establishment (company C) right attitude of managerial staff results from legally imposed systematic controls by the national organs of work condition supervision. Thus quite different motivators for attitudes of managerial staff can give similar results, although activities supporting building high culture – company A, will give in the future better results than obligatory, frequent controls – company B.

Table 1 Comparison of studied groups (the authors' study)

	Company A	Company B	Company C
Employment	80	160	90
Type of activity	Logistic centre	Production of cosmetics	Production of home electronics
Period of activity	0.5 year	10 years	5 years
Number of persons included in the research	30 37%	60 37%	32 36%
Sex: Male Female	20% 80%	62% 38%	59% 415
Education: primary vocational secondary higher	13% 7% 30% 50%	6% 3% 60% 31%	9% 25% 38% 28%
Work position: director manager labourer clerical worker	7% 7% 36% 50%	7% 7% 40% 46%	3% 13% 50% 24%
Occupational seniority: Up to 1 year 1-5 years 5-10 years Above 10 years	0% 23% 17% 60%	0% 44% 31% 25%%	3% 34% 38% 25%

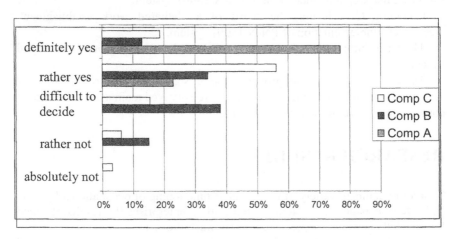

FIGURE 1. Does the top management according to you abide by the occupational hygiene and safety rules? (the authors' study)

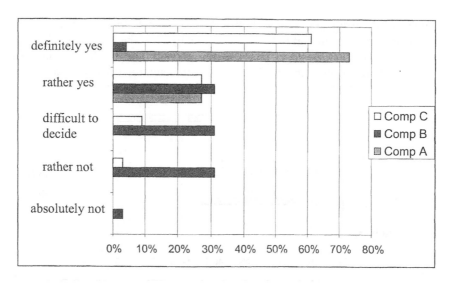

FIGURE 2. Do you know the threats and risk associated with the work you perform? (the authors' study)

Process of occupational risk assessment and associated identification of threads is much better realized in the entities A and C. Company B having no traditions and good models of behavior treats the question of occupational risk only as a legal obligate. It is worth mentioning that in case of the sheltered work establishment (company C) occupational health services are continuously present in the company and engaged in the process of risk assessment and identifying threats, taking into consideration also disabilities of employees.

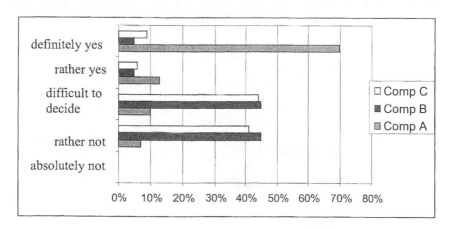

FIGURE 3. Are employees encouraged to report on occupational hygiene and safety (the authors' study)

The aspect of employees participation is not treated in the proper way in B and C entities. The personnel of these companies are treated as partners for discussion and taking decisions, therefore the companies do not use proper forms of motivating, and encouraging them to act also in the area of

524

occupational health and safety. Polish law does not regulate this type of problems, therefore the attitude of the company C – the activities of which concentrate on fulfilling legal requirements and do not result from internal needs of he organization and its high organization culture, is not surprising.

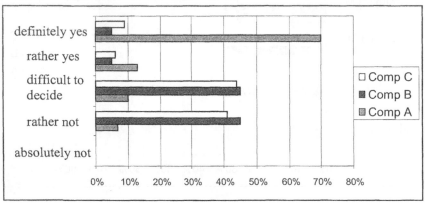

FIGURE 4. Are employees encouraged to take active part in occupational safety training? (the authors' study)

Training methods practiced in company A differ considerably from these in the other two entities. In the entity A a great stress is put on training and improving personnel qualifications. A computer-supported system of the employees self-education in a wide range of problems, from technical issues, through managerial ones, up to human relations has been developed. Decentralization of authority demands from the middle management broadening their knowledge and awareness of subordinate staff, and occupational safety is also included in this process.

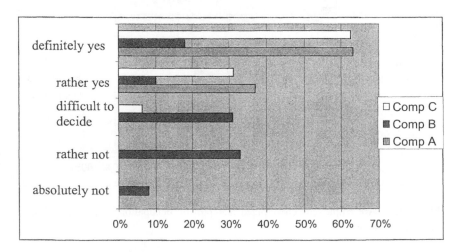

FIGURE 5. Have procedures of alertness and reacting to accidents at work and breakdowns been developed? (the authors' study)

In all the surveyed entities adequate alert procedures have been developed, yet, as the results for company B indicate, only a small number of employees know about the procedures existence. Hence it cannot be expected that in case of critical situations the employees behavior will be adequate. In company A well functioning alert procedures result to a large extent from material responsibility for the clients' goods entrusted to the company. In the entity C, employing disabled persons, mainly with poor hearing and vision, the procedure had to define the range of responsibility for efficient informing or possible evacuation of disabled persons from the danger zone. Therefore the level of the safest system developed in the company C is as high as the level in company B, which is confirmed by the employees.

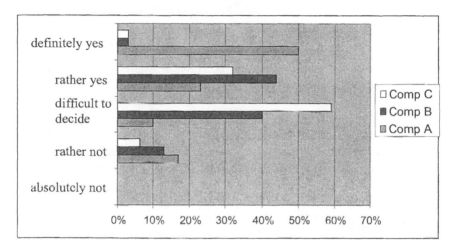

FIGURE 6. Are the employees encouraged to report potentially accidents? (the authors' study)

The key issue from the perspective of accidents prevention are potential accidents. Unfortunately in companies B and C the term is not known to the whole group of employees. Relatively low accident rate and in consequence false sense of security do not encourage employers to run additional documentations of potential accidents. Company A, in which good occupational hygiene and safety practices are used, this element of prophylactics is a priority.

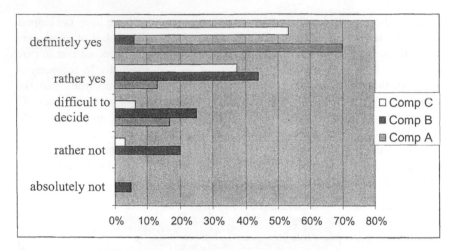

FIGURE 7. Do you feel responsible for occupational hygiene and safety at your workstation and in your company? (the authors' study)

The level of employees awareness concerning their common responsibility for occupational safety is definitely the lowest in the entity B. It is because its employees do not identify themselves with their company, and lack of interest of the managerial staff and their failing to treat occupational safety as a priority influence the attitude of employees and worsen the situation.

SUMMARY

The process of implementation and maintaining of the system of occupational hygiene and safety management does not depend on implemented PN-N 18001 or OHSAS 18001 standard and can process in different way and with different pace.

The fact of having a certificate is a confirmation of having formalized procedures and other required documentation. Still, as it can be seen in the research, it does not guarantee efficient implementation of the requirements of these standards.

Companies have to work out internal mechanisms fostering and supporting development of high culture of safety. The most important problems faced by companies are:

1. Working out good motivators encouraging employees to take initiatives in occupational hygiene and safety, including reporting potential accidents,
2. Developing more efficient activating methods of training fostering?? Personnel responsibility for their own and their workmates safety, and improvement of occupational safety training process as refers to timeliness, reliability, quality, etc. of training.
3. Forming by the highest managerial staff of the rules of safe work through proper behavior and abiding by occupational safety rules as well as assurances of necessary occupational safety means.
4. Motivating personnel to safe working and activating their engagement in

improvement of working conditions at their workstations.
5. Initiating meetings, consulting the personnel, especially the lower rank, active cooperation in the field of occupational safety,
6. Increasing the participation of employees in works associated with identification of threats, occupational risk assessment, studying accidents and potential accidents.

REFERENCES

Boczkowska K., Niziołek K., Bogusiewicz K., *Kultura bhp w małych i średnich przedsiębiorstwach po akcesji do UE*, Politechnika Łódzka 2006
Boczkowska K., Niziołek K., *Kształtowanie kultury bhp skutecznym elementem systemu zarządzania*, red. Jędrych E., Pietras A., Stankiewicz-Mróz A. „Funkcja personalna w zmieniającej się organizacji-diagnoza i perspektywy", Wydawnictwo Media Press Łódź 2008
Boczkowska K., Niziołek K., *Integrated management systems In the context of disabled individuals employment*, red. Lecewicz-Bartoszewska j., Lewandowski J., monografia Ergonomics for the disabled In work organisation and management – results of researches, evaluations and guidelines, Technical Uniwersity of Lodz Press 2008
Milczarek M., *Ocena poziomu kultury bezpieczeństwa w przedsiębiorstwie*, Bezpieczeństwo pracy, 5/2004.
Niczyporuk B., *Motywowanie pracowników do bezpiecznych zachowań jako narzędzie prewencji wypadkowej*, Zarządzanie Bezpieczeństwem i Higieną Pracy w Przedsiębiorstwie, materiały konferencyjne CIOP PIB Toruń 2006
Pawłowska Z., *System zarządzania bezpieczeństwem i higieną pracy w przedsiębiorstwie*, CIOP Warszawa 2004
PN-N 18001:2004 – "Systemy zarządzania bezpieczeństwa i higieny pracy. Wymagania"
Sułkowski Ł.: Kulturowa zmienność organizacji, PWE, Warszawa 2002.
http://pl.osha.europa.eu

Chapter 54

Representing Unstructured User Data for Inclusive Design

Farnaz Nickpour, Hua Dong

Human Centered Design Institute
School of Engineering & Design, Brunel University
Uxbridge, Middlesex, UB8 3PH, UK

ABSTRACT

User data need to be properly communicated to designers when setting up user-centred design specifications. In many cases, first hand field data —mostly unstructured primary data in pictorial, audio, visual or textual formats collected by researchers— needs to be accurately and effectively communicated to designers, which needs structure and organisation. Challenges will arise in the process of structuring such data.

This paper reflects on the opportunities and limitations in the process of representing unstructured user data for designers. A real world design case study is discussed where primary data were collected through observation, video ethnography, interview and questionnaire in hospitals. The researchers applied a database construction tool to create a data structure that could embody the rich primary information collected from various hospital stakeholders (i.e. patients, nurses and visitors).

It is argued that despite major benefits of structuring the unstructured user information (in both visual and textual formats), certain qualities of such rich data are lost in the process of sorting and interpretation. An ideal tool is suggested to enable the designer to see the big picture through structuring the information as well as providing access to the raw data.

Keywords: Inclusive Design, Unstructured Data, User Data, Designers

BACKGROUND

Design, in its conventional practice, has been dealing with humans in their ideal physical, cognitive and sensory performance; i.e. young and able-bodied people. However, this exclusive approach is no longer affordable in our diverse informed societies; it is neither sustainable nor economically, socially or ethically acceptable to design for a generic user with the most typically desirable body and mind.

Practice of user-centered design well addresses these issues, encouraging a more diverse and inclusive approach to designing for people. Inclusive design is an approach to the design of mainstream products and services that are 'accessible to and usable by as many people as reasonably possible, without the need for adaptation or specialist design' (BSI, 2005).

Nevertheless, such good practice needs facilitation and support; designers need to better understand and engage with diverse user groups and their capabilities, limitations and aspirations. Part of this is through access to relevant data on diverse users that not only informs but also inspires designers and resonates with their creative problem solving approach. However, such data are not often included in conventional user data sources such as anthropometric or ergonomic books, handbooks or websites. Bodyspace (Pheasant and Haslegrave, 2006), a familiar book to designers that incorporates UK population data and design guidelines ,is among such conventional user data sources that do not explicitly address special populations. Incorporating the diverse user data into existing and new data sources is one route to be investigated.

However, a lot of user data in design are unstructured; one major source of information on diverse users is the first hand field data. This type of data is mostly primary data collected by researchers, marketers or designers in order to inform different stages of a design process; they come in a variety of audio, visual or textual formats. Such unstructured data are rich and insightful yet mainly unrefined and big in terms of volume, therefore it is hard to manage and go through for designers who constantly face time constraints. Unstructured data needs structure and organisation in order to be more accurately and effectively communicated to designers. Hence, various ways to structure and organize such data should be explored. This has to be done with full consideration of designers' way of engaging with user data.

Designers use various types of user data to inform their design process, these could go under two categories of 'Structured' and "Unstructured" user data. In the context of design, structured user data contains data that have been processed, refined and come in form of tables, charts, graphs or diagrams. Structured user data has are generally quantitative. On the other hand, unstructured user data in design contain any data stored in an unstructured form such as text, audio, graphics or video. They are mainly qualitative and do not follow a data model. Unstructured data communicate information that has not been processed or put into any category/classification.

Table X. 1 shows a classification of user data sources in design on the basis of its structure. The categorization of inclusive design tools developed by Goodman *et al.* (2007) has been used to classify the format and source of structured user data.

Table X. 1 Classification of user data sources in design

	Structured Data	Unstructured Data
Format & Source	1. Paper: Books, booklets, cards, leaflets, etc. 2. Mixed-media: Software, websites, on-line resources and physical kits	1.Textual Reports, interviews, questionnaires 2. Audio Interviews, field recordings, etc. 3. Visual Photos, images, videos, etc.
Routes	Secondary research	Primary research: Observation, video ethnography, interview, focus group, workshop
Provider	Designer	R& D, researcher, marketer, designer

UNSTRUCTURED USER DATA IN DESIGN

In design, unstructured data form a major source for gaining information and inspiration about the users. Unstructured data are an important source of input in various stages of the design process specifically in the 'discover' and 'develop' stages, described in the Double Diamond model (Design Council, 2005); their richness and originality provide inspiration and insights at the discovery stage of the design process and their first-hand specific information provide valuable feedback in the develop stage. However such data are often large in volume and time-consuming to process.

As opposed to structured user data, in most cases unstructured data are collected by people other than designers. As seen in Table X.1, the unstructured user data providers mainly include the marketing team, R&D team, design researchers or third party consultancies. In cases where designers collect the unstructured user data themselves, there is no specific way to manage, organize and store the data and reflect it explicitly; they mainly stay in the designers' mind and turn into an individual-based knowledge. It is therefore critical to effectively and efficiently communicate the collected user data in a manageable way with designers, as well as make explicit designers' retrieved user information.

STRUCTURING UNSTRUCTURED USER DATA

There is a critical need to better communicate, organize, analyse and store unstructured data so that designers can get user information and insights more effectively. Unstructured data, either collected by designers or non-designers, need to be made explicit, manageable and easy to process, as it is an integral part of the knowledge in design. There are different ways to address this need; one potential solution could be to put a structure to the raw unstructured data. This will help overcome the difficulties with large volumes of data and make it more appropriate for procedures such as browse, search or selection.

In order to create a data structure to embody rich primary unstructured data, it is possible to either employ an already existing database construction tool or create a new database tool. However, the potentials and limitations of existing database construction tools in representing unstructured user data needed to be explored and studied before embarking on a journey of creating a totally new database tool, therefore, an already existing relevant database tool was used in our study.

The CES SOFTWARE AND THE HOSPITAL DATA

The Cambridge Engineering Selector (CES) constructor was employed to represent the unstructured data collected in a real world design study. The CES software is typically used for materials information to enable informed material selection in the design process. It has a powerful visualization function and provides three main functions of 'search', 'select' and 'browse' which enable dynamic data integration and manipulation. The CES constructor was selected on the basis of its ability to facilitate the presentation of unstructured data (such as images and quotes).

A real world design project was selected where firstly, a group of researchers from various backgrounds such as anthropology, design and healthcare were brought together to collect primary data about diverse hospital stakeholders (i.e. patients, nurses, porters, and visitors) through observation, video ethnography, interview and questionnaire. These primary unstructured data were then presented to several design consultancies who were commissioned to address specific design briefs regarding patients' dignity in the hospital environment. The rich unstructured data were aimed to provide first-hand insights and understanding for designers and to inform the briefs they were working on.

The unstructured data came in the format of a set of reports (hard copy) in approximately one hundred pages and included a set of observations from different research groups visiting a number of hospitals. Different groups used different formats to communicate their observations and findings; some mainly used text (including their own observations and quotes from various stakeholders) and pictures where relevant, while some also included graphs, journey maps and visual diagrams alongside text, images and videos.

The authors started processing the hospital data report in order to explore potentials for embodying it into the CES constructor. A number of opportunitie and challenges s were observed throughout this process. These issues are briefly discussed below.

STRENGTHS & OPPORTUNITIES

The data available in the report was large in volume, lacked structure and was difficult to browse. Some parts of the information provided lacked relevance, however, the designers had to skim through the whole report to find the relevant and useful bits of information. By categorizing the information and creating tables, sub-categories and forms, it was easy to identify the relevant data.

Following the structural logic of the CES software, the authors were encouraged to think analytically and to develop a 'data model' by break down the chunks of unstructured data into separated sections. This could support developing an information system which would make the data more explicit, accessible and hence usable.

One major benefit of using the CES structure was the opportunity to classify quotes and images from various stakeholders (patients, nurses, porters, doctors, visitors) under various categories at the same time, i.e. the issues each addressed, the stakeholder, the hospital. It was also possible to link the relevant categories; for example linking the quote from a patient regarding lack of hygiene to a photograph from the ward visualizing the same issue.

The 'search' and 'browse' options of the CES software gave fast and easy access to different parts of data provided that the data was efficiently categorized in the first place.

CHALLENGES & LIMITATIONS

By pushing unstructured data under a structure, it is inevitable to limit the analysis of the data to the logic and mental model of the analyzer (here the researcher who populated the CES software). Given the original unstructured report, various researchers would have come with various mental models in terms of how to classify, analyse and structure the data. Therefore the database could look and work totally different based on each developer's way of interpretation.

This becomes critically important when database developers are not necessarily in the same position as the database users and might have different criteria or preferences in terms of how to interpret and analyse unstructured data. In this case, the concern was that the researchers structuring the data may not necessarily share the same mental model with designers who the database was aimed at. It is also worth considering the differences among various designers in terms of approach to working with data and interpreting it.

One limitation was the interface design of the CES software. The CES has been designed and developed by engineers and is mainly aimed at engineering design students, therefore, it has a typical engineering-minded interface. The interface follows a mental logic which is not necessarily compatible with designers' way of processing the data. The current structure also has certain limitations as it is dependent on what data have been collected and is very project-specific.

The other issue is designers' tendency to have access to the original data as well as processed and structured information; designers appreciated structured data in reality but they also wanted to have raw data as a backup.

The transformation of data from 'unstructured' to 'structured', can be clearly shown in Figures X.1 and X.2. Figure X.1 demonstrates one page of the report which included the original unstructured data gathered in hospital. Figure X.2 provides a snapshot of how such data are represented using the CES constructor software.

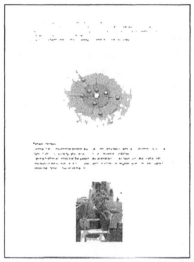

FIGURE 1 One page from the original report including the unstructured data

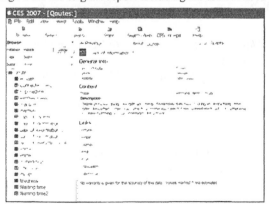

FIGURE 2 Structured data - embodied in the CES constructor software

CONCLUSION AND FUTURE WORK

Structuring the rich and abundant unstructured user data is the way forward to help make design knowledge more explicit, accessible and therefore legitimate. However, the tools implemented for such purposed have to be carefully considered and thoughtfully selected.

Tools such as the CES selector have the capability to represent both structured and unstructured data. In representing the unstructured user data, although the CES constructor is effective in providing a ready structure, it has limitations because of its engineering mental model . The interface design and its incompatibility with designers' preferred visual and graphic representations is another issue to be considered.

Exploring the potential of existing database constructor tools was the first step to understand the scope and possibilities of structuring user data. The next step is to design and develop new database constructor tools that resonate with designers' way of working with rich user data. As well as providing a big picture/outline through structuring the information, the ideal tools should also provide access to the raw data in order to allow designers ownership and authority they need in their creative practice.

ACKNOWLEDGEMENTS

This project is supported by the UK's Engineering and Physical Science Research Council grant, EP/F032145/1.

REFERENCES

BSI. (2005), *Design Management Systems* Part 6: Managing Inclusive Design Guide. British Standards Institute

Design Council, (2005), *The Double Diamond Model*. Available at: http://tiny.cc/fTQcJ. Date viewed: 01.03.2010

Goodman, J. and Langdon, P.M. and Clarkson, P.J. (2007), *Formats for user data in inclusive design* In: 12th International Conference on Human-Computer Interaction (HCI '07), 22-27 July 2007, Beijing, China.

Pheasant, S., Haslegrave, C, (2006), *Bodyspace: anthropometry, ergonomics and the design of work*, Taylor & Francis

CHAPTER 55

Squatting or Sitting: *Alla Turca* or *Alla Franca*?

Oya Demirbilek

Faculty of the Built Environment, Industrial Design Program
University of New South Wales
Sydney, NSW 2052, AU

ABSTRACT

In this paper, the traditional *alla turca* Turkish toilet, or squat toilet, will be compared with the *alla franca* sitting toilet. This will be followed by a discussion about the squat toilet, as well as its more modern sitting version, and the Hygiene etiquette involved, including an overview on modern design solutions. The paper will conclude with the pros and cons of the squatting posture and the squat toilet.

Keywords: squat toilet, sitting toilet, squatting posture.

INTRODUCTION

In most Muslim countries, squat toilets are the norm. These toilets, used by almost two thirds of the world population, may seem archaic and "undignified" to most Westerners, but they have been proven as being much healthier and more hygienic than the sitting ones (Nature's Platform). There are several types of squat toilets (also known as Eastern, *alla turca*, Turkish, or Natural-Position toilet). These all consist essentially of a hole in the ground and places for the feet, with one exception, the "pedestal" squat toilet, which is as high as a standard sitting toilet. Old Turkish squat toilets, as well as the ones found in remote areas in the countryside, have a water tap and/or a container of water for washing the intimate parts with the left hand, and if available, toilet papers (please see figure 1).

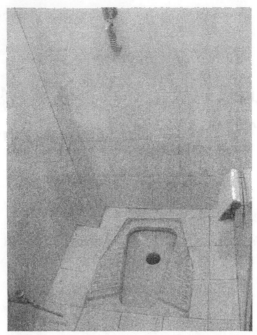

Figure 1. A typical Turkish squat toilet with a water tap on the right, a toilet paper roll holder and a wall mounted flush button.

During the Ottoman period, squat toilets were in private rooms generally located outside of homes for hygienic reasons. This changed with the development and improvement of drainage and sewage systems and these toilets took their place inside the home, in a section called *eyvan* (Genç, 2009). Figure 2 shows the basic types of Turkish squat toilets from the Ottoman period.

In urban Turkish homes, many toilets have 'evolved' to the sitting types (also known as *alla franca*, Flush Toilets or Western Toilets) for most buildings. This was considered as modernisation and many families have opted to the sitting posture, finding it more comfortable and aesthetic, and more Western looking. The hygiene and health aspects of the squat toilet have slowly been ignored. Even after the introduction of toilet paper, water still remained as a cleansing agent, and has also been incorporated to sitting toilets (see figure 7 below), in the form of a nozzle that comes out from underneath the toilet seat, from the back and squirts a jet of water. This is now a common feature in most households.

One could possibly also squat over standard Western sitting toilets, after raising the toilet lid, but this requires extra care, as they are not specifically designed for this purpose. Some retrofitting apparatus and designs are available to facilitate this task. Please see figures 10 and 11 for such examples.

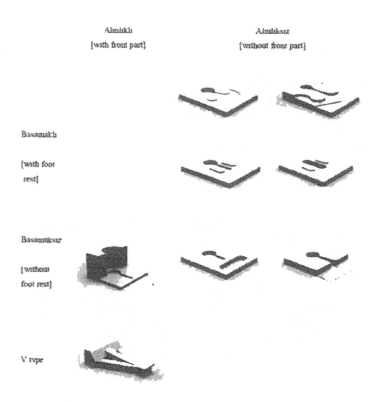

Almlıklı
[with front part]

Almlıksız
[without front part]

Basamaklı

[with foot
rest]

Basamaksız

[without
foot rest]

V type

Figure 2. Ottoman squat toilets (Genç, 2009: page 36).

HYGIENE ETIQUETTE IN USING A TURKISH TOILET

In relation to the grooming activities following the use of a toilet, Gallagher (2008) separates the users in two categories "the wipers and the washers". The Turkish culture belongs mainly to the washers' category. Muslims, Japanese and continental Europeans are all 'washers', mostly using a bidet after passing motion. For the Muslims, this is also a religious requirement, while for the others; washing gives them a greater sense of hygiene. In the Muslim faith, this washing is concerned with cleanliness and purity of body and soul, and can be applied to both sitting and squatting toilet postures. This comes from the fact that the Islamic culture gives an important role to water in praying, to purify the body and the soul. "In the Islamic garden, the water is the mirror of the Heavens and the symbol of life" says Ayşe Birsel, designer of *Zoê*, the washlet designed for Toto (see figure 7 below).

The Islamic faith has a long list of prescriptive rules regarding personal hygiene when it comes to the use of the toilet. This set of rules is known as *Qadaahul Haajah*. It is important to understand that these rules have been established well before the invention of toilet seats and toilet paper. Leaving the religious concerns aside, some of the rules for the hygienic etiquette of using a squat Turkish toilet are

538

as follows:

- One should squat keeping thighs wide apart applying the stress on the left foot.
- After relieving oneself it is essential to perform *Istinja* (washing with water) of the intimate parts with the left hand and water. This has been updated by religious leaders as: "At the beginning of *Istinja*, it is preferable to use toilet paper three times".
- After this process the hands should also be washed thoroughly.

MODERN TOILET DESIGNS

Modern versions of Turkish toilet designs have been created during the years and the following section will discuss examples of these.

Squat toilets

Although it is difficult to find examples of modern squat toilets, there are few designers reinterpreting the squat toilet. Two re-designs of the traditional Turkish squat toilet can be seen in Figure 3, both designed by Inci Mutlu and Gamze Akay for VitrA. In *Sun* (Figure 3, on the right), the foot grid, usually in the shape of two "elephant feet" has been extended all around the recess and the hole.

Figure 3. Squat Toilets by Gamze Turkoğlu Akay and İnci Mutlu for Eczacıbaşı VitrA (sources: Özcan, 2006: page 13, and VitrA Bathroom catalogue, 2008).

The design of the Water Room (Figure 4) designed by Ayşe Birsel is inspired by the beauty of water in nature, and incorporates a minimalist squat toilet which consists of a recess and a hole, with a bar to hold while rising up and a soft rock to lean against (encircled on the left of Figure 4). The water room was awarded first prize at the "Design the Future competition" in Japan, in 1989, and an "ID Magazine Award for Concepts" in 1990.

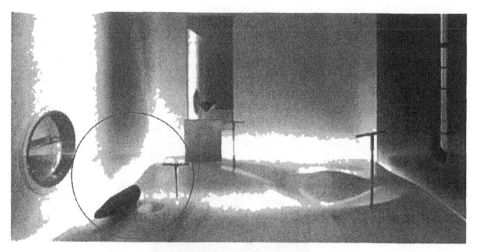

Figure 4. The Water Room incorporating a squat toilet.

Sitting toilet with imbedded washing pipe

As mentioned earlier, most seated type Turkish toilets have a washing copper pipe incorporated. A modern version of this is *Zoê* (Figure 5) designed by Ayşe Birsel for the Japanese company Toto.

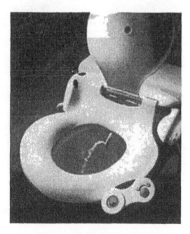

Figure 5. Zoê Washlet

Hybrid toilets, squatting/sitting versions

Flo (Figure 6) was designed by a team of staff and graduate students at Arizona State University, with the aim to design a sustainable, transgenerational toilet that would be usable by toddlers as well as by their grandparents (Christensen et al., 2006). Figure 7 shows *Pinz*, incorporating dual use of sit and squat toilet (http://www.pinz.com.sg). The benefits of *Pinz* are cited as follows: 1) Non-splashing, as the water level is close to he body; 2) Water saving with a pressurized cistern located next to the siphon jet; 3) No blockage with a large trap way (8 cm internal diameter); 4) Safe, as unlike Turkish toilets, this one is above the floor level with a rim around, avoiding slipping into the pan; 5) Hoods on both ends to contain urine spray, like Japanese squat toilets; 6) Easy installation, simply bolted onto the floor, like a normal toilet; and finally, 7) Choice of wet or dry landing.

Figure 6. Flo (Christensen, et al, 2006: page 10)

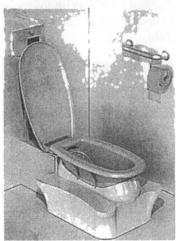

Figure 7. Pinz Ideation 2-in-1 (http://www.pinz.com.sg)

Retrofit squatting devices for sitting toilets

Figure 8 and 9 below show two different temporary retrofit squatting devices easy to install. The first one, *NaturesPlatform™*, provides a platform over an existing seated toilet, enabling the user to squat. This device is manufactured in the UK. The second one, *Lillipad* also allows for a semi-squat position for those with limited flexibility, by raising the feet onto the front step while sitting on the toilet and leaning forwards. A more permanent example, the Toilet Transformer, westernises and converts old-style Japanese squat toilets into the 21st century (please see figure 10 below).

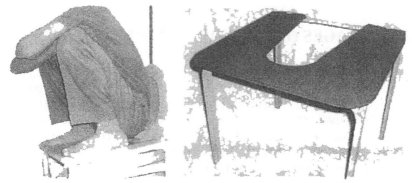

Figure 8. NaturesPlatform™, Nature's Platform toilet converter
(http://www.naturesplatform.com/)

Figure 9. Lillipad: retrofit squatting device for seated toilets (http://lillipad.co.nz/)

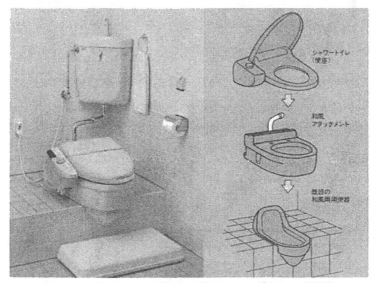

Figure 10. Toilet transformer (Toilet transformer, 2007)

ADVANTAGES OF THE SQUAT TOILET

The use of a squat toilet is said to have many health advantages from a physiological point of view (Aaron, 1938; Bokus, 1944; Hornibrook, 1963; Davenport, 1966; Kira, 1976; Heller and Henkin, 1986; Balaskas, 1991, Christensen et al., 2006). First of all, it is considered hygienic, as it does not involve any contact between the user and a potentially unsanitary surface. There is also no potential splashing as there is no water in the bowl. Ergonomically, the squatting posture provides a natural body posture and is healthier than the sitting one, as it provides for the alignment of the rectum and the anus in a near vertical position. This is facilitating the complete evacuation of bodily waste. Furthermore, elimination of waste in this posture protects the nerves controlling the prostate, bladder and uterus from being stretched and damaged.

Squatting also relaxes the puborectalis muscle and straightens the bend to allow waste to be evacuated easily (see figure 11). It is also said that squatting helps in reducing the occurrence of diseases of the digestive system, such as constipation and hemorrhoids (Dimmer et al, 1996, and Natures Platform, 2002) and other colorectal disorders (such as colitis, diverticulosis and appendicitis). For pregnant women, the squatting posture is also said to be better as it does not apply pressure on the uterus, and daily squatting is reported to help prepare for a more natural delivery (Balaskas, 1991). One other big advantage of squat toilets is that they are very easy to clean. They also consume less water per flush than western toilets and hence are more environmentally friendly.

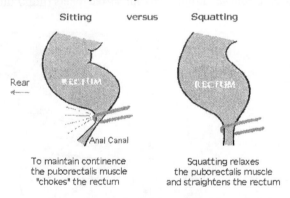

Figure 11. Puborectalis muscle in the sitting and squatting postures. (http://www.toilet-related-ailments.com/colon-cancer.html; and Natures Platform, 2002)

DISADVANTAGES OF THE SQUAT TOILET

From an ergonomic point of view, squat toilets are more difficult to use, requiring careful balancing skills. This is particularly important for people with knee joint problems, limited mobility or recovering from leg injuries. Elderly people may find it very hard to squat and rise back, if they are not used to it. For the large majority of the Muslim population, especially those practicing the religion with a regime of five prayers a day (involving a lot of kneeling down and rising up), squatting would not be a problem.

Another big disadvantage is that squat toilets may often smell bad, as their traps design does not allow for a complete flush. The sitting toilet, due to its bowl design that traps most of the odor under water and to the fact that it is completely flushed after each use, does not retain any odor.

Yet another disadvantage related to hygiene pointed out by Genç (2009) is that the footrest may get dirty and cause the transfer of microbes around. This may not be such a big problem for domestic toilets as most people in Eastern and Asian countries take their shoes off inside their home, and some also have special slippers for using the squat toilet. On the other hand, this is a problem in public restrooms (Genç, 2009).

Finally, squat toilets may also allow splatter to occur on one's own legs and feet, not to mention the potential to lose back pocket belongings into the hole.

CONCLUSIONS

Squat toilets have been designed and used in India, the Far East, Asia and Anatolia since ancient times. The instinctive squatting posture may well have inspired the design of the first squat toilets (Genç, 2009). This posture has advantages and disadvantages. The main advantages can be summarized as follows: considered hygienic, this posture provides a natural body posture that allows better relaxation during body waste evacuation, which prevents diseases in the small intestines. Furthermore, it is eco-friendly as it uses less water.

The disadvantages can be summarised as follows: this posture may not be comfortable for all, especially for the disabled and for elderly people with arthritis in the knee. Another disadvantage is that, in heavily used public restrooms, the footrests may hold dirt and bacteria that could be spread around by users (Genç, 2009).

Despite health and hygiene advantages, the Turkish squat toilet is getting less popular as days go by. The comfort of the sitting posture and the wide range of beautiful sitting toilet designs seem to shadow the advantages of the squat posture. The squat toilet is in desperate need of reinterpretation and innovation by talented designers to return into peoples' daily life. Finally, increased concerns about the environment and irresponsible use of natural resources may well bring squat toilets back as a healthy and sustainable alternative to flushed sitting toilets.

REFERENCES

Aaron, H. 1938. Our Common Ailment. New York: Dodge,

Balaskas, J. 1991. New Active Birth. London: Thorsons.

Birsel, A. On My Way to Water. http://www.core77.com/Reactor/ayse/bio.html [May 5, 2002].

Bokus, H. L. 1944. Gastroenterology. Philadelphia: Saunders.

Christensen, T., Takamura, J., Shin, D., and Bacalzo. D. 2006. Go With The Flo: A report on a collaborative toilet design project that utilized a transdisciplinary approach. International Conference in Lisbon. IADE Design Research Society.

---, Colon Cancer - Why Is It So Common In the West, But Not In Other Places? Toilet related ailments, http://www.toilet-related-ailments.com/colon-cancer.html [October 8, 2009].

Davenport, H. W. 1966. Handbook of physiology, 4 (2nd ed.). Chicago: Appleton-Century-Crofts.

Dimmer, C., Martin, B., Reeves, N, and Sullivan, F. 1996. Squatting for the Prevention of Hemorrhoids? Townsend Letter for Doctors & Patients (159): 66–70. ISSN 1059-5864. http://www.uow.edu.au/arts/sts/bmartin/pubs/96tldp.html.

Gallagher, W. 2008. Bath and Body Works. The Wilson Quarterly. Winter: 32, 1, p. 89

Genç, M. 2009. The Evolution of Toilets and its Current State. Master Thesis. Ankara: Middle East Technical University.

---, 2006. Health Benefits of the Natural Squatting Position". Nature's Platform. http://www.naturesplatform.com/health_benefits.html [October 6, 2009].

Heller, J., and Henkin, W. 1986. Body Wise. New York: J.P Tarcher Inc/St Martin's Press. s

Hornibrook, F. 1963. The Case for the Health Closet. The Architects Journal, July 31, pp. 221-232.

Kira, A. 1976. The Bathroom. New York: Penguin.

---, 2009. Lillipad. http://lillipad.co.nz/ [October 8, 2009].

---, 2002. Natures Platform. A Clinical Study of Sitting vs. Squatting http://www.naturesplatform.co.uk/site/clinical_study_of_sitting_squatting.php [October 12, 2009].

Özcan, C. A. 2006. H2O is Not Water Everywhere: Cultures Evolutionary Design Practices on Water. The 6th International Conference of the European Academy of Design. Bremen. P. 13 (paper 058).

---, 2009. Pinz Ideation. http://www.pinz.com.sg [October 7, 2009].

---, 2009. QuantumFoam. Islamic Toilet Etiquette, QuantumFoam's Diary http://www.kuro5hin.org/story/2009/2/20/15517/5559 [October 10, 2009].

---, 2007. Squat Toilet Totally Explained. http://squat_toilet.totallyexplained.com/ [October 8, 2009].

---, 2007. Toilet Transformer: zaps old-style Japanese squatters into the 21st century. http://www.digitalworldtokyo.com/index.php/digital_tokyo/articles/toilet_transformer_zaps_old_style_japanese_squatters_into_the_21st_century/ [December 5, 2009].

---, 2008. VitrA Bathroom (n. d). VitrA Bathroom Culture. http://enexp.vitra.com.tr/design_culture/overview.aspx [December 3, 2009].

Chapter 56

Doing the Laundry - Different Strokes for Different Folks

Dr. Nicole Busch[a,b] and Henri Christiaans[b]

nbusch@buschwerk.eu and henri.christiaans@io.tudelft.nl
[a]buschwerk - User Experience Research & Design,
Deichstr 39, 20459 Hamburg

[b]Delft University of Technology, Faculty Industrial Design Engineering
Landbergstraat 15, 2628 CE Delft, The Netherlands

ABSTRACT

Globalization and profit maximization force companies to introduce their products in many different regions of the world. Only an awareness of the local circumstances, the needs and desires of the local consumers, their habits and rituals, as well as the local restrictions and limitations will make their efforts a success. This presentation gives an overview of the different washing cultures worldwide to find out if innovations are compatible with existing cultural habits. It appeared that washing machines are used in Europe, Australia and parts of Asia, Africa and America differently. In some cultures the washing machine is placed on wheels to be transportable within the house, sometimes it is positioned on the balcony or in narrow spaces. The content of washing textiles differs and therefore the requirements. Circumstances differ, is washing powder affordable, electricity or water available. This paper shall give an overview about doing the laundry around the world and its results for the design process.

Keywords: Culture, household appliances, everyday, product, design, research.

INTRODUCTION

Today, companies are interested in selling their products worldwide. Different user groups, for example different age groups, use products with different interests. Products have to fit to different abilities and multiple body dimensions as well. Research was done (Busch and Vink, 2006, Busch et al. 2006) and standards were set, to make products as useful and as desirable as possible. For the global market, products have to meet even more requirements, as users in different countries and cultures behave in different ways, depending on their circumstances.

However, many products are designed by European design agencies or produced by European manufacturers. This means that many people have to use products, which were not originally designed for them, their countries or their culture. A company's failure to acknowledge cultural differences often limits its product's marketability (Rutter and Donelson, 2000). Cultural differences were studied by comparing markets and products internationally (e.g. Honhold, 1999, 2000, Ono, 2005). The conclusion from these studies is that products should either be produced and designed locally to satisfy local consumers, or should be equally usable and fit across different countries and cultures. But what does culture mean?

CROSS-CULTURAL RESEARCH

Max Frisch (2005) claims: "We live in a time, in which man are not able to define the meaning of culture anymore." Indeed, the definitions of culture are numerous. As a starting point, most common definitions of culture are described. Kroeber and Kluckhohn (1952) define culture as consisting of patterns of or for behavior, which are transmitted by symbols, constituting the distinctive achievements of human groups, including the design of products. The essential core of culture is made up of traditional ideas and above all of people's values in the present and the future. Anthropologists define culture as a divider between Homo sapiens and animality, i.e. the beginning of hominisation and the development of culture (Grupe et al., 2005). Dahl (2004) states that the word 'culture' is used in "everyday language to describe an abstract entity, which involves a number of usually man-made, collective and shared artifacts, behavioral patterns and values". Hofstede (1994) defines culture as "mental software" – "the collective programming of the mind, which distinguishes the members of one group or category of people from another". In their models, Hofstede (1994) and Trompenaar (1993) provide guidelines to examine differences in culture and how to understand people's own cultural biases. Hall (1977) defines culture as the lifestyle of a group of people, as the sum of their behavioral patterns, attitudes and material things. All definitions state that culture is learned and not inherited. Hoecklin (1995) describes the effect that culture dictates what groups of people pay attention to. The learnability aspect was studied when people were moving from one cultural area to another, taking over habits and behavior of the new culture (see e.g. De Leur et al., 2005). However, it is not always clear at what level this 'new' behavior was internalized: on the level of outward appearance or on deeper levels of beliefs and values?

ASPECTS OF CULTURE

The main global success factor of products is their international usability (Christiaans, 2005). ISO 9241-11 defines usability as the extent to which specified users can use a product to achieve specified goals with effectiveness, efficiency, and satisfaction in a specified context of use. Leonard and Rayport (1997) emphasize the value of ethnographic research to understand how people interact with products, environments, and services. To help companies in foreign markets with various cultural factors and their implications, observational studies on the use of products in different cultures have to be conducted (IDEO, 2003) about various cultural and environmental contexts in which the product might be used. Data can be collected to evaluate perceptions and behaviors to underline similarities or differences of cultures. Intuitive cross-cultural products should reflect the cultural orientation of the users and fit to the users' cultural differences. They have to be able to fulfill people's demands, needs and desires in their environment.

DOING THE LAUNDRY

As white goods companies aim to sell their products, like washing machines, world-wide, it would be smart to know the local needs, rituals, wishes or even limitations. This cultural overview focuses on one household task, which is doing the laundry, an activity that is by its nature a basic need. Washing is a global activity but local circumstances, patterns, automation or washing machine constructions vary. In some areas people still wash manually, in the industrial world the washing process is generally automated. Depending on local circumstances, washing machines work in totally different ways, resulting in very different designs and patterns of usage.

Many companies produce and design washing machines for an international market with diverse cultural differences, which have to be carefully considered (Cushman and Rosenberg, 1991). Kumar (2004) describes how companies underestimate the differences of patterns in daily life in different cultures and therefore fail to meet people's culture-specific needs. For example Whirlpool's "World Washer", which was designed to be a one-size-fits-all machine and was introduced as an important part of the company's global strategy into the Indian market in the late eighties. It failed because designers did not study that most women wore saris, a 30-foot long cloth in India (Honhold, 2000), which became entangled in the machine paddles and caused abrupt water and electrical stoppages. If companies want to expand globally, they need to recognize the differences in people's lifestyles and tailor their offerings according to users' demands and needs. The aim of this study is to give a cultural overview of needs in several countries.

In closer detail, this research answers the following question:

- Do people wash differently when influenced by a different culture?

It is assumed that some of the washing diversity

- may be due to the diversity of environments,
- comes from different types of clothes to wash,
- depends on different histories and backgrounds,
- is driven by economical aspects, or
- could relate to different levels in a social structure.

It is hypothesized that these influences could result in

- different washing strategies with different needs and limitations.

METHODS

In order to answer these research questions about the cultural differences in washing between countries, qualitative research methods were used. For observations countries were selected. It is important to compare leading (Europe, USA and Japan) and upcoming countries (India and China). Countries with additional washing methods were added (Latin America, South Africa, Maldives, Russia). Local information was gathered by asking local researchers. Studies were conducted in America (USA, Venezuela, Argentina, Uruguay, Paraguay, Chile, Brazil), Europe (Germany, the Netherlands, Scandinavia, Spain, Russia), Asia (China, India, Korea, Japan, the Maldives) and Africa (South Africa, the Gambia).

Data were collected through 'cross-cultural observations' (IDEO, 2003). For this method, 11 socio-cultural reports were studied to compare related data about washing in different countries. The relevant comparisons about washing in general, differences in machine systems, loading and unloading, and other environmental, social and economic circumstances were filtered as interface usability was not the focus of this research. To reveal cultural differences in the context of washing reports from Brazil, the Netherlands, Germany, India, China, South Korea, and Australia, general overviews on everyday objects used as well as the habits of carrying loads were researched.

To collect information about latent consciousness, 'cultural probes' (IDEO, 2003) were chosen. Camera journals were assembled and distributed to participants across Asian cultures. Participants were asked to take pictures of their daily routine, with some images and descriptions according to washing ('photo survey', IDEO, 2003). From 20 photo surveys, 10 were returned and 3 (1 from Japan, 2 from China) contained relevant details about washing.

To acquire information about local circumstances for this cultural summary 'foreign correspondents' (IDEO, 2003), i.e. colleagues, friends and their contacts were contacted to learn about local needs, problems and limitations. These contacts sent images and descriptions about the washing process from the USA, Mexico, Brazil, Norway, Spain, Egypt, Russia, China, Taiwan, Japan and Australia.

RESULTS

The results are sorted according to the continents eastbound starting in America. American style washing machines influenced the South American and Asian markets. When presenting cultural habits, first the average user with average income is described and then, if collected data is available, it is pointed out in which way habits differ for people over and under the average income. Differences in washing machines or washing methods, as well as differences in related washing habits are presented.

For a first overview, Proctor & Gamble (2005) compared washing habits for automated washing in the USA, Europe and Japan (figure 1 and 2).

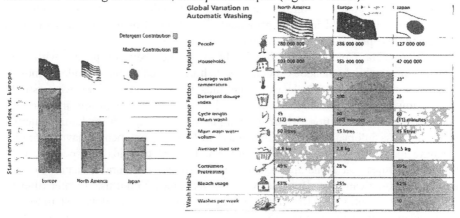

FIGURE 1: Efficiency of washing machines compared by Proctor & Gamble (2005), FIGURE 2: Consumer washing habits analysed by Proctor & Gamble (2005).

NORTH AMERICA

In North America practically every household or apartment building has a washing machine. A shared washing room is common. People mainly use toploaders for washing, described as very comfortable to load and unload. The downside is that the drum rotates on a vertical axis, resulting in laundry pressed to the periphery of the drum. So pieces of laundry do not interfere with each other. This physical rubbing effect is important for the washing result, as it makes clothes cleaner. One remedy is to use stronger washing liquid and bleach to achieve a better result (see figure 2). The effect is that the color of clothes fades quicker, besides stronger environmental issues. Another solution is a mechanical tool to agitate the pieces, with the effect of physical stress on the fabrics and clothes that are worn out sooner. In general, European-style washing machines are becoming more interesting for consumers because they are renowned for their better washing results.

The population structure of the USA with its large proportion of immigrants and people with Latin American background, its many people with an under-average income who cannot afford a washing machine, and those who live in small city apartments without the space for a washing machine, are factors that lead to the popularity of launderettes or shared community washing machines.

LATIN AND SOUTH AMERICA

In many countries of Latin America, people do not have high-tech washing machines, if at all. Old models and basic models are more common. Some of them are only half-automatic, so users have more manual, hard and time-consuming work to do (Clark, 2004) e.g. water has to be filled in by hand, no centrifuge so people, mostly women, have to wring out the wash. Especially in underdeveloped Latin and South American countries, the whole process of laundry is often done manually. Washing powder is sold in small packaged sizes instead of in bulk-size packages mostly because people cannot afford to buy bigger sizes.

In Venezuela, Colombia and Brazil, the difference between people with high and low income is tremendous. People with higher income have modern, mostly American-style washing machines, sometimes used by housemaids who usually have a lower level of education including illiteracy. People with low income still wash with less automation, which is comparable to people from Latin America using old, less automated washing machines. In Brazil, many people have toploader washing machines placed in special cabinets. For the technical perspective, Ono (2005) describes that home appliances are less robust and durable than European and North American ones. Brazilian washing machines do not use a water filter to reduce water pollution. All Brazilian washing machines have a safety system to stop the process when users open the lid, though people like to fool the system, open the lid and see the machine working. As a result, Electrolux integrated a glass lid (in: Ono, 2005). Unlike North America, laundromats are very uncommon and are not used by Brazilians, which is different to many other nations worldwide.

In countries like Uruguay, Paraguay, Argentina and Chile, European-style washing machines are used more often. Many of the people have a middle-sized income. In cities like Buenos Aires or Montevideo, washing machines are placed on balconies. In the countryside, machines stand in garages and the garden as well. As weather conditions show strong variations in temperature and humidity, rust and oxidation are limiting the durability of washing machines.

EUROPE

In Western Europe most households have their own washing machines. Front-loaders are most popular resulting in uncomfortable loading and unloading strategies as analyzed from Busch and Vink (2006). Some people build work-arounds, such as understructures to place their machine on.

In general, only a few differences can be seen between countries, mainly in the circumstances in which a washing machine is used. In Spain, for example, there is no standard for water pressure. Low water pressure is a problem in multi-level apartment buildings. That is why washing machines suffer from a lack of water, causing frustration because predicted washing times cannot be maintained.

German washing machines (e.g. Bosch, Siemens, Miele) are perceived as very robust, solid and conservative (Ono, 2005), compared to other European models.

In Russia the difference between people with high and low income is tremendous. People with higher income have modern, mostly European-style washing machines. In general, though, if at all, people have old, low-tech washing machines of mainly Russian brands. The biggest problem, however, is drying the clothes after the washing procedure. The climate can be very cold in winter or very humid in summer, so clothes stay dank and start to mildew. According to Lindstrom (2006) Russians associate with washing the smell of clothes dried at -20°C.

Northern Europe has the same problems, so dryers are very popular. Where space does not matter, dryers are placed next to the machine. Many people complain that the doors open against each other, to the wrong side and are generally in the way. Where space does matter, dryers are placed on top of the washer and lifting wet clothes becomes an issue.

AFRICA

In most parts of Africa, the laundry is still done manually, mainly because water in the house and electricity are not or only seldom available. In manual washing less water is used than for a washing machine. Most of the washing is done by hand. As firewood is rare and expensive, the laundry is done with cold water, lots of washing powder and scrubbing. Dust is ubiquitous and stains clothes immediately. Washing is done either in plastic washtubs close to the well, pump or directly in the river. After washing white linen is spread out for bleaching. Afterwards all pieces have to be ironed to kills nested bugs. In South Africa big differences can be seen between high and low income. People who own a washing machine (often frontloader) normally have maids to work with it, but most people still wash by hand.

In general, Africa is the continent where clothes are generally washed using pre-industrial resources (Gestora, 2004).

ASIA

In China and Taiwan differences between the newest design of washing machines and manual washing are very large, which is highly related to people's income.

The phrase 'white goods' only works in the Western culture. In China and India, 'white' is perceived as the color of mourning, instead of as the color for cleanliness. Siemens (2001) reports that a Chinese tradition is to receive a washing machine as a wedding present painted pink with hearts stuck on to remind the couple of their wedding gift-giver. Space matters, as people live very closely in China's big cities. While Chinese people in general enjoy noisy conversations, games and music, the noise level of the washing machine, especially when centrifuging, seems to bother people (Siemens, 2001). The washing machine 'Super Silent', produced by BSW (joint venture of Bosch Siemens Home Appliances and Wuxi Little Swan), received an award from the Chinese patent office for washing, rinsing and spinning without producing noise. Moreover, it is a colorful appliance. Not all people in China have washing machines and manual washing in washtubs is still very common.

In India, like in other Southeast-Asian countries, the consumer needs vary dramatically compared to Europe. As garments are very thin, long and colorful, washing by hand or even taking a shower with clothes still on is very common in Southeast Asia. Honold (1999) reports that Indians wash almost once a day. They do not collect soiled clothes for hygiene reasons. Indians use four different methods of washing: manual washing, a washer, semiautomatic and automatic toploaders. In a washer, clothes are rotated in cold water, spinning around a vertical axis without a centrifuge function. Semiautomatic toploaders use two drums, a bigger one for washing, and a small one taking wash portions for centrifuging. The automatic machine combines washing and centrifuging, always around a vertical axis with cold water (Honold, 1999). In cramped living conditions in the cities, washing machines are placed on the balcony, in the hallway or even in the bedroom. Honold reports about a marketing specialist who observed washing machines placed on mobile carts in Mumbai. They draw the conclusion that mobility must be important as place matters in crowded circumstances. But difficulties in cleaning are the main reason for this habit as the climate is hot and humid. It is very dusty, which makes wiping the floor daily necessary, also underneath the machine. If it was not done daily, dirt would collect, mould and fungi would grow and vermin (e.g. rats) would nest. (See Honold, 1999 and 2000.)

In Korea and Japan nearly every household has a washing machine. As America exported their toploaders design with a vertical axis. Washing is mainly a female task in Japan and Korea. People in Japan wash depending on the family size between one and four times a week. They separate colors, whites and delicate fabrics (e.g. silk). In very cramped cities like Tokyo, where it is very common to place the washing machine on the balcony with taps installed for this case. The washing is hung on the balcony as well. In suburban or rural Japan, people have a separate laundry and the washing hangs outside as well. Very hot summers, warm springs and autumns may be the reason that dryers are not that popular in Japan. A new washing machine with angled drum was just recently introduced into the market. The advertising campaign points out that the angle makes the washing machine more usable for the elderly, wheelchair users and even children.

DISCUSSION

Regarding the aim of this study to overview different needs and limitations when washing clothes and to show different strategies of washing in different cultures, the results of this paper are summarized as follows:

Where do people (**who**) wash (**how**) **what**? These attributes phrase the multiple dimensions of washing clothes. Below each attribute is explained:

- **Where** people wash is influenced by various factors. Different climates and environments need to be taken into account. Resources like water or energy are limiting factors. The climate heavily influences the process of washing how often people have to wash. That also depends on how dirty, dusty, sweaty or moldy clothes depending on a specific environment. Furthermore, the climate dictates how drying can be done.

- **What** people wear is also defined by their environment. They adapt their clothes to climate and social circumstances. Depending on their culture people wash different fabrics with different densities, textures and sizes. The differences in colorfulness of clothes, dying technics and how careful they need to be washed are a result of the cultural background. These differences become very important when it comes to defining a product solution.

- **Who** does the laundry is probably the one criterion that most cultures have in common. The laundry, especially at home, is mainly a female task. In some nations, where females do not work outside the house like Arabia or the Maldives, men work as professional washers. In cities with a high percentage of singles or in countries with less gender differences like Northern Europe, men do wash as well. As washing has always been and in some parts of the world still is heavy work, women draw upon their family or other women of the community to help them.

- **How** do people do their washing? All washing methods from manual to automatic washing are used around the world. The automation of this task can be related to how much people can afford to spend for a washing machine and also to whether a machine makes sense, depending on the availability of water and electricity conditions.

Still it is unclear why there are two main systems of washing machines on the market: washing machines spinning on a horizontal or on a vertical axis. It is proven that washing machines spinning on the vertical axis show poorer results, use more water and detergent and wear the fabrics faster. Maybe the main reason lies in the production costs, which are less for a vertical axis machine. This results in a lower consumer price, with more people being able to afford to buy such a machine.

To a great extent, the design of a washing machine depends on the choice of the axis. However, the design impression varies in different countries. Color and finishes are adjusted to local taste or rituals. Depending on the markets interfaces are designed according to local conditions, e.g. certain symbols for regions with a high percentage of illiteracy. The available washing programs are based on local washing habits, e.g. short washing cycles with cold water for saris in India.

Leur et al. (2005) describe an emerging interest in the impact of cultural dimensions on the experience and interaction between people and products. If companies want to design and produce washing machines which will be used in different cultures and environments, they have to study the needs and requirements of their export market, especially when something is designed or produced for a totally different culture. People wash everywhere in this world, but as this paper shows, no global washing culture exist. Local needs vary, even for countries close to each other.

REFERENCES

Busch, N. et al., 2006, *Comfort of washing machines when loading and unloading in a field study,* (Paper accepted), Applied Ergonomics.

Busch, N., Vink, P., 2006, *Comfort, body posture and speed in washing machine loading and unloading,* (Paper accepted), Applied Ergonomics.

Christiaans, H., 2005, *Cultural Differences,* Presentation on internationalisation, TU Delft.

Clark, J., *taken from http://www.geocities.com/jonclark500/weblog/2004/08/ happy-time-cleaning-time.html.*

Cushman, W. and Rosenberg, D., 1991, *Human Factors in Product Design,* Elsevier Science Publishers, Amsterdam.

Dahl, S., 2004, *An Overview of Intercultural Research,* Middlesex University Business School, Middlesex.

Frisch, M., 2005, quote taken from http://de.wikiquote.org/wiki/Kultur.

Gestora, J., 2004, ed., *Equally_Different - Everyday objects from around the world,* FAD, Forum Barcelona 2004.

Grupe, G. et al., 2005, *Anthropologie – Ein einführendes Lehrbuch,* Springer Verlag, Berlin.

Hall, E., 1989, *Beyond Culture,* Anchor Books Editions, New York.

Hoecklin, L., 1995, *Managing Cultural Differences.* Strategies for Competitive Advantage, EIU S, Financial Times Prent. Int, New York.

Hofstede, G., 1994, *Culture and Organizations - Software of the mind,* HarperCollinsBusiness, London.

Honhold, P., 1999, *Interkulturelles Usability Engineering - Eine Untersuchung zu kulturellen Einflüssen auf die Gestaltung und Nutzung technischer Produkte,* Inaugural-Dissertation, Regensburg.

Honhold, P., 2000, *Culture and Context: An Empirical Study for the Development of a Framework for the Elicitation of Cultural Influence in Product Usage,* Journal of Human-Computer Interaction, 12 (3&4), pp. 327-345.

IDEO, 2003, *IDEO Method Cards: 51 Ways to Inspire Design,* Palo Alto.

International Institute for Standardisation, 1994, International ISO DIS 9241-11, Draft, *Ergonomic requirements for office work with visual display terminals* (VDTs) - Guidance on Usability, International Organisation for Standardisation.

Kroeber, A., Kluckhohn, C., 1952, *Culture: A Critical Review of Concepts and Definitions,* in: Adler, N., 1986, International Dimensions of organisational behavior, PWS-Kent Publishing Company, Belmont.

Kumar, V., 2004, *User Insights Tool - A Sharable Database for Global Research,* IWips Conference.

Leonard, D., Rayport, J., 1997, *Sparking Innovation Through Empathic Design,* Harvard Business Review, Nov-Dec 1997, 102113.

Leur, K. de et al., 2005, *Cross-Cultural Product Design: Understanding people from different cultural backgrounds,* IWIPS Conference 2005, Amsterdam.

Lindstrom, M., 2006, *It Simply Makes Sense!* Summary of speeches from the 11th German Trend-day, 11. May, 2006, www.trendbuero.de

Ono, M., 2005, *Design and culture: essentially interlinked, plural, variable, and beyond the predictable*, (Paper submitted), Journal of Design Research, Delft University Press, Delft.

Rutter B., Donelson, T., 2000, *Measuring the Impact of Cultural Variances on Product Design*, Medical Device & Diagnostic Industry Magazine, October 2000.

Siemens, 2001, taken from: *http://w4.siemens.de/Ful/en/archiv/zeitschrift/ heft2_99/artikel02/*.

Trompenaars, F., 1997, *Riding the Waves of Culture*, Nicolas Brealey Publishing, London.

CHAPTER 57

The *Ability/Difficulty Table*: A Tool to Identify the "Limit Target" in the *Design for All* Approach

Giuseppe Di Bucchianico

IDEA Department
University "G. d'Annunzio" of Chieti-Pescara
Viale Pindaro, 42, 65127 Pescara, ITALY

ABSTRACT

One of the major problems for anyone involved in Design for All is identifying the design target. The paper proposes the first results of a research that has developed a tool which eases this task. Starting from a specific activity, the table, termed "A/D Table", enables the identification of groups of individuals who, because of their characteristics, may have the greatest difficulties in carrying it out.

Keywords: Design for All, inclusivity, user limit, abilities and difficulties

INTRODUCTION

"Design for All (DfA) is design for human diversity, social inclusion and equality". This definition, taken from the EIDD "Stockholm Declaration" (European Institute for Design and Disability) represents the fundamental principle underlying the design of environments, facilities, everyday objects and services, usable autonomously by individuals with diversified needs and abilities.

It is a concept, and at the same time, it is a design approach clearly of European character (as such it has been officially acquired by the European Commission). It is certainly tightly connected to the more famous North American "Universal Design", nonetheless differing with regard to the importance given to accessibility design specific for disabled persons.

Currently, the DfA target essentially refers to "all" humanity, with the complex of its diversities, which are not only psychophysical, but also cultural or social, permanent or momentary. Therefore, DfA refers not only to individuals with disabilities, but also to all those who, on different levels, do not fit the psychophysical and sociocultural conditions of "standard" users: adult and young individuals, which are healthy, perfectly able and totally lucid, careful and aware (representing 5% of the population, for which, however, probably more than 95% of products and services are designed).

From this perspective, human diversity can be seen as an enhanceable "resource" rather than a constraint to be considered during the project phase. The concept of "users" is also extended from the "final" players to all the individuals in the–entire product supply chain. So, by definition, DfA " includes" the "whole" inside its target: namely, it pursues the satisfaction of the needs, desires and aspirations of "all" individuals who, for various reasons, want and have a reasonable chance of better "experiencing" the product, or of benefiting from its autonomous use.

However this "field enlargement" may become a problem for designers, because they are in any case obliged to know the user features and needs, even if strictly referred to specific projects.

The paper presents the first results of a research conducted during the development of a Thesis Degree Laboratory in Industrial design on the DfA theme, in the Architecture Faculty of the "G. D'Annunzio" University of Chieti-Pescara[1].

The research objective was to define, through a DfA approach, a design tool useful to simplify the description of the needs reference frame of the "potential" product users, such that it be sufficiently detailed and objective, and easily reset as the project changes.

RESEARCH PURPOSES: DEFINING THE "LIMIT" USERS

In the Design for All approach, the most important and delicate phase of the entire design process is the definition, sufficiently clear and complete, of the so-called "users-system". In other words, the objective is to understand who comprises the "All", that is who are the users that, at the different phases of design development (Accolla, 2009) "desire" ("metaproject" phase) and have a "reasonable probability" ("Project" phase) of enjoying the product. For the designer, this implies: on the one hand the need to understand and interpret the reference scenario, which is the

---

[1] The graduants involved in the research are listed under "Credits"

context in which the product will be used; on the other hand the need to know the psycho-physical characteristics of users interested in its "independent use"[2].

Assuming that the environment, rather than the physical condition of individuals, is responsible for creating the disability, is also true that for every activity it is possible to define the so-called "limit" target, namely the subset of individuals who, in certain contextual conditions and situations of use, represent indeed "the borderline" of the independent use of products and environments. The idea is that, by resolving the project in relation to their characteristics, abilities and needs, it is possible to "include", with good approximation, all the others.

The problem is that, compared to the many possible areas of application of Design for All and to the various activities that characterize them, the "limit" group of individuals (with regard to independent use) is continuously variable and requires redefinition from time to time. In overcoming a height difference, for example, disabled people in wheelchairs have a greater disadvantage with respect to a blind person or an obese person, whereas a blind man expresses greater difficulty, compared to the other two users, in reading a sign along the same path.

The designer, therefore, needs appropriate conceptual and methodological tools with which to identify and define, every time, the "limit" target, to be then able to describe, with the best possible objectivity and completeness, their needs and, based on these, the project requirements.

The research aims to develop a tool that is able to identify the "limit" users in an objective and complete way, for each specific activity or task. The idea is to be able to describe each time the specific needs and to then transform them, eventually, into design requirements. The tool should also facilitate the assignment of a hierarchical order to the different needs, allowing the designer to manage the complex system of constraints arising from the characteristics of users, in complete awareness and with different possible levels of analysis.

METHOD: CONSTRUCTION OF THE "A/D TABLE"

Research activities have focused on the development of the tool now being proposed, defined as "Table of Ability/Difficulty" ("A/D Table"), useful for linking the specific tasks to be performed with a given product or in a given environment and the difficulties expressed from time to time by the various possible users.

In particular, the "Table of Ability / Difficulty" is a rather complex scheme, termed "triple-entry", as it links three groups of variables, which relate to one another two by two.

This tool was synthetically constructed through three different steps of development:

- First step: construction of a matrix relating disabilities reported in scientific

---

[2] Compared to the concept of "self-enjoyment" of the context, please note that the handicap and disability have been defined by the WHO as "a complex interaction between the health of the individual and the influence of the environment surrounding" (ICF, 2001).

literature and their objective and generalized "difficulties", with respect to physical/operational and cognitive/cultural aspects;

- Second step: description, through HTA (Hierarchical Task Analysis) methodologies, of a complete and detailed list of tasks related to the specific activity that will be tested and of the physical/operational and cognitive/cultural "abilities" requested of users to do them;
- Third step: definition of the overall pattern of the "A/D Table" by interrelating the results from the previous two steps.

FIRST STEP

The research began with the establishment of a list, of significant size, of potential physical/operational and cognitive/cultural "difficulties", objective and generalized, expressed by individuals characterized as having the major disabilities reported in scientific literature[3]. The list was also completed, adding to the initial "disabling" conditions, strictly related to human health, some other aspects tied to socio-cultural factors which, if not directly disabling on a health level, may be so with respect to human organizations. The list thus obtained was subsequently briefly reorganized into seven "Macro-classes" of disabling characteristics, that are attributable to: physical disablement, body build, life stages, sensory perceptions, large organ system dysfunctions, social/cultural charactersitics, intellectual difficulties.

The classes, broken down in this way, despite being, in some cases, extremely reductive and generalizing, have nevertheless provided a sufficiently vast description of the possible target-user disabilities. In parallel a list of significant size has been defined, detailing the difficulties expressed in different disabling conditions, characterized by:

- physical/operational difficulties, when the disability leads to difficulties, more or less obvious, of a purely motor type. This applies, for example, to hemiplegia, which expresses an obvious physical "difficulty" if the individual is required to stoop or bend down to the ground independently, and so on. Some operational difficulties were also included in the same category, where these are attributable to procedural activities to be performed in relation to specific spatial contexts: the privation of the use of a lower limb, for example, can lead to operational difficulties, more or less marked, when moving along ramps or gradients, even with the aid of prostheses.
- cognitive/cultural difficulties, when referring to possible perceptual disabilities or to disabilities of semiotic/ cultural nature. An example is given by the possible disability resulting from the difficulty in recognizing

---

[3] In particular, in Italy reference was made especially to a classification of 2008, adopted by the Ministry of Health, Department of health issues, research and organization of the Ministry, entitled "ICD-10. International statistical classification of diseases and related health problems (tenth revision) vol.1, 2,3. IPZS Institute. Rome.
Online: http://apps.who.int/classifications/apps/icd/icd10online/

and interpreting signs and directions provided in an unknown language, or that arising from being able to recognize only clear auditory stimuli, this being the typical difficulty of individuals suffering from hearing loss.

The range of disabilities and of their relative difficulties has allowed the development of a first matrix of values, defined precisely as "Matrix of Disabilities/Difficulties" (figure 1). Here, it has been possible to associate, with each "disability" a set of objective "deficits" or "difficulties", represented in the matrix, as coloured boxes.

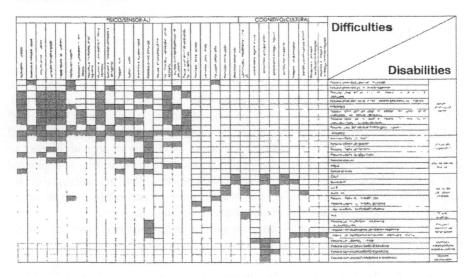

Figure 1: "Matrix of Disabilities/Difficulties"

The so defined table, which directly connects disabilities and deficits, also allows one to identify new and unexpected disabling and discomfort relationships among the different disabilities. In fact, the table shows how each disability may be associated with more than one specific deficit, and how the same deficit may match multiple disabilities.

This is a very useful relation system to handle cases of "temporary" disability (referring to those using bandages, joint braces, etc..), or to suggest at this step some alternative design solutions to offer to users in performing a task.

SECOND STEP

A second matrix, defined as "Matrix of Activities and Skills" (Figure 2) was subsequently developed. The first group of elements of the matrix refers to the activity, or rather the task under analysis whereas the second group takes up the same list of "difficulties" used above, but this time they are considered as "capacities" necessary to carry out such activities.

In particular, with regard to the first group of items, it is extremely variable because it refers to different activities in turn. The idea is to break up the main activity in the different subtasks (Task) and in the individual elementary actions that compose them (Actions), using a classic method of task analysis known as HTA - Hierarchical Task Analysis (Annett and Duncan, 1967). The complete list of elementary actions relates directly with the list of possible skills required to carry them out.

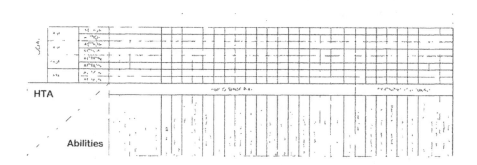

Figure 2: "Matrix of Activities and Skills"

The matrix thus obtained highlights the possibility of identifying a "quantitative" relationship between individual tasks and the skills needed to accomplish them: this allows one to perform an initial assessment of similar activities, distinguishing between the more hazardous ones and those that are more interesting. It is possible to estimate the "quantity" of skills required to perform a given task, by simply "summing" arithmetically the number of skills required for each task (horizontally) and the frequency with which a given capacity is used in the performance of the activity (vertically).

THIRD STEP

The overall scheme that forms the "A/D Table" is constructed by aggregating the two matrices obtained in the–previous two steps. The element of "connection" is given indeed by the two groups of elements, essentially the same but conceptually different, as defined by the list of "difficulties" and list of "abilities" or "skills", in the first and in the second matrix, respectively.

The scheme is completed by inserting a row and two columns showing three different measures of quantitative assessment, referring respectively to:

- "Frequency of HTA Abilities", that is the frequency with which individual skills are required during the conduct of the entire activity in question;
- "Number of skills/task", that is the arithmetic sum of all the physical/operational and/or cognitive/cultural skills that are necessary to perform each single action;
- "Index of the criticality of difficulties", that is a statement of "the weight"

of the disabilities with reference to the activity analyzed.

MAJOR RESULTS: USING THE "A/D TABLE"

Since this is a triple-entry scheme, it is easier to understand the potential of the "A/D Table" (Figure 3) testing its use through the various possible "paths" through which the scheme can be read.

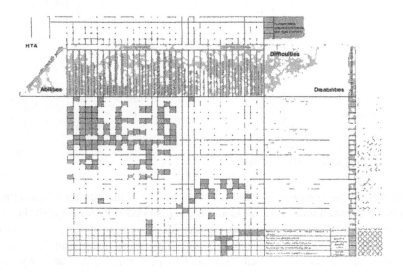

Figure 4: the complete "A/D Table"

As a matter of fact it is possible, for example, to use the table "top-down", that is starting from the abilities required by a specific activity to determine the difficulties and therefore its critical users; or one may want to use the table "bottom-up", starting from given categories of users and their "typical" difficulties, in order to identify their major skills, with the potential aim to intervene in the project by directly changing the hierarchical structure of the tasks related to the activity.

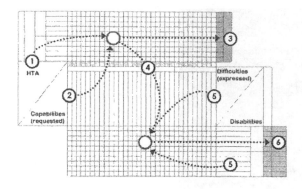

Figure 4: Description of the steps that define the "top-down path"of the "A/D Table".

Among the various possible uses of the table, the following elaborates on the "top-down path" In particular, the most significant points and the possible partial results are shown through six successive passages (Figure 4):

1: starting from the main activity under analysis, it is necessary to break it down into the various tasks and actions, according to the most typical HTA notations, and to then fit the detailed list obtained in the top left column of the scheme.

2: in the top matrix the tasks and the individual actions should be crossed with the list of skills required to carry them out, highlighting among them, those involved.

3: a first result offered by the scheme is shown in the upper right column values in the upper right, which indicates the arithmetic sum of the physical/operational and/or cognitive/cultural abilities necessary to perform each action. This allows the detection of tasks that quantitatively require more skills to be carried out, providing, thus, a first criterion for reading the most critical points of the project.

4: A second result is given in the row "Frequency of HTA Abilities", which expresses another quantitative index of evaluation, as it shows the frequencies with which individual abilities are required during the conduct of the entire activity in question.

5: Writing down the frequency number expressed by the "capacity", in each of the corresponding "colored" boxes in the matrix below (in regard to the respective difficulties), one actually attaches a numerical value to every difficulty. These relate, however, even with the various forms of disability expressed in the matrix below, showing, in fact, an indirect relationship between disability and the activities.

6: Adding the values related to the difficulties shown in each line of disability, it is also possible to define a hierarchy between different types of disability with regard to the analyzed activity, which is useful for determining the "limit" users.

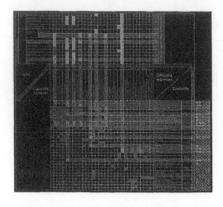

Figure 5: two different examples of "A/D Table" referred to the activities developed in a domestic kitchen (cooking with an oven) and aboard a sailing yacht (being at the helm)

Obviously this result will refer to the entire "All", that is to those who could simply have the "desire" to perform the activity (metadesign phase of target definition).

The designer's actions at this point becomes crucial, for he is the one who, during the design phases, must choose among all the possible users, those who also have a sufficient "likelihood" of carrying out the activities examined in the examined context (oppure "contect of interest") (design phase of target definition).

CONCLUSIONS

The "Table of Abilities/Difficulties" is a flexible instrument of knowledge: it is a useful operational scheme for the description of the needs framework of the "limit" user in regards to the Design for All approach. Sufficiently detailed and objective, conceptually it is based on the idea of being able to analyze "quantitatively" the ability of critical users with respect to a specific activity.

The potentialities offered by the "A/D Table" must be sought first within its relative ease of use, and then within the accuracy of the results which can be obtained: through a few steps, it manages to highlight the real difficulties that variable groups of individuals may face in performing specific tasks. In fact, it shows within an overall activity, the specific tasks or actions that generate physical/operational and cognitive/cultural difficulties, which in fact cause the exclusion of some groups of people.

The "A/D Table," then, is suitable for use as a means of selecting among "all" the possible critical users (metadesign phase), by defining the field of study and consequently facilitating the identification of the "real" design target. Critically identifying the "limit" users by means of an assessment of their difficulties with certain tasks, facilitates the construction of the needs framework: this is indispensable to identify the requirements of the system/product according to the

Design for All approach.

CREDITS

The research described here was conducted under the coordination of Prof. Giuseppe Di Bucchianico within the "Interior design of sustainable living" Degree Laboratory, academic year 2008/2009, in the Faculty of Architecture, University "G. d'Annunzio "of Chieti-Pescara.

In particular, the research was conducted with undergraduates Marco Gregori and Emilio Rossi, who have applied the operating results to their respective Dissertations: "The domestic kitchen for All: enabling solution to facilitate the movement of equipments and food" (relator Prof. G. Di Bucchianico; graduant student Marco Gregori), and "10 meters Daysailer for All: sustainable technological solutions for easy navigation" (relator Prof. G. Di Bucchianico; graduant student Emilio Rossi). Figure 4 shows the "A/D Table" referred to the above Degree Theses.

REFERENCES

Accolla, A. (2009), *Design for All*. Franco Angeli, Milano
Annett, J., Duncan, K. D. (1967), *Task analysis and training design*. Journal of Occupational Psychology, 41, 211-221.
EIDD - European Institute for Design and Disability (2004), *The EIDD Stockholm Declaration*, document of the EIDD Annual General Meeting, Stockholm.
Italian Ministry of Health, Department for the sort of health, research and organization of the Ministry (2008), *ICD-10. International statistical classification of diseases and related health problems (tenth revision) vol.1, 2,3*, IPZS - Istituto Poligrafico e Zecca dello Stato, Roma.
WIIO - World Health Organization (2003), *ICF Checklist. Version 2.1a, Clinical Form for International Classification of Functioning, Disability and Health*, World Health Organization, Geneva.

Chapter 58

Supporting Inclusive Evaluation with User Capability Data

Umesh Persad[1], Patrick Langdon[2], P John Clarkson[2]

[1]Product Design and Interaction Lab,
Centre for Production Systems,
University of Trinidad and Tobago
Trinidad and Tobago, West Indies

[2]Cambridge Engineering Design Centre,
Department of Engineering,
University of Cambridge
Cambridge, United Kingdom

ABSTRACT

Inclusive design is a design philosophy that aims to consider the needs and capabilities of older and disabled users in the design of mainstream products and services. In order to achieve this goal, designers require data on the sensory, cognitive and motor capabilities of disabled populations, together with a supporting framework for evaluating their designs. In order to examine the role of user capabilities in interaction contexts, an empirical study was conducted using four consumer products used in activities of daily living. 19 older and disabled users were recruited, and their sensory, cognitive and motor capabilities were evaluated using objective capability tests. Users then performed a task with each of the products while being videotaped. Difficulty ratings were collected for main actions after task performance. The results were analysed to determine how well the capability measures correlated with rated difficulty, via a series of graphs showing quantified product demands on user capabilities. The results suggest that measures of low-level visual, cognitive, and motor capabilities in general do not correlate strongly with outcome measures such as time, errors and rated difficulty. It is suggested that alternatives to reductionist methods for describing human capability in disabled populations should be investigated and compared.

Keywords: Inclusive Design, Product Evaluation, User Capability Data, Disability

INTRODUCTION

Inclusive design is a design philosophy that aims to consider the needs and capabilities of older and disabled users in the design process. The goal of inclusive design is to design products that are accessible and usable by the maximum number of users without being stigmatising or resorting to special aids and adaptations (Keates and Clarkson, 2003).

In previous work (Persad, Langdon, and Clarkson, 2007), we have examined the possibility of assisting the product design process by providing a framework for analytical assessment that is based on a synthesis of interaction design, human factors and ergonomics and psychology. To do this it was necessary to arrive at a set of requirements for describing in quantitative terms the interaction between the facets of product demand and the statistics of human capability. This was developed in order to encompass the scope of inclusive design that includes the wider capability ranges prevalent in the population, particularly amongst the aging and disabled. The central aim is to evaluate the match between users and the designed product. Thus, we propose that this assessment of compatibility needs to be complete to encompass the whole spectrum of capabilities and should be conducted at a number of levels, based on the distinction between hierarchies of sensory, motor and cognitive levels of human functioning.

BACKGROUND

Research has shown that predictions of real-world interaction problems in disabled populations produce variable results (Kanis, 1993; Steenbekkers and VanBeijsterveldt, 1998; Steenbekkers et al., 1999). In a study of control operation by physically disabled users, Kanis found that he was able to accurately predict users' difficulty in operating controls for a little more than 50% of the cases (after measuring maximum force exertions and the force required by the control) (Kanis, 1993). Steenbeekkers et al. also concluded that laboratory measures have limited predictive value for difficulties experienced in daily life (Steenbekkers and VanBeijsterveldt, 1998). They also mention that it is not clear how individual measures of capability combine to enable an individual to successfully complete a task.

Given this situation, there is scope for further research into understanding how various capability measures of disabled populations interact and relate to the real world performance of tasks with consumer products. If matching demanded capability measures from the product to measured capability measures from users is largely successful at predicting real world difficulty, then it forms a valid base for data collection and supporting inclusive analytical evaluation methods. If capability measures prove incapable of reliably making real world predictions for disabled populations, then the predictive approach to analytical evaluation, with its underlying models of interaction, has to be revised.

Most human factors data tends to be for relatively homogenous populations (Kondraske, 2000a, 2000b). In addition, using capability data to make real world predictions of difficulty and exclusion for disabled people is not well understood (Kanis, 1993; Kondraske, 2000b; Steenbekkers and VanBeijsterveldt, 1998). As a precursor to further developing analytical evaluation approaches and collecting human capability data to support these approaches, this more fundamental problem needs to be addressed. This research effort thus aims to addresses this fundamental problem by investigating the predictive ability of user capability measures in the context of a capability-demand product interaction framework. This should help to determine the type of user capability data that can form the basis of valid and robust analytical evaluation methods.

METHODOLOGY

An empirical study was conducted in the Usability Lab in the William Gates Building at the University of Cambridge. The study design entailed participants using four consumer products chosen to represent activities of daily living: (1) a clock-radio, (2) a mobile phone, (3) a food blender and (4) a vacuum cleaner. Prior to the start of the study, the products were chosen and various characteristics were measured including sizes and colours of text, sizes and colours of interface features (chassis, buttons, handles etc.) and push/pull/rotational forces required for activation. After ethical approval was obtained from the Cambridge Psychology ethics committee, older and disabled users were recruited from organisations in and around Cambridge such as the University of the Third Age (U3A), CAMSIGHT and the Hester Adrian Centre (Papworth Trust). This resulted in the recruitment of 19 participants in total who took part in the study.

Participants first signed a consent form and were given a 10GBP voucher for participating in the study. They were then asked questions to gather demographic, medical and product experience information. Participants were also asked to rate their experience with four consumer products and to describe how they would go about using these products to perform tasks (one task per product). Data was recorded on a questionnaire sheet and via an audio recorder.

Secondly, a series of capability tests were administered using a range of measurement devices. These tests included sensory tests of visual acuity, contrast sensitivity, hearing level; cognitive tests of short term working memory, visuo-spatial working memory, long term memory and speed of processing (reaction time); and motor tests such as push/pull forces exerted by each hand in different positions, walking speed and balance time. Participants had a short break after the sensory and cognitive capability assessment was performed. Some of the participants chose to take breaks during the capability testing session when they became tired. All capability testing data was recorded on a pre-designed testing sheet or a computer database for the computer based cognitive testing (CANTABeclipse from Cambridge Cognition).

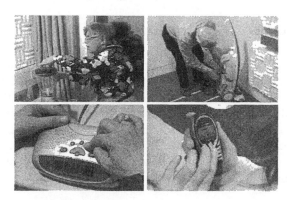

FIGURE 1. Users performing tasks with four consumer products (blender, vacuum cleaner, clock radio and mobile phone).

Thirdly, participants performed one task with each of the products while being videotaped, with tasks randomly assigned to avoid order effects (Figure 1). They were informed that they could stop the task at any time for any reason. The tasks performed were: (a) Clock radio - *setting the time to 4.30 PM*, (b) Mobile phone - *taking the ringer off via the menu*, (c) Blender - *blend a banana and water as if making a smoothie/drink*, and (d) Vacuum cleaner- *vacuum a piece of carpet till clean*. These tasks were analysed for essential constituent actions before the commencement of the study. Therefore, on completion of a task, subjective difficulty and frustration ratings were collected from each participant for these selected actions using a visual analogue scale ranging from 0 to 100. The results were analysed to determine how well the capability measures collected prior to task performance predicted difficulty in actual use, using a capability-demand model. After completing the four tasks, participants were debriefed and thanked for participating in the study.

RESULTS AND ANALYSIS

All collected data was entered into SPSS for analysis and the video data was transferred to a digital format for analysis. The videos of each participant were analysed and task time and errors were recorded. The SPSS data consisted of a dataset of 19 participants (Mean Age=62.68, SD=9.20) with demographic data, capability data, and task outcome measures (times, errors, difficulty and frustration ratings) for each product task. The data was then analysed by graphing scatterplots of rated difficulty versus measured capability for constituent actions and analyzing

570

the strength of linear correlations. Due to space limitations, an overview of the results will be given in the following sections with illustrative examples. The presentation will also be confined to the analysis of capability-demand relationships.

TASK OUTCOMES

Figure 2 shows the proportion of participants who attempted, succeeded or failed the task with each of the consumer products. Of the 16 participants who performed the clock radio task, 56% successfully completed the task, and of the 16 participants who performed the mobile phone task, 19% completed it successfully. Of the 19 participants who performed the blender task, 100% successfully completed the task, and of the 18 participants who performed the vacuum cleaner task, 100% completed it successfully. Thus the mobile phone task and the clock radio task had the highest and second highest failure rate respectively.

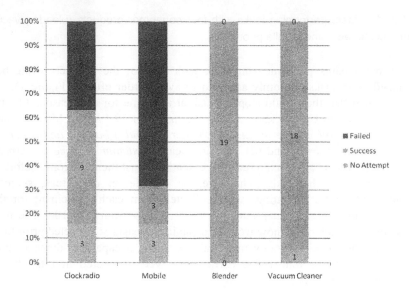

FIGURE 2. Graph of proportion of participants who attempted, failed and succeeded each of the four tasks.

Mean difficulty and frustration ratings were plotted for each product and compared. The mobile phone had the highest mean ratings across all the products for difficulty in starting the task (M=78.24, SD=30.00), difficulty in working out subsequent actions (M=84.59, SD=28.96), and overall mental demand (M=76.47, SD=30.25). The mobile phone also had the highest mean rating for frustration experienced during the task (M=48.89, SD=41.82). These are illustrated in Figure 3.

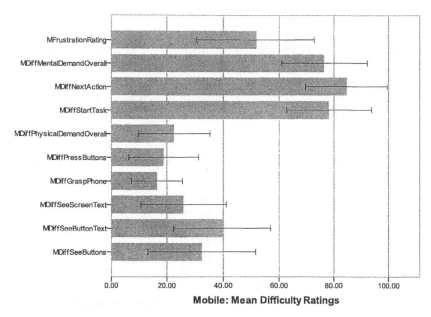

FIGURE 3. Graph of mean difficulty and frustration ratings for the mobile phone task.

In terms of visual demands, the small text on the clock radio display was rated the most difficult to see (M=52.37, SD=34.78), followed by seeing the numbers on buttons (M=46.05, SD=36.84) and seeing the actual buttons (M=39.47, SD=41.16) on the mobile phone. The physical actions of opening (M=47.37, SD=30.25) and closing (M=38.68, SD=26.03) the blender cover and pushing the vacuum cleaner forward (M=28.06, SD=30.69) were also rated as being the most difficult actions on average. In terms of overall mental demands, the mobile phone ranked the highest (M=76.47, SD=30.25), followed by the clock radio (M=42.94, SD=37.54), the blender (M=28.11, SD=32.58) and the vacuum cleaner (M=27.94, SD=27.60). For mean frustration ratings, the mobile phone once again ranked the highest (M=48.89, SD=41.82), followed by the vacuum cleaner (M=28.33, SD=36.22), the clock radio (M=26.39, SD=39.91) and the blender (M=22.11, SD=36.03). In the following sections, the relationships between measured user capabilities and their difficulty ratings would be examined further.

SENSORY CAPABILITIES

Scatterplots between visual capabilities and rated difficulty in visual actions were generated via capability-demand graphs as shown in Figure 4. Similar graphs were also used for physical actions. The plots show an increasing user capability measure on the horizontal axis, while the vertical axis shows the rated difficulty score ranging from 0 to 100. A vertical dashed demand line is plotted on the graph to indicate the specific demand of the product feature being considered. In the case of the left graph in Figure 4, the contrast demand of the clock radio digital display text is 0.85 (in Log Contrast Sensitivity units) at the size of 1.38 LogMar (Log Minimum Angle Of Resolution).

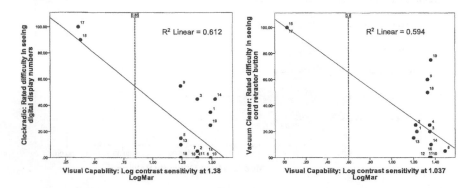

FIGURE 4. Capability-demand graphs of visual capability versus rated difficulty in (a) seeing numbers on the clock radio digital display and (b) seeing the cord retractor button on the vacuum cleaner.

Based on capability-demand theory, as we move toward the demand line from the right of any graph, participant ratings of difficulty are expected to rise till the demand line is encountered. As we cross the demand line and move further to the left of it, participant ratings of difficulty are expected to be near maximum i.e. they should find the action very difficult or near impossible to perform.

A linear model accounted for a significant amount of the variance for actions involving reading textual features on the products. Figure 4 shows a fairly strong negative relationship between reading numbers on the digital display of the clock radio and the contrast sensitivity measured for each participant: $r(16)=-0.782$, $p<0.01$. Some significant linear relationships were also found for actions involving seeing product features, for example seeing the cord retractor button on the vacuum cleaner: $r(14)=-0.771$, $p<0.01$. However, other cases showed no significant linear relationships, for example seeing the buttons on the clock radio: $r(16)=0.051$. This may in part be due to the use of the contrast sensitivity measure for text as an approximation in place of a spatial contrast sensitivity test.

MOTOR CAPABILITIES

In considering fine motor actions and manipulations, no significant linear relationships were found. These actions included pushing buttons, sliding switches and twisting product controls. Figure 5 on the left shows the weak linear relationship between finger push force and difficulty in pushing the clock radio buttons: r(17)=-0.105.

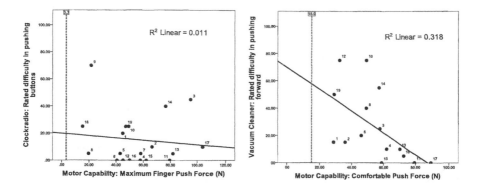

FIGURE 5. Capability-demand graphs of (a) finger push force versus rated difficulty in pushing the clock radio buttons and (b) comfortable push force versus rated difficulty in pushing the vacuum cleaner forward.

Larger push/pull actions such as lifting and opening the blender and moving the vacuum cleaner around showed slightly better linear correlations. An example of this is shown in Figure 5 on the right for pushing the vacuum cleaner forward. In relation to measured comfortable push force, there was a moderate linear relationship with rated difficulty in this action: r(13)=-0.564, p< 0.05.

COGNITIVE CAPABILITIES

In order to investigate the relationships between measured cognitive capabilities and task outcome measures, graphs were plotted of task time, errors, difficulty starting task, difficulty in selecting subsequent actions and overall mental demand against four main cognitive variables: (1) short term working memory (digit span), (2) visuo-spatial working memory (span length), (3) speed of processing (reaction time) and (4) long term memory (GNTpercentcorrect). Significant linear correlations were not found except in the case of errors, where short term working memory and visuo-spatial working memory were found to correlate moderately with errors for the blender r(17)=-0.493, p< 0.05 and vacuum cleaner r(16)=-0.5.16, p< 0.05. These are shown in the top row of Figure 6. Visuo-spatial working memory showed some fairly strong correlations with blender errors r(15)=-0.819, p< 0.01 and vacuum errors r(14)=-0.700, p< 0.01. These relationships are shown in the bottom row of Figure 6.

574

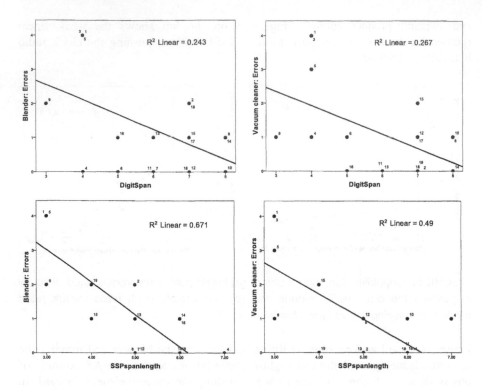

FIGURE 6. Relationships between blender and vacuum cleaner errors versus short term working memory and visuo-spatial working memory.

Long term memory showed a significant relationship with blender errors: r(15)=-0.638, p< 0.01, vacuum errors: r(15)=-0.763, p< 0.01 and clock radio errors: r(14)=-0.502, p< 0.05. These are shown in Figure 7. However, the mobile phone errors had a week linear correlation with long term memory: r(14)=-0.059 as also shown in Figure 7.

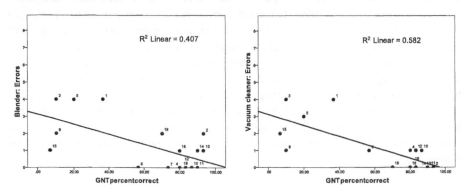

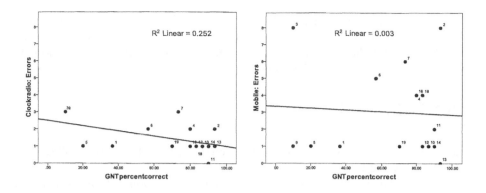

FIGURE 7. Relationships between errors made using the 4 products and long term memory (GNTpercentcorrect).

DISCUSSION

The overview of results presented in the previous sections suggest that, given the limitations of the current study, measures of low-level visual (visual acuity, contrast sensitivity), cognitive (digit span, spatial span, reaction time, long term memory), and motor capabilities (grasp forces, push/pull forces) in general correlate weakly to moderately with outcome measures such as time, errors and rated difficulty. In certain cases, for example vision, correlations were moderate to high indicating that possibly the essential low-level capabilities essential for real world task performance were being captured to some degree.

In striving to achieve a parsimonious model of user capability that can support inclusive design evaluation, the nature of the results indicate that there may be limited value to a linear reductionist model for describing human capability in disabled populations. It could be that the disabled human user taps into multiple low-level capabilities in a non-linear way, relying on a system of accommodation and coping strategies that would be difficult to accurately model with simple linear models. It is suggested that alternatives to a reductionist method for describing human capability in disabled populations should be investigated and compared.

The resulting graphs also show that participants with different capability profiles may be rating their difficulty and frustration differently. Ongoing analysis aims to extract differences between such groups within the study sample.

CONCLUSIONS AND FURTHER WORK

Based on these results, the data is being analysed further to investigate whether non-linear models of fit would better describe the relationship between user capabilities and rated difficulty. In addition, various combination models for user capability variables are being investigated to determine whether higher-level derived measures could show an improvement over low-level measures in predicting task outcomes. Finally, further studies are planned using a similar methodology and a larger sample size to investigate capability-demand interaction in older and disabled populations.

REFERENCES

Kanis, H. (1993), "Operation of Controls on Consumer Products by Physically Impaired Users." *Human Factors*, 35(2), 3.5-328.

Keates, S., Clarkson, J. (2003), *Countering design exclusion - An introduction to inclusive design.* Springer.

Kondraske, G. V. (2000a), "Measurement tools and processes in rehabilitation engineering." In J. D. Bronzino (Ed.), *The Biomedical Engineering Handbook* (2nd ed., Vol. 2, pp. 145-141 - 145-116): CRC Press.

Kondraske, G. V. (2000b), "A working model for human system-task interfaces." In J. D. Bronzino (Ed.), *The Biomedical Engineering Handbook* (2nd ed., Vol. 2, pp. 147-141 - 147-118): CRC Press.

Persad, U., Langdon, P., Clarkson, J. (2007), "Characterising user capabilities to support inclusive design evaluation." *Universal Access in the Information Society,* 6(2), 119-135.

Steenbekkers, L. P. A., VanBeijsterveldt, C. E. M. (Eds.). (1998), *Design-relevant characteristics of ageing users.* Delft, The Netherlands: Delft University Press.

Steenbekkers, L. P. A., VanBeijsterveldt, C. E. M., Dirken, J. M., Houtkamp, J. J., Molenbroek, J. F. M., Voorbij, A. I. M. (1999). "Design-relevant ergonomic data on Dutch elderly." *International Journal for Consumer & Product Safety,* 6(3), 99-115.

Chapter 59

Access 2 All – Innovative HMI Concepts for Barrier-Free Travelling

Katrin Meinken[1], Angelos Bekiari[2]

[1]Fraunhofer-Institute for Industrial Engineering IAO
Stuttgart, Germany

[2]Hellenic Institute for of Transport
Thessaloniki, Greece

ABSTRACT

ACCESS 2 ALL is a European funded project, which aims at encouraging public transport operators, belonging to the project's target group, to adopt innovative technological concepts and mobility schemes that enable mobility and transportation services for all citizens of high quality. In addition the public transport personnel shall be provided with the necessary knowledge on the particularities of specific user groups, such as the elderly or disabled – covered under the term of mobility impaired people. ACCESS 2 ALL aims at defining concrete mobility schemes, guidelines and policy recommendations, ensuring accessibility of public transport to all users through the coordination of current research efforts. Enabling barrier-free transportation does include the improvement of public transport information by means of integrated and personalized travel information systems and novel HMI concepts. The present paper gives a description of the conceptual design of the ACCESS 2 ALL HMI solutions and describes the ACCESS 2 ALL approach of information provision.

Keywords: Public transport, accessibility, MI users, barrier-free travelling

INTRODUCTION

Given that flexible and easy mobility is one of the most important factors affecting the independence and active living, it should be assured that all citizens have the same opportunities regarding travelling. The provision of accessible public transport, especially to users who are handicapped, is therefore a very important matter. In order to allow all people to remain active in society, preferably without the help of a third party, it is mandatory to enhance public transport so that it will be suitable for safe, independent and dignified use by individuals with particular needs. Furthermore information on all areas of public transport must be presented in a clear and simple manner, enabling people to easily obtain the necessary information. As around 13% of the population of Europe is represented by impaired people the need for a barrier-free transportation system must not be underestimated. Approximately 63 million people are affected in Europe alone. The range of disabilities includes mobility and sensory impaired people, people with cognitive or learning difficulties, with communication difficulties such as dyslexia or being illiterate. As most of the times literacy is required in order to use transport facilities about half of these persons concerned can be considered as mobility impaired.

The main aim of the European project ACCESS 2 ALL is a definition of concrete mobility schemes, guidelines and policy recommendations that will ensure the accessibility of public transport. This ambitious goal shall be achieved by encouraging public transport operators as the project's target group to adopt innovative technological concepts as well as mobility schemes that enable high quality transportation services for all different user groups. Furthermore the personnel of public transportation operators need to be provided with the necessary knowledge on the particularities of the different users, such as elderly, wheelchair users or illiterate people. In order to enable accessible public transportation one major objective of the ACCESS 2 ALL project is the improvement of public transport information by means of integrated and personalized travel information systems and novel HMI concepts.

The paper gives a description of the conceptual design of the ACCESS 2 ALL HMI solutions. It takes into account the diverse and sometimes competing needs of mobility impaired users. Considering the different user groups – ranging from wheelchair users to elderly – a lot of heterogeneous requirements have to be taken in to account for information provision. Also the information range has to cover different sections of the journey, such as walking passages or train rides.

ACCESS 2 ALL - PROJECT OVERVIEW

Results from previous European funded projects showed that the demand in the domain of public transport and it's enhancement concerning mobility impaired users is huge. According to COST 335 almost 13% of the population of Europe is represented by disabled people. The range of disabilities include

- Mobility impaired people, including wheelchair users and people being unable to walk far or at all
- Sensory impaired people, including people, who are completely blind or deaf
- People with cognitive and/or learning difficulties
- People with communication problems, including ICT-illiteration, dyslexia, etc.
- Other forms of disabilities, such as asthma or orientation problems.

Amendatory John Gill (2004) estimates the population of Europe with problems using novel technologies. It was defined in terms of the individual functional abilities with specific emphasis on use of information and communication technology systems. The population was classified into groups and does meet the targeted population of ACCESS 2 ALL. Table 1 shows the percentages of occurrences in Europe for each mobility impaired group:

Table 1 Percentages of occurrences of MI groups in Europe

MI group	Percentage of occurrence
Wheelchair users	0.4%
Cannot walk without aid	5%
Cannot use fingers	0.1%
Reduced strength	2.8.%
Reduced coordination	1.4%
Speech impaired	0.25%
Language impaired	0.6%
Dyslexic	1%
Intellectually impaired	3%
Deaf	0.1%
Hard of Hearing	6%
Blind	0.1%
Low vision	1.5%

Furthermore it can be stated that a significant part of people with disabilities or other difficulties is older people, most often defined as 65 years and older. According to the INCLUDE European project about 70% of all disabled people can also be defined as elderly. According to COST 335 the share of elderly in the total population if Europe is expected to rise from 63 million people today to about 75 million people in 2020 and to around 88 million people in 2030. Thus the elderly will be a larger part of the population, but with better health conditions than today and therefore with more requirements for active living and mobility.

Also accompanying persons and able bodied people with temporary mobility restraints, such as parents with strollers or heavy luggage have to be taken into account as people with mobility impairments. As they would also benefit from accessible public transportation, accessibility would then affect 35 to 40% of the population. This does meet the 10/30/100% rule stated in the European initiative "Design for All", which says that accessibility is indispensable for 10%, necessary for 30 to 40% and comfortable for 100% of all people (Neumann & Reuber, 2004).

As society has committed itself to providing all citizens with equal opportunities, the main goal of ACCESS 2 ALL is to better include the named user groups of mobility impaired people into the processes of public transportation. Therefore awareness on the part of public transport providers and their personnel needs to be raised in order to cater to the special needs of MI people. The development and implementation of innovative technological solutions and mobility schemes will help to enable a high quality transportation service for all users. For achieving this goal ACCESS 2 ALL focuses on identifying best practices as well as still-existing deficits, coordinating on-going research activities and promoting novel technology concepts and cooperative systems. Concrete project objectives are thus

- Identification of best practices in all inclusive PT systems, actual problems and real needs of all user groups requiring special attention in transportation
- Definition of specific implementation scenarios covering the targeted user groups
- Promotion of the deployment of innovative technological and service provision concepts based on new vehicle and infrastructure concepts suitable for a barrier-free travel chain
- Improvement of information display and service provision based on integrated, personalized travel information systems and novel HMI concepts
- Promotion of innovative cooperative systems enabling the optimal combination of autonomous and infrastructure based systems
- Defining a concise methodology for the assessment of usability and accessibility of PT services
- Developing guidelines on accessible PT services covering all aspects of transport operations, namely vehicles, information, infrastructure
- Development of a research roadmap towards 2010, 2015 and 2015 on accessibility issues concerning public transport in order to coordinate current and future research efforts.

REQUIREMENTS FOR NOVEL HMI CONCEPTS

One important objective of the ACCESS 2 ALL project is the investigation of innovative HMI concepts for a barrier-free display of information and service provision using the acoustic, visual and/or tactile channel within the infrastructure and vehicles as well as by means of modern equipment like handhelds, cell phones, PDAs or else. The concepts should enable the users to achieve a maximum degree of autonomy throughout the complete transportation process, namely from alternative means of trip planning to accomplish the journey itself, covering the travel chain from walking passages to bus or train rides as well as orientation in and outside the station.

As mentioned above the user group addressed is very heterogeneous as the population of mobility impaired people does include various impairments plus the group of the elderly and parents with strollers or travelers with heavy luggage. Therefore the different users feature a wide variety of needs and requirements that need to be thoroughly addressed in novel concepts for PT information provision.

Basically the requirements of MI people on accessible HMI system concepts can be describes as the following

- All presented information has to be transmitted in a comprehensible and receivable way for all users.
- Information hast to be provided with as little operating steps as possible.
- The user has to be enabled to use the system respectively to place inputs and retrieve information without the help of a third party.
- The system has to be intuitively usable without the need for an introduction to the use.
- The use of profiles or presetting has to be enabled when using mobile services, internet, ticket machines, etc.
- The system has to be accessible for all users or it has to be transportable without effort.
- All supportive items or applications shall be designed in an unobtrusive way in order to not segregate the user as special population.

After interviews with experts and persons concerned two major problems in public transport information and service provision could be identified and addressed:

1. The provision of information in public transport is often lacking important information especially needed by MI users. Not only the lack of this information is problematic but also their timely provision can often not be realized. Platform changes or delays are already a problem for non-impaired travelers concerning schedule or organization, for MI people additional problems in orientation or flexibility can occur. It is therefore of utmost importance to provide all necessary information before the journey and to timely update en-route information of delays

or any other changes during the journey in order to realize a barrier-free travel chain.

2. Orientation in and outside stations was identified as the second difficulty when using public transport. Certain impairments may inhibit the use of main passage-ways if they e.g. include stairs as wheelchair users would then need an elevator. Also signs or directions can be easily overseen when travelling at a sitting position through dense crowds or being visually impaired. Therefore it is crucial to realize a user-friendly and suitable navigation system in order to allow all users to find the easiest and fastest way to the desired destination.

Providing all sub-groups of the user group of mobility impaired people with a unique HMI solution that covers both a constant flow of information and a reliable navigation system considering different input and output modalities as well as individual adaptations to all specific needs of the future users is a challenge we yet have to face. According to the research report of the BAIM project (2008), which aimed at supporting the active and autonomous participation of mobility impaired people in public transport, complex information would be needed in order to realize a barrier-free travel chain for mobility impaired passengers. The target data for an innovative HMI concept would be composed of information on journey details, vehicles used and stations.

Information in journey details would include

- Detailed time-tables with listed places of departure and arrival (stations and tracks), stopovers and connecting times, times of departure and arrival and the used vehicle types
- Fare of the chosen connection.

Information on vehicles would include

- Details on all necessary dimensions, such as door width, access height, dimensions of movement area, steps on vehicle, etc.
- Ramps or lifts for access and egress, including the lifting capacity
- Storing position for wheelchairs, seat for handicapped people, multi-purpose compartments, first class/coach
- Visual displays, announcements, signal of door closing, operation and communication systems

Information on stations would include

- Access
- Stairs, ramps, elevators, escalators
- Parking for cars or bicycles, location of other transportation means, e.g. taxi, bus, etc.
- Restrooms – also handicapped accessible, public telephones/phone booths, emergency call, call number for contact person, waiting area, shelter
- Baggage room, lockers, trolleys

- Ticket machines and ticket window, display of time-tables and fares
- Display panels, dynamic passenger information, destination display
- Kiosks or shops located at the station
- Guidance system for visually impaired people

For complementing the target data additional real time data would be needed in order to announce unscheduled changes during the journey and therewith enabling the user to react timely to unforeseen changes. Real time information would include

- Delays, change of tracks, use of different vehicle types, cancelation of trains/busses/etc.
- Breakdown or malfunction of elevators or escalators
- Closing of routes, relocation of stations

A lot of this information is already provided to the passengers through different systems, such as ticket machines, internet applications or information kiosks. Each of these systems has its own user interface and operating philosophy. Retrieving all necessary data easily can be a challenge, especially when handicapped through any kind of impairment. Therefore the design of a unique HMI application concept which unifies all required information in one single system, providing the user with exactly the data needed for a personal trip planning is aspired in the ACCESS 2 ALL project. In order to integrate the application of the novel HMI solution smoothly in the everyday life an implementation with mobile devices, such as handhelds, smart phones or PDAs would be reasonable. For in- and output different channels, namely acoustic, visual and tactile senses, have to be addressed in order to include as many user groups as possible. The application of existing interaction and communication tools, such as eye tracking or eye following, automated speech recognition (ASR) or text-to-speech (TTS) functions could help to enhance the accessibility of information provision using novel HMI solutions. Also the implementation of less common applications using the tactile channel, such as the "Tactile Vision Substitution System" which creates pictures for blind people via tactile stimulation of the tongue (Goede, 2009) or navigating people via vibration modules, guiding people via simple "taps on the shoulder" (Van Erp & Verschoor, 2004) is entirely conceivable in order to alleviate the travel process for mobility impaired passengers.

CONCEPTS FOR THE ENHANCEMENT OF EXISTING HMI SOLUTIONS

In EU public transportations many accessibility systems do already exist. However certain limitations are given to accessibility as they often do not address all of the identified user groups of mobility impaired passengers. Add-on features or enhancing services used in conjunction with existing systems can solve more specific problems with entirely new features and therewith help to improve these systems and provide a better accessibility. Add-ons are meant to be simple devices or services and are therefore inexpensive to install and to maintain. Furthermore the implementation is easy as add-on functions shall be realized with already existing devices or services the usage will be intuitive and easy to apply.

En-route Headphones

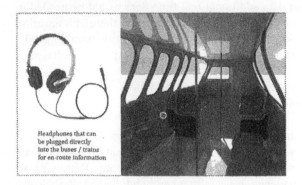

FIGURE 1 En-route Headphones.

En-route Headphone could be available in all public transport means. Being directly plugged-in in designated areas they could provide pertinent travel information, such as departure and arrival times, approximate travel hours or the announcement for next stations, route changes or delays. Therefore acoustic or visual information that was not received can be gained via repetitive listening.

Information via Radio

FIGURE 2 Information via Radio.

Information via radio could be available in all public transport means. Being accessible through any electronics able to receive radio, such as smart phones, mp3 player, etc. this service would provide pertinent travel information, such as departure and arrival times, approximate travel hours or the announcement for next stations, route changes or delays. The service would be easy to implement into existing systems and intuitive to use as applicable on already used devices.

SMS Update via Mobile

FIGURE 3 SMS Update via Mobile.

En-route information on significant travel information, such as delays for more than 1 hr. could be provided to the user's personal mobile devices via SMS alerts. This service would be very flexible and customizable as users could easily sign up or cancel on their own demand. As most passengers would own a mobile the service would be easy to implement and also intuitive to employ as provided on personal devices already in use.

Portable Info Displays

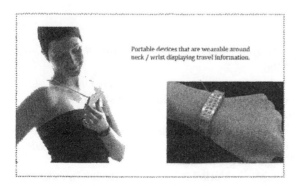

FIGURE 4 Portable Info Displays.

Easy to wear, e.g. as necklace or on the wrist, portable devices could provide short and simple travel information that would be consistently updated throughout the station or in transportation means using technologies such as wireless, IR or Bluetooth. The portable devices should be small and light to carry around, maybe even fancy, and also simple and intuitive to use as no input or programming and therefore no learning of a new system would be necessary.

The introduced add-on concepts to existing information provision systems in public transport do by all means not cover all requirements of the targeted user groups. They are meant as a first approach for the enhancement of existing systems, making them more accessible for a larger number of people. To include all user groups of mobility impaired people further research and development is needed.

CONCLUSION

Accessibility refers to peoples' ability to reach desired goods, services, activities and destinations; hence, quality of accessibility has tremendous direct and indirect impacts. Improving accessibility and reducing accessibility costs can help achieve many economic, social and environmental objectives. As a practical outcome, the ACCESS 2 ALL project would promote the full social and economic integration of mobility impaired people in public transportation processes and would therewith contribute to their independent living.

The requirements identified for the design of novel HMI solutions will help to develop applications and systems considering all users, which will result in benefit for impaired as well as for non-impaired passengers. The concepts outlined in this paper are only a small sample of innovative technologies that can be used. With further research and development, and a consistent contribution and improvement to the current existing system, the goal of the project, to provide indeed, Access to All, can be achieved.

REFERENCES

Gill, J.M. (2004), *Access-Ability: Making technology more useable by people with disabilities*. ISBN 1 86048 0306.

Goede, W. (Ed.) (2009), *Blinde sehen mit einem Zungen-Display*. URL http://www.innovations-report.de/html/berichte/informationstechnologie/bericht-3929.html. – retrieved on 10/12/2009.

Neumann, P., and Reuber, P. (Ed.) (2004), *Ökonomische Impulse eines barrierefreien Tourismus*. Long version of a study on behalf of the Federal Ministry of Economics and Technology, Münster

Rhein-Main-Verkehrsverbund GmbH; Verkehrsverbund Berlin-Brandenburg GmbH; HaCon Ingenieurgesellschaft mbH; IVU Traffic Technologies AG; SemanticEdge GmbH; Forschungsinstitut Technologie und Behinderung (Ed.) (2008), Barrierefreie *ÖV-Information für mobilitätseingeschränkte Personen*. Research report

Van Erp, J.B.F., Verschoor, M.H. (2004), "Cross-modal visual and vibrotactile tracking." *Applied Ergonomics*, 35(2), 105–112.

Chapter 60

Interaction Development for a Public Transit Rider Reporting System

Ellen Ayoob, Timothy Andrianoff, Rafae Dar Aziz, Aaron Steinfeld

Robotics Institute
Carnegie Mellon University

ABSTRACT

End users who encounter an accessibility barrier or a positive experience should have the ability to report the event to the relevant entity and subsequently track the report to completion. Furthermore, users should be able to see all reports submitted by the community, thereby permitting advocacy and transparency. The team is working to develop this type of citizen science system within the realm of public transit. This model will empower riders, resulting in a greater understanding of the transportation system, and improve the feedback loop between rider and provider. While existing citizen science approaches have been successfully deployed for other domains, public transit is particularly complex due to the dynamic nature of the service provided, information required for institutional reasons, and magnitude of many systems. This work describes a user-centered effort to define the website interactions and interfaces for public transit citizen science systems.

Keywords: Citizen science; public transportation accessibility; disabilities; user-centered design; universal design; interaction design

INTRODUCTION

Public transportation systems play a crucial role in our society, particularly for individuals with disabilities and the elderly. In a mobile, global culture, full social participation hinges on accessibility of transportation systems. Approximately 6.5% of the U.S. population is 65 years and older, while more than 20% of the entire population has at least one disability (U.S. Census, 2006). Accessible public transportation in the community allows these individuals, especially those with severe disabilities, to have independent access to works sites, educational programs, health facilities, and social and recreational activities. Unfortunately, more than half a million people cannot leave their homes because of transportation difficulties (BOT, 2003) One-third of people with disabilities have inadequate access to transportation (NOD, 2004). Consequently, four times as many people with disabilities as people with no disabilities lack suitable transportation options to meet daily mobility needs (NCD, 2005).

Accessible public transportation is critical. Current methods and best practices are difficult to adopt for a variety of reasons including budget limitations, scope, complexity, and integration with existing infrastructure. New methods are needed that will be effective given these difficulties and engage *all* riders in a feedback loop with their transportation agencies to effect sustainable change.

Citizen science methods show great promise, using technology to empower end users to contribute data (Paulos, 2008). Universal design offers a promising alternative to conventional accessible design, since it improves usability for everyone, not just people with disabilities (Danford & Maurer, 2005). It also improves public transportation's competitiveness as an alternative to private automobiles for all riders.

It is important to note that without feedback, problem reporting systems can easily appear to be "black holes" where complaints go in and nothing happens (Steinfeld, et al accepted). This black hole problem can lead to low perceived benefit for engaging with the local transit agency. Parallel interviews and interaction concept testing by the team (Yoo, et al 2009) has revealed that riders rarely encounter infrastructure problems that meet the perceived cost-benefit threshold for reporting. As a result, the team's view is that observation reporting should be part of a larger system that includes valuable, frequent information (e.g., arrival times, vehicle fullness, and dynamic route changes), thereby streamlining the infrequent desire to report observations and lowering the cost. This approach gives riders many different reasons to remain engaged with the system and supports pre-loading of important real-time details (e.g., current location, route, etc). Pre-loading is also valuable for riders who have difficulty entering data. This larger context is important when thinking about the mobile reporting case and the demands of on-the-spot reporting.

APPROACH

This paper documents the effort of the team to develop the interaction design of a citizen science transit reporting system. In addition to core team activities, two student teams explored the user-centered website interactions and interfaces needed for such public transit citizen science systems as part of their coursework. While the main focus of their activity was educational in nature, their work supported and informed the design efforts of the core team.

COMPETITIVE ANALYSIS

Students initially surveyed nine existing web-based systems for reporting software bugs and filing complaints. They then surveyed 11web sites designed to foster dialogue between individuals and bureaucracies. Students paid particular attention to the level of interaction required by a user to submit information and the types of information required by a system. Almost all of the issues concern form-based reporting; therefore, it made sense to focus efforts on a series of think-aloud experiments (Lewis, 1994) using paper prototypes (Synder, 2003) of potential forms for the system.

INITIAL THINK-ALOUD SESSIONS

For the initial think-aloud sessions, three scenarios of use were developed by the core team to explore user-system interactions:
1. A bus passes a user by on the street
2. A bus stop has broken glass and the user wants to upload a photo
3. There is a problem with the bus itself.

These scenarios required users to make observations about buses, bus routes, bus stops, and bus drivers and allowed the researchers to test dynamic forms. All think-aloud sessions began with the participant first classifying their observation as a complaint, suggestion, compliment or general discussion then specifying what their observation was related to: bus, bus stop, bus route, bus driver. The experimenters then presented forms depending on the participant's selections. After participants completed a form and said submit, they were presented with a paper prototype screen displaying summaries of other observations and asked to identify like observations.

FIGURE 1. Sample Forms for Creating Observation About a Bus

SUBSEQUENT THINK-ALOUD SESSIONS

In additional think-aloud sessions, students further explored form-based interaction issues, using two task-based scenarios:

- Create an account and observation about a broken bus shelter
- Browse other observations and add a comment to an incident about a bus on a specific route

The sessions used slightly more refined paper prototypes and addressed additional interaction questions, such as motivation for creating a user account.

FINDINGS

COMPETITIVE ANALYSIS

The teams discovered that most systems ask for specifics such as date and time of the occurrence, a subject, and a general description. In addition, they noticed language differences across questions and forms. Finally, sites varied in the transparency of their feedback loop to the user.

One site in particular was seen as a particularly good model for the application of citizen science to public transportation: Parkscan.org. This site was identified early by the core team as a key model for citizen science for an infrastructure-based governmental service. ParkScan.org in San Francisco uses citizen science to improve its public parks through a form interface that allows all users to view all observations, including the status of the observation, at any time. In 2007 alone, ParkScan had 425 registered users, 1,531 observations, and 68% of the issues identified by end users were addressed by the City in a public feedback loop (NPC, 2008).

During an interview with the team, a representative with ParkScan.org stressed the importance of establishing a collaborative relationship between users and service providers. There was an intentional effort to avoid inherently antagonistic interaction. Instead, the organizers focused on helpful outcomes, not looking to create shame, and generation of a team-oriented experience.

Subsequent surveys, contextual inquiries, and interviews revealed issues about high-quality data collection and reporting specific to public transportation, for example:

- Non-fixed locations (unlike a public park)
- Lack of any distinguishable public feedback from the transit agency
- Driver identification and rider privacy issues and challenges with how to collect these descriptions

THINK-ALOUD SESSIONS

Through the rounds of piloting and experimentation, issues with the forms included:

- Confusion with domain-specific terminology
- Ambiguity between bus ID number and bus route.
- Multiple ways to specify a stop resulting in confusion about which stop to select
- Required form fields for information confusing, misleading, or simply ignored e.g., time pertaining to time of report or time of event being reported
- Date ambiguities about whether participant should specify the report date or the date of the item they want to report
- Misinterpreted form fields
- Difficulty distinguishing inbound and outbound routes

Asking participants to choose observations related to their own yielded some unexpected positive results. One first-round participant expressed that seeing other observations at the conclusion of submitting her own gave her a sense that her observation would be read rather than ignored. A participant in the second round remarked that seeing that others had submitted observations similar to his own made him feel more confidence that his observation was worth submitting. Participants also tended to check more than one observation even if the similarities were superficial.

Some participants expressed reluctance to create accounts and simply wanted to make their reports. However, when incentives were added, such as subscribing to a route or multiple routes for email updates, participants reacted more favorably to creating an account.

Bus routes presented an ongoing challenge. For example, participants were asked to input their bus stop or they could click a link "Don't Know Your Bus Stop." If they clicked the link, they received a prompt for their route. The language of the prompts was varied as the think-aloud experiments evolved, yet a common result persisted: some trips can be accomplished using multiple bus route options between the same start and end stops. In this scenario, riders are opportunistic and will take the first available bus.

DISCUSSION

Overall, participants were able to easily submit observations of the situation described in each scenario, even when lacking all details requested by the form. Rather than train the user initially, we will be training our machine-learning equipped system on the backend to parse as much information as possible from whatever input the user provides. The design implications are that:

- a small number of data entry fields will be required but the more completed in the better.
- the date of the report will be prefilled to avoid confusion and this date can be used as part of a search feature.
- the bus id number, which was included in the study, will not be asked for in the initial system release as most users did not know to look for this. It can be reintroduced at a later date.
- a user can provide a guided description of the driver, including an identification number.

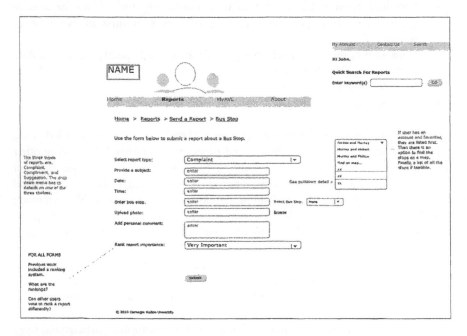

FIGURE 2. Sample revised form embedded in preliminary wireframe

Some problems with the interface stemmed chiefly from issues in language usage; therefore, the language is being streamlined from the think-alouds to a working prototype (Figure 2). The problem of misinterpreted form elements was addressed through changes in the labeling and order of fields on the site. Even though the forms are dynamic, the essential elements appear in the same order.

It would be useful to determine more precisely the impact that seeing other users' reports has on the user's experience of reporting. The unexpected positive response of users to seeing reports related to their own indicates a possibility that being presented with such reports can be perceived by some users as being a form of positive feedback from the system. It may make users more confident in submitting reports and may contribute to users' perception that their reports are usefully communicated through the system rather than being ignored.

Users who choose not to login and/or create an account will still be able to submit reports and see what others have reported simply by providing a valid email address. There is always a risk of abuse of the system, which is why submitting an email address is the minimal requirement. Incentives to create accounts and remain engaged with in reporting activities come from value-added services rather than withheld services. One example of a value-added service is allowing users to subscribe to a route or multiple routes. They can then receive email updates on any observations reported to these routes.

A lingering issue is how users designate routes and stops for inbound and outbound buses if any bus will do. Long drop down menu selections are prohibitive from both a universal design and an accessibility perspective. Several options were tried in the think-alouds and the issue was distilled down to what bus can be caught at what stop. If the person knows their stop or route they can simply input the information. The system will be smart enough to log their favorites and present them to the user in a short drop down alongside the text entry field the next time they issue a report. There are also plans to present routes and stops via Google Maps. This will be the most accurate way to collect this information for sighted users. There is also the potential to leverage short names assigned to stops for rider information purposes (e.g., RouteShout). Finally, a long dropdown will be provided if necessary according to use preference.

NEXT STEPS

As mentioned, the work described here is a portion of a larger effort towards developing, implementing, and evaluating a software infrastructure for this purpose. The project will culminate in a large-scale field test over a whole metro area. Initial testing of the system is scheduled for the near future with iterative development leading up to the deployment.

ACKNOWLEDGMENTS

We wish to thank our student teams whose contributions have been invaluable to our process:
- Team one: JP Arsenault; Lauren Von Dehsen, Priyanka Shetye; Daisy Yoo
- Team two: Jenny Schweers, Meg Davis, Andreia Goncalves, Mark Leung, Nick Matterson

The Rehabilitation Engineering Research Center on Accessible Public Transportation (RERC-APT) is funded by grant number H133E080019 from the United States Department of Education through the National Institute on Disability and Rehabilitation Research.

REFERENCES

Bureau of Transportation Statistics (2003, April). *Transportation difficulties keep over half a million disabled at home, BTS Issue Brief, No. 3*. U.S. Department of Transportation.

Danford, G. S., & Maurer, J. (2005). Empirical tests of the claimed benefits of universal design. *Proceedings of the Thirty-sixth Annual International Conference of the Environment Design Research Association*. Edmond, OK: Environmental Design Research Association, 123-128.

Lewis, C., Rieman, J. (1994). *Task-Centered User interface Design: A Practical Introduction*. Retrieved August 7, 2009, from http://grouplab.cpsc.ucalgary.ca/saul/hci_topics/tcsd-book/chap-5_v-1.txt

National Council on Disability (2000). *The current state of transportation for people with disabilities in the United States*. Washington, DC: National Council on Disability, 2005.

National Organization on Disability (2004). *N.O.D./Harris Survey of Americans with Disabilities*. www.nod.org/index.cfm?fuseaction=Feature.showFeature&FeatureID=1422 (Acc. March 4, 2008).

Neighborhood Park's Council (2008). *2007 ParkScan.org annual report*, www.parkscan.org/pdf/2007/ParkScan_Report_2007_web.pdf (Acc. March 20, 2008).

Paolos, E. (2008). Citizen science: Enabling participatory urbanism, in *Urban informatics: Community integration and implementation, information science reference*, M. Foth, Ed. IGI

Snyder, C. (2003). Paper Prototyping: The Fast and Easy Way to Design and Refine User Interfaces. San Francisco: Morgan Kaufmann Publishers.

Steinfeld, A., Aziz, R.D., Von Dehsen, L., Young Park, S., Maisel, J.L., and Steinfeld, E. (accepted). The value and acceptance of citizen science to promote transit accessibility. Journal of Technology and Disability.

Steinfeld, A., Maisel, J. L., & Steinfeld, E. (2009). The value of citizen science to promote transit accessibility. First International Symposium on Quality of Life Technology.

U.S. Census Bureau. (2006). 2006 American Community Survey.

Yoo, D., Zimmerman, J., Steinfeld, A., and Tomasic, A. (2010). Understanding the space for co-design in riders' interactions with a transit service. *Proceedings of the Conference on Human Factors in Computing Systems*. ACM Press, Atlanta, Georgia.

Chapter 61

Dynamic Analysis of Opportunities and Needs of Persons with Disabilities in the Use of Application Interfaces

Zbigniew Wisniewski, Aleksandra Polak – Sopinska,
Joanna Lecewicz – Bartoszewska

Department of Production Management and Logistics
Technical University of Lodz
Wolczanska 215, 90-924 Lodz, Poland

ABSTRACT

The article presents preliminary results of analyses carried out with the use of the authors' own application developed to study perceptive and motor abilities of various applications users. The aim is the analysis of interface perception due to formants they include, colour palettes used in them and other elements of composition, as well as pointing at relationships between interfaces perception and composition and determining users preferences in terms of interface composition from the perspective of ergonomics. The research was carried out through the Ministry of Science and Higher Education funding under grant N N115 0386 33.

Keywords: elderly, disabled people, digital divide, user interface

INTRODUCTION

It is extremely important to adjust every workstation to psycho-physical features of a man. A working environment should be considered a resultant of a few factors, including technical equipment, furniture, workstation functionality and feelings that the workstation evokes in a worker. Type of works performed at a given workstation plays a very important role here. The following elements influence the comfort of work with a computer:

- Particular elements building up a computer station, directly associated with the computer (keyboard, monitor, printer),
- Shaping spatial structure of a computer workstation,
- Factors of material environment of work (light, microclimate, noise, electromagnetic fields),
- Psychological factors,
- Quality of software.

An interdisciplinary character of ergonomics enables studying all the above problems as indispensable elements of a workstation. When preparing a computer workstation it has to be remembered that everyday work with a computer performed longer than 4 hours in a run can be onerous due to the following reasons:

- Intensive visual work may cause various eye discomforts, headaches, tiredness and weariness,
- Long-term immobility may result in problems with back, shoulders and spinal cord, as well as shallow respiration and slow blood circulation, especially in legs,
- Poor organization of work, fast pace of work, pressure of deadlines, seclusion during work, poor software and lack of support in difficult situations may cause stress leading to tiredness, weariness, frustration and occupational burnout.

In the era of common information and almost unlimited access to computers and internet disabled persons have a possibility to take active part in the so called information society. Hardware and software producers have begun to pay attention to disabled persons, thanks to which more and more advanced applications and tools of human – machine communication are developed.

The greatest problems with the use of a regular equipment of a computer station are faced by persons with such motor dysfunctions as:

- No lower limbs or paralyzed lower limbs,
- Involuntary muscular contractions
- Contractures
- Bones deformations
- Coexisting speech, hearing and visual dysfunctions.

A practical part of the paper is devoted to the analysis of perception of a user working with a computer and the user's (also a disabled one) needs analysis in the area of a computer station. The study was developed in a way enabling checking various factors influencing a user's perception when filing data into a computer as well as when receiving information displayed on the screen.

Recipient's perception when introducing and receiving information has a great influence on the possibility to take right decisions and actions. In the situation when the message transfer is disturbed by certain factors, the time needed for acquiring information gets longer. Situation is similar in case of filing data e.g. with the use of a form. Change of logical order, use of nonstandard formants occurrence of various disturbances limits the ability of an operator to act fast. This results from the fact that many activities are performed intuitively, in a learned way. Introducing modifications into typical and working solutions can have negative influence on users' perception and their ability to take proper actions.

Operator's eyesight and attention are engaged at various degrees. Hence introducing information will be studied with the use of properly designed forms. Use of solutions differing in the way of presenting choice option and filing data enables assessment and pointing at the optimum variation. As a result preferences of users can be checked, and they can be asked to express their opinion.

In case of receiving information users concentrate their eyesight and attention on a display. While reading a content or searching for data operators can be affected by various factors hampering their work. An important role is played here by colours, fonts, contrasts and arrangement of elements on the screen. It has to be examined to what extent these factors change the user's cognitive ability.

Tools used for carrying out the research enabled objective comparison of influence of the described factors and elements of interface on the user. For this purpose the time needed to perform subsequent activities was tested

RESEARCH

The research was run in a few stages. The actual research was started with filling in a questionnaire. With the use of answers obtained in this part of the research it was possible to make a deeper analysis of the results taking into consideration information about the users. Next stages were: filling in the forms and giving answers to the questions concerning these tools. This part of the research was aimed at checking influence of certain factors on a user while introducing data into a computer. Forms, as a very popular and widespread tool, were used for this purpose.

Next stage was the analysis of the influence of elements of interface and composition on the receipt of contents from a computer display. Among others the consequences of placing elements containing information in different way in the visual field of a user were examined.

The surveyed group consisted of 200 persons at the age 18 – 51 (the average age being 26.5 years). Among the survey participants there were secondary schools students, university students and professionally working people. All the participants use internet relatively often.

Most of them can be included into the advanced group because 65% of them use the web everyday and 35% a few times a week. Besides, 95% of the surveyed persons have had contact with internet for at least 5 years (40% for 10 years, 55 % for 5 years). None of the surveyed persons is a computer science engineer or a computer programmer.

The research was made with the use of 17-inch monitor having a resolution of 1024x768 and colour depth of 32 bits/pixel.

Each of the participants received a questionnaire to fill in when realizing particular tasks. After filling in the data in the first part of the questionnaire a user filled in the first form (F1). During filling in of the form the time needed by a participant to introduce all data was measured, and after that the users were asked to assess the legibility of this tool.

Next stage of the research was devoted to the issues associated with influence of different factors on the legibility of a schematic website and its receipt by a user. For the purpose of comparison two different compositions were prepared. The first website was developed in accordance with certain standards, including appropriate fonts (internet dedicated), menu setup and the use of proper colours.

This composition is made upon the model of many popular websites with a characteristic division into sectors. Thanks to the use of various colours the left and the right sector, top panel and the central part were differentiated. In the top part a static commercial banner was placed, on the left side a vertical menu was situated, and in the central panel there were two texts. Above the information a heading was designed introducing the content of a given sector. Both the texts, one 354 signs long and the other 363 signs long with spaces were downloaded from internet portals. After loading them into the application their parameters were settled:

- Font type – Verdana,
- Font colour - white,
- Centered text
- Interline – 4.

The above website composition was presented to the survey participants. Each user was asked to start the clock using the "start" button and to get acquainted with the presented schematic website. It was the participant's decision how much time he/she needs to get acquainted with the website, and necessity to read the texts in the website was not suggested. After getting acquainted with the schematic site the participant stopped the clock.

The next element presented to the participants was the second design of a schematic website. This time with a division into two sectors and different colours of the font and background. The top panel, like previously was highlighted in orange and two texts of the length 442 and 341 signs with spaces were downloaded. After that the following parameters were settled:

- Font type – Times New Roman,
- Text 1 – red font,
- Text 2 – dark blue font
- Aligning: text 1 – to the left, text 2 – to the right,
- Interline – 0.

Two commercial banners, a static and a dynamic one, and a horizontal menu at the bottom were placed in the website. After starting the application by the person coordinating the survey a participant started the clock and began to get acquainted with this schematic website. When the user stopped the time, he/she started to answer questions included in the third part of the survey.

Next part of the questionnaire referred to the presented websites. It included questions comparing both compositions from the perspective of readability and receipt by the participants. At this stage participants were asked to present their subjective evaluations and give answers to questions concerning texts in the websites. The person coordinating the research also recorded the times needed by particular participants to get acquainted with presented contents.

In the end each participant was asked to fill in the second specially designed form (F 2). Times needed to fill in the forms were measured. The forms differed in construction: the first one contained mainly scroll lists, the second contained checkboxes. Because of the difference the first form appeared to be shorter than the second one.

ANALYSIS OF THE RESEARCH RESULTS

It should be noticed here that all participants of the research use computer mainly for their needs and are not specialists in computer science, which means that assessments they make are based on their own subjective feelings and experience. The first information that should be considered is an average time needed by users to fill in particular forms. It is significant that the forms were presented in different order to different participants. Unlike initially expected it turned out that filling in the "Form 2", which was longer because of using smaller number of scroll lists took on average 10 seconds less than filling in the "Form 2". At the stage of forms design it was assumed that the shorter form, with a larger number of scroll lists will be more user-friendly, which influences shortening the time needed to fill it in. Yet it turned out that such a formula is not easier for a user, and the time of filling it in gets longer.

110 (55%) of the surveyed persons chose "Form 2" as the better one. 70 persons chose "Form 1", and for 20 persons there was no difference . Most persons who chose the second form as more friendly justified their decision by the fact that it is more convenient to work with a fill-in form when all choices are visible from the start. Moreover in their opinion it is easier to fill in the forms in which not too many lists have too be scrolled down. Such assessments lead to concluding that the initial assumptions were wrong. The form in which more scroll lists for providing information were used turned out not to be more convenient for the surveyed group. Moreover reducing size by replacing lists with option buttons does not shorten the time needed to fill in the form. Most of the surveyed persons clearly stated that it is more convenient for them to use forms in which all options and answers are visible, and that too many scrolled-down lists makes work a bit more difficult.

Still one more regularity, emerging from the research results, should be noticed. The shortest times in filling the forms in were achieved by the users that could be referred to as the advanced ones. Their advancement in working with computer and internet is the longest. Five of them indicated "Form 1" as more friendly. It could be concluded that their choice results from their experience in working with the web. A great number of forms in internet have the form similar to the one of "Form 1", which is a result of the need to minimize the size. By filling in dozens or even hundreds of forms when opening e-mail accounts, shopping on-line etc. a user gets acquainted with many scrolled –down lists and with the need of searching for information.

SUMMARY

For making an analysis of the influence of various factors on user's perception during receiving information from a computer display two designs of a schematic website were applied. On their basis analysis of objective measurements, such as the time needed to get acquainted with a website as well as subjective evaluations made by particular users were shown.

The first stage of the research was to provide proofs that there exist a number of factors influencing the receipt of information from a computer display. In order to prove the correctness of this thesis two schemes of a website were developed. With their use it was checked how a capability of receiving and remembering information changes in case of alternating the setup of an exemplary website. Both the schematic websites presented to the users contained short texts. Their length was selected so as to exclude the influence of this factor on the ability to remember their content. The comparison was made through using different fonts and changing the scheme of the colours used and setup of the website elements. The research proved that the factors have a great influence on the user's perception.

The research results enable concluding that a poor composition of a website can even prevent users from receiving the information it contains. When preparing services with important information an easy access to the contents should be assured, a good division has to be made to enable to users intuitive moving in the website, and certain standards should be followed. This applies first of all to logical arrangement of a website elements and use of internet dedicated colours and fonts, due to their readability on a computer display.

The most frequently appearing justifications for choosing application 1 as the better one are:

- More readable font used in the texts,
- The colour of background well contrasting with the colour of fonts,
- Buttons arrangement enabling their easy finding,
- Clear division of site with the use of different colours.

As regards application 2, which had been assumed to cause certain problems, the following complaints were made:

- Too garish colours of the site
- No contrast between the background and the text colours is an obstacle for reading,
- Unclear font,
- Disorder on the site (commercial banners on the foreground, menu difficult to find).

Basing on the above statement it can be said that such factors, as colours, fonts, site arrangement etc. have great influence on the receiving of presented contents.

Developing two forms enabled checking preferences of persons introducing data as regards presenting information. Against the initial assumptions in turned out that most users indicated the longer form as easier to fill in. The analysis of time needed to fill in the forms confirmed that the research participants prefer the longer version. This is connected with the way of presenting certain information in the forms. In the first variation the size of the form was reduced by the use of scroll lists. Still this had a negative influence on the convenience of usage, because access to part of information demanded scrolling-down the list with the use of an appropriate button, and searching for the right position in the list. The research proved that filling in of such documents takes more time and requires certain skills. In the second case more option buttons were used (check box, radio button), which caused increase in size, but at the same time enabled fitting of all information directly in the form. The users claimed that such arrangement is more friendly. Moreover such settlement demands less time for introducing data, which is significant from the point of view of a user. It has to mentioned here that the users had no problems with filling in of both the forms. Besides, the information obtained from the survey participants indicate that the fundamental problem for most persons with not agile hands is using a standard keyboard, which is the basic tool of communication with a computer. Other difficulties are associated with manipulating a mouse, switches, using a printer, basic maintenance of the equipment. The buttons and switches of the computer should be accessible in front of it. When arranging a computer station it is worth to apply a popular supply filter with a switch as an element of a central switching-on of supply of the whole computer set. That element should be located in a place enabling efficient and reliable use of the switch, depending on individual abilities of a disabled person. Wires should be arranged so as to prevent hitching with an element of a wheelchair, legs or hands. As needed a keyboard should be stable fixed to the ground, so as to prevents its displacing. Sometimes a proper construction with non-skid backings is sufficient. The desk on which a computer equipment is placed should have a solid construction and not move. A lot of people walking with crutches lean against a table while getting up a chair and carry their whole body weight over the table.

REFERENCES

Mullet, K., Sano, D. (1995). *Designing visual interfaces: Communication oriented techniques*. Sunsoft Press, Prentice Hall

Raskin, J. (2000). The *humane interface: New directions for designing interactive systems*. Addison-Wesley

Wiśniewski, Z., & Polak-Sopińska, A. (2009). HCI standards for handicapped. Paper presented at the *Conference Proceedings: 2009 Human Computer Interaction International (HCI)*, San Diego, CA, USA. , *DVD*

Wiśniewski, Z. (2007a). Ergotronics – object, composition, message, transfer. In L. Pacholski M. (Ed.), *Ergonomics in contemporary enterprise* (J. Grobelny Trans.). (pp. 67-73). Poznań: IEA Press Madison.

Wiśniewski, Z. (2007b). System approach to the process of changes implementation in the organization. In Grudzewski W. M., Hejduk I.,Trzcielinski S. (Ed.), *Organizations in changing environment. current problems, concepts and methods of management* (T. Nowakowski, A Wilk Trans.). (pp. 492-495). Poznań: IEA Press Madison.

Chapter 62

Ergonomic Adjustment of a Selected Workstation to the Function of Occupational Rehabilitation

Aleksandra Polak-Sopinska, Zbigniew Wisniewski

Department of Production Management and Logistics
Technical University of Lodz
Wolczanska 215, 90-924 Lodz, Poland

ABSTRACT

The article presents ergonomic analysis of a selected production line from the perspective of possibility of employing disabled persons, and offers improvement plan. Preliminary research was based on a control list filled in by a manager and on a guided interview with production line labourers. The primary aim of the control list and interview was determining to which disorders particular work stations can be adjusted. In the proper research body position assumed by a worker and problems resulting from poor adjustment of a workstation equipment were evaluated. Also the freedom of movements as well as the seat, physical load and psychical load were evaluated. The attained information enabled identifying barriers that would hinder employment of the disabled, after that preliminary plan of a production line providing for the basic needs of the disabled workers was developed. In case of employing a specific person with particular dysfunctions it will be necessary to adjust the workstation to the person's individual physical abilities.

Keywords: Disabled persons, workstation, occupational rehabilitation

INRODUCTION

Work offered to the disabled is a very important element of their rehabilitation process. It gives to the disabled the possibility to feel needed, to check their abilities and to acquire new skills. Work enables being with persons having similar problems. For most of the disabled the function of work is to satisfy the needs of safety and respect. **Still it has to be remembered that work must not deepen disability**. Therefore the issue of choosing a work and a workstation for the disabled is much more complex than in case of people regarded as healthy.

In practice an employer wishing to employ a disabled person faces a difficult task of selecting a proper workstation and duties to be performed by a specific person. The task gets more difficult as more complex is the dysfunction. Employers do not have at their disposal models of solutions supporting them when taking a decision as well as when employing a disabled person and considering maintaining a productivity of the workstation. The greatest unwillingness to employ he disabled can be noticed among employers in the open market.

RESEARCH AIM AND STAGES

The aim of the research was ergonomic analysis of a selected production line in an enterprise operating in the open job market from the perspective of possibility of employing disabled persons and of improvement plan. The research was also aimed at presenting to the employers the methodology of processing when selecting workstations for persons with disabilities. The research was carried out in six stages:

1. Characteristics of selected workstations.
2. Getting acquainted with general knowledge in the area of diseases. And basic indications and contraindications for work of people suffering from these diseases.
3. Making the analysis of the requirements of work at given workstations from the perspective of possibility of employing the disabled with specific diseases. Choice of workstations at which disabled persons can be employed – **necessary medical consultation**.
4. Getting acquainted with individual abilities and needs of a chosen disabled person.
5. Comparing results of analyses and defining association between skills and abilities of a selected (specific) disabled person, and demands concerning the workstation - **necessary medical consultation**.
6. Adjusting the workstation to individual abilities of a selected disabled person - **necessary medical consultation.**

STAGE I: CHARACTERISTICS OF SELECTED WORKSTATIONS

Ergonomic analysis was carried out for BGR 2 line in BGR department (Figure 1).

The line is one of six lines at which a ready product is finally packed. The analysis and evaluation of every workstation in the production line BGR-2 was aimed at indicating a group of disabled persons that could be employed at them. Making this analysis demanded getting acquainted with tasks performed at particular workstations. Short characteristics of particular workstations is presented in Table 1.

Table 1. Characteristics of workstations at the line BGR2

Workstation name (photo)	Workstation characteristics
Machine operator workstation	1. Work demanding experience, performed by a single person.. 2. Tasks of the worker: current process control , workers training, data collecting.
1 and 2 packer workstation	1. Work is performed for 1.5 hour, in standing or sitting position 2. Clearly defined activities performed all day long. 3. Tasks of the personnel: - picking up packages with products (trays) from the lay down field, - putting products on the belt with the system of followers, - supervision of correct product placing.
3 and 4 packer workstation	1. Work is performed for 1.5 hour, in standing or sitting position 2. Clearly defined activities performed all day long. 3. Tasks of the personnel: - supervision of correct closing of a package (visual standards), - placing products in multipacks, - checking faulty packages .
Material handler workstation 	1. Work is performed for 1.5 hour, in standing or sitting position 2. Clearly defined activities performed all day long. 3. Tasks of the personnel: - preparation of multipacks for station 3 and 4, - receipt of full multipacks from stations 3 and 4 - weighting of full multipacks, - assistance to other workers.

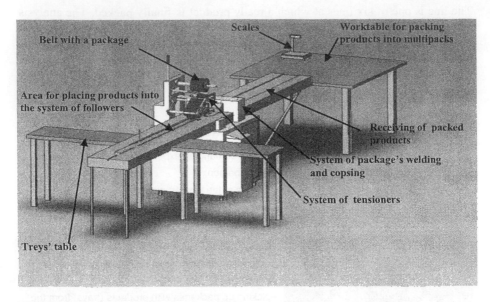

FIGURE 1. Components of Bagger 2 machine

STAGE II: GETTING ACQUAINTED WITH GENERAL KNOWLEDGE ON DISEASES AND BASIC INDICATIONS AND CONTRAINDICATIONS TO WORKING FOR PEOPLE SUFFERING FROM THESE DISEASES

The basic condition for adjusting workstations to the needs of the disabled is knowledge on various diseases and knowledge of the main indications and contraindications for employing persons suffering from them. In Poland for the purpose of occupational rehabilitation the disability classification is used based on the Decree of the Minister of Economy, Labour and Social Policy of 15 July 2003 on certifying disability and its degree. In this classification 11 groups of disability are differentiated:

01-U – mental retardation;
02-P – mental illnesses;
03-L – phonatory, speech and hearing disorders;
04-O – visual changes;
05-R – motor organ dysfunction;
06-E - epilepsy;
07-S – respiratory and circulatory system diseases;
08-T – alimentary system illnesses;
09-M – genitor-urinary system illnesses;
10-N – neurological dysfunctions;
11-I - other, including: endocrinological and metabolic illnesses, enzymatic disorders, infectious and epizootic diseases, disfigurations, hematological illnesses.

For the first nine groups a list of main contraindications and indications for employment has been developed. The purpose of the listings is to make it easier for an employer to decide at which stations disabled persons can be employed. Due to ergonomic range of the article only guidelines for mentally disabled persons are been presented. Basic indications for work for mentally disabled persons:

- works that do not demand continuous tension (e.g. sewing, packing),
- simple, repetitive works (assembling simple elements, tailoring –sewing, cooking, cake making – confectioning),
- works that can be performed without or after a short training (doorman, cloakroom attendant, janitor, cleaner),
- works in which there is a possibility of consultation, and there is no one-person responsibility,
- one- or two-shift works during the day.
- works not demanding using personal protections hampering moving around or distorting receiving information,

Basic contraindications for work for mentally impaired persons:

- exposure to continuous stress (decision process or mental pressure last longer than 50% of working time, e. g. cashier, inspector, engineer of electronics),
- complicated works demanding knowledge in many areas of specialist education,
- only self-reliant, independent works, in which whole responsibility is upon one person. Mentally retarded persons should not work on their own, they should have someone to help them in urgent cases or difficult situations,
- three-shift work in continuous process industry, working time exceeding 12 hours (turns), frequent overtime hours or field work.

STAGE III: ANALYSIS OF WORK DEMANDS AT PARTICULAR WORKSTATIONS FROM THE PERSPECTIVE OF POSSIBILITIES OF EMPLOYMENT OF DISABLED PERSONS. CHOICE OF WORKSTATIONS AT WHICH THEY CAN BE EMPLOYED

In order to assess the analysed workstations from the perspective of their adjustment to various types of disability and pointing at various kinds of dysfunctions which are not an obstacle for employing a person suffering from them at these workstations a preliminary and more profound research was carried out.

Preliminary research

Methods and techniques used in preliminary research

Guided interview with workers, indirect observation and control list were used in the preliminary research.

The interview was developed for the purpose of preliminary analysis diagnosis of a problem area at the analysed workstations. The authors of the questions concentrated mainly on the issues associated with the workstations ergonomics. The workers were asked mainly about their feelings connected with the way of working at the analysed workstations.

The control list was divided into 9 parts corresponding to different types of disability in accordance with the Decree of the Minister of Economy, Labour and Social Policy of 15 July 2003 on certifying disability and its degree. The list did not include questions concerning contraindications for work for persons with neurological and other illnesses, because due to varied complexity of dysfunctions, universal contraindications and indications for workstation organization could not be provided.

Questions in the list were formulated so that answer „yes" confirmed contraindications for employing persons with significant disability, and answer „no" meant that there were no contraindications associated with the content of the question for given disabilities.

In some cases one positive answer could exclude the possibility of employing persons with certain disability if it was impossible to make changes eliminating these contraindications. While in other cases such answer gave a tip concerning the areas which should be adjusted to the needs of persons with a specific dysfunction, so that after introducing changes employing these persons would be possible.

Preliminary research results

Due to extensive character of the problem, only the results for the station of packer 3 and 4 have been presented. The choice was made because this post joins together a number of mistakes common for other workstations.

During a guided interview the workers complained first of all about:
- Pains in the spinal cord (workers are forced to bend over and twist their torso every time they want to put a product into a multipack),
- Tiredness, sleepiness,
- Monotony of work,
- Too fast pace of machine working, which in certain situations is hard to follow by the workers,
- Limited visibility in protective glasses due to their scratching,
- Difficulties with moving with noses put on the tips of boots.

Indirect observation of the production plant proved that the enterprise is not adjusted to employing persons in wheelchairs.

Basing on the interview, observation and control list the disabilities with which persons could work at the given workstations were determined. The results were also consulted with a doctor.

Types f disability accepted at the post of a packer 3 and 4:
- mental retardation – medium or light, with a lower level of development than an average one or an average one;
- phonatory, speech and hearing disorders – light and medium;

- motor organ dysfunction – light and average– selected diseases;
- respiratory and circulatory system disorders – light and medium – selected diseases;
- alimentary and genitor-urinary system illnesses – light and medium – selected illnesses.

In-depth research

Methods and techniques used in the in-depth research

As during preliminary research the workers often complained about afflictions in their motor system, tiredness, sleepiness, monotony of work and too fast pace of machine work, it had to be checked if there are no contraindications for employing disabled persons with the diseases listed above.

The following research methods and techniques were used in the proper research:
- dummies and anthropometrical atlas to analyse and assess working space,
- table-chronometer methods to evaluate energy expenditure,
- Filipkowski-Szumpich method to assess statistical load,
- JSI method to assess monotype character of working movements.

In-depth research results

Anthropometric analysis has proven that the table from which the workers take packed products is either too high or too low, depending on a position taken. In case of a standing position the table is 7.4 cm. too low, whereas in a sitting position it is 10.2 cm too high. Such construction results in assuming enforced body positions. In case of persons with a motor dysfunction any type of enforcing is unacceptable, because of the possibility of fast progressing of the illness. Because of incorrect adjustment of the belt height, when picking up packed products, a worker is forced to slightly rise, At the same time transferring the body weight onto lower limbs. Such a position makes it impossible for persons with motor organ dysfunctions located in toes, tarsus and knee joint to work at the post. Also persons with rigid hip joint, chronic rheumatic arthritis, or ankylosing arthritis, high muscle tone, talipes equinovarus or neck spine or lumbar spine defect may experience discomfort while performing the activity. Intensified physical effort, resulting from enforced body position is not recommended for persons suffering from subacute or chronic repiratory system diseases, circulatory system diseases and is not recommended for persons with neurological dysfunctions, such as multiple sclerosis. Increased working discomfort is caused by inability of sufficient slipping of oneself under the belt, which results from too little space for legs and too small width of the belt. A worker lacks in about 15.5 cm of a distance calculated vertically and 60.3 cm of a distance calculated horizontally (belt width). Construction of the workstation may cause problems with free performing of the activity of inspecting correctness of closing a package and segregating products into these that can be repacked an these that are treated as scrap. Significant bending to reach the sack with faulty packages located under the machine belt causes high tension almost along the whole spinal cord. Hence persons with diseases of this part of the body will have serious problems with performing their activities.

Anthropometric analysis has also proven that a chair with which the stand is equipped is not right. It has no arm-rests and it can be regulated only in two plains – the seat's height and its back orientation. Neither is the station equipped with a separate footrest. At present this role is performed by a rim permanently assembled to the chair.. The worker, when resting his/her feet on the rim experiences tightness in the knee part. Properly selected chair is the basis for assuming correct body position. Its disadvantages may lead in a long run to spinal cord and intervertebral disks degeneration.

Wrong construction of a 3 and 4 packer's station has been confirmed by the result of the assessment of physical loading with work:

- energy expenditure – 3800 kJ/per shift – medium-hardness work for women, can exclude employing persons with disabilities of circulatory and respiratory system,
- static load – medium,
- particularly monotype character of working movements (number of stereotype movements per one shift – 3200), which excludes employing persons with disability of wrist, cubital joint and acromioclavicular joint.

A serious problem at the analysed workstation can be also the rate of machine's work. Due to the fact that the pace cannot be lowered, which is a result of actual production demand - disabled persons may have serious problems with getting adjusted to that pace. The problems gets more serious especially in case of persons with mental disorders, who realising that they cannot keep up with the pace of other workers can feel worse, which in consequence may lead to intensifying their memory, thinking, concentration and situation assessment disorders. A high pace of work may be also unfavourable to persons with circulatory and respiratory system dysfunction.

During the research a physician had to be consulted.

STAGE III, IV, V ADJUSTING WORKSTATION TO INDIVIDUAL ABILITIES AND NEEDS OF A SELECTED DISABLED PERSON

Due to recruitment process being in progress, workstations have not been adjusted yet to individual abilities and needs of a selected disabled person.

Examples of propositions of improvements for a person with motor disability of lower limbs is presented in Figure 2.

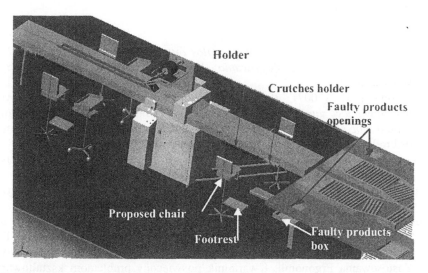

FIGURE 2. Examples of propositions of improvements for a person with motor disability of lower limbs

SUMMARY

Professional work is not only a means of securing financial resources for living, but also a need, social duty and basic condition of personal development. In case of disabled persons professional work performs one more very important function. Work is a critical factor in the progress of the whole rehabilitation process. Thanks to being employed, persons with a specific dysfunction can not only test the skills they already have but also acquire new ones. The very fact of being among other people prevents them from feeling lonely and improves their self-esteem as a result of their contribution to an activities of the enterprise they work for.

In Polish enterprises reluctance towards employing disabled persons can be observed. A conviction of entrepreneurs that employing a person with dysfunction has negative influence on productivity and generates extra costs is a common reason for this situation. Moreover employers do not have at their disposal models of solutions supporting them when taking decision on which disabilities do not prevent candidates for work from being employed in their company and how to maintain productivity of a workstation when employing a disabled person.

The research presented in the article shows practical use of processing method when selecting a workstation for persons with disability. The authors hope that popularising knowledge on employing disabled persons among employers will rise the disabled employment rate.

612

REFERENCES

Batogowska A., Słowikowski J. (1989), *Atlas antropometryczny dorosłej ludności Polski dla potrzeb projektowania*, Instytut Wzornictwa Przemysłowego, Warszawa.

Dziewanowska A. (2007), *Projekt koncepcyjny wybranych stanowisk pracy umożliwiający zatrudnienie osób niepełnosprawnych w przedsiębiorstwie produkcyjnym „X"*. M.A. thesis, scientific care – A. Polak-Sopińśka. Institute of Production Management, Technical University of Lodz, Lodz.

Filipkowski S. (1965), *Kształtowanie warunków pracy*, Państwowe Wydawnictwo Naukowe, Warszawa.

Górska E. (2002), *Projektowanie stanowisk pracy dla osób niepełnosprawnych*, Oficyna Wydawnicza Politechniki Warszawskiej, Warszawa.

Lecewicz-Bartoszewska J., **Polak-Sopińska A.** (2006), *Stanowiska pracy dla niepełnosprawnych – od projektowania indywidualnego do integralnego*. Pod red: J. Charytonowicz, A. Jasiak, L. Pacholski, W. Rybarczyk, E. Tytyk. Zastosowania Ergonomii. Kwartalnik poświęcony problemom kształtowania środowiska egzystencji człowieka. Nr1-3. Poznań-Wrocław-Zielona Góra.

Lecewicz_Bartoszewska J., Polak-Sopińska A. (2008), *Antropometric system approach to designing as a condition of solving the whole problems of persons with disabilities. Introduction of the monograph. Wprowadzenie do monografii.* Ergonomics for the disabled in work organisation and management – results of researches, design,, pod redakcją J. Lecewicz-Bartoszewskiej i A. Polak-Sopińskiej. Wyd. Politechnika Łódzka - seria monografii. Łódź.

Polak-Sopińska A. (2008), *Przystosowanie stanowisk pracy do indywidualnych możliwości osób niepełnosprawnych, podstawą ich aktywizacji zawodowej*, w: Aktywizacja zawodowa osób niepełnosprawnych, Krajowa Izba Gospodarczo-Rehabilitacyjna, Warszawa.

Polak-Sopińska A., Sztobryn-Giercuszkiewicz J., Sznajder D. (2008), *Activity of the Office for the Disabled of the Technical University of Lodz in the aspect of adapting the university to the needs of the disabled*. Ergonomics for the disabled in work organisation and management – results of researchs, evaluations and guidelines, pod redakcją J. Lecewicz-Bartoszewskiej i A. Polak-Sopińskiej. Wyd. Politechnika Łódzka - seria monografii. Łódź.

CHAPTER 63

Evaluation of Methods to Remedy Existing Multi-Family Housing to Maximize Accessibility to Residents

Ilene B. Zackowitz, Alison G. Vredenburgh

Vredenburgh & Associates, Inc.
Carlsbad, CA 92008, USA

ABSTRACT

There is substantial disagreement concerning how developers and owners of existing multi-family housing should remedy properties that have been found by a court not to meet FHA requirements. The Fair Housing Act (FHA) is not a building code; nor does it have any legally required dimensions as to how to design an accessible building. Thus its application tends to be inconsistent. The FHA was enacted to prevent housing discrimination of people in protected classes including race, color, religion, national origin, sexual orientation and people with disabilities. There is disagreement among experts regarding how best to accommodate people with disabilities. This paper addresses two different approaches to remedy existing, already constructed multi-family housing: uniformly applying "safe harbors" retroactively and tailoring units to users' specific needs. Satisfying safe harbor standards during construction ensures avoidance of FHA claims. However, since the capabilities and limitations of people with disabilities differ significantly from one person to the next, we discuss the two approaches in their relative ability to best serve the residents' specific and individual needs after the building has been constructed.

614

Keywords: Fair Housing Act, Accessibility, Disability, Multi-family Housing

INTRODUCTION

The Fair Housing Act (Title VIII of the Civil Rights Act of 1968) was enacted to prevent housing discrimination on the basis of race, color, religion and national origin. In 1977, this list was expanded to include sexual discrimination and the Fair Housing Amendments Act of 1988 sought to cover persons with disabilities by making it unlawful to deny rental or sale of a dwelling unit to people with disabilities. The FHA covers private multi-family housing and requires that all ground floor apartments (including condominiums) and all floors of elevator buildings of four or more units, built for first occupancy after March 13, 1991, comply with the Act.

While the goal of this legislation is to make multi-family housing accessible to people with disabilities, the question is how best to achieve this goal. The FHA has seven design requirements:

1. Accessible building entrance on an accessible route
2. Accessible and usable public and common use areas
3. Usable doors
4. Accessible route into and through the covered unit
5. Light switches, electrical outlets, thermostats and other environmental controls in accessible locations
6. Reinforced walls for grab bars
7. Usable kitchens and bathrooms

Congress recognized six of these requirements to be "features of adaptive design." The Fair Housing Act further incorporates the adaptable/adjustable concept in bathroom walls by requiring that they contain reinforced areas to allow for later installation or grab bars without the need for major structural work on the walls. Thus, incorporating adaptable features into building design during construction can avoid the need for future remedy. Adaptability is "The capability of certain building spaces and elements, such as kitchen counters, sinks, and grab bars, to be altered or added so as to accommodate the needs of persons with and without disabilities, or to accommodate the needs of persons with different types or degrees of disability" (ANSI A117.1-1986, p. 14). Adaptable/adjustable elements and spaces are those with a design, which allows them to be adapted or adjusted to accommodate the needs of different people.

There are many ways to make housing adaptable. The most commonly cited adaptable design feature is a removable sink base cabinet. Since residents often have different needs based on their disability type, many would prefer to keep the standard cabinets and base (e.g. they use a walker or sports chair; they have young children and want to enclose cleaning products; they prefer how it appears). Others would prefer that the base and doors be removed such that they can roll under the sink. Removing a base cabinet typically entails unscrewing 2-4 screws, sliding out the base, and removing hinges from the cabinet doors.

Many other apartment features can also easily be adapted. Outlet locations can be adapted to residents' specific reaches by placing power strips at optimal locations for each residents' reach; people who are quadriplegic often have reaches of 4" or less, or may have use of only one hand, thus locating a power strip in the front of the counter edge or along the refrigerator may be the only usable location for these residents (see Figure 1). These strips can be screwed in, or attached with various adhesives. Other types of easy adaptations include re-hinging doors so that they swing out (such as for bathrooms), adding coat hooks at preferred heights for residents, removing or changing transition strips and adding extensions to thresholds with a gradual slope as needed. If environmental controls are installed with extra wire, they can be moved up or down depending on the residents' needs.

Figure 1. Electrical outlet location can be adapted to the reach of most users.

If a property is already constructed and found by a court not to meet the FHA requirements, there are two principal approaches to "remedy" the properties in this situation. The first approach is based on "safe harbors," or design specifications that, when met, are considered to be in compliance with the FHA. The second method is based on tailoring individual units to the needs of the people who reside there to further the intent of adaptability.

APPROACH 1: SAFE HARBORS

Congress has indicated that the FHA can be met in a variety of ways. One way to satisfy the FHA is through what are called "safe harbors." Safe harbors are not mandatory; they are specifications identified by U.S. Housing and Urban Development (HUD) as fulfilling the FHA requirements, and are used to keep the developers "safe" from accessibility claims.

Safe harbors do not establish minimum requirements and the various identified specifications are not always consistent among each other. There are several sets of specifications that qualify as "safe harbors" as identified by HUD (Paarlberg, 2004) and include: Fair Housing Accessibility Guidelines, Fair Housing Design Manual (both HUD documents), ANSI A117.11986, CABO/ANSI A117.1-1992, ICC/ANSI A117.1-1998 and 2000. Congress has "made it clear that compliance with the Act's accessibility standards did not require adherence to a single set of design specifications" (55 FR 9479).

When retrofitting buildings to the safe harbors as a remedy, advocates contend that all first floor units of multifamily dwellings with stairs and all floors of elevator buildings of four or more units must be modified to comport with the safe harbors.

It is important to note that the safe harbor specifications are *not* the same as *Universal Design*. Universal Design means that buildings are accessible and usable by everyone, including people with disabilities (Steinfeld, 1994). Universal Design is also different from accessible design, which is limited to addressing products and buildings that are accessible and usable by people with disabilities, focusing primarily on wheelchair users. Safe harbors appear to be more consistent with accessible design than universal design because safe harbors are limited to specific populations and are based on sets of specifications, whereas universal design does not currently provide or refer to any specifications. This article addresses the remedies for existing structures and thus while Universal Design is an important topic as a theoretical goal for design and construction, it is not relevant to the topic addressed in this paper.

If individual capabilities and limitations of those with disabilities could be standardized, then the safe harbor approach would be much more effective. However, each person with a disability has a unique set of capabilities, limitations and functional requirements for their home. No single code, specification or guideline could possibly address the huge variations in people with disabilities. This brings us to the second approach: tailoring the environment to the user.

APPROACH 2: TAILOR UNITS TO SPECIFIC NEEDS

A second method of remedy is tailoring multifamily units to the specific needs of the residents who will be living there instead of changing all units (regardless of who lives there) to conform to the safe harbors. This approach focuses on making specific functional accommodations that will benefit all of the actual residents who live in the units. This is not the same as the reasonable modification requirement of the FHA. Whereas the FHA provides that residents may request and pay for reasonable modifications to increase the usability of their own units, if a landlord chooses to use the tailoring approach, they would make all of the residents' accessibility modifications at no cost to the resident. This method applies an ergonomics methodology of evaluating the capabilities and limitations of each user (resident), and designing the environment around their specific needs. This tailoring approach considers the functional analysis of each resident with disabilities within their unit.

Safe harbors are typically based on wheelchair accessibility; however the majority of people with disabilities do not use wheelchairs (Kaye, Kang & LaPlante, 2000). Some features that are inaccessible to wheelchair users may provide greater accessibility than the safe harbor to someone with a different disability; however, the safe harbors do not make this distinction. What is important in this approach is functional usability.

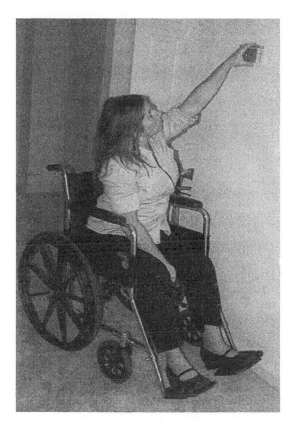

Figure 2. Climate control is a far reach from a wheelchair.

For example, the height of the controls depicted in Figure 2 is high for a wheelchair user, yet well suited for someone who has vision impairment or is ambulatory, but unable to bend (see Figure 3).

Another important consideration is that many people use wheelchairs in part of their unit and a walker or cane in other parts. Therefore, they would want fixtures, controls, etc. higher in the areas where they are ambulatory, and lower in other rooms where they use their wheelchair.

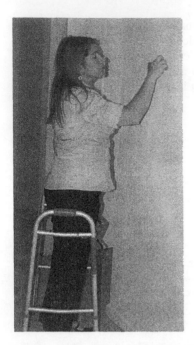

Figure 3. Climate control is accessible when using a walker.

Someone using a wheelchair would probably want their microwave oven lower on the counter, whereas someone with vision impairment or who uses a walker in the kitchen may prefer the cabinet mount design (see Figures 4 and 5).

Figure 4. This microwave is at a good height when used with a walker.

Figure 5. This microwave is inaccessible from a wheelchair.

Figure 6. This dryer is at a good height for someone using a walker, but inaccessible to someone using a wheelchair.

Similarly, a stacked dryer is a good height for someone using a walker (see Figure 6), but inaccessible to someone using a wheelchair. Moreover, the needs of some people who use wheelchairs can be in direct conflict with the needs of ambulatory people with disabilities. For example, since the safe harbors generally cater to people using manual wheelchairs, they require counters, fixtures, appliance and controls to be lower to increase usability for wheelchair users.

However, for residents who are ambulatory, but have a back or neck injury that prevents them from bending, these lower heights may actually be less accessible. Similarly, people with vision impairment may want controls higher, at their eye level.

There is also a lot of diversity among the different models of wheelchairs. For example, the seats of sports chairs are typically lower than standard manual wheelchairs, whereas many motorized wheelchairs are quite a bit higher. The users of motorized wheelchairs often have limited use of their hands. They would optimally want the appliances, outlets, etc. near the edge of the counter, at about the height of the armrests of their chairs.

CONCLUSIONS

There exists a large degree of individual differences in the user needs for accommodations of specific disabilities of apartment residents. Universally changing all covered units to comport with the safe harbor specifications may make units less accessible for more potential residents with mobility impairments than these changes would help.

For example, lowering controls, sinks and counters may make apartments less accessible to residents with multiple sclerosis, walker or cane users, and people with back injuries. Moreover, there are more non-wheelchair users with mobility impairments than wheelchair users (see Table 1).

As indicated in Table 1, the total number of Americans using assistive devices is 6,821,000, the number of wheelchair and scooter users (both power and manual) is 1,679,000 (24.6% of devices used and 0.6 of the population) and the number of users of ambulatory devices (canes, crutches and walkers) is 6,126,000 (89.8% of devices used and 2.4% of the population). Therefore, the usage of canes, walkers and crutches is about 3.6 times the amount of the usage of wheelchairs and scooters.

When selecting a solution to accessibility problems for existing multi-family housing properties, tailoring individual dwelling units to accommodate, to the extent reasonably possible, all residents' specific disabilities would provide greater accessibility to residents with disabilities than universal changes to all covered units regardless of the occupant. When addressing accessibility of individual residents in their housing, it is critical to evaluate their specific needs and tailor their living environments to these limitations. Empirical research indicates that the best way to meet this goal is to adapt the environments at the time of rental to the specific needs of the residents since making global changes as per specifications may reduce accessibility for more residents than it helps (Vredenburgh, Williams, Zackowitz, & Welner, in press).

TABLE 1. Number of people using mobility devices by age, device used and gender. (*1000s)

Device	All people	Under 18	18-64	65 and over	Male	Female
Any device	6821	145	2310	4366	2832	3989
Wheelchair or scooter	1679	88	658	933	692	987
Wheelchair	1599	88	614	897	658	941
Manual	1503	79	560	864	606	897
Power	155	18	90	47	84	71
Scooter	142	0	78	64	60	82
Other mobility device	6126	73	1987	4065	2502	3624
Cane	4755	19	1535	3200	2014	2741
Crutches	566	36	375	155	328	238
Walker	1820	27	373	1421	508	1312

*The numbers total more than 100% because some people use more than one device (Source: Kaye, Kang and LaPlante, 2000).

Safe harbors do not take into account whether residents have disabilities, or what types of specific accessibility issues that they may experience in their units. For example, a bathroom can conform to the safe harbors but be completely inaccessible to a user who only has use of his left hand in a "right handed" bathroom (wall to the left and fixtures to the right of the wheelchair). Thus, meeting safe harbors does not necessarily result in accessibility. Thus, for existing units covered by the FHA, it would be more effective to tailor units to the individual and specific needs of residents with disabilities (e.g. raise or lower counters and controls, re-hinge bathrooms and refrigerator doors, change fixtures and appliances, add grab bars at locations selected by resident, replace carpet with vinyl floors, etc.) than to make universal changes consistent with a safe harbor specification that may increase accessibility for no one.

The tailoring approach can be achieved by building owners establishing an account to be used by apartment management to tailor units to accommodate the specific disabilities of the occupants. This can occur during the signing of a lease by asking new residents with disabilities to complete a checklist of needed design modifications, policies and services to accommodate their individual usability requirements. It should be noted that the potential tenant must disclose their disability in order for the landlord to make the necessary changes.

Cost is a significant issue in retrofitting existing multi-family housing communities. Arguments could be made against the practicality and expense of making universal changes to all covered units in a building. If a covered unit that has been retrofitted is never used by a person with disabilities or if the safe harbor changes made do not increase the accessibility of that unit to the person who lives there, that expense would have no positive effect. Instead, if this money were used to make specific modifications to units, these changes will effectively impact 100% of residents with disabilities. Moreover, these modifications and changes to policies and services will typically exceed the specifications of safe harbors. Whereas the FHA specifies that residents may make reasonable modifications at their own expense, with the tailoring approach, these modifications will be performed and paid for by the developers.

An additional issue to consider is the validity of the specifications accepted as safe harbors. Certain safe harbor specifications and building codes are based on empirical research, which has been used to determine the environmental design elements that best meet the accessibility needs of people with disabilities. For example, stairway and handrail designs are based on research by John Templer (1974) and Jake Pauls (1984), and reach ranges are based on research by Steinfeld, et al (2005). However, this is not true for other safe harbor specifications, which are consensus standards. A strict adherence to safe harbors does not address the practical accessibility of units to actual or future residents.

As consultants who have worked extensively with people with disabilities across the country and conducted accessibility studies regarding threshold detection of deviation from safe harbors, and usability of a variety of environmental designs (Vredenburgh, Hedge, Zackowitz & Welner, (2009); Vredenburgh, Williams, Zackowitz, & Welner, (in press)), we have found that since disabilities are not standardized, individual residential units cannot best accommodate the maximum number of people with all of their unique needs based on their individual capabilities and limitations without tailoring the environmental design to the needs of each person. While common areas should still conform to ADA specifications, the users of private housing areas can have their needs best met by adapting their environments to their specific usability requirements.

The authors are currently consulting for developers on cases in several regions of the United States.

REFERENCES

ANSI A117.1-1986. Providing Accessibility and Usability for Physically Handicapped People.

CABO/ANSI A117.1-1992. Accessible and Usable Buildings and Facilities.

Fair Housing Accessibility Guidelines, (1991), FR Vol. 56, No. 44.

Fair Housing Act Amendments, 42 USC 3604(f).

Fair Housing Design Manual (Rev. April, 1998). US Department of Housing and Urban Development.

Fair Housing Act Regulations, 24 CFR 100.200.

ICC/ANSI A117.1-1998. Accessible and Usable Buildings and Facilities.

2000 Code Requirements for Housing Accessibility.

Kaye, Kang & LaPlante (2000). Mobility Device Use in the United States. *Disability Statistics Report 14.* Disability Statistics Center, University of California, San Francisco.

Paarlberg, (2004). Safe harbor documents. *Building Safety Journal.* International Code Council.

Pauls, J.L. (1984). Stair Safety: Review of Research. *Proceedings of the 1984 International Conference on Occupational Ergonomics,* (171-179).

Steinfeld, E. (1994). The Concept of Universal Design. Center of Inclusive Design and Environmental Access. New York: University of New York at Buffalo.

Steinfeld, Schroeder, S. Bishop, M. (2005). *Accessible Buildings for People with Walking and Reaching Limitations.* U.S. Department of Housing and Urban Development.

Templer, J.A. (1974). *Stair Shape and Human Movement.* Doctoral dissertation, Columbia University, New York.

Vredenburgh, A.G., Hedge, A. Zackowitz, I.B. & Welner, J.M. (2009). Evaluation of Wheelchair Users' Perceived Sidewalk and Ramp Slope: Effort and Accessibility. *Journal of Architectural and Planning Research, 145-158.*

Vredenburgh, A.G., Williams, K. Zackowitz, I.B. & Welner, J.M. (in press). Evaluation of Wheelchair Users' Perceived Kitchen and Bathroom Usability: Effort and Accessibility. *Journal of Architectural and Planning Research.*

CHAPTER 64

A Program to Help Persons with Disabilities Become Entrepreneurs

Peter McAlindon[1], Clayton Mayo[2], Peter Schoemann[3]

[1]Blue Orb, Inc.
Orlando, FL 32751, USA

[2]University of Central Florida
Orlando, FL 32816, USA

[3]Central Florida Disability Chamber
Orlando, FL 32801, USA

ABSTRACT

A step-by-step process involving technology and human expertise provides persons with disabilities an opportunity to independently develop their own ideas as entrepreneurs. Many persons with disabilities have a propensity to become entrepreneurs but often times have difficulty accessing and participating in traditional entrepreneurial activities. An online environment that allows persons with disabilities to participate more fully in the process of becoming and being an entrepreneur is the aim of this research. In addition to providing these individuals with a tool set to improve and enhance entrepreneurship among them, the collaborative environment will be a very powerful research tool to test hypotheses about the cognitive, affective, and inter-personal constructs and mechanisms involved in entrepreneurship and among persons with disabilities. For example, the environment would allow both formal and informal firsthand inquiry of entrepreneurial and technology questions. Additionally, the environment would allow research on entrepreneurship and on cognitions of persons with disabilities in non-laboratory and non-classroom situations. Authentic field and informal learning experiences are regarded by educators as important for advancing learning and motivation to pursue career paths in entrepreneurship. Unlike other laboratory-based research on these topics, the environment would be able to obtain relatively large and potentially quite representative samples of individuals engaging in real-life problem solving and assessment. Thus, in addition to allowing the testing of new mechanisms/hypotheses, the environment also allows for generalized studies from prior laboratory-based research into the "real world" applications.

Keywords: Entrepreneurship, Persons with Disabilities, Support Systems, Systems Modeling

INTRODUCTION

A number of persons with disabilities have what it takes to become successful entrepreneurs. Persons with disabilities are consistently underrepresented in entrepreneurship (Weiss, 2009). Given a proper environment and access to information necessary to thrive, problems amongst persons with disabilities could be solved in greater numbers and more quickly. Many have ideas for products and processes to not only improve their own life, but the lives of many other people, whether able-bodied or disabled. A system is required to support entrepreneurs who wish to identify, develop, and test special population products used by the people who could benefit from them most, namely persons with disabilities. Our aim is to identify and use best practices of entrepreneurship programs and adapt them for use with persons with disabilities.

Recently there has been a strong push for persons with disabilities to more actively participate in science and technology professions, including entrepreneurship. Legislation such as The Individuals with Disabilities Education Act, or IDEA, which mandates that all students with disabilities shall have a free and appropriate education (Federal legislation, Public Law 94-142) and the Americans with Disabilities Act has helped create an awareness of these professions to persons with disabilities. Persons with disabilities have developed strong problem solving skills as a result of their disabilities, and these problem-solving skills may have the potential of leading to significant contributions to the scientific and/or entrepreneur community. A specific aim of the research was to develop a collaborative environment for facilitating educational, motivational, and supportive processes for entrepreneurial development among persons with disabilities. These aims are crucial for persons with disabilities to access research and business information, providing them with the capacity to function independently as entrepreneurs, a potentially crucial factor for fostering their own ideas and for owning their own business.

Without such an environment, persons with disabilities are relegated to passive learning conditions in becoming entrepreneurs. Passive learning is not conducive to developing self-confidence and may lead to learned helplessness (Pintrich, 1992, Schunk, 1991). Lack of self-confidence with regard to independent functioning and learned helplessness may be key reasons for persons with disabilities' under representation in entrepreneurial ventures (Weiss, 2009). The research also involves the efficacy testing of the new system to measure learning, as well as the development of self-confidence and the desire to pursue entrepreneurship as a career. Minorities and women are also underrepresented as entrepreneurs (Entrepreneurship, Self Employment and Disabilities, 2007, Landmark Disability Survey Finds Pervasive Disadvantages, 2004). It is likely that a similar underlying mechanism contributes to under representation of these groups — hence, the research provides insights into strategies for increasing representation of minorities and women, as well as that of persons with disabilities. The research collaborative consists of organizations including, but not limited to: Central Florida Disability Chamber (CFDC), the Center for Independent Living in Central Florida, Rollins College, University of Central Florida, local companies including Blue Orb, Caxiam Group, and the law firm Broad and Cassel.

IDENTIFICATION OF THE NEED

In 2007, 12.1% of persons of working age (from 21 years of age to 64 years of age) in Florida (1,241,000 of the 10,247,000 working-age individuals in Florida) reported one or more disabilities (Disability Status Report Florida, by Cornell University, 2007). 42.4% of persons with disabilities in Florida have a Bachelor's degree or at least some college education (Disability Status Report Florida, by Cornell University, 2007). Yet, in July 2009, the employment-population ratio (i.e., the ratio of working-age persons who are employed to the total of all working age persons) for persons with a disability was 19.5%, compared with 65% for persons with no disability (US Census Data, 2009). In the United States, 22,295,000 persons of working age are disabled and the employment-population ratio of such persons is 18.4% (Disability Status Report Florida, by Cornell University, Bureau of Labor Statistics, 2007). Therefore, what is true in Florida is also true throughout our country. The reason for the low employment ratio of the disability population likely includes a combination of following barriers: the inability of employers to accommodate persons with disabilities, the inability of persons with disabilities to obtain reliable methods of transportation to the employer's workplace, and the limitations of public benefits to persons with disabilities. Persons with disabilities are also significantly underrepresented in STEM-related fields (National Science Foundation, 2000, Broadening Participation in Science, Technology, Engineering, and Mathematics (STEM) A Summary of Opportunities through the National Science Foundation, 2006). Similar under representation is apparent in schools, where less than 2% of graduate students in STEM-related fields are persons with disabilities (National Science Foundation, 2000).

Most unemployed persons with disabilities are underemployed and also receive aid (such as Supplemental Security Income, Social Security Disability Insurance, Medicare, Medicaid, and/or vocational rehabilitation services) under both state and federal programs. In 2007-2008 the Florida Department of Vocational Rehabilitation assisted 12,458 persons with disabilities in finding gainful employment within the state, many of whom were employed on a part time basis (Florida Vocational Rehabilitation Statistics, 2009). This underemployment is a huge governmental and taxpayer burden. According to U.S. Census figures, persons with disabilities comprise 10.4% of the overall workforce, but only 2.7% of the sciences, engineering, or entrepreneurial workforce. This gap is not necessarily indicative of a lack of interest in entrepreneurial careers by disabled people. According to the American Council on Education, college freshmen with disabilities are equally interested in majoring in science as are freshmen without disabilities. However, this initial interest is rarely realized as an actual career in a STEM-related field. Further evidence of the shortage of individuals with disabilities in STEM fields is offered by a National Science Foundation study finding that fewer than 275 persons with disabilities received Ph.D.s in either science or engineering in 2006, compared with the more than 15,000 non-disabled individuals who received Ph.D.s that year in science or engineering (Committee on Equal Opportunities in Science and Engineering, 2001).

While the employment rate of persons with disabilities is low, persons with disabilities have exhibited entrepreneurship. Approximately 40% of all home-based businesses are owned and operated by persons with disabilities (Weiss, 2009). Twice as many persons with disabilities are starting their own business as people who are able-bodied (Weiss, 2009).

The question is: How can we help disabled persons of working age to become more independent in the workforce and to become independent of the need for government assistance? The answer is: We need a system and process that will help would-be-entrepreneurs who are disabled gain self-confidence and the knowledge necessary to realize their dreams of independence.

DEVELOPING ENTREPRENEURIAL INTEGRATION TOOLS TO SUPPORT SYSTEMS DESIGN

We propose to develop a highly structured, step-by-step process via a collaborative environment both online and face-to-face to help foster and support the entrepreneurial spirit of persons with disabilities. With each step in the process, entrepreneurs will be connected to other persons with disabilities, as well as able-bodied business people and entrepreneurs, to share their ideas, brainstorm new ideas, develop products, and launch businesses. The collaborative online environment has to be fully accessible. Persons whom are visually impaired, have motor control issues, etc. will aid in developing and testing the system and will be accommodated in using it. For each step, people will be paired with the best support person(s) for that particular step. Support for the entrepreneurs comes from personal expert help, libraries of information, and organizations and systems already in place that are tasked with helping form ideas and grow businesses. Locally, these organizations include the Disney Entrepreneur Center, the University of Central

Florida Venture Lab, Small Business Incubator, Small Business Development Center, the Senior Corp of Retired Executives (SCORE), and online, they include LinkedIn (www.linkedin.com), Growthink (www.growthink.com), Kauffman Foundation (http://www.kauffman.org/), and the Florida High-Tech Corridor (http://www.floridahightech.com/).

STEPS OF THE ENTREPRENURIAL LEARNING SYSTEM

The development of a step-by-step online learning system will enable persons with disabilities to pursue their entrepreneurial ambitions. Realistic hands-on, inquiry-based learning is important for teaching and developing entrepreneurial skills, self-confidence, and the desire to succeed in business. Entrepreneurs are increasingly turning to handheld devices such as the iPhone to obtain on-demand, real-time instruction and support. However, few of these handheld devices, including the iPhone, provide the means of operation to persons with disabilities. This creates an access barrier for the use of these devices and thus for complete involvement in the educational entrepreneurial experience by persons with disabilities. The proposed innovation will be the first software implementation to allow greater access to becoming an entrepreneur via a handheld device. The system consists of the following 16 steps from business basics to successful growth or sale of a business and each one of these steps operates in the process shown in Figure 1.

1) **Business basics overview** (e.g., how to form a legal entity, tax considerations, etc.)
2) **Selection of a mentor or coach**
3) **Determination of roles:** inventor, concept person, marketing
4) **Identify organizations that can help formulate ideas** (e.g., Central Florida Disability Chamber, Disney Entrepreneur Center, Nexus, University of Central Florida Incubator, University of Central Florida College of Engineering, Venture Lab, Center for Independent Living, Edward Lowe Foundation, etc.)
5) **Develop plan for "fail fast" testing of product or service.** What can cause an idea to fail fast? It's been done; it's not practical or profitable to make; it doesn't meet/satisfy customer needs. So, check patents, do thorough market research, and talk to customers!
6) **Pursue preliminary patent protection and prepare business plan framework**
7) **Obtain seed funds**
8) **Design service/product**
9) **Develop and test prototype**
10) **Obtain input from customers, groups, potential partners and investors**
11) **Launch viable product/service**
12) **Pursue intellectual property protection**
13) **Sell to first customers and potential partners in 10 above.**

14) **Begin to incorporate all function of the business plan** (marketing, web strategy and plan, etc.)
15) **Refine as necessary**
16) **Grow, Innovate, Exit**

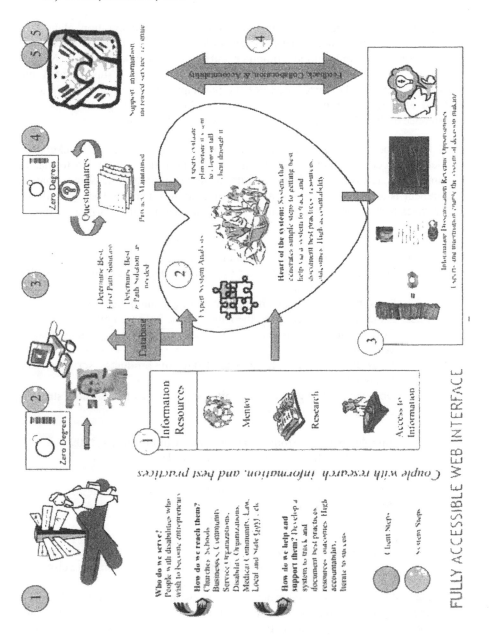

Figure 1. An entrepreneurial process for persons with disabilities.

ASSESSMENT

An important component of the human/computer systems integration plan is to verify and validate the process that provides a clear way to evaluate the success of mentor, participant, and online computer curriculum integration. The systems integration team has developed a battery of tests that can easily be incorporated into the assessment plan. The performance of the participant in the system needs to be validated as part of the overall system. Using this methodology addresses the performance of the participant with respect to the overall system.

It is important to ascertain to what extent a person has the background knowledge to become an entrepreneur. This will be performed through a formative assessment. An assessment will also be administered to learn more about what an entrepreneur is and does. It is only after understanding a persons background and what it takes to become an entrepreneur that the decision is made to begin the 16 step progam.

Formative assessment example questions include:
1. Does the applicant have the requisite background, education, skills, and work experience?
2. Does the applicant have management or accounting experience?
3. Is the applicant clear as to the nature of the proposed business?
4. Does the applicant have an understanding of marketing and estimating sales potential?
5. Does the applicant have knowledge of products, services, and location?
6. Does the applicant have the capital requirements for business start-up, expansion, or acquisition?

Some example responses from the assessment (from Isenberg ,2009):
1. I always look for new and better ways to do things.
2. I like to question conventional wisdom.
3. I like to get people together in order to get things done.
4. People get excited by my ideas.
5. I am rarely satisfied or complacent.
6. I can't sit still.
7. I can usually work my way out of a difficult situation.
8. I would rather fail at my own thing than succeed at someone else's.

Each of the 16 steps in the process will be evaluated in much the same way. For example, in step 1, business basics overview, information will be presented using accessible videos, podcasts, books, and articles to highlight principles of getting started with an idea. Again, we will be using best practices from centers across the country to develop the information most pertinent to a person's background and entrepreneurial needs. To illustrate:

Who's involved: The participant. Human experts in each business basics topic covered (e.g., Marketing; Corporate Structure and Taxes; Product Design and Development; Networking and Resources). A local agency (i.e., Center for Independent Living) would act as the support personnel for the participant using the system and the process.

Assessment: A summative online assessment that requires a > 90% pass rate on each of the business basic topics. Each topic will be developed with text, videos, audio playback, and offer links to definitions and related principles. Feedback on the assessment is examined and offered to the participant from the participating agency and through an on-line tool. The participant will also provide feedback on how the information helped increase their knowledge of business basics and in becoming an entrepreneur. And in subsequent steps, tell us how well it prepared them for the next step in the process.

Milestone(s): Acquiring basic knowledge of the process of becoming an entrepreneur. Summative assessment scores will determine whether or not a participant can advance to the next step in the process.

A primary goal of the assessment is to develop a profile of applicants in the program. We began by exploring a number of research questions, many of which will be investigated in subsequent studies. Example research questions include:

- What types of entrepreneurial activities are of interest to applicants?
- What are the characteristics of applicants in the program, and what motivates them to opt for self-employment? How is this group of self-selecting individuals different than others with and without disabilities?
- Is there a relation between the severity or type of disability and other personal and economic characteristics of applicants?
- What are the economic and employment backgrounds of applicants in the program? And, what can we learn about the characteristics of applicants that will assist in the assessment of successful employment outcomes?
- To what extent are applicants receiving federal and state support for workforce development activities?

To begin to address the research questions listed above, the data were organized in five general categories:

1) Disability measures;
2) Background measures;
3) Prior employment and economic measures;
4) Prior public and private assistance measures; and
5) Proposed self-employment measures.

CONCLUSIONS

This project provides technology to persons with disabilities that will allow them to experience entrepreneurial education activities from a hands-on perspective. This will promote learning and may enhance these participants' self-confidence in their capacity to function independently in starting their own businesses. This could lead to an increase in the number of individuals with disabilities entering the workforce and supporting related activities of other disabled or able-bodied entrepreneurs. Persons with disabilities are frequently unemployed and draw taxpayer monies for support. Increased representation of persons with disabilities in entrepreneurship, by its very nature, will reduce their unemployment numbers and the need for taxpayer support for not just the founder of the effort, but perhaps several employees who also have disabilities.

REFERENCES

Broadening Participation in Science, Technology, Engineering, and Mathematics (2007) from http://opas.ous.edu/Committees/OPASS/NSFPrimer.pdf

Committee on Equal Opportunities in Science and Engineering. (2000). Graduate education: A declining share for SMET from http://www.nsf.gov/pubs/2001/ceose2000rpt/congress_5.pdf

Cornell Disability Statistics Online Resource for U.S. Disability Statistics. http://www.ilr.cornell.edu/edi/disabilitystatistics/

Entrepreneurship, Self-Employment and Disabilities from http://www.diversityshop.com/store/entre.html

Florida Vocational Rehabilitation Statistics (2009). http://www.rehabworks.org/Files/DocumentsLinks/2009%20Annual%20Repo rt%20Email_smFinal.pdf

Landmark Disability Survey Finds Pervasive Disadvantages. From http://www.nod. org/index.cfm?fuseaction=Feature.showFeature&FeatureID=1422

National Science Foundation (2000). Women, Minorities, and Persons with Disabilities in Science and Engineering: 2000 (NSF-0327); Arlington: VA.

Pintrich, P.R., & Schrauben, B. (1992). Students' motivational beliefs and their cognitive engagement in classroom tasks. In D. Schunk & J. Meece (Eds.), Student perceptions in the classroom: Causes and consequences (pp. 149-183). Hillsdale, NJ: Erlbaum.

Schunk, D.H. (1991). Self-efficacy and academic motivation. Educational Psychologist, 26, 207-231.

Stoddard, S., Jans, L., Ripple, J. & Kraus, L. (1998). Chartbook on Work and Disability in the United States, 1998. An InfoUse Report. Washington, D.C.: U.S. National Institute on Disability and Rehabilitation Research.

U.S. Census Data. (Researched September 2, 2009) from http://www.census.gov

Weiss, T. (2009). Persons with Disabilities and Entrepreneurship. Disability World.
    http://www.disabled-world.com/editorials/disablity-entrepreneurship-tips.php
Isenberg, D. (2010). Should you be an entrepreneur? Harvard Business Review
    from http://blogs.hbr.org/cs/2010/02/should_you_be_an_entrepreneur.html

CHAPTER 65

Situational Disabilities in Firefighting - Impact on Communications

Jari Järvinen

Motorola
Plantation, FL 33322-9947, USA

ABSTRACT

When thinking about fit and able firefighters, one might not see a connection to design for disabled access. Due to their demanding and high stress work environment, firefighters can temporarily experience functional limitations from all main categories of disabilities. These "situational disabilities" have a detrimental effect on firefighters' ability to communicate to each other and to the command post. The principles of design for disabled access can and should be applied to the design of communication devices for firefighters.

The findings and insights presented in this paper are based on contextual and participative design research with firefighters as well as literature reviews on disabilities, FCC guidelines and principles of Design for Disabled Access.

Keywords: Situational disabilities, Design for Disabled Access, Firefighters, Communications, User Research, Usability, Two-way Radios

INTRODUCTION

In 1996, the U.S. Congress passed the Telecommunications Act to ensure that manufacturers of telecommunications products and providers of telecommunications services meet the needs of people with disabilities. Section

255, "Access by Persons with Disabilities" mandates that manufacturers make telecommunications equipment accessible and usable by individuals with disabilities, if it is readily achievable. Due to these Federal Communications Commission (FCC) regulations, there was a lot of emphasis on design for disabled access for consumer communications devices in the late 90's and early 2000's.

Public safety two-way radios do not need to comply with the same FCC rules as consumer telecommunication devices. However, some similarities were discovered between the functional limitations of persons with disabilities and the temporary or "situational" disabilities experienced by firefighters during their work.

This paper describes how two initially separate exploratory research projects (Design for Disabled Access and Design Research for Firefighter Communications Devices) found a common ground and resulted in a design approach that can be used to design more usable and effective communications devices for first responder firefighters.

OBJECTIVES

The objectives of this exploratory study were to:
- Learn about challenges faced by firefighters in their communications and current communication devices
- Explore how principles of design for disabled access can be expanded to the design of public safety first responder communication devices.

METHODS

The following methods were used in Design for Disabled Access Research:

- Literature review on disabilities and their classification.
- Literature review on Design for Disabled Access.
- Research on FCC regulations related to disabled access of telecommunications devices.

Firefighter user research at Motorola is an on-going effort aimed at learning about the challenges of firefighters' work environment that are relevant to their communication needs. The following methods were used:

- Contextual research in the form of firefighter training. Several members of the cross-disciplinary design team have participated in firefighter training in North-America and Asia. The training included putting out fires and rescuing "victims" from the burning training facility wearing the full protective gear. Radio communications were part of the training.
- Interviews with firefighters in group sessions.

- Participatory design sessions with groups of firefighters.

FUNCTIONAL LIMITATIONS

Just being human means that we all have physical or mental limitations to some extent. These limitations may be *permanent*, such as severe arthritis; *temporary*, such as a broken arm in a cast; or *situational*, such as trying to dial a wireless phone using one hand as you wheel your suitcase through the airport.

In the United States about 20 % of the population has some level of disability and about 10% have a severe disability (McNeil, 1995)

There are five main categories of disabilities that were relevant to this study:

- *Visual Impairments*: Degradation of visual capability usually ranges from 20/20 vision to wearing glasses to low vision to blindness.
- *Hearing Impairments*: Degradation of hearing ability can range from hearing whispers to difficulty understanding conversation to deafness.
- *Speech and language impairments*: Degradation of language capabilities can range from clear speech to difficulty being understood to lack of speech
- *Perceptual and cognitive impairments*: Degradation of perceptual and cognitive abilities can range from average IQ to reduced memory and sequencing ability to no interpretive skills.
- *Physical impairments*: Mobility and dexterity can range from, for example, being physically fit to minor arthritis to total paralysis.

VISUAL IMPAIRMENTS

☐Four categories of uncorrectable vision loss are commonly identified. These overlap to some extent, as explained below:

- *Legal blindness* – The principal criterion for legal blindness is that the vision in the better eye is 20/200 or poorer even with corrective glasses/contact lenses OR that the person's field of vision is no greater than 20 degrees (Elkind, 1990). Approximately 85% of people who are "legally blind" have some residual vision. The remaining 15% are "totally blind" (Corn and Keonig, 1996).
- *Total blindness* – An individual who is totally blind has no vision at all (Corn and Keonig, 1996).
- *Functional blindness* – A person who is functionally blind can perceive light; however, the shape and direction of the light cannot be determined. People who are totally blind are also included in this category (Corn and Keonig, 1996).

- *Low vision* – Low vision means that the person's vision loss is severe enough to affect the person's performance of everyday tasks, but the person still has some useful visual discrimination. Individuals with low vision may make use of specialized magnifying technologies and lenses (Corn and Keonig, 1996).

HEARING IMPAIRMENTS

About 37 million adults in the United States had a hearing loss ranging from a little trouble to being deaf (Pleis and Lethbridge-Cejku, 2007).

A person is considered to have *mild* hearing loss when sound must be presented at a level between 25-45 dB SPL above the average hearing level threshold for individuals with "normal" hearing. *Moderate* hearing loss requires levels at 65-85 dB SPL above the average hearing threshold. Individuals are considered *profoundly* deaf when levels in excess of 85 dB SPL above the average hearing level threshold are required (i.e. the sound must be at least 5 to 10 times louder than normal speech to be heard (van Cleve, 1987)

SPEECH IMPAIRMENTS

In the United States there were 2.5 million people whose speech is difficult to understand. Of that number, about 10% had a severe speech disability (McNeil, 1993).

Speech is a muscular process linked to a person's oral motor skills and is not connected to the person's capacity to produce language. People who have speech disabilities can vary greatly in their ability to produce speech. Speech disabilities can be temporary or permanent, chronic, or stable, fluctuating or improving (Simpson, 1997).

COGNITIVE IMPAIRMENTS

☐Cognitive disabilities can range from severe retardation to memory limitations to the impairment of a particular cognitive function, such as language.

Learning disabilities affect people's ability to either interpret what they see and hear or to link information from different parts of the brain. These limitations can appear in various ways, such as specific difficulties with spoken and written language, coordination, self-control, or attention (LD OnLine).

PHYSICAL IMPAIRMENTS

There were 13.6 million people that had limited use of their hands (National Center for Health Statistics, 1994). Approximately 5.7 million people had arthritis, which ranks as the third most common health condition that causes a person's activities to be limited (National Center for Health Statistics, 1992).

FIREFIGHTER RESEARCH

Deaths from fires are the fifth most common cause of unintentional injury deaths in the United States (CDC, 2006). The loss of life, destruction and property damage highlight the value of the firefighters in any society. Firefighters are perceived as rescuers.

The firefighters have their own unique culture, which influences the fireground communications. The main goals of firefighters are to save lives and put out fires. The firefighter culture emphasizes aggressiveness, action and the ability to overcome obstacles. They are under pressure to demonstrate bravery. There is a stigma attached to calling for reinforcements and fear of being judged as incompetent, slow and non-aggressive.

MAIN FIREFIGHTER TASKS

Main firefighter tasks during fire incidents include 1) rescuing people and animals, and 2) extinguishing and controlling fire. The rescue tasks start by a search in most likely locations, followed by the second patterned search. An important part of rescue tasks is orienting in the building with low or no visibility. Extinguishing and controlling fires include activities related to hose transfer and manipulation.

FIREGROUND COMMUNICATIONS

Coordination of the fireground tasks and several firefighter teams (or "companies" usually consisting of four firefighters) requires effective communications. There are three levels of communications during a fire incident: 1) Within a company, 2) between companies, and 3) between a company and incident commander.

The communications *within a company* are related to coordinating the immediate rescue or fire control activities between the team members. The two-way radios are rarely used in company's internal communications in order to reduce the radio traffic on the fireground channel. Moreover, using the radios would require the team member to interrupt the on-going task.

The main task in *between companies* communications is the coordination of assignments between the companies. Most company-to-company two-way radio communications happen through incident commander.

The radio communications *between a company and incident commander* include transmitting progress reports on assignments and receiving instructions and new assignments.

The volume of radio traffic in complex multi-alarm incidents may become overwhelming and all messages do not always get through. A fire department may also face difficulty in communicating with neighboring departments due to different radio frequency allocations. The incident commanders may not always have all the necessary information about the status of the fireground activities. The decision making may be adversely affected due to insufficient information.

The firefighters have their hands full during a fire incident. They have to grasp equipment and make sure that breathing gear works properly. They have to make sure that they understand the instructions, and they have to avoid myriad of hazards. As far as radio communications go, they don't want to have to think about them at all.

The user interface of the radio has to be very intuitive, and require minimal attention and effort to use. Ergonomics and usability issues with consumer devices are an annoyance and cost the user just a few extra minutes and some frustration. But when first responders, such as firefighters, have difficulty using their devices, those few extra minutes could cost them their lives.

SITUATIONAL DISABILITIES IN FIREFIGHTING

The harsh and high stress work environment can temporarily impair almost all of the firefighter's senses. The usually abled and fit firefighters can experience the following situational disabilities:

- *Blindness or low vision* due to smoke. In a typical house fire the smoke is black and acrid when carpeting, drapes and furniture burn. They may not be able to see their hand or even a backlit LCD display in front of their eyes. The Self-contained Breathing Apparatus (SCBA) mask is often fogged.
- *Deafness or hearing loss* due to noise caused by fire and firefighting activities. The noise can be created by chain saws and axes breaking doors and walls, fire truck engines pumping water, etc. Also, the SCBA low oxygen alert or the Personal Alert Safety System (PASS) both create loud audible alerts that can mask the radio reception or interfere with the transmission. Firefighters also face radio interference issues such as acoustic feedback ("squealing noise") when two radios get too close to each other.
- *Speech limitations* due to SCBA mask. The SCBA is critical for firefighter safety, but it interferes with both face-to-face and radio communications by muffling the speech.
- *Perceptual and cognitive limitations* due to high stress, and high sensory and cognitive load. The stress levels elevate even for experienced firefighters in a burning structure, unable to see their hands in front of their faces and forced to find their way out solely by feel.
- *Mobility and dexterity limitations* due to thick fire gloves and other fire gear. They typically carry 40-50 lbs of equipment, such as SCBA tank, axe, saw, and hoses. Use of water for fire control results in additional

dexterity issues, because wet radios, tools and other objects become slippery. The wet fire gloves become even stiffer making it harder to grasp objects or manipulate radio controls.

DESIGN FOR DISABLED ACCESS

Accessible design refers to maximizing the number of potential customers who can readily use a product (Scadden, 1993). Product features that make products usable by people with disabilities or functional limitations normally make them convenient for everyone else.

The FCC rules (FCC, 1999) define how manufacturers must comply with the requirements to make products (1) accessible to people with disabilities when readily achievable. The following FCC rules provide specific direction for the accessibility and compatibility of telecommunications equipment.

INPUT, CONTROL AND MECHANICAL FUNCTIONS

It is required, when readily achievable that input, control and mechanical functions be locatable, identifiable, and operable in accordance with specific criteria to be assessed independently.

- *Operable without vision*. Provide at least one mode that does not require user vision.
- *Operable with low vision and limited or no hearing*. Provide at least one mode that permits operation by users with visual acuity between 20/70 and 20/200, without relying on audio output.
- *Operable with little or no color perception*. Provide at least one mode that does not require user color perception.
- *Operable without hearing*. Provide at least one mode that does not require user auditory perception.
- *Operable with limited manual dexterity*. Provide at least one mode that does not require user fine motor control or simultaneous actions.
- *Operable with limited reach and strength*. Provide at least one mode that is operable with limited reach and strength.
- *Operable with a Prosthetic Device*. Controls shall be operable without requiring body contact or close body proximity.
- *Operable without time-dependent control*. Provide at least one mode that does not require a response time or allows response time to be by-passed or adjusted by the user over a wide range.
- *Operable without speech*. Provide at least one mode that does not require user speech.

- *Operable with limited cognitive skills.* Provide at least one mode that minimizes the cognitive, memory, language, and learning skills required of the user.

INFORMATION NECESSARY TO OPERATE AND USE THE PRODUCT

It is required, when readily achievable that all information necessary to operate and use the product, including but not limited to, text, static or dynamic images, icons, labels, sounds, or incidental operating cues, must comply with specific criteria to be assessed independently.

- *Availability of visual information.* Provide visual information through at least one mode in auditory form.
- *Availability of visual information for low vision users.* Provide visual information through at least one mode to users with visual acuity between 20/70 and 20/200 without relying on audio.
- *Access to moving text.* Provide moving text in at least one static presentation mode at the option of the user.
- *Availability of auditory information.* Provide auditory information through at least one mode in visual form and, where appropriate, in tactile form.
- *Availability of auditory information for people who are hard of hearing.* Provide audio or acoustic information, including any auditory feedback tones that are important for the use of the product, through at least one mode in enhanced auditory fashion.
- *Prevention of visually-induced seizures.* Visual displays and indicators shall minimize visual flicker that might induce seizures in people with photosensitive epilepsy.
- *Availability of audio cutoff.* Where a product delivers audio output through an external speaker, provide an industry standard connector for headphones or personal listening devices (e.g. phone-like handset or ear cup) which cuts off the speaker(s) when used.
- *Non-interference with hearing technologies.* Reduce interference to hearing technologies (including hearing aids, cochlear implants, and assistive listening devices) to the lowest possible level that allows a user to utilize the product.
- *Hearing aid coupling.* Where a product delivers output by an audio transducer that is normally held up to the ear, provide a means for effective wireless coupling to hearing aids.

DISABLED ACCESS DESIGN STRATEGIES AND GUIDELINES

Examples of additional design guidelines and strategies for design for disabled access are presented below:

- Provide discrete buttons to allow a person to locate them tactilely (Vanderheiden, 1997).
- If the device has a standard number pad arrangement, putting a nib on the "5" key helps to orient oneself to the numbered keys on the keypad (ETSI, 1998).
- Ideally, no key should be more than one key away from a tactile landmark, such as a corner, a uniquely shaped key, nibbed key, or easily identifiable home key.
- Providing distinct shapes for keys can either indicate their function or make it easy to tell them apart.
- Providing a rotational or linear stop and tactile or audio detent is one strategy that can be used. If the product has an audio system and microprocessor, audio feedback of the setting may be used.
- Controls can be shaped in a fashion that they can easily be tactilely read. For example, a twist knob shaped like a pie wedge (Vanderheiden, 1997). If the key is a two-state key (on/off), use a key that is physically different in these states, so the person can tell what state it is in by feeling it. If using keys that do not have any physical travel, some type of audio and tactile feedback should be provided so that the individual knows when the key has been activated.
- Avoid controls that require simultaneous activation of multiple keys or buttons, or provide an alternative method to achieve the same result that does not require simultaneous actions.
- Symbols can sometimes be more legible and understandable than fine print.
- A judicious use of color- and shape-coding and following standard conventions and stereotypes can be used to reduce the need to read labels (Sanders and McCormick, 1993; Cushman and Rosenberg, 1991).

CONCLUSIONS

Due to their demanding and high stress work environment, firefighters can temporarily experience functional limitations from all main categories of disabilities. These situational disabilities have a detrimental effect on firefighters' ability to communicate to each other and to the incident command post. The principles of design for disabled access can be applied to the design of communication devices for public safety first responders. Eventual usability improvements can save lives of both firefighters and victims of fire, and prevent property damage.

REFERENCES

Centers for Disease Control and Prevention (2006). Web-based Injury Statistics Query and Reporting System (WISQARS) [Online]. Website: www.cdc.gov/ncipc/wisqars

Corn, Anne and Keonig, Alan, ed. (1996). Foundations of Low Vision: Clinical and Functional Perspectives, American Foundation for the Blind Press, New York

Cushman, W.H. and Rosenberg, D. J. (1991). Human Factors in Product Design, Elsevier, New York.

Elkind, J.I. (1990). The Incidence of Disabilities in the United States, Human Factors 32(4), 1990, pp. 397-405.

European Telecommunications Standards Institute, (1998). ES 201 381, Human Factors; Telecommunications keypads and keyboards; Tactile identifiers.

FCC Report and Order, (September 29, 1999). Docket No. 96-198, Section 6.3(a)(1).

LD OnLine. Website: http://www.ldonline.org

McNeil, John M. (1993). "Americans with Disabilities: 1991-92", Data from the Survey of Income and Program Participation, Bureau of the Census Current Population Reports, Household Economic Studies, US Department of Commerce, Economics and Statistics Administration.

McNeil, John M. (1995). "Americans with Disabilities: 1994-95", Data from the Survey of Income and Program Participation, Bureau of the Census Current Population Reports, Household Economic Studies, US Department of Commerce, Economics and Statistics Administration.

National Center for Health Statistics, (1992). National Health Interview Survey.

National Center for Health Statistics, (1994). National Health Interview Survey.

Pleis JR and Lethbridge-Cejku M. (2007). Summary health statistics for U.S. adults: National Health Interview Survey, 2006. National Center for Health Statistics. Vital Health Stat 10(235).

Sanders, M.S. and McCormick, E.J. (1993). Human Factors and Engineering Design, 7th edition. McGraw-Hill, Inc, New York.

☐Scadden, Lawrence (1993). "Design for Everyone – A Look at Features That Affect Usability of Consumer Electronics, CE Network News, Consumer Electronics Manufacturers Association.

Simpson, Jenifer J., Principal Investigator. (September, 1997). Telecommunications Problems and Strategies of People Who Use Augmentative and Alternative Communication Devices, A report of United Cerebral Palsy Associations for the Rehabilitation Engineering Center on Universal Telecommunications Access, p. 1.

Van Cleve, J.V., ed. (1987), Gallaudet Encyclopedia of Deaf People and Deafness. McGraw-Hill, New York.

Vanderheiden, Gregg C. (1997), Design for People with Functional Limitations Resulting from Disability, Aging or Circumstances. In G. Salvendy (ed.), Handbook of Human Factors and Ergonomics, pp. 2010-2052. New York: John Wiley & Sons, Inc.

Chapter 66

Ethnicity, Aging, and the Usability of Blood Glucose Meters

Shadeequa D. Miller[1], Alicia N. Nolden[2]
Tonya L. Smith-Jackson[2], Karen A. Roberto[3]

[1]Department of Industrial and Systems Engineering
University of Wisconsin-Madison
Madison, Wisconsin, 53706, USA

[2]Grado Department of Industrial and Systems Engineering
Virginia Polytechnic Institute and State University
Blacksburg, Virginia 24061, USA

[3]Center for Gerontology
Virginia Polytechnic Institute and State University
Blacksburg, Virginia 24061, USA

ABSTRACT

The successful use of blood glucose meters (BGMs) in diabetes self-management amongst older adults is becoming increasingly important. Type 2 diabetes, which accounts for approximately 90% -95% of all diagnosed cases of diabetes in the U.S., is a common problem in older adults and minorities. There is sufficient evidence that BGMs are not easy to use. Although there are culturally-centered diabetes education programs, limited research has assessed the relationship between ethnicity and usability of BGMs. This pilot study explored how the usability of BGMs varied among European-Americans and African-Americans. In two product interactive focus groups with older adults, two BGMs were compared on three tasks to evaluate usability. In addition, the Medical Technology Innovativeness (MTI) questionnaire was administered. Results indicated that ethnicity may be a significant factor in the usability of BGMs. However, there were no significant differences in perceptions of MTI. Understanding the association between ethnicity and usability of blood glucose meters is fundamental to addressing the technology-based factors that contribute to cultural disparities in diabetes self-management.

Keywords: usability, ethnicity, aging, blood glucose meters, Type 2 diabetes

INTRODUCTION

Diabetes mellitus is a chronic disease whose prognosis is highly dependent upon consistent management of diet, physiology, complications, co-morbidities, and psychosocial factors. A major goal of clinical management of diabetes is to maintain glucose and lipid/triglyceride levels as close to normal as possible (Winter & Signorino, 2002). Poor management of diabetes can lead to a host of complications that undermine the quality of life and may even lead to death. In fact, diabetes is the seventh leading cause of death in the United States. In addition to its impact on mortality, diabetes mellitus continues to challenge U.S. healthcare system capacity. The number of persons with diabetes has increased each year for the past 30 years, with 1.6 million new cases reported in 2007. Ninety to ninety-five percent of people in the U.S. diagnosed with diabetes have Type 2 diabetes (Centers for Disease Control & Prevention, 2007).

Several disparities in diabetes care have been identified, however the most significant are associated with ethnicity and socioeconomic status. According to the Agency for Healthcare Research and Quality (AHRQ, 2001), almost all minority groups in the United States have a prevalence of Type 2 diabetes that is two to six times greater than that of the European-American population. More than half of the 24 million Americans estimated to have diagnosed and undiagnosed diabetes are over 60 years old (CDC, 2007).

Although income is one barrier to adoption of technologies, the extent to which the technology is culturally valid and meaningful may be a more substantial barrier. A recent study showed that age, gender, income, and insurance status explained 13% to 38% of observed differences in diabetes care (Neale, 2008). Ethnicity has been shown to introduce system barriers, which undermine the quality of care received by minority users (Mull, 1993; Quill, 1989). Current interface design models for self-management technologies such as blood glucose meters (BGMs) are both 'gendered' and 'ethnocentric' (Green et al. 1993; Wyer & Adam, 1999). This pilot study was an effort to address the need for better technological designs to reduce health disparities in diabetes by exploring how usability of BGMs varied among European-Americans and African-Americans.

LITERATURE REVIEW

Individuals continue to be challenged by the demands of managing Type 2 diabetes. Diabetes self-management technologies such as BGMs are commonly recommended to patients that have Type 2 diabetes, yet a large number do not conduct the minimum recommended daily blood glucose checks (Hankó et al., 2006; Vincze et al., 2004). The use of BGMs has made self-monitoring of BGM levels easier by enabling people with diabetes to monitor their glucose any place at any time. The BGM was designed to be used as a management tool to support self-management. Patients are expected to understand high or low test results and make specific life-style adjustments or change their medication regimen without assistance from a health care professional (Paulshock & Johnson, 1988). Diabetes self-management using BGMs has helped to reduce long-term health costs and improve convenience by minimizing the number of doctor and hospital visits. Consequently, the BGM has shifted the responsibilities of glucose monitoring, glucose testing quality, and decision-making regarding behavioral adjustments from healthcare providers to patients (Paulshock & Johnson, 1988; Rogers et al., 2001). Since treatment is often determined by the glucose level readings, errors in the use of BGMs can have serious consequences for the short- and long-term health of diabetics. Although BGMs have migrated into the home, the knowledge required for effective use has not always followed.

A study conducted by the Center for Devices and Radiological Health in 2001 found that the top-rated features for BGM use were accuracy and ease of use (FDA, 2005); however, there is sufficient evidence that BGMs are not easy to use (Colagiuri, R. et al., 1990; Rogers et al., 2001). One study showed that, of the persons diagnosed with diabetes, 70% had trouble with the BGMs while first learning how to use them (Rogers et al., 2001). Operation of BGMs can often be misleading. They are often advertised as easy to use, but the reality is that it can take over 50 steps to check blood glucose levels and each task and subtask must be successfully completed before moving on to the next (Rogers et al., 2001). Given that both technical and cognitive skills are required for effective self-management of blood glucose (Skelly et al., 2005), older adults experience more problems with using BGMs due to functional limitations associated with aging. Based on feedback from users who are 65 and over, the two major challenges with BGMs are learning how to use it and understanding how to interpret the results (Skelly et al., 2005).

The U.S. population is aging and medical devices are becoming an integral component of daily life. These changes have led to the need for human factors engineers to focus on two important parameters - design and usability. Problems with medical devices are often attributed to the user; however it has been more evident that the device could be at fault (FDA, 1996; Winter, 2007). Additional features are added to medical devices to improve usability but these features only help the medical device to be more aesthetically pleasing and often at the expense of safety and effectiveness (Kaye & Crowley, 2000). A medical device is not usable if a design feature jeopardizes the usability attributes of the device. Sawyer (1996) stated that a BGM can only be used safely and effectively if the interaction between the physical and social environments, user capabilities, and device design is considered during the design process. In general, a good design is one that accounts for the ergonomic needs of a diverse population.

Fleming et al. (1997) identified several weaknesses in usability engineering for older adults (and consequent problems with usability and safety) that stem from the exclusion of older adults in user-centered design activities. This exclusion tends to produce products or systems that are difficult to use, hazardous, and sometimes ineffective. When specific groups that have traditionally been disenfranchised on the basis of cultural factors such as ethnicity, age, or socioeconomic status continue to experience inequities in technology benefits, the negative impacts accumulate and perpetuate further inequities (i.e., cumulative disadvantages). Thus, the developed technologies begin to solely reflect the needs and capabilities of the dominant groups and exclude the marginalized groups. Literature supports the presence of disparities due to technology design as well as provider interactions and quality of care. The purpose of this study was to determine whether such inequities exist by exploring users' perceptions of BGMs.

METHODS

PARTICIPANTS

The research involved a total of nine participants. All participants were 60 years of age or over, diagnosed with Type 2 diabetes; had used a blood glucose meter for at least three months; and self-identified as African-American (n=4) or European-American (n=5). Participants were recruited via flyers, newspaper ads, and word of mouth. Participants were chosen after successful completion of an initial phone screening in which they were asked questions related to type of diabetes, age, ethnicity, experience with blood glucose meter, education level, and residence. The number of participants recruited for each ethnicity was justified based on the findings of (Virzi, 1992) who stated that 80% of the usability problems are detected with four or five subjects.

QUESTIONNAIRES AND EQUIPMENT

Demographic Questionnaire: Participants were asked to complete a demographic questionnaire before performing the three BGM tasks (e.g., gender, ethnicity, occupation, adherence to glucose monitoring, health insurance). There were also two questions related to comprehension of glucose levels.

Medical Technology Innovativeness (MTI) Questionnaire: This questionnaire was adapted from Goldsmith and Hofacker (1991) and Groeneveld et al. (2006). The MTI measures innovativeness within a specific domain of interest familiar to the consumer. The MTI was administered after the demographic questionnaire and again after all BGM tasks were completed.

Post-Task Questionnaire: Upon completion of each of the three tasks for each BGM model, participants were given a usability questionnaire to complete. Table 2 provides a breakdown of the usability questions per task. A Likert-type scale with five response levels was used. The scale ranged from 1 (*very hard*) to 5 (*very easy*) as the anchor points. The questionnaire was designed to elicit the users' ratings of the ease-of-use, accessibility of desired functions, and comprehension of symbols.

Equipment: Two different blood glucose meters were used to investigate usability. The manufacturer's labels on the blood glucose meters and accessories were covered and not visible to the participants. The models were labeled as 'A' and 'B' corresponding to the brands respectively. In addition to the meter, each participant was given a lancing device, lancets, test strips, and the control solution. The blood glucose meters and other equipment that were not in use were kept out of the participants' view.

Table 1. Demographics Characteristics of Study Sample

	European –Americans n=5 Mean(SD)/n	African-Americans n=4 Mean(SD)/n
Age (years)	67.8 (9.73)	71 (7.96)
Gender	Female = 3 Male = 2	Female = 2 Male = 2
Duration of Diabetes (years)	3*	3.5*
Blood Glucose Meter Experience (years)	2*	3.5*
Education	High School = 0 Some College= 1 Bachelor's Degree = 2 Graduate Degree =2	High School = 1 Some College= 1 Bachelor's Degree = 2 Graduate Degree =0
Health Insurance	No Insurance = 1 Medicare/Medicaid = 0 Private Insurance =0 Both=4	No Insurance = 0 Medicare/Medicaid = 0 Private Insurance =2 Both=2

* Median reported

Table 2. Post-Task Usability Questions by Task Scenarios

Task Scenarios	Questions
Task 1: Turn on Blood Glucose Meter	**Q1.** How easy was it to turn the blood glucose meter on? **Q2.** How easy was it to find the correct way to turn on the blood glucose meter? **Q3.** Once you understood how to turn it on, how easy was it to operate, push, or insert something in order to make it work?
Task 2: Change Setting	**Q1.** How well did the symbols match the intended function? **Q2.** How easy was it to get to the set-up menu? **Q3.** How easy was it to change the time and date? **Q4.** How easy was it to change the unit measurements? **Q5.** Once you understood how to get to the set-up menu, how easy was it to conduct the necessary tasks?
Task 3: Measure Glucose Level (simulated)	**Q1.** How easy was it to get to a glucose reading? **Q2.** How understandable/comprehensible was the information that you saw on the screen? **Q3.** How well did the information you saw match what you would have like to have if you had to do the task again? **Q4.** Overall, how would you rate the ease-of-use of this BGM for checking your glucose level?

PROCEDURE

Product interactive focus groups, which allow participants to examine and use products by completing several tasks, were used in this study. The product interactive focus group involving European-Americans was conducted in a conference room at Virginia Tech; the product interactive focus group involving African-Americans was conducted in a conference room at a local church. Participants gave consent by signing an informed consent document and retained one copy for themselves. After completing the demographic questionnaire and the first MTI questionnaire, participants were asked to examine the BGM hardware interface, which consisted of the keys, the display, shape and weight of the BGM, and how the BGM felt in their hands. For the software interface examination, the moderator instructed the participants to complete three task scenarios using the two different BGMs. The three task scenarios were:

1. Turn on
2. Change Setting
3. Measure Glucose Level (simulated)

During the performance of a task no additional instructions or assistance was given to participants. Participants were not timed during the task, however they were asked to stop after three minutes had elapsed. After each task scenario, participants completed a post-task questionnaire and participated in open discussions. Upon completion of the three task scenarios and discussions, the MTI was administered a second time. Each focus group session lasted 90 minutes and participants were compensated at the end of the session.

RESULTS

The usability ratings from the post-task questionnaire were analyzed for each model using ethnicity as the predictor variable. As data distribution was not normal, the Mann-Whitney U non-parametric test was used. The significance levels was set at *alpha* = .10. The two groups differed significantly for Model A, Task 2, Change Setting, Z (7) = -2.44, p < .01; European-Americans experienced more difficultly with the Change Setting task than African-Americans for Model A. In addition, the two groups differed significantly for Model B, Task 2, Change Setting, Z (7) = 2.02, p < .04; African-Americans experienced more difficultly with Change Setting (Task 2) than European-Americans for Model B. The median usability ratings for Change Setting (Task 2) for each model by ethnicity can be seen in Figure 1.

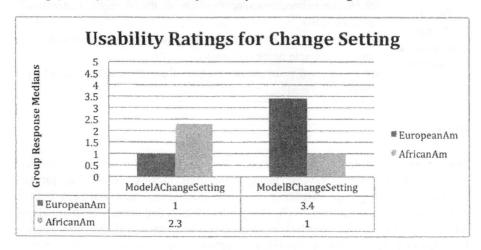

	ModelAChangeSetting	ModelBChangeSetting
EuropeanAm	1	3.4
AfricanAm	2.3	1

FIGURE 1. Median Usability Ratings by Model

There were no other statistically significant differences for usability ratings. Although from the data it cannot be conclusively inferred that the variation in the median usability ratings for the Change Setting task for each model is due to ethnicity, it may be noted as a significant factor and the trend should be investigated further. A more specific comparison between groups for each item is shown in Figure 2. We used paired sample Wilcoxon signed-rank test to analyze the responses from the medical innovativeness questionnaire, since it was administered twice to each participant. No statistically significant difference was found between the two groups on medical innovativeness.

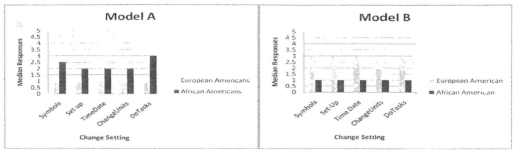

FIGURE 2. Post Task Questionnaire Medians by Item

DISCUSSION AND CONCLUSIONS

A central assumption of this exploratory research was that ethnicity is associated with some of the differences in perceptions of usability of BGM technologies. One contributor to disparities in use of disease management technologies is the usability of a technology, which is influenced by the relevance and validity of a technology from the perspective of the users. In this study, ethnicity played a role in the usability of a blood glucose meter for a particular task. Specifically, European-American participants found the Change Setting task to be more difficult compared to African-American participants for Model A; while African-American participants found the Change Setting task to be more difficult compared to European-American participants for Model B.

During the open discussion for all three tasks using Model B, the African-American participants made several references about how <m> and <c> buttons were not intuitive. One participant mentioned how he had tried a different button to complete each task. Another participant said she guessed that <m> was the button to turn on the meter because she thought <m> stood for meter. The correct way to insert the test strip was also a major topic of discussion among the African-American participants. In addition to experiencing difficulties with completing the Change Setting (Task 2) on Model B, the reasons listed above may have also impacted the African-American participants' preference for Model A over Model B.

On the other hand, the European-American participants stressed the shape and the ability to hold the BGM comfortably saying that Model A was not as comfortable to hold as Model B. One participant, referring to Model B, made the comment "It fits perfectly in my hand." Another concern of the European-American participants was that symbols on Model A did not match what the user was doing. The participant asked, "How would I have known that I had to press and hold the <S> button to change the date and time and what do the <+> and <-> buttons mean?" The shape, symbols, and the inability to complete the Change Setting (Task 2) on Model A may have influenced the European-American participants' preference for Model B over Model A.

Further studies related to this project need to be conducted to discover other tasks that show differences among different ethnic groups. Given that ethnicity may influence the perceptions of usability of BGMs among older adults with Type 2 diabetes, it is important for designers to include older and minority users early in the system design cycle. A more usable BGM can play a major role in the adherence to self-monitoring of blood glucose.

Currently, HF/E is operating within a more inclusive paradigm, which assures that research and practices account for the needs of all users, including those who were ignored in the past (based on ethnicity, gender, nationality, ability/disability, age, etc.). The HF/E focus on cultural ergonomics (Kaplan, 2004) and such efforts to redefine how biomechanics and anthropometric research is conducted (Kroemer, 2006) demonstrates an increasingly apparent shift toward inclusion as the driving foundation for current and future research efforts.

REFERENCES

Agency for Healthcare Research and Quality (2001). *Diabetes Disparities Among Racial and Ethnic Minorities: Fact Sheet*. Rockville, MD: U.S.Department of Health and Human Services.

Centers for Disease Control & Prevention. (2007). *National Diabetes Fact Sheet: United States.*

Colagiuri, R., Colagiuri, S., Jones, S. & Moses, R.G. (1990). "The quality of self-monitoring of blood glucose." *Diabetics Medicine, 7,* 800-4.

U.S. Food and Drug Administration (1996). Review Criteria Assessment of Portable Blood Glucose Monitoring In Vitro Diagnostic Devices Using Glucose Oxidase, Dehydrogenase or Hexokinase Methodology (Draft Document).

Fleming, T.E., Morrissey, S.J., and Kinghorn, R.A. (1997). *Subjects in human factors: Who should they be? Designing for an Aging Population: Ten Years of Human Factors/Ergonomics.* Human Factors and Ergonomics Society. Santa Monica, CA: Human Factors and Ergonomics Society.

Goldsmith, R.E. & Hofacker, C.S. (1991). Measuring consumer innovativeness. *Journal of Academy of Marketing Science, 19(3),* 209-221

Green, E., Owen, J., and Pain, D. (Eds.) (1993). *Gendered by Design? Information Technology and Office Systems.* London, UK: Taylor and Francis.

Groeneveld PW, Sonnad SS, Lee AK, Asch DA, Shea JE. (2006). Racial Differences in Attitudes Toward Innovative Medical Technology. *Journal of General Internal Medicine.* 21, 559–563.

Hankó B., Kázmér M., Kumli P., Hrágyel Z., Samu A., Vincze Z., and Zelkó R. (2007). "Self-reported medication and lifestyle adherence in Hungarian patients with Type 2 diabetes." *Pharmarcy World & Science*, Apr: 29(2):58-66.

Kaplan, M. (2004). E. Salas (series ed.), "Advances in Human Performance and cognitive engineering research." *Cultural Ergonomics. 4.* Amsterdam: Elsevier.

Kaye, R. and Crowley, J. (2000). US Department of Health and Human Services, Food and Drug Administration, Center for Devices and Radiological Health. Washington, DC, USA

Kroemer, K. H. E. (2006). *"Extra-ordinary Ergonomics."* HFES Issues in Human Factors and Ergonomics Series, *4*. Santa Monica, CA: Taylor and Francis.

Mull, J. D. (1993). "Cross-cultural communication in the physician's office." *Western Journal of Medicine, 159*, 609 – 613.

Neale, T. (2008). "Racial Disparities in Diabetes Care Rooted in Physician Performance." *MedPage Today* Retrieved June 11, 2008.

Paulshock, B. Z., & Johnson, L. H. (1988). "Monitoring techniques in non-insulin-dependent diabetes mellitus." *Prim Care, 15*(2), 389-408.

Quill, T. E. (1989). "Recognizing and adjusting to barriers in doctor-patient communication." *Annals of Internal Medicine, 111*, 51 – 57.

Rogers, W. A., Mykityshyn, A. L., Campbell, R. H., & Fisk, A. D. (2001). "Analysis of a Simple Medical Device." *Ergonomics in Design, 9*(1), 6-14.

Sawyer, D. (1996), *Do It By Design: An Introduction to Human Factors in Medical Devices*. Food and Drug Administration, Bethesda, MD.

Skelly, A. H., Arcury, T. A., Snively, B. M., Bell, R. A., Smith, S. L., Wetmore, L. K., et al. (2005). "Self-monitoring of blood glucose in a multiethnic population of rural older adults with diabetes." *Diabetes Educator, 31*(1), 84-90.

Vincze G, Barner JC, and Lopez D (2004). "Factors associated with adherence to self-monitoring of blood glucose among persons with diabetes." *Diabetes Educator, 30*(1), 112-25.

Virzi, R. A. (1992). *Refining the test phase of usability evaluation: How many subjects is enough?* Human Factors, 34, 4, 457-468.

Winter, W.E. and Signorino, M.R. (2002*). Diabetes Mellitus: Pathophysiology, Etiologies, Complications, and Lab Evaluation – Special Topics in Diagnostic Testing*. Washington, DC: AAAC Press.

Wyer, M., and Adam, A. (1999/2000). "Gender and computer technologies." *IEEE Technology and Society Magazine 18* (4), 4-6

Chapter 67

Lessons Learned from Blind Individuals on VideoDescription

Claude Chapdelaine [1] Anne Jarry [2]

[1] CRIM, [2] School of Optometry-University of Montreal
Montréal (Québec), Canada

ABSTRACT

Audio-visual contents are made accessible to blind individuals with added videodescription (VD) that translate the visual information into auditory information. We interviewed ten legally blind individuals (with and without residual vision) to monitor the quantity and frequency of information needed in VD. It was found that residual vision and the complexity of the content have a significant impact of the level of VD needed. This suggests that a tool for the dissemination of VD must provide a basic level of information and also offer enough flexibility to confirm some information on demand.

Keywords: Blindness and Visual Impairment, Videodescription, Verbal Protocol Analysis

INTRODUCTION

Our research is concerned by the processing of audio-visual information done by blind and visually impaired individuals when viewing television and film for learning or entertainment purposes. Audio-visual content such as television and film are made accessible to them by producing additional audio information which describes the relevant visual information. This added information is called audio description or video description (VD). In the five past years, CRIM has been developing tools for VD production and dissemination that aims at minimizing production time while integrating a comprehensive understanding of the cognitive and memory capacities of the intended population through many users testing.

Our actual study aims to establish from a diverse population of congenitally and late blind individuals the level of VD needed. We want to identify 1) what is the information needed, 2) how much and 3) when. In short, we are seeking their definition of an effective VD to achieve an understandable and enjoyable experience.

We report on the method and the results of phase I of our study involving blind individuals classified as level 3, 4 and 5 by the World Health Organization (WHO). In the coming year, a second group of participants of level 1 and 2 will be interviewed to complete the entire study with intended audience for VD.

AUDIO-VISUAL INFORMATION PROCESSING BY BLIND INDIVIDUALS

The WHO classifies visual deficiency into five levels. Level 1 is for degree of acuity lesser than 20/70 or a field of view (FOV) less than 60 degrees after correction. Level 2 is for acuity lesser than 20/200 or a FOV less than 20 degrees; level 3 is for acuity less than 20/400 or a FOV less then 10 degrees; level 4 is for acuity less than 20/1200 or FOV less then 5 degrees. Level 5 means no remaining vision (Mergier, 1999).

Levels 1 and 2 are persons with remaining functional vision, often referred to as persons living with low vision. When it comes to maximizing the use of their remaining vision, the most important variables to adjust are illumination and contrast (Ponchillia, 1996). As for persons in the level 3, they have a slight remaining functional vision. They have severe low vision and are sometimes classified as blind or visually impaired. For levels 1, 2 and even more for level 3, a combination of visual, auditory and tactile system may be needed to successfully complete a task. For example, with a computer, a large screen with magnification and a voice synthesizer will often be needed depending on different factors such as document length, fatigue level, etc. (Presley, 2009). For watching TV, sitting close to a high quality LCD or plasma screen, with reduced illumination and good contrast is a must for people with mild to severe vision loss. Good and easy to follow auditory content with little auditory conflicting messages is also very useful to confirm what is partially seen for person in the level 3 category.

Levels 4 and 5 bring together persons using only auditory or tactile mechanism to access information. Level 4 means light perception and seeing forms with high contrast. Level 5 means no vision. For these two levels, a combination of auditory and tactile system may be needed to successfully complete a task. A voice synthesizer and/or a Braille display will be necessary when using a computer (Presley, 2009). For watching TV, good and easy to follow auditory content with little auditory conflicting messages is almost mandatory.

Classification based on FOV and acuity results are useful for statistical purposes. However, there is great variability within the levels since visual impairment takes many forms and can be seen in many various degrees. Other factors have to be taken into account such as when the deficiency occurred if not congenital. Sadato (2002) found that individuals who lost their sight before the age of sixteen could redirect their primary visual cortex from processing visual input to tactile input.

The degree of visual residue and the adaptation of the blind individual will have a tremendous impact on his ability to process information since the human sensory system has limitations when processing quantity and quality of the information. It also have a preference, Colavita (1974) found the visual dominance effect, proving that humans show a strong tendency to rely more on visual information in a multichannel environment. However, when this predominance is lost, the individual must rely on the auditory channel which is omnidirectional and transient (Wickens, 2000). This very nature of the audio modality has a huge impact on the cognitive capacities of the blind people. Research show that blindness modifies the way information is processed and that auditory, tactile and kinesthetic channel will be central (Gouzman, 2000). Rokem and Ahissar (2008) found that congenitally blind individuals have higher auditory and memory capacities than sighted people. They were more resilient to noise and better at frequency discrimination which allow them to reach better speech perception. This advantage also fosters a short-term memory advantage. So, in order to make sense of the world in which we live in, individuals with little or no vision will listen carefully to auditory information and they need a form of translation for the visual information which is the role of VD.

ON CREATING VD

The production of VD requires that videoscripters insert description of relevant visual elements in the gaps between dialogues. This subjects VD to complex timing constraints since it is dependent on the frequency and length of available gaps. All relevant elements cannot be described, thus a choice must be made and is usually guided by the story line or can be inspired by the existing guidelines. However, adding VD of a visual cue among an already existing audio track is not without hurdle. The VD may not be synchronized without masking other relevant sounds (such as a door opening) and not covering too much of the music to eliminate all ambiance. Furthermore, the VD should at best be synchronized with the image that is being described and this could be impossible if it occurs during a dialogue.

Scripting VD is a challenging task with only a few emerging guidelines (Ofcom, 2000) (ADS, 2009) (Morisset, 2008). These guidelines are often based on intuition or convention which does not provide indications why some VD may be more effective than others (Braun, 2008). Therefore, there is a need for user-oriented research to clarify guidelines. Research on VD is fairly recent and is mostly conducted on VD done by professionals seeking to understand how visual cues could be translated into words. Turner (1998) proposed a VD typology to enrich film indexing and help to automate VD production (Turner, 2008). He analyzed 11 productions of various types and found that most of the information types given in VD are composed of action/movement, character identification, description of the surroundings, expressions of emotion, and textual information included in the image. Piety (2004) demonstrated how the constraints imposed on VD production create a distinctive usage of language that has its own form and function. Salway (2007) showed the linguistic relation of frequent VD words to the characters, the action and the scene. Peli (1996), Pettit (1996), Schmeidler (2001) and Ely (2006) proved that VD is valuable and appreciated by the visually impaired. Up to now, those researches give many insights on the nature of the VD but little is known on the effectiveness of VD to convey meaning to the intended audience (Piety, 2004).

METHOD

In our prior work (Gagnon, 2009), the feedback of blind and visually impaired individuals to whom films with VD were presented, suggested that various VD levels must be offered to accommodate a broad range of vision problems and individual differences. In a more recent work (Chapdelaine, 2009), we presented films with two levels of VD (standard and extended). The visually impaired and the congenitally blind individuals stated that they required less VD and preferred the standard level while the late blind individuals preferred the extended version. Furthermore, individuals with residual vision reported that they found annoying or confusing the VD when it was not synchronized with the image. Those results indicated that a more user-oriented study was needed to identify what would be an efficient VD for them.

Procedure

We conducted a user study using verbal protocol analysis. The design scenario aimed a reproducing the context of watching television for a legally blind individual (participant) with a sighted person (experimenter). A brief synopsis of the video was read to the participants before viewing. Participants could ask questions before and during viewing. After the viewing, participants were asked to summarize what they remember as if telling the story to a friend. This was used to build the mental representation of their comprehension. If concepts were omitted, the experimenter could ask a related question to know if the concept was understood or not.

The verbal protocol was analyzed to extract all the requests made by the participants and they were classified into two groups. A first group included all requests that were made to confirm information, for example: "It is Marc speaking?" The second group included all the requests that were an inquiry on something in the content, such as: "Where is this happening?", "What is the man doing?" Both groups were further classified into six types of information: who, where, action, facial expression, description, explanation of sound or speech.

The summary made by participants and their answers to the questions of the experimenter were used to build a mental representation. We favored the landscape model approach for which Roskos-Ewoldsen et al. (2003) confirmed its adequacy to describe the mental representation of TV series. The concepts extracted from the model of each participant were compared to the concepts collected from the models of four sighted persons (control group) who summed up what they remembered from their viewing. The inter-reliability among sighted viewers on concept identification was 94% after discrepancies were resolved.

Participants

The individual sessions were done with ten participants divided into two groups. Group A was composed of five individuals who could be classified WHO's levels 4 and 5 (2 congenitally blinds, 2 late blinds before than age of 16 and 1 late blind for more than 10 years). Group A was between 36 and 65 years old and they all reported listening to less than five hours of television per week. All of them except one stated that they rarely watch television alone. Three of them wished that the level of VD would be less while the remaining two were satisfied with the actual level.

Group B was composed of five legally blind individuals with some residual vision (level 3 of WHO). They reported being able to detect a human face, some of them could identify a movement done by one person but would be confused if it was a group. They all stated that they needed to be very close to the screen and that image contrast played a very important role in their ability to identify anything. They are all between 46 and 65 years old. Three of them often watch television alone while the remaining two rarely did so alone.

Participants of both groups had experiences with VD and they all preferred VD to a human reporter. Everyone stated that the more important information that should be described in VD was: who is talking, what is the action and when relevant where is the action taking place, the facial expression of the actors and the description of specific objects.

Corpus

Two videos were shown to all participants. One was a short film telling the story of a man meeting a woman on a beach. The second was an excerpt of a TV drama about the life of three doormen.

The film had 4 main actors with no secondary actors and one scene in a public place. It contained a long scene without speech and was almost without background noise except music. In that scene, action was happening and we expected that it would be difficult for blind people to imagine the action without any residual vision. The TV drama had three main actors with four secondary actors and two scenes with a crowd. All the scenes had many dialogues often set in very noisy environments which added strong background noise. It contains very few non-speech segments.

Table 1 Description of Corpus

Description	Film	TV Drama
Length (mm:ss)	08:49	07:35
Nb. Scenes	4	5
Nb. Actors	4	6
Nb. Speech Units	7	10
% Speech in Video	27%	71%
Nb Non-speech Units	10	14
% Non-speech in Video	73%	29%

RESULTS

We report on the analysis of requests made by Group A and B classified either as confirmation or as inquiry. We also present results on the distribution of these requests among the different information types. Finally, we compare the concepts found by the control group against the concepts stated by each participant.

Analysis of Requests for Confirmation versus Inquiry

As shown in Table 2, Group A made the majority of request while viewing the film (60.4%) as opposed to Group B who made only 39.6% of the request. This was a predictable outcome since the film had a long scene without speech that was expected to be less accessible to individual without residual vision. This is further confirmed by the fact that 60.7% of the requests made by Group A were to inquire about information as opposed to the requests of Group B which were mostly to confirm information (67.3%). Indeed, a request to confirm is also an indication of a lesser need for information since the person knows but may need reassurance to avoid confusion, as opposed to an inquiry that is needed to avoid misunderstanding or being loss.

Table 2 Percentage of requests per film per group

	Group A			Group B		
	% of request	% to confirm	% to inquire	% of request	% to confirm	% to inquire
Film	60.4	39.3	60.7	39.6	67.3	32.7
TV drama	48.4	55.7	44.3	51.6	49.2	50.8

The behavior of both groups is much more similar when we look at the results obtained for the TV drama. Indeed, Group B made almost as much requests as Group A (51.6% and 48.4% respectively). This is probably caused by the greater complexity of the video itself. The TV drama had more actors than the film and dialogues were long with noisy background. Also, some scenes were happening at night which meant less contrast in the images and therefore less discriminating elements were available to Group B. This explains why Group B made as much requests to confirm as they made inquiries (49.2% and 50.8% respectively). As for Group A, they made significantly more requests (55.7%) to confirm information rather than to inquire (44.3%) indicating that with complex content they also need more feedback.

Analysis of Requests Distributed among Information Types

Table 3 shows the requests percentage made per information type. For the film, we observed that the majority of requests were about the action for both groups (Group A with 52.4% and Group B with 60%). This indicates that for both groups, some cues help them identified the actors since this is their most desired information. In this case, it seems the small number of actors made identification easy for individual who needed less indications about who was in the scene (11.9% and 14.5%) and also where the scene was taking place (11.9%, 7.3%). A more detailed analysis of the request on action revealed that for Group A, 68.2% of these were inquiries and in Group B, 66.7% were confirmation of action.

For the TV Drama, the requests on action were not as numerous as for the film (37.7% in Group A, 41.5% in Group B). We observed that requests are distributed among the other types (by order of importance): who, where, description and sounds. The larger number of actors in the TV Drama had a significant impact since the request for identification is more than double for Group A (11.9% for the film versus 26.2%) and slightly less for Group B (14.5% for the film versus 23.1%).

Table 3 Percentage per Information Type in Requests

Information Types	Group A		Group B	
	Film	TV Drama	Film	TV Drama
Who	11.9%	26.2%	14.5%	23.1%
Where	11.9%	13.1%	7.3%	13.8%
Action	52.4%	37.7%	60%	41.5%
Description	7.1%	11.5%	9.1%	12.3%
Expression	8.3%	3.3%	7.3%	-
Sound	8.3%	8.2%	1.8%	9.2%

The noisy background had a stronger impact on Group B than on Group A. Indeed, requests for what has been said or to identify a sound (for example: hair dryer, opening of white cane) has been constant in Group A for both video (8.3% for the film and 8.2% for the TV Drama). However, we observed in Group B that those requests were only of 1.8% in the film but they increased significantly during the TV Drama (9.2%). Those results indicates that Group B adopted a behavior similar to Group A when confronted to a complex content that would not provide them with discriminating visual elements they needed.

Analysis of Mental Representations

The control group had identified 12 concepts for the film and 14 for the TV drama (Table 4). For both groups, we observed that stated concepts are about the same on average for both video (8.2 and 8.1 respectively) even if the TV Drama had two more concepts than the film. This again points to the more complex content of the TV Drama compared to the film.

Table 4 Percentage of found concepts.

	Control group	Both Groups	Group A	Group B
	Total nb. of concepts	Average	Average	Average
Film	12	8.2	9.0	7.4
TV Drama	14	8.1	8.6	7.6

The results revealed that Group B omitted on average more concepts than Group A for both films (7.4 over 9.0 for the film and 7.6 over 8.6 for the TV Drama). To assume that this indicates that Group A had a better understanding of the video than the Group B would be a hasty interpretation of the data. Indeed, a correlation of this data with the confirmed requests done by Group B indicates that most of omitted concepts (62%) were in the confirmation request. This indicates that Group B did understand the concepts but simply did not state them in their summary. Further study is needed to understand why those concepts were omitted from their summaries, could it be that they were judge irrelevant or that they were forgotten.

On average, Group A stated more concepts that Group B and this could partially be explained by their better memory capacities. Indeed, as mentioned earlier (Sadato, 2002) there is evidence that suggests that congenital blinds and potentially the late blinds before the age of 16 could store more information in their memory than late blind after the age of 16. Since four of the five persons in Group A meet this criteria than these individuals would have been able to recall more concepts.

DISCUSSIONS

How much information is enough so that a blind individual can enjoy watching a movie or television? This is the basic question this study has addressed. First, we found that the individuals without residual vision and who preferred to have less VD were the ones who asked the greater number of questions. This implies that they need a certain quantity that is more than what they expect. Their better performance in the mental representation analysis suggests that they have better memory capacities which probably required a lot more attention. So their claim may not be about getting less VD but more a preoccupation that VD should not require more attention and thus changing an entertainment into an ordeal.

The basic quantity of VD needed would probably be less for individuals with residual vision. However, we found that when viewing conditions are not optimal, the advantage of individuals with residual vision is quickly overturned. In this case, they will need about the same amount of information as individuals without residual vision. Another interesting outcome of our study is the high percentage of request to confirm information made by the individuals with residual vision. It demonstrates perfectly the statement of Dr. Colenbrander (2006) that a person will use his vision and remaining vision until there is no vision left. Furthermore, a request to confirm is an indication of a lesser need for information since the person has perceived or deducted the information but still is seeking reassurance to avoid confusion.

Based on those results, we concluded that any tool that would disseminate VD should not only provide a basic quantity of information but also that it must offer the possibility to confirm information on demand.

Our next step is to conduct other interviews with individuals with low vision that are classified level 1 and 2 by the WHO. Our aim is to gain a comprehensive understanding of the cognitive and memory capacities of a large spectrum of individuals with visual impairments to assess their need for VD and to design tools that are truly accessible.

REFERENCES

ADS (2009). Guidelines for audiodescription (initial draft of May 2009). http://www.adinternational.org/ad.html

Braun, Sabine (2008). Audiodescription research: state of the art and beyond. Translation Studies in the New Millennium, Vol. 6, 14-30.

Chapdelaine, C., Gagnon, L. (2009). Accessible Videodescription On-Demand. In Eleventh International ACM SIGACCESS (ASSETS'09). Pittsburgh, PA, USA, October 26-28.

Colavita, F.B. (1974) Human sensory dominance. Perception and Psychophysics, 16, 409-412.

Colenbrander, A. (2006) How blind is blind? Smith—Kettlewell Eye Research Institute. http://www.mdsupport.org/presentation-howblind1/index.html

Ely R., Emerson R. W., Maggiore T., O'Connell T., & Hudson L. (2006). Increased content knowledge of students with visual impairments as a result of extended descriptions. Journal of Special Education Technology, 21(3), 31-43.

Gagnon L., Foucher S., Héritier M., Lalonde M., Byrns D., Chapdelaine C., Turner J., Mathieu S., Laurendeau D., Nguyen N.T., Ouellet D. (2009) Towards Computer-Vision Software Tools to Increase Production and Accessibility of Video Description to Visually-Impaired People, Universal Access in the Information Society, Springer-Verlag, Vol 8, no. 3, 199-218.

Gouzman, R. and Kozulin A. (2000) Enhancing Cognitive Skills in Blind Learners. The Educator: 20-29.

Mergier, J. (1999) Le classement OMS des déficiences visuelles, http://www.irrp.asso.fr/articles/article007.html Davis, G.A., and Nihan, N.L. (1991), "Nonparametric regression and short-term freeway traffic forecasting." *Journal of Transportation Engineering*, ASCE, 177(2), 178–188.

Morisset L., Gonant F. (2008). Charte de l'audiodescription. http://www.travail-solidarite.gouv.fr/IMG/pdf/Charte_de_l_audiodescription_300908.pdf

Ofcom (2000). ITC Guidance on Standards for Audiodescription. http://www.ofcom.org.uk/tv/ifi/guidance/tv_access_serv/archive/audio_description_stnds

Peli E., Fine E. and Labianca A. (1996). Evaluating visual information provided by audio description. JVIB 90:5. 378-385.

Pettitt B., Sharpe K. and Cooper S. (1996). AUDETEL: Enhancing television for visually impaired people. BJVI 14:2. 48-52.

Piety, P. (2004). The language system of audio description: an investigation as a discursive process. JVIB 98:8. 453-469.

Ponchillia P.E., Ponchillia, S.V. (1996) Foundations of Rehabilitation Teaching with Persons Who Are Blind or Visually Impaired, AFB Press, 432 pp.

Presley, I., D'Andrea, F.M. (2009) Assistive Technology for Students Who Are Blind or Visually Impaired: A Guide to Assessment, AFB Press, 500 pp.

Rokem, A. and Ahissar M. (2008) Interactions of cognitive and auditory abilities in congenitally blind individuals. Neurophychologia, 47, 843-848.

Roskos-Ewoldsen B., Roskos-Ewoldsen D. R. and Yang M., (2003). Testing the Landscape Model of text comprehension. Paper presented at the annual meeting of the International Communication Association, San Diego, CA.

Salway, Andrew. 2007. "A Corpus-based analysis of the language of Audio Description". In Media for All, Díaz Cintas, Jorge, Pilar Orero and Aline Remael, eds. 151-174.

Turner, J., and Mathieu S. (2008). Audio description text for indexing films. International Cataloguing and Bibliographic Control 37, no. 3 (July/September), 52-56.

Turner, J. (1998). Some Characteristics of Audio Description and the Corresponding Moving Image. Proceedings of 61st ASIS Annual Meeting, vol. 35, 108-117. Medford, NJ: Information Today.

Sadato, N., Okada, T., Honda, M., and Yonekura, Y. (2002) Critical period fro cross-modal plasticity in blind humans : A functional MRI study. Neuroimage, 16, 389-400.

Schmeidler E. and Kirchner C. (2001). Adding audio-description: does it make a difference? JVIB 95:4. 197-212.

Wickens, C.D., Holland J.G. (2000) Engineering Psychology and Human Performance, 3rd Ed. Upper Saddle River, NJ, Prentice-Hall, 572 pp.

Chapter 68

Application of the Critical Incident Technique to Evaluate a Haptic Science Learning System

Kyunghui Oh[1], Swethan Anand[1], Hyung Nam Kim[1], Na Mi[1], Heidi Kleiner[1], Takehiko Yamaguchi[2], Tonya Smith-Jackson[1]

[1]The Grado Department of Industrial and Systems Engineering
Virginia Polytechnic Institute and State University
Blacksburg, VA 24061, USA

[2] Department of Industrial Engineering
University of Arkansas
Fayetteville, AR 72701, USA

ABSTRACT

Assistive technologies have become increasingly widespread, yet many developers have been criticized for a lack of usability evaluation. This paper presents a research study where usability evaluation was done on a haptic science learning system using the critical incident technique. The system was evaluated by seven teachers experienced in working with students with visual disabilities. The Retrospective Think Aloud (RTA) method was conducted as teachers used the system. Two researchers analyzed and elicited critical incidents from video recordings of teachers' interaction with the system. Results from the verbal protocol analysis revealed several areas of concern regarding system usability.

Keywords: Assistive Technology, Usability Evaluation, Critical Incident Technique, Inclusive design

INTRODUCTION

Learning systems have experienced rapid growth as technological advancements to enhance learning within and beyond classroom environments. There has also been an increasing need to make these advancements available and accessible to everyone including students with disabilities. Assistive technology, in particular, plays as important role in providing equitable opportunities for students with disabilities (Riemer-Reiss, 1999).

Even though these technologies are currently used in educational environments and can help students with disabilities (Johnson et al., 1997), many students have discontinued the use of devices; ultimately resulting in their abandonment (Riemer-Reiss & Wacker, 2000). According to Scherer (1996), in an amount ranging from 8% for life saving devices to 75% for hearing aids, nearly one third of assistive technologies are abandoned. However, there are few existing studies examining the abandonment rate across all kinds of assistive technologies (Magiera & Goetz, 2001).

Many studies have been conducted to identify factors related to the acceptance or abandonment of assistive technology (Batavia et al., 1990; Phillips & Zhao, 1993). For various reasons (e.g., poor device performance, device unreliability, environmental factors, fear of technology), the principal reason leading to the abandonment of assistive technology is the inability of the technology to meet users' needs and requirements (Scherer, 1996). To ensure successful usage of assistive technology, it is important to design, develop and implement systems that comply with the needs and desires of users including learners, teachers, friends, and family members (Kintsch & Depaula, 2002).

Researchers in the Human-Computer Interaction (HCI) field have also been trying to address problems in assistive technology (Barnicle, 1999, Wehmeyer, 1999). In particular, usability evaluation has been employed as one of most important methods to verify effectiveness, efficiency, safety and some degree of comfort and satisfaction among users. However, studies show that evaluation of assistive technology has been insufficient and irrelevant in many cases (Aquilano et al., 2007; Stevens & Edwards, 1996). One of the key factors is the deployment of improper evaluation techniques for evaluation of assistive technology. It is therefore imperative to evaluate the system using techniques such that address users' perspectives.

The purpose of this study was to identify usability problems and design requirements in a haptic science learning system using the critical incident technique. The haptic science learning system mentioned here is a prototype designed to help students with visual impairments, ranging from those with partial sight to those with total blindness, and to understand fundamentals of scientific concepts via physical and virtual interactions with the system. The paper also aims to explore the effectiveness of the critical incident technique in evaluating assistive technology.

CRITICAL INCIDENT TECHNIQUE

The critical incident technique developed by Flanagan (1954) is a method used to identify behaviors that contribute to the success or failure of individuals or organizations in specific situations. Flanagan defined an incident as "any observable human activity that is sufficiently complete in itself to permit inferences and predictions about the person performing the action" (Flanagan 1954, p.335). The term 'critical' refers to the fact that the behavior described in the incident plays an important role in determining an outcome.

The technique has been used widely in diverse areas such as aviation, the service industry, HCI and many others. In the context of usability evaluation, this method has been applied to identify common features of critical incidents, such as classifying usability problems during formative evaluation (del Galdo et al., 1986). This technique has been used as either an expert-based or user-based usability evaluation in laboratory or remote evaluations (Thompson & Williges, 2000). In addition, it has been used as a collaborative critical incident procedure for improving the usability problem specification that collects dialogue between the user and the expert evaluator (Neale et al., 2000). With regard to assistive technology, usability evaluation using the critical incident technique has been applied to assess usability problems (Sutcliff et al., 2003); in particular, as a basis for constructing studies such as surveys, interviews, and observations.

HAPTIC SCIENCE LEARNING SYSTEM

The haptic science learning system is designed to help students with varying degrees of visual impairment obtain a better understanding of essential concepts via physical and virtual interaction with the system. The haptic system was developed by Novint Falcon. This paper focuses on the Molecular Properties Module (MPM), where students can learn about molecular concepts of basic chemistry. The MPM is composed of three main sections: Menu Selection, Structure Mode, and Force Mode. Table 1 shows the descriptions of each section.

Table 1. The Molecular Properties Module (MPM)

Sections	Descriptions	Types of Interfaces	Components
Menu Selection	The Main Menu is MPM's central hub to enter Force Mode or Structure Mode respectively with a haptic device.	• Matrix Selection • Circular Selection • Linear Selection	
Structure Mode	User can feel 2-Dimensional (2D) and 3-Dimensional (3D) molecular models with a haptic device	• 2D space filling model • 3D space filling model • 2D ball & stick model • 3D ball & stick model	- Visual Components - Sound Components - Haptic components
Force Mode	User can explore intermolecular force mode and intramolecular force mode for a particular molecule	• Intermolecular force mode • Intramolecular force mode	

EXPERIMENTAL DESIGN

Semi-structured interviews were conducted in laboratory usability evaluations. A total of seven teachers with experience teaching students with visual impairments participated. Six of the seven were sighted; one had a visual impairment. Four interview sessions were conducted with seven teachers. The composition of teachers in each interview is presented below:

- Interview 1: Three teachers (three females)
- Interview 2: One teacher (one female)
- Interview 3: Two teachers (one female, one male)
- Interview 4: One teacher with visual impairments (one female)

Interviews were semi-structured to allow for investigation of diverse responses. Structured questions were used to provide the boundaries of topics discussed while unstructured probes and follow-up questions were used for clarification and specificity. The interview consisted of 16 questions with constructs including Efficiency, Effectiveness and Satisfaction as main properties of usability suggested by ISO (ISO 9241-11). In particular, questions considering use in context such as student capabilities, learning situations and system implementation were included. The RTA method (Ericsson & Simon, 1993) was administered during usability evaluation, where teachers' thoughts were captured during interaction with the system.

Prior to beginning the interview, each teacher completed an informed consent document. Teachers were asked to explore and perform tasks specified by the researcher in each mode. Upon completion, teachers' thoughts were gathered through RTA sessions. Structured questions were administrated at the end of each mode. Interviews were scheduled to last 60 minutes. Each session was videotaped using a stable camera, a Sony VHS camcorder.

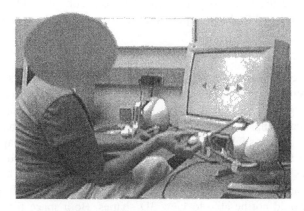

FIGURE 1. Experimental Setup

Two researchers individually analyzed interviews using the critical incident technique. Two researchers analyzed all four interviews for the entire duration observing teachers' interaction with the system for critical incidents. As a result, each researcher identified the characteristics and frequency of critical incidents (CIs).

In order to improve the reliability of the result, a reconciliation meeting was conducted. During the reconciliation meeting, each researcher presented their findings to the other researcher. The researchers compared their results to gather the number of agreements and disagreements among each other. A final set of CIs emerged during the discussion and percent agreement (Hartmann, 1977) calculated (i.e., agreements divided by agreements plus disagreements). Prior to the reconciliation meeting the average percent agreement between the two researchers was 54.5%. The average percent agreement between the two coders was 92.5% (after the reconciliation meeting).

RESULTS

A total of forty-seven (47) critical incidents were identified and twenty-three (23) suggestions were collected (Table 2). The results are categorized into four groups (three sections in system: menu selection, structure mode, force mode, and overall system).

Menu Selection

Nineteen (19) critical incidents and two (2) suggestions were collected in the menu selection section. The Matrix Menu was the most preferred menu option as it matched the teaching and students' learning patterns. The Circular Menu was also preferred, because teachers thought it would be easier for students to move around the edges rather than moving up and down. The Linear Menu was the least preferred because teachers thought it would be difficult for students to navigate through the system and find more options rather than having them all laid out.

Structure Mode

In the structure mode section, sixteen (16) critical incidents and ten (10) suggestions were collected. The responses about structure mode were complicated as teachers frequently navigated in and out of the different interfaces (2D space filling, 3D space filling, 2D ball & stick and 3-D ball & stick). 2D shapes (2D space filling and 2D ball & stick) were easier to figure out than 3D shapes (3D space filling and 3D ball & stick). However, teachers were very excited about the 3D shapes and the feedback they provided, but felt it would take a longer time for students to learn them. Teachers also stated that students with partial visual impairments would not be able to view 3D shapes. Both space filling and ball & stick shapes provided important meanings to teachers, which are sizes, angles and bonds relationships.

Force Mode

In the force mode section, two (2) critical incidents and two (2) suggestions were collected. Among three sections, force mode was the one where teachers had difficulties with awareness of the functions and operations of the device. One reason was that the system did not give enough information about this section and its operation in an auditory format. Therefore, the researcher had to give details of information presented on the screen frequently.

Overall System

Six (6) critical incidents and nine (9) suggestions that were related to the overall system were collected. Teachers mostly preferred the auditory information at each section, however they argued that it needed improvement to reflect students' independent use and learning environments. In reference to the boundary, teachers preferred the boundary and reported that the boundary was an absolute necessity to guide the user. Although the auditory feedback with the boundary was appreciated by the teachers, they also suggested stronger haptic feedback to be incorporated in the boundary. In the matter of joystick design, a tension control system was suggested so that students can physically change the tension as required.

In general, the teachers were excited about the prototype as a tool that could enhance learning for students with visual impairments. Teachers stated that this system would greatly help learning and maintain a competitive and interactive learning atmosphere. However, several areas of concern were expressed. Teachers suggested a training session be given prior to implementing the system. They also suggested that add-on features such as ZoomText would enable students to enlarge images on the screen. Having the learning modules in a game format was also suggested.

Table 2. Critical Incidents and Suggestions

Critical Incidents & Suggestions				Frequency
Menu Selection	Positive		(Common feature: sound) give sound information for skipping one	3
			(Circular menu: joystick movement) easy to move the position/identify the current position	2
			(Matrix menu: joystick movement) able to move in different ways-diagonal direction	2
			(Common feature: color) need to have solid color and distinct contrast	2
	Negative		(Main menu: joystick movement) high possibility of error during navigation	1
			(Circular menu: joystick Movement) cannot go through middle area (no providing short cut)	2
			(Linear menu: menu structure) simplest, but lose the sense of position	4
			(Matrix menu: joystick movement): difficult to identify the position (Especially, middle one)	2
			(Matrix menu: force feedback) easy to move too far (skipping the middle one)	1
	Suggestions		(Common feature: navigation/magnifier function) if there is a function magnifying the screen for students with severe visual impairment (e.g., ZoomText), it will be helpful	1
			(Common feature: navigation/joystick movement) using keyboard is easier and faster than using joystick	1
Structure Mode	Positive		(Common feature: color/contrast) good contrast	2
			(2D Space filling: haptic feedback) able to feel the difference of roughness in shape	2
			(3D Space filling: haptic feedback/Joystick movement) good to know the size of molecule/able to feel the molecule	2
			(2D Ball & Stick: joystick movement) give angle information	1
			(2D Ball & Stick: auditory information) good to know what the user needs to look for (e.g., bonds, angle)	1
	Negative		(Common feature: shape) need to reflect the actual ratio in molecule	1
			(2D Space filling: haptic feedback) cannot feel the shape of object	1
			(2D Space filling: shape) too short distance from molecule to molecule/hard to detect the center molecule	1
			(3D Space filling: haptic feedback/Joystick movement) hard to feel the shape/the edge of object	3
			(3D Ball & Stick: haptic feedback/Joystick movement) hard to find the whole object (especially, z axis boundary problem)	2
	Suggestions		(Common features: sound) If system give sound when users are contacting the molecule, it will be helpful	2
			(Common feature: sound) If system give sound about user's position where they are in the space (left, right, top, down), it will be helpful	1
			(Common feature: joystick movement) automatically guided navigation	3
			(Common feature: haptic feedback/joystick movement) it will be easier if users use Touch pad rather than 3D object joystick	1
			(2D Ball & Stick: haptic feedback) across molecules, consistently distinct haptic feedback in each part of molecule (e.g., main	1

		component, bond, other components)	
		(3D Ball & Stick: sound) If system give sound for z axis boundary, it will be helpful	1
		(3D Ball & Stick: sound) If system give sound when users are contacting on the middle of the frame, it will be helpful	1
Force Mode	Negative	(Common feature: Auditory information) no auditory information about learning contents	2
	Suggestion	(Common feature: sound) if system gives sound when one molecule is getting closer to other molecule, it will be helpful for users to find their orientation	2
Overall System	Positive	(Boundary) good about giving boundary information	2
		(Human Voice) Using human voice is good	2
		(Sound) most sounds are discrete and clear	1
		(Force feedback) good for operation	2
	Negative	(Auditory information) need more descriptive auditory information for considering independent use	2
		(Pointer) need to have distinctive color/shape	1
	Suggestions	(Boundary) haptic/force feedback would be helpful	
		(Auditory information) customized auditory information	2
		(Auditory information/color) different color or auditory information according to region (left, right, up, down) will be helpful	1
		(Force feedback) adjustable force feedback using wheel adjustment	1
		(Error Recovery) Error recovery functions (ex: ESC key)	1
		(Training/Instruction) Training Tape that have been used to teach a blind person to use new piece of technology will be helpful (e.g., aph.org, bookport, victor reader stream)	1
		(Fun) gaming contents will be helpful	1

DISCUSSION

In this study, we identified usability problems in a haptic science learning system using the critical incident technique. Results indicated several areas of concern with regard to usability of the system, including navigation, haptic feedback, and auditory feedback. Suggestions by teachers that presented alternative or supportive solutions were also collected. The researchers will further analyze collected critical incidents and discover redesign solutions, which enhance the haptic science learning system prototype. In addition, further analysis of these CIs and suggestions can lead to generation of design guidelines in a haptic science learning system.

This study also explored the effectiveness of the critical incident technique in evaluating assistive technology. The technique was successful in the collection of critical incidents and gave important information to researchers who are interested in designing assistive technology. First, it is becoming evident that before actual implementation, collecting users' perceptions and examining the system are critical in the design of assistive technology. Therefore, studies using the critical incident technique can offer valuable insights about users' experiences with system. Second, through analyzing critical incidents, the researcher identified several factors besides the interface itself that impact users' experiences, which include learning environment, learning context, nature of task, and users' abilities. Considering the importance of interactions between users and other persons (e.g., teacher, other students) in adopting and utilizing assistive technology (Kintsch & Depaula, 2002), this technique presented opportunities to understand and analyze users' experience from diverse points of view.

This study used the critical incident technique through observations of semi-structured interviews. Therefore, it had the advantages of thorough analysis and reliability using two analyzers. However, if the components of acquiring critical incident logs including critical characteristics of assistive technology are further developed, user-based and remote-based evaluation with lower intrusiveness and demand effects would be applied in this area.

CONCLUSIONS

Deployment of relevant evaluation techniques plays a vital role in successful implementation and acceptance of assistive technologies. The critical incident technique used for usability evaluation of the new haptic science learning system helped generate various points of interest for design modifications. These points of interest if analyzed further with other relevant evaluation techniques can lead to generation of design guidelines for further changes to the system to enhance the learning experience.

ACKNOWLEDGEMENTS

This study is funded by the National Science Foundation, Grant DRL-0736221. The authors wish to thank all the teachers for their time and valuable insights.

REFERENCES

Aquilano, M., Salatino, C., and Carrozza, M. C. (2007), Assistive Technology: a New Approach to Evaluation. *Proceedings of the 2007 IEEE 10th International Conference on Rehabilitation Robotics*, 809-819.

Barnicle, K.A. (1999), *Evaluation of the interaction between users of screen reading technology and graphical user interface elements:* PhD dissertation. Graduate School of Arts and Sciences, Columbia University, New York.

Batavia, A. I., Dillard, D. and Phillips, B. (1990), *How to avoid technology abandonment.* Washington, DC: National Institute on Disability and Rehabilitation Research.

del Galdo, E. M., Williges, R. C., Williges, B. H., and Wixon, D. R. (1986), An Evaluation of Critical Incidents for Software Documentation Design. *Proceedings of Thirtieth Annual Human Factors Society Conference.* Anaheim, CA: Human Factors Society, 19-23.

Ericsson, K. and Simon, H. (1993), *Protocol analysis: Verbal Reports as Data.* Cambridge, MA: MIT Press.

Flanagan, J. C. (1954), *The Critical Incident Technique.* Psychological Bulletin, 51, 327-359

Hartmann, D. P. (1977), Considerations in the choice of interobserver reliability estimates. Journal of Applied Behavioral Analysis, 10, 103-116.

ISO 9241-11 (1998), *Ergonomic requirements for office work with visual display terminals (VDT)s - Part 11 Guidance on usability.*

Johnson, K. L., Dudgeon, B., and Amtmann, D. (1997), Assistive technology in rehabilitation. *Physical Medicine and Rehabilitation Clinics of North America,* 8(2), 389-403.

Kintsch, A. and DePaula, R. (2002), A Framework for the Adoption of Assistive Technology. *SWAAAC 2002: Supporting Learning Through Assistive Technology.* E3 1-10.

Magiera, J., and Goetz, J. (2001), *Achieving New Heights with Assistive Technology.*

Neale, D. C., Dunlap, R., Isenhour, P., and Carroll, J. M. (2000), Collaborative critical incident development. *In Human Factors and Ergonomics Society 43rd Annual Meeting,* 598–601.

Phillips, B. and Zhao, H. (1993), Predictors of assistive technology abandonment, *Assistive Technology,* 5(1), 36-45.

Scherer, M. J. (1996), *Living in the State of Stuck: How Technology Impacts the Lives of People with Disabilities* (Second ed.). Cambridge: Brookline Books.

Stevens, R. D. and Edwards, A. D. N. (1996), An approach to the evaluation of Assistive Technology. *Proceedings of ASSETS 1996,* ACM Press, 64-71.

Sutcliffe, A., Fickas, S., Sohlberg, M. M., and Ehlhardt, L. A. (2003), Investigating the usability of assistive user interfaces. *Interacting with computers,* 15(4), 577-692.

Thompson, J. A., and Williges, R. C. (2000), Web-based collection of critical incidents during remote usability evaluation. *Proceedings of Human Factors and Ergonomics Society 43rd Annual Meeting,* Santa Monica. 602-605.

Riemer-Reiss, M. (1999), Assistive technology use and abandonment among college students with disabilities. *International Electronic Journal for Leadership in Learning* 3(23).

Riemer-Reiss, M. L. and Wacker, R. R. (2000), Factors associated with assistive technology discontinuance among individuals with disabilities. *Journal of Rehabilitation,* 66(3), 44-50.

Wehmeyer, M. L. (1999), Assistive technology and students with mental retardation: utilization and barriers. *Journal of Special Education Technology* 14 (1), 48.

Chapter 69

Analyzing the Behavior of Users with Visual Impairments in a Haptic Learning Application

Steve Johnson[1], Takehiko Yamaguchi[1], Yueqing Li[1],
Hyung Nam Kim[2], Chang S. Nam[1]

[1]Department of Industrial Engineering,
University of Arkansas
Fayetteville, AR 72701, USA

[2]Department of Industrial & Systems Engineering,
Virginia Tech
Blacksburg, VA 72701, USA

ABSTRACT

This paper presents empirical methods and techniques for performing haptic system user behavior analysis, specifically of target users with visual impairments, in an effort to improve the development and design of future haptic interfaces. A haptic learning application was developed entitled the Molecular Properties Module (MPM) supporting the Novint Technologies' Falcon haptic device (3 Degrees of Freedom). A usability test was conducted at the Arkansas School for the Blind in Little Rock, AR with two independent variables: (1) five key tasks and (2) thirteen user interfaces, resulting in thirteen total scenarios. A Cursor Trajectory Management System (CTMS) was developed to accurately and empirically capture a user's cursor behavior throughout each scenario. Statistical analysis techniques are considered for analyzing haptic user behavior. Such techniques could aid in the development of future haptic interfaces, the development of haptic interface guidelines, and provide meaningful insight into the behavior of users with visual impairment. Likewise, relationships could be established between user behavior and haptic applications' interface objects and interactions.

Keywords: Haptic, Sensorial, Modality, Tactile, Interface Development, User Behavior

INTRODUCTION

Haptic interfaces offer interesting and exciting sensorial experiences through tactile and force feedback. This ever-growing dynamic technology has been applied in surgical simulations (Guthart & Salisbury, 2000; Madhani, Niemeyer, & Salisbury, 1998), training environments (Basdogan, Ho, & Srinivasan, 2001; Tendick, et al., 2000), scientific visualizations (Avila & Sobierajski, 1996; Iwata & Noma, 1993), and assistive technologies for the visually impaired (Grabowski & Barner, 1998). In addition, it has been found that haptic stimuli and sensorial feedback enrich students' understanding of scientific ideas (Brooks, Ouh-Young, Batter, & Kilpatrick, 1990; Kilpatrick, 1976; Sauer, Hastings, & Okamura, 2004). For example, combining haptic perceptions with hearing has been shown to improve students with visual impairments' ability to understand scientific concepts (McLinden, 2004; Yu & Brewster, Evaluation of multimodal graphs for blind people, 2003). Student's critical and problem solving ability is aided by investigating and experimenting with concrete objects (Paulu & Martin, 1991). Finally, users' task performance and sense of copresence have been shown to be affected by the influences of haptic force feedback (Basdogan, Ho, & Srinivasan, 2001; Hubbold, 2002; Sallnas, Rassmus-Grohn, & Sjostrom, 2000).

This study seeks to develop a haptic media that supports the systematic and empiric evaluation of user behavior in a haptic system with the purpose of identifying and improving key design elements and guidelines for haptic interface development. Although a number of graphical user interface design guidelines exist, their compatibility to haptic systems is questionable. Haptic researchers, utilizing regression-based techniques, have analyzed Time Lag and Visual Force Fields using Fitt's Law (Li, 2007; Ahlström, 2005), Haptic Delay (Jay, Glencross, & Hubbold, 2007), Behavioral Force adaptation in relation to spring stiffness (Sulzer, Salamat, Chib, & Colgate, 2007; Nisky, Mussa-Ivaldi, & Karniel, 2008), Wavelet Networks (Miller & Colgate, 1998), and Haptic-BCI Object Recognition (Grunwald, et al., 2001) in a variety of haptic systems. However, these statistical methods generally analyze short, controlled, and repetitive cursor movements or utilize simple haptic devices with 1 Degree of Freedom (DOF).

This paper provides: first, a haptic learning application entitled the Molecular Properties Module (MPM) is presented. Second, results are presented from a usability test conducted at the Arkansas School for the Blind in Little Rock, AR with two, within-subject, independent variables: (1) five key tasks and (2) thirteen user interfaces, resulting in thirteen total scenarios. Third, user behavioral data from the Cursor Trajectory Management System (CTMS) is presented and discussed.

Fourth, proposed empirical and statistical behavioral paradigms are brought forth as a means to systematically evaluate haptic user behavior. Finally, this paper concludes with a general discussion of all topics presented therein.

MOLECULAR PROPERTIES MODULE

The Molecular Properties Module (MPM) is a science learning application that supports Novint Technologies' Falcon haptic device (3 DOF). The MPM teaches molecular concepts through three interactive modes: *Menu Selection, Structure Mode,* and *Force Mode.* In essence, nine molecules are presented as selectable buttons in the *Main Menu.* A molecule can then be selected and interacted with in *Structure Mode*, which presents the user with tangible, tactile and/or audio based feedback, or *Force Mode*, which presents the user with intermolecular and intramolecular forces between two molecules of the same type. FIGURE 1 illustrates some sample interfaces developed for the MPM.

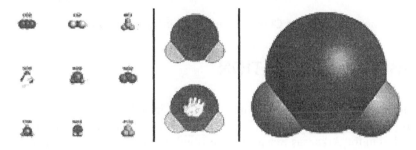

FIGURE 1 Example MPM Interfaces from *Main Menu Selection* (Left), *Intermolecular Force Mode* (Middle), and *Structure Mode* (Right)

CURSOR TRAJECTORY MANAGEMENT SYSTEM

In order to accurately and empirically capture user behavior for further analysis, a Cursor Trajectory Management System (CTMS) was developed and incorporated into the MPM (See Yamaguchi et al., 2010 for details). FIGURE 2 illustrates the theory behind CTMS for an example interface from MPM's *Structure Mode.*

The CTMS is a C++ based algorithm that calculates, stores, and plots the user's cursor activity throughout the duration of the MPM application. The CTMS devises the user's cursor information by (1) Dividing the environmental space into a two-dimensional grid where each R×C cell, or element, within the grid is approximately X by Y pixels and is described as an (x,y) coordinate within the screen's resolution. (2) An array is initialized to hold an increment counter for each element (i.e. Array[3][5] corresponds to the element in Row 3, Column 5 in the onscreen grid). (3) During the duration of the MPM, a cell location is incremented within the array every time positional data is captured. Therefore, the higher the increment within a particular array location, the longer the user's cursor remained within a particular element.

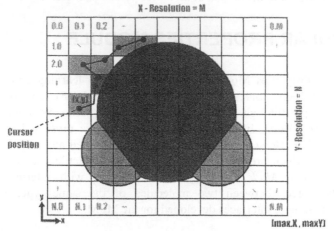

FIGURE 2 Theory Behind the Cursor Trajectory Management System

USABILITY EVALUATION

A usability evaluation of the MPM was conducted in order to obtain user behavior data from target participants with visual impairments. Fourteen participants were recruited from the Arkansas School for the Blind in Little Rock, AR. There were 6 female and 8 male participants whose mean (M) age was 14 years (SD, = 1.88). All participants had little or no experience with Novint Technologies Falcon haptic devices or haptic application interfaces.

Two independent variables were manipulated in the study: five key tasks and thirteen user interfaces, resulting in thirteen total scenarios. Five key tasks were evaluated in the usability study: *Menu Selection* (Task 1), *Space Filling Model Structure Recognition* (Task 2), *Ball & Stick Model Structure Recognition* (Task 3), *Intermolecular Force Recognition* (Task 4), and *Intramolecular Force Recognition* (Task 5). Tasks contained haptic elements such as Gravity Wells, Haptic Boundaries, and Tactile Feedback for interface ease of use and navigation. Tasks also employed navigation and feedback audio to provide users with structure recognition features. NASA TLX questionnaires were administered to assess users' cognitive workload. Additionally, questionnaires were administered to obtain user preference and comments.

CTMS USABILITY DATA

Raw usability data captured by the CTMS was subdivided into three zone types in order to quantify proportions of the user's total cursor activity/behavior. Zone types included (1) TR: Region(s) closest to haptic object(s), (2) SR: Region(s) surrounding Trace Region(s), and (3) ER: Region(s) surrounding *Stability Region(s)* or inside a 2D haptic object. Essentially, TR signified that the user was interacting with a haptic object, SR signified that the user's cursor was very close to a haptic object, but may not be touching it, and ER signified the user's cursor was not near a haptic object.

A **Positional Data Analysis** (PDA) of the data resulted in representational proportions of the user's total actions for a particular scenario. PDA proportions include (1) ERP(%): Portion of Total User Behavior in ER for a particular interface, (2) SRP(%): Portion of Total User Behavior in SR for a particular interface, and (3) TRP(%): Portion of Total User Behavior in TR for a particular interface. Correspondingly, for a particular scenario, the summation of the three PDA proportions accounts for 100% of the user's behavior.

Based on the PDA, a **Transitional Data Analysis** (TDA) was conducted to analyze user transitions between zone types. Transitions are determined by comparing subsequent cursor positions (P_i) within the data collected. Transitions are defined as $\sum_{i=1}^{Total\ i} P_{i-1} > P_i$ for each of the nine possible transitions between zone types TR, SR, and ER, where *Total i* represents the total number of transitions for a particular scenario.

USER BEHAVIOR ANALYSIS TECHNIQUES

In order to analyze haptic user behavior, especially of target users with visual impairments, statistical and empirical analysis techniques can be applied to haptic CTMS usability data. In this paper, Regression and Markov Chains are presented as potential analysis techniques. These techniques, though presented in the context of the MPM, will be presented in such a manner that they may be extended to future haptic applications and interfaces.

REGRESSION

A logistic regression (LOGIT) technique similar to that presented by (Ginsberg, et al., 2009) produces a transformation over the response Y, reducing data variability and producing more meaningful results, especially for usability data that is not constrained by a time cutoff. To extrapolate, the LOGIT technique, in the case of the MPM, assimilated modified usability data such that Trace Region and Stability Region Data were combined into a new Stability-Trace Region (STR). Using the LOGIT technique, the response variable Y is described as $LOGIT(Y) = ln\ (STR/ER)$. For each task, first order regression models can be developed to account for dependent measurements such as *Total Task Time*, *Zone Type Transitions*, and *Interface Type*. Table 1 illustrates resulting regression statistics for MPM Tasks 1-3 (i.e. Y_1, Y_2, and Y3, respectively). Such models can be examined to determine how much data variation is accounted for in the data (i.e. R^2 and Adjusted R^2) as well as how factors correlate with regression trends.

Table 1 Regression Statistics for First Order Models in MPM Tasks 1-3

	Y_1	Y_2	Y_3
R^2	0.775	0.860	0.836
ADJUSTED R^2	0.729	0.831	0.803
STD. ERROR	0.957	0.422	0.442
F VALUE	16.75	29.84	24.84
P VALUE	0.000	0.000	0.000
OBSERVATIONS	42	42	42

MARKOV CHAINS

Markov chains can provide useful insight into the probability of future user behavior(s). Figure 3 shows an example Markov Model constructed from obtained MPM Task 1 usability data.

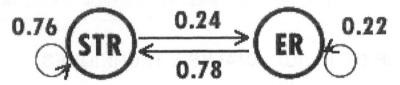

FIGURE 3 Example Markov Model constructed from MPM Task 1 usability data

Markov chains can first be constructed for each zone type within a haptic application. For the MPM, STR (combined TR and SR regions) and ER zone types from each task were used to represent a state (q). State probabilities can be determined using obtained usability data, more specifically data resulting from a TDA. Because TDA transition data describes the proportion of transition types for each subject and each task, transition data can be averaged across tasks for all subjects to represent state change probabilities. From this model, a transition matrix can be constructed in order to predetermine the probabilities associated with future user behavior. For example, assuming the user is currently in zone type STR, their behavior can be predicted three "moves" from now utilizing matrix algebra:

$$[1 \quad 0] \begin{bmatrix} 0.76 & 0.24 \\ 0.22 & 0.78 \end{bmatrix}^3 = [0.6304 \quad 0.3696]$$

Therefore, the user has a 63% chance of remaining in STR and a 37% chance of moving into ER in three moves.

DISCUSSION AND CONCLUSION

Using statistical techniques such as Logistic Regression and Markov Chains in the realm of Haptic System Usability can help to further aid in the assessment and evaluation of current generation haptic interfaces. A statistical and empirical approach to analyzing haptic user behavior can help to establish coherent guidelines and practices for the future of haptic interface development. Such guidelines and practices could also influence the future development of haptic Integrated Development Environments (IDEs), attributing the spread of interactive, user-friendly haptic media. Catering haptic interfaces to their target audiences, such as users with visual impairments, should be a key focus for haptic developers in regards to future haptic interface development.

This paper described a haptic learning application entitled the Molecular Properties Module (MPM) and introduced a Cursor Trajectory Management System (CTMS) as a method of accurately capturing raw user behavior data. Empirical and statistical behavioral paradigms are brought forth as a means to systematically evaluate haptic user behavior. Though sophisticated behavior analysis techniques exist such as Hidden Markov Models, Fast Fourier Transformations, or Mel-frequency cepstral coefficients (MFCCs), this paper presented logistic regression and Markov Chains as an introduction to analyzing haptic end user behavior. These methods, combined with more sophisticated techniques, could potentially aid in the understanding the behavior of users with visual impairments as well as the establishment of haptic design guidelines for the future of haptic technology.

ACKNOWLEDGEMENT

This research was in part supported by the National Science Foundation (NSF) under Grant No. DRL-0736221. Any opinions, findings, and conclusions or recommendations expressed in this material are those of the authors and do not necessarily reflect the views of the NSF.

682

REFERENCES

Ahlström, D. (2005). Modeling and Improving Selection in Cascading Pull-Down Menus Using Fitts' Law, the Steering Law and Force Fields. *Proceedings of CHI 2005* (pp. 61-70). Portland, OR: ACM Press.

Avila, R. S., & Sobierajski, L. M. (1996). A Haptic Interaction Method for Volume Visualization. *Proceedings of IEEE Visualization*. San Fransisco.

Basdogan, C., Ho, C., & Srinivasan, M. A. (2001). Virtual environments for medical training: Graphical and haptic simulation of laproscopic common bile duct exploration. *IEEE/ASME Transactions on Mechatronics , 6* (3), 269-285.

Brooks, F. P., Ouh-Young, M., Batter, J., & Kilpatrick, P. (1990). Project GROPE - Haptic displays.

Ginsberg, J., Mohebbi, M. H., Patel, R. S., Brammer, L., Smolinski, M. S., & Brilliant, L. (2009). Detecting influenza epidemics using search engine query data. *Nature , 457.*

Grabowski, N. A., & Barner, K. E. (1998). Data Visualisation Methods for the Blind Using Force Feedback and Sonification. *The SPIE conference on Telemanipulator and Telepresence Technologies V*. Boston.

Grunwald, M., Weiss, T., Krause, W., Beyer, L., Rost, R., Gutberlet, I., et al. (2001. 11). *Theta power in the EEG of humans during ongoing processing in a haptic object recognition task*. Research Report, Cognitive Brain Research.

Guthart, G. S., & Salisbury, J. K. (2000). The IntuitiveTM telesurgery system: Overview and application. *Proceedings of IEEE Conference on Robotics and Automation*, (pp. 618-621).

Hubbold, R. (2002). Collaborative stretcher carrying: A case study. *Proceedings of 2002 EUROGRAPHICS workshop on virtual environments*. Barcelona, Spain.

Iwata, H., & Noma, H. (1993). Volume Haptization. *Proceedings of the IEEE 1993 Symposium on Research Frontiers in Virtual Reality*.

Jay, C., Glencross, M., & Hubbold, R. (2007). Modeling the effects of delayed haptic and visual feedback in a collaborative virtual environment. *ACM Trans. Computer-Human Interaction , 14(2)* (8).

Kilpatrick, P. (1976). *The use of kinesthetic supplement in an interactive system*. Ph. D dissertation, University of North Carolina at Chapel Hill, Computer Science Department.

Li, Y. (2007). *Modeling the Effects of Time Lag in Virtual Reality (VR)-Based Haptic Surgical Simulator*. Master's Thesis, NCSU, Industrial Engineering, Raleigh, NC.

Madhani, A., Niemeyer, G., & Salisbury, J. (1998). The Black Falcon: A teleoperated surgical instrument for minimally invasive surgery. *IEEE/RSJ International conference on Intelligent Robotic Systems*, *2*, pp. 936-944.

McLinden, M. (2004). Haptic exploratory strategies and children who are blind and have additional disabilities. *Journal of Visual Impairment and Blindness , 98* (2), 99-115.

Miller, B., & Colgate, J. (1998). Using a Wavelet Network to Characterize Real Environments for Haptic Display. *Proceedings of the 7th Annual Symposium on Haptic Interfaces for Virual Environment and Teleoperator Systems.* Anaheim, CA: ASME/IMECE, DSC.

Nisky, I., Mussa-Ivaldi, F. A., & Karniel, A. (2008). A Regression and Boundary-Crossing Based Model for the Perception of Delayed Stiffness. *IEEE Transactions on Haptics , 1,* 73-82.

Paulu, N., & Martin, M. (1991). *Helping your child learn science.* U.S. Department of Education Office of Educational Research and Improvement.

Sallnas, E. L., Rassmus-Grohn, K., & Sjostrom, C. (2000). Supporting presence in collaborative environments by haptic force feedback. *ACM Transaction on Computer-Human Interaction ,* 461-476.

Sauer, C. M., Hastings, W. A., & Okamura, A. M. (2004). Virtual environment for exploring atomic bonding. *Proceeding of Eurohaptics,* (pp. 232-239).

Sulzer, J., Salamat, A., Chib, V., & Colgate, J. (2007). A Behavioral Adaptation Approach to Identifying Visual Dependence of Haptic Perception. *Symposium on Haptic Interfaces for Virtual Environment and Teleoperator Systems, World Haptics 2007, Second Joint,* (pp. 3-8).

Tendick, F., Downes, M., Goktekin, T., Cavusoglu, M., Feygin, D., Wu, X., et al. (2000). A virtual environment testbed for training laparoscopic surgical skills. *Presence , 9* (3), 236-255.

Yamaguchi, T., Nam, C. S., Johnson, S., and Smith-Jackson, T. L., (2010) Co-Touch: A Component-Based Framework for Creating Haptic Applications in Flash Media. *IEEE Multimedia* (under review).

Yu, W., & Brewster, S. (2003). Evaluation of multimodal graphs for blind people. *Universal Access in the Information Society , 2* (2), 105-124.

Chapter 70

Comparison of Hand Grip Strength Between Wheeled Mobility Device Users and Non-Disabled Adults

Caroline Joseph[1], Clive D'Souza[1], Victor Paquet[1], David Feathers[2]

[1] Department of Industrial and Systems Engineering
Center for Inclusive Design & Environmental Access
University at Buffalo
Buffalo, NY 14260, USA

[2] Department of Design and Environmental Analysis
Cornell University
Ithaca, NY 14853, USA

ABSTRACT

Human factors engineers often use anthropometric data to design and develop specifications for products and features of the environment. Often this data is limited to studies of adults who are reported as healthy and those with functional limitations may not be accommodated. The purpose of this paper is to compare the peak isometric hand grip strength measured on the dominant hand of wheeled mobility device users to the peak isometric hand grip strength of the reported for the U.S. adult population. Three hundred and twenty six individuals out of the 369 users of manual wheelchairs, power wheelchairs and scooter recruited were able to demonstrate a successful power grip that could be recorded in a posture in agreement with the recommendations of the American Society of Hand Therapists.

Means, minimum and maximum values across gender (males and females), age groups (20-24, 25-29, 30-34, 35-39, 40-44, 45-49, 50-54, 55-59, 60-64, 65-69, 70-74 and 75+ years of age) and type of wheeled mobility device (manual wheelchairs, power chairs, and electric scooters) are provided. Also presented are graphical comparisons showing the mean and 95% confidence intervals between grip strength measures from the current study and equivalent measures from the adult U.S. population without disabilities available in the literature. Analysis reveal important differences in power grip strength between healthy ambulatory adults vs. users of mobility devices, with the mean hand grip of wheeled mobility device users often being lower than the 95% confidence interval of the healthy ambulatory adult population. These differences suggest the need for careful consideration of potential end user groups that vary in physical and functional abilities when designing products and environments. In particular, for products and environments intended for wheeled mobility device users, the hand grip force requirement should be much lower than would be expected to accommodate the vast majority of healthy ambulatory adults. Also, the inability of many wheeled mobility device users to form a functional power grip suggests that tasks should be designed to afford multiple/alternate grip configurations, or incorporate technologies that would eliminate the need for a power grip. By doing so, products and environments would likely be easier to use and more accessible to the entire population.

Keywords: Hand Power Grip Strength, Wheeled Mobility Device Users

INTRODUCTION

Human factors engineers and product designers often use anthropometric data to develop technical specifications and design criteria for products and features in the environment. Hand grip strength measurement is important to the design of different hand-operated products and environmental features. A majority of the literature on hand anthropometry of adults considers only a population that is considered healthy and able (Bohannon, Peolsson, Massy-Westropp, Desrosiers, & Bear-Lehman, 2006; Crosby, Wehbé, & Mawr, 1994; Mathiowetz et al., 1985), with only a smaller proportion of studies considering an elderly population (Bohannon, Bear-Lehman, Desrosiers, Massy-Westropp, & Mathiowetz, 2007; Desrosiers, Bravo, Hebert, & Mercier, 1995). If human factor engineers and products designers base their designs solely on these data, it is possible that a large proportion of adults with functional limitations would be unable to use such products. For instance, hand grip strength data on wheeled mobility device users is still hard to find. With the population of wheeled mobility device users in the U.S. estimated at approximately 1.7 million individuals and expecting to increase dramatically (Kaye, Kang, & LaPlante, 2000), it is important to ensure that the design needs of mobility device users are being met. A better understanding of the differences in hand grip strength across different user groups is also important in order to emphasize any differences in functional abilities that may exist, and towards developing more inclusive design criteria for designing accessible products and environments.

The purpose of this paper is to compare the peak isometric hand power grip strength measured on the dominant hand for a diverse sample of wheeled mobility device users with normative hand grip strength data for the U.S. adult population.

METHODOLOGY

PARTICIPANTS

Participants were recruited from the local community and with the help of different organizations in the Buffalo, NY and Pittsburgh, PA areas. A total of 369 individuals participated in this study, but only a subset 326 individuals of this group was able to form a functional hand grip with recordable strength due their inability to hold the prescribed posture with their arm and/or hand, to lack of dexterity, lack of hand strength, pain while performing the hand grip, spasticity, etc. Of the individuals with recordable strength, 146 were females (77 manual wheelchair users, 56 power wheelchair users, and 13 scooter users) and 180 were males (101 manual wheelchair users, 64 power wheelchair users, and 15 scooter users). The overall sample was not representative of the U.S. population of wheeled mobility device users (Jones & Sanford, 1996) in terms of age, device type and disability, but allowed grip strength analysis for some of the more prevalent subgroups of users within our sample (e.g., power wheeled mobility device users).

PROCEDURE

Hand power grip was only recorded for the participant's dominant hand. Participants were asked to perform their maximal grip strength with their dominant arm by their side and with their elbow at 90 degrees as recommended by the American Society of Hand Therapists (Fess, 1992). Participants were also instructed not to brace themselves on the arm rest of their wheeled mobility device when performing the hand grip. Data from three trials were measured with sufficient rest breaks between trials. However, some individuals could only complete one or two trials due to discomfort or were unable to exert any recordable strength in the subsequent trials. When two or three trials were completed, the mean of the trials was taken to represent the maximum grip strength of the individuals (Mathiowetz, Weber, Volland, & Kashman, 1984), and when only one trial was done, the value for the unique trial was taken as the maximum grip strength of the individual.

ANALYSIS

All the hand grip strength data collected duri9ng this study was stratified by age (20-24, 25-29, 30-34, 35-39, 40-44, 45-49, 50-54, 55-59, 60-64, 65-69, 70-74 and 75+ years of age), gender (males and females), and wheeled mobility type (manual wheelchairs, power chairs, and electric scooters). The means, standard deviation and the range were calculated for each subgroup. Graphical comparisons between the current sample of wheeled mobility device users and the adult U.S. population without disabilities was done by plotting the means and 95% confidence intervals for grip strengths obtained from the current study sample versus the normative power grip strength data for the adult U.S. population reported by Bohannon, et al. (2006).

RESULTS

Dominant hand grip strength for wheeled mobility device users stratified by age, gender and mobility device type are presented in Table 1 (see Appendix). Mean grip strength for females was always lower than mean grip strength for males across all age groups. The variation of mean grip strength in females (range from 14 kg (32 lbs) to 22 kg (49 lbs)) was lower than in males (range from 23 kg (50 lbs) to 41 kg (91 lbs)). Differences in mean grip strength were also observed between wheeled mobility device types, with manual wheelchair users having greater mean power grip strength both than power wheelchair users and scooter users.

Graphical comparisons between the current sample of wheeled mobility device users and the adult U.S. population without disabilities are presented in Figures 1 and 2, for females and for males, respectively. For both females and males, the mean grip strengths for the adult U.S. population without disabilities was greater than the current study sample of wheeled mobility device users. Also in most cases, mean grip strength for wheeled mobility device users was less than the lower bound of the 95% confidence interval grip strength for the adult U.S. population without disabilities. Greater variability in power grip strength was also found in the current sample of wheeled mobility device users as compared to the values reported by Bohannon et al. (2006).

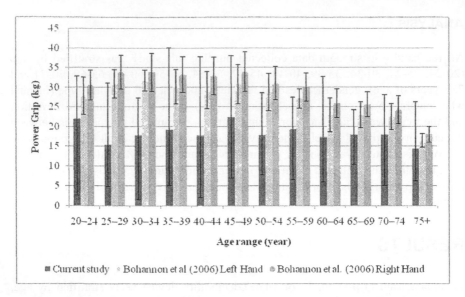

FIGURE 1 Comparison of hand power grip of dominant hand of female wheeled mobility devices users and the power grip of a female non-disabled population (Bohannon, et al., 2006).

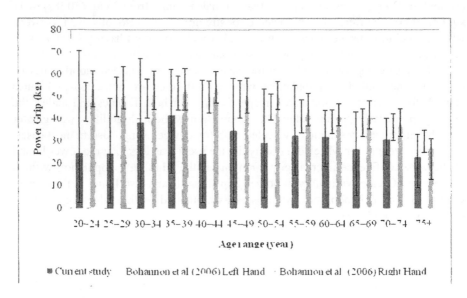

FIGURE 2 Comparison of hand power grip of dominant hand of male wheeled mobility devices users and the power grip of a male non-disabled population (Bohannon, et al., 2006).

DISCUSSION

Graphical comparisons revealed important differences in hand power grip strength between healthy ambulatory adults versus users of mobility devices. For most of the age groups, the mean hand grip of wheeled mobility device users was lower than the 95% confidence interval of the healthy ambulatory adult population. Grip strength also varied substantially across users of manual wheelchairs, power wheelchairs and scooters. This is particularly true for users of power wheelchairs that often times have more severe medical conditions and impairments in the upper extremities.

However, care should be taken when applying the overall hand grip strength data to design, as the current study sample was not representative of the U.S. wheeled mobility device user population in terms of proportion of representation by gender, wheeled mobility type and age. Power wheelchair users were intentionally over-sampled to get a better understanding of their physical and functional abilities in the built environment. Also, scooter users are not adequately represented in the study sample to-date as the proportion of scooter users in the wheeled mobility device population is lower in comparison to manual and power wheelchair users. This lower level of representation of scooter users was then presenting a challenge during participant recruitment.

Data collection for this study is still on-going and efforts are being devoted to making the sample more representative of the U.S. wheeled mobility device user population. In this regard, the current analysis provides insights into the sub-groups within the study sample that need additional representation, e.g. scooter users, and requiring special attention during participant recruitment in the future. Enhancing the geographical diversity of the sample is also an important consideration for improving the generalization of the data. Plans for data collection in other U.S. cities, particularly regions with more tropical climates, are also underway.

CONCLUSION

Findings from the current study suggest the need for careful consideration of potential end user groups when designing products and environments. In particular, for products and environments intended for wheeled mobility device users, the hand grip force requirements should be much lower than the values expected of a healthy ambulatory population. The use of lower grip strength requirements in design would also be more inclusive of a healthy ambulatory population. Also, the inability of many wheeled mobility device users to form a functional power grip suggests that tasks should be designed to afford multiple/alternate grip configurations, or incorporate technologies that would eliminate the need for a power grip. By doing so, products and environments would likely be easier to use and more accessible to a larger user group.

ACKNOWLEDGEMENTS

This research was supported with funding from the National Institute on Disability and Rehabilitation Research (NIDRR), U.S. Department of Education through the Rehabilitation Engineering Research Center on Universal Design at Buffalo (Grant #H133E990005) the U.S. Access Board (Contract # TPD-02-C-0033). The opinions expressed in this paper are those of the authors and do not represent the policy of the National Institute on Disability and Rehabilitation Research, Department of Education, or the U.S. Access Board.

REFERENCES

Bohannon, R. W., Bear-Lehman, J., Desrosiers, J., Massy-Westropp, N., & Mathiowetz, V. (2007). Average grip strength: A meta-Analysis of data obtained with a Jamar dynamometer from individuals 75 years or more of age. *Journal of Geriatric Physical Therapy, 30*(1), 28-30.

Bohannon, R. W., Peolsson, A., Massy-Westropp, N., Desrosiers, J., & Bear-Lehman, J. (2006). Reference values for adult grip strength measured with a Jamar dynamometer: a descriptive meta-analysis. *Physiotherapy, 92*(1), 11-15.

Crosby, C. A., Wehbé, M. A., & Mawr, B. (1994). Hand strength: Normative values. *Journal of Hand Surgery, 19A*, 665-670.

Desrosiers, J., Bravo, G., Hebert, R., & Mercier, L. (1995). Normative data for grip strength of elderly men and women. *American Journal of Occupational Therapy, 49*, 637-644.

Fess, E. E. (1992). Grip Strength *Clinical assessment recommendations* (2nd ed., pp. 41-45). Chicago, IL: American Society of Hand Therapists.

Jones, M. L., & Sanford, J. A. (1996). People with mobility impairments in the United States today and in 2010. *Assistive technology, 8*(1), 43-53.

Kaye, H. S., Kang, T., & LaPlante, M. P. (2000). Mobility Device Use in the United States. Washington, D.C.: U.S. Department of Education, National Institute on Disability and Rehabilitation Research.

Mathiowetz, V., Kashman, N., Volland, G., Weber, K., Dowe, M., & Rogers, S. (1985). Grip and pinch strength: Normative data for adults. *Archives of Physical Medicine and Rehabilitation, 66*, 69-72.

Mathiowetz, V., Weber, K., Volland, G., & Kashman, N. (1984). Reliability and validity of grip and pinch strength evaluations. *Journal of Hand Surgery, 9A*(2), 222-226.

Age range (year)	Overall		Manual			Power			Scooter		
	N	Mean (SD)	N	Mean (SD)	Range (Min-Max)	N	Mean (SD)	Range (Min-Max)	N	Mean (SD)	Range (Min-Max)
Females											
20-24	4	22 (13)	1	24		2	15 (18)	3-28	1	33	
25-29	5	15 (10)	2	18 (19)	5-31	3	14 (2)	12-15	0		
30-34	5	18 (23)	3	21 (6)	15-27	2	13 (17)	2-25	0		
35-39	8	19 (10)	4	24 (11)	17-40	4	14 (7)	5-21	0		
40-44	17	18 (11)	6	27 (9)	16-39	11	13 (9)	1-34	0		
45-49	17	22 (8)	11	24 (7)	16-42	6	18 (9)	4-27	0		
50-54	13	18 (7)	7	16 (8)	8-29	5	18 (5)	11-22	1	21	
55-59	18	19 (6)	10	22 (4)	17-28	4	16 (8)	6-25	4	18 (6)	7-20
60-64	14	17 (8)	6	16 (5)	8-23	5	21 (12)	6-34	3	14 (7)	6-18
65-69	15	18 (4)	6	18 (3)	14-24	7	19 (4)	16-24	2	13 (6)	9-17
70-74	9	18 (6)	6	19 (5)	14-28	2	14 (12)	5-20	1	22	
75+	21	14 (6)	15	13 (5)	5-25	5	18 (7)	9-28	1	20	
Males											
20-24	11	24 (19)	5	36 (21)	20-73	6	14 (9)	2-25	0		
25-29	10	24 (16)	4	40 (8)	29-49	6	14 (10)	2-29	0		
30-34	11	38 (17)	10	40 (17)	6-68	0			1	23	
35-39	12	41 (13)	9	42 (15)	14-64	2	40 (12)	32-49	1	40	
40-44	14	24 (17)	4	34 (18)	18-59	9	19 (17)	2-51	1	31	
45-49	25	35 (17)	14	37 (19)	1-67	8	35 (17)	5-52	3	22 (5)	17-27
50-54	25	29 (14)	14	33 (13)	15-55	10	24 (15)	0-44	1	31	
55-59	23	33 (13)	15	36 (11)	15-57	5	25 (17)	15-54	3	31 (7)	23-36
60-64	9	32 (8)	5	32 (5)	27-39	4	32 (11)	19-44	0		
65-69	7	26 (14)	3	39 (4)	36-43	3	15 (12)	5-28	1	24	
70-74	7	31 (7)	4	28 (3)	24-30	2	40 (1)	39-40	1	24	
75+	26	23 (8)	14	24 (8)	9-34	9	23 (8)	9-33	3	16 (3)	13-19

Table 1 Dominant hand grip strength for females and males wheeled mobility device users in kg (to convert in lbs multiply by 2.2)

Chapter 71

Enhancing the Wheelchair Transfer Process to Elements in the Built Environment

Alicia M. Koontz, Maria L. Toro, Padmaja Kankipati, Rory A. Cooper

Human Engineering Research Laboratories
Department of Veterans Affairs
Pittsburgh, PA 15206
University of Pittsburgh
Pittsburgh, PA 15260, USA

ABSTRACT

In 2002, the United States Access Board published specific guidelines concerning transfers to amusement park rides however they were not based on evidence and only provide very limited specifications concerning transfers (ADAAG, Section 15). The purpose of our study is two-fold 1) gather expert opinion on the relevance and strength of the scientific literature to date related to the performance of independent transfers among individuals who have lower limb dysfunction and 2) determine the impact of various environmental factors on transfer performance. Data generated in this study will be useful for developing new assistive technologies and making changes in the environment for the purposes of enhancing the transfer process.

Keywords: Environmental Accessibility, Wheelchair, Mobility, Disability, Accessibility Guidelines

INTRODUCTION

For individuals who rely on wheeled mobility devices, performing transfers is essential to achieving independence with activities of daily living (ADL) inside the home and for participating in activities in the community. For example, transfers are required for getting to and from the mobility device to bed, bath tub/shower seat, commode seat, motor vehicle seat and so on. Individuals who perform routine transfers independently (e.g. no human or technological assistance) are predisposed to developing overuse related injuries due to large weight-bearing loads borne by the upper limbs (Boninger et al., 2005). In addition, the built environment can pose difficulties to individuals who would like to be independent with transfers but require assistance because the surface they need to transfer to is not easily assessible (e.g., too high, too far away).

The United States Access Board develops and maintains design criteria for the built environment to maximize accessibility to public places. The Access Board produced the Americans with Disabilities Act Accessibility Guidelines (ADAAG) which includes general recommendations on transfer heights and clear space for a limited number of elements where transfer is expected (Department of Justice, 1994). In 2002, the Board published specific guidelines concerning transfers to amusement park rides however they were not based on evidence and only provide very limited specifications concerning transfers (ADAAG, Section 15) (US Access Board, 2002). Little research is available on the issue of transferring from one's mobility device to another element. We have identified several aspects of a transfer where data are needed. These include the following:

1) use and placement of handhelds or grab bars for facilitating transfers;
2) the ranges of heights that individuals can realistically transfer up and down to;
3) how close the transfer surfaces need to be to each other;
4) transfers where space for positioning one's mobility device is limited;
5) transfers into confined spaces;
6) how obstacles in between the device and target surface affect the transfer process.

This kind of information is essential to refining the guidelines related to transfers and enabling designers and engineers to create an environment that is more accessible to individuals who independently transfer.

A two-phase study is currently underway to collect data that can be used to revise guidelines concerning transfers in built environment. In Phase I, the purpose of the research is to gather expert opinion on the relevance and strength of the literature to date related to the performance of independent transfers among individuals who have lower limb dysfunction (e.g. due to spinal cord injury (SCI), amputation or other disability). Phase II of the study will determine the impact of setup (e.g. environmental factors) on transfer performance and provide data that the Board can use to modify equipment and/or environment for the purposes of enhancing the transfer process.

EXPERT LITERATURE REVIEW (PHASE I)

A systematic review of the literature was conducted using techniques consistent with those reported in the literature (Levy and Ellis, 2006). Four scientific and medical databases were searched using keywords: wheelchair + activities of daily living; biomechanics; efficiency; electromyographic; force; force plate; function; functional electrical stimulation; gait; isokinetic; kinematics; kinetics; measurement system; moment; motion analysis; movement; muscle balance; muscular demand; orthosis; paralysis; paraplegia; rehabilitation; scapula; shoulder; SCI; stroke; SCI patient; shoulder impingement; standing up; task performance and analysis; technology; tetraplegia; torque; torque ratio; transfer; transfer motion;   transfer strategy; transfer movement strategies; upper extremity; upper limb; weight-bearing; weight bearing; and three dimensional kinematics.

After excluding duplicates 339 articles were initially identified and internally reviewed by three experts.   Forty-one peer-reviewed full journal articles were determined to be related to the performance of independent wheelchair transfers. These articles were sent to 12 external reviewers with clinical experience and/or expertise on transfers. They completed a scoring sheet (one for each article) which was comprised of questions related to the relevance and strength of the evidence. Details concerning how the articles were scored are under review elsewhere (Toro et al., *in review*). Twenty-six of these articles met the minimum cut-off score for 'relevance' (primary criteria for inclusion) and a minimum cut-off score for 'strength of the evidence' (secondary criteria for inclusion). Results showed that there were critical aspects of the transfer process that were not addressed in the literature including space available to place and maneuver the mobility device, availability of supports (i.e. grab bars), number of the transfers to go from the initial location to the final destination, use of a transfer assistive device, constrained space available for transfers, and physical obstacles or barriers present while transferring.

THE IMPACT OF TRANSFER SETUP ON THE PERFORMANCE OF INDEPENDENT TRANSFERS (PHASE II)

In order to improve the transfer process for individuals with disabilities, the standards concerning transfers to elements in the built environment must be updated and expanded. Data on transfers from a broad spectrum of community-dwelling mobility device users is necessary to inform engineers, architects, and designers who design public and private spaces about how to modify the environment to enable the highest degree of independence. A custom-built modular, transfer station was designed and fabricated in house to systematically investigate the impact of various setup parameters on transfer performance. The station is designed to mimic potential transfer scenerios that may be experienced in the real world and consists of a height adjustable platform with a detachable backrest and allows for attaching/detaching side guards and grab bars of varying heights (see Figure 1).

The design criteria for the transfer station were:

- Platform adjustable in vertical height from 10 to 28 in with respect to the floor
- Adjustable in horizontal gap (e.g. distance between the mobility device and the platform)
- Adjustable in the xy plane (e.g. 'clear space' where the mobility device is positioned with respect to the platform)
- Detachable physical obstacle/barrier to transfer
- Constraint for foot placement
- Flexibility in the side of approach (e.g. direction of transfer is unrestricted)
- Detachable backrest
- Detachable frontal grab bar (height variable) and side grab bar of a height consistent with current accessibility guidelines
- Portable
- Dimensions: Seat: 18x18 in, Backrest: 18x18 in, Overall length: 57 in

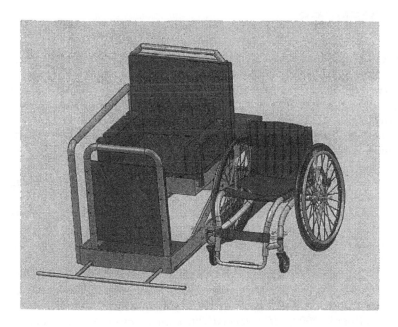

Figure 1 SolidWorks rendering of the transfer station.

We designed a test protocol using the station to evaluate the influence of setup on transfer performance. For the Initial Setup of the transfer station, the platform will be adjusted to be level with the subject's seat, with no side guard in place, and no frontal grab bar in place. The subjects will be asked to position themselves next to the platform as they normally would to prepare for a transfer. Wheeled device position (e.g. angular orientation and linear distances from the target) will be recorded. The subject will be asked to transfer from their own wheelchair to the platform. Hand placement, method of transfer, and use of a transfer board will be noted. Their perceived level of exertion will be recorded using the OMNI Scale (Robertson et al., 2003). Next, they will be asked to perform several test protocols in random order. Each protocol tests a specific environmental factor such as change in vertical height. After each transfer in each protocol, their perceived level of exertion, changes made to device positioning (if any), use of a board, or multiple transfer (e.g. transfer occurs in smaller steps versus one fluid motion) will be recorded. The main outcome measures will be the maximum/minimum attainable vertical and/or horizontal gap distances with/without presence of a side guard barrier and/or frontal grab bar. Descriptive statistics (e.g. means, medians, frequencies, standard deviations) of the data will be used to examine results of the first 50 community dwelling mobility device users tested. Additional subjects will be recruited as necessary until results reach a point of saturation. Data collection for this phase of the project is slated to begin in April 2010.

REFERENCES

Boninger M.L., Waters R.L., Chase T., Dijkers M.P.J.M., Gellman H., Gironda R.J., Goldstein B., Johnson-Taylor S., Koontz A., and McDowell S. (2005), *Preservation of upper limb function following spinal cord injury: a clinical practice guideline for healthcare professionals.* Consortium for Spinal Cord Medicine.

Department of Justice . (1994), *ADA Standards for Accessible Design.* Code of Federal Regulations. 490-580.

Levy Y. and Ellis T.J. (2006), "A systems approach to conduct an effective literature review in support of information systems research." *Informing Science Journal,* 9, 181-212.

Robertson,R.J., Goss,F.L., Rutkowski,J., Lenz,B., Dixon,C., Timmer,J., Frazee,K., Dube,J., and Andreacci,J. (2003), "Concurrent validation of the OMNI perceived exertion scale for resistance exercise." *Medicine and Science in Sports and Exercise,* 35, 333-341.

Toro M.L., Koontz A.M., Kankipati P., Naber M., and Cooper R.A. (in review) "Independent wheelchair transfer: a systematic literature review." *Proceedings of the Annual RESNA Conference, Las Vegas, NV, June 26-30, 2010.*

US Access Board. (2002), *ADA Accessibility Guidelines for Buildings and Facilities (ADAAG),* http://www.access-board.gov/adaag.

ACKNOWLEDGEMENTS

Funding for this study was provided by the Department of Education (NIDRR) H133E070024 and United States Access Board Project #84.133E. This material is the result of work supported with resources and the use of facilities at the Human Engineering Research Laboratories, VA Pittsburgh Healthcare System. The contents of this paper do not represent the views of the Department of Veterans Affairs or the United States Government.

Chapter 72

Clear Floor Area for Wheeled Mobility Users

Edward Steinfeld, Clive D'Souza
Victor Paquet, Jonathan White

University at Buffalo
State University of New York

ABSTRACT

Many provisions in accessibility standards are based on assumptions about the sizes of wheeled mobility devices. Thus, the "clear floor space" (CFS) of an occupied wheelchair is a basic "building block" of standards. The CFS is used to establish an "accessible" approach area to fixtures, equipment and controls, like water fountains, paper dispensers, electrical controls, and ATMs. It is also used to establish the space needed in seating areas reserved for people who use wheeled mobility devices, both in buildings and in transportation vehicles. Since 1980, several research studies on the anthropometry of wheeled mobility devices and their users in Canada, the UK and Australia, have demonstrated that many wheeled mobility devices and their users are larger than those on which accessibility standards and codes are based. In this article we describe a study in the U.S. that verified these findings and provides new information that can be used to improve standards and codes. We also discuss the importance of how research information is communicated to designers and standards developers and present a design resource that can be used both by both groups to address the needs of contemporary wheeled mobility users from a universal design perspective.

Keywords: accessibility, architecture, transportation, disability

INTRODUCTION

Many provisions in accessibility standards are based on assumptions about the sizes of wheeled mobility devices. The "clear floor space" (CFS) of an occupied wheelchair is a basic "building block" of standards. In particular, the CFS is used to establish an "accessible" approach area to fixtures, equipment and controls, like water fountains, paper dispensers, electrical controls, ATMs and service counters. It is also used to establish the space needed in seating areas reserved for people who use wheeled mobility devices, for example, the space needed to provide a reserved area for wheelchair and scooter users on vehicles, the size of viewing areas for wheeled mobility users in stadia and other public assembly areas and clearance needed for positioning a wheelchair at work stations .

The research on wheeled mobility users that created the knowledge base for the technical requirements of the accessibility standards used in the U.S., including ICC/ANSI A117.1 (2003) Accessible and Usable Buildings and Facilities (ICC/ANSI), the Fair Housing Accessibility Guidelines, (FHAG) and the ADA Accessibility Guidelines (ADAAG) was completed during 1974 -1978 using a research sample that included about 60 individuals who used wheelchairs (see Steinfeld et al., 1979). In the ensuing years, many changes have occurred in the body sizes of the U.S. population, the demographics of people who use wheeled mobility devices and the characteristics of equipment that they use. Yet, the standards have not changed. In fact, until recently, a newer anthropometric data set on wheeled mobility users in the U.S. was not available. Yet, research on the anthropometry of wheeled mobility devices and their users conducted since 1980 in Canada, the UK and Australia, demonstrates that many contemporary wheeled mobility devices and their users are larger than those on which accessibility standards and codes are based. These studies, however, have many differences in methodology as well as sample recruitment and size; furthermore, there may be differences in the types of devices available in different countries as well as the policies that third party payers use to purchase devices which could make those findings less than ideal for application in the U.S. (Steinfeld et al, 2010).

In response to the lack of current data on U.S. populations, the Center for Inclusive Design and Environmental Access (IDeA Center) has been developing a comprehensive data set with a high level of accuracy and methodological rigor (Steinfeld et al, 2004; Feathers et al, 2004; Paquet and Feathers, 2004, Steinfeld et al., 2010). Although data collection is ongoing, we have now achieved a sample size and breadth that we believe is sufficient to start a dialogue about the needs for revision to current U.S. standards and to provide guidance to designers who wish to exceed the current standards and accommodate a wider range of people. This paper provides a brief overview of our research activities, reports the findings on CFS and provides some resources that can be used to revise standards as well as design guidance. In the process of completing the research, we also identified differences between research practices and standards development that impedes the development of evidence based design practice in this field. The need to reconcile research methods with design standards is discussed. Recommendations and examples are provided as to how this might be accomplished.

METHODS

At the time this paper was written, data from 369 participants with a wide range of chronic conditions had been analyzed. The participants were recruited through outreach efforts with several organizations in Western New York and mass media. Fifty-eight percent of the sample was male and 42% female. The mean age of the sample was 52.4 with a range of 18-94 years of age. Fifty-three percent used manual wheelchairs, 39% used power wheelchairs and 8% used scooters.

We collected structural anthropometric measurements of people and devices using an electromechanical probe (Paquet and Feathers, 2004). Three-dimensional coordinates of the landmarks were then used to derive estimates of widths, heights and depths of key mobility device characteristics and body dimensions. This method of obtaining anthropometric dimensions provides reliable measurements but with differences from those obtained with more traditional anthropometric measuring devices (Feathers et al, 2004). Two types of landmarks were used in the data collection: body landmarks based on anatomical reference points and device landmarks corresponding to defined planes or points on the wheeled mobility device or accessory object.

The digital data collection and recording method allowed us to project dimensions of the body and device onto a virtual "floor" plane. In this way, we were able to describe the occupied length and width that accommodated each individual. Occupied dimensions refer to the dimensions of the person and device together. We believe that these are the data that are most appropriate for computing CFS. We developed definitions of the body and device landmarks used to compute these measurements. Our definitions clarify the concept of occupied width and length for design purposes. We discovered that establishing standardized landmarks required precise definitions that could address the variation across different types of devices. Thus landmarks for occupied devices could be the outside edge of the device housing or armrests, the participants' elbows, arms or legs, extending beyond the outside edges of the device housing. Unlike previous studies, we included accessory devices mounted to the device such as backpacks and baskets. We also measured people in their "normal resting posture" rather than a standardized position. This provides a more realistic perspective on space needs, especially for situations when an individual may remain in position for an extended period of time, e.g. in a public assembly space or transit vehicle.

Percentile data was computed for occupied width and length. A composite measure of occupied floor area for each participant, defined as the product of the occupied length and occupied width, was also computed and used to represent the estimated rectangular floor area required for each occupant-mobility device in the study sample. The results were illustrated in diagrams that can be used to make improvements in standards and also to practice universal design with respect to wheeled mobility needs (Steinfeld et al, 2010). Universal design seeks to accommodate the broadest range of abilities and characteristics possible, something that accessibility standards at present do not generally address.

We collected data to study many other design issues as well, including knee and toe clearance, reach envelopes, grip strength and size, and maneuvering clearances, although the other results are not reported here. For more details on the methodology used and findings, see Steinfeld et al. (2010), Paquet and Feathers (2004) and D'Souza et al. (2009).

RESULTS

The results confirmed findings from research in other countries that current dimensions for "clear floor space" prescribed in U.S. accessibility standards, and adopted by other countries as well, are inadequate for accommodating many users of wheeled mobility devices, especially those that use power chairs and scooters.

We also found inconsistencies in the way CFS is defined that can lead to confusion about what standards are supposed to achieve. In the U.S. standards, illustrations of CFS show the width dimension the CFS based on the width of an unoccupied device. On the other hand, they show length based on an occupied device. Only a manual wheelchair is shown in the standards yet they are used to address the needs of power wheelchair and scooter users as well. And, no accessories like baskets, backpacks, tip protectors, etc. are included in the dimensions. Thus, not only do the values used in the standards need to be revised to accommodate contemporary needs but the dimensions (variables) used in the standards need to be more carefully defined to ensure that research findings can be applied effectively. Moreover, the standards need to identify what types of devices must be accommodated for various uses.

Another important gap between research findings and standards is the relationship of a reaching target in relationship to the location of the CFS. Functional reach should be measured along an axis with the shoulder point, since this reflects the actual distance of reach. Thus, in analyzing our research results we measured reaching ability with the target positioned along an axis through the shoulder both for lateral and forward reach D'Souza et al, 2009). The methods we used allowed us to compute the location of each individual's shoulder in relationship to the boundaries of the occupied wheeled mobility device and aggregate that data for sub-samples and the total sample by aligning all shoulder points in virtual space.

Two other factors that come into play in the use of CFS are handedness and the orientation of the CFS with respect to the target (e.g. lateral or forward approach). The most functional reach is achieved when an individual uses the hand that is preferred for reaching. Our findings showed that about 26% of our sample preferred to use their left hand for reaching tasks, compared to a prevalence of 10% maximum in the general population (D'Souza, et al, 2010); the difference between our sample and the general population is probably due to impairments that limited use of right hands, even if an individual was naturally right handed. Our findings on reaching ability demonstrated that lateral reach was much more effective than forward reach

(D'Souza, et al, 2009). In fact, the findings show that only about 62% of our sample could reach a target on a plane across the anterior most point of device or toes as shown in the current standards. Thus, the ideal approach to CFS would include both the space needed for a forward and lateral approach. For a forward approach our data demonstrates that a knee and toe clearance would increase the accommodation level for forward reach significantly for any target height (D'Souza, et al, 2009).

DISCUSSION

Current standards do not take the axis of reach, handedness or approach direction into consideration. The provision of space for either a forward or lateral approach is generally allowed and some standards require the CFS to be centered on the target (U.S Department of Housing and Urban Development, 1991; ICC/ANSI, 2003). Centering essentially ignores performance related to handedness and reduces performance for all wheelchair users. Our findings suggest a very different conceptual model of CFS, one that would provide enough space for both a right-handed and a left-handed approach and that would line up the target on axis with the shoulder point of the user rather than the center of the CFS. It would also require a lateral orientation of the CFS to the target wherever possible and knee and toe clearance for any situation that required a forward reach.

The U.S. design standards for accessibility are referenced by building codes and other regulations. This means that they are intended for use in a legal framework. In the United States, the inclusion of additional information that could help designers achieve a higher level of accessibility than provided by minimum regulatory requirements is rare. Such information is usually provided in separate guidance documents to avoid confusion between the legal requirements and best practices (for example, see Levine, 2003). However, designers often use the standards as their only reference and do not know that they could, often with little impact on the cost of construction or technical difficulty, achieve a higher level of accommodation. Moreover, standards developers are often unaware of the research available. Thus, it often takes a long time before the results of research find their way into standards, particularly if they are part of building codes, which are notoriously difficult to change. Thus, there is often a lag between the completion of research and incorporation of findings into standards and codes.

To address the knowledge utilization gaps above, we propose a new approach for adoption in both standards and best practice documents, an "accommodation model." Using graphics that are similar to those that are currently incorporated in standards and codes, the full range of human characteristics and abilities can be conveyed and the level of accommodation required by standards and codes can then be specified within the model. In this way, designers can have ready reference to best practice information and they will understand that minimum compliance does not necessarily mean that the entire population is accommodated. Standards developers also will see the implications of their decisions and thus be more informed. The decision to exclude various groups of people will be transparent and thus lead to a more democratic approach to decision making. Figure 1 (see Appendix) is an example of an accommodation model for CFS based on our research findings.

CONCLUSION

The research described here has many implications for design and standards development. Not only do the findings suggest that revisions to standards are necessary but also that new methods be used to communicate findings and apply them in design. While we have used CFS to illustrate a new approach, it is only one example drawn from our research. We have also developed similar models for reach (D'Souza et al, 2009), knee and toe clearance, and maneuvering clearances.

The definition of clear floor area in the existing U.S. accessibility standards (ICC/ANSI, 2003) prescribes a minimum rectangular floor space of 760mm x 1220mm. Fig. 1 suggests that this space should be increased in width and length to accommodate 95% of our sample of manual chair users (dimensions A and B in Fig. 1). Increases of 117 mm (5 in.), 178 mm (7 in.), and 217 mm (8 in.) would be needed to accommodate 95% of the manual chair users, power chair users and scooter users, respectively, in our sample. The CFS should also be wider (forward approach) or longer (lateral approach) than the actual space needed for the occupied wheeled mobility device, although not necessarily as wide as the 2D dimension shown in Fig. 1. This would provide more space to allow both right-handed and left handed people to position their preferred hand closer to their target. The target should be specified for each application, e.g. the centerline of a sink, the handle or push plate of a drinking fountain, the control panel of a range, etc. The clear floor space should align with the target to be used by the hand rather than a fixture or appliance. Finally, a lateral approach to the target should be required unless a forward approach is needed to use a device, e.g. a computer workstation. In that case, a knee and toe clearance should be required as well. The knee space would allow individuals to slide in close to the target improving functional reach greatly.

Taking a universal design perspective, however, more flexibility should be provided. For example, the CFS could be increased even more, to accommodate 95% of the sample in both left and right handed reach and either a lateral or forward approach by using the full T-shaped space in Fig. 1. This clearly presents a challenge where space is limited, however. There are levels of accommodation in between minimum standards and the greatest degree of accommodation that could include reducing the percentage of people accommodated and/or accommodating only one approach instead of both. The decision on how many people should be accommodated should be made with participation of all stakeholders. By providing the research information as in the accommodation model in Fig. 1, all stakeholders involved are aware of the implications. The limitations of any approach can then be understood and other strategies employed to improve access such as redesign of equipment, automation and remote controls.

Using a transparent approach could require some difficult policy decisions to avoid a significant impact in cost. Social science research can help us understand patterns of utilization of wheeled mobility devices and would provide useful information for such decisions. For example, knowing more about the types of environments in which scooters are used would be helpful. If scooter users, for example, tend not to use their device at home, it may not make sense to base minimum standards for CFS within dwelling units on that group except for providing a space to park and recharge the device.

To address concerns about affordability and technical issues, developers of minimum accessibility standards may choose not to accommodate all mobility device users, but they should make decisions on CFS issues based on the latest knowledge available. If standards do not accommodate the entire population of wheeled mobility users, information should be available to inform designers of those limitations and provide design guidance for addressing them when desired.

It is important that the building and transportation industries, citizens with disabilities, design professionals, mobility device manufacturers, vendors and prescribers understand the limitations of current standards and become involved in the dialogue about how to address the need for improving them. Further analyses will investigate other issues such as unoccupied width of devices at different measurement planes above the floor, which is important for establishing the width of ramps and pathways, and compare the data from structural measurements with actual performance in activities such as passing through doors and entering transportation vehicles. The accommodation model will also be used to development universal design standards that provide guidance in exceeding minimum standards to achieve a higher level of inclusion.

REFERENCES

D'Souza, C., Steinfeld, E. and Paquet, V. (2009), "Functional reach abilities of wheeled mobility device users: toward inclusive design." *Proceedings of the 2009 International conference on Inclusive Design, INCLUDE 2009*, ISBN 978-1-905000-80-7, Royal College of Art, London, UK, April 2009.

Feathers, D., Paquet, V., and Drury C. (2004), "Measurement consistency and three-dimensional electromechanical anthropometry." *International Journal of Industrial Ergonomics, 33*(3): 181-190.

ICC/ANSI. (2003). ICC/ANSI A117.1-2003 Standard on Accessible and Usable Buildings and Facilities. ISBN: 9033S03. Washington, DC: ANSI, ICC.

ICC/ANSI. (1998). ICC/ANSI A117.1-1998 Standard on Accessible and Usable Buildings and Facilities. Washington, DC: ANSI, ICC.

Levine, D. (2003). *Universal Design New York 2*. New York: Department of Design and Construction in participation with the Mayor's Office for People with Disabilities, The City of New York.

Paquet, V. and Feathers, D. (2004). "An anthropometric study of manual and powered wheelchair users." *International Journal of Industrial Ergonomics, 33*(3), 191 - 204.

Steinfeld, E., Schroeder, S., & Bishop, M. (1979). *Accessible buildings for people with walking and reaching limitations*. Washington, D.C.: U.S. Department of Housing and Urban Development..

Steinfeld, E., Paquet, V., and Feathers, D. (2004). "Space requirements for wheeled mobility devices." *In: Proceedings of the Human Factors and Ergonomics Society 48th Annual Meeting*.

Steinfeld, E., Maisel, J., Feathers, D. & D'Souza, C. (2010, forthcoming). *Standards and anthropometry for wheeled mobility. Report prepared for the U.S. Access Board*. Buffalo, NY: IDEA Center. Retrieved from http://www.udeworld.com/anthropometrics.

U.S Department of Housing and Urban Development. (1991). Fair Housing Accessibility Guidelines. 24 CFR Part 100 [Docket No. N-91-2011; FR 2665-N-06] Retrieved from http://www.hud.gov/offices/fheo/disabilities/fhefhag.cfm on March 1, 2010.

706

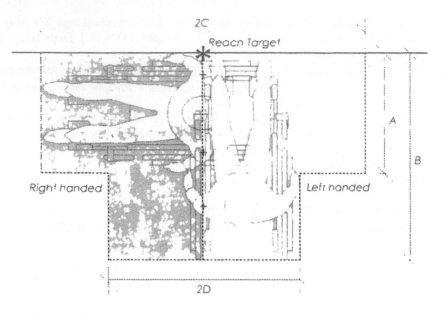

Dimension values in mm [n]

Manual chair users (n =194)

aim max % occ.	50%	75%	90%	95%
A	675 [27]	712 [28]	753 [30]	780 [31]
B	1145 [45]	1213 [48]	1295 [51]	1337 [53]
2C	1671 [66]	1823 [72]	1955 [77]	2088 [82]
2D	1030 [41]	1108 [44]	1175 [46]	1229 [48]

Power chair users (n =147)

aim max % occ.	50%	75%	90%	95%
A	691 [27]	747 [29]	796 [31]	822 [32]
B	1184 [47]	1283 [51]	1352 [53]	1398 [55]
2C	1644 [65]	1811 [71]	1985 [78]	2127 [84]
2D	1041 [41]	1135 [45]	1213 [48]	1272 [50]

Scooter users (n =28)

aim max % occ.	50%	75%	90%	95%
A	622 [24]	722 [28]	818 [32]	840 [33]
B	1203 [47]	1275 [50]	1378 [54]	1435 [56]
2C	1940 [76]	2052 [81]	2135 [84]	2304 [91]
2D	1020 [40]	1081 [43]	1208 [48]	1274 [50]

FIGURE 1.1 Accommodation Model for Clear Floor Space

Chapter 73

Anthropometric Data Visualization Tools to Improve Accessibility of Built Environments

Clive D'Souza[1], Victor Paquet[1], Edward Steinfeld[1], David Feathers[2]

[1]University at Buffalo, State University of New York

[2]Cornell University

ABSTRACT

The Center for Inclusive Design and Environmental Access (IDeA Center) at the University at Buffalo has been developing an anthropometry database of wheeled mobility device users in the U.S. in a multi-year study. The objective of this paper is to demonstrate data visualization tools and methods that are being developed to help inform the design of environments and products. One tool is an interactive software application that is used to visualize body size and shapes, three-dimensional (3D) reach envelopes, and distributions of structural and functional anthropometric dimensions. The tool also allows the user to select subsets of wheeled mobility user groups from the database. Other design tools developed at the IDeA Center are performance-based graphical charts. The tools are described in the context of designing for physical accessibility and inclusive design for mobility device users in the built environment. There is an important need for communicating ergonomics research findings in forms that can be readily applied to the task of designing for accessibility.

Keywords: Accessibility, Wheeled Mobility Users, Anthropometry, Visualization

INTRODUCTION

Research on the anthropometry and functional abilities of wheeled mobility has important implications for the improvement of accessible design practice. First, it provides the necessary data to develop tools for designing accessible spaces. Anthropometry and ergonomics data on people with disabilities also serves as an important resource for federal agencies when developing accessibility guidelines and standards (e.g. ICC/ANSI, 2003; U.S. Access Board, 2004). In fact, much of the technical requirements in the existing standards for accessibility are based on research completed in the 1970's at the Syracuse University that involved, from today's perspective, a very limited sample of manual chair users (Steinfeld et al., 1979). The lack of available knowledge about the size and functional abilities of wheeled mobility device users makes it difficult for standards developers and law-makers to develop and enforce standards that are accommodating.

THE ANTHROPOMETRY OF WHEELED MOBILITY PROJECT

In 2002, the IDeA Center initiated a comprehensive anthropometric study of wheeled mobility users in the U.S. This is an on-going study with a current sample size of close to 400 mobility device users. Comprehensive data is collected in this research, including measurements of mobility device dimensions, body sizes of individuals in 3D, functional reaching abilities in a pick-and-place task, maneuvering abilities, and grip and pinch strengths. Structured interviews to obtain information on participant demographics and mobility device attributes and usage, and visual inspection of the mobility device for specifications and optional device accessories are also conducted (Paquet & Feathers, 2004; Steinfeld et al., 2005).

Findings from this study are being used to provide technical assistance to designers by the identification of best practices as well as to standards developers in revising the technical requirements in current accessibility standards (Steinfeld et al., 2005). Two types of design tools are being developed to support these activities. First, an interactive software program to visualize anthropometry information. And second, a set of printable graphical performance-based design tools.

INTERACTIVE ANTHROPOMETRIC DATA VISUALIZATION

The process of measuring body and mobility device size made use of an electromechanical device for digitizing a selection of landmark points on the individual and device in 3D. The measurements for each person yield a rich set of 3D landmark coordinates, and represent a significant improvement over previous anthropometry studies on mobility device users that used conventional two-dimensional (2D) measurement methods. The 3D coordinate data for each individual are archived as individual datasets in a relational database. This provides the ability to retrieve and analyze the data in many ways, including the ability to construct 3D static digital representations of the individual and mobility device (D'Souza et al., 2007), as well as generate summary statistics on a number of different body and device dimensions (Paquet & Feathers, 2004).

Visual interface software to the database was developed using Microsoft Visual C++ and OpenGL. The software application generates in runtime either graphical and numeric displays such as histograms, summary statistics, percentile values, etc. for a sub-group of individuals based on user-selected demographic and anthropometric variables (Figure 1.1), or digital human models of specific individual cases. These individual cases can be selected from a sub-sample using an interactive histogram that helps identify individuals who possess extreme or 'outlier' values for a particular univariate dimension i.e. starting at the tails of the distribution. This provides designers with in-depth information on the characteristics and functional abilities of certain individuals who present design needs beyond that of the typical wheelchair sub-sample.

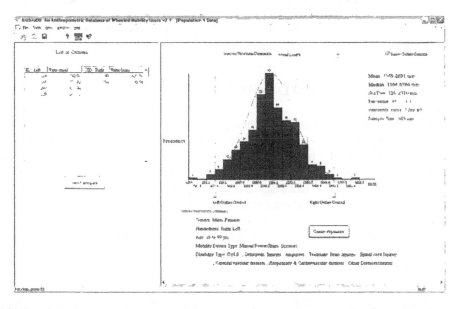

FIGURE 1.1 Example interactive histogram of an anthropometric dimension that allows individuals to be selected for additional analysis.

Figure 1.2 shows a photograph of a female power wheelchair user in the database along with the corresponding digital model and superimposed reach envelopes color-coded for different object weight conditions. Using 3D coordinate data for constructing the model helps to create digital models that reflect the relative size, position and spatial orientation of individual body size and postures, and device size and shape. Reach envelopes are constructed using maximum reach distances recorded electromechanically in an object transfer task. These reach distances were measured in 3D at five normalized shelf heights in three different directions (lateral, forward and an intermediate 45 degrees) resulting in 15 reach data points for each of four different weight conditions (0.45 kg, 1.36 kg, 2.27 kg and 0 kg) to simulate reach and object placement conditions one might attempt during typical activities of daily living.

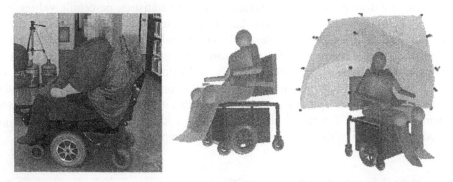

FIGURE 1.2 3D human model and visualization of reach data of a female power wheelchair user.

PERFORMANCE-BASED DESIGN TOOLS

"Performance-based design tools" are graphical tools that demonstrate how key environmental design parameters relate to human performance variables that are empirically measured. In terms of physical accessibility, relevant "performance-based design tools" might depict the proportion of individuals expected to successfully complete a reaching task that is parameterized in some way by the physical characteristics of the environment, or the proportion of wheelchair users capable of maneuvering in a tight space.

For example, we have developed graphical tools that illustrate the effect of reach distance, height of the target from the floor, reaching direction (forward vs. sideways), and object weight on the percent of wheeled mobility users in our sample that would be able to reach a target location during an object transfer task (D'Souza et al., 2009). And an "accommodation model" that depicts the relationship between occupied device width and length, choice of reach direction (forward vs. sideways) and hand use (left vs. right), and the corresponding floor space required for wheelchair users to position themselves when reaching (Steinfeld et al., 2010a).

It is important to note that this focus on performance to define environment accessibility is very different from a prescriptive approach to accessibility, such as that followed in the current U.S. accessibility standards and guidelines. The Americans with Disabilities Act Architectural Guidelines (ADAAG), for example, prescribes limits or thresholds for design dimensions of elements in the built environment to ensure access for wheelchair users (U.S. Access Board, 2004). These include the height ranges between which hand-operated controls should be placed, or the minimum rectangular floor area that needs to be provided adjacent to elements like ATM's and vending machines to provide space for wheelchair users. Using a prescriptive approach provides for a more straightforward evaluation of an environment for compliance with standards by simply comparing whether minimum dimensional criteria are met or not. However, the drawback is that this approach does not allow one to predict how accommodating a design may be, and seldom encourages designers to go beyond minimum criteria (Salmen, 2001).

While we do not necessarily advocate "performance-based accessibility standards", we are of the opinion that "performance-based design tools" could assist both, standards developers and designers. These tools can help standards developers the ability to evaluate and revise the prescriptive criteria in existing accessibility standards and guidelines in order to make them better representative of contemporary user needs. Second, these tools can serve as a supplementary design resource to designers and architects by helping them to understand the potential impacts of their design decisions on physical accessibility and encourage them to adopt best practices in accommodation beyond the minimum criteria (D'Souza et al., 2009). Currently the adoption of universal design, a more inclusive design approach, is emerging around the world as an alternative to minimum standards (Steinfeld, 2010).

RESEARCH CONSIDERATIONS

In this section we will highlight some of the research considerations that can hopefully improve the research and practice of inclusive design and environmental accessibility. These are described in terms of methodological concerns and are related to our experiences in collecting and analyzing anthropometric data on wheeled mobility users.

METHODOLOGICAL CONCERNS IN DATA COLLECTION AND VISUALIZATION

A number of methodological and measurement considerations arise when studying samples that are anthropometrically very diverse and with varied levels of disability. For instance, the inability of participants to maintain standard postures or maintain postures for prolonged durations greatly limited the choice of reliable and consistent measurements tools for use and the data that could be collected. Hence, structural body and device measurements were measured while participants were seated in their own wheelchairs in a comfortable posture, and wearing light everyday clothing.

Given the lack of established measurement procedures in wheelchair anthropometry, standardization of measurements is also a critical component to ensure that data can be pooled for statistical analysis. Considerable efforts went into deciding upon the landmark locations on the person and wheelchair that needed to be digitized (Feathers, 2004) and in establishing consistency and reliability of measurement methods (Feathers et al., 2004). It is imperative that research methods get documented thoroughly. A review of wheeled mobility anthropometry studies conducted internationally demonstrated that good documentation is not always provided (Steinfeld et al., 2010b).

Unlike anthropometry studies involving persons without disabilities, the body and device characteristics in the current sample were not homogenous. Providing data on participants that did not meet the measurement criteria, yielded missing data, or dropped out prior to completion of data collection due to medical reasons or fatigue can also be valuable information. Some examples include the proportion of the sample that could not form a functional grip or show some functional grip capability (Joseph et al., 2010), the prevalence of persons with amputations making it unable to obtain leg or foot measurements, or missing hand dimensions due to involuntary contracture or spasms in the upper extremities. Mobility device characteristics can also vary extensively in size, design, user customizations, optional devices and postural support accessories such as footrests, legrests, and arm-rests.

Such information can make designers aware that certain segments within the population may still not be accommodated despite the designers' best intentions. For example, roughly 21% of our sample was unable to reach above shoulder height. These users may be unable to open a door or operate a light switch, irrespective of the door handle or switch location that a designer opts for. It is very likely that providing this information would encourage designers to adopt universal design strategies to accommodate these users, such as technologies based on motion or proximity sensors to open doors or turn on lighting.

The focus of the Anthropometry of Wheeled Mobility Project has been on person-environment fit for wheeled mobility users in environments such as buildings and architectural spaces. Additional research is still needed to investigate human performance, usability, safety and accessibility for other user groups and environments to enable universal design. Design changes and improvements tailored to one user group can sometimes create barriers for other users. An emphasis on studying different environments is necessary since behavior and performance may vary considerable across different settings. Related to this is the need to establish consensus on how to define accessibility in terms of human performance, and define minimum levels of accessibility in different environments and for different user groups. Full-scale mock-ups provide a useful way to simulate different environments, while providing an opportunity to test different conditions rapidly, at a relatively low cost with a high level of control (Steinfeld, 2004).

DEVELOPMENT OF DESIGN TOOLS FOR ACCESSIBILITY

It is critically important that ergonomics research findings are communicated to designers in forms that can be readily applied in the process of design. This includes the development of tools that convey information about anthropometric variability and multivariate accommodation, and tools that provide designers the ability to anticipate or evaluate the impact of potential design decisions.

Digital human models of individuals provide a highly visual, interactive tool to evaluate physical accessibility and ergonomics problems relating to physical fit, reach and posture. Such tools enable designs to be evaluated against the body sizes and capabilities of 'whole' individuals rather than individual dimensions or human models constructed by scaling manikins using percentile data. HADRIAN is an example of a human modeling and task analysis software developed in the U.K. with data from individuals with disabilities (Marshall et al., 2009; Porter et al., 2004). The HADRIAN database contains detailed structural, kinematic and behavioral data on 100 people, most of who are older and have some type of disability. Also noteworthy is a software tool developed using movement and joint force data collected from 84 older adults in Scotland (Loudon & Macdonald, 2009; Macdonald et al., 2007). The tool depicts stick-figures with representations of the functional demand on the hip and knee during tasks such as object transfer, stair climbing, and when sitting or rising from a chair.

Despite the value of human modeling software, the use of seemingly less sophisticated 2D charts or visual aids should not be ignored (see examples by D'Souza et al., 2009; Steinfeld et al., 2010a). Our experience has shown that 2D representations can be used to depict multivariate data (the simultaneous consideration for the effects of more than one anthropometric or design dimension), are easy to comprehend by a broader audience, and can be a very effective decision-aiding tool during design and standards development.

Finally, the presentation of research findings and the design tools developed should be relevant or contextualized to the information needs of the various stakeholders that rely on such data e.g. architects, product designers, work place ergonomists, policy- and law-makers, mobility device manufacturers, clinicians, etc. For instance, complying with standards such as the ANSI (2003) is an important aspect for designers of the built environment. Hence, our research findings on functional reach (D'Souza et al., 2009) were presented in terms of environment design dimensions that the designer is interested in, but using a presentation scheme that is similar to the illustrations in the U.S. standards and guidelines, allowing easy reference to ensure code compliance.

CONCLUSIONS

This paper provides an overview of the anthropometric data visualization tools developed as part of the Anthropometry of Wheeled Mobility Project at the IDeA Center, and focus on physical accessibility in the built environment for mobility device users. To enable the advancement of universal design a number of such data sources and analysis tools that give consideration to other user groups, including people with mobility, sensory, cognitive, and communications impairments, the elderly, children and other groups are needed. Methodological considerations of data presentation that relate to anthropometry research involving persons with disabilities are discussed. It is important that research findings get presented to designers in forms that focus on maximizing accessibility in design. Although minimum accessibility can be achieved by compliance with regulations and codes, further information is needed to accommodate a larger proportion of the population. Developing a better evidence base can also provide standards developers and policy-makers the ability to evaluate and create more effective accessibility standards.

ACKNOWLEDGEMENTS

This research was supported with funding from the U.S. Access Board (contract # TDP-02-C-0033) and the Department of Education, National Institute on Disability and Rehabilitation Engineering Research through the Rehabilitation Engineering Research Center on Universal Design at Buffalo (Grant # H133E990005). The opinions expressed in this paper are those of the authors and do not represent the policy of the U.S. Access Board, the U.S. Department of Education nor the National Institute on Disability and Rehabilitation Research.

The authors wish to acknowledge the contributions of the numerous researchers and graduate students at the University at Buffalo that assisted with data collection during different phases of the research. The authors are also grateful for the co-operation of Rory Cooper, Alicia Koontz, Padmaja Kankipati, and the research staff at the University at Pittsburgh, where a portion of these data were collected.

REFERENCES

D'Souza, C., Feathers, D., & Paquet, V. (2007). *Constructing Three-Dimensional Models of Individuals and Their Wheeled Mobility Devices from Landmark Data*, Technical Paper 2007-01-2494. Warrendale, PA: SAE Inc.

D'Souza, C., Steinfeld, E., & Paquet, V. (2009). "Functional reach abilities of wheeled mobilty device users: towards inclusive design", *Proceedings of the 2009 International Conference on Inclusive Design, INCLUDE 2009*, ISBN 987-905000-80-7, Royal College of Art, London, UK, April 2009.

Feathers, D. (2004). *Digital Human Modeling and Measurement Considerations for Wheeled Mobility Device Users*, Technical Paper 2004-01-2135. Warrendale, PA: SAE Inc.

Feathers, D., Paquet, V., & Drury, C. (2004). "Measurement consistency and three-dimensional electromechanical anthropometry", *International Journal of Industrial Ergonomics*, 33, 181-190.

International Code Council/American National Standards Institute. (2003). *ICC/ANSI A117.1-2003, Accessible and usable buildings and facilities.* New York: International Code Council.

Joseph, C., D'Souza, C., Paquet, V., & Feathers, D. (2010). "Comparison of hand grip strength between wheeled mobility device users and non-disabled adults", *Proceedings of the 3rd International Conference on Applied Human Factors and Ergonomics, 2010 AHFE International,* Miami, Florida, July 2010.

Loudon, D., & Macdonald, A. S. (2009). "Towards a Visual Representation of the Effects of Reduced Muscle Strength in Older Adults: New Insights and Applications for Design and Healthcare". In V. Duffy (Ed.), *Proceedings of the Second International Conference on Digital Human Modeling, ICDHM 2009,* (Vol. 5620/2009, pp. 540-549). Berlin: Springer Berlin / Heidelberg.

Macdonald, A., Loudon, D., Rowe, P., Samuel, D., Hood, V., Nicol, A., et al. (2007). "Towards a design tool for visualizing the functional demand placed on older adults by everyday living tasks", *Universal Access in the Information Society,* 6(2), 137-144.

Marshall, R., Case, K., Summerskill, S., Sims, R., Gyi, D., & Davis, P. (2009). "Virtual Task Simulation for Inclusive Design". In V. Duffy (Ed.), *Proceedings of the Second International Conference on Digital Human Modeling, ICDHM 2009,* (Vol. 5620/2009, pp. 700-709). Berlin: Springer Berlin / Heidelberg.

Paquet, V., & Feathers, D. (2004). "An anthropometric study of manual and powered wheelchair users", *International Journal of Industrial Ergonomics,* 33, 191-204.

Porter, M. J., Case, K., Marshall, R., Gyi, D., & Sims, R. (2004). "Beyond Jack and Jill: designing for individuals using HADRIAN", *International Journal of Industrial Ergonomics,* 33(3), 249-264.

Salmen, J. P. S. (2001). "U.S. Accessibility Codes and Standards: Challenges for Universal Design". In W. Preiser & E. Ostroff (Eds.), *Universal Design Handbook* (pp. 12.11-12.18). New York, NY: McGraw-Hill Companies, Inc.

Steinfeld, E., Schroeder, S., & Bishop, M. (1979). *Accessible buildings for people with walking and reaching limitations.* Washington, D.C.: U.S. Department of Housing and Urban Development.

Steinfeld, E. (2004). "Modeling spatial interaction through full-scale modeling", *International Journal of Industrial Ergonomics,* 33(3), 265-278.

Steinfeld, E., Maisel, J., & Feathers, D. (2005). *Standards and anthropometry for wheeled mobility.* Report prepared for the U.S. Access Board. Buffalo, NY: IDeA Center.

Steinfeld, E. (2010). "Universal Design". In J. H. Stone & M. Blouin (Eds.), *International Encyclopedia of Rehabilitation.* Available from http://cirrie.buffalo.edu/encyclopedia/article.php?id=107&language=en

Steinfeld, E., D'Souza, C., Paquet, V., & White, J. (2010a). "Clear floor area for wheeled mobility users", *Proceedings of the 3rd International Conference on Applied Human Factors and Ergonomics, 2010 AHFE International,* Miami, Florida, July 2010.

Steinfeld, E., Maisel, J., Feathers, D., & D'Souza, C. (2010b). "Anthropometry and Standards for Wheeled Mobility: An International Comparison", *Assistive Technology*, in press.

U.S. Access Board. (2004). *Americans with Disabilities Act and Architectural Barriers Act Accessibility Guidelines for Buildings and Facilities*. Available at http://access-board.gov/ada-aba/index.htm

Chapter 74

Mobility Impaired Users and Usability Requests in a Context of Public Transport

Anabela Simões

High Institute for Education and Science (ISEC)
UNIVERSITAS
Al. Linhas de Torres, 179
1750-142 Lisboa, Portugal

ABSTRACT

Public transportation authorities and operators should be required and directed to provide accessible public transport to all potential users. For this, they should make necessary adjustments in vehicles, infrastructure and information systems so that they will be suitable for safe, independent, and dignified use by individuals with mobility impairments. Modifications must be made to ensure that there is no need for assistance from others in climbing stairs, maneuvering, locating access aides, etc. Furthermore, information regarding public transportation must be presented in a clear and simple manner, enabling people to easily obtain the necessary guidance and fact.

The ACCESS2ALL European Coordination Action (FP7) aims at defining concrete mobility schemes, guidelines and policy recommendations, ensuring accessibility of PT to all users, through the coordination of current research efforts, the production of common research roadmaps, the identification of best practice models and the appropriate use of information and communication technology (ICT) aids and networks. Introducing the concept of Mobility Impaired people, which is a very heterogeneous population, the first step in this project was the clustering of the target users and the identification of mobility needs and activity limitations of each category.

This paper is centered on methods and tools developed for the usability evaluation in the field of PT to be disseminated among PT operators in order to promote the accessibility and so the quality of their delivered service.

Keywords: Public transport; Mobility; Mobility impaired people; Accessibility; Usability

INTRODUCTION

Public transportation authorities and operators (local, regional and interregional buses, trams, light rail systems, regional and interregional trains, terminals and airports, ports, etc.) should be required and directed to provide accessible public transport (PT) to all potential users. For this, they should make necessary adjustments in vehicles, infrastructure and information systems so that they will be suitable for safe, independent, and dignified use by individuals with mobility impairments. Modifications must be made to ensure that there is no need for assistance from others in climbing stairs, maneuvering, locating access aides, etc. Furthermore, information regarding public transportation must be presented in a clear and simple manner, enabling people to easily obtain the necessary guidance and fact.

The ACCESS2ALL European Coordination Action (FP7) aims at defining concrete mobility schemes, guidelines and policy recommendations, ensuring the accessibility of PT to all users, through the coordination of current research efforts, the production of common research roadmaps, the identification of best practice models and the appropriate use of information and communication technology (ICT) aids and networks. With this aim, methods and tools for usability evaluation of PT were developed to be applied in every PT system.

MOBILITY IMPAIRED PEOPLE

Mobility impaired (MI) users were defined as the target population for this project. The concept of MI extends beyond the traditional definitions of disability and ageing, referring to any activity limitation that prohibits the free movement of a person. This is a widely used term that encompasses a very heterogeneous group of the population. In the frame of the ACCESS2ALL project and having this heterogeneity in mind, MI users were clustered into different categories according to each type of limitation. This classification is composed of thirteen groups containing different levels of limitations and totalising 55 sub-categories. Considering the demographic projections and the increasing number of elderly people in EU, a category of healthy elderly PT users was incorporated in the users groups clustering. Thus, the category of age-related declines involves psychomotor, visual and cognitive limitations, characterising the elderly and reported to the related effects on activity (Eby, Molnar and Kartje, 2009). Furthermore, it was decided to incorporate two more categories regarding the users' groups classification developed for the ASK-IT project1: (1) anthropometric features addressing extra high stature, extra low stature (dwarf) and obesity; and (2) a category addressing problems that lead to social exclusion, like economic problems (e.g. long term unemployment, other), language, cultural differences, etc. Stakeholders that were considered to be relevant bodies for discussing needs and requirements of MI users are clustered into seven categories: PT operators, municipalities, transport authorities, MI organisations (NGO), knowledge providers, industry and infomobility services providers.

The users clustering represents an important basis to identify and collect the specific needs of each category of users in the context of PT, aiming at developing a holistic and modular model to be considered at the planning and operation of PT services, to enhance MI users "transportability"2. This new concept of continuous support of different needs by modular and easily personalised PT elements will allow for PT services that are as comfortable and as individualised as private transport. It should be stressed that the application of this model will make any PT much user friendly, accommodating as well the mobility needs of temporary (pregnant women, injured people) and occasional (adults with children or luggage) MI users.

MOBILITY AND PT ACCESSIBILITY

The ability to move around at will to be engaged in social and recreational activities when desired and to reach business and social services when needed are key components of quality of life. Transportation services and related facilities are part of a package that allows MI people to perform these important activities. However, the existence of facilities and services is meaningless without access to them. Mobility is a key function in determining the degree of matching needs and resources for meeting them. Therefore, mobility becomes fundamental regarding the participation in social relations and activities for older and disabled people (Mollenkopf et al., 1997). Outdoor mobility could be seen as a prerequisite, not only to be able to consume services and products, but also to participate in society generally (Mollenkopf et al., 2005). The qualities of mobility that facilitate the meeting of users' needs and thereby support well-being are:

- Feasibility for the user, which includes his or her abilities to perform the related activities;
- Safety;
- Personal control of mobility (Carp, 1988).

---

[1] ASK-IT Project – ASK-IT Project (Ambient Intelligence System of Agents for Knowledge-based and Integrated Services for Mobility Impaired users) aimed at developing an Ambient Intelligence space for the integration of functions and services for Mobility Impaired people across various environments, enabling the provision of personalised, self-configurable, intuitive and context-related applications and services and facilitating knowledge and content organisation and processing. ASK-IT was a FP6 Project from DG InfoSO (2004-2008).

[2] Transportability model – A new model that is being developed in the frame of ACCESS2ALL. Resulting from the decomposition of the transportation task of MI users from origin to destination in its subtasks, recognising the key contributors to it on behalf of user resources and PT attributes.

These mobility qualities can be further impacted by factors including transportation technology and infrastructure, as well as the user's socioeconomic status, gender, culture and geographical setting (Lukas, 2002).

Although the great majority of European PT operators are nowadays committed in providing fully accessible transportation, there are still barriers in different steps of the travel chain, which highlight the need for a holistic approach in the design and deployment of a PT system. In the narrow sense people who have activity limitations could also have mobility limitations resulting from sensory, physical or learning impairments. In the broader sense anyone who is over-weighted, old, small, tall, pregnant, illiterate, suffering temporarily from a broken limb in an accident, carrying heavy luggage or people carrying small children or people who cannot understand the local language, can have mobility limitations. In all cases, mobility limitations arise from an occasional, temporary or permanent inability to interact with or make use of a transport system, its infrastructure and its component parts. This inability also extends to the built environment, which is part of the transport infrastructure.

"Accessibility is the encounter between the functional capacity of the person or the group and the design and demands of the physical environment", and as such, it "refers to compliance with official norms and standards" (Iwarsson and Ståhl, 2003). In the context of PT, accessibility is the concept the most used to refer the barrier-free features of the entire system. Schmöcker (2009) discusses accessibility and mobility in terms of land-use and public transport planning. In contrast to mobility, a term often used to describe actual or realised travel, accessibility refers to potential travel or the possibility of reaching a destination. Unfortunately, accessibility is not equally distributed across society. Persons with mobility impairments often find it difficult or impossible to reach their desired destinations. Schmöcker points out the fact that in many policy documents, measures aimed at improving accessibility are presented as synonymous with "universal" or "barrier-free" design methodologies. Policy questions are complex because the concept of accessibility is tied to the characteristics and circumstances of individuals – improving accessibility for some might imply worsening it for others. For example, on a very local level, raised kerbs are an important amenity for visually impaired persons but can be problematic for wheelchair users; on a larger scale, both societies that emphasize private automobile travel and those with transportation systems dominated by public transportation impose accessibility limitations on some part of their populations.

Many accessibility indicators, however, are not suitable for older people as they do not reflect the types and characteristics of journeys older people actually make and aspire to make. The missing element is the importance of mobility and independence for this group – the ability to just get out and about, the ability to meet people, to partake in social interactions (Titheridge & Solomon, 2007).

PT operators (intra and extra-city buses, trains, light rail systems, airplanes, ships, ports, etc.) should be obliged and instructed to make the necessary adjustments in vehicles, so that they will be suitable for safe, independent, and dignified use by MI people. Modifications must be made to ensure that there is no need for assistance from others in climbing stairs, maneuvering, locating access aides, etc. Furthermore, information regarding PT must be presented in a clear and simple manner, enabling people to easily obtain the required information.

USABILITY PRINCIPLES

"Usability refers to the extent to which a product or a service can be used by specified users to achieve specified goals with effectiveness, efficiency and their satisfaction in a specified context of use" (ISO DIS 9241-11, in Jordan 1998). The system effectiveness refers to the extent to which a goal or task is achieved (e.g. a PT user succeeds to get a ticket from an automatic ticket machine, so the machine has accomplished its function); the system efficiency refers to the amount of effort and/or time required to accomplish a goal or task (e.g. a PT user succeeds to get a ticket easily from an automatic ticket machine, which means a high level of efficiency); the users' satisfaction refers to the level of comfort felt by users when using the system and how acceptable the system is to them as a means to achieving their goals (e.g. the user is pleased to use the ticket machine as it is easy to use and allows for saving time). Satisfaction is an important aspect of usability for systems involving voluntary and generalized use.

Usability is therefore an important request for the success of any product or service and its criteria should be incorporated since the early phases of a product or service design and deployment. Designing user-friendly products or services means also designing with people in mind, particularly those who will use them. Ensuring an easy, effective and efficient use of a system requires the assessment of each one of the above-referred usability criteria (effectiveness, efficiency and users' satisfaction) in a specified context and taking into account specified groups of potential or actual end users. Actually, usability should be viewed as a property of a product or service in relation to the end user, the related tasks and the environment of use. Usability started to be seen as a bonus; nowadays it is rapidly becoming an expectation, being users disappointed with products or services that do not support an adequate quality of use. Products and services are advertised as being "user-friendly" and "ergonomically designed".

Broader than "user's satisfaction", Nielsen (1993) introduces the concept of "system acceptability", as a combination of social and practical acceptability. The social acceptability refers to how the system responds to the needs of users (ethics and legal) and practical acceptability is established by usefulness, cost, reliability, etc.

Although following usability requests in the design of products or services, that is to say, considering the needs and limitations of those who will be using the system, usability evaluations are required to ensure that users' expectancies were taken into account on the design process. By evaluating usability, problems and difficulties met by the users will be identified and recommendations and guidelines for re-designing the system or service will be provided. Therefore, the levels of effectiveness and efficiency, as well as the users' satisfaction should be assessed.

METHODS AND TECHNIQUES FOR USABILITY EVALUATION

Usability evaluations should be carried out to identify problems and difficulties experienced by MI users in using public transport. The complexity of a PT system, usually integrated in a local transport network, requires taking into consideration the different elements of the travel chain in any project towards accessibility. Thus, each element of the PT system corresponding to a specific step within the travel chain should be designed and assessed according to usability principles and criteria (effectiveness, efficiency and users' satisfaction). As the travel chain involves different technologies, the usability evaluation will address separately the different elements of the PT system; however, these elements will be grouped according to their relevance to the proposed assessment. Therefore, the usability evaluation will address each step of the travel chain and the corresponding technology separately using different evaluation tools.

According to Oppermann and Reiterer (1997), usability evaluation methods can be classified as:

- Subjective methods based on user's judgment, after the system's use:

  - Questionnaires, providing data on the end user's views about the system's usability in a real or reference situation; however, the selection of samples must be cautious in order to ensure that subjects represent the actual or intended group of end-users;
  - Interviews, requiring careful pre-planning and a good degree of expertise on the part of the interviewer; they consume more time than questionnaires and must be structured with a pre-determined set of precisely phrased questions in order to allow for an accurate analysis.

- Objective methods based on observational methods involve real users performing relevant tasks:

  - Direct observation is the simplest form of observational methods and requires a human factors expert to observe the user performing the requested tasks;
  - Video recording allowing data gathering to be analyzed later using accurate data processing computer-based methods; they increase costs and time but provide accurate and liable results.
  - Interaction monitoring requiring capture tools to gather data about people using the system; they can be replay-based or logging-based;
  - Co-operative evaluation involving actively the users in the process of evaluation by means of asking the user some questions during the task performance or to think aloud or a video self-confrontation after the system's use; an alternative method is the constructive interaction method involving two users performing the tasks together and telling each other what they feel, do or intend to do.

- Expert evaluation method drawn upon expert knowledge to make judgments about the usability of the system for particular tasks and end-users:

  - Specialist reports centered on a critical evaluation based upon expert knowledge; the expert walkthrough is a more methodological variation of the specialist report generated by the expert on the basis of "walking trough" representative tasks the system has been designed to support;

- Cognitive walkthrough based on a theory of learning by exploration and on modern research on problem solving, gives expert evaluation of usability a more accurate theoretical basis;
- Checklists, guidelines and principles providing a heuristic evaluation by means of examining the system's compliance with the defined checklist items, guidelines or principles.

- Experimental evaluation methods: usually carried out in controlled conditions with precise hypothesis and a correct definition of dependent and independent variables, as well as the selection of the adequate environment for testing; however, it is sometimes difficult to find theoretical framework on human-machine interaction.

USABILITY EVALUATION TOOLS FOR ACCESS2ALL

Apart the method to be chosen, the evaluation should address the usability criteria: effectiveness, efficiency and users' satisfaction. All methods involve users at the exception of an expert evaluation; for this reason, this method does not allow for the assessment of users' satisfaction, being applied at an early stage of the design process to test early prototypes.

Therefore, the evaluation tools to be used in the ACCESS2ALL project consist of:

- Checklists, which are useful to assess the system effectiveness and efficiency;
- A questionnaire to assess the user's satisfaction regarding each group of elements of PT systems.

Field observations of users performing the PT related tasks could provide a more accurate evaluation of the PT system usability.

In this context, the evaluation tools should be adapted to the different elements of the travel chain to be addressed: interactive information and ticketing systems, PT infrastructure, vehicles and walking environment.

Therefore, a checklist composed of the following parts has been developed to be used in the assessment of each category of the system in terms of effectiveness, efficiency and users' satisfaction:

- Pedestrian accessibility (addressing walking environment to reach the PT network and inside it);
- Information and ticketing systems user-friendliness;
- Vehicle accessibility, safety and comfort;
- Infrastructure accessibility;
- Service availability and affordability;
- Social inclusion.

A questionnaire was developed being composed of the following parts: user profile (city/country, age, gender, type of mobility limitation), mobility habits (questions concerning different types of public transport, ticket and information) and information and communication technologies (use of internet and phone).

CONCLUSION

In the frame of the ACCESS2ALL project, the developed tools for assessing PT accessibility will be available to any European PT operator. As each component of the travel chain is technologically different, the usability assessment should involve the use of specific tools and be carried out separately. Aiming at ensuring the accessibility of PT to any user, PT operators should carry out usability assessments with samples of MI people. Actually, being accessible to them the system will be much easier to any potential user.

Furthermore, policies for the environment protection include the promotion of PT use, but the success of these policies will depend on the quality of the delivered PT service to all citizens. Therefore, the usability of each component of the travel chain should be a main request in the evaluation process of a PT operator quality of service. In this perspective, a PT system will fulfill the needs and expectations of MI people if any potential gaps between the required service and the delivered service will not exist.

REFERENCES

Carp, F. (1988), Significance of Mobility for the Well-Being of the Elderly. In Transportation in an Aging Society – Improving Mobility and Safety for Older Persons, Volume 2.TRB, Washington, D.C.

Eby, D.; Molnar, L. and P. Kartje (2009), Maintaining Safe Mobility in an Aging Society. CRC Press – Taylor and Francis, London.

García-Pastor, A. and López-Lambas, M.E. (2005), Quality Issues In Transport Operation Tenders. The Difficult Equilibrium Between Price And Service Level. European Transport Conference 2005 Proceedings. Strasbourg, France.

Iwarsson, S., & Ståhl, A. (2003), Accessibility, usability and universal design; positioning and definition of concepts describing person-environment relationships. Disability and Rehabilitation, 25 (2), 57 – 66.

Jordan, P. (1998), An Introduction to Usability. Taylor & Francis, London.

Lucas, K. (2002), Transport and Social Exclusion: The UK Perspective, Presented at Cities on the Move Transport Seminar in Paris, December 5-6, 2002.

Mollenkopf, H., Marcellini, F., Ruoppila, I., Flaschentrager, P., Gagliardi, C., & Spazzafumo, L. (1997), Outdoor mobility and social relationships of elderly people. Archives of Gerontology and Geriatrics, 24(3), 295-310.

Mollenkopf, H., Marcellini, F., Ruoppila, I., Széman, Z., & Tacken, M. E. (2005), Enhancing Mobility in Later Life. Amsterdam: IOS Press.

Nielsen, J. (1993), Usability Engineering. Morgan Kaufmann.

Oppermann, R., Reiterer, R. (1997), Software Evaluation Using the 9241 Evaluator Usability Evaluation Methods. Behaviour and Information Technologyv.16 n.4/5 p.232-245.

Schmöcker, J.D. (2009), Access, Aging, and Impairments Part A: Impairments and Behavioral Responses. Journal of Transport and Land Use 2(1), pp. 1–2.

Titheridge, H., & Solomon, J. (2007), Benchmarking accessibility for elderly persons. Paper presented at the 11th International Conference on Mobility and Transport for Elderly and Disabled people (TRANSED), Montreal, Canada, June 18-22, 2004.

Chapter 75

Priority Implementation Scenarios for Accessibility of Public Transport for Mobility Impaired Users

Evangelia Gaitanidou, Evangelos Bekiaris

Centre for Research and Technology Hellas/
Hellenic Institute of Transport
6th Km Charilaou-Thermi Rd
57001, Thermi, Thessaloniki, Greece

ABSTRACT

The issue of accessibility of public transport (PT) for mobility impaired (MI) people is a major concern in modern societies, within the concept of assuring equal opportunities for all citizens. The mobility impairment may be related either with a temporal or a permanent disability. The disabilities in question could be of different types (i.e. lower limb impairment, vision impairment, age-decline disabilities, etc.), thus covering a significant percentage of the population. The ACCESS2ALL FP7 Coordination Action project aims at defining concrete mobility schemes, guidelines and policy recommendations, ensuring accessibility of public transport to all users, through the coordination of current research efforts, the production of common research roadmaps, the identification of best practice models and the appropriate use of ICT (Information & Communication Technologies) aids and networks. Within this concept, one of the targets is to issue a set of priority implementation scenarios to promote the uptake of MI users needs and wants into the planning and operation of PT. Towards this, the first step has been to clearly cluster the addressed user groups and perform literature review on their needs, the current situation and existing good practices in the area. Then, the definition of implementation scenarios followed, taking into account the needs of the users as well as the technological potential and organizational structures. In ACCESS2ALL, a set of 18 scenarios has been defined by the project experts, structured in a special template, including the PT mode, the relevant issues addressed, the related user groups, the type of operation (i.e. urban, interurban, periurban, etc.), the business chain actors involved, the most adequate aids and processes to be followed, an estimation of the relevant cost, maintenance and service update issues, etc. These scenarios involved most of the issues of interest of the ACCESS2ALL project, i.e. information (e.g. adapted pedestrian route guidance system for elderly and disabled (E&D), dynamic route accessibility), safety in transportation (e.g. special dummies of E&D for crash tests, modular and adjustable tie-down systems and anchorage points in PT vehicles), vehicle and/or transportation hub ingress/egress (e.g. standardisation of automatic and appropriate ingress/egress lifts, automatic lift at trains for wheelchair users ingress/egress) use of accessible aids (e.g. modelling Accessibility Technology devices in mainstream products design), comfort issues and service facilities (e.g. modelling E&D users in ergonomic design of transportation systems and stations), emergency support (e.g. personalized guidance in mobile nomad device of MI travellers in case of emergency in transportation hubs, personalized models of emergency handling, taking into account E&D behaviour and presence), ticketing/booking/cancellation (personalized smart card or mobile phone application for ticketing/booking/cancellation taking into account user profile, Automatic info-kiosk interface with mobile), multimodality of transport chain (e.g. selection of preferred modes and mode changes based on user's profile, selection of walking distance between modes based on user's profile). In order to prioritize these scenarios, the Analytical Hierarchy Process Multi-Criteria Analysis has been applied. According to this methodology, a set of criteria should be originally defined, which will be the factors for prioritization. The criteria defined for the needs or this analysis were: range of affected users, maturity of technology, European dimension, cost efficiency, range of application/transportation means covered, impact to safety and impact to the environment. In order to perform a concise analysis, these criteria needed to be prioritized, so as to assign a weight to each criterion. For this reason, a pairwise comparison matrix has been composed, in which each criterion is rated in comparison to each of the rest of them, following a pre-defined range of values. The second step was to create a second matrix, in which a mark is assigned for each scenario – criterion combination, again according to a pre-defined scale. A dedicated interactive session has taken place during the first project workshop, for the application of this methodology by internal and external to the project experts. The analysis of the results followed, from which the scenario priorities derived. This prioritization exercise has been performed in the first year of the project; it will also be performed later in the project, including more scenarios and more and/or different criteria, which will derive from the work performed within the several tasks of the ACCESS2ALL project, as well as from other related projects in cooperation with ACCESS2ALL. The final list of priority implementation scenarios will as well contribute in the definition of further research roadmaps, which also lies among the aims of the ACCESS2ALL project.

Keywords: public transport, accessibility, disabled, scenarios, ICT, ACCESS2ALL

INTRODUCTION

Accessibility in public transport (PT) for all the members of the society is an objective undertaken by most of the modern societies. By the term "all" it is implied that the transportation systems should provide the appropriate facilities so as to be able to accommodate any person that wishes to use public transport, regardless of the disability they may have (either physical or mental, temporary or permanent). This might seem to be an easy task; however, in practice, the difficulties and obstacles that many people face while trying to move in the cities are numerous and of great variety. Worldwide, this problem has been recognized and many initiatives have been undertaken. More specifically, the Americans with Disabilities Act (ADA, 1990), states that no individual with a disability shall be denied the opportunity to use the public transportation system. Additionally, where necessary or upon request, the personnel of both public and private transportation services must assist a disabled person with the use of ramps and lifts enabling the individual to enter or exit the transportation service. The personnel may have to leave their seats in order to provide such assistance. Another example is the Disability Discrimination Act (DDA, 1995&2005) in the UK, where again specific regulations are foreseen for the unobstructed and by equal terms use of transportation by people with disabilities of any kind.

In this view, the research community has also been active in the field during the last decades, striving to identify the needs, locate the problems and, most notably, provide the solutions to them, either with legislative and organizational or technological means. ACCESS2ALL project, funded by the European Commission within the 7th Framework Programme, is a research initiative that "aims at defining concrete mobility schemes, guidelines and policy recommendations, ensuring accessibility of public transport to all users, through the coordination of current research efforts, the production of common research roadmaps, the identification of best practice models and the appropriate use of Information Communication Technologies (ICT) aids and networks" (ACCESS2ALL Consortium, 2008). Within the context of this work, the definition of a set of implementation scenarios is foreseen, which would imply the priority application areas and emerging technologies that can be used in the service of PT accessibility, guided by the actual and most outstanding needs of the users.

BACKGROUND

In order to reach the stage of defining implementation scenarios and prioritising them, a series of steps have been realised in the work undertaken within the project. First, the user groups to be addressed have been identified and clustered. After that, their needs in terms of using PT systems and the problems they are confronted with have been investigated, while best practices of applied systems and/or measures have been collected and a relevant database created. This work has set the basis for defining and prioritising implementation scenarios.

CLUSTERING OF USER GROUPS

The first task aimed at describing the different groups of public transport mobility impaired users, as well as the related stakeholders involved in creating the required conditions and adopting the necessary measures for an inclusive PT system. The ICF (International Classification of Functioning, Disability and Health codes) (WHO, 2001) were adopted to classify PT MI users according to the different activity limitations.

The classification that has derived has been based to previous related work undertaken within the framework of different research initiatives like ASK-IT Integrated Project (Simoes, 2005), adapted and extended to cover the needs of the work within ACCESS2ALL. In this sense, more user groups have been added, in order to include also users with age-related declines in abilities, social exclusion parameters, as well as the affected stakeholders. The main categories of this clustering can be seen in Table 1. Each of them is divided in several sub-categories so as to incorporate and specify the variations of each user group (Alauzet et al, 2009).

Table 1. User groups clustering and stakeholders categories

User Groups	Stakeholders groups
Lower limb impairment	PT operators
Wheelchair users	Municipalities
Upper limb impairment	Transport Authorities
Upper body impairment	MI Organisations (NGO)
Physiological impairment	Industry
Psychological impairment	Knowledge providers
Psychological impairment	Infomobility services providers
Cognitive impairment	
Vision impairment	
Hearing impairment	
Communication producing and receiving difficulties	
Age-related decline in abilities	
Anthropometric features	
Factors leading to social exclusion	

REVIEW OF LITERATURE, NATIONAL SURVEYS AND BEST PRACTICES

After identifying the affected user groups, the next step has been to investigate their needs and problems regarding the use of PT. For this reason, extensive literature survey has been undertaken, reviewing more than 50 sources as well as incorporating relevant results coming from earlier research initiatives. Moreover, existing national surveys on accessibility issues have been reviewed and interviews have taken place with transportation providers and representatives of disabled organizations in different EU countries. The results of this work are presented in detail in (Alauzet et. al., 2009).

Within the context of this work, best (or bad, as examples to be avoided) practices have also been identified, which refer to services, measures, technologies, etc. that have already been applied, aiming to facilitate the accessibility of mobility impaired groups in public transport systems. The collection of best practices is an ongoing work throughout the duration of the ACCESS2ALL project and the collected best (and bad practices) are structured in an online database which will be publicly available upon verifying and assessing all gathered data (Gaitanidou et. al., 2009).

IMPLEMENTATION SCENARIOS

DEFINITION AND COLLECTION

After identifying the user groups and investigating their needs and problems, as well as finding best practices in the area of accessible mobility, the next step has been to define implementation scenarios for the application of relevant measures, systems and technologies. For the systemization and facilitation of this work, a template was created, where each implementation scenario was described in terms of: user characteristics (primary, secondary and business chain actors involved), operative characteristics (PT mode addressed, type of operation, mobility issues addressed, devices/accessibility aids used), cost estimation, maintenance and service, scenario description. This template was distributed to the experts participating in the project, in order for them to propose implementation scenarios addressing to the different accessible mobility issues, defined since the beginning of the project, which are: pre-trip and on-trip information, safety on transportation, ingress/egress, use of accessible aids, comfort issues and service facilities, emergency support, ticketing/booking/cancelation and multimodality of transport chain.

Initially, more than 20 scenarios suggestions have been gathered which, after removing duplications and the ones lacking important data, resulted in a list of 18 scenarios, covering all of the above mentioned issues (Table 2). These were the scenarios that were subject to the preliminary prioritization round, as can be seen below.

Table 2. Selected list of implementation scenarios per PT transport accessibility issue

Issues	#	Scenario
Information	1	Adapted pedestrian route guidance system for elderly and disabled
	2	Dynamic route accessibility
	3	Dynamic POI accessibility
Safety on transportation	4	Special dummies of E&D for crash tests
	5	Modular and adjustable tie-down systems and anchorage points in PT vehicles.
	6	Rear Facing wheelchair position for buses
Ingress/ Egress	7	Automatic lift at trains for wheelchair users ingress/egress.
	8	Standardisation of automatic and appropriate ingress/egress lifts.
	9	Talking doors for blind.
Use of accessible aids	10	Modelling Accessibility Technology devices in mainstream products design.
Comfort issues and service facilities	11	Modelling E&D users in ergonomic design of transportation systems and stations.
Emergency support	12	Personalized guidance in mobile nomad device of MI travellers in case of emergency in transportation hubs.
	13	Personalized models of emergency handling, taking into account E&D behavior and presence.
	14	Emergency module for mobility impaired people
Ticketing/ booking/ cancelation	15	Automatic info-kiosk interface with mobile.
	16	Personalized smart card or mobile phone application for ticketing/booking/cancellation, taking into account user profile.
Multimodality of transport chain	17	Selection of preferred modes and mode changes based on user's profile.
	18	Selection of walking distance between modes based on user's profile.

EXTRACTING PRIORITIES

PRIORITIZATION METHODOLOGY

For the prioritisation of the preliminary selected list of implementation scenarios the AHP/MCA (Analytic Hierarchy Method/Multi Criteria Analysis) methodology has been used [Saaty, 1995]. In short this is a method to derive ratio scales from paired comparisons. Following this methodology for the needs of ACCESS2ALL, a set of criteria is primarily defined, according to which the scenarios will be prioritized. These criteria are defined according to the nature and content of the scenarios as well as their targeted effect. In the specific case, a set of seven criteria was selected, trying to cover different aspects of the use of the scenarios in the context of public transport and accessibility of users with mobility impairment. These criteria are: a) Range of affected users, b) Maturity of technology, c) EU dimension, d) Cost efficiency, e) Range of application/ transportation means covered, f) Impact to safety, g) Impact to environment. Each of these criteria may have a different and of varied significance effect on each of the implementation scenarios and their intended impact; thus a different weight factor should be assigned to each criterion. To do this, pairwise comparisons are performed, comparing each criterion with all the rest (in a 7x7 matrix), using the scale seen in Table 3.

Table 3. Rating scale for criteria pairwise comparisons (Saaty, 1995)

Intensity of importance	Definition
1	Both elements have equal importance
3	Moderately higher importance of row element (RE) as compared to column element (CE)
5	Higher importance of RE as compared to CE
7	Much higher importance of RE as compared to CE
9	Complete dominance in terms of importance of RE over CE
2, 4, 6, 8 (Intermediate values)	
1/2, 1/3, 1/4, ... 1/9 (reciprocals)	

Further on, an overall 7x18 matrix is being created, where the first column includes the criteria and the first row the scenarios. This matrix is used in order to give a rate for each scenario according to each criterion. The rating scale used in this case is: 1-2: irrelevant, 3-4: relevant, 5-6: moderately important, 7-8: important, 9-10: absolutely important.

ORGANISATION OF PRIORITISATION PROCESS

The prioritisation of the selected implementation scenarios is planned in more than one stages throughout the duration of the project. The first stage has been the prioritisation of the preliminary selected scenarios by external to the project experts. This has been realised during the 1st ACCESS2ALL Workshop that was held in Porto, Portugal, on September 2009. During an especially dedicated interactive session, the preliminary list of scenarios and related criteria were presented in detail and afterwards the participants were asked to use the comparison matrices and provide ratings for the criteria and the scenarios, according to the above presented methodology, which was explained to them beforehand. The results of this session are analysed in the present paper. The next stage of prioritisation will be to perform this exercise also with other experts representing as much as possible the different affected user groups (project participants, members of the project user forum, etc.) with the addition of more scenarios. Another prioritisation round may be performed also during the 2nd ACCESS2ALL Workshop and the final priorities will emerge by the combination of results of the different stages.

RESULTS

After gathering the input from the participants (10 assessments), analysis of the data has been performed according to the applied methodology, adapted for the needs of the specific research. First, the relevant priorities (weights) of the criteria were calculated, according to which the final scenario priorities were extracted.

Weights of criteria

In order to synthesize the various pairwise comparisons given by each representative, the geometric mean was calculated, as suggested by (Saaty, 1995). The final weights that were calculated can be seed in Table 4:

Table 4. Relative priorities of criteria (weights)

Criterion	Relative priority
Range of affected users	0,16751
Maturity of technology	0,149831
EU dimension	0,09602
Cost efficiency	0,143328
Range of application/ transportation means covered	0,10997
Impact to safety	0,209041
Impact to environment	0,124301

734

Scenarios priorities

Taking into account the relevant priorities of the different criteria and the ratings given by the participants to each scenario in relation to each criterion, the final ratings of the scenarios were calculated, by applying the weighted mean values. The top five rated scenarios according to this prioritisation are: 1) Modelling E&D users in ergonomic design of transportation systems and stations, 2) Modelling Accessibility Technology devices in mainstream products design, 3) Special dummies of E&D for crash tests, 4) Personalized guidance in mobile nomad device of MI travelers in case of emergency in transportation hubs, 5) Dynamic POI accessibility (Figure 1). It should of course be noted that most of the scenarios have very little difference in their obtained ratings, which indicates that most of them are of comparable importance and should, according to the opinion of the participants, be considered for application.

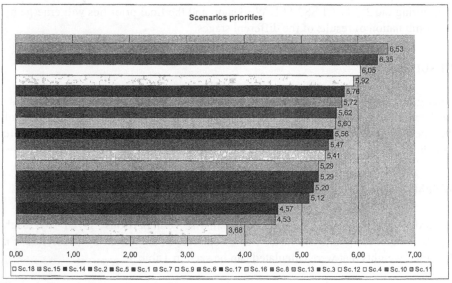

Figure 1: Extracted scenarios priorities

DISCUSSION AND CONCLUSIONS

After selecting a preliminary list of 18 scenarios and 7 criteria by which these would be prioritised and upon defining the prioritisation methodology and tools, the first stage of prioritising these scenarios was performed. As can be concluded from the results of this stage, the scenarios that received the higher ratings are referring to design and emergency procedures, which implies that there should be given special focus on designing the transportation systems in such a way so as to be accessible and safe for all users (including mobility impaired ones) and, at the same time, transportation infrastructure (stations, hubs, etc.) should foresee the adoption of special procedures in order to handle emergency situations in an effective way, also in the case of mobility impaired people that may be present on site at the time of emergency. Moreover, the use of new technologies, and mostly nomadic devices, in terms of guidance and identification of accessible routes and associated points of interest (POIs) is a very promising area, as it can provide assistance to all users in a personalized and independent way. These research domains have a lot to present in the future and the scenarios that will be finally proposed as priority implementation scenarios from ACCESS2ALL (upon completion of the different prioritization stages mentioned) is aiming to indicate the main areas where special focus should be put on in researching and developing of services and applications, towards the creation of a transportation environment that would be usable and friendly to all the members of the society.

REFERENCES

ACCESS2ALL Consortium (2008), "Mobility Schemes Ensuring Accessibility of Public Transport for All Users", Grant agreement No: 218462, Annex I – Description of Work.

ADA (1990), "Americans with Disabilities Act of 1990, as amended in 2008", US Department of Justice, Online, Available from: http://www.ada.gov/pubs/adastatute08.htm, Reached on 25 February 2010

Alauzet, A., Dejoux, V., Simoes, A., Rocci, A. (2009), "User needs and preferences per user group", Deliverable 1.1, ACCESS2ALL project.

DDA, (1995), "Disability Discrimination Act 1995 – Chapter 50", Online, Available from: http://www.opsi.gov.uk/acts/acts1995/ukpga_19950050_en_1 Reached on 25 February 2010.

DDA, (2005), "Disability Discrimination Act 2005 – Chapter 13", Online, Available from: http://www.opsi.gov.uk/acts/acts2005/ukpga_20050013_en_1 Reached on 25 February 2010.

Gaitanidou, E., Panou, M., Chalkia, E., Kesidou, E., Agnantis, K. (2009), "Best practices database and Transportability s/w tool", Deliverable 1.2, ACCESS2ALL project.

Saaty, T.L. (1995), "Decision Making for Leaders. The Analytic Hierarchy Process for Decisions in a Complex World", RWS Publications, Pittsburgh.

Simões, A. and Gomes, A. (2005), "User groups classification and Potential demography of MI in Europe", ASK-IT IP – Internal Document "ASK-IT SP1 User Groups.doc" (A1.1.1)

WHO (2001), "International Classification of Functioning, Disability and Health: ICF", World Health Organization.

Chapter 76

Handwriting Starts at Childhood: a Systematic Review of Variations on Handwriting Grip

Marianela Diaz Meyer, Luciana Sica,
Andrea Sinn-Behrendt, Ralph Bruder

Institute of Ergonomics
Darmstadt University of Technology
Petersenstrasse 30, 64287 Darmstadt, Germany

ABSTRACT

One activity that children spend a lot of time on is the usage of a handwriting instrument for different tasks, such as scribbling, drawing or writing. Although there are well-known and accepted handwriting grips, such as the dynamic tripod grip, a large variation of functional and dysfunctional grips can be observed in adults and children. A literature review and systematic classification of the variations of handwriting grips, according to stages of development, were carried out. A total of 34 different handwriting grips were identified in the literature. Additional findings from an experimental analysis with 23 children aged 3 to 14 years are discussed. The results of this study will benefit the ergonomic design of the grip zone of writing instruments for children.

Keywords: Handwriting grip, writing instrument, children ergonomics, ergonomic product design

INTRODUCTION

Classical ergonomic product design guidelines have been conceived and used for, in particular, improving hand-guided tools that require high force exertions. Nowadays, nevertheless, occupational ergonomics and human factor knowledge is being applied both to the design for all, as well as, to the design of user-oriented products for special groups. That is, the case of ergonomics for children (e.g., Lueder and Rice, 2007).

One activity that children spend a lot of time on is the usage of a handwriting instrument for different tasks, such as scribbling, drawing or writing. Young school children spend 30 to 60 percent of school days learning to write and improving other fine motor skills (McHale and Cermak, 1992). The stabilization of the writing instrument during these tasks is done through the handwriting grip.

Although there are well-known and accepted handwriting grips, such as the dynamic tripod grip, a large variation of functional and dysfunctional grips can be observed in adults and children. For this reason as well as to support the further development of ergonomic writing instruments, a critical review of the literature and systematic classification of the variations of handwriting grips was done. The developmental process of children was considered by comparing the well-known and accepted dynamic tripod grip against other modified or less conventional grips. Additional influence factors, such as the writing task (e.g., scribbling, drawing, coloring, shading, writing, etc.) and different features of writing instruments were analyzed in an experiment.

Inadequate handwriting grips can produce negative effects, such as body/muscle strain due to the exertion of too much pressure and awkward postures over extended periods of time. Measures to alleviate such negative effects for young children can be achieved by means of an adequate behavior and ergonomic writing instruments specifically designed for children. It can be assumed that the handwriting grip of children is influenced by the design of the writing instrument together with the interaction between the writing instrument and the writing surface.

Designing ergonomic writing instruments for an early childhood stage can provide children the tools they need to become proficient at writing while improving fine motor control and postural stability. Furthermore ergonomic writing instruments can ensure skill acquisition, and reduce the negative effects of inappropriate postures over extended periods of time.

At present, a large number of writing instruments with innovative and so-called ergonomic designs are offered. However, their ergonomic advantages remain questionable in many cases. There are unexplored ergonomic criteria for the design of writing instruments, as well as, a lack of specifications for special target groups such as children. Suitable evaluation methods should help to scientifically define and prove the above mentioned aspects. Ergonomists, psychologists, occupational physicians and designers should work together in order to design appropriate products and tools for children, society's future workforce.

Proposing guidelines for the ergonomic design of handwriting instruments should help to:

- improve handwriting (i.e., performance and quality)
- reduce children's difficulties and time for developing an appropriate handwriting grip
- reduce the problems and negative effects over extended periods of time
- fit the writing instrument to the child, i.e.,
  - meet the needs of a variety of children in different developmental stages
  - consider the changes during growth and developmental processes.

CLASSIFICATION OF HANDWRITING GRIPS

A total of 34 different handwriting grips were identified in the literature and systematically classified into subgroups according to stages of development (see Table 1).

Rosenbloom and Horton (1971) researched on the maturation of grips in children and identified two stages of development: tripod posture and dynamic tripod. Schneck and Henderson (1990) extended this classification to three stages of development: primitive, transitional and mature grips. Selin (2003) employed this developmental scale and included additional types of grips. In this study, we added further grips following the work of Schneck and Henderson (1990) and Selin (2003) and extended the classification with descriptors.

There are four groups of descriptors: grip configuration, body posture and movement, interaction within writing system's elements, and human characteristics, skills and capacities (see Figure 1). Whereas early definitions of the dynamic tripod were created in the 1960s (e.g., Winn-Parry, 1966, cited by Rosenbloom and Horton, 1971), Ziviani (Ziviani, 1983, Ziviani and Elkins, 1986) was one of the pioneers, who discriminated between different handwriting grips using elements related to the grip configuration (e.g., finger contact, finger position, finger flexion, etc.). These and other grip configuration elements (e.g., Blöte and Dijkstra, 1989, Carlson and Cunningham, 1990, Braswell and Rosengren, 2007) describe the aspects concerning the contact of the fingers with the writing instrument. Examples of these aspects are the number of fingers and the part of the finger in contact with the writing instrument (pad contact, tip contact, side contact, no contact), the finger used to hold the shaft, thumb-index position, index flexion of proximal and distal joints, location of shaft-rest (open web, sticking up radial to the index finger), finger intrinsic/coordinated movement and proximity of grip to the pen-point (called longitudinal grip position of lowest finger tip). Blöte and Dijkstra (1989) determined that the optimal distance between the point of the pen (pen-point) and the lowest fingertip is 2.5cm. Other researchers (e.g. Braswell and Rosengren, 2007) employed qualitative criteria for this aspect, i.e, low, middle, high. The grip configuration group of descriptors was used to define the grips contained in Table 1.

Primitive grips are characterized by a larger quantity of fingers in contact with the shank and a larger flexion of finger joints (see Figure 2). On the contrary, mature grips tend to use fewer fingers while increasing finger dexterity. The finger-to-finger position (e.g., thumb-index position) is an important criterion that shows the advance in development. However, this criterion cannot be explicitly defined. In the case of the dynamic tripod, Callewaert (1963, cited by Bailey, 1988) and Schenk et al. (2004) defined an alternative mature grip to the traditional dynamic tripod, called the modified handwriting grip. This type of grip does not require pad-to-pad opposition between the thumb and index finger, but rather the shaft of the instrument should be held between the middle and index fingers. Concerning the movement, Blöte and Dijkstra (1989) reported that coordinated finger movement is not exclusively restricted to the dynamic tripod, it can also be found in other mature

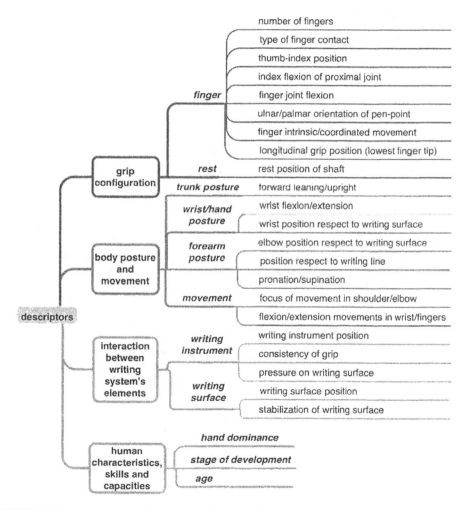

FIGURE 1 Descriptors used for the classification of grips.

Table 1 Systematic classification of writing grips according to the stages of development (Diaz Meyer et al., 2010). (a) Blöte and Dijkstra (1989), (b) Braswell and Rosengren (2007), (c) Burt and Benbow (2007), (d) Koziatek and Powell (2003), (e) Schneck (1991), (f) Selin (2003).

Stages of development	**Mature grips**	quadrupod	dynamic quadrupod (c)				
		tripod / tripod-like	dynamic tripod (usual type) (a,c,e,f)	combined type (c,f)	modified tripod (b)		
	Transitional grips	quadrupod / quadrupod-like	static quadrupod (c)	lateral quadrupod (d)	quadrupod (broken type) (d)		
		four-finger	four finger (e,f)	four finger (broken type) (d)			
		tripod / tripod-like	lateral tripod (e,f)	lateral pinch (c)	lateral tripod (d)		
			static tripod (c,e,f)	static tripod (a,b)	fixed tripod (f)	solitary tripod (f)	broken tripod (f)	dynamic tripod (broken type) (d)
	Primitive grips	other	supinate (f)	with extended fingers (e,f)	index (c,f)	brush (f)	transpalmar inter digital brace (f)
		thumb	cross thumb (f)	cross thumb (e)	thumb wrap (f)	thumb wrap (c)	thumb tuck (f)	thumb tuck (c)
		digital	tong or ulnar digital (a)	digital or radial digital (b)	digital pronate, only index finger extended (e,f)		
		palmar	radial cross palmar (e,f)	palmar supinate or ulnar palmar (e,f)			

grips. Primitive grips are characterized by rather static holding compared to dynamic holding. The main criterion for the distinction between transitional and mature grips is the development of well-coordinated and intrinsic movements.

The second group of descriptors is the body posture, which also includes hand-arm movement. Blöte and Heuden (1988) and Blöte and Dijkstra (1989) observed distinctive body postures of the trunk, wrist/hand and forearm, as well as, characteristic movements (i.e., flexion/extension of wrist and fingers) during handwriting. Rosenbloom and Horton (1971) discussed the change of the focus of movement respect to developmental stages, from focus in the shoulder, elbow and wrist to focus in the fingers. These aspects were operationalized and included with the descriptors in Figure 1.

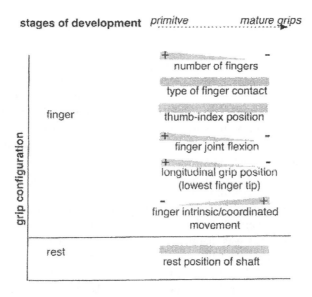

FIGURE 2 Variation of values of the grip configuration descriptors according to the stages of the development.

The third group of descriptors illustrates the interaction between the writing system's elements, which can be used to estimate the body force exertion (e.g., pressure on the writing surface), as well as influence factors that are external to the hand-tool system (e.g., position and orientation of the writing surface). Parush et al. (1998) employed these descriptors within the Hebrew Handwriting Evaluation (HHE), in order to collect data concerning speed of handwriting, ergonomic factors and handwriting quality. These hand-tool-system descriptors are in accordance to Kao (1979), who defined the handwriting system by three major components: the hand, the writing instrument and the writing surface. The three components are closely related and interact with each other through visual, tactile, kinesthetic and auditory feedback. Receiving the movement of the pen shank provides kinesthetic feedback. Viewing the pen-point and the strokes in writing letters provides visual feedback. Kao (1979) proposed an off-centered pen-point design in order to improve the visual feedback. The smoothness of the point of the pen (e.g., ballpoint smoothness depends on the type of ink, size of the ball, etc.) and the writing surface characteristics provide auditory feedback. Noise can be produced between the writing surface and the writing instrument due to friction.

The fourth group of descriptors describes human characteristics, skills and capacities such as hand dominance (e.g., Burt and Benbow, 2007), stage of development and age (e.g., Rosenbloom and Horton, 1971, Schneck and Henderson, 1990, Ziviani and Elkins, 1986). Ziviani and Elkins (1986) stated that based on the 'normal' grip development timeline, children start to learn the dynamic tripod grip at the age of seven. However, there are many adults who do not achieve a mature grip.

EXPERIMENTAL ANALYSIS

The handwriting grips of 23 children aged 3 to 14 years were observed und analyzed in experiments. The task consisted of writing for the older children, as well as, drawing and coloring for the youngest. The handling of different types of writing instruments was observed, using photo and video analysis during the tasks. How to use new types of writing instruments was not explained.

In this section, we summarize three additional observations, captured during the experimental analysis of test subjects.

Firstly, despite being offered pencils with different grip zone types, the children's handwriting grip did not completely change. An "automatic" correction of the handwriting grip only occurs in some of the experiments. We observed that some children tried several different positions of the fingers in order to adapt to the pen grip. Our hypothesis was that the handwriting grip for children is influenced by the design of the writing instruments together with the interaction between the writing instrument and the writing surface. A possible cause for not observing this correction can be due to our short observation periods and the long times required in the handwriting, grip-modifying learning process. To validate this, we suggest the execution of long-term experiments both with young children, as well as with adults who have a primitive handwriting grip.

Secondly, we observed that the handwriting grip did not only depend on the skills and developmental process of the person, but also on the writing task. Concretely, during hachuring, it was frequently noticed that the pen's lead wore down to one side and the distance of the lowest fingertip to the pen-point increased, compared to drawing fine lines (see Figure 3 and Figure 4). This led the test subjects to adopt other postures to achieve the desired painting or drawing effect.

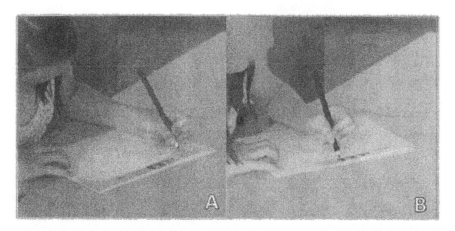

FIGURE 3 Increment of distance of the lowest fingertip to the pen-point. (A) Drawing fine lines, (B) hachuring. Subject Nr. 11, girl, 11 years old.

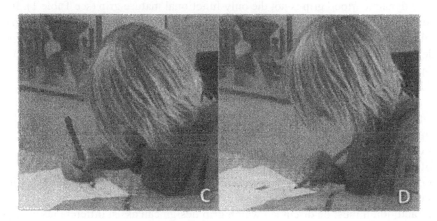

FIGURE 4 Wear of pen's lead to one side. (C) Drawing fine lines, (D) hachuring. Subject Nr. 8, boy, 9 years old.

Thirdly, we observed that the handwriting grip was adjusted depending on the characteristics of the writing instrument (e.g. features of the pen-point). To avoid having to sharpen the pencil continuously, test subjects rotated it towards the less worn side. Symmetric pencils facilitated a uniform utilization of the lead, and were intuitive and simple to grab –especially for small children-. This seemed to be of importance for (colored) pencils and crayons, but less relevant for pens, fountain pens and quills. Another example concerns the flow of ink, which can also influence the grip. This is due to the required higher inclination of the shaft, which is needed in order to achieve a fluent, soundless writing.

CONCLUSIONS AND FUTURE WORK

The objective of this paper was the systematic review of variations on handwriting grips. Designers and scientists can use the broad compendium of grips and descriptors as a basic tool for the design and evaluation of ergonomic products, as well as for the ergonomic assessment of handwriting grips.

The goal should be to develop a mature grip, which is characterized by

- a relaxed-muscle body posture
- well-coordinated and intrinsic finger movement.

Although it is known that there is a relationship between the stage of development and the age (Ziviani and Elkins, 1986), there are many adults, who do not achieve a mature grip. Further research regarding the frequency and the reasons for this is required.

Writing instruments for young children, who have not yet developed a mature pen grip, should offer an optimal ergonomic pen grip. Designers should be aware that the dynamic tripod grip is not the only functional mature grip (see Table 1). It is however not clear, whether the answer favors a pen design with multiple grip possibilities or whether it is more convenient to define preference and rigorous guidelines for a single grip. Investigations on adequate teaching material and guidance regarding ergonomic handwriting grips are necessary.

The experiments described in the literature, including our experimental analysis, only consider short-term experiments. The investigation of the human response to long-term practice is necessary, i.e., the familiarization with an unusual but recommended handwriting grip and/or with a new handwriting instrument.

Further research is also needed regarding the recommended ergonomic design for older children and adults, who practice a dysfunctional pen grip. Should the pen grip design be adapted to frequently found dysfunctional grips? Will this adaption help reduce the finger pressure and muscle cramps?

In addition it must be discussed which grip design can assist different tasks (e.g., scribbling, drawing or writing) and which grip design can assist different characteristics of the writing instrument (e.g. features of the pen-point). The experimental analysis of this study strongly suggests a dependency between the adopted handwriting grip and the characteristics of the writing instrument, as well as, between the adopted grip and the writing task.

The results of this work on writing instruments can also be applied to the investigation and design of other precision tools, which do not require high force exertion (e.g., chirurgical instruments, paint brush, etc.). Adaption of the instrument's grip to the anthropometry of the hand requires the analysis of the human grip and the research of its influencing factors.

ACKNOWLEDGEMENTS

The authors would like to thank to Yangpeng Zhao for his valuable contribution of drawing the handwriting grips and collecting literature.

REFERENCES

Bailey, C.A. (1988), "Handwriting: Ergonomics, Assessment and Instruction." *British Journal of Special Education*, 15(2), 65–71.

Blöte, A. W., and Dijkstra, J.F. (1989), "Task effects on young children's performance in manipulating a pencil." *Human Movement Science*, 8, 515–528.

Blöte, A.W., and Heuden, P.G.M. v.d. (1988), "A follow-up study on writing posture and writing movement of young children." *Journal of Human Movement Studies*, 14, 57–74.

Braswell, G.S., and Rosengren, K.S. (2007), "Task constraints on preschool children's grip configurations during drawing." *Developmental Psychobiology*, 49, 216–225.

Burt, C., and Benbow, M. (2007), "Children and handwriting ergonomics." In: *Ergonomics for children: Designing products and places for toddler to teens.* R. Lueder, and V.J.B. Rice (Ed.), Taylor & Francis, London and N.Y., 690–720.

Callewaert, H. (1963), "For easy and legible handwriting". *New Horizons for Research in Handwriting*, University of Wisconsin Press.

Carlson, K., and Cunningham, J.L. (1990), "Effect of pencil diameter on the graphomotor skill of preschoolers." *Early Childhood Research Quarterly*, 5, 279–293.

Diaz Meyer, M., Sica, L., Sinn-Behrendt, A., and Bruder, R. (2010), "Klassifikation von Schreibhaltungen: Unterstützung für eine ergonomische Schreibhaltung bei Kindern." In: *56th spring conference of Gesellschaft für Arbeitswissenschaft (GfA)*, March 24 – 26, Darmstadt. GfA-Press, Dortmund.

Kao, H.S.R. (1979), "Conventional and cybernized writing instruments." *IPSI Conference*, Montenegro.

Koziatek, S.M., and Powell, N.J. (2003), "Pencil grips, Legibility, and speed of fourth-grader's writing in cursive." *The American Journal of Occupational Therapy*, 57(3), 284–288.

Lueder, R., and Rice, V.J.B. (2007), *Ergonomics for children: Designing products and places for toddler to teens.* Taylor & Francis, London and N.Y.

McHale, K., and Cermak, S.A. (1992), "Fine motor activities in elementary school: preliminary findings and provisional implications for children with fine motor problems." *The American Journal of Occupational Therapy*, 46, 898–903.

Parush, S., Levanon-Erez, N., and Weintraub, N. (1998), "Ergonomic factors influencing handwriting performance." *Work*, 11, 295–305.

Rosenbloom, L., and Horton, M.E. (1971), "The maturation of fine prehension in young children." *Developmental Medicine & Child Neurology*, 13, 3–8.

Schenk, T., Bauer, B, Steidle, B., and Marquardt, C. (2004), "Does training improve writer's cramp? An evaluation of a behavioral treatment approach using kinematic analysis." *Journal of Hand Therapy*, 17, 349–363.

Schneck, C.M. (1991), "Comparison of pencil-grip patterns in first graders with good and poor writing skills." *The American Journal of Occupational Therapy*, 45, 701–706.

746

Schneck, C.M., and Henderson, A. (1990), "Descriptive analysis of the developmental progression of grip position for pencil and crayon control in nondysfunctional children." *The American Journal of Occupational Therapy*, 44(10), 893–900.

Selin, A.S. (2003), *Pencil grip: a descriptive model and four empirical studies*. Abo Akademi University Press. Doctoral thesis, 140.

Winn-Parry, C.B. (1966), *Rehabilitation of the Hand* (2nd edn.). Butterworth, London.

Ziviani, J. (1983), "Qualitative changes in the dynamic tripod grip of seven to fourteen year olds." *Developmental Medicine and Child Neurology*, 25, 778–782.

Ziviani, J., and Elkins, J. (1986), "Effect of Pencil grip on handwriting speed and legibility." *Educational Review*, 38 (3), 247–257.

Chapter 77

Ergonomic Student Workstation as a Condition for Proper Shaping of Body Posture

Joanna Lecewicz-Bartoszewska, Aleksandra Polak-Sopinska,
Zbigniew Wisniewski

Department of Production Management and Logistics
Technical University of Lodz
Wolczanska 215, 90-924 Lodz, Poland

ABSTRACT

In recent years there has been a sudden increase in the number of people with body posture defects, whereas the average age of the affected has been constantly lowering. The young, not completely shaped organism of a child, is especially susceptible to the influence of physical factors. Incorrect classroom design, not adjusted to rapidly changing features of adolescent organism, poses a serious threat to its proper development. The analysis of the sitting positions adopted by children during classes at school, disclosed the need for better adjustment of student workstations to anthropometric requirements, for increased ergonomic awareness of work organization, as well as the need for considering changes in arrangement of workstation, depending on the objectives and forms of didactic activities. Based upon the research results, an ergonomic example design was developed for a school desk for the youngest students.

Keywords: Student workstation, anthropometric requirements, body posture

INTRODUCTION – SPACE MANAGEMENT DETERMINANTS

Managing space of any human activity is predominantly identified through formal regulations. The above incorporates definite area or cubature per one person depending on the length of time spent in the room, number of persons and harmfulness of the environmental conditions. The data provided belongs to the quantity category and, what is even worse, it is used also numerically, without taking into consideration geometrical setup of space and the whole range of interactions between particular stations.

It is obvious that the area of education space first of all, directly or indirectly, depends on anthropometrical dimensions of its users. These determinants refer to all elements of space, such as convenient entering and getting out of the room, moving around, accessibility of regulating, control and operation elements, equipment and, the last but not least, stations of direct work, which to a large extent determine other ergonomic relations in education area. The workstations, due to for example frequency of usage, influence postural, musculoskeletal load of persons using classrooms.

Besides anthropometrical conditions and their derivational – physiological conditions also psychical factors are important. These factors provide for needs, abilities and limitations of users and they refer also to the closest environment (colour climate, good frame of mind, full audiovisual comfort, free choice of colleagues in the nearest neighbourhood).

Pupil's workstation as the most important element of teaching space, also from the perspective of ergonomics should be adjusted to changing work conditions resulting from various forms of classes demanded by the teaching programme.

Only certain, basic ergonomic factors have been mentioned above. At their full-scale they are a significant carrier of functions of proper body posture and consequence good health prophylactics.

REASON FOR TAKING UP PUPIL WORKSTATION DESIGN

In recent years there has been a sudden increase in the number of people with body posture defects, whereas the average age of persons affected with them has been constantly lowering. A young, still forming organism of a child is particularly susceptible to influence of physical factors. Incorrect classroom design, not adjusted to the rapidly changing features of the adolescent organism, poses a serious threat to its proper development.

This situation prompted the authors to carry out a research among the youngest pupils (from first to third grade) in five primary schools in the Łódź region and in three primary schools in the city of Łódź [Glinka and Lecewicz-Bartoszewska, 2002; Lecewicz-Bartoszewska, 2005; Lewandowska and Lecewicz-Bartoszewska, 2002]. In the objects selection the rule of their accessibility was used. The research included:

- observations of body positions assumed by pupils during classes in classrooms, determined mainly by dimensions of their workstations,
- guided interviews with pupils on the sensations they experience during classes and their satisfaction with the workstations they use,
- surveys among teachers on problems of matching workstations with anthropometric features of pupils, on their knowledge of the rules of matching and organisation of the teaching space.

The observations were carried out in 13 groups, for 65 pupils (random choice of 5 from each grade) and 20 teachers (form masters of all the observed classes and representatives of school authorities).

Classrooms for the youngest pupils in the surveyed schools were equipped with two-persons desks and with chairs with not regulated height. They all had certificates of being in accordance with children's anthropometric dimensions upon the standard PN-EN 1729:2007 „Furniture. Chairs and desks for education institutions. Functional dimensions".

Differences in height between children in the same age group – first grade (separately for boys and girls) is presented in Figure 1. Lack of adjustment of pupils' workstations to their anthropometrical dimensions, and in consequence assuming incorrect body positions when writing and listening, as well as attempts of "compensating" for these problems is presented in photos under Figure 2.

FIGURE 1. Differences in height of boys and girls in first grade [Glinka and Lecewicz-Bartoszewska, 2002; Lewandowska and Lecewicz-Bartoszewska, 2002]

The photos disclose the maladjustment of the desks height to children's height and of the seats height to the children's popliteal height. They also show wrong matching of desks and seats.

A short girl and a tall boy seat together at one desk. The desk is adjusted to the boy's height. When writing, the girl has to raise her arms too high. Boys sitting at too low desks bend over too much when writing. They even push back their chairs far from the desk.

FIGURE 2. Body positions assumed by pupils during writing and listening and positions "compensating" for static load fatigue. [Glinka and Lecewicz-Bartoszewska, 2002; Lewandowska and Lecewicz-Bartoszewska, 2002]

The length of the desks does not allow for leaning forearms at their whole length (both elbows) upon them, which is especially inconvenient when the desk is pushed to the wall. While listening the pupils cannot find enough room under the desktop for their legs. They often twist their bodies and sit sidelong to the working surface of the desk. Sitting at a high desk on a low chair is compensated by putting legs under buttocks. When sitting on a chair which is too high and has too deep seat it is impossible to rest the both feet completely upon the ground. If a pupil leans his/her back against the back-rest of such a chair while listening, the pupil's spine is excessively deflected to the back. When writing a child would move towards the desk and sit at the front edge of the chair.

The surveys among children prove that almost half of the pupils in first grade and 2 remember instructions given by the teacher concerning the choice of desks and chairs to sit at. Unfortunately they do not seem to understand the purpose of the rules. Relations with their friends and peer-colleagues are the most important for them, also when choosing or changing a place to sit. Easy to move chairs foster changing places. The situation in third grade is better.

The situation presented above indicates the need of "connecting" the seat for the youngest pupils with the desk.

The questionnaires and interviews with teachers prove that all of them have knowledge on adjustment of desks chairs to pupils' dimensions. Yet, unfortunately two out of the interviewed group of teachers expressed their doubt about purposefulness of this adjustment, motivating them by lowering the aesthetic properties of the classroom arrangement.

The knowledge and awareness of the need of taking up the enumerated adjustment activities has led to their almost common usage. Still in this area the teachers pointed at certain problems. Children, in particular the first and second grade pupils, could not remember the colours of signs of chairs and desks assigned to them or did not pay attention to them, as they did not understand the sense of such instructions. Very often pupils when seated at a proper chair joined a chosen colleague at a desk not adjusted to their specific height. Controlling if the children abide by the rules of using the furniture assigned for them is very difficult in terms of organisation. In practice control should be performed at the beginning of every lesson. Hence the notion to restrict the possibility of free matching of a desk with chairs and introducing for the youngest pupils the solutions referring to a traditional desk, but not a single-person one. The teachers were sceptical about regulation of the angle of the seat lean, even a two-position one, motivating their doubts with the statement that "the more regulation options, the more difficult t is to control". Small children are eager to make changes without any need, just for fun. The teachers perceive the possibility of solving the problem of reducing postural load of pupils in changing organisation of classes for the youngest of them, shortening the classes and introducing more dynamic activities.

EXAMPLE OF ERGONOMIC DESIGN OF A SCHOOL DESK

ERGONOMIC-APPLICATION GUIDELINES

The results of the observations, interviews with pupils and surveys among teachers were the basis for developing ergonomic-application guidelines for design project of a pupil workstation. The most important of them include:
- single-person workstation,
- anthropometric adjustment to the youngest pupils,
- possibility of creating various arrangement of workstations,
- lightness, mobility and strength of workstation construction,
- connecting the seat with the desk,
- permanent height of the working surface to assure possibility of setting group workstations,
- range of seat depth regulation corresponding to 5. and 95. percentile of users.
- height of the seat adjusted to the height of the desktop assuring correct body position during work,
- shape and size of the desktop providing support for elbows,
- regulation of the lumbar spine support,

- user-friendly seat and desktop material,
- application of a movable seat in order to liquidate the differences between the seat height for the shortest and the tallest pupils,
- planning a place for keeping a pupil's belongings,
- neutral colour of pupil's working surface, assuring a small contrast (of intensity and colour) with the background for a book or a copybook,
- convenience of changing body position – sitting down, standing up,
- assuring correct position of a pupil while writing and listening,
- minimising the risk of an injury – eliminating sharp edges, protruding elements,
- possibility and easiness of exchange of specific destroyed elements of the workstation – minimising purchase of new equipment, lowering repair costs.

DESIGN DESCRIPTION

Pupil's workstation is a combination of a desk and a seat, a sort of reminiscent of a traditional school desk (Figure 3). The desk has been designed as a single– person piece, to be used by pupils in their first to third grade. The height of its desktop and seat is settled. The parameters to be regulated are: the seat's depth, height of lumbar spine support and height of footrest.

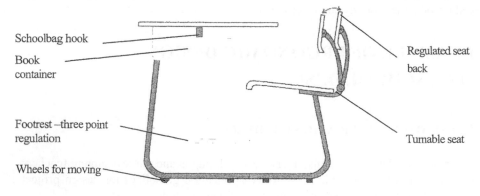

FIGURE 3. Concept design of a school desk for the youngest pupils [Glinka and Lecewicz-Bartoszewska, 2002]

The above mentioned regulations, and in particular the shape of the desktop assure correct position of pupils during writing (in accordance with the requirements of Phelps rehabilitation method). Both forearms of a child can be completely rested on the desktop next to the torso being maximum close to the front edge of the desktop. At the desktop, with every setup, there is enough room for a book, a copybook or A3 sheet of paper (Figure 4.).

These solutions and regulation ranges enable adjusting the desk to anthropometric dimensions of pupils at the age of 7 to 10 and representing 5. to 95. percentile (Figure 5. and 6.).

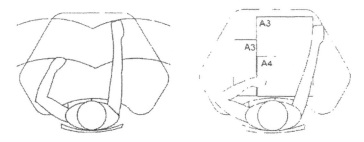

FIGURE 4. Shape of desktop, arranging it's area, reach of forearms [Glinka and Lecewicz-Bartoszewska, 2002]

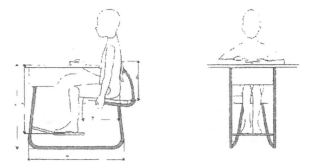

FIGURE 5. Adjusting the desk to a seven years old 5. percentile girl [Glinka and Lecewicz-Bartoszewska, 2002]

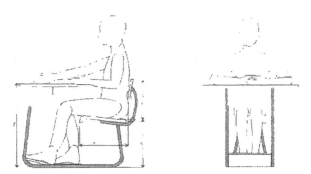

FIGURE 6. Adjusting the desk to a ten years old 95. percentile girl [Glinka and Lecewicz-Bartoszewska, 2002]

In the concept design of the desk a turnable seat is planned with set height and angle of the sitting surface inclination. A turnable seat enables free change of body position from the seated to the standing one and the other way round, both ways in relation to the desk. Change of the seat depth is possible thanks to application of spine support "following the user" and is associated with changing the height of support of the lumbar spine (Figure 7. and 8.).

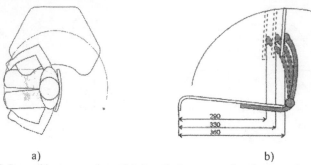

a)   b)

FIGURE 7. a) Preparing to stand up; b) Regulation range for the seat depth [Glinka and Lecewicz-Bartoszewska, 2002]

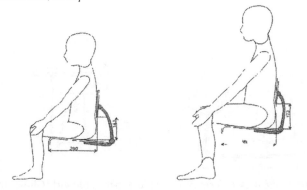

FIGURE 8. Change of the seat depth and back support height for girls: seven years old 5. percentile and ten-year old 95. percentile [Glinka and Lecewicz-Bartoszewska, 2002]

Following the assumption referring to the desk mobility, it has been equipped with wheels situated at the front of the support frame. Together with the lightness of construction they enable free and comfortable moving of the desk after raising it minimally from the seat side. The way of using the desk in this aspect is presented in Figure 9.

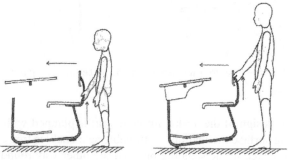

FIGURE 9. The way of moving the desk by a seven years old 5. percentile and ten-year old 95. percentile girl [Glinka and Lecewicz-Bartoszewska, 2002]

Figure 10 presents possibilities of creating free arrangements of desks depending on the needs of the teaching programme, its aims and forms. Various configurations of desks enable wider realisation of the postulate of integration of healthy children and children with disabilities.

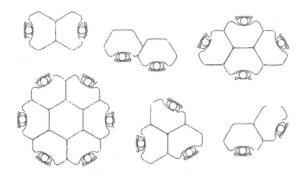

FIGURE 10. Possibilities of the desks arrangement [Glinka and Lecewicz-Bartoszewska, 2002]

DESIGN SOLUTION – DESK PROTOTYPE

Bearing construction of the desk has been made of metal closed profiles with square diameter (20x20mm), coloured surface (powder coating, in case of aluminium – anodizing). The construction consists of elements, stable joins, easy to assembly and disassembly, which assures convenient transport, storage and changing destroyed elements. The surfaces in direct contact with the user – the desktop, the seat and its back are made of a natural "friendly" material – a 10 mm thick profiled leafy plywood. The edges of the plywood are rounded. External surfaces of the plywood are bright, matt, assuring small contrast with the sheets in a book or a copybook, and causing no light reflexes. The wood pattern, especially on the desktop is amorphous (no visible rings) to avoid putting a pupil off stride.

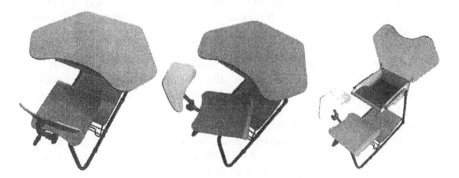

FIGURE 11. Desk prototype – general view, turned seat and raised desktop [Glinka and Lecewicz-Bartoszewska, 2002]

Under the desktop a container for a pupil's belongings, like pens, pencils and other "treasures" has been designed. The container is covered inside with a soft, pliable fabric, facilitating gripping of small objects.

CONCLUSION

The above results of estimate research of the state of adjustment of basic equipment of classrooms and ergonomic design of pupil workstations proves the necessity and possibility of ergonomic, complex managing of teaching areas for the youngest pupils.

Proper, anthropometric design of a pupil workstation is the essential element of heath prophylactics.

The desk can be a specific carrier of rehabilitation functions performed outside medical units, and because of its mass character an extremely important element in the chain of activities for rehabilitation of children with motor disabilities.

Still, developing an ergonomically correct design solution itself cannot be sufficient activity in heath prophylactics. It is necessary to promote knowledge on basic rules of ergonomics, which are recommendations of healthy, correct behaviours and ways of using products and objects of everyday usage. Also more awareness in ergonomics concerning organisation of pupils' work process and programming conditions of teaching areas should be stimulated.

BIBLIOGRAPHY

Glinka M. (2002), *Analiza ergonomiczna stanowiska pracy ucznia klas wstępnych szkoły podstawowej. Założenia ergonomiczne i projekt koncepcyjny stanowiska.* M.A. thesis, scientific care – J. Lecewicz-Bartoszewska. Institute of Ergonomics WFP ASP w Łodzi.

Glinka M. (2002), *Projekt wzorniczy ławki szkolnej dla uczniów klas pierwszych.* Design, scientific care – J. Ginalski. Institute of Design WFP ASP w Łodzi.

Lecewicz-Bartoszewska J. (2005), *Wzornicze projektowanie ergonomiczne pomocy rehabilitacyjnych dla dzieci z dysfunkcją ruchu.* Przegląd Pediatryczny. Suplement 1, Wyd. Cornetis, Wrocław.

Lewandowska M. (2002), *Analiza i ocena ergonomiczna stanowisk pracy uczniów w szkołach regionu łódzkiego.* M.A. thesis, scientific care – J. Lecewicz-Bartoszewska. Institute of Production Management, Technical University of Lodz, Lodz.

Polak-Sopińska A. (2008), *Analysis and ergonomic assessment of a selected production line and improvement plan.* in W. Karwowski & G. Salvendy (Eds.), Proceedings of the 2008 AHFE International Conference, Las Vegas, Nevada, USA, 14-17 July 2008. [CD-ROM]. USA Publishing (ISBN 978-1-60643-712-4)

CHAPTER 78

Of Age Effects and the Role of Psychomotor Abilities and Practice when Using Interaction Devices

Christine Sutter[1], Michael Oehl[2]

[1]Department of Work and Cognitive Psychology
RWTH Aachen University
Germany

[2]Institute of Experimental Industrial Psychology
Leuphana University of Lüneburg
Germany

ABSTRACT

Interacting with technical environments often challenges users in terms of cognitive and psychomotor requirements. Considering the users' developmental changes over the lifespan, technical devices do often not satisfactorily meet the demands of children or older adults. Recent evidence (e.g., Sutter and Ziefle, 2006) suggests that these so-called age-effects do not originate from age in general, but from a lack of practice in particular. In the present study we focused on fine psychomotor abilities as a further factor that might modulate age effects. We surveyed twenty-eight children and teenagers between 9 and 18 years of age. Fine psychomotor abilities were assessed with a standardized motor performance test battery. Then, participants performed point-click and point-drag-drop tasks with either a touchpad or a mini-joystick. Results showed that fine psychomotor abilities were less matured in children and that this was associated with a less efficient handling of the interaction devices compared to the older group. However, taking mouse practice into account age effects disappeared. Developmental changes over the lifespan and their implications for an efficient use of interaction devices will be discussed.

Keywords: Touchpad, Mini-joystick, Children, Teenager, Development

INTRODUCTION

Interacting with technical environments often challenges users in terms of cognitive and psychomotor requirements. Considering the users' developmental changes over the lifespan, technical devices do often not satisfactorily meet the demands of children or older adults. Consequently, performance is inferior within these two age groups compared to those of young and middle-aged adults (e.g., Armbruester et al., 2007; Inkpen, 2001; Smith et al., 1999). Recent evidence (e.g., Sutter and Ziefle, 2006) lets us assume that these so-called age-effects do not originate from age in general, but from a lack of practice in particular. Further findings (Barker et al., 1990; Smith et al., 1999; Sutter and Ziefle, 2005) give reason that fine psychomotor abilities are associated with input device performance, too. Both former studies assessed manual and finger dexterity with the pegboard test (Tiffin, 1968). Participants had to insert as many pegs as possible into the pegboard in a given time interval. It was found that high finger and manual dexterity was associated with a fast and accurate mouse performance. In a more recent study in our lab we used a touchpad and a mini-joystick. Psychomotor abilities were assessed with the "Motorische Leistungsserie" (= MLS by Schuhfried, 2002), a standardized motor performance test battery (Figure 1). According to Fleishman's structure of fine psychomotor abilities (1972) the MLS evaluates six factors (Table 1): aiming, steadiness, precision of arm-hand movement, manual dexterity and finger dexterity, rate of arm and hand movements, and wrist-finger speed. Our results corroborated once more, that manual/finger dexterity was highly associated with input device performance ($r = 0.51$, $p < 0.05$). Wrist-finger speed also correlated with time of cursor control ($r = -0.53$, $p < 0.05$) that means high wrist-finger speed was associated with fast input operations.

We can conclude that fine psychomotor abilities are associated with successfully using interaction devices. Furthermore, psychomotor abilities change over the lifespan. We find worse performance in children in whom fine psychomotor abilities are not fully matured, and in older adults in whom they are declining (e.g., Teipel, 1988). In sum, the motivation of the present study is to examine whether and how fine psychomotor abilities modulate age-effects. According to previous studies we will focus on manual/finger dexterity and wrist-finger speed as crucial factors that are associated with interaction performance. We hypothesize a strong correlation with movement time in that way, that children, whose psychomotor abilities are less matured need more time for cursor operations, whereas teenagers, whose psychomotor abilities are fully matured show faster cursor operations.

METHOD

In the present study we surveyed twenty-eight children and teenagers from 9 to 18 years of age. Fine psychomotor abilities were assessed with a standardized motor performance test battery. Then, participants performed point-click and point-drag-drop tasks with either a touchpad or a mini-joystick. The experiment lasted 1 h.

FINE PSYCHOMOTOR TASKS

We assessed the *fine psychomotor abilities* with the MLS and evaluated the factors *aiming, steadiness, precision of arm-hand movement, manual dexterity and finger dexterity, rate of arm and hand movements*, and *wrist-finger speed* in terms of errors, error time, total time and number of hits (Table 1). Participants performed the following five tasks with the right and the left hand in succession. They were instructed to work as fast and accurate as possible.

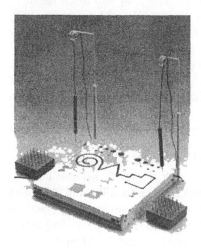

- Steadiness: Insert and hold the pen inside the smallest borehole (diameter 5.8 mm) for 32 s.
- Line tracking: Insert the pen inside the groove and follow it from "start" to "end" without any contact to the panel.
- Aiming: Hit each of the circles with the pen.
- Tapping: Tap the pen as fast and often as possible onto the square.
- Peg inserting: Take pegs and insert them into the boreholes on the right (right hand condition) or left side (left hand condition) of the panel, respectively.

FIGURE 1.Motor performance test battery (MLS[1]).

[1] http://www.schuhfried.at/de/produkte/wiener-testsystem-wts/testverfahren/spezielle-leistungstests/mls-motorische-leistungsserie.html, 12.01.2010

Table 1 Psychomotor abilities with MLS tasks and dependent variables

Psychomotor ability	MLS task	Dependent variable
Aiming	Aiming	Error, error time, hit
Steadiness	Steadiness	Error, error time
Precision of arm-hand movement	Line tracking	Error, error time
Manual dexterity and finger dexterity	Peg inserting	Total time
Rate of arm and hand movements	Aiming, Line tracking	Total time
Wrist-finger speed	Tapping	Hit

TASKS FOR INTERACTION DEVICES

Participants used a laptop computer (Dell Inspiron 8100), connected to an external 15" TFT flat screen (Iiyama TXA 3841J) with a 1024 x 768 resolution. The input devices integrated in the laptop were a touchpad, a flat 60 x 44 mm touch sensitive panel and a mini-joystick, a small force-sensitive joystick placed between the "G", "H" and "B" keys on the keyboard. Two mouse buttons were arranged horizontally in the wrist rest.

Participants were randomly assigned to operate either touchpad or mini-joystick. The touchpad is an isotonic device. Its gain is a function of finger motion to cursor motion. The mini-joystick represents an isometric device, i.e., the gain is a function of finger force to cursor speed.

The *point-click* and the *point-drag-drop task* represent typical demands of computer work. To meet demands of ecological validity, cursor actions varied from easy (big and near target) to difficult (small and far target). A trial started with a self-paced press on the space bar. The point-click task (Figure 2, subtask 1 alone) comprised a movement of the cursor inside the target box and a following click with the left mouse button. The point-drag-drop task consisted of several single actions that were executed one after another (Figure 2, all subtasks). In a point-drag action the centrally placed symbols were highlighted (subtasks 1 and 2). Then the object was picked up and dragged and dropped inside the square target on the right side (subtasks 3 and 4). For the drag actions participants were instructed to drag the cursor by pressing the left mouse button. For every successful action a visual feedback was given. Releasing the left mouse button at the end of subtask 4 completed the trial.

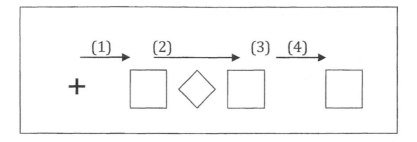

FIGURE 2. Task types: point-click task (subtask 1 alone) and point-drag-drop task (subtasks 1-4).

Participants were instructed to move the cursor as fast and as accurate as possible, and to operate the input device with their dominant hand, the left mouse button with their non-dominant hand. They worked throughout a block of point-click tasks and a block of point-drag-drop tasks. Each block consisted of 96 trials and additional 5 training trials in advance of each block. *Performance* was assessed in terms of movement time and errors. *Movement time* was defined as the interval from the onset of cursor movement to the final button press/release. *Error* percentages were calculated on the basis of trials where the left mouse button was pressed / released when the cursor was not inside the target box.

PARTICIPANTS

Twenty-eight children and teenagers ($N = 28$; 15 male) volunteered for the study. The children's group consisted of participants aged between 9 and 13 years ($M = 11$ years of age). The age of the teenager group ranged from 14 to 18 years ($M = 15$ years of age). In each age sample 14 participants were included. All participants were regular computer users with a contact time of approximately $M = 1$ h per day ($SD = 0.64$ h). The teenagers showed a significant longer contact time than the children (1.32 vs. 0.72 h/d; $F(1,20) = 9.41$, $p < 0.01$). Neither of them had any or only less experience with touchpad or mini-joystick.

RESULTS

EFFECTS OF AGE AND FINE MOTOR ABILITIES

Errors, error time, total time and number of hits observed in the MLS-tasks were analyzed with two-tailed t-tests for independent groups (11 vs. 15 years old).
    Right hand: For the *factor aiming* errors (0.6 vs. 0.9 errors), error time (each 0.03 s) and hits (each 20 hits) did not differ significantly between the 11 and 15 years old ($p > 0.1$). For the *factor steadiness* errors (10 vs. 5 errors; $t(20.54) = 2.36$,

$p < 0.05$) and error time (1 vs. 0.4 s; $t(26)=1.87$, $p < 0.1$) were higher for the 11 years old than for the 15 years old. The same pattern of results was found for the *factor precision of arm-hand movement*. Errors (30 vs. 20 errors; $t(26) = 2.17$, $p < 0.05$) and error time (2.7 vs. 1.7 s; $t(21.93) = 2.13$, $p < 0.05$) were significantly increased for the 11 years old. For the *factor manual dexterity and finger dexterity* total time was also higher for the 11 years old than for the 15 years old (51 vs. 47 s; $t(26) = 2.83$, $p < 0.01$). For the *factor rate of arm and hand movements* we found marginally significant differences between the 11 and 15 years old in total time of the aiming task (12 vs. 9 s; $t(26) = 1.84$, $p < 0.1$), but no differences in total time of the line tracking task (42 vs. 49 s; $p > 0.1$). At least, the analysis of the *factor wrist-finger speed* revealed a significantly lower number of hits for the 11 than for the 15 years old (166 vs. 186 hits; $t(26) = 2.94$, $p < 0.01$).

Left hand: For the *factors aiming, steadiness, and wrist-finger speed* a similar pattern of results was observed for the left hand. Again, age differences were not apparent in the *factor aiming* (each $p > 0.1$). However, for the *factor steadiness* children showed a less accurate performance in terms of errors (20 vs. 6 errors; $t(15.42) = 2.58$, $p < 0.05$) and error time (2 vs. 0.4 s; $t(14.04) = 2.96$, $p < 0.05$). And also for the *factor wrist-finger speed* children were outperformed by teenagers (143 vs. 160 hits; $t(26) = 2.39$, $p < 0.05$). For the *factor precision of arm-hand movement* we found marginally significant differences between the 11 and 15 years old in errors of the line tracking task (44 vs. 35 errors; $t(26) = 1.87$, $p < 0.1$), but no differences in error time in the same task (3.8 vs. 3.4 s; $p > 0.1$). The *factor manual dexterity and finger dexterity* did not reveal any differences between age groups (53 vs. 50 s; $p > 0.1$). However, the *factor rate of arm and hand movements* was significantly worse for the 11 than for the 15 years old in total time of the aiming task (12.6 vs. 9.7 s; $t(20.31) = 3.09$, $p < 0.01$), but it was significantly better in total time of the line tracking task (36.1 vs. 51.3 s; $t(20.06) = 2.26$, $p < 0.05$). Please note that the better performance of the 11 years old in total time of the line tracking task was due to a speed-accuracy trade-off, since at the same time errors and error times were substantially higher in the 11 than in the 15 years old.

EFFECTS OF AGE AND INPUT DEVICE PERFORMANCE

Movement times and error percentages observed in the point-click and point-drag-drop task (Figure 3)were analyzed with a 2 x 2 x 2 analysis of variance with the *between-subject factors age* (11 vs. 15 years of age) and *sensorimotor transformation* (motion vs. force transformation of touchpad and mini-joystick respectively), and the *within-subject factor task type* (point-click vs. point-drag-drop).

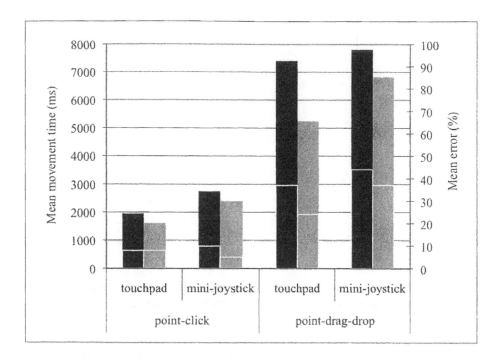

FIGURE 3. Movement times (ms) and errors (%) for point-click and point-drag-drop task (black bars = children, grey bars = teenagers).

For movement times the ANOVA revealed significant main effects for the *factors age* ($F(1,24) = 4.40$, $p < 0.05$) and *task type* ($F(1,24) = 375.25$, $p < 0.01$). The *factor sensorimotor transformation* was marginally significant ($F(1,24) = 3.75$, $p = 0.06$). Movement times were generally 965 ms higher in children than in teenagers (4981 vs. 4016 ms). In point-drag-drop tasks movement times substantially rose by 4641 ms. This increase was more pronounced in the 11 year old users than in the 15 year old users, yielding a significant *interaction between age and task type* ($F(1,24) = 6.65$, $p < 0.05$). However, these age-related effects totally diminished when the *factor computer practice* was included as a covariate in the ANOVA. At the same time the *factor sensorimotor transformation* gathered weight and revealed a significant impact on movement times ($F(1,24) = 6.31$, $p < 0.05$).

The analysis of the error percentages showed only a significant effect for the *factor task type* ($F(1,24) = 64.81$, $p < 0.01$). Other effects or interactions were not significant. Errors remarkably increased by 28% from the point-click to the point-drag-drop task (8% vs. 36%).

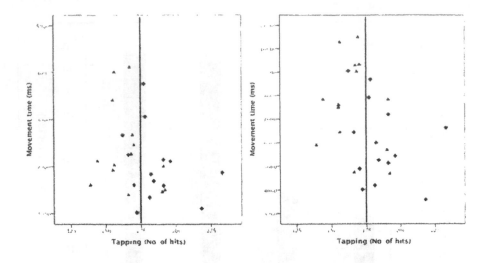

FIGURE 4. Wrist-finger speed of children (triangle) and teenagers (squares) as a function of movement time in the point-click task (left) and point-drag-drop task (right).

In the following section we examined whether and how fine psychomotor abilities - especially manual/finger dexterity and wrist-finger speed - modulated age-effects. We used rank correlations (two-tailed) between errors, error time, total time and number of hits observed in the MLS-tasks (right hand alone), and movement times and errors in the point-click and point-drag-drop task. For wrist finger speed we found high correlations with movement time in the point-click ($r = -0.38$, $p < 0.05$) and in the point-drag-drop task ($r = -0.41$, $p < 0.05$). The data are depicted in Figure 4. The faster the wrist-finger speed was, in terms of many hits in the tapping task, the faster was the cursor movement. Children were mostly represented in the lower half of the distribution of hits, and teenagers in the upper half. All other correlations did not reach significance.

DISCUSSION

The present study examined whether and how fine psychomotor abilities modulate age effects when using interaction devices. With regard to previous studies (Barker et al., 1990; Smith et al., 1999; Sutter and Ziefle, 2005) finger and manual dexterity and wrist-finger speed are associated with input device performance. Consequently, we assumed that if psychomotor abilities in children are less matured, this is associated with a less efficient handling of the interaction devices compared to the older group, in which psychomotor abilities are fully matured. The main finding confirmed our hypothesis: Wrist-finger speed was mostly lower in children and associated with slower input operations, and vice versa for teenagers. In general, the significant correlation between wrist-finger speed and movement time when using interaction devices corroborates findings from previous studies in our lab (Sutter and Ziefle, 2005). Surprisingly, we could not find any significant association with manual/finger dexterity in the present study. However, a closer look into the data showed the same pattern of association, although considerably weaker (point-click task $r = 0.26$, $p = 0.17$ and point-drag-drop task $r = 0.33$, $p = 0.08$), as observed in earlier studies (Barker et al., 1990; Smith et al., 1999; Sutter and Ziefle, 2005) for mouse, touchpad and mini-joystick.

A second finding was that age effects in the input device tasks diminished when we took computer and mouse practice into account. This was also found in one of our earlier studies for the computer mouse (Sutter and Ziefle, 2006). Teenagers were mostly more experienced in interacting with mouse and computers than children, but all participants were totally inexperienced with touchpad and mini-joystick. The fact that computer/mouse practice benefited the use of a new interaction device is remarkable. However, what cannot be answered at this point is, if this is due to a transfer of general task and computer knowledge and/or to specific motor skills.

As outlined in the introduction, ergonomic literature gives strong evidence for distinct age-related declines in children and older adults using interaction devices. These age differences were mostly observed to be stronger in complex and difficult tasks than in simple tasks (e.g., Armbruester et al., 2007; Inkpen, 2001; Joiner et al., 1998; Oehl et al., 2007). This was also the case in the present study. The performance of children substantially decreased in the point-drag-drop task, and this effect was stronger than for teenagers. As a reason for this we already discussed the development of fine psychomotor abilities. The ability to fine motor adjustment and to motor coordination becomes relevant for actions like double-/clicking, dragging and dropping, in particular.

At least we observed a worse performance for the mini-joystick compared to the touchpad. This is an already well-known finding when comparing force- versus motion-transforming input devices (an overview is given by Sutter, 2007). The less efficient handling of force-transforming input device seems to be totally independent from fine psychomotor abilities. And it is recently more and more discussed in the context of cognitive psychology and the cognitive processes underlying the use of tools with sensorimotor transformations (e.g., Lukas et al., 2009; Sutter et al., in press).

In sum, the present experiment showed that the observed age effects were moderated by computer/mouse practice and the fine psychomotor ability wrist-finger speed. Further studies should address the question of practice transfer and in particular examine whether general task and computer knowledge or specific motor skills benefit the use of a new input device.

ACKNOWLEDGEMENTS

We wish to thank Kathrin Wendler for supporting this research.

REFERENCES

Armbruester, C., Sutter, and C., Ziefle, M. (2007), Notebook input devices put to the age test: The usability of trackpoint and touchpad for middle-aged adults. *Ergonomics*, *50* (3), 426-445.

Barker, D., Carey, M.S., and Taylor, R.G. (1990), Factors underlying mouse pointing performance. In E.J. Lovesey (Ed.), *Contemporary Ergonomics* (pp. 359-364). Taylor & Francis, London, UK.

Fleishman, E.A. (1972), Structure and measurement of psychomotor abilities. In R.N. Singer (Ed.), *The psychomotor domain* (pp. 78-196). Lea & Febiger, Philadelphia, Pennsylvania.

Inkpen, K.M. (2001), Drag-and-drop versus point-click mouse interaction styles for children. *ACM Transactions on Computer-Human Interaction*, *8*, 1-33.

Joiner, R., Messer, D., Light, P., and Littleton, K. (1998), It is better to point for young children: A comparison of children's pointing and dragging. *Computers in Human Behavior*, *14*, 513-529.

Lukas, S., Brau, H., and Koch, I. (2009), Anticipatory movement compatibility for virtual reality interaction devices. *Behaviour & Information Technology*, First published on: 24 February 2009 (iFirst).

Oehl, M., Sutter, C., and Ziefle, M. (2007), Considerations on efficient touch interfaces – How display size influences the performance in an applied pointing task. In M.J. Smith & G. Salvendy (Eds.), *Human Interface, Part I, HCII 2007, LNCS 4557* (pp. 136-143). Springer, Berlin, Germany.

Schuhfried, G. (2002), Motorische Leistungsserie [Motor performance test battery MLS]. Version 23. Schuhfried, Mödling, Austria.

Smith, M.W., Sharit, J., and Czaja, S.J.(1999), Aging, motor control and the performance of computer mouse task. *Human Factors*, *41*, 389-396.

Sutter, C. (2007), Sensumotor transformation of input devices and the impact on practice and task difficulty. *Ergonomics*, *50* (12), 1999-2016.

Sutter, C., Müsseler, J., and Bardos, L. (in press), Effects of sensorimotor transformations with graphical input devices. *Behavior & Information Technology*.

Sutter, C., and Ziefle, M. (2005), Psychomotor user characteristics as predictors for a successful usage of small input devices. In *Proceedings of the HCI International 2005*. Mira Digital Publishing, St.Louis, MO.

Sutter, C., and Ziefle, M. (2006), Impact of practice and age on motor performance of mouse users. In *Proceedings of the International Ergonomics Association 16th World Congress*. Elsevier, Amsterdam, Netherlands.

Teipel, D. (1988), Diagnostik koordinativer Fähigkeiten. Eine Studie zur Struktur und querschnittlich betrachteten Entwicklung fein- und grobmotorischer Leistungen [Evaluation of coordination skills. A study for the structure and cross-sectional observed development of fine and gross psychomotor performance]. Profil, München, Germany.

Tiffin, J. (1968), *Purdue pegboard: Examiner manual*. Science Research Associates, Chicago, IL.

CHAPTER 79

Autism and the orbiTouch Keyless Keyboard

Peter McAlindon[1], Clayton Mayo[2], Breson DeLater[3]

[1]Blue Orb, Inc.
Orlando, FL 32751, USA

[2]University of Central Florida
Orlando, FL 32816, USA

[3]Colorado State University
Ft. Collins, CO 80521, USA

ABSTRACT

This research project investigates (a) the relationship between communication and motor skill deficits among individuals with Autism Spectrum Disorders (ASD) with their acquisition of specific computing skills and (b) the utility of assistive technologies to aid individuals with ASD in interacting with computer-based instruction tools that focus on language skills, such as vocabulary-building, typing, etc. An evaluation of the communication and motor skill development of individuals with ASD is timely as it provides the basis for the development of assistive technologies. The results will (a) advance the knowledge and understanding of ASD and its impairments and (b) provide further benefit by studying and developing improved input devices and protocols for users with ASD. This project contributes to the areas of human computer interaction for special populations and universal access to computer technology.

Keywords: Human Computer Interaction, Autism, Persons with Disabilities, orbiTouch, Ergonomic Keyboard

INTRODUCTION

There is an increasing population of children with developmental disorders, including ASD. In fact, ASD directly affects approximately 300,000 children in the U.S. population aged 6 to 22, and it indirectly affects (i.e., parents, caregivers, siblings) in much larger numbers. At the same time, progress in both computer technology and assistive technologies for the disabled has created opportunities for improving many areas of daily living, including communication and education. Thus, an extensive evaluation of the communication and motor skill development of individuals with ASD is timely, as it provides the basis for the development of assistive technologies. Furthermore, an investigation into the interaction of individuals with ASD with computers and assistive technology is warranted.

The project sets the foundations for improving the ability of individuals with ASD to access computer-based learning tools by studying the differential and combined effects of their motor and linguistic deficits on the use of computer input devices. Specifically, individuals' progress in interactions with vocabulary learning and typing software as they use traditional and alternative input devices (i.e., keyboards) will be studied. One of the alternative assistive technologies available, the orbiTouch Keyless Keyboard by Blue Orb, Inc. was developed primarily for people with cumulative stress injuries (CSIs) through a NSF Small Business Innovative Research (SBIR) grant. In recent months, however, the orbiTouch Keyless Keyboard has shown to be effective in improving access to computers for users with ASD (http://www.nsf.gov/news/news_summ.jsp?cntn_id=115476& org= NSF&from=news).

Motor Deficits in Autism Spectrum Disorders

The focus of this section is the effect of ASD on motor skill development and consequently the differences in motor skill functioning between autistic children versus normal functioning children. It should be noted, however, that the deficits interact with one another, and thus, some of the other characteristics of ASD can play a role in the delayed development of motor skills.

Motor skills can be divided into two groups: gross motor skills and fine motor skills. Gross motor skills are comprised of the larger movements of the entire body or the major limbs, including walking, jumping, and riding a bicycle. In contrast, fine motor skills consist of the smaller movements of the body, particularly those involving the hands (fingers and wrists), toes, and mouth (lips and tongue). Thus, fine motor skills include handwriting, speech, or using scissors. Gross and fine motor skills develop in conjunction with each other, as many activities rely on the coordination of these skills (Hames, 2003). A deficit in gross or fine motor skills will most likely impact the development of the accompanying skill as well.

Gross Motor Skills

One of the common characteristics of ASD is uneven gross motor movements (Autism Society of America, n.d.). Specifically, the gross motor movements such as crawling and walking may develop later than normal. These movements may appear to be clumsy or awkward due to their delayed development (HSDC, 1999). In addition to development delays, individuals with ASD may have difficulty walking through a revolving door or guiding a shopping cart through a store (Downey, 2001).

Fine Motor Skills

As previously mentioned, deficits in gross motor skills can also lead to deficits in fine motors skills. Consequently, children with ASD may also exhibit deficiencies in fine motor skills as well. Delayed development of fine motor skills can lead to difficulties with activities such as speech and the grasping of objects (HSDC, 1999). Additionally, children with ASD may experience apraxia of motors skills, i.e., difficulty carrying out voluntary motor skills. Similar to gross motor skills, fine motor skills in children with ASD may appear to be clumsy, awkward, or uneven (Autism-PDD Resources Network, n.d.).

Deficits in the sensory processing of touch may also affect fine motor skills. For example, individuals with ASD may not be able to discriminate between objects merely by touch (CT Autism Spectrum Resource Center, n.d.). Moreover, many children with ASD experience hypersensitive or hyposensitive senses (Edelson, 1999) which can contribute to the inability to process incoming stimuli and consequently react to those stimuli. Thus, a heightened sensitivity to touch may increase a child's difficulty to withstand normal stimulation, resulting in an aversion to grasping or manipulating foreign objects.

Potential Solution: Assistive Technology for Persons with ASD

Assistive Technology (AT) can be broadly conceptualized as any technology with the potential to enhance the performance of persons with disabilities. As defined by the Individuals with Disabilities Education Act Amendment of 1997, assistive technology is "any item, piece of equipment, or product system... that is used to increase, maintain, or improve functional capabilities of individuals with disabilities" (Part A, Sec. 602 (1)). Each year, tens of thousands of people are confronted with some form of disability as a consequence of an accident, disease, or other causes—and they wonder how they will cope (U.S. Department of Commerce, 2003). Assistive technology offers a wide range of alternatives. It includes both "low tech" and "high tech" devices, and it incorporates technologies designed specifically for people with disabilities as well as generic technologies developed for use by the general public (Langdon, Clarkson, & Robinson, 2007). AT holds great promise for bridging the digital divide between persons with disabilities and our information society.

Autism Spectrum Disorders provide a unique challenge to AT because, as a diagnosis, ASD is much less defined and understood than most other AT applications. It is further complicated by its multi-factor nature. There are few set solutions for AT assessments and applications as they often vary greatly case by case. Nowhere is this variable more pronounced than in working with persons with ASD. In cases of arthritis, for example, AT solutions are fairly well established and can be isolated to primarily motor control types of AT. In cases of ASD, by contrast, an AT professional needs to consider many more aspects of use including sensory, cognitive, communication, socialization, and motor abilities. This is further compounded by AT products becoming more complex, requiring users to integrate a variety of engineering, information technology, and human resource processes and components.

Individuals with ASD may particularly benefit from the use of technology because many technologies allow individual customization, flexibility, predictability, and clarity. These technologies also emphasize visual stimulation. Assistive Technologies include both simple and complex tools that enhance learning and cognitive skill, communication, manual dexterity, and social skill. Individuals with ASD demonstrate significant adaptation to a variety of skills derived from AT. AT can provide increasingly flexible and variable accommodations in helping a child with ASD develop various skills and abilities.

Because AT implementation is continually updated and reviewed, once technologies are identified and processes are implemented it is imperative that the team members examine the specific characteristics of the particular child with whom they work, and the way in which ASD manifests itself to understand how best the AT works for the child. ASD is a disorder that affects many abilities. Although many people with ASD demonstrate common characteristics, they vary greatly by degree and type. Moreover, the characteristics that an individual demonstrates can change dramatically over time. Thus, many specific considerations related to both technology and ASD must also be taken into account for the AT and the child to be successful. These specific considerations include communication issues, motor issues, input considerations, output considerations, software considerations, and troubleshooting strategies (Harden, 2001).

Moore and Calvert (2000) found that students with ASD appeared to be more motivated and more attentive, in addition to learning more vocabulary, while using a computer software program as opposed to traditional methods. All of the above citations provide evidence for the beneficial effects of using computer-based assistive technologies to aid in the development and acquisition of communication and language skills for individuals with ASD.

In the following, we summarize some of the key issues involved in the development and testing of specific assistive technologies for individuals with ASD.

- Proper evaluation and assessment not only occurs at the AT level but must also occur at the level of the individual with ASD. An evaluation should begin with an emphasis on the individual's functional capabilities prior to the examination of component skill definition and AT system construction.

- Students with ASD may have upper extremity motor control issues that affect their ability to appropriately use AT. Fine motor control issues can be overcome by use of alternative input devices including large key keyboards, head tracking devices, large buttons and switches, and the like. Thus, because gross motor skills in children with ASD are sometimes better developed, and therefore can be better adapted, this knowledge can drive the technology support that is selected and how well the child uses and benefits from the technology.

- When instructing a student with ASD on how best to use AT, the visual and physical setup should clearly indicate what item does which task. The function of each specific task and technology is not always readily apparent. When the student responds favorably to a particular aspect of the system, it should be explained and demonstrated and the function of each as it relates to the tasks to be completed should be reviewed.

Why the orbiTouch Versus Other Keyboards?

As indicated, children with autism oftentimes have limited fine motor skill (Kurtz, 2007; Green, et al, 2002) as well as visual perception problems (Kurtz, 2006). As a keyboard that lends itself well to gross motor skill (vs. fine motor skill in using a regular keyboard), the orbiTouch accommodates a physical need of the child with ASD. Moreover, the orbiTouch utilizes an alphabetical, color coded, character arrangement that aids in remedying visual perception problems. The orbiTouch also has mouse capability. However, perhaps its greatest advantage over other keyboards is that it removes a major first hurdle of perception in using a keyboard -- "it doesn't scare me", "the keys are in order" and "I operate the orbiTouch like I would two joysticks when playing a video game". This video illustrates these principles through the eyes of a young man with autism: http://www.nsf.gov/ news/News_ summ.jsp?cntn_id=115476&org=NSF&from=news. A story on how the orbiTouch Keyless Keyboard is being used by persons with autism can also be found at: http://www.nod.org/index.cfm?fuseaction=feature.showFeature&FeatureID=1184& C:\CFusion8\verity\Data\dummy.txt

Objectives

In this research, we investigated the effects of differences in motor skills among autistic children on the acquisition of basic skills in human-computer interaction. We achieved this by studying the children's progress in interactions with typing software as they use traditional and alternative input devices (i.e., keyboards). Our premise is that the existing, standard QWERTY-keyboard has two characteristics that make it particularly difficult to use for children with ASD. First, the keyboard requires fine motor skills for the actuation of the keys. Second, the presentation of all letters simultaneously, one on each key, with most keyboards having between 75 and 101 keys, presents a particular challenge to the attention requirements for children with ASD, a difficulty also associated with limitations in language development among autistic children.

One of the alternative technologies that we propose to test is the orbiTouch Keyless Keyboard by Blue Orb, Inc. (see Figure 1). The orbiTouch is a unique alphanumeric AT device that offers individuals a means of typing without the use of finger or wrist motion. It is a gross-motor oriented, full-featured keyboard and mouse. The orbiTouch is comprised of two domes where the hands rest comfortably. The typist creates a keystroke by sliding the two domes into one of their eight respective positions. Typists can achieve adequate typing speeds very quickly and with very little training.

This ergonomic keyboard may be used by children with ASD. In fact, the orbiTouch Keyless Keyboard has demonstrated promise for users with ASD (Williams, 2003). A user with ASD was able to learn and effectively use the orbiTouch, typing at a speed of 33 words-per-minute. We believe the orbiTouch may be particularly useful for users with ASD as it addresses two critical impairments that can be hypothesized to prevent the effective use of the regular keyboard and mouse combination: The orbiTouch Keyless Keyboard (a) requires more gross rather than fine motor skills for actuation and (b) being chorded, the orbiTouch does not present the same high-density display as the regular keyboard. Our purpose was to identify if, and how, these differences between the orbiTouch and the regular QWERTY keyboard affect interactions with typing skill acquisition software.

Figure 1. The orbiTouch Keyless Keyboard by Blue Orb, Inc.

METHOD

In a 2 x 2 between-subjects factorial design, participants were assigned to high and low ability groups in motor skills. Then, one-half of the participants in each cell will be randomly assigned to interact with the QWERTY keyboard; the other half will interact with the orbiTouch Keyless Keyboard. Using measurements of motor skills we determined the differential effects keyboard type has on performance.

	High Motor Skill	Low Motor Skill
orbiTouch Keyboard	22 wpm (n=4)	14 wpm (n=4)
QWERTY	8 wpm (n=4)	7 wpm (n=4)

Hypothesis – Effect of keyboard on performance: A significant difference in performance will be observed between the groups using the QWERTY and orbiTouch keyboards.

Participants

In this research, caregivers of current students in the several Central Florida autism schools were solicited for voluntary screening and potential participation in the research. Initial selection criteria for inclusion of a child in the screening will be:

- Child's age is between 8 and 15 years old
- Telephone screening indicates a minimum level of functioning necessary for inclusion in the study
- Parents/guardians consent to participate in the study

A review of the records indicated that, of the 1,500+ students in the database, more than 60% (i.e., 600+) were eligible for detailed screening using the criteria stated above.

Based on the assessment of the children's motor during the detailed screening (see below), 16 participants with ASD were selected for participation, fully crossing high and low ability groups in motor and language skills, respectively. One-half of the participants in each of the four (2 x 2) ability-based cells (i.e., 4 per cell) were assigned to interact with the QWERTY keyboard; the other half will interact with the orbiTouch Keyless Keyboard.

Each person's participation lasted approximately 10 hours. The participants were asked to complete ten (10) sessions of approximately 1 hour duration twice per week, for a total study participation of approximately 5 weeks.

Screening and Selection Evaluation

In order to better assess the efficacy of the orbiTouch, it will be important that we have a comprehensive overview of each child's overall functioning. Specifically it would be important to be able to identify, at a minimum, the level of motor skill, language competency, and social adaptability. Consequently, evaluation of children for inclusion in this study included a functional assessment of gross and fine motor skills and attending skills that are thought to be useful in distinguishing both high and low levels of motor skill competency.

Motor skills assessment

Motor skills will provide a measure of the degree of motoric facility needed to control or manipulate the orbiTouch during a given task. We will use a combination of items created from the following two detailed assessments:

- The Beery-Buketenica Developmental Test of Visual-Motor Integration (VMI) is suitable for ages 3 years old to adult. There is both a full format version and a separate motor coordination version. While this test covers the entire age range we are interested in, it mostly covers fine motor skills, and it requires the child to cooperate in an activity.

- The McCarthy Scales of Children's Abilities contains both fine and gross motor skills tests. The sections used for motor skills include items such as right-left orientation, leg coordination, arm coordination, imitative action, draw-a-design test, etc. While this test covers both fine and gross motor skills, the child has to be involved and follow directions, and the administration will take longer than in other tests.

In addition, in cooperation with occupational therapists and with ASD specialists, we have developed a pre-screening questionnaire for parents, caregivers, and guardians. This tool was to be used to determine minimum capabilities for inclusion in the detail screening described above.

RESULTS

The orbiTouch typists typing speeds averaged 18 wpm when compared to the QWERTY typist's average typing speed of 8 wpm. Character analysis revealed that the eight positions of dome movement on the orbiTouch were for the most part proportionally balanced. This finding indicates that no one position was more or less difficult to actuate than any of the other positions.

The main effect motor skill was found to be statistically significant, $F(1,14) = 19.67$, $p < 0.001$. Gaining proficiency was much more pronounced in the orbiTouch group than in the QWERTY group. Figure 2 indicates the non-linear performance for the 10 hours for this research study and beyond of both the orbiTouch and

QWERTY keyboards when used by children with autism. Also shown in the graph are the average typing speeds and times of non-autistic typing performances.

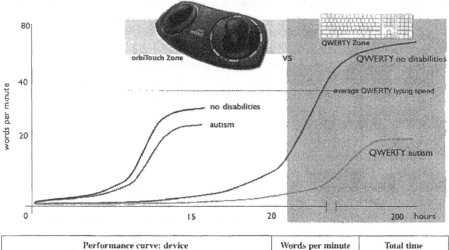

Performance curve: device	Words per minute	Total time
No disabilities: orbiTouch	32	15
Autism: orbiTouch	22	15
AVERAGE TYPING SPEED" QWERTY	39	100+
No disabilities: QWERTY	>70	200+
Autism: QWERTY	18	100

Figure 2. Learning curves of children with autism using the orbiTouch Keyless Keyboard and the QWERTY keyboard

CONCLUSIONS

The project sets the foundations for improving the ability of individuals with Autism Spectrum Disorder (ASD) to access computer-based learning tools by studying the differential and combined effects of their motor and linguistic deficits on the use of computer input devices. There is an increasing population of children with developmental disorders, including ASD. In fact, ASD directly affects approximately 0.4% of the population, and it indirectly affects (i.e., parents, caregivers, siblings) a much larger number. At the same time, progress in both computer technology and assistive technologies for the disabled has created opportunities for improving many areas of daily living, including communication and education. Thus, an investigation into the interaction of individuals with ASD with computers and assistive technology is warranted. Understanding the barriers to computer use for individuals with ASD will help to provide new tools to harness the power of computing.

REFERENCES

Autism Society of America (n.d.). *Common characteristics of Autism.* Retrieved on January 3, 2004 from http://www.Autismsociety.org/site/PageServer? pagename=AutismCharacteristics

Autism-PDD Resources Network (n.d.). *Autism checklist.* Retrieved on January 4, 2004 from http://www.Autism-pdd.net/checklist.html

CT Autism Spectrum Resource Center (n.d.). *Sensory processing issues in Autism.* Retrieved on January 4, 2004 from http://www.ct-src.org/doc/Sensory%20Processing%20Issues%20 in%20Autism.doc

Downey, M. K. (2001). *List of possible characteristics of a person with Asperger's Syndrome (AS), High Functioning Autism (HFA) or Pervasive Developmental Disorder-Not Otherwise Specified (PDD-NOS).* Retrieved on January 4, 2004 from http://www.mkdowney.com/characteristics.html

Edelson, S. M. (1999). *Overview of Autism.* Retrieved on January 4, 2004 from http://www.Autism.org/overview.html

Green, D., Baird, G., Barnett, A., Henderson, L., Huber, J., & Herderson, S. E. (2002). *The severity and nature of motor impairment in Asperger's syndrome: a comparison with Specific Developmental Disorder of Motor Function. Association for Child Psychology and Psychiatry, 43*(5), 655-668.

Hames, P. (2003). *What's the difference between fine motor and gross motor skills?* Retrieved on January 3, 2004 from http://www.babycentre.co.uk/expert/6562.html

Harden, B. (2001). Assistive technology for student with Autism. In H. Miller-Kuhaneck (Ed.), *Autism: A Comprehensive Occupational Therapy Approach* (pp. 201-224). Bethesda, MD: American Occupational Therapy Association.

Hearing, Speech, and Deafness Center (HSDC) (1999). *Communication update: Autism, what is it?* Retrieved on January 3, 2004 from www.hsdc.org

Kurtz (2007). *Understanding Motor Skills in Children with Dyspraxia, ADHD, Autism, and Other Learning Disabilities: A Guide to Improving Coordination.* Jessica Kingsley Publications.

Kurtz (2006). *Visual Perception Problems in Children With AD/HD, Autism, And Other Learning Disabilities.* Jessica Kingsley Publications.

Langdon, P., Clarkson, J., & Robinson, P. (2007). Designing accessible technology. *Universal Access Information Society 6:*117-118.

Moore, M., & Calvert, S. (2000). Brief report: Vocabulary acquisition for children with Autism: Teacher or computer instruction. *Journal of Autism and Developmental Disorders, 30*(4), 359-362.

U.S. Department of Commerce. (2003, February). *Technology Assessment of U.S. Assistive Technology Industry.* Washington: Office of Strategic Industries and Economic Security, Bureau of Industry & Security (in cooperation with National Institute on Disability and Rehabilitation Research, U.S. Department of Education and Federal Laboratory Consortium).

Center for Disease Control (2010). From http://www.fightingautism.org/idea/autism.php

Chapter 80

Using Evidence Based Research Method to Design Inclusive Learning Environment for Autism

Rachna Khare[1], Abir Mullick[2], Ajay Khare[3]

[1]National Institute of Design
Ahmedabad, India

[2]Georgia Institute of Technology
Atlanta, USA

[3]School of Planning and Architecture
Bhopal, India

ABSTRACT

Autism is a developmental disorder that leads to a different and characteristic pattern of perceiving, thinking and learning. To design a supportive learning environment for autism, it is necessary to understand behavioral pattern of children and its relation with the physical environment. The present study employs an evidence based research method to explore the impact of environmental settings on education for children with autism. There are several stages to this study; identify, refine and validate. First, an analysis of children's behavior in educational spaces identified 'eighteen environmental design parameters' that are enabling for autism. These design parameters are tested in the subsequent stages to provide evidence based body of knowledge to design autism friendly spaces that are inclusive of other children. The present study examines many inter-related aspects, education across ages, disability and cultures. This paper discusses the research process as well as some results from educational settings in the USA and India.

Key words: Inclusive Design, Learning environment, Evidence based Research

INTRODUCTION

Children with autism are found in all countries; they represent a range of cultural expectations, attitudes to education and disability, and resources (Jordan 1997). It is a lifelong developmental disorder that affects communication and social abilities of an individual. Other characteristics often associated with autism are engagement in repetitive activities and stereotyped movements and unusual responses to sensory experiences (APA 2000). It is not an uncommon disorder, according to the data released by the Center of Disease Control's (CDC 2001) Autism and Developmental Disabilities Monitoring (ADDM) Network in 2007, about 1in150 eight-year-old children in multiple areas of the United States had Autism Spectrum Disorder.

Educating children with autism is a challenge for both parents and teachers as they have unique strengths and weaknesses (Hodgdon 1995). Some may be of average to above average functional capabilities, while others may be of below average, but all children with autism can make significant progress if the intervention is appropriate and consistent. Educational programming for students with autism often addresses a wide range of skill development including academics, communication and language skills, social skills, self-help skills, leisure skills and behavioral issues (Schopler, Lansing & Waters 1983, Maurice, Green & Luce 1996). Today with dawn of inclusive education in the world, it has become vital to explore the influence of environmental design for autism. The present research study develops a framework to design user-friendly and high performance educational spaces for children with autism. Because of the inclusive and cross-cultural nature of the study, the results are beneficial globally and to a wide range of children.

METHODOLOGY

The methodology followed relies on evidence based practices in design that not only adds precision to the study but also helps in building a realistic and convincing argument in favor of the developed enabling aspects for autism. The research process mainly derives from Environment-Behavior research methods, discussed by Zeisel (2006), Preiser (2001), Steinfeld & Danford (1999) and Cherulnik (1993). It employs several research approaches. Sequentially starting with a concept, it draws from accumulated knowledge, existing theories and preliminary field survey to formulate the hypothesis. The hypothesis is then tested to verify the concept for the purposes, those can be generalized. Largely, the aim of the study is to recognize the environmental aspects effecting performance of children with autism, measure the environmental impact on learning and then develop a set of guidelines for architects and designers to design autism friendly educational settings. The study also makes an effort to explore the effect of the identified environmental aspects on able-bodied children and in cross cultural settings to establish a global perspective and a base for universal design.

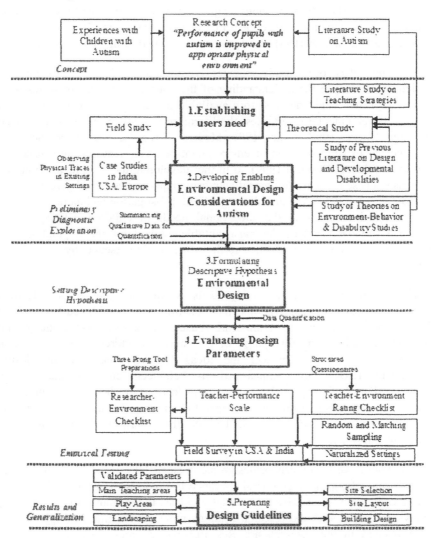

Figure 1 Five Phase Research Process adopted in the Study

The concept that propels the entire study is 'Performance of pupils with autism is enhanced in appropriate physical environment'. This concept originates from authors' experiences with children with autism, and the literature on autism. Acting as a reference for all future observations, it sets the objectives and raises the research questions. These research questions are systematically analyzed in five phases of the study (refer figure 1). These five phases are: (1) Establish relation between environment and the needs of children with autism (2) Develop Environmental Design Considerations (EDC) to address these needs (3) Derive Design Parameters (DP) from previous stages to present tangible and testable ideas (4) Conduct evaluations to validate identified design parameters (5) Prepare autism friendly Design Guidelines (DG) based on these evaluated design parameters.

In the *first phase* of the study, an extensive preliminary diagnostic exploration is carried out to deepen the understanding of the concept. This elaborate research stage in the present study derives from available literature on previous works on design and developmental disabilities, the literature on autism and strategies used for teaching (Siegel 1998, Jordan 1997, Maurice, Green & Luce 1996, Schopler 1983), and the existing theories of environment-behavior research and disability studies (Steinfeld and Danford 1999), to further develop unfocussed ideas about the users' needs. These ideas are further refined through preliminary field investigations carried out in Europe, USA, and India. The field investigation technique in natural settings adopted to establish the educational needs of children with autism is unobtrusive and described as the most suitable method for the population that cannot be interviewed and is sensitive to the researcher's presence, by John Zeisel (2006).

The *second phase* of the study leads to the development of detailed environmental design considerations that enable pupils with autism. These environmental design considerations are based on the elaborate diagnostic exploration carried out in first phase of the study to address the deficits and conditions in autism. To present the ideas tangibly, the enabling design considerations (EDC) are summarized as 'eighteen environmental design parameters' (DP) (refer figure 2). These design parameters help to organize and simplify the extensive information generated during preliminary diagnostic exploration. This is the *third phase* and it sets the descriptive hypothesis in the present research study.

The descriptive hypothesis or design parameters (DP) are tested in the *fourth phase* subsequently, to have empirical evidence in favor of the concept. Empirical testing is carried out to quantify the qualitative ideas; it contributes to the accuracy of knowledge and makes the research findings more convincing. Multiple sets of tools developed in the research are Researcher-Environment Checklist (REC), Teacher-Performance Scale (TPS) and Teacher-Environment Rating Checklist (TEC). The first two tools present sequential evaluations that assess environment and measures performance of the pupils in an environment. The third tool rate the importance of the developed parameters for the educational environment. The design parameters are tested in the existing educational setups based on the post occupancy evaluation in environment-behavior studies as discussed by Zeisel (2006) and Preiser (2001).

An extensive survey is carried out in the fourth stage of research and Researcher-Environment (REC) and Teacher-Performance (TPS) data are collected from sixteen educational spaces in USA and six in India. Teacher-Environment Rating (TEC) data is collected from eighteen experts working with low functioning children with autism and also from fourteen education experts working with able bodied children. The data is collected from teachers and therapists working with the children with autism, in a naturalized setting that is familiar and comfortable for children. The samples although selected randomly, represent all age groups in elementary, middle and primary schools; different type of educations settings, inclusive (children with and without disabilities) and specialized (only for children with disabilities); different education experts, working with children with autism and able-bodied children; and different countries, developed and developing. The empirical data is then structured, compared and analyzed both intimately and distantly at the same time to study similarities, differences and trends. Manifest and latent inferences from observations are drawn to answer the research questions formulated in the beginning of research.

To present a successful research in environmental design, it is necessary to fill the 'gap' between environment-behavior research, and design and planning practice (Reizenstein 1975, Cherulnik 1993). The *fifth and final stage* addresses that gap and the presents the research findings in a way, which can be used in applied design situation. It presents highly rated universally applicable parameters for children with autism those are also beneficial for able-bodied children and in different cultures. The information is presented in the form of architectural design guidelines for use by teachers, therapists, designers and evaluators. The guidelines are prepared for site selection, site layout, building design, main teaching areas, play areas and landscaping. These guidelines provide a framework to design high performance educational spaces for children with autism. The guidelines are expected to be universally beneficial and supplement existing design standards for schools to provide equal educational opportunity for everyone.

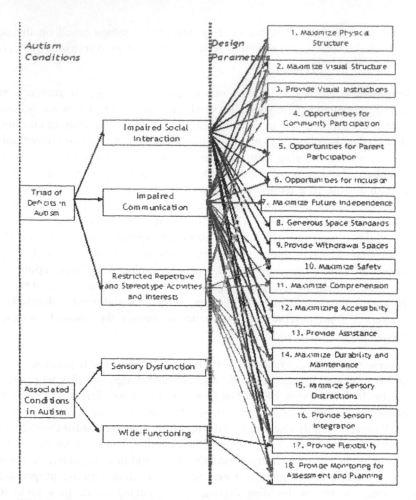

Figure 2 Design Parameter derived from Autism Conditions

RESULTS AND DISCUSSIONS

As described earlier, Researcher-Environment (REC) and Teacher-Performance (TPS) data are collected from sixteen educational spaces in USA and six in India. The Teacher-Environment Rating (TEC) data is collected from eighteen experts working with low functioning children with autism and also from fourteen regular education experts. The collated data is systematically analyzed to validate the eighteen design parameters in the study. To begin with, existing environment in the USA and India (representing a developed and a developing country) is assessed using Researcher-Environment (REC) tool. It is found that the existing environment in both responds to the identified environmental design parameters (DP) and thus ascertain cross-cultural validity of the parameters (refer figure 3). The teachers in both have arranged their classrooms according to the knowledge and resources available to them. The other areas in the school buildings designed by architects and designers did not respond to the needs of children with autism. It is also observed that the existing environment in the USA respond to the environmental design parameters better than in India. This is because of higher level of awareness, better resources, and stringent laws to support autism education, accessibility and inclusion in the school environment.

The empirical testing of environmental design parameters in elementary, middle and high schools settings for autism demonstrate intense association between enabling environments and educational performance of children with autism at all age levels (refer figure 4). Some variations in the graph profile illustrate that the performance is not exclusively dependent on the environment. There are many other factors affecting it, ranging from educational to social.

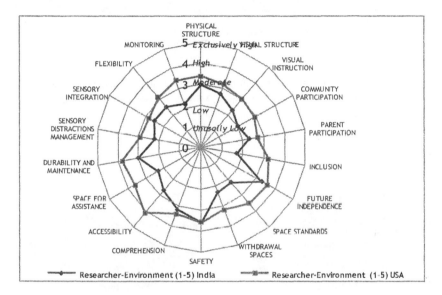

Figure 3 Graphical Representation of Comparative Analysis of Existing Environment in India and in the USA for Children with Autism

784

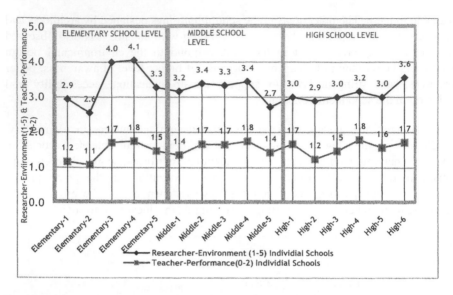

Figure 4 Graphical Representation of Relation between Educational Environment and Performance of Children with Autism in Primary, Middle and Secondary Schools

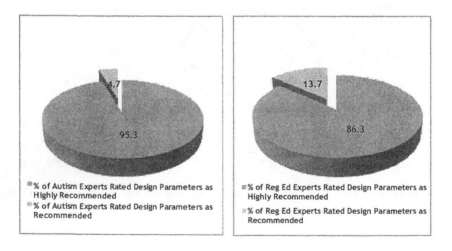

Figure 5 Overall Teacher-Environment Rating for Children with Autism and able-bodied kids

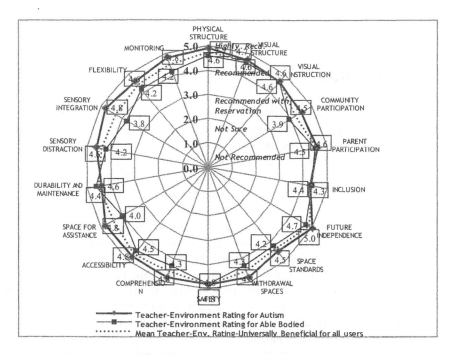

Figure 6 Teacher-Environment Rating for Children with Autism and Able-Bodied Children for Individual Parameters, and Mean Rating Universally Beneficial for All Users

The identified enabling design parameters (DP) are reviewed by autism experts and education experts for able-bodied children (figure 5). All teaching experts rated the environmental design parameters very high on Teacher-Environment Rating Checklist (TEC). Out of all experts, 95.3% experts rated design parameters as highly recommended for children with autism, whereas 86.1 % experts rated design parameters as highly recommended for able-bodied children. This confirms that the eighteen design parameters formulated for children with autism are not only favorable for kids with autism but are also beneficial for able-bodied kids.

Some thought provoking results are generated when Teacher-Environment ratings (TEC) of autism experts for children with autism and education experts for able bodied-children are compared for individual parameters (figure 6). It is found that environmental design parameters (DP) that are generally expected to be helpful for conditions like autism and other developmental disabilities are actually more beneficial for able bodied children. For example inclusion, expected to be in favor of children with autism, is rated higher for able-bodied children. Durability and maintenance is also rated as a bigger issue for able-bodied children than children with autism. Universal consequence of the design parameters to the educational environments is furthermore defined by the mean values of 'Teacher-Environment rating' by autism and regular education experts, that establishes universally beneficial inclusive environment for all users (figure 6).

CONCLUSIONS

The present research is inter-related, multi-disciplinary and multi-stage approach to achieve complex research objectives dealing with education, environment, autism, ability and inclusion. It tries to incorporate inter subjectivity, reliability, validity, tenability, testability and specifiability and generalizability, the seven indicators of quality research as defined by Zeisel (2006) and Steinfeld & Danford (1999). To incorporate inter-subjective opinions, data is presented and published at every stage, in different forums, at different countries. The subject matter is shared with educators, autism experts, researchers and designers to have diverse viewpoint. The research uses multiple techniques, draws from theory and employs literature surveys, field surveys, and multidisciplinary inputs, adding to its reliability. A testable hypothesis and the multi-layered evaluation tools, augment the validity of the research, and, the empirical testing of the design parameters to answer the research questions, assures the testability and tenability in the research findings. The specific results in the research are in the form of highly rated, inclusive, environmental design parameters, enabling children with autism. Finally, the amplified breadth and depth of the study in cross cultural settings developed generalizable design guidelines that have universal acceptability and global relevance.

The research is expected to draw attention of multidisciplinary team of researchers and has direct application in the field for all stake holders. The developed enabling environmental parameters may prove valuable to the researches in the field of autism, design, education and environmental-behavior research. The design guidelines provide a framework that may be used by designers, architects, teachers, educators, therapists, experts, school administrators and all those who wish to provide a successful environment for children with autism. A new understanding is developed through research, and practitioners may like to implement the findings to design effective schools for children with autism that will have positive educational impact on other children.

To conclude, the present research is a sequential progression that stands on accumulated body of knowledge, to produce environmental design guidelines that enable children with and without autism. Using a research process, with pre-established foundations, it generates new evidence based knowledge, to design supportive, accessible and inclusive learning environment for all in international context.

REFERENCES

APA (2000), Pervasive Developmental Disorders, *Diagnostic and Statistical Manual of Mental Disorders*. Fourth edition-text revision (DSM-IV-TR). American Psychiatric Association, Washington DC, pp. 69-70.

CDC (2001), Autism Community Report by Autism Developmental Disabilities Monitoring Network, Centers for Disease Control and Prevention, Department of Health and Human Services, USA.

Cherulnik, Paul D. (1993), *Applications of Environment-Behavior Research; Case-Studies and Analysis*. Cambridge University Press, New York.

Hodgdon, Linda A. (1995), *Visual Strategies for Improving Communication*. Quick Roberts Publishing.

Jordan, R. (1997), *Education of Children and Young People with Autism*. UNESCO.

Maurice, C., Green, G. & Luce, S. C. (1996), *Behavioral Intervention for Young Children with Autism: A Manual for Parents and Professionals*. Pro-ed, Austin, USA.

Preiser, W.F.E. (2001), "Towards Universal Design Evaluation", In W.F.E. Preiser, and E. Ostroff (Eds.) *Universal Design Handbook*. McGraw-Hill, New York, USA, pp.9.1-9.18.

Reizenstein, J. (1975 December), "Linking Social Research and Design", *Journal of Architectural Research*. No.3, pp. 26-38.

Schopler, E., Lansing, M. & Waters, L. (1983), *Individualized Assessment and Treatment for Autistic and Developmentally Disabled Children: Teaching Activities for Autistic Children*. TEACCH, NCSU, USA.

Siegel, B. (1998), *The World of the Autistic Child: Understanding and Treating Autistic Spectrum Disorders*. Oxford University Press.

Steinfeld, E., Danford, Scott G. (1999), "Theory as a basis for research on Enabling Environments", In E. Steinfeld & G.Scott Danford (Eds.), *Enabling Environments*. Plenum Press, New York, pp. 11-32.

Zeisel, J. (2006), *Inquiry by Design*. W. W. Norton and Company, New York, USA.

Chapter 81

The Older Drivers' Adaptive Strategies: Some Factors of Interest

Catherine Gabaude*, Ladislav Moták*, Jean-Claude Marquié**

*INRETS-LESCOT
Bron, France

**CNRS-CLEE-LTC
Toulouse, France
catherine.gabaude@inrets.fr

ABSTRACT

Four experiments are presented to describe some factors influencing the adoption of adaptive strategies by older drivers. (1) A case control study with 40 senior drivers (mean age 66 years) has been conducted to evaluate, in real road setting, their driving performance. Results showed that, cognitive and perceptive declines only explain a small part of the variance of the driving errors and that case group (volunteers having at least 3 accidents during a three year period) commits more violations. To better understand the influence of aberrant driving behavior a complementary approach has been developed. (2) We studied, in a sample of 568 older drivers (mean age 71 years), the impact of self-evaluation of functional abilities and declared aberrant driving behavior on the self-reported driving avoidance through questionnaires. Sequential multiple linear regressions revealed that 8 variables accounted for 49% of the variance in self-imposed driving limitations in the whole sample. The study suggests that perceived abilities, especially self-assessed driving-related processing speed and attention abilities play a major role in the decision to avoid difficult driving situations. (3) Then, we explored the impact of the awareness of functional abilities on driving avoidance. Seventy-six drivers took part in the experiment (mean age 76 years). We measured the differences between the subjective and objective evaluations of skills in three areas: vision, speed of processing and selective attention. A strong heterogeneity has been observed between participants in their functional abilities awareness. This study highlights the fact that people underestimating their attentional abilities reduce their mobility while people overestimating the same abilities are less avoiding difficult driving situations and probably are more exposed to accident risk. (4) In a last experiment we explored how older drivers cope behaviorally with difficult driving situations. Twenty-one younger drivers (mean age 30 years) and sixteen older drivers (mean age 72 years) drove on simulator through four circuits of increasing difficulty. In spite of the older drivers' lower level of chosen difficulty, both younger and older adults proceeded from easier to harder items during their training which refers to a well adapted self-regulation. These results are jointly presented to illustrate the influence of self-evaluation and self-regulation on the driving activity of older drivers so as to conclude on future research.

Keywords: older driver, cognitive abilities, driving behaviors, self-evaluation, self-regulation

INTRODUCTION

Allowing seniors to continue to drive safely and thus preserve their mobility is a strong social issue. Main part of research on aging and driving focused so far on cognitive abilities and on high-risk sub-groups of drivers (Rogé et al., 2009). In the driving assessment literature then, behavioral regulations are often mentioned. However, as noted in many studies, the observed perceptive and cognitive declines of older drivers often do not impair their driving performance as much as one would expect (e.g., Schlag, 1993), and this result leads different authors to account also for a possible higher-order cognitive (i.e., metacognitive) adaptations of older drivers (e.g., Kostyniuk & Molnar, 2008; but see also Marquié & Isingrini, 2001, for an overview of forms of adaptation to cognitive aging).

In this vein, we try to demonstrate through our four experiments exposed in this chapter that the efforts to accompany older drivers must be global and multifaceted. In accordance with other recent authors (see above), we claim that one's functional abilities self-awareness affects the use of adaptive strategies, and that it will probably enhance one's safety if accurate.

Our chapter is then progressively structured around these different adaptation-related factors as follows: perceptive and cognitive factors (Exp. 1), aberrant driving behaviors and self-assessment of functional abilities (Exp. 2), awareness of one's functional abilities (Exp. 3) and behavioral self-regulation (Exp. 4).

FIRST EXPERIMENT[1]

Objective: The objective of this first on road experiment was to better understand the impact of visual and cognitive declines on driving performance (Gabaude & Paire-Ficout, 2005).

---

[1] on road evaluation of the driving performance

Materials and methods: participants were recruited using two regional databases obtained from an insurance company (members having or not declared an insurance claim during the last three years). Forty senior drivers (mean age 66 years) were recruited for the real road experiment, 20 participants were case volunteers who had 3 or more at-fault accidents during a three year period and 20 were control volunteers with no accident record during the same period. The two groups were matched on age, gender and study level.

The assessment procedure included a medical exam with visual tests (static visual acuity in far vision, low-contrast vision and movement perception), neuropsychological tests (Mini-Mental Status Evaluation, Zazzo crossing-out test, Wechsler digit symbol substitution test) and finally a road test. The driving performance evaluation (quantifying a penalty score) was conducted with an instrumented vehicle along a fixed route (25 km) including urban, rural and motorways sections. Participants were assessed at 45 pre-designated check-points on the test route following or not the instructions of the experimenter (guided versus non guided condition). A road test observation grid was filled out in real time by an experimenter. Eight dimensions (mirror check, visual search, indicator use, speed, direction following, lane choice, violation and positioning at intersection) were coded and then a scoring grid was completed using weightings according to the characteristics of each situation. At the end, a penalty score was calculated.

Results: no significant difference was observed on the penalty score between men and women and the study level was not correlated with the penalty score. The penalty points were attributed as followed: 28% of penalty point for visual search, 24% for mirror checking, 23% for indicator use, 14 % for lane choice, 6% for violation and 5 % for positioning at intersection.

Regarding case and control group comparisons: significant differences are observed between case and control group for the total score and for the guided score: case individuals having a higher mean penalty score (table 1). An analysis of the subtotals of the eight rated dimensions showed only a significant difference between case and control group on violation criteria. Twenty violations were observed in the case group and 5 in the control group. For both groups, violations were observed more frequently for left turn intersections.

Table 1 Comparison of non-driving and driving tests scores for case and control group

	Case (n=20)	Control (n=20)	p<
Visual tests m (sd)			
Movement perception	10,05 (1,10)	10,35 (1,09)	0,39
Contrast vision m (sd)	3,90 (2,08)	3,25 (2.57)	0,38
Cognitive tests m (sd)			
MMSE	28,30 (1,42)	27,85 (1,39)	0,32
Zazzo time	56,05 (9,29)	49,80 (12,98)	0,09
Digit symbol test	44,65 (9,08)	50,90 (8,26)	**0,03**
Road test score m (sd)			
Total score	38,87 (14,55)	29,55 (10,80)	**0,03**
Guided score	20,57 (8,78)	14,24 (6,16)	**0,01**
Non guided score	17,98 (7,84)	15,62 (6,92)	0,32
Violation score	2,93 (3,72)	0,85 (1,32)	**0,03**

Regarding road test performance prediction: all variables (age, visual and cognitive variables) were significantly correlated with the penalty score excepted MMSE score. Forty three percent of the variation of the penalty score could be predicted based on age, group and scores obtained with 2 non-driving tests: movement perception and time for Zazzo's test (R^2= 43 %, R^2 adjusted= 36 %, F= 6.55, p<0.001).

Discussion: this study confirmed the relationship between perceptive and cognitive functions and the driving performance. The result from multiple regression analysis underlines the relevance of two tests which can be used easily. The time for Zazzo's test was the best cognitive measure to detect unsafe drivers. It is not surprising as it is a composite test including speed of information processing and selective attention (more particularly inhibition). Some authors have showed that these mechanisms play an important role in driving activities (Lafont et al, 2010).

Even if participants were observed by an experimenter we noticed that case group commit violations. It might be interesting to find a tool to consider this personality treat. Within this framework, the driving behaviour questionnaire (DBQ) might be interesting (Reason et al. 1990).

Finally, results showed that, cognitive and perceptive declines only explain a small part of the variance (43%) of the driving errors. Some other factors must therefore be at play when committing these errors. The adoption of aberrant driving behaviors is a first avenue worth exploring and the discrepancy between perceived abilities and objective ones can be another one. It seems interesting to develop questionnaires prospecting these issues.

SECOND EXPERIMENT[2]

Objectives: the most common adaptive strategies adopted by older drivers is the avoidance of dangerous driving situations (Vance et al., 2006; Kostynuik & Molnar, 2008). Then, to better understand the limits of these adaptive strategies we decided to explore the impact of self-evaluation of functional abilities and declared aberrant driving behaviors on the self-reported driving avoidance through questionnaires (Gabaude et al., 2010). We examined the relative predictive validity, with respect to avoidance by older drivers, of the DBQ and of material addressing driving-related perceptual and cognitive abilities more directly. Behavioural scales such as the DBQ may help identify the psychological factors that lead to the adoption of self-regulation designed to compensate for cognitive aging.

Materials and methods: a questionnaire was mailed to 1,500 men and women retirees owning a car. Among the returned questionnaires, 568 were properly completed and could be included in the analyses (mean age of the participants 71 years).

The questionnaire was divided into 4 sections: general information on participants (age, sex, health status and recent accidents), self-reported aberrant driving behaviors (French version of the DBQ, Gabaude et al, 2010), self-evaluation of three perceptual and cognitive abilities known to be important in driving (visual efficiency, speed of information processing, and attentional abilities) and avoidance of eleven driving situations reputed difficult for older drivers (bad weather, driving alone, heavy traffic, slippery roads, foggy weather, unknown route, night driving, roundabouts, left-turn situations, high speed roads and glaring sunlight).

Results: three types of self-reported driving errors measured by DBQ were studied (inattention errors, serious errors and violations).
Entering the above mentioned variables in three steps (table 2: model 1, 2 and 3), sequential multiple linear regressions revealed that 8 variables accounted for 49% of the variance in self-imposed driving limitations in the whole sample. By decreasing importance, avoidance was more frequent in drivers with less efficient self-assessed cognitive abilities, older subjects, females, reported smaller annual mileage, no recent crash experience, fewer violations and poorer health.

---

[2] influence of self-evaluation of functional abilities and declared aberrant driving behavior on driving avoidance

Table 2 Predictors of driving avoidance (sequential multiple regression analyses)

(n =568)	model 1 (β std)	model 2 (β std)	model 3 (β std)
Age	.198**	.199**	.180**
Females (vs males)	.133**	.165**	.127**
Annual mileage (> 8,000km vs ≤ 8,000km)	-.226**	-.124**	.115**
Health status	-.225**	-.076*	-.075*
Accident / 3 years (vs none)	-	-.124**	-.116**
Perceived abilities:			
Sight		-	-
Processing speed		-.286**	-.260**
Attention		-.292**	-.275**
Aberrant driving behaviour:			
Inattention errors (F1)			-
Serious errors (F2)			-
Violations (F3)			-.098**
R^2	.22	.48	.49

p≤.01**, p≤.05*

Discussion: findings show that the DBQ is not as good a predictor of the driving avoidance as self-evaluation using a few items of the cognitive abilities involved in driving such as processing speed and attention. The study suggests that perceived abilities, especially self-assessed driving-related processing speed and attention abilities play a major role in the decision to avoid difficult driving situations. It can also be concluded that such self-efficacy beliefs are a stronger predictor of avoidance than the DBQ for the older drivers.

In a third time, to better understand the limits of these adaptive strategies, we want to explore the influence of the discrepancy between subjective and objective evaluation on the strategic choices done by the drivers. Then, we invited some participants to another experiment.

THIRD EXPERIMENT[3]

Objectives: then, we sought to specify the conditions associated with activation of this regulation form (Gabaude et al., 2007). We examine the role the awareness of changes affecting its own functional abilities in the implementation of avoidance in the older. To date, the role of awareness deficit has been studied and demonstrated for the visual abilities (Holland & Rabbit, 1992). We sought here to extend the study to other aspects of the skills involved in driving (speed of information processing and attention), with the assumption that an accurate awareness is more favorable to the establishment of regulations of avoidance type.

---

[3] Impacts of subjective and objective evaluations on driving avoidance

Materials and methods: this experiment included 76 drivers (mean age 76 years). We first measured the differences between the subjective and objective evaluations of skills in three areas: vision, speed of processing and selective attention. For each area, the subjective evaluations (self-evaluations) were explored using 4 questions related to the driver's abilities in specific situations. Participants responded on a 10 points scale between "very good" and "very bad" and a subjective evaluation score was calculated. The objective evaluations were obtained with the measure of the visual acuity in far vision, the Wechsler digit symbol substitution test score, and the Stroop test score of interference. For each area, participants were allocated in a double entry table (objective and subjective measures) according to three skill levels (good, medium, bad – using 33th and 66 th percentiles). Then, individuals were grouped into three classes either reflecting an underestimation, a good accuracy, or an over-estimation of their subjective assessments (evaluation bias).

Participants should also indicate the frequency of avoidance of each of 10 difficult driving situations previously cited (all excepted foggy weather). The question was worded as follows: "Because in these situations you feel uncomfortable, you avoid getting behind the wheel." Responses were made on a 10-point scale ranging from "never" to "Always" and an avoidance score was then calculated by summing the 10 responses.

Results: correlations between objective and subjective evaluations scores are not significant. To go further, using the 33th and 66th percentiles of the distributions of subjective and objective functional abilities we define three modalities as independent variables (good, medium, or low) and we conducted analyses of variance with the avoidance scores as dependant variables. Only one significant effect was observed for the subjective evaluation of attention ($F[2,71]=8.83$, $p<0.001$), the avoidance score increases when the subjective score of attention decreases.

Regarding the influence of the evaluation bias on avoidance score, analyses of variance were also conducted. For vision and processing speed we did not find significant relationships between evaluation bias and the avoidance score; nevertheless, for attention, people who underestimated their attentional abilities reported more avoidance than those who overestimated (figure 1).

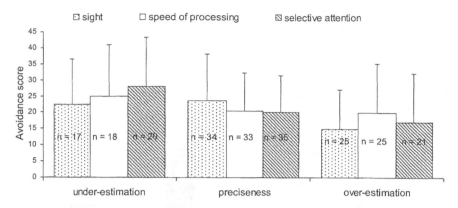

Figure 1 variation of the driving avoidance score according to evaluation bias (mean and standard deviation)

Discussion: concerning functional abilities awareness, a strong heterogeneity has been observed between participants. This study highlights the fact that people underestimating their attentional abilities reduce their mobility (possibly wrongly) while people overestimating their attentional abilities are less avoiding difficult driving situations and probably are more exposed to accident risk. We notice that avoidance is a self-regulatory behavior depending on high-level strategic choices like self-evaluation or deficit awareness. To better understand how these adaptive strategies are implemented it seems interesting to really observe them in a context allowing a good control on the environmental constraints (i.e. a driving simulator).

FOURTH EXPERIMENT[4]

Objectives: Almost all of the recent studies discussed the elderly drivers' adaptation through the analysis of self-declared avoidance. No behavioral data are available, however, to describe the way older drivers really deal with experienced difficulties while driving. Therefore, inspired by metacognitive theories (e.g., Metcalfe & Kornell, 2005), we tried to propose a new experimental design to explore (a) the way older drivers adapt their behavior when faced with situations of variable difficulty (see Metcalfe, 2002 for a similar experimental paradigm), and (b) the difference between older and younger drivers in this adaptation.

Materials and methods: twenty-one younger drivers (mean age 30 years) and 16 older drivers (mean age 72 years) drove on simulator through four circuits of increasing difficulty, lasting 5 minutes each. Immediately after, during a training phase of only 8 minutes, participants chose in which circuits they wanted to train themselves. In order to allow a fine-tuned behavioral adaptation, the time left was indicated on a special screen embedded in the cabin during the whole training

---

[4] Behavioral adaptation to difficult driving situations while driving a simulator

session. In accordance with a model of self-regulated learning proposed by Metcalfe and Kornell (2005), our statistical analyses focused then on time spent in each circuit and the order of choices during a training phase.

Preliminary results: older drivers spent more time by training themselves in the easiest circuit than in the hardest one, $t(15) = 2.58$, $p <$ .05 (Figure 1). The contrary was the case for younger drivers, $t(21) = 3.38$, $p < .01$. Importantly, both younger and older adults proceeded from easier to harder items during their training, $F(1, 35) = 6.56$, $p < .05$ (Mean levels of difficulty for the first two choices: $M = 1.81$, and $M = 2.56$ for older; $M = 2.76$ and $M = 3.14$ for younger drivers).

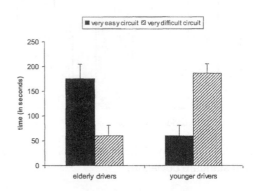

FIGURE 1. Time spent by two age groups during the training session in the easiest and the hardest circuits (mean and mean standard errors).

Discussion: As both age groups started their training by some easier items and only after moved to some harder ones, our preliminary results indicate that both age groups evaluate well what is difficult when driving, and more importantly, can adapt accurately to this perceived difficulty. Yet, even if older drivers' adaptation is of comparable accuracy and has the same form as that of younger drivers, it differs in its content as they spent more time by training themselves on easier circuits.

At the moment, we admit two concurrent hypotheses to explain this last result. First, it is possible that the choice to train themselves mainly on easier circuit is a consequence of the older drivers' propensity to avoid some difficult driving situations in a real life (Vance et al., 2006; Ball et al., 1998). This choice would then mirror, and possibly prove predictive of the real-life older drivers' self-regulation.

The second, maybe more realistic hypothesis is that the older drivers' choices were task adapted as they involved the most profitable items. The training time being very short and the elderly people's learning rates obviously slower than these of younger ones, our older drivers turned spontaneously to those items where they might expect some training benefits, instead of wasting their time in excessively hard circuits with only low probability to drive through safely. This result is completely in line with predictions of the Metcalfe and Kornell's model of an accurate self-regulated learning (Metcalfe & Kornell, 2005).

CONCLUSION

Through our four experiments, we tried to offer a better understanding of the older drivers' adaptive behaviors from different, but highly complementary points of view. Our point was here to indicate varied factors to be emphasized in the development of more efficient training procedures, i.e., the factors of potential benefit for elderly drivers.

First, on the level of functional abilities, we confirmed an effect of some perceptive and cognitive abilities, and above all that of selective attention. As expected, however, the eventual declines in this ability cannot account for all drivers' behaviors, and we turned therefore to other, higher-order cognitive abilities. These latter, namely the self-assessment and the awareness of one's functional abilities and self-regulation, are supposed to compensate for the older drivers' functional declines if accurate.

Until now, one of the most widespread operationalizations of how to investigate the older drivers' adaptation was by means of comparison between their self-declared and/or objectively measured physical and cognitive status at one hand, and the driving situations they affirm to avoid at the other (e.g., Gabaude et al. 2010, Charlton et al., 2006; Ball et al., 1998). Our final experiment, however, provides evidence that (a) it is possible – and worthy – to observe the real older drivers' behavioral adaptations, and that (b) the measures of this adaptation on different levels (self-declared avoidance or observable strategic choices), are absolutely complementary and should be taken into account when deciding about the older drivers' functional abilities. In this vein, our results proved rather positive: elderly drivers of our experiments can efficiently adapt their behaviors provided they are accurately aware of their functional abilities, i.e., they are aware of difficulties they may encounter when driving.

In all industrialized countries with a growing population of elderly drivers, relevant tools and methods should be developed in the near future to find the right balance between strict control of driving abilities and possible accompaniment programs to preserve safe mobility for older drivers. This kind of programs should not be based solely on training of specific skills but also on knowledge and awareness acquisition when driving. Moreover, future research should include the factors we mentioned above so as to better conceive the accompaniment program. Some statistical analyses allow to prospect the status of these different factors to distinguish main factor, moderator or mediator to better understand how we can operate to improve the elderly drivers' performance.

Acknowledgement: authors wish to thank Frédérique Obriot-Claudel for her help (data from experiment 2 and 3 were collected during her PhD work). Data from experiment 4 are collected by Ladislav Moták within his PhD work.

REFERENCES

Ball, K., Owsley, C., Stalvey, B., Roenker, D.L., Sloane, M.E., and Graves, M. (1998), "Driving avoidance and functional impairment in older drivers." *Accident Analysis and Prevention,* 30 (3), 313-322.

Charlton, J.L., Oxley, J.A., Fildes, B., Oxley, P., Newstead, S., Koppel, S., and O'Hare, M. (2006), "Characteristics of older drivers who adopt self-regulatory driving behaviours." *Transportation Research Part F: Traffic Psychology and Behaviour,* 9 (5), 363-373.

Gabaude, C., Marquié, J.-C., and Obriot-Claudel, F. (2010), "Self-regulatory driving behaviour in the elderly: relationships with aberrant driving behaviours and perceived abilities." *Le Travail Humain,* 73(1), 31-52.

Gabaude, C., Obriot-Claudel, F., and Marquié, J.-C. (2007), "Conduite automobile et vieillissement : relation entre prise de conscience de ses capacités visuo-attentionnelles et évitement de situations difficiles." *Congrès national de la Société Française de Psychologie* (p. 66). Nantes, France.

Gabaude, C., and Paire-Ficout, L. (2005), "Toward a Driving Competency Assessment Encouraging Elderly's Automobility: a french point of view." *Third international driving symposium on human factors in driver assessment, training and vehicle design,* (pp. 325-334). Rockport, June 27-30, Maine, USA.

Holland, C. A., and Rabbitt, P. M. A. (1992), "People's awareness of their age-related sensory and cognitive deficits and the implications for road-safety." *Applied Cognitive Psychology, 6,* 217-231.

Kostyniuk, L.P., and Molnar, J. (2008), "Self-regulatory driving pratices among older adults: health, age and sex effects." *Accident Analysis and Prevention,* 40 (4), 1576-1580.

Lafont, S., Marin-Lamellet, C., Paire-Ficout, L., Thomas-Anterion, C., Laurent, B., & fabrigoule, C. (in press), "The Wechsler Digit Symbol Substitution Test as the best indicator of the risk of impaired driving in Alzheimer disease and normal aging. " *Dementia and cognitive geriatrics disorders.*

Marquié, J.-C., and Isingrini, M. (2001), "Aspects cognitifs du vieillissement normal. " In E. Aubert & J.-M. Albaret (Eds.), *Vieillissement et Psychomotricité* (pp. 77-113). Marseille: Solal Editeur.

Metcalfe, J. (2002), "Is Study Time Allocated Selectively to a Region of Proximal Learning?" *Journal of Experimental Psychology: General,* 131(3), 349-363.

Metcalfe, J., and Kornell, N. (2005), "A Region of Proximal Learning model of study time allocation." *Journal of Memory and Language,* 52, 463-477.

Reason, J., Manstead, A., Stradling, S., Baxter, J., and Campbell, K. L. (1990). "Errors and violations on the roads: A real distinction?" *Ergonomics,* 33(10/11), 1315-1332.

Rogé, J., Paire-Ficout, L., Gabaude, C., Motak, L., and Marin-Lamellet, C. (2009). "New trends in road traffic safety research on senior drivers." In Paterson S, Allan L (Eds.): *Road Traffic: Safety, Modeling, and Impacts,* (pp. 321-345). New York: NOVA publishers.

Schlag, B. (1993), "Elderly drivers in Germany – Fitness and driving behavior." *Accident Analysis and Prevention*, 25(1), 47-55.

Vance, D. E., Roenker, D. L., Cissell, G. M., Edwards, J. D., Wadley, V. G., and Ball, K. K. (2006). "Predictors of driving exposure and avoidance in a field study of older drivers from the state of Maryland." *Accident Analysis and Prevention,* 38(4), 823-831.

CHAPTER 82

Evolution of Mobility by Public Transport and of Drivers' Working Conditions: New Relationships Between Work and Aging

Liliana Cunha, Marianne Lacomblez

Centro de Psicologia da Universidade do Porto
Rua do Dr. Manuel Pereira da Silva
4200-392 Porto

ABSTRACT

In most European countries, the evolution of the utilization rate of public transport has involved frequent oscillations (Eurostat, 2006). In Portugal, and specifically regarding collective road passenger transportation, it is common to discuss the (un)reliability of this kind of transportation, in terms of schedule and frequency, due to external factors (crashes; traffic jams; road works;...) (Antrop, 2002), as well as factors resulting from changes in working conditions and work organization options in the sector (Cunha & Lacomblez, 2007). Several measures for the improvement of existing infrastructure and accessibility, as the construction of bus corridors or paid parking lots in bigger urban centers, have been taken in order to counter that tendency, even if their contribution has not been decisive in the increase of users for this mode of transportation. Recognizing this was determinant in guiding our research towards the analysis of working conditions and work organization options which regulate the real activity of those providing the service – as it is assumed by our scientific tradition, that of Activity Ergonomics – given that the point of view of the driver's activity is not being emphasized on mobility through public transportation debates, and changes taking place aiming at mobility improvement do not happen without costs to these workers. The guiding thread of this article is based on the idea that the intervention in the sector cannot, therefore, be made without a simultaneous focus on what happens *inside* it, this is, without considering the working conditions and their impact on health and on drivers' aging quality.

Keywords: Mobility by bus; driver's working conditions; aging *by* work.

INTRODUCTION

Nowadays, it is clear the statement that aging does not start at retirement but is a process that begins early in life (Robert, 1983). Aging is accompanied by several organism changes with some negative effects on physical and cognitive abilities, which may change the way of working. Thus, it is called by Teiger (1989) as "aging *in* work". On the other hand, the age advance is also made of positive changes, related to experience, maturity and regulation processes, developed to compensate the deficit of some functions. Work and its characteristics intervene in this evolution (Laville, Teiger & Wisner, 1975), accelerating, in some cases, the aging process, due to an early deterioration of certain health dimensions. In this case, we refer to "aging *by* work" (Teiger, 1989; Ramos & Lacomblez, 2008a), resulting from medium and long-term effects of working conditions on workers' health. In these situations we state that worker addressed demands have gone beyond his/her limits (BTS, 2001) and the conditions in which he/she ended up performing his/her professional activity have not allowed the development of compensatory mechanisms (Ramos & Lacomblez, 2008b).

Thus, if it is true that the advance on age may allow an individual the construction of experience – namely by balancing professional maturity and health preservation – one cannot neglect that activity constraints on work context may difficult and even preclude such compromise, with obvious negative effects on health.

This kind of analysis and reflection is a surplus in comprehending and framing the relations between work and aging (Cassou & Laville, 1996; Derriennic, Touranchet & Volkoff, 1996; Laville, Teiger & Wisner, 1975; Marquié, Paumès & Volkoff, 1998; Teiger, 1989), an issue that we assume as particularly pertinent face to the current Portuguese conjuncture in collective road passenger transportation sector.

Two important questions will support our research presentation: (i) even in the absence of an explicit age management policy, can the way how career paths are managed reveal different choices according to age? (ii) how are career paths affected by work organization changes and what is its relation with the sector's current situation?

DRIVERS' CAREER(S) AND MOBILITY(IES)

Among researches on bus drivers' activity, we can enhance some, focused on their working conditions and consequences on their health. In the UK, Tse, Flin & Mearns (2006) trace drivers' occupation health research evolution, beginning at the mid twentieth century. These authors analyse several studies in this scope and conclude that drivers' health can be affected by working in specific conditions, as inadequate workplaces, shift work, continuous time pressure, traffic or risk of passengers violence, which may end up turning into both physical (such as cardiovascular, gastrointestinal and muscular-skeletal problems) and psychological problems (such as depression or anxiety). Drivers must build a commitment between these sometimes incompatible activity demands, concerning safety, client service and company goals. Meijman & Kompier (1998), in a study conducted in Holland, also refer the contradictory activity demands, analysing the way how drivers establish priorities in face of them and how those strategies impact their health. For example, in face of the impossibility of keeping up with the schedules while performing a safe driving, the drivers who consider more important to keep the schedules reported more health problems than those who privileged safety, even if it could lead to delays. Rydstedt, Johansson & Evans (1998) identified some activity's characteristics, like negative interactions with passengers or strictly predicted schedules as responsible for the stress that drivers refer to.

Whilst, some researches we have been developing in this sector of activity in Portugal (Cunha & Lacomblez, 2007; 2008), led us to admit that referring to bus drivers' activity, the conditions under which they work and the kind of constraints with which they are confronted necessarily implies a concrete reference to their specificities, considering the definition of new goals in providing this type of transport service. In this sense, it seems important to begin by approaching the relation between the drivers' career paths and their exposure to specific working conditions (Ramos & Lacomblez, 2008b).

From the activity analyses conducted in 2003 we were able to conclude that, in many companies, remains a model of career path management which contemplates the drivers' evolution according to three professional ranks. It is then possible to distinguish, among drivers, those who found themselves in the categories "outside-the-rules driver", "scaled driver" or "assigned driver". The transition from one professional category to the following is mainly determined by the drivers' seniority and by the "opportunity" that arises from retiring of elder drivers or exiting of ones in upper categories.

Thus, in a first moment, of career beginning at the company, drivers have irregular schedules and only know their schedule and time of driving within two days of advance – what gives them the status of "outside-the-rules driver". The fact that they can perform services in any route and in any vehicle justifies the way they are designated, since their activity is not guided by the same rules as his/her colleagues. These are the drivers who end up "solving" unexpected situations (accidents, other drivers' absences or delays, for instance), guaranteeing the service. This professional category can be found at the bottom of the career and is mostly occupied by young drivers and the newly admitted at the company. Later, these drivers may be integrated in a scale of services ("scaled driver") that, even though rotating, are known earlier, which facilitates somehow planning life beyond working time. Nevertheless, these drivers' activity is still characterized by irregular schedules, which is often translated in beginning the service very early and ending it very late, even if there are some breaks in the middle of the driving period.

Finally, the status of "assigned driver" corresponds to the highest category in the career, more associated to senior drivers. This category is underlined by the performance of a "complete schedule" (a complete morning or a complete afternoon), this is, with no breaks besides the small ones between the "turns"[1]; and also by driving in a pre-defined group of routes and in the same vehicle. In this sense, the transition to this category is perceived as the access to better working conditions.

Despite the fact that the first two categories are seen as "transitional" until the effective integration in regular schedules and routes and even if they are frequently occupied by young drivers, one cannot neglect the hardship associated with the working conditions that characterize them, and we must question the influence of this career evolution in the (in)visibility of risks that drivers are exposed to (Volkoff, Laville & Maillard, 1996; Marquié, Paumès & Volkoff, 1998).

Nevertheless, changes were recently introduced in the sector, related namely with another conception of mobility. Sustained by the articulation between different means of transportation and by a regular service offer throughout the day, it is intended to make public transport more attractive. These changes were determinant to redefine drivers' working conditions and the work content itself, endowing some other specificities to their professional path.

THE PROJECT OF AN "INTEGRATED MOBILITY" AND ITS IMPACT ON DRIVERS' PROFESSIONAL PATH

The adhesion of Portugal to the European Community influenced the definition of transport policies that privilege *intermodality* (INOFOR, 2000), this is, the integration of several means of transportation in the mobility system, favoring the use of public transport.

This aim, defined national wide, became imperative in the metropolitan area of Porto, when it introduced a new mean of transportation, in December 2002 – the subway. This evolution towards creating an intermodal transport system, demanded collective road passenger transportation companies to re-conceive their routes and ticketing scheme, avoiding overlapped offers of these two means of transportation, as well as scenery of competition.

The activity analyses conducted recently, since 2007, have revealed new demanding made to drivers as a result from these changes.

In the current context, the bus as a mean of transportation assumes, more than ever, an adjustment role in the mobility system, considering that it is sufficiently

---

[1] Expression commonly used by drivers to designate the end of a trip and the beginning of the following.

"flexible": it does not require the construction of exclusive paths, as the subway does; and it is possible to reconfigure its routes according to demand dynamics and schedules of other types of transportation.

More than ever, scheduling compliance is a guarantee to mobility system functioning, despite this mean of transportation being particularly likely to have delays, related to traffic congestion.

This requirement over determines the activity of every driver, regardless of his/her category, presenting itself as a double constraint: having to comply with the schedule prescribed by the company and with the compromise assumed with the other transport companies. Besides, the re-conception of most bus lines ended up questioning the strategies drivers had already developed, by their experience, to comply with former schedules, since they had to adapt to new routes and new schedules.

These changes, that affected every driver, put out for debate the differentiation that once existed between different professional categories. Nowadays, referring the category doesn't necessarily reveal, by itself, differenced requirements, which makes us question the impact that hardship extension, associated to its working conditions, may have.

METHOD

We already made reference to one of the privileged methods on our research, identity of our scientific tradition, Activity Ergonomics: the analysis of work activity, in real context, in its human, social and organizational dimensions and in its complexity (Lacomblez & Melo, 1988).

Besides, and because it was our goal to have access to a global picture of drivers' current working conditions and their impact on health, research was developed using an epidemiological questionnaire – INSAT, Health and Work Survey - (Barros-Duarte, Cunha & Lacomblez, 2007) in six public and private companies, with a sample of 161 drivers, with an average age of 43 years (M=43,42; SD=10,20) and an average professional seniority of 14 years (M=14,14; SD=9,48).

INSAT is thus organized into 7 main areas: (I) The work; (II) Conditions and characteristics of the work; (III) Conditions of life outside work; (IV) What bothers me most at work; (V) Training and work; (VI) My state of health; and (VII) Health at work.

For INSAT's data analysis, we recurred to *SPSS* software (Statistical Package for the Social Sciences). We analyzed some survey items that had revealed themselves more relevant to the thematic in study, this is, those related to working schedules; to activity demands; and to health, thus focusing mainly on the following lines of the survey: (I) The work; (II) Conditions and characteristics of the work; and (VII) Health at work. The information obtained through the activity analyses constituted, finally, a way to enrich and explore INSAT data.

RESULTS AND DISCUSSION

Considering the importance of the variable "age" in the analysis, four age groups were constituted, presented on Table 1, which also demonstrates the sample distribution in each one of these groups.

Table 1: Drivers' distribution per age group

	Age group 1	Age group 2	Age group 3	Age group 4
Age	≤ 34 years	35 – 42 years	43 – 52 years	≥ 53 years
Frequencies	39	43	38	41
Percentage	24,2%	26,7%	23,6%	25,5%

Specifically, taking into account variables such as work schedules (Table 2), we can note that the drivers in the first two age groups are the ones mainly exposed to working conditions involving irregular schedules; rotating shifts; working at night; work schedules which "frequently make one go to bed after midnight" or "getting up before 5 in the morning". This means that despite all drivers being subject to schedules changes in almost every services, it seems to remain the option to assign the most irregular and atypical schedules to the youngest ones.

Table 2: Characteristics of drivers' work schedules at different age groups

	Age group 1	Age group 2	Age group 3	Age group 4
Irregular schedules	100%	90,7%	42,1%	9,8%
Working at weekends	79,5%	74,4%	73,7%	7,3%
Rotating shifts	92,3%	86,0%	23,7%	2,4%
Working at night	64,1%	55,8%	31,6%	0,0%
Schedules which make one "go to bed after midnight"	61,5%	39,5%	18,4%	0,0%
Schedules which make one "get up before 5 in the morning"	71,8%	55,8%	26,3%	9,8%

However, this is not the case when analyzing variables more related to work content. Actually, certain constraints that we would easily associate to "outside-the-rules" categorized workers, whose experience would not have allowed them to develop strategies in face of these situations, nowadays seem to be part of every drivers' activity, even those with great experience, as we can see by their responses, referred by more than half the sample (Table 3). One aspect that should be underlined is that all data presented refers to the "current exposure" and not to the past exposure, as we would expect in the case of age group 3 and 4.

Table 3: Exposure to certain activity constraints, associated with rhythm of work and contact with the public

	Age group 1	Age group 2	Age group 3	Age group 4
Depending on direct requests of costumers	87,2%	60,5%	71,1%	63,4%
Frequent interruptions	90,3%	79,1%	68,4%	65,9%
Need to rush	92,3%	84,0%	78,9%	76,9%
Solving unexpected situations or problems without help	87,2%	81,4%	76,3%	73,2%
Withstand demands of public	94,9%	86,0%	84,2%	36,6%
Moments with multiple demands	87,2%	83,0%	86,8%	56,1%
"Work that I will be able to perform when I am 60 years old"*	10,3%	18,6%	7,9%	41,5%**

* Here are included the answers to the statement "My work is/was a kind of work that I will be able to perform when I am 60 years old", considering that it corresponds to an important indicator of workers' perception on the hardship associated to their work.

** In the amount of drivers referring that they can perform that work activity at the age of 60, approximately 30% had, at the moment of data collection, 60 years or more.

Concerning workers' perceived impact on health, it is possible to verify (Table 4) that most answers correspond to "affirmative" options, this is, more than half the sample agrees with the statement "I consider that my health and safety are or were affected by my work".

Table 4: Degree to which one agrees with the sentence "I consider that my health and safety are or were affected by my work"

	Age group 1	Age group 2	Age group 3	Age group 4
Totally agree	35,9%	39,5%	42,1%	51,2%
Stongly agree	10,3%	27,9%	21,1%	22,0%
Agree	33,3%	16,3%	15,8%	12,2%
Undecided	5,1%	4,7%	2,6%	0,0%
Disagree	2,6%	2,3%	7,9%	4,9%
Strongly disagree	7,7%	4,7%	7,9%	2,4%
Totally disagree	5,1%	4,7%	2,6%	7,3%

Nowadays, we note an extension of activity's associated hardship – namely, as a result of exploring the articulation with other means of transport, which translates into additional constraints for bus drivers. But we also understand that it corresponds to a cumulative process, evolving throughout one's life and accelerating in the last years of activity, as we can note from drivers' reported health problems at different ages (Table 5).

Table 5: Main health problems referred by drivers, as well as reported to work

	Age group 1	Age group 2	Age group 3	Age group 4
Hypertension	7,7%	11,6%	26,3%	34,1%
Sleep disorders	30,8%	37,2%	39,5%	41,5%
Nervous problems	46,2%	48,8%	42,1%	58,5%
Limited movements at the cervical area of the back	33,3%	37,2%	44,7%	53,7%
Limited movements at the low back	30,8%	34,9%	42,7%	63,4%
Fatigue	59,0%	65,1%	47,4%	70,7%
Sleep problems	20,5%	37,2%	34,2%	46,3%

In fact, it's the elder drivers who report more complaints regarding the above mentioned health problems. Nevertheless, one cannot neglect that, even though with a lowest percentage, the youngest drivers also refer having these health problems and relate them to work, which denounces their potential contribution to a premature aging.

CONCLUSIONS

In the last two decades, the sector's working conditions were under deep transformations, as a result of social and macroeconomic changes. On one hand, the aim was to promote *intermodality* and, subsequently, the increase of public transport demand; on the other hand, it was expected that bus transportation would improve the integrative potential of the system, acting as a reliable adjustment amongst other means of transportation. In this sense, companies, workers, skills and work itself gained new shapes (Lacomblez, 2000). However, in the specific case of this mean of transportation, reliability is not intrinsic, considering that at each moment several factors interfere with scheduling compliance: traffic, conditioned by the area's level of urbanization and the moment of the day; driving in a route with great or low demand; general conditions of circulation...

Thus, if it is important to produce new ways of mobility, it is also fundamental to comprehend the circumstances under which those services are produced. This means that it is necessary to characterize the activity's contribution in face of requirements made by this "productive space" and their impact on the workers.

The analyses, conducted in real context, allowed to conclude that drivers represent the "adjustment variable", in a system whose integration results from a compromise that is often precarious. And since this is common for all of them, regardless the professional path or category, it seems that there have been created conditions that may accelerate the aging process, due to an early deterioration of certain dimensions of health. This is also an activity result that must be known and discussed in the scope of "sustainable mobility" projects.

REFERENCES

ANTROP (2002). *Linhas de orientação estratégica para o sector de transportes colectivos rodoviários de passageiros*. Vol.1 Porto: Associação Nacional de Transportadores Rodoviários de Pesados de Passageiros.

Barros-Duarte, C., Cunha, L. & Lacomblez, M. (2007). "INSAT: uma proposta metodológica para análise dos efeitos das condições de trabalho sobre a saúde". *Laboreal,* 3, (2), 54-62. http://laboreal.up.pt/revista/artigo.php?id=37t45nSU5471123111:499682571

BTS (2001). "Le travail sans limites? Réorganiser le travail et repenser la santé des travailleurs". *Numéro spécial – Bulletin d'Information du Bureau Technique Syndical Européen pour la Santé et la Sécurité*.

Cassou, B. & Laville, A. (1996). "Vieillissement et travail : cadre général de l'enquête ESTEV". *In* Derriennic, Touranchet & Volkoff (Eds.). *Âge, travail, santé. Études sur les salariés âgés de 37 à 52 ans. Enquête ESTEV 1990*. Paris : INSERM, 13-31.

Cunha, L. & Lacomblez, M. (2007). "Market and regulation of general interest in the passenger land transport sector: a debate renewed by drivers' activity". *@ctivités, 4 (1)*, pp. 141-148, http://www.activites.org/v4n1/v4n1.pdf

Cunha, L. & Lacomblez, M. (2008). "A influência do traçado de mobilidade na noção de território e nas oportunidades de desenvolvimento local". *Laboreal, 4, (1)*, 56-67.
http://laboreal.up.pt/revista/artigo.php?id=37t45nSU5471123285483322811

Derrienic, F., Touranchet, A. & Volkoff, S. (1996). "ESTEV, une méthode d'enquête". *In* Derrienic, Touranchet & Volkoff (Eds.). *Âge, travail, santé. Études sur les salariés âges de 37 à 52 ans. Enquête ESTEV 1990*. Paris: INSERM, 33-55.

Eurostat (2006). *Le transport de voyageurs dans l'Union européenne. Statistiques en bref*. Communautés européennes.

INOFOR (2000). *O sector de transportes em Portugal*. Lisboa: INOFOR.

Lacomblez, M. & Melo, A. (1988). *Informatisation des petites et moyennes entreprises et conditions du travail – Analyse de cas au Portugal*. Porto: Comissão da Comunidade Europeia, D.G.5.

Lacomblez, M. (2000). *Factores psicossociais associados aos riscos emergentes. Riscos emergentes da nova organização do trabalho*. Lisboa: IDICT.

Laville, A., Teiger, C. & Wisner, A. (1975). *Âge et contraintes de travail*. Jouy-en-Josas: N.E.B. Éditions Scientifiques.

Marquié, J-C., Paumès, D. & Volkoff, S. (1998). *Working with age*. London: Taylor & Francis.

Meijman, T. F. & Kompier, M. A. J. (1998). "Busy business: How urban bus drivers cope with time pressure, passengers and traffic safety". *Journal of Occupational Health Psychology, 3*(2), 109-121.

Ramos, S. & Lacomblez, M. (2008a). "Soi-même comme un « vieux »: variations dans les regards sur les fins de vie au travail", (La fabrique des vieillissements, sous direction de Thibauld Moulaert et Madeleine Moulin), *Revue de l'Institut de Sociologie – Université Libre de Bruxelles, 1-4, 21-38*.

Ramos, S. & Lacomblez, M. (2008b). "L'âge : évolutions des approches d'un marqueur des histoires de travail". *Les Politiques sociales*, 3&4, 14-27.

Robert, L. (1983). *Mécanismes cellulaires et moléculaires du vieillissement*. Paris : Masson.

Rydstedt, L. W., Johansson, G. & Evans, G. W. (1998). "A longitudinal study of workload, health and well-being among male and female urban bus drivers". *Journal of Occupational and Organizational Psychology*, 71, 35-45.

Teiger, C. (1989). "Le vieillissement différentiel par et dans le travail: un vieux problème dans un contexte récent". *Le Travail Humain*, 52 (1), 21-56.

Tse, J. L. M., Flin, R. & Mcarns, K. (2006). "Bus driver well-being review: 50 years of research". *Transportation Research Part F, 9*, 89-114.

Volkoff, S., Laville, A. & Maillard, C. (1996). "Age et Travail : contraints, sélection et difficultés". In Derrienic, F., Touranchet, A. & Volkoff, S. (Eds.). *Age, travail, santé. Études sur les salariés âges de 37 à 52 ans*. Enquête ESTEV 1990. Paris : INSERM : 57-77.

Chapter 83

Ergonomic Workplace Design for the Elderly: Empirical Analysis and Biomechanical Simulation of Information Input on Large Touch Screens

Sebastian Vetter, Jennifer Bützler,
Nicole Jochems, Christopher M. Schlick

Institute of Industrial Engineering and Ergonomics
RWTH Aachen University
Bergdriesch 27, D-52062 Aachen

ABSTRACT

When designing computer workstations in times of demographic change, it is especially important to satisfy the ergonomic requirements of elderly employees. Large touch screens seem to have ergonomic benefits for many work tasks. However, when using a touch screen, the muscles of the shoulder, upper arm, forearm and index finger are repetitively activated which is especially critical for elderly users. In an empirical study a sample of 11 younger and 11 older subjects were tested in a pointing task on a large format touch screen mockup. Body postures, reaching areas, and subjective comfort were measured at three tilt angles of the mockup surface (0°, 8°, 16°). The AnyBody Modeling System was used for a biomechanical investigation. A typical body posture when pointing to a screen object in the upper left corner was used to evaluate the muscular strain at different tilt angles. The subjective comfort rating indicates a preferred tilt angle of 16°. However, the biomechanical model indicates lowest muscular strain at 8°.

Keywords: ergonomic workplace design, human-computer interaction, aging, body postures, reaching areas, biomechanical modeling

INTRODUCTION

The touch screen as an input device is growing quickly in popularity. It is a direct input device that requires neither spatial nor spatial-temporal transformation. As shown by Rogers et al. (2005) direct input devices are superior for long, ballistic movements. In a comparative study of direct and indirect input devices – mouse, touch screen, and eye-gaze– the authors found that age-related differences between execution times in pointing tasks are compensated by touch-input. When using a touch screen, elderly computer users are as fast as their younger counterparts using a computer mouse (Jochems, 2010). The impact of direct information input on computer work will rise as multiple simultaneous touches can be discriminated, and even larger touch screens become affordable. Large touch screens seem beneficial for many work settings (e.g. project management, plant design, architecture). However, when using a touch screen, the muscles of the shoulder, upper arm, forearm and index finger are activated. Thus, longer work periods involving a touch screen often lead to significant arm fatigue. This is ergonomically critical especially for older users (Ahlström et al., 1992).

A promising approach in the ergonomic evaluation of human-machine systems is the biomechanical modeling of skeletal and muscular dynamics. A widely used system is the so called AnyBody Modeling System for developing musculoskeletal models (Rasmussen et al., 2001; Rasmussen et al., 2003; Rausch et al., 2006). AnyBody allows for modeling of various body parts with different anthropometrics and muscle parameters.

In the present paper we will address the question of how to design large touch screens workplaces regarding: (1) participant age, body postures, reaching areas and subjective comfort; (2) the simulated muscular strain at different tilt angles.

METHOD

To answer these research questions, a sample of younger and older subjects was tested empirically in a pointing task. Furthermore, a typical body posture when pointing an object in an extreme target position in the upper left corner was reconstructed in the AnyBody Modeling System to compute the muscular strain at different tilt angles.

EMPIRICAL STUDY

A mockup with the exact physical dimensions (working area: 99x77cm) of the "Diamond Touch" screen was built for the pointing task (Dietz & Leigh, 2001). The tilt angle of the mockup was designated as an independent variable and was varied at three angles (0°, 8°, 16° degree). Body postures, maximum reaching areas and subjective comfort were dependent variables.

The participants' body postures were video recorded. The maximum reaching areas of the right and left arm were marked by the participants at mockup angles of 0° and 16°. The subjective comfort was measured with the help of the ZEIS-Scale (Pitrella & Käppler, 1988) on a 14 level rating scale (0 = very uncomfortable to 14 = very comfortable) for the three angles.

11 younger (20-32 years, M=23.82, SD=3.516) and 11 older (57-69 years, M=64.36, SD=3.443) subjects participated in the experiment. Relevant anthropometric variables (e.g. body height, arm length) of the participants were determined according to DIN EN ISO 33402-2:2005.

The participants were seated in front of the touch screen mockup. The experimental task was to point small plastic pads at 8 different target positions (Ø=3cm) equally distributed over the working area. The video recorded body postures were analyzed descriptively for the two age groups.

The maximum reaching areas and the level of comfort were analyzed by a mixed design ANOVA with age-group as a between-group factor, and the mockup angle as a within-subjects factor. The level of significance for each analysis was set to $\alpha=0.05$.

BIOMECHANICAL SIMULATION

The maximum muscle activity resulting from work on a large touch screen was analyzed on the basis of the AnyBody system. AnyBody uses an inverse dynamic technique to drive the body by a given movement. The forces that are used to produce the movement are calculated by the software. To do so a minimum fatigue criterion (min/max criterion) that minimizes the maximum relative muscle load is used (Rasmussen et al., 2001).

As in the empirical analysis, the tilt angle of the simulated screen was varied in three levels (0°, 8°, 16°). The muscle activity is defined as the muscle force in relation to the theoretical maximum force of the muscle. As dependent variable the maximum muscle activity is calculated from all muscles involved in the movement. A pointing activity on the large touch screen was simulated to detect the maximum muscle activity. Therefore, pointing with the right hand to the upper left corner of the touch screen was modeled. The anthropometrics of the model were defined as 50th percentile male. The body posture and the extremity angles of the AnyBody Model were analyzed and matched to the subjects of the empirical analysis. For the three tilt angles the maximum muscle activity was calculated and compared.

RESULTS

EMPIRICAL STUDY

Body Postures

The descriptive analysis of participants' body postures from the video recordings shows large between-subjects differences in extremity angles (see Figure 1). This effect is even more pronounced in the older age group because of a greater variability in anthropometrics.

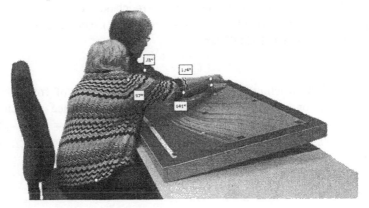

FIGURE 1. Between-subjects body postures during the pointing task. The participant in the front has extremity angles of 97° (shoulder-arm angle), and 141° (upper arm-forearm angle). A person of the same age group with different anthropometric variables has extremity angles of 78° and 135°, respectively.

Reaching Areas

The reaching areas in square centimeter (cm²) of the subjects also show large between-subjects differences. The individual areas also depend on the particular tilt angle (0° and 16°). Figure 2 shows an illustration of between-subject differences in the reaching areas.

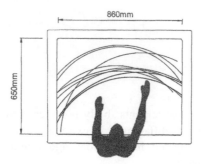

FIGURE 2. Between-subjects differences in the reaching areas (exemplarily shown for the younger age-group and for a mockup angle of 0°). The reaching areas range between 2701.83 cm² and 4629.66 cm².

As expected the reaching areas are larger for a tilt angle of 16° (M=3649.40, SD=567.98) than for 0° (M=3456.93, SD=591.56). The difference in reaching areas is not significant ($F_{(1,19)}$=4.088, p=0.058) with an effect size of ω^2=0.108. Age does not have any significant influence on the size of the reaching area ($F_{(1.19)}$=1.530, p=0.700).

Subjective Comfort Rating by Means of the ZEIS-Scale

The subjective comfort rating shows differences between the angles 0° and 16° ($F_{(2,40)}$=5.909, p=0.006) with an effect size of ω^2=0.182 (see Figure 3). The angles 0° and 8° as well as 8° and 16° do not differ significantly. The main effect of age-group on the subjective comfort rating is non-significant ($F_{(1,19)}$=0.040, p=0.843). The angle 16° is rated best and 0° worst regardless of age.

BIOMECHANICAL SIMULATION

The inverse kinematic of the AnyBody Model results in a "natural" body posture. Figure 4 show that the model predicts extremity angles, that are very similar to an examined subject with comparable anthropometrics (50th percentile male).

The weighted maximum muscle activities for the three tilt angles are depicted in Figure 5. One can see that the muscle activity is highest for the angle 0° (59%) and lowest for the angle 8° (51.1%). Pointing tasks carried out at a tilt angle of 16° evoke 52.9% muscle activity.

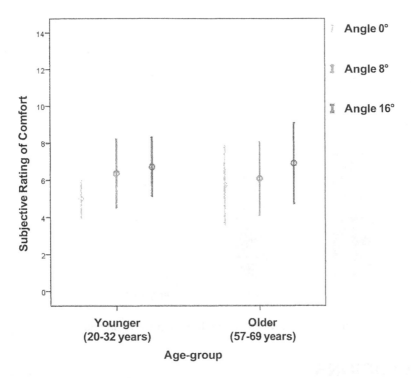

FIGURE 3. Subjective comfort rating by means of the ZEIS-Scale depending on angle and age

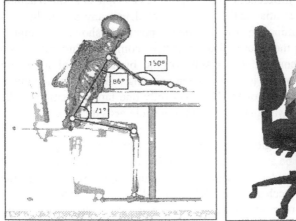

FIGURE 4. Body Posture Analysis of the AnyBody Model (left) in comparison to the subject (right)

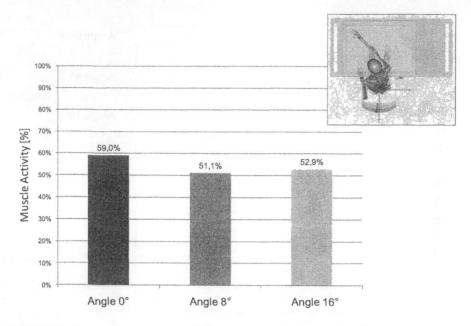

FIGURE 5. Maximum muscle activities for the three examined angles

CONCLUSIONS

Execution of a pointing task on a mockup of a large touch screen showed considerable between-subject differences in body postures as well as in the reaching areas of the participants. Inferential-statistics could not detect any age influences though the descriptive examination of the body postures showed a higher heterogeneity in the group of the elderly. The subjective comfort rating indicates a preferred angle of 16°. In contrast, the results of the biomechanical modeling showed that the load of the muscles is lowest when pointing tasks are carried out at a simulated screen with a tilt angle of 8°. However only one target position in the upper left corner of the screen was analyzed and the muscle activity for a 16° angle is only slightly higher.

Clearly, the question which angle is ergonomically "optimal" for computer work on large touch screens cannot be answered in general. It has to be taken into account that the current study does consider the effect of muscular fatigue over a long time period. Furthermore, the biomechanical model included only one extreme target position in the upper left corner, so the results of the muscular strain cannot be generalized for other target positions. In summary, the results show that the biomechanical model can be successfully applied to assess pointing activities in human-computer interaction to gather not only time and comfort data but also additional information about muscular dynamics.

ACKNOWLEDGEMENTS

This research was funded by the German Research Foundation according to the priority program no. 1184, Age-differentiated Work Systems. The authors extend their gratitude to all participants taking part in the experiment.

REFERENCES

Ahlström, B., Lehman, S., Marmolin, T. (1992), "Overcoming touchscreen user fatigue by workplace design." *Conference on Human Factors in Computing Systems*, Monterey, California, 101-102.

Dietz, P. & Leigh, D. (2001), "Diamondtouch: a multi-user touch technology." *Proceedings of the ACM Symposium on User interface Software and Technology*, USA, 14.

DIN 33402-2: Ergonomie Körpermaße des Menschen Teil 2: Werte, Deutsche Fassung EN ISO 33402-2:2005. (in german)

Jochems, N. (2010), *Altersdifferenzierte Gestaltung der Mensch-Rechner-Interaktion am Beispiel von Projektmanagementaufgaben*. Dissertation RWTH Aachen, Shaker-Verlag, Aachen. (in german)

Pitrella, F.D., Käppler, W.D. (1988), *Identification and evaluation of scale design principles in the development of the extended range, sequential judgement scale*. Research Report No. 80, Forschungsinstitut für Anthropotechnik, Wachtberg.

Rasmussen, J., Dahlquist, J., Damsgaard, M., de Zee, M., Christensen, ST. (2003), "Musculoskeletal modeling as an ergonomic design method". *IEA Proceedings*.

Rasmussen, J., Damsgaard, M., Voigt, M. (2001), "Muscle recruitment by the min/max criterion – a comparative numerical study". *Journal Biomechanics*, 34, 409-415.

Rausch, J., Siebertz, K., Christensen, S.T., Rasmussen, J. (2006), "Simulation des menschlichen Bewegungsapparates zur Innenraumgestaltung von Fahrzeugen". *VDI-Berichte 2006*, VDI, Düsseldorf. (in german)

Rogers, W. A., Fisk, A. D., McLaughlin, A. D. and Pak, R. (2005), "Touch a screen or turn a knob: choosing the best device for the job". *Human Factors*, 47 (2), 271-288.

Chapter 84

Computer Workstation Ergonomics for an Aging Workforce

Neil Charness

Psychology Department and
Pepper Institute on Aging and Public Policy
Florida State University

ABSTRACT

A rapidly aging workforce coupled to ubiquitous computing presents challenges to employers. I report some results from a study of workstation ergonomics for younger (< 40 yr) and older (> 50 yr) workers, conducted at a large southeastern university in the United States. The sample was a representative one (n=206) of workers who spent at least part of their day at a computer workstation. A NIOSH ergonomics checklist and self-report comfort measures were taken initially in the office and a subset of the sample permitted us to measure these variables at their home workstations. We found greater comfort on most measures at home, and superior ergonomics at work than at home. Employers need to be encouraged to support better computer workstation setups for teleworkers.

Keywords: laptop, desktop, work, home, workstation, ergonomics, pain, older worker, aging, age, telework

INTRODUCTION

World populations are aging both in developed and developing countries due to declining birth rates coupled with declining death rates (http://www.un.org/esa/population/publications/worldageing19502050/ accessed 2/25/2010). To cope with the increasing costs to support burgeoning retired populations, many developed countries are adopting policies of encouraging workers to remain employed longer. For instance, countries are increasing the age of entitlement for public pension benefits. As an example, in the United States the retirement age for full Social Security pensions has risen from age 65 for those born in 1937 or earlier to age 67 for those born in 1960 or later and there are incentives to delay retirement to age 70 in exchange for higher monthly pension payments. As the general population ages, so too does the work force population.

The median age of a worker exceeded 40 years of age in 2006 in the United States (Toosi, 2007) and in Canada (http://www.agingworkforcenews.com/2008/03/canada-median-age-of-workforce-goes.html). Aging is associated with a variety of normative changes in perception, cognition, and psychomotor performance that have implications for design of technology (Fisk et al., 2009) and awareness of these issues has led to the newly developing field of gerontechnology (Charness & Jastrzembski, 2009). My goal in this paper is to explore the implications of an aging workforce for computer workstation ergonomics, particularly in the context of two important recent trends: increased telework, which may be an attractive form of work for aging adults (Sharit et al., 2009), and increased use of laptop computers.

COMPUTER USE

Computers are becoming ubiquitous in workplaces given the rapid expansion of service sector jobs and the computerization of many other work sectors from farming to manufacturing. By 2002, about half of all jobs made use of computers in the EU15 countries (http://www.eurofound.europa.eu/ewco/reports/TN0412TR01/TN0412TR01.pdf). Although there are ergonomic guidelines for computer workstations (ANSI/HFES 100-2007), they are based mainly on data from young adult populations and pertain mainly to the use of desktop computers. With telework (work from home or satellite offices) increasing in frequency, it appears likely that workers will be increasingly responsible for setting up their own home workstations, rather than relying on human factors and ergonomics experts to design their workspace. Further, since 2005, laptop computers have outsold desktop computers in the United States (Datapoints, 2007). Laptops are an attractive alternative to desktops for companies when considering equipping an employee who is likely to be working at least part-time from home. If one examines the US population's use of computers at home, a recent study (Lenhart et al., 2009) indicated that for the 18-29 year old age group, laptops dominated desktops (66% vs 53%), though for those age 65+ the reverse was true (37% desktops vs 18% laptops).

STUDY OF WORK AND HOME COMPUTER WORKSTATIONS

In our study we investigated a variety of features of workstation ergonomics for a representative sample of younger (age < 40 years) and older (age > 50 years) workers from a large Southeastern US University who were tested at work and then one year later at work following feedback about their initial workstation setup. A subset permitted us to evaluate home computer workstations. Details of the sampling and procedures are given in Weaver et al. (2008) which dealt with arm/shoulder and wrist pain reports for work settings, and in Charness et al. (submitted) which dealt with postural and pain variables at work and at home, focusing on laptop versus desktop use. Here I emphasize the pattern of results for variables assessing workstation ergonomics, comfort with noise, lighting, text size, and temperature, as well as overall workstation satisfaction.

Method

Participants (N = 206; 103 Age > 50, M = 58 yr and 103 Age < 40, M =31 yr) were assessed at work. A total of 64 were later assessed at home (29 Age > 50, M = 57 yr; 25 Age < 40, M = 32 yr). All were asked to assume their usual posture in front of their computer workstation. Base posture was assessed while viewing a document on the computer via the NIOSH Tray 5-G Computer Workstation Checklist (NIOSH, 1997) and from photos and videos taken of the workstation. Checklist items were pooled initially into 11-item posture and ergonomics scales and also 1) a Posture scale: 4 items tapping horizontal thighs, neutral wrists, etc., 2) a Chair scale: 6 items tapping lumbar support, arm rests, etc., 3) an Environment scale: 9 items tapping detachable keyboard, glare avoided, space for knees & feet, etc., and 4) a Training scale: 8 items tapping trained in proper postures, how to adjust workstation, etc. Comfort was assessed using 0-10 rating scales for temperature, glare, fonts for text, noise, light levels, eye strain, and pain reported from arm/shoulder and from wrist. Demographics were also assessed via a questionnaire.

Results

Comfort Variables. Repeated measures ANOVA with age group (younger, older) as a between subjects factor and type of measure (temperature, noise, text, light, overall workstation comfort) and location (work, home) as within subjects factors showed only main effects of type of comfort variable, $F_{(4, 204)} = 6.09$, MSe = 1.69, $p < .01$, and a location by comfort type interaction, $F_{(4, 204)} = 6.5$, MSe = 1.69, $p < 01$. Age group was not a significant factor ($p > .11$) and did not interact with the other factors (though there was a marginal 3-way interaction ($p < .10$). Figure 1.1 shows the means with standard error bars. Comfort with temperature and noise levels showed a home advantage. Actual temperature readings did not vary significantly by age group, location, or their interaction, with mean level being 24 C. Comfort ratings for temperature were not significantly correlated with measured temperature. Similarly, ambient sound levels did not vary significantly between home and office, between age groups, or show an interaction between these factors, though there was a trend ($p < .08$) for sound levels to be lower at home (42 db) than at work (43.5 db). Again, there was no significant correlation between comfort with noise levels and measured sound levels.

As would be expected from a prior survey of actual light levels in Tallahassee offices and homes (Charness & Dijkstra, 1999), lighting comfort was better in the office than home. There was no difference in self-reported comfort ratings for office and home computer workstations or for comfort with text on the screen at work and at home.

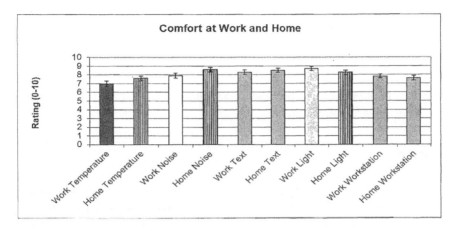

FIGURE 1.1 Comfort ratings at work and at home. Error bars represent 95% confidence intervals for the means.

Posture Variables

Proportion correct ergonomic score (all NIOSH items) for computer workstation was better at work (M = .81, SE = .019) than at home (M = .76, SE = .017). Total score was partitioned into subscales and a mixed ANOVA with a between groups factor (age group: younger, older) and 2 within subjects factors (4 NIOSH scale types; 2 locations: work, home) showed main effects of NIOSH scale (F (3, 129) = 5.1, MSE = 256, p < .01), location (F (1, 129) = 8.63, MSE = 256, p < .01), and a scale by location interaction (F (3, 129) = 4.56, MSE = 256, p < .01). The interaction seen in Figure 1.2 shows that work chairs were superior to home chairs. There was a trend for people to adopt better postures at work (perhaps in part to due to having a better adjustable chair).

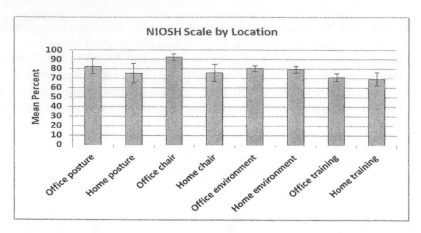

FIGURE 1.2 NIOSH subscale by location interaction. Error bars represent 95% confidence intervals for the means.

Conclusions

For this particular workforce, office and home computer workstation ergonomics were generally quite good, though there was a significant difference favoring the workplace over the home. The advantage to work for ergonomics was primarily attributable to use of a less adjustable chair at home than at work. Light levels were also better at work than at home. However, people were more comfortable at home, particularly for ambient characteristics such as temperature, noise, and eye strain. This was true despite minimal relationships between self-reports of comfort and actual values of temperature and noise. The home advantage may be due to better perceived control over many variables (e.g., temperature, noise) in that setting. Nonetheless, such findings suggest that telework at home may be associated with higher levels of worker satisfaction.

Generally, there were few significant age differences observed between younger and older workers, with the exception that despite reports of generally low levels of arm/shoulder and wrist pain, somewhat more pain was reported by younger workers who used laptops at home than those who used desktop computers at home (Charness et al., submitted). This finding suggests that younger workers need to be educated to adopt better home workstation setups in order to prevent future musculoskeletal injuries. However, the main recommendation from this study would be to advise employers and computer equipment manufacturers to instruct teleworkers on how to achieve better home workstation ergonomics. Within this sample, the advice would be to invest in a better chair for home use. Employers should consider supporting the purchase of better office furniture, not just computer and telecommunications equipment, when assigning employees to telework.

REFERENCES

ANSI/HFES 100-2007 *Human Factors Engineering of Computer Workstations.* Published by the Human Factors and Ergonomics Society. P.O. Box 1369, Santa Monica, California, USA 90406-2410.

Charness, N., & Dijkstra, K. (1999). Age, luminance, and print legibility in homes, offices, and public places. *Human Factors, 41*(2), 173-193.

Charness, N., Dijkstra, K., Jastrzembski, T. S., Weaver, S. J., & Champion, M. (submitted). Are Laptop Computers a Health Risk for an Aging Population?

Charness, N., & Jastrzembski, T. S. (2009). Gerontechnology. In P. Saariluoma & H. Isomäki (Eds.). *Future Interaction Design II* (pp. 1-29). London: Springer-Verlag.

Datapoints (2007). 53% of computers sold in May 2005 were laptops, the first time that laptops have surpassed desktops in market share. Newsweek, 145 (June 27): E2.

Fisk, A. D., Rogers, W. A., Charness, N., Czaja, S. J., & Sharit, J. (2009). *Designing for older adults: Principles and creative human factors approaches* (2nd Ed.). Boca Raton: CRC Press.

Lenhart, A., Purcell, K., Smith, A., & Zickuhr, K. (2009). Social media & mobile internet use among teens and young adults. Pew Internet & American Life Project, 1615 L St., NW – Suite 700, Washington, D.C. 20036. Accessed 2/17/2010 from http://www.pewinternet.org/~/media//Files/Reports/2010/PIP_Social_Media_a nd_Young_Adults_Report.pdf

National Institute for Occupational Safety & Health (NIOSH) (1997). Toolbox-tray 5-G: Computer workstation checklist. Elements of Ergonomics Programs: A Primer Based on Workplace Evaluations of Musculoskeletal Disorders. NIOSH Publication No. 97-117. Cincinnati: DHHS.

Sharit, J., Czaja, S. J., Hernandez, M. A., & Nair, S. N. (2009). The employability of older workers as teleworkers: An appraisal of issues and an empirical study. *Human Factors and Ergonomics in Manufacturing, 19*, 457-477.

Toosi, M. (2007). Labor force projections to 2016: more workers in their golden years. Monthly Labor Review, November, 2007. http://www.bls.gov/opub/mlr/2007/11/art3full.pdf accessed 7/15/08.

Weaver, S. J., Charness, N., Dijkstra, K., & Jastrzembski, T. (2008). Musculoskeletal pain prevalence in randomly sampled university employees: Age and gender effects. *Gerontechnology, 7*, 279-292.

Chapter 85

Work System Design with the Ergonomic Harmony Framework for Aging Workforce

L.Y. Wang, Henry Y.K. Lau

Department of Industrial and Manufacturing Systems Engineering
The University of Hong Kong
Hong Kong, P.R. China

ABSTRACT

The general workforce is aging, especially in the Asian countries where most of the world's production operation is taking place, and it has indeed become one of the dominant issues in various sectors of industry all around the world. This paper develops a theoretical ergonomic evaluation and design model named "Ergonomic Harmony Framework" for the aging workforce based on four existing approaches which are separately used to evaluate and solve the physiological, psychological, psychosocial and behavioral problems of human factors in a work system. A descriptive work system model and a mathematical framework specifically assessing this model are integrated for obtaining the harmony of the entire work system.

Keywords: Aging workforce, Work system design, Ergonomic Harmony Framework

INTRODUCTION

Aging workforce is a fierce challenge in the context of unprecedented demographic changes that much of the world is facing an increasingly aging population (McQuaid, 2007). On one hand, aging workforce is a rich resource in the work system due to their accumulated knowledge, extensive experience, skills, loyalty and reliability. On the other hand, age-related loss of physiological and psychological functional capacities which including declines of strength, flexibility, sensory acuity, attention and memory are obstacles for creating high performance work system (Ilmarinen, 2001). Ergonomics is the art and science of designing the work to fit the worker (Manuele, 1993) instead of forcing the worker to fit the work. Thus ergonomics measures should be taken seriously for the sake of optimizing aging workforce and overall work system performance by providing well designed equipment, workplace layout, process, work organization and management so as to achieve optimum efficiency, accuracy, productivity, safety, health and comfort.

Worker and machine are two complementary elements in work system, the human element is the most difficult part to quantize and control due to its inherent uncertainties and variations. Aged worker can be particularly exposed to excessive physical and cognitive stress if ergonomics is neglected. Thus, the use of an ergonomics approach in early stages of work system design is an effective and essential step to accommodate aged workers' capabilities and limitations to work and obtain healthy and sustainable work systems in order to improve the overall producing capability. In terms of the physical aspect, appropriate work facilities should be widely used to prevent aged workers from undue physical strain, stress, heavy lifting, repetitive motion, etc. Consequently, less reworking and fewer errors will be achieved to improve production activity. At the same time, physical sense alone that without much consideration of topics and issues in mental area is not enough to satisfy comfortable work while extensive knowledge of human factors is required. As such, undue attention should be paid to the fatigue, health, safety and comfort aspects. In relation to psychosocial working conditions, combinatorial elements including flow, layout and organization should also play an important role in establishing a reliable work system which exists and operates in space and time for the aging workforce. Particularly, theories and models that addressed only some parts of the work system are not holistic and sufficient.

BACKGROUND

AGING WORKFORCE

The graying of workforce is becoming progressively serious along with the population aging as a whole on account of the declining fertility rates and delaying retirement (Rix, 2009). As workers age, their functional capacities inevitably show a declining trend in a number of functions, in which physical characteristics are mainly taken into account. The weakening of physical work ability in relation to aging which has primarily concentrated on strength, flexibility, balance, range of motion, speed of movement, motor skills and vision (Haight, 2003) is associating with reductions in musculoskeletal, metabolic, cardiovascular, respiratory and nervous systems (Kenny et al., 2008). These weaknesses impact work performance and productivity of older workers through forcing them to the individual maximal capacities, especially injuring those who are engaged in the physically demanding occupations like construction workers, manual assembly workers, firefighters and so on (de Zwart et al., 1995). Meanwhile, changes in cognitive capabilities are related to decreased tactile feedback, reaction times and information processing precision. These weakening of actual age-related changes can indisputably be compensated by the experience-based knowledge and skills to a great extent.

WORK SYSTEM

A work system is a system in which human participants and/or machines perform work using information, technology, and other resources to produce products and/or services for internal or external customers (Alter, 1999). The role ergonomics plays in work system design has been well documented. Das (1999) developed a comprehensive ergonomics model in consideration of various relevant factors such as job satisfaction, worker productivity, health, safety and so on to test the expected work design attributes and outcomes. Later, Parker et al. (2001) proposed an elaborated model of work design after examining a greater range of work characteristics like physical and cognitive demands, role ambiguity and conflict, group and social contact, etc. They also examined work outcomes such as individual creativity, job motivation, and organizational performance. Shikdar and Das (2003) considered worker participation and performance feedback as an optimum management strategy to improve job attitudes and worker satisfaction in a repetitive industrial production task. Subsequently, Cho and Eppinger (2005) presented an ergonomic process analysis and modeling technique for managing complex work system design using advanced simulation. Their model is useful for identifying strategies of process improvements and evaluating different system design plans. Furthermore, Kleiner (2006) investigated ergonomic problems of work systems within a holistic systems context. He scanned the work environmental and organizational sub-systems, defined production system type and unit operations process, then created the key variance control table and role network for performing joint design and function allocation, thus to implement, iterate and improve the work system performance. A recent study about work organization constructs and ergonomics by Hanse and Winkel (2008) showed that job rotation contributed significantly in all experiment models. This rotation rule could be used as a source for ergonomics improvements to work organization design. The problems of work system design are complex and multidimensional, and accordingly both physical ergonomics and mental ergonomics are essential to simplify the complexity and optimize the work system.

OBJECTIVE

The main aim of this paper is to develop a fully integrated and harmonized framework for designing work system especially for aging workforce based on an ergonomics approach which covers four relevant fields: (i) physiological factors, (ii) psychological factors, (iii) psychosocial factors, and (iv) behavioral factors. It is critical to identify the basic obstacles faced by aged workers in a work system with claims resulting from poor designed equipment and tools, as well as inappropriate work processes and methods, and to use efficient, fledged, and valuable ergonomics principles to provide potential solutions at a top priority.

The first step in our research is to reduce work hazard and improve aging workforce protection by investigating the impact of work-related musculoskeletal disorders among aged workers while carrying, pushing, pulling, lifting, lowering and reaching. These impact factors comprise human factors and technical elements which are related to work stations, fixtures and material handling tools. The other aspect of an ergonomics improvement is setting up appropriate system arrangements. Frequent rotation of employees from high risk operating posts to low risk operating posts can be helpful to reduce exposure of risk factors. Proper training, educations and awareness programs will also fulfill the requirement related to advance aged workers' morale, quality of work life and desirable worker values.

METHODOLOGY

For establishing the integrated framework of ergonomics work system design for aging workforce, well consolidated techniques, theories and evolutionary models about the design of work system are discussed and developed in each of these four areas, and the proposed research investigate and apply the following five main methods for a harmony work system on a theoretic as well as a practical level.

Firstly, pertinent background information from literatures at relevant areas in work system design and aging workforce is gathered to identify the issues.

Secondly, a digital human modeling (DHM) tool (Helin et al., 2007) is adopted and used in modeling and analyzing physiological factors such as vision, height, weight, forward arm reach and such anthropometric elements. Powerful digital human modeling and simulation tools use digital humans as representations of the workers inserted into a simulation or virtual environment to facilitate the prediction of performance and safety. It includes visualizations of the human body shapes, dimensions, posture, motion constraints, reachability, operation sequences and field of view with the mathematics and science in the background to minimize redundant changes by eliminating ergonomics related issues (Sundin and Örtengren, 2006). Also, It contributes to the efficiency of iterative work system design process involving evaluation, diagnosis and revision (Chaffin, 2001). The tool which relies on a database of recorded and measured real human motions is used to replicate the ongoing operation of complex systems for purposes of analyzing and evaluating work condition quickly and objectively in order to identify and avoid potentially harmful working postures, forces and durations which can lead to work-related musculoskeletal disorders, these disorders in turn can result in company costs for worker replacement, compensation and rehabilitation.

Thirdly, in view of psychological factors like fatigue, stress, boredom, satisfaction and so forth, participatory design principles (Schuler and Namioka, 1993) are enhanced and extended to involve workers at the beginning of work system design process including work processes, procedures, and environment components, in order to actively ensure that the work environment is usable and meets aged workers' mental needs. These needs consist of emotional, spiritual and cultural satisfaction.

Fourthly, the cognitive work analysis approach (Rasmussen et al., 1994) is adopted to cope with psychosocial factors including work content, role ambiguity, employment relationship, community and all that. This formative and constraint-based approach is benefit for the analysis, design, modeling, development and evaluation of complex, dynamic and high-risk socio-technical work systems by aiding in distinguishing between workers' essential goal-directed behaviors. It guides the analyst through the process of answering the question of why the system exists, what activities are conducted within the domain, as well as how this activity is achieved, and who is performing it (Jenkins et al., 2008). Analysis of aged workers' behaviors competencies can guide development of support tools, programs training and system design.

Fifthly, task analysis measure (Stanton, 2006) is illustrated to quantify and manage the response accuracy, reaction time, adaptation, endurance and such behavioral factors. It is the analytical process of how a task is carried out to achieve by analyzing system function in terms of the system goals and sub-goals inherent in performing the task (Diaper, 1990). A comprehensive consideration of work durations, allocation, frequency, complexity and any other unique necessary factors is adopted for the determination of the instructional work goals, describing the detailed work flow and evaluating the work performance.

ERGONOMIC HARMONY FRAMEWORK

An integrated theoretical ergonomic evaluation and design framework named "Ergonomic Harmony Framework" (see Figure 1) is developed for the aging workforce following the research proposed in the previously described.

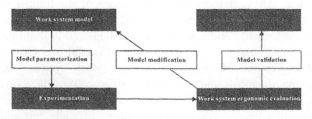

Figure 1. The ergonomic harmony framework.

Work on ergonomics harmony modeling including the definition of the work system model and the mathematical harmony evaluation has been underway.

A work system model is constructed through the integration of four major related elements based on the balance theory (Carayon, 2009), while each element incorporates multiple factors and parameters for both internal and external interactions of the work system. A simplified descriptive model comprising these important and related subsystems named personnel subsystem, technology subsystem, environment subsystem and organization subsystem is proposed to describe the internal dynamics of the integral work system (see Figure 2). It provides a visual way of picturing all the elements of work system that affect aged workers and the entire outcomes. It is assumed that whenever there is a change introducing in any element of the work system, other elements will change to seek an equilibrium between the work demands (task, technology, environment and organization) and the individual aging workers' capacity (physiological, psychological, psychosocial and behavioral).

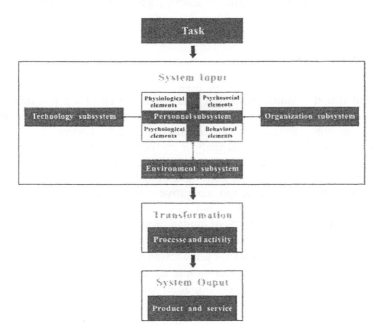

Figure 2.   The proposed work system model for aging workforce within the proposed ergonomic harmony framework

In general, traditional work system models are descriptive models, which are unable to provide measurable and accurate evaluation and design guidance for a work system. Therefore, a mathematical framework specifically simulating and formulating this model is combined for evaluating the numerical harmony of the entire work system. This framework that incorporates the insight into two layers of nodes with distinct semantics, with the upper layer representing the elements of the work system inputs, while the lower layer denotes the work system outputs.

A harmony function $H(r,a)$ is defined to evaluate the state of work system harmony in the form of a quantitative value (Hara and Mogi, 1993). It is given by:

$$H(r,a) = \sum_i \sigma_i a_i h(r,k_i) \tag{1}$$

where a_i is the value of element of work system input, σ_i is a strength weighting for each element, r_j is the value of work system output, h is a degree of harmony contributed by activating an element of input a_i for an output r_j, k is a parameter which determines a matching rate between input and output.

With the proposed work system model introduced based on the ergonomic harmony framework, we are undertaking subject test and simulation studies with local manufacturing industries in the Greater Pearl River Delta (GPRD) region where a very large labor workforce is deployed to carry out production of commodities, to collect definitive data for analyzing the value and practicality of the framework.

CONCLUSIONS

It is expected that the contributions and finding of the proposed research would ensure harmony among the key factors of the work system by establishing an integrated ergonomic framework for instructing work system design. This framework consists of four enhanced analysis methods for designers to identify the main factors and behavior measures for work system design. The framework is extended and integrated based on the existing approaches generally known as digital human modeling, participatory design, cognitive work analysis and task analysis which are separately used to evaluate and solve the physiological, psychological, psychosocial and behavioral problems of human factors in a work system.

The benefit of using this ergonomic approach in designing a new work system for aging workforce is that it allows for a reduction in overall work time as well as cost effectiveness, improved communication of both ergonomic concerns and design alternatives, much more comfortable and safe workstation. For the producers, the new work system organization means reduced stimulation from work, less opportunity to influence their workers' health and also decreased musculoskeletal disorders. As a bonus, the aging workforce will develop higher work morale and better control costs of the work.

In general, to ensure a highly usable and truly effective work system for aging workforce, the most effective means to accomplish this is through ergonomic approach. However, the explicitly linking of this ergonomic harmony framework to a specific work system design remains a great challenge for industrial engineers and system designers. This challenge concerns the integration of the separate personal, technical, environmental and organizational subsystems into one integral model. More significantly, the study will broaden the field of study with the introduction of a generalized framework.

REFERENCES

Alter, S. (1999), "A general, yet useful theory of information systems", *Communications of the Association for Information Systems*, 1(13), 1–69.

Carayon, P. (2009), "The balance theory and the work system model . . . twenty years later", *International Journal of Human-Computer Interaction*, 25(5), 313–327.

Chaffin, D.B. (2001), *Digital Human Modeling for Vehicle and Workplace Design*. Society of Automotive Engineer, Warrendale, PA.

Cho, S.-H., and Eppinger, S.D. (2005), "A simulation-based process model for managing complex design projects", *IEEE Transactions on Engineering Management*, 52(3), 316–328.

Das, B. (1999), "Development of a comprehensive industrial work design model", *Human Factors and Ergonomics in Manufacturing*, 9(4), 393–411.

de Zwart, B.C., Frings-Dresen, M.H., and van Dijk, F.J. (1995), "Physical workload and the aging worker: a review of the literature", *International Archives of Occupational and Environmental Health*, 68(1), 1–12.

Diaper, D. (1990), *Task Analysis for Human-computer Interaction*. Prentice Hall PTR, Upper Saddle River, NJ.

Haight, M.J. (2003), "Human error and the challenges of an aging workforce: considerations for improving workplace safety", *Professional Safety*, 48(12), 18–24.

Hanse, J.J., and Winkel, J. (2008), "Work organisation constructs and ergonomic outcomes among European forest machine operators", *Ergonomics*, 51(7), 968–981.

Hara, F., and Mogi, S. (1993), "A computational model of artificial emotion by using harmony theory and genetic algorithm", *Proceedings of the 2nd IEEE International Workshop on Robot and Human Communication*, 414–419.

Helin, K., Viitaniemi, J., Aromaa, S., Montonen, J., Evilä, T., Leino, S.-P., and Määttä, T. (2007), "OSKU digital human model in the participatory design approach: A new tool to improve work tasks and workplaces", *VTT Working Papers*, VTT-WORK-83.

Ilmarinen J.E. (2001), "Aging workers", *Occupational and Environmental Medicine*, 58(8), 546–552.

Jenkins, D.P., Stanton, N.A., Walker, G.H., Salmon, P.M., and Young, M.S. (2008), "Applying cognitive work analysis to the design of rapidly reconfigurable interfaces in complex networks", *Theoretical Issues in Ergonomics Science*, 9(4), 273–295.

Kenny, G.P., Yardley, J.E., Martineau, L., and Jay, O. (2008), "Physical work capacity in older adults: Implications for the aging worker", *American Journal of Industrial Medicine*, 51(8), 610–625.

Kleiner, B.M. (2006), "Macroergonomics: Analysis and design of work systems", *Applied Ergonomics*, 37(1), 81–89.

Manuele, F.A. (1993), *On the Practice of Safety*. Van Nostrand Reinhold, New York.

McQuaid, R.W. (2007), "The aging of the labor force and globalization", in: *Globalization and Regional Economic Modelling*, Cooper, R.J., Donaghy, K.P., and Hewings, G.J.D. (Ed.). Springer-Verlag, Heidelberg, pp. 69–85.

Parker, S.K., Wall, T.D., and Cordery, J.L. (2001), "Future work design research and practice: Towards an elaborated model of work design", *Journal of Occupational and Organizational Psychology*, 74(4), 413–440.

Rasmussen, J., Pejtersen, A.M., and Goodstein, L.P. (1994), *Cognitive Systems Engineering*. Wiley Interscience, New York.

Rix, S.E. (2009), "Employment at older ages", in: *International Handbook of Population Aging*, Uhlenberg, P. (Ed.). Springer, Dordrecht, The Netherlands, pp. 445–470.

Schuler, D., and Namioka, A. (Ed.) (1993), *Participatory Design: Principles and Practices*. Lawrence Earlbaum, Hillsdale, NJ.

Shikdar, A.A., and Das, B. (2003), "The relationship between worker satisfaction and productivity in a repetitive industrial task", *Applied Ergonomics*, 34(6), 603–610.

Stanton, N. (2006), "Hierarchical task analysis: Developments, applications, and extensions", *Applied Ergonomics*, 37(1), 55–79.

Sundin, A., Örtengren, R. (2006), "Digital human modeling for CAE applications", in: *Handbook of Human Factors and Ergonomics*, Salvendy, G. (Ed.). John Wiley & Sons, Hoboken, New Jersey, pp. 1053–1078.

CHAPTER 86

Website Usability Design Principles for the Elderly

Rashaad E.T. Jones, Mary E. Mossey, Mica R. Endsley

SA Technologies, Inc.
Marietta, GA 30066

ABSTRACT

This paper reviews and establishes usability design principles for websites that provide services and information to the elderly populace. This paper addresses usability design principles in three core areas: visual, motor, and cognitive. Specifically within the visual area, this paper establishes guidelines for the proper use of font, color contrast, and text spacing to support older populations. Motor-related issues that impact speed and accuracy of selecting the correct interface component (e.g., button, textbox, etc) and effective scrolling are discussed. Lastly, cognitive issues are discussed that include supporting working memory, navigating through websites, and proper utilization of error messages. Finally, the paper concludes with how these usability design principles are being leveraged for the development of an online dispatch, coordination, communication, and support system to support Red Cross volunteers, which is primarily comprised of seniors.

Keywords: Usability Design, Websites, Elderly, Human Factors

INTRODUCTION

The "baby boomer" generation, those born approximately between 1946 and 1964, have begun approaching senior or "elderly" status. By 2000, it was reported that more than half the adult population was 45 years old or older, and the 85 and older age group became the fastest growing in the country (Zhao, 2001; Fisk, Rogers, Charness, Czaja, & Sharit, 2009). Accompanying this increase in the senior population is the use of technology, specifically the internet, which has become more omnipresent. Today, people of all ages use the internet in every aspect of their lives, including work, personal, social, and recreational activities (Fisk, Rogers, Charness, Czaja, & Sharit, 2009). Consequently, larger percentages of seniors are online now than in the past (Jones & Fox, 2009). Seniors, particularly, have demonstrated great interest in the internet because it gives them another method to communicate and stay informed (Nielsen, 2002).

As the adult population ages and the internet pervades more aspects of life and culture, the need to effectively design websites to accommodate older adults increases. Beginning in the mid forties, the physical and cognitive signs of aging begin to be noticeable, making interface design aimed at the 45 and older age group challenging and important to successful interaction with technology (Hawthorn, 2000). Many seniors begin using internet technology later in life and are not as proficient as younger generations who began using computers at a relatively young age (Nielsen, 2002). For example, scrolling is a concept that younger generations readily understand, but seniors have been found to struggle with (Usability for older web users, 2006). Due to their unfamiliarity with modern interface metaphors, seniors need more time to understand pages, scan text, and extract information (Watanabe, Yonemura, & Asano, 2009). Seniors' inexperience combined with declining physical and cognitive abilities exacerbates their frustration with trying to learn new technology. Unfortunately, the elderly tend to blame themselves for their inability to maneuver through web interfaces rather than poor interface design (Usability for older web users, 2006). However, despite their frustration, seniors still have an interest in technology and recognize its benefits to themselves and society (Nielsen, 2002).

Despite the impact of the aging population, very little research has been done to identify clear design standards for interfaces backed by specific research. In 2000, Hawthorn noted that no research was done on what makes an interface usable for older adults. Towards the end of the last decade, it was reported no Web design methods that are appropriate for elderly users have still been published (Watanabe, Yonemura, & Asano, 2009). Furthermore, real world applications do not have proper design guidelines for seniors. For example, (Lamendola, 2008) found that medicare websites, which are aimed at seniors, are too difficult to be used by them. Effective approaches to deciding what is important or what is not important in website designs for seniors have longed been sought (Watanabe, Yonemura, & Asano, 2009). Thus, the objective of this paper is to address this issue by providing website design principles for the senior population.

A notable set of work has been accomplished by Fisk and colleagues (2009), who has recently discussed human factors principles for seniors for various applications, including interface design. This paper continues this discussion by identifying critical design guidelines for websites targeted for seniors and briefly discussing how it has been applied in an web-based application targeted for seniors that volunteer in the American Red Cross. The guidelines presented in this paper are established from a methodology that includes a comprehensive literature review and an examination of common practices that have been found to be effective in current websites aimed for seniors. The organization of this paper is as follows: the analysis section outlines the impacts of aging as it relates to using a website application and how proper design principles address those issues. The practice section highlights our work in leveraging these principles to design and develop a web-based application targeted for seniors. The paper concludes with a summary of our research and how our efforts are significant to the human factors community.

ANALYSIS

Prior research indicates three common weakness areas for the elderly that require proper user interface (UI) design considerations: declining vision, limited short-term memory, and insufficient conceptual understanding (Zajicek, 2001; AgeLight, 2001; Akutsu, Legge, Ross, & Schubel, 1991; Nielsen, 2002; Ownby & Czaja, 2003; Hart & Chapparro, 2004). Visual hindrances for the elderly consist of physical conditions that affect their vision, which ultimately impacts their online experience. Motor hindrances have been found to decrease web performance. For example, motor issues, such as shaky fine movements, can make selecting small buttons difficult. Lastly, cognitive hindrances arise from aging issues, such as decreased short term memory, and unfamiliarity with technical web terms, due to inaccurate mental models. The following sections highlight the research completed in each area and concludes by identifying critical design principles.

VISION

As people age, their vision declines for a variety of reasons including stroke, glaucoma, presbyopia, cataracts, and the shrinking of the pupil (AgeLight, 2001). Theses disadvantages lead to the importance of proper contrast for web pages (Hawthorn, 2000; Watanabe, Yonemura, & Asano, 2009). Contrast includes colors, patterns, and designs for the background and the foreground (which includes font color). For example, designers should avoid style techniques like shading text and patterned backgrounds, as these can be distracting and create eye fatigue (Zhao, 2001). Additionally, text should be dark with a light background rather than light with a dark background. Bright or light colors, especially on dark backgrounds create a halo that make letters appear to blur into themselves (AgeLight, 2001; Zhao, 2001). Furthermore, color contrast must be considered when placing colored elements on top of one another. Choosing complimentary colors or colors from the top of the color wheel to utilize with colors on the bottom of the wheel generally results in good contrast (AgeLight, 2001; Zhao, 2001). Colors should be bright and bold without being florescent; other color properties, such as hue, lightness and saturation should also be considered (AgeLight, 2001). Colors such as black, white, blue and yellow are more easily visible than red, green, brown, grey, and purple (AgeLight, 2001). Lastly, seniors generally are more sensitive to glare and changes in illumination, so abrupt changes in screen illumination should be avoided as well as distracting flashing graphics (AgeLight, 2001; Zhao, 2001).

Seniors have a harder time reading small fonts, which most websites typically utilize to maximize their content per page area. Thus, proper usage of font size and type should be considered to provide quality usability for a webpage. While there is a discourse as to whether larger font actually improves performance, research has noted that seniors generally prefer larger text (Chadwick-Dias, McNulty, & Tullis, 2003). Nayak and associates (2006) reports that webpages with large, easy to read text with high contrast is essential for seniors. For example, 12 point font should be the minimum size and headings should be 2 points larger (Usability for older web users, 2006; Nielsen, 2002; AgeLight, 2001; Zhao, 2001). Simple and common fonts should be used for their legibility (AgeLight, 2001; Zhao, 2001). Researchers contend that san serif should generally be used (Hawthorn, 2000). Additionally, websites must provide an easy and simplistic manner for users to increase the font size to meet their needs in order to ensure all users can read the text (AgeLight, 2001; Zhao, 2001). Designers should use bold text only in appropriate places, particularly bold text should not be used when it is surrounded by text that is immediately above and below it (as in the case of a paragraph). This general rule is advisable because while bold text may seem larger, line spacing has been reduced, making text harder to read (AgeLight, 2001; Zhao, 2001). Text should never be placed on a patterned background because it causes eye fatigue when the letters blur with the background. Finally, text should be left aligned for readability (AgeLight, 2001; Zhao, 2001).

MOTOR

Motor issues have been found to impact web performance. Motor activities include physical movements that are required by the user to interact with a webpage. An example include using a peripheral device, such as a mouse to select a button. Motor activities require proper hand-eye coordination and for seniors, physical issues, such as shaky fine movements, can make selecting small buttons difficult. Additionally, since speed of physical movements and accuracy of hand eye coordination generally decrease with age, research suggests that buttons and links should be large and readily selectable (Zhao, 2001; Hawthorn, 2000; Hollinworth, 2009). While the size of the button is a relative term, a "large" button should appear to be bigger than normal to help seniors recognize it quickly and select it easily. Furthermore, there should be sufficient space between buttons and hyperlinks to prevent accidental selections of the wrong button. These principles should also apply to text and text boxes. Navigational activities, such as scrolling is impacted by motor hindrances. Scrolling has been found to be cumbersome for seniors that do not have a scroll wheel on their mouse device or alternatively, or a computer with touchpad scrolling (AgeLight, 2001). Older users have been found to accidentally make large jumps in scrolling that cause them to lose their place on the webpage (Hawthorn, 2000). Finally, drop down menus and nested menus have been found to cause frustration in seniors when the menus disappear upon mouse movement or if the menu items are too close together (Nielsen, 2002).

COGNITIVE

Certain cognitive processes deteriorate with age and should be taken into consideration when designing webpages. For example, memory issues impact web navigation. Seniors with poor memory have been found to lose their sense of "internet orientation and direction" and forget the specific webpage they had recently visited or the hyperlinks that they had clicked (Nielsen, 2002; AgeLight, 2001; Zhao, 2001). A clear indication of where users currently are and where they just were can also aid in navigation. Consequently, a sound design principle that has been practiced is to change the color of the hyperlink text, preferably from blue to purple, once it has been clicked. To address reduced working memory in seniors, information should be presented in a simplified manner, so that seniors have only the information they need to process at any moment in time and the information should be presented in an instructional manner to assist in comprehension. For example, instructions that use action words, such as "Go to My Profile" has been found to guide seniors to the appropriate page rather than one word descriptions, such as "Profile" (Usability for older web users, 2006; Chadwick-Dias, McNulty, & Tullis, 2003).

As many seniors have learned to use computers later in life, they struggle with technical terminology that younger groups readily understand. Terms, such as "homepage" and "URL" are basic internet terms, but are difficult for seniors to grasp (Usability for older web users, 2006; Lamendola, 2008; Chadwick-Dias, McNulty, & Tullis, 2003). Additionally, seniors have been found to have poor mental models of how webpages are structured (Chadwick-Dias, McNulty, & Tullis, 2003). Avoiding complex, nested menus have been found to help seniors avoid getting lost in a task (Tang, 2005). Furthermore, many seniors do not know understand scrolling and subsequently miss information at the bottom of the page without realizing it (Usability for older web users, 2006; Lamendola, 2008; AgeLight, 2001; Wantanabe, 2009). Consequently, designing webpages that do not scroll and minimizing navigation to other pages will ensure that they can access all of the information available (AgeLight, 2001). Tabs are a popular UI component that have been used to divide a webpage into sections and eliminate scrolling (Sanchez & Wiley, 2009). When unable to find information, many seniors resort to the "Search" function of the website, so an effective search engine that examines the entire website will help make up for shortcomings in navigational capabilities (Usability for older web users, 2006; AgeLight, 2001).

To properly support senior comprehension, webpages should provide clear, simple, and instructional feedback. For example, errors and error messages should not frustrate the user, but rather simply explain the issue and offer solutions (Nielsen, 2002; Watanabe, Yonemura, & Asano, 2009). Seniors have been found to be slower at understanding and recovering from errors, so error messages that help highlight the next steps will assist seniors to recover from the error faster (Hawthorn, 2000). Seniors tend to read all of the text on a page before making a selection which can double their time to complete a task (Usability for older web users, 2006; Chadwick-Dias, 2003). Consequently, a large amount of white space and use of short sentences or bullets makes reading text more manageable and less time consuming (AgeLight, 2001; Zhao, 2001).

Visual	Motor	Cognitive
Colors and Illumination • Use complimentary colors • Colors should be bright and bold • Colors, such as black, white, blue, and yellow are preferable • Avoid flashing graphics **Text** • Avoid shading text and placing text on patterned backgrounds • Text color (foreground color) should be dark with a light background • Use 12 point font for regular text • Headlines should be 2 points larger • Use San Serif font • Provide easy way to increase font size • Only use bold text in appropriate places	**Buttons** • Use buttons that are relatively larger than the normal size typically used • Buttons and hyperlinks should be large and readily selectable • Use sufficient space between buttons and hyperlinks	**Terminology** • Use instructional action phrases for buttons • Avoid technical terms • Use forgiving and corrective feedback messages • Use short phrases **Search** • Use effective search engines that examines the entire website
	Navigation • Eliminate scrolling by placing all information on the visible portion of the screen • Avoid using drop-down and nested menus • Use blue color and underline hyperlink text and set the color to purple if the user has already clicked it • Use tabs, when appropriate	

Table 1: Design guidelines

PRACTICE

The design principles outlined in this paper were put into practice to inform the design a web-based solution for the American Red Cross (ARC). Our research with the ARC indicated that they have a critical need to support their Disaster Action Teams (DAT) in coordinating and dispatching resources. A single DAT is responsible for a geographical area and DAT team members travel to locations of emergencies, such as a fire at a residence, to provide assistance to victims. Each

DAT team has a assistant captain that dispatches DAT team members. A DAT captain is responsible for several DATs and must assign volunteers to a specific DAT. The DAT captain and assistant captain both must utilize DAT team members strategically so that all areas are covered without over-expending their team members.

Members of a DAT are primarily volunteers, with a majority of them being in the senior age group. Consequently, a design requirement involved developing a solution that can be easily used by seniors. The design principles established in this paper, such as utilizing action words for buttons and minimizing scrolling, guided the design and development of PinPoint™, which is a technological solution that supports ARC volunteers preparation and response operations by pinpointing resources, locations, activities, and information.

PinPoint is a web-based solution that effectively enhances coordination, supports information sharing, and facilitates communication, while requiring little training from its users. One of the major features of the PinPoint system is its capability to support the multiple roles of the DAT. It includes Google Mapping Services to display geospatial information, such as the location of events, that supports the DAT Captain's goal to manage members strategically (Figure 1). It also supports navigational tasks, such as providing directions to an ongoing emergency. Additionally, humanitarian efforts like those of the Red Cross provide a unique design challenge due to its constantly changing personnel and need to immediately share critical information. As such, PinPoint addresses this design challenging by providing a system that is robust, streamlined, and flexible. It allows for resource coordination in constantly changing roles by providing only the critical information when needed, and facilitating communication between resources, all while maintaining a high ease of use. While there has not been usability studies to measure the efficacy of PinPoint, there has been overwhelming positive feedback from members of the American Red Cross, specifically in the simplicity and ease of use of the design of the PinPoint application.

CONCLUSION

This paper provides a documented set of guidelines for an underexplored research area . The usability design principles established by this paper can be used as a template to guide future website designs. Additionally, though the 65 and up age group require more assistance in the above areas, those in their mid 40s and up can also benefit from these assistive designs, as many of the ailments of the 65 and up age group begin in the mid 40s. Furthermore, the guidelines highlighted in the following report parallel standard human factors practice. Consequently, these guidelines that are specifically targeted for the elderly will not inhibit the usability of other age groups.

840

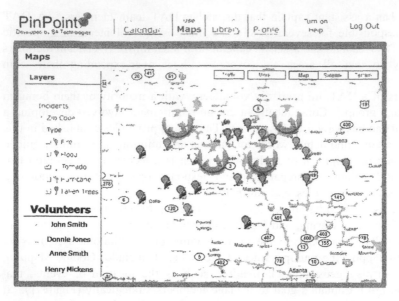

Figure 1: Sample screenshot of PinPoint™ application

REFERENCES

AgeLight. (2001). Interface Design Guidelines for Users of All Ages. AgeLight.

Akutsu, H., Legge, G. E., Ross, J. A., & Schuebel, K. J. (1991). Psychophysics of reading: X. Effects of Age-Related Changes in Vision. Journal of Gerontology: Psychological Sciences , 325-331.

Chadwick-Dias, A., McNulty, M., & Tullis, T. (2003). Web Usability and Age: How Design Changes can Improve Performance. ACM SIGCAPH Computers and the Physically Handicapped (pp. 30-37). New York: ACM.

Charness, N. (2008). Aging and Human Performance. Human Factors: The Journal of the Human Factors and Ergonomics Society , 548-557.

Fisk, A. D., Rogers, W. A., Charness, N., Czaja, S. J., & Sharit, J. (2009). *Designing for Older Adults: Principles and Creative Human Factor Approaches.* Boca Raton: CRC Press.

Gribbons, B. (2006, October 24). Functional Illiteracy and the Aging Population: Creating Appropriate Design Support. Bentley College.

Hawthorn, D. (2000). Possible implications of aging for interface designs. Interacting with Computers , 507-528.

Hollinworth, N. (2009). Improving computer interaction for older adults. ACM SIGACCESS Accessibility and Computing (pp. 11-17). New York: ACM.

Jones, S., & Fox, S. (2009, January 28). PEW Internet Project Data Memo. Retrieved from Pew Internet & American Life Project: http://pewinternet.org

Lamendola, B. (2008, August 20). Medicare Web Site Just Doesn't Click with Users, Report Shows. Retrieved from AARPBulletintoday: http://bulletin.aarp.org

Making Your Website Senior Friendly. (2009, February). Retrieved from National Institute of Aging: http://nia.nih.gov

Nielsen, J. (2008, March 31). Middle-Aged Users' Declining Web Peformance. Retrieved from Useit.com: http://www.useit.com

Nielsen, J. (2002, April 28). Usability for Senior Citizens. Retrieved from Useit.com: http://www.useit.com

Ownby, R. L., & Czaja, S. J. (2003). Healthcare Website Design for the Elderly: Improving Usability. American Medical Informatics Association Annual Symposium Proceedings, (p. 960).

Sanchez, C. A., & Wiley, J. (2009). To Scroll or Not to Scroll: Scrolling, Working Memory Capacity, and Comprehending Complex Texts. Human Factors: The Journal of the Human Factors and Ergonomics Society , 730-738.

Sibley, C. (2008, December). Web Usability and Aging. Retrieved from Usability.gov: http://www.usability.gov

Usability for older web users. (2006, February). Retrieved from webcredible: http://www.webredible.co.uk

Watanabe, M., Yonemura, S., & Asano, Y. (2009). Investigation of Web Usability Based on the Dialogue Principles. In M. Kurosu, Human Centered Design (pp. 825-832). Berlin: Springer-Verlag.

Zajicek, M. (2001). Supporting older adults at the interface. In C. Stephanidis, Universal Access in HCI: Towards an Information Society for All (pp. 454-457). CRC Press.

Zhao, H. (2001, April). Universal Usability Web Design Guidelines for the Elderly (Age 65 and Older). Retrieved from Universal Usability in Practice: http://www.otal.umd.edu

Chapter 87

OASIS HMI: Design for Elderly – A Challenge

Vivien Melcher[1], Jan-Paul Leuteritz[2], Maria Panou[3], Asterios Leonidis[4]

[1]Fraunhofer Institute for Industrial Engineering IAO
Stuttgart, Germany

[2]Institute for Human Factors and Technology Management IAT
University of Stuttgart, Germany

[3]Centre of Research and Technology
Hellas, Greece

[4]Foundation for Research and Technology
Hellas, Greece

ABSTRACT

The older population is increasing very rapidly in the western world. This fact leads to new requirements of ICT with the focus on HMI design for the elderly. In order to ensure the independence and quality of life of elderly citizens, it is important to remove obstacles and avoid hassles that prevent them from having optimal access to information technologies. The aim of the OASIS project is to support elderly people in their daily activities by a number of web based services. These services are interconnected using an open reference architecture and they will be available at different end devices like PC, PDA, and even mobile phone.

The user group of the OASIS applications has different preconditions for using ICT. They can differ in aspects of mobility, vision, computer literacy or background knowledge. Furthermore different context characteristics simplify or handicap the usage of different end devices. For this reason a self-adaptive user interface technology will be used to automatically adapt different parameter of the graphical user interface and the end device to the user requirements. Additionally, in order to give OASIS a common look-and-feel and to assure overall usability, a style guide is used to help developers of new web based services fit their application into OASIS and its concept of an elderly friendly interface. The development of the adaptation rules and the OASIS style guide will be described with focusing on a cultural adaptation.

Keywords: HMI for elderly, senior users, user interface design, self-adaptive interfaces, culture

INTRODUCTION

The societies of European countries will grow older significantly in the next few decades. For instance the population in Germany will have decreased by about 6,4 percent in 2030 in comparison to 2005 (Statistische Ämter des Bundes und der Länder, 2007). The decrease of the population goes along with a structural change of the population's composition. In contradiction to the decreasing population the number of elderly over 65 will have increased by about 40 percent in 2030. Additionally, the expectancy of life will increase by about 4 years from 2005 to 2030. An appropriate medical care and an integration into the social system becomes more and more difficult. Help from health institutions or the government will be increasingly limited, due to the fact that the ratio of retired people to working people will shift dramatically during the coming decades (e.g., Statistisches Bundesamt, 2006).This demographic change requires new solutions towards improving the independence, the quality of life, and the active ageing of older citizens.

Elderly people are willing to use new technologies to improve their living situation and to be independent until a high living age. Therefore it gets more and more important to remove obstacles that prevent elderly users from having optimal access to information technologies.

OASIS is an Integrated Project of the 7th FP of the EC in the area of eInclusion that aims at increasing the quality of life and the autonomy of elderly people by facilitating their access to innovative web-based services. OASIS stands for "Open architecture for Accessible Services Integration and Standardisation", which hints at the project's way towards making this vision a reality: OASIS aims at creating an open reference architecture, which allows not only for a seamless interconnection of Web services, but also for plug-and-play of new services. In order to give the OASIS architecture a critical mass for widespread implementation, the project consortium will make the reference architecture in question, and the related tools, available as open source. 12 initial services have been selected for prototype development in the project's lifetime. They are classified into three main categories considered vital for the quality of life enhancement of the elderly: Independent Living Applications, Autonomous Mobility, and Smart Workplaces Applications (OASIS Consortium, 2007).

DESIGN FOR THE ELDERLY

An application's usability depends on its appearance and behaviour and on the way the appearance and behaviour provide access to the application's functionalities. In turn optimal access to ICT depends on the user requirements. The primary problem when designing for the elderly is the heterogeneity of the user group and therewith a diversity of requirements to the system or application. The "generation 55+" is one of the largest user groups possible. First of all, there are huge differences in age. A person at the age of 55 certainly has user needs that differ a lot from those of a person that is 85 years old. Additionally, different kinds of deteriorations can be expected, which are not predictable by age only. These include deficits in cognition or attention, learning and memory, partial loss of sight or hearing and psychomotor deteriorations.

The vision of the OASIS project is to provide optimal ICT access to users worldwide. This means next to user needs based on age, gender, deteriorations or impairments additional user requirements have to be taken into account, for instance the user's cultural background.

Reeves (1998) showed that humans are using the same rules for communicating with computers as with other humans. Both are processes where information is exchanged. This leads to the fact, that problems in communication between persons with different cultural backgrounds derived from cultural misunderstandings would also appear in the communication between a user of a specific culture and an application developed on the basis of another cultural knowledge (Xie et al., 2008).

THE USER INTERFACE ADAPTATION METHOD

All the above mentioned makes it clear that there is no "one size fits all" solution to this problem. The OASIS project is approaching a new concept to bridge the gap between universal UI design and user needs. The OASIS project is making first steps in the area of automated user interface adaptation and approaching those high goals using a concept called "Unified User Interface Design Method" (Savidis & Stephanidis, 2004). This method helps developers of services and applications by providing developer tools and internet-based infrastructure to create self-adapting applications. In the context of the OASIS project, developers will be provided with an *Adaptive Widget Library*. At runtime, the infrastructure automatically does the adaptation for any application that has been programmed using these widgets. The end device sends important adaptation triggers, such as the user profile, to the *DMSL Server* (Savidis et al., 2005; Antona et al., 2006) (*DMSL* means "Decision Making Specification Language"). This server then applies its predefined adaptation rules to the interface on the user end device. The adaptation rules were derived from a literature review. Special design guidelines for elderly users were researched in order to find or create reasonable adaptation rules (e.g. Fisk et al., 2004). The result of this research is a huge number of "if, then" rules that define which adaptation should be done.

DMSL-Server   **User end device**

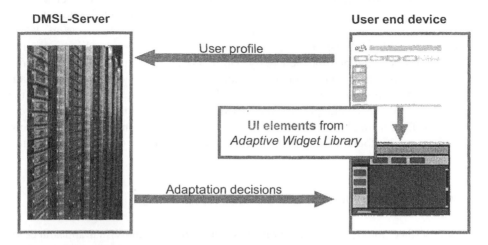

Figure 1: OASIS adaptation approach

THE OASIS STYLE GUIDE

A consistent look-and-feel helps users learn an application faster and use it more efficiently. In addition, a consistent look-and-feel will increase the memorability of an application and helps users learn to navigate through other applications that share the same look-and-feel ISO 9241, 2007). Based on the usability principles of the European standard 9241 a style guide has been defined to create a unified look & feel of all applications associated with the OASIS project. The main objective of this OASIS style guide is to phrase a set of design principles and guidelines, which will enable the UI developer to conceive and implement the identified user requirements, corresponding to the various alternative modalities, interaction methods applicable to end-user devices, and the widgets of the adaptive library.

The style guide can be distinguished into two parts. The first part describes aspects, which cannot be influenced by the adaptation tool. These aspects include basic guidelines concerning the layout of the screen, the navigation principles, menu structure and main metaphors. The layout copies a website look-and-feel because this is considered to be best known among the end users. The layout is separated in three distinct zones. The first one serves for the personalisation of the interface, in case the user is not content with the automatic adaptation and wants to override some settings. The user can e.g. switch on and off audio features, change the audio volume and chose a different colour profile. The second area is dedicated to the navigation. It divides into a horizontal and a vertical tab menu, as Gao et al. (2007) have shown that the concept of tabs is very well understood by elderly users. The horizontal tabs are the main navigation elements, the vertical ones second level of menu options. The third zone is the content area, which has only few restrictions of design.

The second part of the style guide describes all basic UI elements which are part of the adaptive widget library. Two colour profiles can be adapted automatically. The first colour profile is for the elderly with no impairments in vision. It's a coloured profile with the green as the main colour theme. A second less coloured profile has been defined for the elderly with decreased seeing ability. This profile fulfils the request for a maximum colour contrast by using black and white as main colours. A third colour profile with inverted colours can be set manually by the user. Next to the colour profiles the OASIS style guide specifies four different profiles for font and UI element sizes. The size of UI elements increases from profile one to four. The profile selection is handled by the decision making adaptation engine.

IMPACT OF CULTURE ON USER INTERFACES

As described above, the OASIS project is going to adapt the visual appearance of the user interface to the characteristics of the user (e.g. age, deteriorations) and the context of usage (e.g. on the way, at home). A new concept within the OASIS project is a cultural adaptation of the graphical user interface to the user's cultural background. This concept will be described in the next paragraphs.

Sir Edward B. Taylor (1920, p.1) defines culture as "That complex whole which includes knowledge, belief, art, morals, law, custom, and any other capabilities and habits acquired by man as a member of society."

The Dutch cultural anthropologist Hofstede (1984) made the groundbreaking contribution to defining culture. He defines culture as „...the collective programming of the mind which distinguishes the members of one group or category of people from another". Hofstede announced that culture is a learned construct, not an inherited one. He identified five dimensions of culture: Power Distance (PD), Uncertainty avoidance (UA), Masculinity (MAS), Indiviualism (IDV), and Long- term vs. short-term orientation (LTO) (Hofstede, 1984, 1997, 2006)

Power Distance (PD): Power distance can be defined as the extent to which less powerful members of institutions and organizations within a culture expect and accept that power is distributed unequally (Hofstede, 1997). High PD countries tend to have a centralized political power and exhibit tall hierarchies in organizations with large differences in salary and status (Marcus & Gould, 2000).

In contrast low PD countries tend to view subordinates and supervisors as closer together and more interchangeable, with flatter hierarchies in organizations and less differences in salaries and status (Hofstede, 1997; Marcus & Gould, 2000).

Uncertainty avoidance (UAI): Uncertainty avoidance is the extent to which the members of a culture feel threatened by uncertain or unknown situations and try to avoid such situations (Hofstede, 1997).

Marcus and Gould (2000) noted that high UA cultures may have a business with more formal rules, a longer career commitment and a focus on tactical operations. They expect structure in organizations to help make events clearly interpretable and predictable. Business in low UA cultures may be more informal and focus more on long-range strategic matters.

Masculinity (MAS): Masculinity and femininity are two ends of a continuum. In cultures with a high masculinity index gender roles are important and clearly distinct. Certain values such as assertiveness, toughness, focus on material success, and being unemotional are linked to masculinity. In contrast values like tenderness, caring, quality of life and emotional factors are linked to femininity. In cultures with a high femininity index gender roles overlap (Hofstede, 1997).

Individualism (IDV): Individualism is the degree to which a culture reinforces individual achievement and relationships. Cultures in which the ties between individuals are loose have a high individualism index. Members of those cultures tend to look after their family and themselves but no one else. Collectivism is the contrary of individualism. Members of cultures with a high collectivism index are strongly bounded from birth into cohesive groups that protect them in unquestioning loyalty (Hofstede, 1997).

Long-term vs. short-term orientation (LTO): This dimension was not included in the first approach to culture dimensions of Hofstede and has been added afterwards. It is related to the choice of focus for people's efforts: the future or the present and past. Values associated with Long-term orientation are thrift and perseverance; values associated with Short-term orientation are respect for tradition, fulfilling social obligations, and protecting one's 'face' (Hofstede, 1997, 2006).

Cultural impact on UI design has become a main topic of research in the last few years. Due to the distribution of the Internet people get more and more connected. Customer satisfaction now depends on how seriously service providers take into account regional user needs combined with culturally-suitable user interface design (Rau & Liang, 2003).

The representation of the real world in a graphical user interface (for instance an Internet website) is likely to be judged by the user in reference to the surrounding environment. Therefore in addition to usability, cultural aspects of interface design have to be considered. Allwood and Wang (1990) found significant differences between cultures in attitudes towards computers. Marcus (2000, 2002) demonstrated the influence of cultural dimensions on the design and layout of web sites. Rau, Liang (2003) and Choong and Salvendy (1997) found clear cultural differences in terms of time orientation and communication style. This research points out, that cultural differences do indeed impact attitudes towards and acceptance of technology.

Cultural differences manifest themselves in several ways – symbols, heroes, rituals, and values (Hofstede, 2006). Culture can have an impact on different components in user interfaces (Smith & Salvendy, 2001). It influences the metaphors and the mental models used in User Interfaces. Furthermore Navigation structure and interaction techniques can depend on cultural aspects. At least the whole appearance of a UI, including colour choices or UI elements, is defined by the cultural background of the designer.

In a first step the aim of the OASIS HCI concept is to provide a culturally adaptable appearance of the OASIS User Interface based on Hofstede's dimensions of culture.

APPROACH TO GUIDELINES DERIVED FROM HOFSTEDE'S CULTURE MODEL

Power distance (PDI): Marcus and Gould (2000) defined aspects of a user interface that can be influenced by Hofstede's power distance index. A cultural impact exists for the emphasis on the social and moral order and its symbols. High PD cultures use more frequently national and religious picture motives than low PD cultures. Marcus and Gould (2000) announced that web pages from high PD cultures give more prominence to leaders than low PD cultures where citizens, customers or employees are more prominent.

Based on Hofstede's power distance index and the theory of Marcus and Gould, the following adaptations within the OASIS applications could be applied: Pictures and icons showing men or national symbols could be set automatically for cultures with high PDI. For countries with a low PD index values pictures with elderly of both gender as motives could be set as default by the adaptation tool.

Uncertainty avoidance (UAI): Marcus and Gould (2000) believe that cultures with a high UA would emphasise simplicity on user interfaces with clear metaphors, limited choices, and restricted amounts of data. Navigation schemes would intend to prevent users from becoming lost. Redundant cues like colour, typography, and sound are used to reduce ambiguity. The theory of Marcus and Gould notes that low UA cultures would emphasise the reverse. Low UAI culture would pay attention to complexity with maximal content, they would accept wandering and risk and they would have less control of navigation. The coding of colour, sound, and typography would occur to maximize information.

Based on Hofstede's uncertainty avoidance index and the theory of Marcus and Gould, the following rules for a culturally adaptive user interface in the OASIS framework are possible: For cultures with a high index in uncertainty avoidance changes in the application could be announced very early on the start page to avoid confusion. Cultures with a high UA index value may feel secure in using the OASIS application in case experience reports or statements of other users regarding functionality of the application and their satisfaction are arranged on the start page of the application or the OASIS main application.

Masculinity (MAS): For design in feminine cultures poetry and unifying values, images of nature, and traditional arts used to generate emotional or aesthetic appeal, might play a more important role than practical, strictly goal orientated organization, navigation and use of graphics (Smith & Salvendy, 2001).

Regarding the MAS a cultural adaptation can occur for colours and picture themes. For cultures with a high MAS index an interface layout with cold colours like blue as dominant colour theme could be set as default. Warm colours like orange, light red or yellow could be used in interface design as default for cultures with a high femininity index. Another possibility to adapt the interface design to cultures with a high femininity index is to use kittenish symbols and icons like comic figures.

Individualism (IDV): Regarding Hofstede's culture dimension 'Individualism' the following guidelines could apply:

Cultures with a high individualism index value may prefer pictures with individual active elder persons as default. Cultures with a high collectivism score may feel closely connected with pictures showing groups of elderly as default.

For cultures with a high individualism index a newsflash with the latest news of the OASIS applications (e.g. new embedded service providers, news that might be important for elderly) could be displayed at the start screen of the regarding application or the main application. For cultures with a low individualism score this newsflash could be hidden in an option of the main menu (for example the help menu). In contradiction cultures with a low IDV index may prefer experience reports or statements of other users at the start page regarding functionality of the application and the user's satisfaction.

For cultures with a high individualism score more opportunities for interface personalization may be provided within each application. In this case the automatic adaptation will be limited to some aspects.

BOUNDARIES OF THE ADAPTIVE UI APPROACH

Not all of the differences Marcus and Gould (2000) found within web site design for different cultures could be turned into adaptation guidelines. The difficulty for culturally adaptive interfaces is to get the balance between cultural adaptation and usability. Hofstede's concepts of Power distance and Uncertainty avoidance are difficult to realize without breaking the rules for designing intuitive, comprehensible and predictable interfaces. For example, distinguishing between a simple and complex layout and between a clearly controllable and a less controllable navigation structure for cultures of different PD indexes will not be feasible within the OASIS applications, because the aim of the OASIS project is to provide an intuitive and predictable system to the end user. Making the interface design too complex or providing a less controllable navigation would thwart this aim.

Another problem derives with the definition of the user groups. What characteristics should a user have to assign him/her to a definite user group? Should the adaptation be in regard to the preselected country, religion, peer group, specified ages, gender, etc..?. Some characteristics like age, gender or the degree of age related deteriorations are ascertainable. However, cultural characteristics of a user are more difficult to measure. User and their cultural knowledge can be distinct into two groups: single-culture user and multi-culture user (Röse, 2001). Single-culture users do only have knowledge about the representation of meanings in their own culture. In contrary multi-culture user have the knowledge of different representations of a meaning - they discovered different cultures. What about those multi-culture users, who grow up and live as foreign nationals in another country? What culture do they have and to what culture related layout should the UI be adapted? At our current state we are not able to answer those questions. A concept of assignment of users to user groups with defined characteristics (e.g., value of culture dimension, level of deteriorations) has to be developed at first.

CONCLUSION AND FUTURE PERSPECTIVES

This paper has presented the OASIS approach to user interface adaptation in the context of applications for older users. The approach of a self-adaptive user interface has been complemented with an approach for a culturally adaptation. The OASIS style guide and the adaptive UI approach are still concepts in their beginning. They have to be evaluated to ensure their feasibility and to avoid a decrease in user satisfaction. The OASIS project tries to identify the best fitting UI solution for every user. From the huge number of possible trigger combinations for the adaptation arise a huge number of possible interface alternatives. These interface alternatives have to be validated in regard to the user requirements. This will be done in a next step. For this purpose a defined number of interface alternatives will be presented to users of different user groups within usability laboratories across Europe. If the results indicate that a defined user interface layout based on a set of rules improves interaction for certain user characteristics, then this set of rules can be consolidated and re-used. If usability problems result from the UI design or the adaptation is not accepted by the user group, then the rules have to be revised.

The approach for a culturally adaptation should be understood as a first idea. Further research is necessary to consolidate this concept, taking into account technical and ethical aspects and the compliance with general usability standards. The outcomes of a first testing with users will give more insight in the feasibility of this approach.

REFERENCES

Allwood, C. M., & Wang, Z. (1990). Conceptions of computers among students in China and Sweden. *Computers in Human Behavior, 6*(2), 185-199.

Antona M., Savidis A., & Stephanidis C. (2006). A Process–Oriented Interactive Design Environment for Automatic User Interface Adaptation. *International Journal of Human Computer Interaction,* 20 (2), 79-116.

Choong, Y.Y. & Salvendy, G. (1998). Design of icons for use by Chinese in mainland China. *Interacting with Computers,* 9, 417-430.

Fisk A., Rogers W., & Charness N. (2004). *Designing for older adults: Principles and creative human factor approaches.* London: Crc Pr Inc.

Gao Q., Sato H., Rau P.-L. P., and Asano Y. (2007). *Design Effective Navigation Tools for Older Web Users.* In: Jacko, Julie A. (ed.) HCI International 2007 - 12th International Conference - Part I July 22-27, 2007, Beijing, China. pp. 765-773.

Hofstede, G. (1984). *Culture's consequences: International differences in work-related values,* Newbury Park, CA: Sage.

Hofstede, G. (1997). *Cultures and Organizations: Software of the mind,* McGraw-Hill, New York.

Hofstede, G. (2006): Dimensionalizing cultures: The Hofstede model in context. In W. J. Lonner, D. L. Dinnel, S. A. Hayes, & D. N. Sattler (Eds.), *Online Readings in Psychology and Culture* (Unit 2, Chapter 14)

ISO 9241 (2008). *Ergonomics of human-system interaction.* International

Leuteritz, J.-P., Widlroither, H., Mourouzis, A., Panou, M., Antona, M., Leonidis, A. (2009): *Development of Open Platform Based Adaptive HCI Concepts for Elderly Users.* HCI (6) 2009: 684-693Organization for Standardisation.

Marcus A., Gould E. W. (2000): *Cultural Dimensions and Global Web User-Interface Design: What? So What? Now What?* Proceedings of the 6th Conference on Human Factors and the Web in Austin, Texas, 19 June 2000.

OASIS Consortium (2007). (OASIS) *Grant Agreement no 215754 – Annex I - Description of Work.* European Commission, Brussels, Belgium.

Rau P.-L. P., Liang S.-F. M. (2003): A Study of the cultural effects of designing a user interface for web-based service. *In International Journal of Services Technology and Management* - Vol. 4, No.4/5/6 pp. 480 - 493.

Reeves, B.; Nass, C.(1998) *The Media Equation: How People Treats Computers, Televisoin, and New Media Like Real People and Places.* Campridge, August (22-26), pp.793-797.

Röse, K. (2001). : Kultur als Variable des UI Design. In H. Oberquelle, R. Oppermann, J. Krause (Eds.): *Mensch Computer 2001*: 1, Fachübergreifende Konferenz. Stuttgart: B.G. Teubner; 2001, S. 153-162

Savidis A., Antona M., Stephanidis C. (2005). A Decision-Making Specification Language for Verifiable User-Interface Adaptation Logic. *International Journal of Software Engineering and Knowledge Engineering,* 15 (6), 1063-1094.

Savidis A., & Stephanidis C. (2004). Unified User Interface Design: Designing Universally Accessible Interactions. *International Journal of Interacting with Computers,* 16 (2), 243-270.

Smith M. J., Salvendy G. (2001): *Cross-Cultural User-Interface Design.* Proceedings, Vol. 2, Human-Computer Interface International (HCII) Conference.

Statistisches Bundesamt (2006). *Bevölkerung Deutschlands bis 2050.* Wiesbaden.

Statistische Ämter des Bundes und der Länder (2007): *Bevölkerungs- und Hausentwicklung im Bund und in den Ländern. Heft 1.* Wiesbaden.

Tylor, E. (1920). *Primitive Culture.* New York: J.P. Putnam's Sons.410.

Xie, A.; Rau P.-L., P; Tseng, Y.; Su, H.; Zhao, C. (2008), Cross-cultural influence on communication effectivenes and user interface design. *International Journal of Intercultural Relation*

CHAPTER 88

Collecting Data About Elderly and Making Them Available for Designers

Johan FM Molenbroek[1], Bea (LPA) Steenbekkers[2]

[1]Delft University of Technology
Faculty Industrial Design Engineering
Delft, The Netherlands

[2]Wageningen University and Research Centre, Wageningen
The Netherlands

ABSTRACT

This paper shows an overview of how data about 80 characteristics of older people are collected in the period 1993-1998 in the Netherlands. These characteristics were physical, psychomotor, sensory and some cognitive from origin. The sample consists of 600 healthy 50-plus people and an additional sample of 150 people of age 20-30 to compare the results and to show an age effect, if existent. The first result was a book in 1998 and several papers in journals and conferences.

Because the data are mainly used by designers, also a digital interface with an interactive tool was made. This tool is in ongoing development and is on a continuous basis improved with feedback from users. This tool is now used for more than ten years by thousands of students from all over the world. Attention is given to the usefulness and limitations of this digital design tool.

INTRODUCTION

Are consumer goods and interiors of houses sufficiently adapted to the usage by the elderly? When designing products to be handled at home, in a professional situation or in the public domain, designers ought to base their choices for technical properties on the capabilities, habits and preferences of the user group. Although the 'grey sector' of society is continuously increasing, design-relevant data on elderly users are almost nonexistent. The Delft Geron project was initiated in 1993. It ended in 1998 when the first book (Steenbekkers and van Beijsterveldt, 1998) about the project was published. This laborious project was an attempt to narrow the gap in gerontechnology: product design for the elderly.

During daily-life activities in and around the house individuals often interact with products varying from a hammer to a television and from a microwave oven to a bicycle. Often these interactions occur rather easy and without thinking, because it has become automatic. In some cases these interactions are more troublesome (Norman, 1988). In the occupational environment, during leisure activities and in public transportation difficulties also seem to exist. Sometimes the cause of the difficulties can be related to either the product or the environment - maybe even the user; in other instances the reason for these difficulties might be a concurrence of factors or circumstances.

Some user groups appear to have more difficulty using products than others (Czaja et al., 1993). Sometimes these difficulties are so severe that accidents occur. The elderly are frequently victims of accidents, especially falls (Consumer Safety Institute, 2009). In these cases characteristics of the individual, the environment and/or product might cause accidents. It seems plausible accidents of elderly persons might be attributed to ageing, especially the change in functional capabilities related to the ageing process. The interaction between a senior citizen and products might also differ from the pattern adopted when they were young. The changing physical, psychomotor, sensory and cognitive capabilities of the elderly might influence their interaction with products.

When growing older, the ability to learn (and to unlearn) decreases, which implies that it becomes more difficult to adapt to new ways of operating consumer goods. This implies for example for those persons that have grown up in the 'mechanical period' with handles and wheels and in those times being in the age of about 10-25 years, this metaphor should or could be used when creating virtual controls for those same people when they are in the age of 75-100 years.

The growing number of older people is a extra reason why the elderly are so important to be taken into account in the design process. In the Netherlands in 1900 we had 6% elderly over age 65 and now we have 15% elderly. This number is predicted to rise to 25% in 2050 (van Duin, 2009)

Furthermore, governmental policies encourage a larger proportion of the population to live independently for a longer period of time: 'bad' products might hamper this policy, whereas 'good' products could facilitate independent living (de Klerk, 1997).

These considerations led to the initiation of this Geron-project to investigate various capabilities of users which are supposed to be relevant to man-product interaction and to assess whether and how age influences the level of functioning.

METHODS

RESEARCH QUESTIONS AND GOALS OF THE GERON-PROJECT

This study was set up in order to be able to answer the following research questions:

1. Which physical, psychomotor, sensory and cognitive capabilities of the Dutch elderly are relevant to the design of daily-life durable products?

2. Is it possible to quantify these characteristics and if so, what level do they have and how much variation do they exhibit?

3. Are there mechanisms which compensate for the decreasing or changing capabilities?

Figure 1: The cover of the book about the Delft Geron-Project

In daily life it is not always required or even necessary and desirable to use only one capability during interaction. Due to changing capabilities it might be common practice to use the best available capability in order to achieve a comparable and successful interaction or to investigate how to exert less effort. Older users might set lower standards for themselves or might adopt a strategy of neglecting products (e.g. using mobile phones or internet) or their effects. Such way of 'interacting' might not or probably has not been foreseen by the designer of the product. By means of a written questionnaire and an interview in this Geron-project, insight into compensation mechanisms is obtained.

When these research questions are answered the following goals of the project will be met:

I. To describe and quantify physical, psychomotor, sensory and cognitive characteristics of users of consumer products in the age groups 20 to 30 and over 50 years of age, resulting in a database of human characteristics relevant to the design and innovation of daily-life equipment.

2. To compare the capacities of different age groups of adults, this might enrich various theories on the ageing of capabilities.

3. To generate design guidelines for designers of durable daily-life products based on the measurement of capacities.

These design guidelines should be applied during the conceptualization phase and development phase of daily-life products, in order to ensure a more useful, safe, comfortable and efficient interaction with new and redesigned products. This could very well mean that the elderly can live independently for a longer period of time.

The design of the research

In a national study 750 subjects, who lived independently, were assessed. In total about 80 variables, all more or less important for product use, were measured. The sample consisted of four age groups ranging from 50 to over 80 years of age: a group of young people (20 - 30 years) was also studied for the purpose of comparison. Women and men participated in about equal numbers. The variables covered a variety of human characteristics, such as body and limb measurements, maximum forces exerted, speed of movements, eye-hand co-ordination, etc., as well as their seeing, hearing and feeling capabilities and certain aspects of memory. In addition, a questionnaire was used to probe the problems experienced with various products.

RESULTS

Book

In the scientific part of this study is assessed how characteristics differ between the age groups, i.e. generations, studied. Some variables proved to deteriorate earlier and some later in life, and some hardly at all. The theory that the ageing process means a slow increase in the differences between people, i.e. individualization, is confirmed. The ageing process also implies a decrease in the level of capabilities: older people exhibit a growing tendency to resemble weaker young people. In some respects, however, differences between people are influenced more by sex than by age. These general rules can provide inspiration for product innovators.

The design data of this study is found mainly in the `yellow pages' of the book. This part contains the graphs and tables, the statistical parameters and some hints on how

856

to use them. See Figure 2. Each of the 80 pages for 80 different variables includes Design guidelines and examples of products or components and their ergonomic or gerontechnological points of interest. This part is meant especially for professionals who innovate, design or judge durable goods for daily life. These data can serve as checks during product development in the phase of analysis and conceptualization and can provide inspiration for new 'transgenerational designs', i.e. products fit for the younger and the older people in our society or special products for the elderly. The Section Applied Ergonomics of the Faculty Industrial Design Engineering of the Delft University of Technology offers this 'database' to designers and evaluators of products for contemplation, application, improvement and / or augmentation.

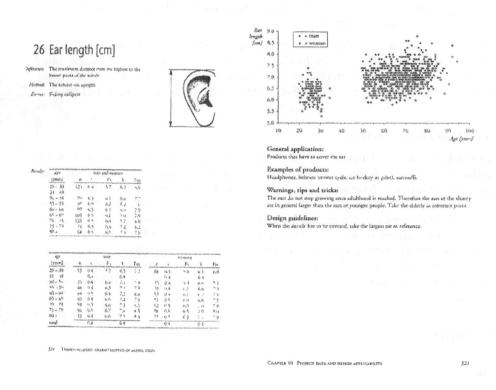

FIGURE 2. Examples of 2 pages of the data presentation in the Delft Geron Book

This book is a first step in the dissemination process. The following is the interactive website www.DINED.nl.

Interactive website

The main reason for making an interactive website is the apparent way industrial designers work. Currently designers use computers and internet as a great part of their working knowledgebase during their design process. Therefore it makes much sense to provide them the data easy to use through the internet.

Furthermore designers mostly appear to have a gap in their knowledge concerning anatomy and statistics. So if 'popliteal height' is mentioned a lot of designers don't know which body part is meant. Therefore clear graphical support from human figures is necessary. Also about percentiles and other statistics their knowledge seems to be limited. Most users of anthropometric data are not aware what the consequences are when relevant human dimensions are correlated and how to design for this in the 2D or 3D space. Percentiles are only meaningful in 1D space. In 2D we need scatterplots to understand the problem and in 3D the problem should be split in several 2D views, but not many researchers or designers do so.

Around 1995 the digital Dined-tool was initiated with very primitive visualisation means. Before that time it was just a percentile calculation tool and a presentation of one table with Dutch data of adults between 20-60 years of age. Since 1982 this was set up to help students Industrial Design Engineering to make their anthropometrical calculations. With the increasing interaction design techniques this digital Dined-tool became a platform independent interactive design tool and is currently made in JavaScript©.

Design Process

When designing a product anthropometry is only one of the many aspects the designer has in mind, but it is continuously essential to know whether the product in progress fits the user group (User Centred Design). An impression of that anthropometric design process is shown as flowchart in Figure 3 when it was applied during the development of a smart toilet for elderly: the Friendly Rest Room Project (FRR project). The objectives were 'to carry out the necessary research and design, the engineering and evaluation of prototypes for a more user friendly restroom for elderly and persons with disabilities' (Molenbroek et al., 2010).

The target group of the FRR consists of elderly and people with disabilities. Though differences in behaviour and preferences are without doubt to be expected, no specific differentiation was made in gender; men and women were equally subject of study. With regard to ethnicity and type of disability initially no distinction was made either.

What we learned from this project is that the observational research in the beginning of a design process decides which anthropometric variables are relevant and critical.

'Relevant' means the following dimensions for example for a product to be seated upon:

- lower leg length, that gives an indication for seat height
- buttock-popliteal length, that gives an indication for seat depth

- hip width, that gives an indication for seat width
- elbow-rest height, that gives an indication for the arm rest height

So for each product a typical series of relevant dimensions can be described.

Then something about the critical parts within the relevant dimensions: e.g. for the height of the control buttons in an elevator the reach height is relevant but the lowest percentiles are critical, because the small people should also to be able to touch the controls from all the floors and not only the lowest floor as happens if the highest floors have buttons that are mounted too high.

The ceiling from the elevator however is critical for the highest percentiles, because otherwise they can't enter the elevator without bending; the same yields for people with big hands in relation to tools, or people with wide hips in relation to airplane passenger chairs.

Each dimension has a certain variability which has to be taken into account during product development. Several possibilities exist:

1 "One size fits all" is the easiest and the cheapest, e.g. a chair in a restaurant. There will be lots of users that will not fit, but the usage is only during a relatively short period of time and in special cases like a child a special chair is arranged.

2 Adjustability seems ideal, but don't forget the lower and the upper boundary have to be subject of careful study. Also the control that holds the adjustability in place. It should be easily accessible and should give feedback and feed forward to the user about how and how far to adjust the dimension. Also the to be exerted force on the control should not be too high for the weakest user. Otherwise the strongest user can put the adjustment on hold but the weakest user can't open it again.

3 Several product types seems ideal: but in shoe or clothing design, the extreme users might be excluded because the retailer only meets the users that fit, and mostly not the one that are excluded.

Another important issue is "how many sizes".

4 Made to measure seems very desirable, but is it viable and feasible?

From this analysis we can learn that a sizing system depends on the product, the context and the market.

5 Universal design seems also very desirable and is often more feasible and viable than one thinks on beforehand. Universal design means to design for the widest possible audience and does often imply that less potential users are excluded.

859

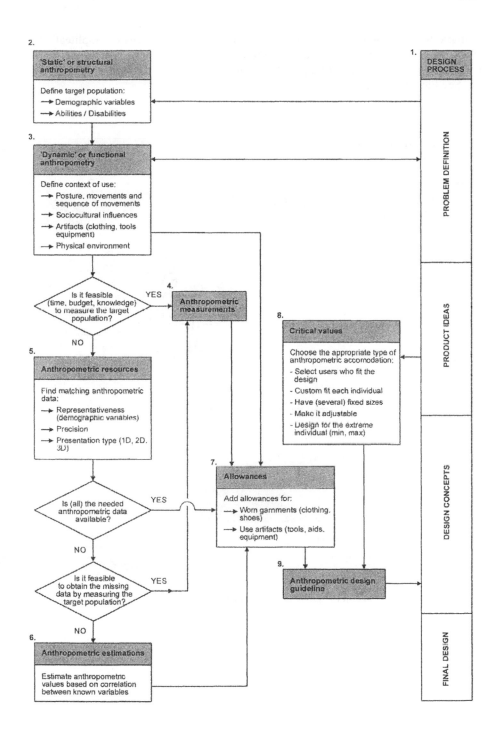

FIGURE 3. Flow diagram to develop an anthropometric design guideline

For example for the product dimension seat height the human dimension popliteal height (lower leg length) is relevant. With the interactive Dined-design tool designers can play with the data of a population or a human dimension or with a set of both. See Figure 4 and 5

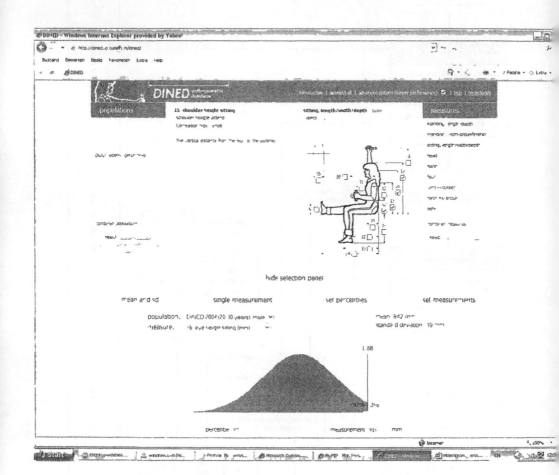

FIGURE 4. Example of a page in the DINED design tool when looking for the percentage to exclude in the next design decision.

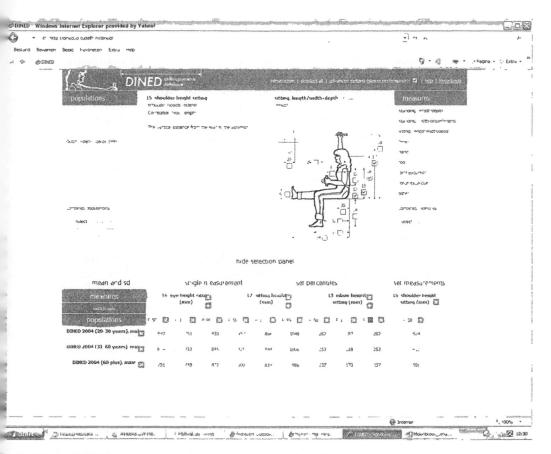

FIGURE 5. DINED shows an overview of percentiles in sitting heights

For application in 2D a special tool was developed: ELLIPS. This tool can be downloaded for free from the website www.dined.nl . ELLIPS supports the designer in thinking about the number of required sizes of a product like school furniture or shoes to accommodate the target group (Molenbroek et al., 2003).

For the 3D level the CAESAR-project delivered in 2002 the first good quality dataset: www.sae.org/caesar . It shows 4000 full body scans from USA and Europe, but the details from the ears, noses and fingers are too vague to be used for design purposes (Robinette and Daanen, 2006)

The Sizechina project followed in 2005 with data from Asian people and focused on the form of the Chinese head. In this project 3D measurements of the heads of 2000 Chinese adults of the main land of China and also a group of children were assessed Ball (2009). After the project these data were made available for designers via www.certiform.org in physical head forms or as digital head forms.

DISCUSSION

An ideal ergonomics tool for designers would be within their design environment and will support the designer during the making of the design decisions in the whole design process from the first thought until the testing of the series production. The most common design CAD software in the concept phase is ProEngineer, Cathia, Solid Works and Rhinoceros. It is not easy for developers to create a tool that shows human diversity during the whole design process and in all the three areas of ergonomics the International Ergonomics Association defines: physical, cognitive and organization ergonomics. Currently there is no tool that covers all these areas. The Dined-tool covers currently only a selection of anthropometrical measurements, but yet no cognitive or organizational aspects.

Moreover DINED is focused on Dutch data, but has a section of global anthropometric data.

More global data are for free available on www.humanics-es.com and through WEAR. WEAR is a legal organization and was founded in 2000 by 15 experts in anthropometry: see www.wearanthro.org . WEAR strives for a web portal with anthropometrical data of the whole world. A pilot information system from WEAR was launched at the IEA Beijing Conference in 2008. Also around this matter the IEA Technical Committee on anthropometry was started, see www.iea.cc . This committee gathers the people that are experts or are just interested in anthropometry in the whole world and will show on the next IEA conference the state of the art of the current knowledge in anthropometry.

CONCLUSION

Collecting anthropometric data has been done seriously since the second halve of the past century. A good overview of data is given in the Anthropometric Source Book published in 1978 in which 900 dimensions are defined in words and clear illustrations. The disadvantage is that almost no designer will know this valuable work. Moreover the data might be less accurate for current application due to secular changes in the populations. Most designers however know the human scales of Diffrient et al. (1974), which has a first graphical and adjustable interface. It is easy to use but the data are limited useful.

WEAR has a great potential to serve in this field but has to create an economic feasible organization around this matter to be able to pay the programmers of the interface and to satisfy the sources of the data. In the mean time DINED is just a simple and free tool for design students and designers.

REFERENCES

Ball, R. (2009), 3D Design tools from the SizeChina Project. *Ergonomics in Design, Summer 2009, 8-13*

Bruin, R de, Molenbroek, JFM, Groothuizen, T & Weeren, MH (2003). *On the development of a friendly rest room.* In R Coleman & J Myerson (Eds.), Include 2003: inclusive design for society and business (pp. 14: 570-14:576). London: Helen Hamlyn Research Centre.

Buzink, SN, Molenbroek, JFM, Haagsman, EM & Bruin, R de (2005). Falls in the toilet environment: a study on influential factors. *Gerontology, 4(1), 15-26.*

Consumer Safety Institute, *Factsheet ongevallen 55+.* Retrieved at www.veiligheid.nl, d.d. 10-02-2010

Czaja, SJ, RA Weeber, SN Nair (1993), A human factors analysis of ADL-activities: a capability-demand approach. *The Journal of Gerontology, 48 (special issue), 44-48*

Dekker, D, Buzink, SN, Molenbroek, JFM & Bruin, R de (2007). Hand supports to assist toilet use among the elderly. *Applied ergonomics, 38(1), 109-118.*

Diffrient et al., (1974) *Human Scales.* MIT Press .

Duin, C van (2009) *Bevolkingsprognose 2008-2050: naar 17,5 miljoen inwoners.* CBS bevolkingstrends, 1ᵉ kwartaal 2009. DenHaag/Heerlen, CBS.

Klerk, MMY de (1997), *The use of technical aids by elderly people, a study of determinants and possibilities for substitution (in Dutch),* Erasmus University, Rotterdam.

Molenbroek, JFM (2000). *Making an anthropometric size system interactively.* In: Proceedings of the XIVth triennial congress of the International Ergonomics Association and 44th annual meeting of the Human Factors and Ergonomics Society (pp. 6-766 - 6-769) Santa Monica CA (USA): Human Factors and Ergonomics Society.

Molenbroek, JFM & Steenbekkers, LPA (2003). *Age effects in body joint motion of older and younger adults.* In s.n. (Ed.), IEA 2003 (pp. 1-4). s.l.: s.n..

Molenbroek, JFM, Kroon-Ramaekers, YMT & Snijders, CJ (2003). Revision of the design of a standard for the dimensions of school furniture. *Ergonomics, 46(7), 681-694.*

Molenbroek, JFM, J Mantas and R de Bruin (Eds.) (2010), *Developing adaptable toilets of the future for disabled and elderly people,* IOS-Press, Amsterdam.

Norman, DE (1988), *The psychology of everyday things.* Basic books Inc, NewYork.

Robinette, K. and H. Daanen, Precision of the CAESAR scan-extracted measurements. *Applied Ergonomics 37 (2006) 259-265*

Steenbekkers, LPA and CEM van Beijsterveldt (Eds.) (1998), *Design relevant characteristics of aging users.* Faculty Industrial Design Engineering, Delft University of Technology.

Acknowledgement: Student assistant Tjeerd Ijtsma who created the last version of DINED in Javascript during the last few years.

Chapter 89

Design of a Mobile Social Community Platform for Chinese Older People

Daniel Ebert, Xing Chen, Qin Gao, Pei-luen Rau

Institute of Human Factors and Ergonomics
Dept. of Industrial Engineering, Tsinghua University
Beijing, 100084, China

ABSTRACT

The increasing proportion of the older population and the change of Chinese family structure make Chinese older people more vulnerable to social isolation and exclusion than ever. To support older people's socialization activities, we designed a mobile social community platform (SCP) which facilitates organization of real life encounters and leisure time activities between adjacent older people who have similar interests. Users with this SCP will be able to search for nearby users to arrange activities with, to search for activities around to join in, or to rate activities according to their own interests. The implications of aging for interface design were collected and reviewed from literature, and Chinese older people's socialization needs and patterns were collected through interviews. Findings helped build the set up of the SCP, which was embodied by an entirely developed prototype application for a PDA via a user-centered design approach.

Keywords: Older People, User-centered Design, Mobile Application, Usability

INTRODUCTION

Chinese family structure has undergone a big change due to the rapid economic development and the strict population control in the past 20 years. Nuclear family has become the most common family arrangement instead of extended family.

These changes, in addition to the fast population aging speed, lead to increasing social isolation and exclusion for older people. According to China National Committee on Aging 2009, the average rate of "empty nest" family has reached 49.7% in China urban area, which means almost half of the older people in cities live alone without company of young ones. Especially older people whose spouses have died and who completely live alone report the highest rates of social exclusion (Demakakos, 2008). The spread of computing has the potential to extend older people's access to social contacts, entertainments, and other activities. But the impact of age-related capability impairments on interface design and older people's preference must be taken into considerations to make the technology accessible and useful for them.

This research aims to design a mobile communication platform which helps older people stay connected with friends via a user-centered design (UCD) approach. We first reviewed limitations and characteristics of older people and their impacts on system design from literature, and then conducted a series of semi-structured interviews to gather older people' socialization requirements. It is noteworthy that Chinese older people show different socialization patterns compared with young people or western older people. They are more likely to take part in collective activities with other older people, like Tai-Chi, chorus singing, or Mahjong. A former study conducted Institute of Human Factors and Ergonomics of Tsinghua University (2008) found that the Chinese older people emphasize the importance of communication but prefer the manner of face-to-face. And the motivation of making new friends is quite low. All the findings were concretized and integrated into a social community platform aiming to facilitate real life encounters between the older people with similar interests.

LITERATURE REVIEW

To gather design requirements of older users, we began with obtaining usability concerns from literature review. After understanding the special requirements of older people, implications can be derived and transformed into clear and easy-to-follow design guidelines.

LIMITATIONS OF OLDER PEOPLE AND THEIR IMPLICATIONS FOR INTERFACE DESIGN

Numerous research reported age-related decline in vision, hearing, psychomotor abilities, and intellectual skills. The effects of these impairments on older people's usage of computers should be considered to make the application accessible.

An obvious obstacle for older people to use computers is the decline of vision beginning in the middle age (Fozard 1990, as cited in Hawthorn, 2000). The eye's ability of near focus declines and due to changes in the cornea, pupil, iris, and lens less amount of light is transmitted to the retina (Morrell & Echt, 1997). Comparing

50 year olds to 20 year olds, Owsley et al. (1983, as cited in Hawthorn, 2000) reported a significant decline in contrast sensitivity, which increases again by the age of 80. From around 60, older people may show a reduction of the effective vision field, which means that peripheral stimuli must be stronger and/or closer to the center of the visual field to be detected (Cerella, 1983, as cited in Hawthorn, 2000). Helve and Krause (1972, as cited in Hawthorn, 2000) detected a loss of ability of distinguishing color, especially in the range of blue-green wave lengths. Also the capability to identify figures that are embedded within other figures is reduced (Capitani et al., 1988, as cited in Hawthorn, 2000). Morrell and Echt (1997) also reviewed several studies on text preferences of older readers on screen display and suggest that older adults may benefit from sans-serif fonts like Helvetia or Arial in a 12-14 point range. In our system, we tried to enlarge letters, icons, as well as other symbols as possible, with a set of colors of relatively high contrast ratio.

Older people also experience a loss of ability to detect high pitched tones. Laboratory tests showed that older people miss attention getting sounds with peaks over 2500 Hz (Rockstein & Sussman, 1979, Schieber, 1992, as cited in Hawthorn, 1998a). This should be considered when the system is designed to use sound to get users' attention.

Aging is associated with decline in psychomotor abilities. The response times in motor tasks of older people become longer. This limitation should be considered for interactions depending on speed (e.g., double clicks with a short interval to trigger an action). Charness and Bosman (1990, as cited in Hawthorn, 1998a) reported that older people have problems to position the cursor if the target has the size of letters or spaces in text. To meet the special requirements of older people Welford (1985, as cited in Hawthorn, 2000) recommended adding an age correction factor to Fitts' law. These researches suggest that bigger font sizes, buttons and icons should be used for older users.

There is general agreement in literature that the ability to concentrate on relevant information in the presence of distracting stimuli declines with age (Connelly & Hasher, 1993, Kotary & Hoyer, 1995, as cited in Hawthorn, 1998b). This bases the design guideline that graphics for decoration and animations should be avoided, as well as multiple overlapping windows (Kurniawan & Zaphiris, 2005). According to Vercruyssen (1996, as cited in Hawthorn, 2000) another problem of older adults is to maintain attention over a long time. Especially tasks requiring rapid or continuous scanning are fatiguing for them. Hence, too dazzling or complex interfaces should be eliminated if possible. Tasks that require multiple continuous operations should also be avoided.

MOBILE APPLICATIONS AND OLDER PEOPLE

Design mobile applications for older people is particularly challenging due to the small screen size versus the poor vision of older people. In a study using a PDA as mobile device the participants had problems with text reading and recognition of smaller icons and buttons. There were difficulties with handwriting caused by problems with hand stability and motion, given the space for input of Chinese characters was small (C.-F. Lee & Kuo, 2007). Older people were found preferring bigger buttons compared to standard mobile phones.

A team from Korea (H. Kim et al., 2007) conducted qualitative and contextual research on older users' experience and interaction difficulties in using their mobile phones in everyday life. In the study, older people think that their mobile phone offers too many and complex functions. They don't understand them and have problems finding them in the deep structured menu. These findings are consistent with Melenhorst et al. (2001) and Lee's study (2007), both finding that older people use mobile phones for very limited purposes, mainly for communication and security. Melenhorst et al. (2001) also suggested that the main benefit for telephones is to keep in touch with someone emotionally close, and other benefits are seen in setting leisure time activities with friends or relatives, and immediately sharing exciting news.

The design guidelines derived from perceptional, psychomotor and cognitive limitations of older people users can be applied for user interfaces of our design.

REQUIREMENTS GATHERING

To understand how Chinese older people socialize with others and what they really need regarding to communication facilities, we conducted interviews.

APPROACH

We conducted semi-structured interviews with eight Chinese older people aged from 60 to 83. All the four female interviewees were widowed and the four males lived with their wives. Two male interviewees lived with their children. Many interviewees have brothers or sisters, but in most cases they live in other towns or even provinces. So do their children.

Older people were asked about their use of mobile phone and computer, leisure time activities they usually have, their perception of social situation and how they dealt with loneliness.

RESULTS

All the participants use their mobile phone in daily life for connecting family and relatives, while calling friends was mentioned by four of them. However, they seldom use short messages (SMS). The two main reasons reported are too small characters and complicated operations. No participant uses the computer for Email, but two men have QQ, the most popular instant messenger in China. Five of the participants have a quite positive attitude towards new technology, though they emphasize the difficulty of learning it.

Nearly all of the interviewees mentioned that they would like to join in collective leisure time activities (LTAs), which range from playing games like Mahjong or Chinese Chess with friends or neighbors, painting or chorus to doing morning exercises or talking a walk. Majority of these activities would be planned in some previous face-to-face gathering. But they had difficulties when they wanted to immediately find someone sharing the same interest to play together occasionally without previous arrangements. So, they also mentioned watching TV, reading newspaper, magazines, or books sometimes, alone, instead.

It is also noteworthy that two interviewees are reserved regarding meeting new people since "strangers are not reliable".

As we have discovered so far from interviews, functions of mobile devices have to be accessible in an easy way and not be hidden in a deep structured menu, where every step has to be remembered. One aid will be cues that require recognition rather than recall. Moreover, the complexity of the whole application has to be reduced to a minimum. This ranges from a simple and consistent layout design to processes with least pages, steps and options. The layout also has to be taken account of the older people's visual impairments as well as difficulties with inputting text, especially Chinese characters.

PLATFORM DESIGN

The interview results are analyzed and transformed into the design of the social community platform (SCP) with the focus on the group of friends or neighbors who live in the interviewee's neighborhood. Here, the functional and technical design of the SCP is derived to describe the functions of the system, followed by discussion of graphical design to address the concern in usability of the system for older people.

FUNCTIONAL DESIGN

The central idea behind the social community platform is to utilize information and computer technologies in order to prevent older people from loneliness, stimulate their social behavior, and foster their emotional well being. Users can create their own profiles in which LTA preferences are included, make friends with users around, arrange activities, and make phone calls as well. In this first draft of the platform with limited functions the focus lies on a software program for a Personal Digital Assistant (PDA) that looks a little similar to an instant messenger like MSN or QQ. The main differences are the implementation on a mobile device instead of a desktop computer and the purpose of arranging a LTA when using the system. Additionally the application includes a phone book with calling function. Also a map function is integrated, where the position of oneself and friends is visualized.

Similar to an instant messenger each user has the possibility to choose between three modes, *offline*, *online*, or *invisible*, called status. The relationship between two members of the SCP can be a *friend* relationship or a *non-friend* relationship. If User A and B are friend, they can mutually see their status even in *invisible* status.

When *invisible* User A and *online* User C are not friends, A can see that C is *online* but is indicated as *offline* towards C.

The purpose of the different statuses and the kind of relationship between users is the arrangement of LTAs. When members of the social community feel lonely or just want to undertake some activities, they can switch their statuses from *offline* to *online* respectively *invisible* and search for users who also want to do something with other members of the platform. Thereby the system helps through an intelligent sorting of search results and suggestions when and where to meet. The users can decide whether they accept the suggestion or not, and they can also call each other at any point of the process. However, due to the wide application of mobile phones, voice communication and phone book are omitted in the prototype.

The start page is shown as Fig 4.1a, presenting portals to five main functions, which are *Search for Users*, *Search for Activities*, *Show Map*, *Phone Book*, and *Preferences*. The flag in the start page stands for the status of the user, thereby green stands for online, grey for invisible and red for offline. The first three functions cannot be used when the status is offline. A user can find out other nearby users by selecting *Search for Users*. When this function is utilized, the system automatically finds users who share the same LTA, see Fig 4.1b. The user with the result can decide whether to arrange activities with nearby users to take part in the activities favored by the both sides. This works because of the preference list that each user has saved as we assumed in the sample database. Friend relationship is indicated by a figure with a red heart. When a user wants to know activities happening or going to happen around, this user should select *Search for Activities*. When this function is utilized, similarly, possible activity arrangement suggestions based on the preference of this user with other member willing to take part in the activities will be shown on the screen, as shown in Fig 4.1c. When the user selects *Show Map*, other friends around will be on the map, but no suggestion will be shown compared to the result of *Search for Users*. See Fig 4.1d. Notice that the user himself/herself is shown by a red dot, and all online and invisible friends are displayed by a blue dot. Non-friends will not be displayed due to security concern.

FIGURE 4.1 Interfaces of the system, (a) Start Page, (b) Search for Users, (c) Search for Activities, (d) Show Map

As previously described, our SCP tries to help older people to develop new friendships, both in the social network and in real life. However, as we found through the interviews, older people have reservations regarding getting known new people. Thus we offer *invisible* status when they want to avoid inquiries by strangers and also offer friend relationship for trusted friends to know that the user is actually online even the user is *invisible* for non-friends.

Users will manage their preferences in *Preferences* option. In this system the user can rate seven LTAs on a scale from 0 to 5. A LTA that is rated with 0 does not appear on one's own preference list. Following LTAs can be chosen: go to the cinema, go to the park, watch TV, play Chess, play Mahjong and take a walk.

TECHNICAL DESIGN

The three central elements of the social community platform are users, activities, and locations. See Fig 4.2.

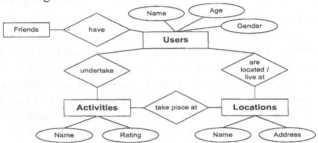

FIGURE 4.2 Entity-Relationship Diagram of the SCP

The main function of the SCP is the arrangement of activities. The system contains a list with activities that probably matches the members' favorite LTAs, although they cannot directly manipulate the list. But they have the possibility to rate each activity that is offered by the system and generate an own LTA preference list. Activities take place at definite locations, whereas the same activity can be executed at various locations and at the same location can be executed several activities. Not only activities but also users are connected to locations.

A possible residential neighborhood for implementation of the SCP is sketched up in Figure 4.3a. The houses surrounded by an orange circle represent locations where a LTA can be executed. They are located within walking distance from the user's home, the third house with a circle. All locations have in common that they are equipped with W-LAN (Wireless Local Area Network). As soon as one user of the system enters this physical activity space, his/her PDA automatically connects to the wireless network and thereby to the SCP.

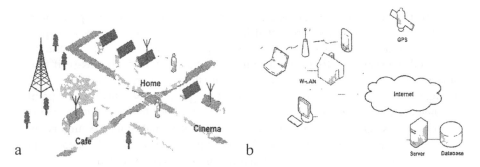

FIGURE 4.3 Neighborhood Sketch and Architecture of SCP. (a) Sketch. (b) Architecture

The system is consisted of client part, server part, and connection part. PDA client or PC client will act as the client part directly sending or receiving information for the users. A server and a database where all data like user information and lists of activities and locations are stored consist of server part. GPS (Global Positioning System) and W-LAN devices are the intermediate that helps information exchange under Hyper Text Transfer Protocol (HTTP). Core of this SCP is a PDA with touch screen and integrated GPS receiver. Especially here a PDA as mobile client of the system connects to the server of the SCP via W-LAN. The PDA constantly receives its current position via GPS and sends it to the server of the SCP. The SCP server communicates with all kinds of clients and provides them access to the database. Beside the PDA the second kind of client is a conventional mouse driven desktop PC or Laptop, which was ignored in the first version of the prototype. See Fig 4.3b.

In order to meet these special requirements of older users, the PDA application will completely go without text input. Administrators thus are required to input text information for them ahead of use. After users logging in, the users can manage their accounts and change their personal profiles by simple selecting, not typing.

A prototype was designed on a PDA. The model is an ASUS Mypal A639 with 3.5" display and a resolution of 240x320 pixels on a Windows Mobile 5.0 operating system integrated with Microsoft Visual Studio 2008 as the IDE. Furthermore, Microsoft .NET Compact Framework 3.5 and C# as coding language was used.

GRAPHICAL DESIGN

The main finding of the literature review requirements specification was that the complexity of the whole application and its graphical user interface must be reduced to a minimum. Low complexity is a very widespread term and ranges from a simple and consistent layout design to processes with least pages, steps and options. One major problem for older people is a deep structured menu, which was mentioned in studies as well as in the interview. Thus, our system goes without a menu and all main functions are directly accessible from the start page, while additional functions can be reached on a logical way.

A further way to reduce the application's complexity is the keeping of a consistent layout design. This is reflected in repetitive usage page layouts, shapes, text fonts and sizes, colors, icons and process steps. One outcome of the literature review was that a sans serif font should be used. The system uses the sans-serif font Tahoma in an 11, 12 and 14 point range. The labeling of buttons has always a 12 point range, important titles or notices a 14 point range and longer texts an 11 point range when the screen size does not allow a 12 point range. Also the button size is bigger compared standard buttons that are suggested by Visual Studio. Sound is not widely used in the system due to the detectability concern for older people as previously reviewed. Only alarm for arrangement is conveyed by sound.

In order to keep the layout design consistent two general methods to display information were developed. In the following they are called page and popup window. As it can be seen in Fig 4.4 the difference lies in the size of the window that contains for the user relevant controls and information. While a page uses the whole available screen size of 240x320 pixels, a popup window is content with 206x234 pixels. It always appears centered and the button controls of the main page become disabled. Furthermore, it owns a grey frame to increase the contrast towards the main page and is only displayed when necessary. The user should not be forced to confirm every selection that he made. Popup windows often come along with a notice like *Meeting arranged* or *Suggestion Denied* in combination with a meaningful icon and offer alternatives in appropriate situations.

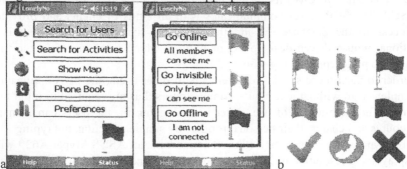

FIGURE 4.4 Graphical results, (a) Page and Popup Window, (b) Use of Icons

SUMMARY

The entire work is established upon a UCD approach, since literature review and interview build the base for SCP and PDA application. By this way the older people's requirements were found out and transformed into the current status of work. The core of the SCP, a prototype application running on a PDA, was successfully finished. During the development process several expert tests showed that although all factors seemed to be considered improvements and changes became necessary. Unfortunately, due to time limitations the logical next step in the

UCD process could not be applied. The finished application needs to be tested and evaluated by real users in order to validate if it functions as supposed and fulfils their requirements, since in theory good approaches must not automatically lead to in practice successful results. Real users need to evaluate the application and its interface with regard to usability and user friendliness. But even more important, they need to give some feedback about the underlying idea of the SCP and the arrangements of LTAs using a mobile device application.

REFERENCES

Demakakos, P. (2008), *Being socially excluded and living alone in old age,* Finding from the English Longitudinal Study of Ageing (ELSA). Age Concern Englando, London.

Hawthorn, D. (1998a), *Psychophysical Aging and Human Computer Interface Design.* Paper presented at the Computer Human Interaction Conference, Adelaide, SA, Australia.

Hawthorn, D. (1998b), *Cognitive Aging and Human Computer Interface Design.* Paper presented at the Computer Human Interaction Conference, Adelaide, SA, Australia.

Hawthorn, D. (2000), "Possible implications of aging for interface designers." *Interacting with Computers*, 12(5), 507–528.

Institute of Human Factors and Ergonomics of Tsinghua University (2008), *Investigation of the requirements of older people related to the OASIS use cases.* Tsinghua University, Department of Industrial Engineering.

Kim, H., Heo, J., Shim, J., Kim, M., Park, S., & Park, S. (2007), "Contextual Research on Elderly Users' Needs for Developing Universal Design Mobile Phone." In C. Stephanidis (Ed.), *Universal Acess in Human Computer Interaction*. Coping with Diversity (pp. 950-959). Springer Verlag, Berlin Heidelberg

Kim, S., Symons, M., & Popkin, B. M. (2004), "Contrasting Socioeconomic Profiles Related to Healthier Lifestyles in China and the United States." *American Journal of Epidemiology*, 159(2), 184-191.

Lee, C.-F., & Kuo, C.-C. (2007), "Difficulties on Small-Touch-Screens for Various Ages." In C. Stephanidis (Ed.), *Universal Acess in Human Computer Interaction*. Coping with Diversity (pp. 968–974). Springer Verlag, Berlin Heidelberg

Lee, Y. S. (2007), *Older Adult's User Experience with Mobile Phones: Identification of User Clusters and User Requirements.* Faculty of the Virginia Polytechnic Institute and State University, Blacksburg, Virginia.

Li, J. (2008), "Association of Education with the Longevity of the Chinese Elderly" In Z. Yi, D. L. Poston, Jr. & D. A. Vlosky (Eds.). *Healthy Longevity in China* (pp. 149-156), Springer Netherlands.

Melenhorst, A.-S., Rogers, W. A., & Caylor, E. C. (2001), *The Use of Communication Technologies by Older Adults: Exploring the Benefits from the User's Perspective.* In Proceedings of the Human Factors and Ergonomics

Society's 45th Annual Meeting, Minneapolis.

Morrell, R. W., & Echt, K. V. (1997), "Designing Written Instructions for Older Adults: Learning to Use Computers." In A. D. Fisk & W. A. Rogers (Eds.), *Handbook of Human Factors and the Older Adults* (pp. 335-362). San Diego: Academic Press, Inc.

Zajicek, M. (2001), *Special Interface Requirements for Older Adults*. In Proceedings of the 2001 EC/NSF Workshop on Universal Accessibility of Ubiquitous Computing, Alcácer do Sal, Portugal.

Zimmer, Z., & Kwong, J. (2004), "Socioeconomic Status and Health among Older Adults in Rural and Urban China." *Journal of Aging and Health*, 16(1).

CHAPTER 90

Embodied Cognition and Inclusive Design: Using Gestures to Elicit User Requirements for Interactive Systems

Young Sam Ryu[1], Tonya Smith-Jackson[2], Katherine Carroll[3], Si-Jung Kim[2], Minyoung Suh[3]

[1]Texas State University-San Marcos
USA

[2]Virginia Polytechnic Institute and State University
USA

[3]North Carolina State University
USA

ABSTRACT

This paper presents a research effort to determine the needs of older adults with severe visual impairments (SVIs) based on a wearable system to support wayfinding and object recognition in a social environment. The research is predicated on the premise that gestures are manifestations of embodied cognition in the form of thoughts, ideas, creative discourse, and reasoning. Individuals with SVIs participated as design team members to design a prototype of a wayfinding system consisting of orientation and navigational components. The design sessions were recorded and gesture analysis was conducted, using categories of gestures identified in the literature, to capture critical incidents that reflect user requirements.

Keywords: inclusive design, participatory design, usability, gesture analysis, severe visual impairment

INTRODUCTION

Social interaction refers to relationships between two or more individuals influencing each other. Individuals with Severe Visual Impairments (SVIs) face problems in activities of daily living as they lose vision and this may hinder their social interactions (Hatlen, 1996). A prototype of a garment-based wearable environment awareness system, called Near and Far Environmental Awareness System (NaFEAS), was developed to facilitate users' social interactions by supporting wayfinding and object recognition, which in turn underwent iterations to develop a low fidelity prototype (Kim, Smith-Jackson, Carroll, Suh, & Mi, 2009). NaFEAS can identify objects tagged by a radio frequency identification (RFID) system, information from sensors, and conditions in the near space, which is defined as 1.22 meters (4 feet) around the human envelope. The system can also identify the current location of the user and provide near-space auditory and haptic feedback through user interfaces embedded in clothing.

The basic design features and functions of NaFEAS were identified and applied to concept and prototype design by our participatory design team, which included individuals with SVIs (Kim et al., 2009). Participatory design meetings were used to access user requirements (Maiden, 2008) for NaFEAS functionality and garment design. The participatory design activities included attending design sessions, brainstorming, critiquing, and reviewing design concepts to ensure the needs of the target users. However, more requirement gathering and analyses are needed to improve the current design for use.

Contemporary research has shown the importance of somatosensory processing of environmental cues and cognitive processing. Embodied cognition is used to describe this reciprocal relationship between the body, senses, and mind. Based on embodied cognition, mind elements such as thoughts are embodied through physical and sensory acts and not as abstract representations (Lakoff, 1980; Varela, Thompson, & Rosch, 1991). Thus, physical acts such as gesturing are direct representations of abstract cognitions. Speech-linked or conversational gestures have been recognized as either complementary to speech or serve as supplements to speech, expanding the content expressed in verbal statements (Kendon, 1987; McNeill, 2005). The cognitive-motor perspective on gestures views gestures as codified representations of concrete objects. The use of gesture-based analysis is increasing in the research domain as a means to expand user-centered design efforts (Beattie & Shovelton, 2005; Cassell, Bickmore, Campbell, Vilhjálmsson, & Yan, 2001). Therefore, we used gestures as a means to extract additional data from the requirements discussions.

This paper presents lessons learned from a participatory design session with consultants with SVIs to elicit additional requirements for NaFEAS that were specific to social interaction within leisure settings. The primary pieces of information acquired from the participatory design session were gestures and verbal responses. We explored gestures as an effective manifestation of design inputs by target groups. Just as discourse or verbal expressions communicate design ideas, we propose that gestures are an equivalent manifestation of design ideas and can be used as "data" by designers.

METHOD

PARTICIPANTS

Five individuals with SVIs were recruited via the Roanoke Alliance for the Visually Enabled (RAVE) for a participatory design session as consultants The participants' ages varied according to the sociological categories: Generation Y (1982-2003, at least above 18), Generation X (1961-1981), Baby Boomers (1943-1960), and Silent Generation (1925-1942). The participants' age distributions were as follows: one in Generation Y, one in Generation X, two in Baby Boomers, and one in Silent Generation. The first and second participants were completely blind. The remaining three participants have different eye levels in Snellen: the third participant - worse than 20/800 (left), between 20/600-20/800 (right); the fourth participant - between 20/400-20/600 (left), between 20/600-20/800 (right); and the fifth participant: 20/200 (left), completely blind (right). All the participants except one had experience with the white cane. The participants' experiences with the cane include that it is helpful because it helps them avoid obstacles and prevents them from tripping or falling. They commented that it has limitations in finding objects at near eye level sometimes resulting in collisions.

DATA COLLECTION INSTRUMENTS

The participants provided user-requirement data by discussing the topic of social interaction within a party setting and the extent to which the current NaFEAS prototype (FIGURE 1) would be useful.

Informal questions listed below were used to prompt a design discussion:

1. What are some of the problems you encounter (or think you might encounter) when interacting with people?
2. What are some of the problems you encounter (or think you might encounter) when you are trying to mingle by moving around the room?

This portion of the design discussion was videotaped to capture both verbal and nonverbal data for analysis and extraction of gestures (FIGURE 2).

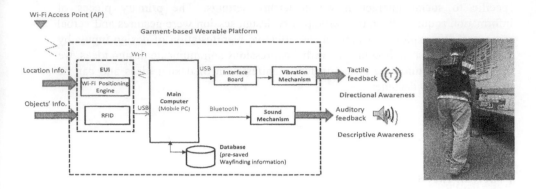

FIGURE 1. A prototype of NaFEAS (Near and Far Environmental Awareness System)

FIGURE 2. A participatory design session with five consultants with SVIs

Users with SVIs employ gestures in patterns comparable to users with full vision (Iverson, Tencer, Lany, & Goldin-Meadow, 2000; Magnusson & Karlsson, 2008). Thus, we used gestures as a means to extract additional data from the requirements discussions. In addition to gestures, we also included lexical affiliates; the verbal content that corresponds to each gesture (Kendon, 1987; McNeill, 2005).

A gesture coding matrix was developed based on current gesture research, namely McNeil and Kendon's taxonomies. We selected gesture categories that were most likely to be used to represent concrete design concepts. The types of gestures that were coded are shown in Table 1.

Table 1. Gesture codes and definitions

GESTURE (CODE)	DEFINITION
Mimicry (M)	One person copies another person's gestures.
Appropriation (A)	One person takes over the gesture of another person to continue the idea.
Abstract (Abs)	Common gestures with no specific or concrete meaning.
Referent/Reference/ Deictic (R)	A gesture that is directed at something or someone.
Iconic (I)	A gesture that refers to a tangible or concrete object.
Metaphoric (M)	A gesture that refers to an idea while using a concrete object to express the idea.
Agreement	One person expresses agreement with another using their body (not their voice).

PROCEDURE

Design Session

The participatory design session was centered on the low-fidelity NaFEAS prototype, and some of the features of the prototype were demonstrated in that session. At the end of the session, participants were asked to use a party scenario to discuss the usefulness of NaFEAS to support social interaction. The following script was used:

Scenario: Imagine yourself at a cocktail or other type of party that is not business-related. You want to chat with people and walk around the room to meet different people as well as find the all-important refreshments table. The goal is for you to do all of this independently without relying on a cane, a seeing-eye dog, or an assistant.

The discussion was videotaped with the permission of the consultants to assist in extracting as many user requirements data as possible. The discussion lasted 25 minutes. Consultants were compensated for their time.

Gesture Coding

A Youtube video (2 minutes and 12 seconds long) was selected to use as a training tool to train an independent coder to a criterion of 75% accuracy. To develop the

880

criterion set of gestures, two of the researchers coded the training video independently and then reviewed the codes until a consensus was reached on the number and types of codes in the training video. The independent coder was given a training session in which codes were defined and examples were provided. After the training session, the coder analyzed the training video. After completion, codes were checked for accuracy, and an accuracy level of 77% was achieved, which met the criterion. The coder was then given access to the participatory design session video and used the same matrix for gesture coding.

RESULTS

A total of 33 codes were extracted from the video (FIGURE 3). Most of the gestures, 16 out of 33, were classified as abstract. Eight were coded as metaphoric, four as agreement, three as referents, and one each were classified as mimicry and iconic.

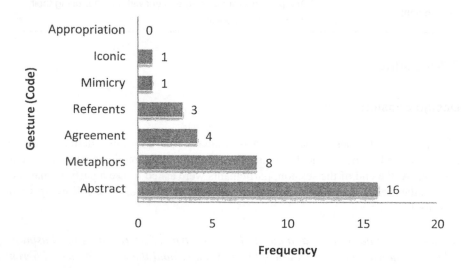

FIGURE 3. Gesture code classification from the participatory design session video

Some of the abstract gestures and metaphoric gestures were matched with expressions that directly related to user requirements for a system that would support social interaction. These topics were:

- Navigation through the social space:
  a. including going up and down stairs;
  b. walking on balconies;
  c. entering and exiting doors;
  d. getting help from other people.

- Independent wayfinding:
     a. fear of being abandoned by friends;
     b. moving from a door to a specific area;
     c. determining the gender of others during approach.

DISCUSSION

The gesture data helped us to capture concepts that may not have been isolated through analyzing the verbal content alone, but the data did not provide the level of detail we expected. The complexity of spatial navigation, however, was illustrated in more detail through analyzing the gestures, and we may not have been able to acquire spatial complexity information using alternate methods such as discourse analysis. Negotiating space during a party or social gathering is not simply a factor of the design of the space, but also includes the spatial configurations formed by balconies, stairs,, doorways and other architectural elements. Users with SVIs perceived these spaces to be a necessary part of social engagement in this type of setting. Since the mental models seem to be based on the most difficult navigation elements (i.e., balconies), these requirements will be given high priority in the next iteration.

The gesture data also revealed issues regarding problems with abandonment in social situations by friends or assistants. Older adults with SVIs reported instances in which they were abandoned in social gatherings and had difficulty finding their assistants, leaving them to negotiate the environment on their own. These experiences support the need to ensure that NaFEAS provides collaborative components or tags that can be worn by others who provide assistance when needed.

Gender was identified as an important consideration. Interestingly, the female participant who discussed the need for the system to provide information regarding gender used a "stop" hand gesture while speaking about it. The hand was held vertically with the palm facing out. This gesture may signify the value placed on interaction with different genders based on different goals. It may also reveal a hesitation about selecting people with which to interact without first knowing their gender.

CONCLUSION

This effort was focused on determining the needs of older adults with SVIs when coupled with a garment-based wearable system to support wayfinding and object recognition in a party environment. Upon the participatory design session to elicit the requirements necessary to iterate the existing prototype, analyzing the gestures of the older adults with SVIs participated in the design session was attempted based on gesture analysis literature. The gesture analysis could capture concepts that may not have been isolated through other methods although it did not provide the level of detail we expected.

For future work, questions prompting garment design discussion such as

- If you could wear something (like a watch or something on your clothes) that would help you recognize people and objects in the room, what would you want it to be and how would you want it to work?
- If you wore a garment to hide this device, what would you want it to look like? (Pockets? Shirt/Blouse? Tag?).

will be used as data collection instruments in the iterative participatory design sessions.

ACKNOWLEDGEMENTS

We are grateful to the five consultants with SVIs for sharing their time and giving us valuable feedback. We thank the Roanoke Alliance for the Visually Enabled (RAVE) for their support. This project was funded by the National Science Foundation (NSF) and the State Council for Higher Education of Virginia.

REFERENCES

Beattie, G., & Shovelton, H. (2005). Why the spontaneous images created by the hands during talk can help make TV advertisements more effective. *British Journal of Psychology, 96*(1), 21-37.

Cassell, J., Bickmore, T., Campbell, L., Vilhjálmsson, H., & Yan, H. (2001). More than just a pretty face: conversational protocols and the affordances of embodiment. *Knowledge-Based Systems, 14*(1-2), 55-64.

Hatlen, P. (1996). The Core Curriculum for Blind and Visually Impaired Students, Including Those with Additional Disabilities. *RE:view, 28*(1), 25-32.

Iverson, J. M., Tencer, H. L., Lany, J., & Goldin-Meadow, S. (2000). The relation between gesture and speech in congenitally blind and sighted language-learners. *Journal of Nonverbal Behavior, 24*(2), 105–130.

Kendon, A. (1987). On gesture: Its complementary relationship with speech. In A. Siegman & S. Feldstein (Eds.), *Nonverbal behavior and communication* (2nd ed., pp. 65-97). Hillsdale, NJ: Erlbaum.

Kim, S. J., Smith-Jackson, T., Carroll, K., Suh, M., & Mi, N. (2009). Implications of Participatory Design for a Wearable Near and Far Environment Awareness System (NaFEAS) for Users with Severe Visual Impairments. In *Proceedings of the 5th International Conference on Universal Access in Human-Computer Interaction. Addressing Diversity. Part I: Held as Part of HCI International 2009* (p. 95).

Lakoff, G. (1980). *Metaphors we live by*. Chicago: University of Chicago Press.

Magnusson, A., & Karlsson, G. (2008). The Body Language of Adults Who Are Blind. *Scandinavian Journal of Disability Research, 10*(2), 71.

Maiden, N. (2008). User Requirements and System Requirements. *IEEE SOFTWARE*, 90–91.

McNeill, D. (2005). *Gesture and thought.* Chicago: University of Chicago Press.

Varela, F., Thompson, E., & Rosch, E. (1991). *The embodied mind : cognitive science and human experience.* Cambridge Mass.: MIT Press.

Chapter 91

Iterative Ergonomic Design and Usability Testing for Elderly Citizens Applications

Katerina Touliou, Maria Panou, Evangelos Bekiaris

Hellenic Institute of Transport (HIT)
Centre of Research and Technology (CERTH)
6th Km Charilaou Thermis, Thessaloniki, 57001, Greece

ABSTRACT

Usability/iterative tests were planned and are being conducted within the European Integrated project OASIS which introduces an innovative, ontology-driven, open reference architecture and system enabling and facilitating interoperability, seamless connectivity and sharing of content between different services and ontologies, in all application domains relevant for the elderly and beyond. The OASIS System is open, modular, holistic, easy to use and standards abiding. It includes a set of novel tools for content/services connection and management, for user interfaces creation and adaptation and for service personalization and integration. Through this new Architecture, over 12 different types of services are connected for the benefit of the elderly, covering user needs and wants in terms of Independent Living Applications (nutritional advisor, activity coach, brain and skills trainers, social communities platform, health monitoring and environmental control), Autonomous Mobility (elderly-friendly transport information services, elderly-friendly route guidance, personal mobility services, mobile devices, biometric authentication interface and multimodal dialogue mitigation) and Smart Workplaces Applications. Applications are all integrated as a unified, dynamic service batch, managed by the OASIS Service Centre and supporting all types of mobile devices (tablet PC, PDA, smartphone, automotive device, ITV, infokiosk) and all types of environments (living labs, sheltered homes, private homes, two car demonstrators, public transport, DRT, etc.). The key aim is the empowerment, independence and productivity of the target population. The iterative tests are based on mock ups, in the four pilot sites of OASIS, in order to ensure high user involvement since the early stages of HCI concepts and modules design and development. The application of ergonomic design methodology facilitates the iterative testing procedure (Nielsen, 1993). Mock ups were created with appropriate software and presented as pen and paper procedure. The Brain Trainer service was tested at CERTH/HIT usability lab in Thessaloniki.The brain trainer module aims at cognitive exercise and training of elder users by offering a game-like experience. Specific testing plans were developed in order to accommodate the elder people identified needs and requirements. Usability was a major objective for the mock ups creation and it was essential to be attained prior the iterative procedure. As user friendliness and acceptability is a top priority for the project, a user centred- design approach is followed along the service and application development. Task analysis procedure facilitated the mock ups creation according to the project Use Cases and the respective scenarios. Additionally, the test design was based on important considerations such as the acceptance of ergonomics standards, avoidance of complexity of tasks assigned, close observation of mental workload. The aforementioned assumptions reflect the apprehensive attitude towards end-users' needs without removing the affect of their capacities (mental, physical, and psychological). Mock ups will be evaluated by 40 elderly users across sites. The OASIS main end-user group, were broken down into 3 sub-groups according to their age classification scheme. "Young" elderly (55-65 years) are users who still live a healthy and active life with minor decline. The elderly group (65-75 years) are still healthy but with a higher risk, compared to the previous group, to experience mild cognitive and physical problems. The "elder" elderly group (>75 years) are individuals who are more likely to present health and cognitive deterioration due to aging. It is evident from the above that within the iterative tests, the specific needs of the users for the stepwise design of the provided services (e.g. in terms of functionality, UI, etc.) will be investigated in depth and beyond experts' analysis. The evaluation process comprises questionnaires completion and focus groups sessions. In addition, during the tests, users are recorded while carrying out certain tasks and notes are kept by the moderators in order to track all users' reaction and feedback. These results will be the intermediate step in the development cycle of the OASIS System and applications, aiming to be optimized before the final evaluation tests.

Keywords: Iterative design, elderly, usability tests

INTRODUCTION

Ageing has for many become a synonym of leading a passive life, being confined to one's home, and dependent on others, ultimately spending the rest of your life in a residential care home. OASIS offers elderly the opportunity to continue to live independently at home, to travel and even to continue to work.

Although, other tests have been contacted for many of the OASIS modules, this paper focuses on iterative tests conducted for the brain trainer module at the Greek pilot site. Before moving on discussing the iterative process, it is helpful to briefly discuss cognitive decline in elderly.

A great body of research has indicated that cognitive aging is accompanied by reduction of memory capacity and the decline in suppressing the acquisition of necessary information. The relation between testing and everyday functioning and activities has been shown to be weak. Cognitive skills may be defined as the mental processes that individuals apply in order to acquire and process new information (i.e., to learn to live in this world).

The main scope was to focus mainly on these cognitive skills on which decline affects their daily functioning. A great deal of research has shown that practicing cognitive skills encourages their preservation and development at all ages.

The underlying principle is to actively engage the elderly in a series of mind activities with aiming at reduced stress levels; emotional frustration may lead elder users to respond as they are being testing. The latter is a main concern in this attempt as the longevity of the brain trainer application relies on its integration to daily activities by embedding them into existing neuropsychological tests. It is evident that every day activities incorporate complex coordination of motor, psychological, emotional and mental processes and activities.

Various cognitive tests have been developed and validated so as to reinforce and enhance elderly agility and health promotion. However, the actual integration of daily activities within the training curriculum is the aim and purpose of this tool. Furthermore, the materialisation process entails the incorporation of current findings in cognitive research and mental training product development. Thus, the rationale of the trainer's development is based on the well-known screening tool "Instrumental Activities of Daily Living Scale" (IADL) (Lawton & Brody, 1969).

The application of simulated experiences isolates to a substantial degree the cognitive element so as to investigate also its role to the mental-motor coordination. Nonetheless, applying a verisimilitude approach could augment its ecological validity.

More specifically, the objectives of the mock ups were to provide with paper and pencil reproductions of the expert team's design for both PC and mobile phone to be tested by elderly participants. The interactive nature of brain games is strong, thus support, feedback and story boards were, also, used to fill in the gaps of mock ups.

In brief, the content of the brain training module may be divided in three major sections. Firstly, a profile is created based on the OASIS user profile with information on basic demographics. Secondly, two tests are applied to create an assessment of brain level or brain power, which is a term widely used in brain games. Thirdly, allocation to cognitive exercise level will be based on a consolidated score. Scalable features are embedded in the development of the respective cognitive exercises. Finally, elders will be able to play a fun game, which connects face recognition and short term memory; a direct link to their own memories and experiences.

TESTED FUNCTIONALITIES IN MOCK UPS

The mock ups were created with the Balsamiq application. Paper and pencil conceptual model mock ups were created. Iterative evaluation will lead to successful modifications towards developing prototypes. The "trick" for prolific usability testing was to impose the feeling of real testing and, of course, to get the basis for conceptually sound and relatively stable design. Real testing is imperative for elder users as they are identified as a group with limited ICT literacy and special related needs. Mock ups were created and tested for both desktop and mobile phone and are presented in separate sections.

DESKTOP ENVIRONMENT

The brain exercises module was given the name "COGNISIS" derived from COGNItion and oaSIS. The name was used so as to add to the feeling of playing a game. The proposed functionalities in mock ups were the following:
- Profile creation
- Brain power estimation (baseline assessment)
- Cognitive exercises
- Fun game (edutainment) with dynamic elements (fig. 2)

The paper and pencil versions of the desktop environment appeared as a desktop window with its basic characteristics. Users were provided with scenarios to complete. Scenarios included a short introduction followed by an analytic task completion list. A facilitator accompanied participants in their step-by-step conduction of the scenario. The selection of an action (pressing of a button, back or forward) initiated the transition to another A4 page of the application window. The facilitator took down notes on duration (minutes to complete an action), any aloud thinking processes (comments, failures, questions, etc.).

The scenarios and tasks were personalised according to gender and profile. Nevertheless, the scenario steps were clear but if needed further information was provided. The procedure was identical for the mobile phone.

MOBILE PHONE

The content and functionalities remain the same. The GUI differences will be evaluated and compared, although, the tested environments are inherently different. It was decided not to compare two different versions of mock ups at this point with and emphasize on differences in sizes and colours but implement the heuristic evaluations to the mock ups and allow users to evaluate them with regard UI elements, but most importantly on content and functions included as it is the scope of such type of evaluation. Again, content and functionalities are very close in design. Still their variability due to different devices' architecture and programming options enables us to look at other perspectives on GUI.

ITERATIVE TESTING PROCESS

The iterative evaluation procedure comprises two cycles. The first one targeted the evaluation of the brain trainer (cognisis) design for a desktop environment. The second one was created for mobile phone.

SAMPLE

Eleven (65±7.21 years old) users participated in the iterative evalution. Users from all OASIS elder user groups were included. However, two elder users (>75 years old) were excluded from the study as their ICT literacy level was restricting. It was very difficult to grasp basic ideas, thus could not fully comprehend the objectives and tasks involved. All users were good English speakers and had fairly good PC knowledge. Users were recruited by an ICT for the Elderly county programme. Hitherto elder users were acquainted with technology and, thus, deriving participants from special computer training programmes for elder people, allow at least controlling for enthusiasm and motivation as possible.

Description of mock ups tested

As mentioned above two sets of mock ups were presented but not distinguished on differences in GUI rather than in target application. It seemed important not to confuse users with numerous mockups as the tests aimed at evaluation for both desktop and mobile phone, but rather focus on one set for each application. The abstract thinking required for the comprehension of games is usually high and demands a built up a gaming knowledge and/or experience, thus emphasis was given on comments and productive discussions in focus groups in one set, rather on alternatives. All participants were good English speakers, however, guidelines and explanations were given in native language. The presentation of mock ups followed the sequence of application usage.

In order to be allocated in levels, users have to complete two tests (baseline assessment). Both scores will be consolidated in order to create one score, which is translated into a level (beginner, intermediate, advanced). The first test is a word recognition test based on an everyday shopping list. 15 words are flashed for a second, and then a list of 30 words is presented and users have to decide if the word belongs to the initial list (shopping list) or not (fig. 1). Similarly, a score is calculated. The second test is based on the universally standardised stroop test. Users have to find which rule applies at the given situation (either two features have the same colour or shape) as fast as possible. A consolidated score of the two aforementioned tests will allocate users to one of the three levels. Cognitive games focus on daily activities. Their philosophy is based on finding the right items, their position; differentiate from non-shown items depending always on the level of difficulty of allocated level. Games are scalable aiming at enhancing cognitive functioning. Returning to the foundation of the developed exercises, it is essential to note that they represent everyday activities (e.g. bathroom items and every day grooming, which are core functions of IADL). It is not possible to present all mock ups from the iterative process, hence examples for both desktop and mobile phone are shown (fig.1-4).

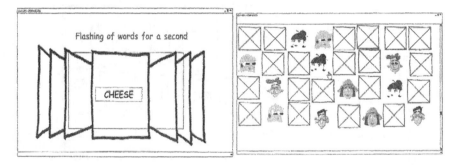

Figure 1. Word recognition test   Figure 2. Photo album fun game

The mock ups created for the mobile phone are similar to the desktop ones, but adjusted to the requirements of a mobile phone.

Description of the testing procedure

Iterative tests were carried out according to the following programme:
- Welcome/introduction
- Information sheet/consent form completion
- ICT literacy questionnaire
- Demonstration of mock ups /power point presentation
- Mock ups handling
- Questionnaires
- Debriefing

890

This procedure lasted approximately two and a half hours. Coffee and snacks were available to elderly. Confidentiality of data was attained by applying a coding scheme to gathered data.

Mock ups were presented in paper and pencil version. Users were allowed to make notes in their provided paper versions. Facilitators kept notes on their diaries on errors, understanding hindrances, questions, when (interaction step) and where (scenario step) help and guidance was asked and/or needed and required level of assistance. Proposed alterations, additions, user acceptance and perceived usability were, also, recorded. The think-aloud protocol (TAP) (Lewis, 1982) was applied in order to efficiently and fruitfully gather data. The design of mock ups was a pluralistic walkthrough process with diverse input. However, the actual usability and efficiency of the proposed brain training application was acquired by the participants' both filled in questionnaires, think aloud notes, focus groups and related gathered data.

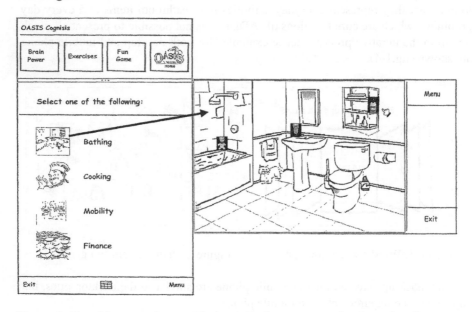

Figure 3. Cognitive exercises   Figure 4. Bathroom game (beginner level)

Results

The main areas of investigation focus were: a) questionnaires, b) think aloud processes notes (both participants and facilitators), and c) focus groups themes for both types of mock ups (incorporated in Table 1).

The System Usability Scale (Digital Equipment Corporation, 1986) was applied. Usability encompasses *effectiveness, satisfaction*, and *efficiency*. In addition, a quantified user acceptance questionnaire was applied.

Desktop

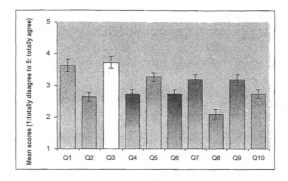

Figure 5. SUS mean score (SE)

As shown above participants rated quite highly the consistency and coherence of the mock ups for the brain trainer application (7/11). However, users believe that if they were to test the actual application, they would need assistance, regardless of instructions. These comments reflected more elderly users (2/11), whose digital literacy was lower compared to the rest of the participants. The average SUS score after transformation was 67%. In addition, user acceptance of the different areas of UI was investigated. Mean user acceptance was quantified into 58.6%. Elder users reacted positively in the possibility of applying the shown cognitive exercises. Overall, mock ups experience was described as easy. However, the absence of information on errors was thought to be high (8/11). Thus, a more elaborate help function is needed.

Mobile phone

The same questions were asked for the mobile phone mock ups, which content-wise were identical to the desktop with slight but required interface changes.

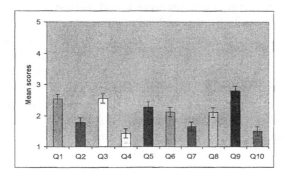

Figure 6. Mean usability scores (SE) for mobile phone

According to the above graph usability was moderately rated, especially on interaction and complexity levels. Users were, also, hesitant regarding the applicability of such an idea to the elderly. They assumed the fonts and the size to be relatively small for elder users. It is important to bear in mind that, mobile phones are rarely used for anything than calling by elder people.

Think aloud findings

While users carried out the scenarios, they were prompted to vocalise (think aloud) all steps taken, and the accompanying thoughts and feelings. The recurrent themes are shown in tables categorised with respect the undertaken task. These are the qualitative analysis recurrent themes. However coding may provide quantitative findings in the future pilots.

The following table provides comments made by the participants during testing procedure. Most common comments are presented. These comments were quantified and are presented in the following table according to the evaluation objectives.

CONCLUSIONS

In conclusion, the above findings (Table 1) and the previously described results may be summarised in the following recommendations for the developers' team.
- Constant zoom should available
- Colourful interface but more emphasis on distinctiveness
- Buttons instead of wording
- Instructions always available
- At times pause function should be an option
- Touch screen necessary for mobile phone
- Less fonts and more pictures for mobile phone

Overall brain training exercises were found interesting and fun. The latter is major objective not only for the mock up creation but, also, for the development of the application. These findings (remarks, suggestions, acceptance levels, satisfaction, and usability percentages) are the basis for the next part of the development cycle. Next step encompass a second round of iterative tests with demos of developed exercises. Gathered data will be enhanced by video material and reaction measurements. Finally, two rounds of iterative testing will assist and allow a smoother coordination for the final pilot testing of the developed application within the OASIS system architecture.

Table 1. Consolidated user satisfaction

Mockup characteristics	User group satisfaction (%) and corresponding comments	
Colours	67%	Need for real colours to clearly get the feeling (72%)
		Should not have loads of colours, distracting (54%)
Lay out	78%	Less words, more buttons (63%)
		Buttons to give instant explanation or pop ups (82%)
Wording	61%	Comprehensability is quite high but should be in English (79%)
		Font size should be altered or scalable (63%)
Buttons/functions	58%	More colourful (57%)
		More realistic (64%)
Instructions	69%	Always available (81%)
Coherence in lay out	73%	More colours (63%)
Coherence in presentation of content	58%	Detailed description in instructions (24%)
Alterations	73%	As shown above
Additions	29%	Chess games (19%), connect with friends (41%)

REFERENCES

Brooke, J. (1996), "SUS: a "quick and dirty" usability scale". in P. W. Jordan, B. Thomas, B.A. Weerdmeester, & A.L. McClelland. *Usability Evaluation in Industry. London*: Taylor and Francis.

Lawton, M. P., and Brody, E. M. (1969), "Assessment of Older People: Self-maintaining and Instrumental Activities of Daily Living." *Gerontologist,* 9, 179-186.

Lewis, C. H. (1982), "Using the "Thinking Aloud" Method In Cognitive Interface Design." *Technical Report IBM* RC-9265.

Nielsen, J. (1993), "Iterative User Iterface Design". *IEEE Compute,* .26(11), 32-41.

Stroop, J. R. (1935), " Studies of Interference in Serial Verbal Reaction." *Journal of Experimental Psychology,* 18, 643-662.

Treisman, A. & Fearnley, S. (1969), "The Stroop Test: Selective Attention to Colours and Words." *Nature,* 222, 437-439.

http://www.oasis-project.eu/

http://www.usabilitynet.org/trump/documents/Suschapt.doc.

Printed and bound by CPI Group (UK) Ltd, Croydon, CR0 4YY

18/10/2024

01776255-0001